AF865149

Navigating the EU AI Act: A Practical Guide for Global Manufacturing and Engineering Leaders

Marcos Kauffman

Navigating the EU AI Act: A Practical Guide for Global Manufacturing and Engineering Leaders

Springer

Marcos Kauffman
Research Centre for Manufacturing
and Materials
Institute for Advanced Manufacturing
and Engineering
Coventry University
Coventry, UK

ISBN 978-3-031-99612-2 ISBN 978-3-031-99613-9 (eBook)
https://doi.org/10.1007/978-3-031-99613-9

© The Editor(s) (if applicable) and The Author(s), under exclusive license to Springer Nature Switzerland AG 2025

This work is subject to copyright. All rights are solely and exclusively licensed by the Publisher, whether the whole or part of the material is concerned, specifically the rights of translation, reprinting, reuse of illustrations, recitation, broadcasting, reproduction on microfilms or in any other physical way, and transmission or information storage and retrieval, electronic adaptation, computer software, or by similar or dissimilar methodology now known or hereafter developed.
The use of general descriptive names, registered names, trademarks, service marks, etc. in this publication does not imply, even in the absence of a specific statement, that such names are exempt from the relevant protective laws and regulations and therefore free for general use.
The publisher, the authors and the editors are safe to assume that the advice and information in this book are believed to be true and accurate at the date of publication. Neither the publisher nor the authors or the editors give a warranty, expressed or implied, with respect to the material contained herein or for any errors or omissions that may have been made. The publisher remains neutral with regard to jurisdictional claims in published maps and institutional affiliations.

This Springer imprint is published by the registered company Springer Nature Switzerland AG
The registered company address is: Gewerbestrasse 11, 6330 Cham, Switzerland

If disposing of this product, please recycle the paper.

"The impediment to action advances action. What stands in the way becomes the way."
(Marcus Aurelius—Meditations, Book V, 20)

Regulatory hurdles, including rigorous risk management and conformity assessments, can feel like obstacles. But by embracing these requirements, organisations discover new and stronger business propositions. The compliance journey becomes a source of genuine progress and competitive advantage, rather than a mere roadblock.

Introduction

Artificial intelligence (AI) has become a defining force of the twenty-first century, reshaping virtually every sector from manufacturing and engineering to healthcare, finance, and beyond. As the technology progresses, it brings forth not only tremendous opportunities but also pressing questions around governance, ethics, and societal impact. In the European Union (EU), these questions have culminated in a landmark piece of legislation known as the European Union's Artificial Intelligence Act (EU AI Act), formally Regulation (EU) 2024/1689, the EU AI Act, a comprehensive legal framework that seeks to regulate AI systems in a manner that safeguards public safety, fundamental rights, and ethical standards, while also supporting technological innovation.

Why This Book Matters

In 2018, the introduction of the General Data Protection Regulation (GDPR) set new global standards for data privacy and personal data handling. Organisations worldwide scrambled to meet GDPR requirements, realising that compliance with EU regulations could not be brushed aside merely because they were not physically located in Europe. The so-called Brussels Effect demonstrated that EU regulations often become de facto global standards because businesses outside the EU adapt to maintain their market access.

In a similar vein, the EU AI Act heralds a new era in AI governance, and its scope extends far beyond European borders. Any organisation that markets AI-related products or services to EU clients or whose AI systems affect EU-based individuals will be subject to these rules. Consequently, manufacturers, engineers, and business leaders around the world must prepare for the Act's implications if they hope to remain competitive in the European market. However, compliance need not be seen solely as a burden; it can be a chance to refine processes, embed ethical considerations from the outset, and build a reputation for responsible AI innovation.

Who Should Read This Book

This book is tailored primarily for professionals and organisations operating in manufacturing and engineering. However, it will also prove valuable for:

- Business leaders and executives: Seeking to understand the strategic implications of AI regulation.
- Compliance officers and legal teams: Charged with aligning company practices to meet the EU AI Act's requirements.
- AI developers and data scientists: Involved in the design, deployment, and monitoring of AI systems that may be impacted by the Act.
- Researchers and academics: Interested in the interplay between AI governance and industrial innovation, particularly in the European context.

Objectives and Structure

The content is structured to provide you with both a high-level understanding of the EU AI Act and a detailed roadmap for achieving compliance. Each chapter focuses on a particular aspect, building a coherent narrative that moves from foundational knowledge to practical implementation strategies and future outlook.

Chapter 1: The EU AI Act—Foundations and Framework

This chapter lays the groundwork by examining why the EU decided to develop AI-specific legislation, how it compares to previous regulations such as the GDPR, and what the EU AI Act seeks to achieve in terms of trust, innovation, and protection of fundamental rights.

Chapter 2: Core Definitions and Risk Classifications

Here, we delve into the definitions set by the Act. What qualifies as AI under this legislation, and how does the risk-based approach categorise systems into unacceptable risk, high risk, limited risk, or minimal risk? This chapter provides clarity on which rules apply to which AI systems.

Chapter 3: Global Implications for Non-EU Manufacturers

While the regulation originates in Europe, its extraterritorial reach means that organisations around the globe must pay heed. This chapter discusses how non-EU manufacturers can navigate compliance and manage divergent regulatory frameworks in different jurisdictions.

Chapter 4: AI in Manufacturing and Engineering

Moving into the sector-specific focus, we look at how AI is revolutionising manufacturing and engineering—from predictive maintenance and supply chain optimisation to robotics and the unique challenges and responsibilities these applications bring.

Chapter 5: Compliance Requirements

This chapter breaks down the Act's legal obligations. What are the rules on data governance, conformity assessments, transparency measures, and human oversight? Here we provide a checklist to ensure a thorough understanding of what meeting the Act's standards entails.

Chapter 6: Impact on Innovation

One common concern with new regulations is whether they stifle innovation. In this chapter, we investigate how businesses can leverage the EU AI Act not merely as a compliance exercise, but as a catalyst for better design, enhanced safety, and stronger consumer trust.

Chapter 7: Steps to Compliance

Building on the requirements, we outline a pragmatic roadmap that takes your organisation from initial AI inventory to post-market monitoring. This chapter offers step-by-step guidance and best practices.

Chapter 8: Collaborating with EU Partners

Compliance is not a task undertaken in isolation. Especially in manufacturing, supply chains and partnerships are complex and often global. We examine how to engage proactively with EU-based distributors, suppliers, and regulators to streamline cross-border compliance.

Chapter 9: Leveraging Technology for Compliance

Technology itself can aid compliance. This chapter highlights various auditing tools, model management platforms, and compliance dashboards that streamline and automate many of the tasks required by the EU AI Act.

Chapter 10: Case Studies—Lessons from the Field

Real-world illustrations often shed more light than theoretical discussions. Here we present case studies demonstrating how different manufacturing and engineering firms adapted to, and sometimes benefited from, aligning with the EU AI Act's standards.

Chapter 11: Insights from Industry Leaders

This chapter compiles key perspectives from executives, engineers, and compliance officers who have had hands-on experience with AI governance. Their insights provide both cautionary tales and success stories.

Chapter 12: Looking Ahead

As technology rapidly evolves, so too will the regulatory environment. We assess potential amendments or expansions of AI law, explore the growing drive for global harmonisation, and outline how organisations can remain future-ready.

Chapter 13: The Ethical Imperative

Finally, we close by addressing the ethics dimension in AI development. Going beyond mere compliance, how can manufacturers and engineers embed fairness, transparency, and accountability into their AI systems from the ground up?

Key Themes

Throughout this book, certain themes recur:

Risk-Based Regulation

The Act's proportional approach ensures that the stringency of requirements aligns with the risk posed by an AI system. Understanding where a system falls on this spectrum is crucial.

Data Quality and Governance

Sound data practices underpin trustworthy AI. From ensuring data representativeness to documenting its origin, the EU AI Act repeatedly emphasises robust data management.

Transparency and Human Oversight

New rules around informing users they are interacting with AI, allowing for human overrides, and disclosing how decisions are reached aim to maintain public confidence in AI systems.

The Brussels Effect

Reflecting the EU's clout in global markets, the Act's reach extends outside Europe. This extraterritorial dimension mirrors the precedent set by the GDPR, underscoring that organisations worldwide must prepare.

Collaboration and Continuous Monitoring

Compliance isn't a one-off milestone. The legislation encourages robust post-market monitoring and alliances with relevant stakeholders, ensuring AI remains safe and reliable throughout its lifecycle.

How to Use This Book

You need not read the entire text in strict linear order, though the structure is designed to guide you from foundational knowledge to more advanced insights. If you are already familiar with the basics of the EU AI Act, you may wish to jump straight to the chapters dealing with your immediate concerns, be it technical documentation, risk classification, or forging partnerships with EU entities. However, reading from start to finish will provide a cohesive understanding of the Act's rationale, provisions, and practical implications.

In each chapter, you will find:

- **Clear Headings and Subheadings**
  Organised and consistently numbered, making it easy to locate relevant material.
- **Practical Examples**
  Illustrating abstract regulations with tangible scenarios drawn from manufacturing and engineering contexts.
- **Guidance Actions**
  Concluding tips or bullet points that summarise essential steps or reflections for the reader to apply in their own organisational context.
- **References and Glossary**
  A unified references section appears at the end of the book, ensuring consistent citation. A glossary covers key terms and abbreviations, helping readers navigate the sometimes complex terminology of AI regulation.

Limitations and Acknowledgements

While this book aims to present a thorough analysis of the EU AI Act, legal texts are subject to change and interpretation. Always check the latest version of the legislation and consider seeking professional legal counsel for complex compliance matters. Moreover, this text is tailored primarily for the manufacturing and engineering industries, though many of the principles and steps can be adapted across diverse sectors.

Final Thoughts

The EU AI Act is not just a new rulebook; it is a significant milestone in the global progression of AI governance. By placing accountability, transparency, and fundamental rights at the forefront, it encourages manufacturers and engineers to develop technologies that are as trustworthy as they are innovative. Organisations that embrace compliance as an opportunity, rather than a hurdle, will be well-placed to benefit from the next wave of AI-driven transformation building solutions that are safer, sustainable, and ethically sound.

With that, let us embark on this journey together, beginning with a deeper look into the foundations and framework of the EU AI Act.

A Note on Moving Goalposts

Regulators, like engineers, work with tolerances, but the political and technological tolerances around AI are currently measured in weeks, not microns. As we wrap up this manuscript (mid-2025), Brussels is still debating delegated acts, Washington is as unpredictable as ever and at least three large language models have broken the FLOP (floating point operations per second) speed record since you started this paragraph. In other words, the insights and technical information you will find on the pages ahead are accurate at the time of printing, yet despite being fundamental principles enshrined in law, they sit on shifting foundations. Think of them as a well-calibrated digital twin of today's landscape rather than a static CNC drawing guaranteed to fit tomorrow's parts.

Why mention this? Because good engineering practice demands documented assumptions. The core assumption here is that Regulation (EU) 2024/1689 remains materially unchanged and that the accompanying ISO/IEC standards progress along their current trajectories. Should a geopolitical tremor, a breakthrough in neuromorphic chips, or an unexpected parliamentary amendment alter that path, you may need to redline a section or two, preferably before your next conformity assessment.

So, please read the book with a screwdriver in one hand and a version-control tag in the other. We have done our best to future-proof the guidance, but just like any predictive-maintenance model, it will require fine-tuning as new data arrive. Until then, torque these insights to spec, keep an eye on the dashboard, and enjoy the build.

Contents

Chapter 1
The EU AI Act—Foundations and Framework

Abstract This chapter explains why the European Union introduced a dedicated Artificial Intelligence Act and situates the regulation within the EU's broader policy landscape. It contrasts the Act's aims: Trust, innovation, and protection of fundamental rights, with earlier cornerstone legislation such as the GDPR, highlighting both continuities and novel elements (e.g. the risk-based tiers for AI systems). Readers gain a concise overview of the Act's structure, key articles, and implementation timeline, together with an introduction to the "Brussels Effect," which helps EU norms shape global practice. By the end of the chapter, manufacturers and engineers understand the legal foundations they must work with and the strategic rationale that underpins the new rules.

1.1 Introduction

The European Union (EU) has taken decisive steps to regulate the rapidly evolving field of artificial intelligence (AI). In doing so, it seeks not only to protect fundamental rights and consumer safety but also to establish clear rules that encourage responsible innovation. This chapter explores the foundational rationale behind the EU AI Act, its overarching objectives, and how it aligns with or diverges from previous legislative efforts such as the General Data Protection Regulation (GDPR). We will also examine its extraterritorial reach, timeline, and its significance for both EU-based and global organisations.

This journey begins with understanding the motivations of the EU in embarking on such a pioneering legislative effort. As with the GDPR, the EU AI Act embodies a vision in which regulatory oversight goes hand in hand with industrial competitiveness and protection of human values. For businesses, especially those in manufacturing and engineering, gaining clarity on why the EU enacted this law is a valuable first step towards effective compliance. By the end of this chapter, readers will grasp how the EU AI Act lays the groundwork for a new era in AI governance, outlining both responsibilities and opportunities for those looking to harness the power of AI responsibly.

© The Author(s), under exclusive license to Springer Nature Switzerland AG 2025

M. Kauffman, *Navigating the EU AI Act: A Practical Guide for Global Manufacturing and Engineering Leaders*, https://doi.org/10.1007/978-3-031-99613-9_1

1.2 The Case for AI Regulation

AI has gradually permeated almost every sector of the economy, reshaping processes in ways that were unthinkable a decade ago. Manufacturers deploy machine learning models to streamline supply chains and automate quality control; financial services use AI-driven tools for fraud detection and credit assessment; healthcare systems rely on predictive analytics for diagnosis support. With this widespread adoption, however, has come a parallel rise in concerns: questions about fairness and bias, the opacity of so-called black-box models, the reliability of safety–critical AI applications and the broader ethical implications of widespread AI-based automation in decision systems.

Evolving Public Perception

Public opinion on AI has fluctuated between excitement and wariness. On one hand, AI is championed for its potential to boost productivity and spur economic growth. On the other, stories of job displacement, biased facial recognition systems, and intrusive surveillance have fuelled scepticism and calls for stronger oversight. The EU, noting these developments, resolved that it was time to set out a unifying set of rules that would guide AI development in a manner deemed safe, trustworthy, and respectful of fundamental rights [1].

A key inspiration in the EU's approach to AI regulation has been its own GDPR. Introduced in 2018, the GDPR rapidly became a global benchmark for data privacy, with businesses worldwide scrambling to align their data protection practices to the new standard. The EU AI Act, although broader in scope, mirrors the GDPR in its extraterritorial reach and risk-based philosophy. Much like the GDPR, it establishes specific obligations and potential penalties, incentivising developers and deployers of AI to incorporate compliance considerations from the earliest phases of design. In essence, the success of the GDPR in raising global awareness around data privacy reassured EU policymakers that well-calibrated regulation could shape international norms and support innovation within the EU market [2].

Upholding Fundamental Rights

The EU places great emphasis on fundamental rights such as human dignity, privacy, and non-discrimination. Throughout the drafting stages of the EU AI Act, policymakers repeatedly stressed that AI must operate in ways that uphold, rather than undermine, these values. Issues such as social scoring, the use of manipulative AI in public spaces, and the deployment of real-time facial recognition systems have all been highlighted as areas too sensitive or risky to leave unregulated. The Act thus includes provisions explicitly prohibiting certain uses of AI that are deemed incompatible with the EU's ethical and social values [3].

Despite concerns that regulatory controls might stifle innovation, the EU's stated aim is to encourage responsible AI development. The Act endeavours to provide a stable, predictable environment in which companies can confidently invest in AI solutions without the threat of unpredictable legal complications. This sense of legal certainty, it is argued, will bolster the EU's competitiveness on the global stage, much as the GDPR arguably enhanced consumer trust in digital services. For manufacturers and engineering firms alike, having consistent rules across the EU can simplify compliance compared to managing a patchwork of country-specific requirements across the EU 27 members [4].

1.3 Objectives of the EU AI Act

Given this backdrop, the EU AI Act can be seen as resting on three primary objectives, each of which underpins the central provisions:

1. **Fostering Trustworthy AI**

The legislation introduces a structured approach to risk categorisation, ensuring that systems most likely to cause harm, be it physical, psychological, or societal, face the strictest requirements. By focusing on transparency, accountability, and data governance, the Act aims to promote responsible design choices and mitigate the risks of untested or unscrutinised algorithms.

2. **Safeguarding Rights and Well-Being**

In line with the EU's human-centric ethos, the Act safeguards citizens from opaque, manipulative, or dangerous AI deployments. The explicit prohibition of certain practices like subliminal manipulation and social scoring, signals that not all technological capabilities are acceptable, regardless of their economic potential. Similarly, high-risk applications (for instance, in employment or healthcare) must meet stringent guidelines to ensure individuals' rights and well-being remain paramount.

3. **Championing Innovation, Competitiveness, and Growth**

The EU AI Act is not conceived merely as a protective measure; it also aims to create the conditions in which innovation can flourish. Clear responsibilities and shared standards offer a more predictable environment, encouraging investment. It is believed that a robust regulatory framework will assure consumers and business partners that AI-driven products are safe and reliable, leading to higher adoption rates.

1.4 Scope and Applicability

One distinctive aspect of the EU AI Act which mirrors the GDPR is its extraterritorial scope. The EU AI Act's reach extends beyond the geographical boundaries of the European Union. It applies to providers who place AI systems on the EU market, irrespective of their establishment or location. Furthermore, the Act is applicable to providers and deployers located outside the EU if the output produced by their AI systems is used within the EU. This broad scope explicitly includes product manufacturers based outside the EU who place AI systems on the EU market together with their products. Therefore, non-EU engineering firms and manufacturers intending to offer AI-powered products or whose AI-driven processes generate outputs used in the EU must comply with the regulations outlined in the EU AI Act.

Who Must Comply?

In essence, three categories of actors are expected to comply with the EU AI Act:

1. **Provider**

The entity (e.g. a manufacturer or developer) that places an AI system on the EU market or puts it into service under its name or trademark. For a manufacturing company, if you develop or integrate an AI component into your product and sell it, you are the provider.

2. **Deployer (User)**

The organisation or person using an AI system under their authority. In manufacturing, a deployer might be a factory owner using an AI tool in production.

3. **Importer**

An EU-based intermediary that places a non-EU provider's AI system on the EU market. For example, if a non-European company's AI-equipped machine is sold in Europe via a local agent, that agent is the importer with compliance responsibilities.

4. **Distributor**

Other intermediaries in the supply chain making the AI system available in the EU, who are not the provider or importer.

These definitions matter because obligations under the Act differ for providers, deployers, importers, and distributors. A single manufacturing firm might even play multiple roles (e.g. developing an AI system as a provider and also deploying AI internally as a user). Understanding these roles clarifies who must do what to comply.

Territorial Extent

The territorial scope covers all EU Member States. Additionally, if a non-EU company places AI technology on the EU market, either in physical goods or as software services, it also falls under the Act. This means a robotics firm in Japan, a healthcare AI company in the United States, or a predictive analytics vendor in India must ensure their offerings meet the Act's obligations if they have EU customers.

Material Scope: Defining AI Systems

The Act adopts a broad definition of AI, covering an extensive range of software that can infer, predict, or decide based on input data. Whether an algorithm employs classical statistical methods, machine learning, or symbolic reasoning, it can qualify as AI under this law. This inclusivity ensures that new, emerging AI approaches do not slip through regulatory gaps simply because they do not fit a narrow description.

Manufacturers utilising automation, data analytics, or algorithmic decision-making in their production lines must evaluate whether they fall within the Act's domain.

1.5 Key Provisions and Structure

At a structural level, the EU AI Act is divided into several chapters and articles, each addressing different aspects:

- **Risk Classification (Articles 5–8)**

This section outlines the four risk tiers (unacceptable, high, limited, and minimal), detailing which AI applications are banned outright and which require higher levels of scrutiny.

- **Obligations for High-Risk AI (Articles 9–16)**

Here, the Act sets requirements around data governance, transparency, human oversight, accuracy, and robustness. Any system deemed high risk must undergo conformity assessments before it can be deployed.

- **Transparency Obligations for Certain AI (Article 52)**

For systems interacting directly with humans, such as chatbots or AI that generates synthetic media, transparency mandates ensure users are informed they are engaging with AI.

- **Governance and Enforcement Mechanisms (Articles 60–71)**

These articles detail how national authorities, notified bodies, and EU institutions will coordinate to enforce the Act, including penalties for non-compliance.

Manufacturers need to be particularly mindful of the high-risk provisions. AI systems in assembly lines, industrial robots, or safety-monitoring solutions often carry potentially serious implications for worker safety and product integrity. Consequently, such systems are frequently designated as high risk, bringing extra responsibilities around data validation, documentation, and monitoring.

1.6 The Timeline for Implementation

Much like the GDPR's staggered approach, the EU AI Act foresees a phased implementation to give organisations time to adjust. Although precise dates may shift slightly based on legislative processes and final approvals, the general schedule includes:

1. **Official Publication and Entry into Force**

Shortly after the text is published in the Official Journal of the European Union, the regulation formally "enters into force," generally a few weeks from publication.

2. **Grace Period**

There is an interval (often ranging from 12 to 24 months) before many of the Act's provisions become fully enforceable. This grace period allows businesses time to understand the Act and begin aligning systems and procedures.

3. **Full Applicability**

By a specific date, all or most obligations come into effect. High-risk AI systems will typically have earlier deadlines for conformity assessments, given their potential impact on safety and rights.

4. **Further Stages**

Certain provisions may come into play over additional time frames—for instance, rules specific to "general-purpose AI," if included in final texts, or new appendices that list specific technologies subject to mandatory risk assessments.

For manufacturers, this timeline underscores the importance of proactive efforts. Waiting until the last moment could result in non-compliant products, market entry delays, or the forced withdrawal of AI systems from critical industrial settings.

The EU AI Act's implementation timeline is designed to give businesses time to adapt:

July 2024: Published in the Official Journal.
August 2024: The Act enters into force, establishing its legal framework.
February 2025: Obligations for prohibited AI systems and basic compliance requirements begin.
August 2025: General-purpose AI models must comply with specific provisions.
August 2026: Most obligations, including those for high-risk AI systems, become enforceable.
August 2027: Final provisions for certain applications, including high-risk systems in Annex I, take effect.

While the staggered timeline provides some flexibility, businesses must act early to address high-risk systems, which often require detailed assessments, documentation, and in-market monitoring [5–7].

1.7 Comparison with the GDPR

Drawing a parallel between the EU AI Act and the GDPR is inevitable, as the latter has already transformed how the world approaches data privacy and protection. While both pieces of legislation share a similar ambition to protect fundamental rights and encourage responsible innovation, they differ in their scope and operational details.

Similarities

- **Extraterritorial Reach**—Both laws apply to organisations outside the EU if they handle the data of EU citizens or, in the case of the AI Act, provide AI systems that affect EU individuals or markets.
- **Risk-Based Approach**—The GDPR's emphasis on data impact assessments parallels the AI Act's classification of AI systems by risk, each requiring proportionate controls.
- **Strong Penalties**—Non-compliance with GDPR can incur fines of up to 4% of global annual turnover. The AI Act similarly proposes significant fines, reflecting the EU's willingness to enforce compliance via financial and reputational repercussions.

Differences

- **Subject Matter**—The GDPR focuses primarily on personal data, whereas the AI Act addresses the broader functioning of AI systems, including safety, transparency, and ethical considerations.
- **Technology Coverage**—The GDPR remains technology-agnostic regarding data processing methods, while the AI Act specifically outlines AI techniques (machine learning, logic-based systems, etc.) in its annex.
- **Levels of Requirements**—The GDPR imposes a single set of obligations on all data controllers/processors, albeit with variations for SMEs or special categories of data. The AI Act, by contrast, introduces four tiers of risk classification, each with distinct obligations.

Overall, for manufacturers that have already navigated GDPR compliance, the experience may offer valuable lessons in risk assessment, documentation, and dealing with EU supervisory authorities. However, the AI Act adds new layers of technical and ethical considerations that go beyond mere data handling [7].

1.8 The Brussels Effect on AI Governance

The term "Brussels Effect" describes how the EU's large market effectively compels multinational corporations to adopt its regulatory standards. Once a company modifies its AI systems to comply with the EU's rules, it may choose to apply the same standards globally rather than maintaining multiple, often conflicting, sets of practices. This phenomenon could spread the EU AI Act's influence in regions far beyond Europe, making it a de facto global standard for AI governance.

From a practical standpoint, manufacturers worldwide might find it more efficient to design AI solutions from the outset to meet the highest regulatory bar. This ensures a smoother entry into EU markets and can serve as a badge of credibility for customers and partners elsewhere. Indeed, companies that excel at conforming to stringent EU regulations often enjoy heightened global trust, as they can demonstrate robust processes and ethical considerations integrated into their product lifecycle.

1.9 Real-World Implications for Manufacturers

For manufacturing and engineering, the implications are multifaceted:

Integration with Existing Product Directives

Many manufacturers already comply with EU product safety directives, such as the Machinery Directive or Low Voltage Directive. The EU AI Act adds another layer to these obligations. A factory robot or industrial control system that uses AI may require both compliance with relevant machinery safety standards and the new AI-specific conformity checks.

Supply Chain Collaboration

Manufacturers often source components from a network of global suppliers. Ensuring that each supplier meets the Act's standards, particularly for high-risk components, requires contractual clauses, data-sharing agreements, and in some cases, audits.

Market Differentiation

Compliance can become an advantage. For instance, a British firm specialising in AI-driven predictive maintenance solutions that meet the EU AI Act's data governance standards can market its products as trustworthy, secure, and future-proof. This credibility may be valued not only in Europe but also by international clients who view the EU's regulations as a benchmark [7].

Innovation Catalyst

Rather than stifling creativity, the Act's clarity around regulatory expectations could spur R&D in safer, more reliable AI systems. Manufacturers that invest early in compliance may find their innovations rewarded by faster time-to-market and lower long-term risk.

1.10 Conclusion

The EU AI Act is poised to shape the trajectory of AI governance both in Europe and worldwide. For manufacturers and engineers, understanding the foundations of this legislative framework is critical. Grounded in an ethos that seeks to balance innovation with the protection of fundamental rights, the Act adopts a risk-based approach that imposes the strictest requirements on those AI systems most likely to impact safety and well-being.

From its parallels with the GDPR to its overarching ambition to foster responsible AI innovation, the Act signals an era in which AI development must incorporate not merely technical expertise but ethical and regulatory considerations too. Compliance, therefore, is not merely a matter of ticking legal boxes; it has profound implications for product design, supply chain relationships, and brand reputation.

Subsequent chapters will delve deeper into the specifics of risk classification, global ramifications, sector-specific applications, and practical strategies for meeting the Act's obligations. By understanding the rationale behind the EU AI Act, as outlined in this chapter, manufacturers, engineers and business leaders are better positioned to engage with its requirements constructively. In doing so, they can transform regulatory compliance from a challenge into a strategic enabler, fuelling progress while maintaining the trust of stakeholders around the world.

Key Takeaways

I. The EU introduced the AI Act to safeguard fundamental rights, ensure safety, and provide a clear framework for AI developers and users.
II. The Act aims to build trustworthy AI, protect societal interests, and foster innovation.
III. Organisations worldwide that provide AI to EU citizens or the EU market must comply.
IV. AI systems are categorised into risk tiers, with stricter rules for those posing higher potential harm.
V. Implementation is staged, giving organisations time to adapt, but proactive steps are strongly recommended.
VI. Similar extraterritorial reach and strong enforcement mechanisms, but the Act focuses on AI system governance rather than purely data handling.
VII. Likely to shape global AI standards, prompting businesses to design AI products that meet EU requirements by default.
VIII. From re-aligning supply chains to integrating AI-specific checks, compliance can also serve as a market differentiator.

With this foundational understanding in place, we now progress to Chap. 2, where we examine the Act's core definitions and risk classifications in detail. This next step is vital, for it tells us precisely how various AI systems, ranging from minimal to high risk, are evaluated and regulated under this new framework.

References

1. European Union (2024) Regulation (EU) 2024/1689. https://eur-lex.europa.eu/eli/reg/2024/1689/oj/eng
2. White and Case LLP (2024) Long-awaited EU AI act becomes law after publication in the EU's Official Journal. https://www.whitecase.com/insight-alert/long-awaited-eu-ai-act-becomes-law-after-publication-eus-official-journal
3. European Commission (2024) European artificial intelligence act comes into force. Press Release IP_24_4123. https://ec.europa.eu/commission/presscorner/detail/en/IP_24_4123
4. Matheson (2024) EU AI Act finalised. https://www.matheson.com/insights/detail/eu-ai-act-finalised
5. Skadden (2024) The EU AI Act: what businesses need to know. https://www.skadden.com/insights/publications/2024/06/quarterly-insights/the-eu-ai-act-what-businesses-need-to-know

6. Hunton Andrews Kurth LLP (2024) AI act published in the Official Journal of the EU. https://www.hunton.com/privacy-and-information-security-law/ai-act-published-in-the-official-journal-of-the-eu
7. European Commission (2025) Regulatory framework for AI. Dig Strat Portal. https://digital-strategy.ec.europa.eu/en/policies/regulatory-framework-ai

Chapter 2
Core Definitions and Risk Classifications

Abstract This chapter clarifies the Act's terminology and the four-tier risk taxonomy that determines regulatory obligations. We dissect the legal definition of an "AI system," explain the criteria for unacceptable, high, limited, and minimal risk, and explore the emerging focus on general-purpose AI. Practical manufacturing examples illustrate how identical technologies may fall into different tiers depending on context. The chapter closes with guidance on common risk-classification pitfalls and the importance of aligning internal assessment methods with Annex III and Article 6.

2.1 Introduction

Understanding the European Union (EU) Artificial Intelligence (AI) Act 2024 begins with defining what the legislation means by "artificial intelligence" and how it classifies AI systems according to their potential for harm or abuse. This tiered, risk-based framework is a keystone of the Act, establishing precise obligations proportional to the risks posed. For manufacturers and engineers, it is essential to grasp these definitions and categories, because they determine everything from data governance requirements to post-market monitoring expectations.

In this chapter, we explore the EU AI Act's broad and inclusive definition of AI, examine each risk category in detail, and highlight how these risk classifications apply in real-world scenarios, particularly in manufacturing contexts. We will discuss prohibited systems (unacceptable risk), the obligations for high-risk systems, and the relatively lighter requirements for limited-risk and minimal-risk AI. By chapter's end, you should have a clear grasp of where different AI technologies fall within this continuum and the steps necessary to remain compliant.

© The Author(s), under exclusive license to Springer Nature Switzerland AG 2025

M. Kauffman, *Navigating the EU AI Act: A Practical Guide for Global Manufacturing and Engineering Leaders*, https://doi.org/10.1007/978-3-031-99613-9_2

2.2 The EU AI Act's Broad Definition of AI

One of the Act's most striking features is how comprehensively it defines "artificial intelligence." This approach reduces the possibility that new or evolving AI methodologies might slip through regulatory gaps simply because they do not neatly fit a narrow definition.

Article 3(1): A Wide-Reaching Definition

Article 3(1) of the Act describes an AI system as:

> Software developed with techniques such as machine learning, logic- and knowledge-based reasoning, or statistical methods, which, for a given set of human-defined objectives, can generate outputs such as predictions, recommendations, or decisions that influence real or virtual environments [1].

This definition is intentionally technology-agnostic. Whether you are working with deep neural networks, symbolic AI, knowledge graphs, expert systems, or simpler rule-based algorithms, you may well be subject to the Act if your system makes autonomous inferences or decisions. The scope extends to:

- **Machine Learning Algorithms**: Ranging from linear regressions and decision trees to complex reinforcement learning models.
- **Expert Systems**: Utilising if-then rules or knowledge bases to mimic human decision-making in specific domains.
- **Statistical Models**: Traditional data-driven methods that produce predictions or classifications based on correlations.

Why Such a Comprehensive Scope?

The EU recognises that AI evolves rapidly. Once, technologies like large language models or deep reinforcement learning might have seemed theoretical or niche, yet they are now deployed in real-world systems at scale. By specifying a broad coverage, the Act ensures that new AI developments no matter how unexpected are regulated under its umbrella.

For manufacturers, this means even seemingly simple algorithms used for production planning or stock management could come under the Act if they produce actionable recommendations. It is therefore crucial to assess whether existing software solutions qualify as AI and, if so, whether they fall into higher- or lower-risk categories.

2.3 Four Tiers of Risk

A central principle of the EU AI Act is its risk-based approach. Rather than imposing uniform obligations on all AI applications, the legislation recognises that some uses are inherently more hazardous or ethically sensitive than others. Accordingly, it segments AI systems into four risk levels, each with distinct legal consequences:

1. **Unacceptable Risk** (*Prohibited Systems*).
2. **High Risk.**
3. **Limited Risk.**
4. **Minimal Risk.**

We will now examine each category, the rationale behind it, and its implications.

2.4 Unacceptable Risk: Prohibited AI Systems

Unacceptable-risk AI covers those applications deemed so detrimental to fundamental rights, public safety, or ethical standards that they are outright banned within the EU. These include systems that deploy subliminal techniques to manipulate people, exploit vulnerable groups, or facilitate social scoring by governments. For example, an AI that covertly influences factory workers' behaviour in a way that could cause harm, or a state-run system ranking citizens (social credit scoring), would fall in this banned category. Such systems cannot be developed or used at all within the EU. (Fortunately, most manufacturers would not contemplate these use cases in legitimate operations.) Article 5 of the Act enumerates examples that include [1]:

Subliminal Manipulation: AI used to manipulate human behaviour beyond their conscious awareness. This might involve hidden advertising cues that exploit vulnerabilities.

Social Scoring: Systems that rank or evaluate individuals based on personal data or observed behaviour, in ways reminiscent of certain authoritarian regimes.

Real-Time Remote Biometric Identification in Public: Facial recognition systems used by law enforcement in publicly accessible spaces, subject to very narrow exceptions (e.g. a terrorist threat).

Implications for Manufacturing

Although the above examples may not seem immediately relevant to a typical factory setting, manufacturers must still guard against inadvertently deploying or supplying technology that could be used for these prohibited purposes. For instance, an industrial camera system with facial recognition features might not be a direct violation if used for safety alerts in a closed environment. However, selling the same system with software enabling real-time public surveillance without user consent could place it under the banned category.

Because these systems are "prohibited," there is no compliance procedure or conformity assessment that could render them lawful. For a manufacturing or engineering firm, awareness is key. You must ensure that none of your AI-enabled solutions can be applied or repurposed into these disallowed functionalities, either by design or inadvertent misuse.

2.5 High-Risk AI Systems

For many organisations in manufacturing, this is the most critical classification. High-risk AI encompasses those applications whose malfunction or misuse could seriously endanger personal health, safety, or fundamental rights. These are allowed on the market only if they comply with strict requirements (detailed in Chap. 5) before and after deployment. High-risk examples relevant to manufacturing include AI components of safety–critical machinery (e.g. a crane's autonomous control system), AI used in medical devices, or AI for recruitment/human resources decisions. The AI Act's Annex III provides a detailed list of use cases automatically deemed high risk [1, 2] (we discuss manufacturing-relevant items in later chapters). If your AI system is high risk, you face the most extensive obligations under the Act.

The official listing of high-risk use cases includes applications in sectors such as:

- Medical devices and healthcare (e.g. AI-driven diagnostic tools).
- Transport and automotive (e.g. advanced driver assistance systems, autonomous driving software).
- Public infrastructure (e.g. AI coordinating utilities or traffic systems).
- Employment and workforce management (e.g. AI used for recruitment or performance monitoring).
- Essential private services (e.g. systems affecting credit approval).
- Law enforcement and border control.

Safety Components in Regulated Products

One especially relevant category for manufacturers is the intersection of AI with already regulated products such as machinery, transport vehicles, lifts, or pressure vessels under EU harmonisation legislation. If the product in question requires a conformity assessment for safety and it incorporates AI in a safety–critical manner, it automatically becomes high risk.

Example: A robotic arm is used on an assembly line to handle hazardous materials. If the AI controls critical safety features such as collision detection or emergency shutdown, and a malfunction could injure workers, the system is likely to be classified as high risk.

Obligations for High-Risk Systems

The Act imposes a range of obligations on high-risk AI providers and deployers, as specified in Articles 9–16:

- **Quality and Representativeness of Data (Article 10)**

High-risk AI systems must be trained, validated, and tested on datasets that are complete, representative, and free from errors likely to cause significant bias or safety issues. For manufacturers, this might involve collecting operational data from diverse factory environments and thoroughly documenting its origin.

- **Technical Documentation and Record-Keeping (Article 11)**

Organisations must maintain detailed records describing how the AI was designed, tested, and deployed. This includes information about algorithms, model performance, and version histories. Inspectors or notified bodies can request this documentation to verify compliance.

- **Transparency and Provision of Information to Users (Article 13)**

The Act demands that manufacturers provide clear instructions for use, warnings about limitations, and information on how the system processes data or reaches decisions. In a manufacturing environment, this could mean user manuals detailing how an AI-based quality control system flags defects, the reliability rates, and scenarios in which operators must intervene.

- **Human Oversight (Article 14)**

High-risk AI systems must be designed so that human operators can oversee their operations and override automated decisions if necessary. In a factory setting, this might entail an emergency stop mechanism or a dashboard enabling staff to monitor real-time AI analysis and intervene when anomalies appear.

- **Accuracy, Robustness, and Cybersecurity (Article 15)**

The system must meet essential performance metrics in real-world conditions, remain resilient against adversarial attacks, and incorporate cybersecurity measures to prevent unauthorised alterations.

- **Post-market Monitoring (Article 61)**

Even after deployment, organisations are responsible for tracking the performance of high-risk AI, collecting user feedback, and promptly addressing any risks or malfunctions that might emerge [1, 3].

Conformity Assessments

High-risk AI often requires a conformity assessment prior to market release [4]. Depending on the specific use case, this may be an internal assessment (self-assessment) or a third-party assessment by a notified body designated by the EU.

In many manufacturing contexts, the higher the potential harm, the more likely an external, independent review is mandated.

The assessment scrutinises design documentation, testing reports, and the adequacy of risk mitigation. Once validated, the AI system can be affixed with a CE marking (if part of a wider product requiring CE marking) or other relevant compliance indicators before being placed on the EU market.

2.6 Limited-Risk AI Systems

Not all AI is high stakes. Many applications pose relatively lower risks to health, safety, or fundamental rights. This category covers AI systems that are not high risk but still warrant some transparency. Limited-risk AI is permitted with relatively lighter requirements, and the Act mandates that users be informed they are interacting with an AI when it might not be obvious. For instance, an AI chatbot interface for equipment support should disclose that it is AI-driven. In manufacturing, an AI-driven analytics dashboard for managers might be limited risk; it doesn't directly impact people's safety, but transparency (e.g. labelling AI-generated forecasts) is required by Article 52 of the Act. No rigorous pre-approval is needed for limited-risk AI, but good practice calls for responsible design and some user instructions. Typically, systems in this category involve:

- AI chatbots providing general customer support.
- AI systems that interact with users but do not make critical decisions impacting safety or essential rights.
- AI-based productivity tools that simply recommend, rather than execute, tasks.

Transparency Obligations (Article 52)

The most notable requirement for limited-risk systems is transparency. Providers and deployers must ensure users are aware they are interacting with an AI system. For instance:

AI chatbots: Users should be clearly informed that they are conversing with AI rather than a human agent.

AI-Generated Content: If the system produces synthetic text, images, or other media, it must be disclosed as such.

These disclosures build trust and help users make informed choices, particularly in consumer-facing contexts. In manufacturing, although user interactions are often restricted to trained operators, there may be scenarios such as AI-based software assisting engineering decisions where explicit disclaimers remain necessary.

Potential Upgrade to High-Risk AI

It is important to keep in mind that a limited-risk system can move into a high-risk category if its deployment environment changes. A visual inspection AI might be considered limited risk if used to detect cosmetic flaws in consumer goods. However, if, for example, the same technology is repurposed to identify structural defects in aircraft parts, it likely becomes high risk due to the safety implications.

2.7 Minimal-Risk AI Systems

This category includes AI applications deemed so unlikely to harm individuals or society that the Act imposes no specific legal obligations beyond general principles of transparency and ethical design. At present it is likely that the vast majority of AI systems used in manufacturing operations will fall here, posing only negligible risks. These face no specific legal obligations under the Act aside from general EU laws. Examples might include AI that sorts incoming raw materials for inventory (with no safety impact) or AI in office productivity tools, such as invoice matching or automatic email responses. While legally there's no action required, manufacturers are encouraged to still follow ethical AI guidelines (transparency, data privacy, etc.) as a matter of corporate responsibility. Examples might include:

AI-Driven Spelling and Grammar Checkers

Automated Personal Assistants

Simple Data Visualisation Tools

While these systems remain subject to the overarching spirit of ethical AI ensuring no clandestine manipulation or unlawful data processing, the EU AI Act does not impose formal risk assessments, conformity procedures, or heightened transparency rules on them.

Maintaining Good Practices

It's important to note that the classification depends on the use case, not just the technology itself. The very same AI algorithm could be minimal risk in one context and high risk in another. For example, an image recognition AI is low risk when sorting parts on a conveyor, but if it's used as a quality gate for detecting defects in airplane engine components, it could become high risk due to safety implications (lives might depend on those decisions). Therefore, manufacturers must evaluate each specific application of AI in their products or processes.

Even when the law does not mandate thorough documentation, businesses are encouraged to maintain good development and quality control practices. Doing so may prevent liability if a system unexpectedly transitions to a higher-risk context in future. In the dynamic world of industrial automation, an apparently mundane AI tool could be later integrated into critical workflows, raising its risk profile.

2.8 General-Purpose AI: An Emerging Focus

In addition to the four risk tiers, there is growing attention on general-purpose AI (GPAI). Large language models, generative AI systems, and other multi-use technologies can be applied in various sectors with varying risk outcomes. While not formally categorised as a separate tier, the EU AI Act has opened discussions on how to handle systems whose ultimate applications may be beyond the direct control of their original developers.

For instance, a general-purpose vision model might be fine-tuned for use in either benign inventory management or invasive facial recognition. The legislation may add further clarifications or obligations, prompting developers of GPAI to share usage guidelines, maintain traceability of training data, or implement usage restrictions [5]. Manufacturers relying on GPAI for tasks like predictive maintenance must monitor how updates or modifications to underlying models affect risk classification.

2.9 Practical Examples in Manufacturing

The following list provides a series of 10 practical, real-world AI scenarios identified from recent empirical research. These are grouped below into two main categories: (1) manufacturing-based embedded AI use cases and (2) product-based embedded AI use cases.

Group 1: Manufacturing Embedded AI

1. **Predictive Maintenance in High-Precision Manufacturing**

AI algorithms continuously monitor real-time sensor data from high-precision manufacturing equipment to identify potential failures before they occur. These systems directly impact machinery used in critical sectors, such as aerospace or medical device production, where errors can lead to safety concerns or regulatory violations.

2. **Autonomous Robotics for Assembly Lines**

AI controlled robots execute repetitive assembly tasks, such as welding, painting, or assembling components, in a controlled manufacturing environment. These systems are designed with minimal human interaction and risk of harm, relying on adaptive decision-making to respond to anomalies in real time and correct errors autonomously.

3. **AI-Driven Quality Control in Medical Device Production**

Computer vision-powered AI systems are deployed to detect micro-defects in medical device production. These systems ensure compliance with stringent quality standards to prevent defective products from entering the market, which could pose risks to patient health and safety.

4. **AI for Energy Optimisation in Smart Factories**

AI algorithms dynamically manage energy consumption by monitoring and adjusting machinery and HVAC systems in real time to reduce waste and improve energy efficiency. These systems indirectly contribute to sustainability goals without impacting human safety.

5. **Digital Twin for Factory Operations**

AI-driven digital twins replicate physical manufacturing processes in a virtual environment. This allows manufacturers to test workflows, predict bottlenecks, and optimise production plans without disrupting the actual factory floor or exposing workers to physical risks.

Group 2: Product Use Cases with Embedded AI

6. **AI for Autonomous Navigation in Vehicles**

AI systems control braking, steering, and navigation in semi-autonomous vehicles. These systems rely on real-time data from road and the driver collected via sensors and cameras to make split-second decisions in dynamic environments. Some implementations, however, may use subliminal techniques to influence driver behaviour.

7. **AI-Powered Battery Management for Electric Vehicles**

AI systems monitor and optimise battery usage, predict energy needs, and prevent overheating or failure in electric vehicles. Failures in these systems could lead to catastrophic events, such as battery fires, posing risks to vehicle occupants and surrounding infrastructure.

8. **In-Cabin Monitoring Systems in Autonomous Vehicles**

AI systems analyse driver and passenger behaviours to enhance safety, such as detecting drowsiness, distraction, or improper seatbelt use. These systems provide real-time alerts and can influence vehicle operation based on their findings.

9. **AI-Driven Collision Avoidance in Automotive Systems**

AI-powered collision avoidance systems use advanced sensor fusion and predictive modelling to detect and respond to potential on-road collisions in vehicles. These systems are critical for ensuring safety in dynamic environments.

10. **AI in Consumer Robotics for Home Automation**

AI-powered consumer robots perform home automation tasks, such as vacuuming. They primarily enhance convenience and have limited safety or ethical implications.

2.10 Challenges in Determining Risk Levels

While the Act aims to provide clarity, grey areas remain. Determining the boundary between limited risk and high risk can be difficult, particularly in industrial contexts where the same AI model can be used in multiple ways. The following challenges often arise:

- **Context Dependence**

A single AI product's risk level can change based on operational deployment. The same image recognition system might be minimal risk in a benign environment yet become high risk in safety-critical applications.

- **Evolving Systems**

AI that continues learning post-deployment (online learning models, adaptive systems) could initially pass compliance checks but later drift into riskier territory if updates alter its decision-making.

- **Supply Chain Complexities**

An AI subsystem integrated into a broader product might appear low risk in isolation, yet the entire product's risk could be elevated. Collaboration and due diligence with suppliers are crucial to maintain compliance.

- **Unforeseen Emerging Risks**

Rapid advancements or novel use cases can lead to previously unconsidered risks. For instance, a data aggregator tool might inadvertently collect sensitive personal data from employees, raising privacy and ethical concerns.

To navigate these complexities, manufacturers should adopt conservative approaches, consult frequently with legal and technical experts, and remain prepared to reclassify AI systems if new information suggests a higher-risk category.

2.11 Aligning Risk Categories with Compliance Strategy

The main takeaway for organisations is that every AI system should undergo an internal review or "risk classification check." This process could be implemented into the engineering process flow for developing new products or processes and should typically involve:

- **Mapping Functionalities**

Document the AI system's purpose, scope, and control over critical processes or decisions.

- **Assessing Potential Harms**

Consider worst-case scenarios, paying attention to safety, health, privacy, or fundamental rights. Where these harms appear severe, a higher-risk tier may apply.

- **Reviewing Annexes and Guidelines**

Consult the official annexes (particularly Annex III) and any sector-specific regulatory guidance to clarify categories.

- **Determining Compliance Measures**

High-risk systems necessitate thorough documentation, external audits (if required), and post-market checks. Limited-risk and minimal-risk systems have fewer obligations but still benefit from good design practices and ethical considerations.

By methodically implementing this classification step, manufacturers can avoid last-minute surprises. A well-structured risk analysis also lays a solid foundation for subsequent chapters' topics, such as data governance, transparency, and post-market monitoring obligations.

2.12 Conclusion

Risk classification is the linchpin of the EU AI Act, guiding organisations in how intensively they must prepare for compliance. Whether an AI tool is deemed prohibited, high risk, limited risk, or minimal risk depend largely on the potential impact on individual safety, fundamental rights, and societal well-being. Manufacturers operating in or selling to the EU must be thorough in categorising their AI systems, for misclassification could result in significant legal liabilities or the forced withdrawal of non-compliant products from the market.

A few major points bear repeating:

- **Definition of AI**: Incredibly broad, capturing nearly any software that infers or decides, irrespective of the underlying technique.
- **Unacceptable/Prohibited Systems**: No scope for legal compliance; these are simply banned.
- **High Risk**: The most stringent requirements, covering critical applications that could severely affect safety or rights.
- **Limited Risk**: Lighter obligations, mainly transparency, though systems can be reclassified if circumstances change.
- **Minimal Risk**: Few or no explicit requirements, yet best practices remain advisable.
- **General-Purpose AI**: Though not a separate tier, concerns around multi-use or large-scale models could expand obligations for developers and adopters alike.

In the chapters that follow, we will delve deeper into how these risk tiers influence the specific compliance obligations, starting with a global perspective in Chap. 3, where we discuss how non-EU manufacturers can manage cross-border compliance challenges under the EU AI Act. Through this lens, we will also explore how different regulatory frameworks around the world affect (and are affected by) the EU's leadership role in AI governance.

Key Takeaways

I. The EU AI Act covers most AI techniques, ensuring new technologies do not sidestep regulatory oversight.
II. From prohibited to minimal, the Act enforces proportionate obligations based on potential harm.
III. High-risk systems, common in manufacturing, involve strict requirements on data, documentation, human oversight, accuracy, and post-market monitoring.
IV. Lower tiers face fewer legal obligations, though transparency and good design practices remain essential.
V. An AI's classification depends on its use case and environment; the same system can move from limited to high risk if deployed in safety–critical scenarios.
VI. High-risk systems may need thorough internal or external audits.
VII. The classification landscape may evolve as AI technologies develop, so ongoing vigilance is required.

Armed with this foundational knowledge, manufacturers can more effectively chart their compliance strategies under the EU AI Act and avoid pitfalls that arise from misunderstanding the Act's definitions and risk categories.

References

1. European Union (2024) EU AI act, art. 6 and Annex III. https://artificialintelligenceact.eu/article/6/ and https://artificialintelligenceact.eu/annex/3/
2. Holistic AI (2024) Identify high-risk AI systems according to the EU AI act. https://www.holisticai.com/blog/identify-high-risk-ai-systems-according-to-eu-ai-act
3. Holistic AI (2024) Penalties of the EU AI act. https://www.holisticai.com/blog/penalties-of-the-eu-ai-act
4. Security Compass (2024) Understanding EU AI act risk categories. https://www.securitycompass.com/blog/understanding-eu-ai-act-risk-categories/
5. WilmerHale (2024) What are high-risk AI systems within the meaning of the EU's AI act?. https://www.wilmerhale.com/en/insights/blogs/wilmerhale-privacy-and-cybersecurity-law/20240717-what-are-highrisk-ai-systems-within-the-meaning-of-the-eus-ai-act-and-what-requirements-apply-to-them

Chapter 3
Global Implications for Non-EU Manufacturers

Abstract Although the EU AI Act is European legislation, its extraterritorial reach compels non-EU firms to comply whenever their AI-enabled products or outputs reach the Union. This chapter analyses the Act's global ripple effects, compares it with parallel frameworks in the United States and Asia, and outlines practical strategies for navigating overlapping regimes. Case vignettes demonstrate how supply chain partners, distributors, and local subsidiaries can trigger EU jurisdiction. Readers learn to anticipate compliance pressures, leverage early alignment as a market differentiator, and prepare for future interoperability or divergence across regions.

3.1 Introduction

The European Union (EU) has a storied history of setting influential regulatory precedents, and the EU Artificial Intelligence (AI) Act is no exception. Although this legislation originates within the EU, its effects extend far beyond European borders. Any manufacturer whose AI systems reach the EU market or impact EU-based individuals fall under the Act's jurisdiction, regardless of where that manufacturer is headquartered. In today's interconnected world of global supply chains and cross-border data flows, the practical consequence is that countless non-EU businesses are now compelled to adopt EU standards for their AI products and services.

This chapter delves into what the EU AI Act means for manufacturers and engineers located outside the EU. We begin by exploring the Act's extraterritorial scope, which underpins its global reach. We then examine the compliance pressures driving non-EU entities to align with EU rules, often independently of any formal enforcement mechanism in their home countries. Following that, we discuss cross-border compliance strategies and the steps non-EU manufacturers can take to adapt their processes, supply chains, and governance frameworks. We also explore how the Act interacts with other regulatory frameworks worldwide, from the United States to China, highlighting both divergences and areas for possible convergence.

By the end of this chapter, you should understand not only why non-EU manufacturers need to worry about EU AI law but also how they can proactively meet its

© The Author(s), under exclusive license to Springer Nature Switzerland AG 2025

M. Kauffman, *Navigating the EU AI Act: A Practical Guide for Global Manufacturing and Engineering Leaders*, https://doi.org/10.1007/978-3-031-99613-9_3

demands. This knowledge is crucial for any enterprise keen on retaining access to the lucrative European market and fostering a reputation for responsible AI innovation on the global stage.

3.2 The Extraterritorial Scope of the EU AI Act

A hallmark of modern EU regulations, starting perhaps most prominently with the General Data Protection Regulation (GDPR), is their reach beyond the borders of EU Member States. The EU AI Act follows this blueprint. Article 2(1) stipulates that the Act applies to:

1. **Providers** offering AI systems in the EU, whether they are established within the EU or in a third country.
2. **Users** of AI systems within the EU.
3. **Providers and users** located in a third country, if the output produced by their AI system is used within the EU.

For example, a Canadian firm producing AI-enabled robotics distributed to Germany must ensure its products comply with the Act because they are ultimately utilised in an EU context.

This extraterritorial application drives non-EU manufacturers to integrate EU standards into global product designs, reducing complexity and mitigating legal conflicts, the phenomenon known as the "Brussels Effect" [1].

Who Is Affected?

From a manufacturing standpoint, any company producing AI-enhanced machinery, robotics, or software that is sold or utilised in EU Member States must comply. Similarly, if the organisation sells its products to a third party who then imports those products into the EU, the Act's obligations also filter through the supply chain.

Example: A Canadian firm produces AI-enabled robotics for industrial warehouses. Although the robots are assembled in Canada, they are sold to distributors worldwide, including in Germany. Because these robots are eventually deployed in an EU warehouse environment, the Canadian firm's AI components must meet the EU AI Act's requirements.

Implications for Corporate Strategy

The extraterritorial application prompts many non-EU manufacturers to revise their internal AI governance models. Compliance can no longer be an afterthought or an "EU-only" affair. Instead, integrating the EU AI Act's principles into global product design ensures efficiency, consistency, and reduced liability, rather than juggling multiple versions of the same product for different markets.

Moreover, non-EU businesses may find themselves compelled to adopt EU standards universally, a phenomenon termed the "Brussels Effect." By standardising on the EU's benchmark, global manufacturers can reduce complexity and potential legal conflicts across multiple jurisdictions.

3.3 Compliance Pressures Beyond Regulatory Enforcement

A common question is how the EU AI Act can practically enforce compliance against an enterprise situated thousands of miles away, outside EU boundaries. While direct enforcement might pose certain challenges, several indirect but potent forces ensure compliance.

The Role of Market Access

The EU's single market is one of the largest and most lucrative in the world. Many global manufacturers cannot afford to lose access. If the EU AI Act bars non-compliant products, or if fines and injunctions threaten a manufacturer's dealings with EU partners, the potential financial damage is considerable [2]. Consequently, fear of losing a significant market share incentivises alignment with EU rules.

Example: A Brazilian manufacturer of AI-powered diagnostic equipment may have a small but growing client base in Spain. Non-compliance with the EU AI Act could lead to product bans, forfeiture of potential sales, and reputational harm, thus undermining the company's European ambitions.

Supply Chain Requirements

Even if a manufacturer has no direct contracts in EU states, it might be part of a broader supply chain that ends up selling AI-integrated products in Europe. In such cases, downstream compliance obligations transfer upstream. Distributors, importers, or EU-based firms may insist that suppliers certify their AI modules meet EU standards. Failure to provide documentation such as risk assessments, data governance proof, or conformity certificates may result in lost business opportunities [3].

Reputational Considerations

Increasingly, stakeholders and customers worldwide are conscious of ethical and safety standards. Aligning with the EU AI Act, given its focus on transparency and risk mitigation, can enhance a manufacturer's global reputation. Conversely, being publicly found non-compliant can tarnish a brand's credibility, even in regions without stringent AI laws [4].

3.4 Navigating Cross-Border Compliance Challenges

Understanding the rationale behind the EU AI Act is one thing; translating that into day-to-day operations across different continents is another. Non-EU manufacturers often grapple with logistical, legal, and cultural hurdles.

Data Transfers and Governance

One frequent issue is where and how data is stored, processed, or analysed. The EU AI Act dovetails with broader EU data privacy and security norms, including the GDPR. This interplay becomes particularly relevant if an AI system requires data from EU-based users or sensors. Manufacturers must ensure that:

1. Data is collected lawfully.
2. Adequate measures are in place to avoid bias, as mandated by Article 10 (data governance).
3. Personal data, if any, is protected in line with GDPR or equivalent frameworks.

Cross-Border Implications: A factory in South Korea capturing real-time performance data from machines operating in Poland must ensure that not only is the machine's AI module compliant under the EU AI Act but that data handling meets relevant EU data protection standards.

Conformity Assessments from Abroad

High-risk AI systems may require formal conformity assessments (Article 43) to demonstrate compliance before being placed on the EU market. For non-EU manufacturers, this typically involves working with a notified body recognised by EU authorities. The complexity here lies in:

Logistics: Arranging for prototypes or documentation to be reviewed, potentially involving the shipping of sample products or remote auditing.

Documentation Gaps: Non-EU firms may lack the precise format or level of detail demanded by EU authorities.

To streamline the process, many businesses designate an "Authorised Representative" within the EU who liaises with notified bodies and regulators, ensuring local presence and quicker response times [5].

Language and Cultural Barriers

EU regulations often require user manuals, risk assessments, or disclaimers in one or more official EU languages. A manufacturer in Thailand exporting AI-integrated machining tools to Germany and France must provide accurate translations and ensure that the instructions meet local language and cultural expectations [6]. Miscommunication can lead to user misinterpretation of safety procedures, which in turn could violate the Act's transparency or oversight clauses.

Legal Overlaps and Conflicts

Some countries outside the EU may have their own AI-related legislation, or sectoral laws that conflict with certain EU provisions. In the United States, for example, no single overarching federal AI law exists, but numerous agencies provide guidance, creating a patchwork of norms. A non-EU manufacturer may need to reconcile:

- EU demands for human oversight with local norms that emphasise a "light-touch" approach to AI.
- EU emphasis on ethical design with local or regional regulations that lack explicit ethical guidelines.

The challenge is more acute for multinational firms operating in multiple jurisdictions. They might need to create multi-tiered compliance structures or, conversely, adopt the strictest common denominator, often the EU's standard.

3.5 Comparing the EU AI Act with Other Regulatory Frameworks

To better appreciate the global implications, it helps to contrast the EU AI Act with other prominent regulatory approaches around the world.

The United States

Regulatory Landscape: The US primarily relies on sector-specific guidelines, e.g. the Food and Drug Administration (FDA) for medical devices or the National Highway Traffic Safety Administration (NHTSA) for autonomous vehicles.

Federal Initiatives: There have been discussions around an AI Bill of Rights and voluntary frameworks led by the National Institute of Standards and Technology (NIST). However, no single federal law comparable to the EU AI Act exists.

Key Differences: US regulators tend to focus on innovation and competitiveness, with an emphasis on self-regulation. The EU AI Act, by contrast, is more prescriptive, mandating audits and risk mitigation steps [1].

For a non-EU manufacturer aiming to sell AI-powered medical robots both in the United States and Europe, reconciling the US's less centralised approach with the EU's structured, risk-based framework can be challenging. Nevertheless, meeting EU standards often positions the product well for acceptance in many US contexts, though additional clarifications might still be required for FDA approval.

China

Centralised Oversight: China's government actively shapes AI development, emphasising national security, social stability, and economic competitiveness.

Recent Regulations: Provisions around algorithmic transparency, data security, and content management have been introduced, often focusing on social media and data privacy.

Key Differences: While the EU approach centres on fundamental rights and risk mitigation, China's AI governance is shaped heavily by public security objectives. For example, social credit systems are widely used there, a practice the EU AI Act explicitly prohibits.

A non-EU manufacturer dealing with China faces quite a different environment from that of the EU. In effect, it must navigate two distinct sets of rules, leading some global enterprises to adopt separate product strategies per region.

The United Kingdom

Post-Brexit Divergence: The UK is formulating its own AI regulatory approach, which is generally more "light-touch" and principle-based compared to the EU's prescriptive model.

Sector Regulators: The UK government envisages that existing regulators—such as those in health, finance, or transport—will provide AI guidance within their domains, rather than introducing a single AI law.

Key Differences: While many UK rules may remain compatible with or influenced by the EU approach, there is potential for divergence over time.

UK-based manufacturers selling to the EU must still comply with EU rules for their exports. Conversely, domestic UK regulations might differ, creating dual compliance burdens or, in some cases, potential advantage if the UK offers less stringent compliance obligations.

3.6 Strategies for Non-EU Manufacturers

Despite these complexities, non-EU manufacturers can implement several proven strategies to align themselves with the EU AI Act.

Conduct a Comprehensive AI Inventory

Start by identifying which AI systems within your product lines interact with EU-based users or markets. For each system, document its function, data sources, and potential risks (as per the risk classification framework discussed in Chap. 2). This step underpins all subsequent compliance decisions.

Appoint EU-Focused Compliance Leads

It often helps to have dedicated staff members versed in EU regulations. Large multinationals may form cross-functional teams: combining legal, engineering, and operational expertise. Even smaller organisations can benefit from designating a compliance officer responsible for understanding the EU AI Act and coordinating with third-party specialists or notified bodies.

Embed Compliance Early in Product Development

Adopt a compliance-by-design mentality. Rather than retrofitting AI systems at the final stage, incorporate the Act's requirements around data governance, transparency, and risk mitigation from day one. Doing so reduces redesign costs and shortens time-to-market.

Use Local Partnerships

Forming partnerships with EU-based distributors or technology partners can be beneficial. Such collaborators may already have experience with CE marking and EU conformity processes, making it easier to secure market entry. However, ensure that roles and responsibilities (e.g. who handles the conformity assessment) are explicitly defined in contracts.

Seek Certification and Third-Party Audits

For high-risk systems, a successful external audit by a recognised notified body reassures EU regulators and customers that you meet the Act's standards. Proactively seeking such certification can also serve as a market differentiator, highlighting your organisation's dedication to safety and ethics.

Maintain Post-Market Vigilance

The AI Act's obligations do not end at the point of sale. Implement robust post-market monitoring structures. Collect performance data, user feedback, and incident reports systematically. This not only keeps you prepared for random inspections or audits but also allows you to refine and improve AI systems over their operational life.

3.7 Opportunities and Competitive Advantages

Although it is tempting to view the EU AI Act solely as a regulatory challenge, it can also confer significant strategic benefits upon proactive non-EU manufacturers.

First-Mover Advantage: Early adopters of the Act's high standards can secure a reputation for responsibility and quality. EU clients, especially those in sectors like automotive or healthcare, increasingly seek partners who can demonstrate robust compliance.

Globally Consistent Standards: Designing AI systems to meet stringent EU obligations often results in better overall quality, safety, and reliability. This can make products more attractive not just in Europe but in other regions where consumer expectations are rapidly evolving.

Reduced Liability Risks: Aligning with best-practice guidelines (e.g. data governance, transparency, oversight) helps mitigate the risk of lawsuits or public outcry if AI systems behave unexpectedly. This fosters resilience and longevity for your brand.

Enhanced Investor Confidence: In a world attuned to corporate social responsibility and ethical considerations, demonstrating compliance with tough AI laws can attract socially conscious investors. Your AI solutions are less likely to face ethical controversies, thereby reducing reputational risk.

3.8 Case in Point: A Hypothetical Non-EU Manufacturer's Journey

To bring these considerations to life, imagine a fictitious American company, **TechTronics Inc.**, that specialises in automated production line robotics. TechTronics has traditionally sold its products in the Americas and Asia, but it is now eyeing expansion into the EU market.

1. **Discovery**: Upon investigating potential distributors in Germany, TechTronics learns of the EU AI Act and realises its robotics' AI modules likely qualify as high risk under the EU AI Act due to the potential for workplace accidents if the AI malfunctions.
2. **Gap Analysis**: The compliance officer at TechTronics conducts an AI inventory and discovers the existing documentation is incomplete and does not address EU requirements around data governance or human oversight. This prompts internal discussions about how to embed safety features and thorough monitoring capabilities.
3. **Collaboration**: TechTronics partners with a notified body in the Netherlands. The organisation reviews TechTronics' design, testing protocols, and post-market monitoring plans. The process unveils the need for a more transparent user interface, enabling operators to intervene or override AI-based decisions in real time.
4. **Implementation**: Engineering teams add robust logging systems, user notification dashboards, and multi-tiered fail-safes. Legal teams revise user manuals and ensure data protection standards align with GDPR. After repeated iterative testing, TechTronics achieves conformity for its robotic arms, gaining CE marking status and documentation completed.
5. **Outcome**: TechTronics successfully launches in the EU with products that meet or exceed local safety benchmarks. Soon after, they market this compliance worldwide, distinguishing themselves from less-prepared competitors. Sales

grow not just in Europe but in Asia and Latin America, where industrial clients see TechTronics' EU-grade certification as a mark of higher reliability.

While hypothetical, this scenario is based on interviews with non-EU manufacturers as part of an empirical study carried out by the author and underscores how the EU AI Act pushes non-EU manufacturers to refine their products, streamline documentation, and enhance safety, ultimately producing benefits that transcend European markets.

3.9 Future Outlook: Towards Global Harmonisation?

With AI governance rising as a hot topic worldwide, there is a chance that the EU AI Act's risk-based approach could form part of a broader movement towards global regulatory convergence. Various international standard-setting bodies like the International Organization for Standardization (ISO) are already working on AI-related standards. Meanwhile, the G20, OECD, and other multinational forums consistently discuss AI regulation principles. Although each jurisdiction sets its own laws, a consensus may eventually emerge around core tenets of safe, transparent, and fair AI.

For non-EU manufacturers, early compliance with the EU AI Act can put them on the cutting edge of this potential harmonisation, freeing them to devote more energy to product improvement rather than regulatory patchwork. Nonetheless, businesses should keep tabs on developments in other key markets, particularly the United States, China, and the UK, to ensure that changes there do not catch them off guard.

3.10 Conclusion

The EU AI Act's extraterritorial scope and potential for global influence have profound implications for manufacturers operating outside Europe. Far from being a matter for EU-based businesses alone, the legislation compels non-EU organisations to examine how their AI-driven products interface with European customers, supply chains, or data. But while compliance may appear daunting, it is best viewed through a strategic lens:

Market Access: Maintaining and expanding EU market share often hinges on meeting EU regulatory standards.

Supply Chain Alignment: Non-EU suppliers might find themselves contractually bound to comply, even if indirect sales routes funnel their products into Europe.

Reputation Building: Aligning with robust EU rules can strengthen a brand's global standing, signalling a commitment to responsible innovation.

Product Enhancement: Integrating thorough data governance, risk assessments, and transparency from the start can yield better, safer, and more reliable AI solutions.

In an era of increasingly globalised AI development, the EU AI Act embodies a major step towards harmonised governance. Non-EU manufacturers need to grasp its demands, adapt their processes, and embrace the momentum for ethical and trustworthy AI. By doing so, they protect themselves from enforcement risks and tap into new commercial and reputational opportunities, cementing their place in a rapidly evolving marketplace.

Key Takeaways

I. Even if you are based outside the EU, the Act applies if your AI products or services touch the EU market or its citizens.
II. Market access, supply chain requirements, and reputational concerns strongly incentivise alignment with EU standards.
III. Legal complexities arise in data governance, conformity assessments, and language requirements, but partnerships with EU-based notified bodies can mitigate these hurdles.
IV. The EU's approach is more prescriptive than many others, yet meeting EU standards can give you an edge in multiple regions.
V. Appoint dedicated EU compliance teams, embed risk analysis early in product development, and foster close collaboration with EU partners.
VI. As global AI regulations proliferate, the EU AI Act's provisions may become a blueprint for future laws, underscoring the long-term value of proactive compliance.

With a deeper grasp of how the EU AI Act affects non-EU entities, we turn next to the heart of industry practice, Chap. 4: AI in Manufacturing and Engineering. There, we will examine how AI is shaping modern factories and engineering processes, and how the Act's requirements fit into that technological evolution.

References

1. White and Case LLP (2025) AI watch: global regulatory tracker—European Union. https://www.whitecase.com/insight-our-thinking/ai-watch-global-regulatory-tracker-european-union
2. Holistic AI (2024) Implications of the EU AI act for non-EU leadership teams. https://www.holisticai.com/blog/implications-of-the-eu-ai-act-for-non-eu-leadership-teams
3. KPMG (2024) What non-EU companies need to know about the recently passed EU AI act. https://kpmg.com/xx/en/our-insights/transformation/what-non-eu-companies-need-to-know-about-the-recently-passed-eu-ai-act.html
4. Orrick (2024) The EU AI act: who and what is covered?. https://www.orrick.com/en/Insights/2024/09/The-EU-AI-Act-Who-and-What-Is-Covered
5. Baker Donelson (2024) EU AI act tightens grip on high-risk AI systems. https://www.bakerdonelson.com/eu-ai-act-tightens-grip-on-high-risk-ai-systems-five-critical-questions-for-us-companies
6. CMS Law (2024) Guide to the EU AI act for businesses outside the EU. https://cms.law/en/col/publication/guide-to-the-eu-ai-act-for-businesses-outside-the-eu

Chapter 4
AI in Manufacturing and Engineering

Abstract Focusing on shop-floor realities, this chapter surveys today's most prevalent AI use cases such as predictive maintenance, computer vision quality control, collaborative robotics, digital twins, generative design, and supply chain optimisation. Each example is mapped to the Act's risk tiers, revealing when additional safeguards (data governance, human oversight) become mandatory. The discussion extends to organisational change, workforce impacts, and environmental considerations, showing how AI is reshaping engineering lifecycles and operational culture. Actionable checklists help readers link technical deployments to regulatory obligations from the outset.

4.1 Introduction

The world of manufacturing and engineering is undergoing a profound transformation, largely driven by developments in AI. What began as a handful of isolated applications such as basic automation or rudimentary predictive tools has evolved into comprehensive AI-driven solutions reshaping entire production lines, supply chains, and product lifecycles. From robotic arms making split-second decisions on the factory floor to advanced analytics optimising energy usage, AI has become central to modern industrial strategy.

This chapter explores how AI is integrated within the manufacturing and engineering sectors, highlighting the practical processes it enhances, and examining specific use cases where AI adds tangible value. We then connect these real-world innovations back to the EU AI Act's regulatory framework, paying close attention to how risk classifications (see Chap. 2) and global compliance concerns (see Chap. 3) shape the responsibilities of manufacturers. By understanding both the potential of AI-driven manufacturing and the attendant obligations, businesses can capitalise on technology's benefits while remaining aligned with the evolving regulatory landscape.

© The Author(s), under exclusive license to Springer Nature Switzerland AG 2025

M. Kauffman, *Navigating the EU AI Act: A Practical Guide for Global Manufacturing and Engineering Leaders*, https://doi.org/10.1007/978-3-031-99613-9_4

4.2 The Ascendancy of AI in Industrial Settings

Catalysts for AI Adoption

For decades, manufacturers have sought ways to reduce costs, improve quality, and minimise waste. Historically, these objectives were pursued through lean manufacturing methods, Six Sigma initiatives, and rigid automation processes. However, a confluence of factors, particularly the explosion of available data, the affordability of powerful computing resources, and breakthroughs in machine learning algorithms, has empowered a new generation of AI technologies that can tackle challenges previously deemed too complex [1].

Data Ubiquity

Manufacturing environments generate torrents of data, from machine sensors logging vibrations and temperatures to enterprise software tracking inventory and order fulfilment. AI systems thrive on such data, unveiling hidden patterns and anomalies.

Computational Advancements

High-performance computing, edge computing devices, and cloud-based architectures have slashed the cost of processing vast datasets. This accessibility allows even mid-sized firms to experiment with AI solutions once reserved for multinational corporations.

Algorithmic Maturation

Modern machine learning techniques, including deep learning and reinforcement learning, excel at tasks like image recognition, anomaly detection, and real-time decision-making. Manufacturing contexts benefit directly from these capabilities.

Competitive Pressures

Globally, manufacturers face fierce competition, compelling them to streamline operations, speed up innovation, and rapidly adapt to shifting consumer demands [2]. AI-enabled systems can enhance flexibility and responsiveness in complex supply chains.

The Evolving Role of Robotics and Automation

When many people imagine AI in manufacturing, they picture robots on assembly lines. Historically, these robots were scripted to perform repetitive tasks: welding car frames, packaging goods, or stacking pallets. AI changes the paradigm. Instead of programming each action meticulously, robotics can now leverage computer vision and machine learning to adapt dynamically, identifying parts regardless of orientation, adjusting movement if a human enters its workspace, or learning from real-time feedback to reduce errors. This evolution has spurred the concept of collaborative robots (cobots), which work alongside human operators, supplementing their capabilities with speed, precision, and strength.

Yet as noted in Chap. 2, any AI-empowered robot carrying out tasks that could harm human operators or produce dangerous defects may qualify as high risk under the EU AI Act. This classification obliges manufacturers to implement human oversight mechanisms, robust cybersecurity defences, and thorough documentation. Thus, while AI-driven robotics can significantly boost productivity, they also bring heightened accountability.

4.3 Real-World AI Applications in Manufacturing

AI's impact is not confined to robotic arms alone. It permeates nearly every level of manufacturing and engineering, from design and prototyping to supply chain logistics and after-sales services. Below, we highlight several core AI applications, showing how each aligns with or poses challenges under the EU AI Act.

Predictive Maintenance

Predictive maintenance exemplifies AI's potential to reduce downtime and extend equipment lifespan. Rather than relying on routine check-ups or waiting until a machine fails, AI systems analyse sensor data such as temperature, pressure, vibration, acoustic signals, to predict when a part is likely to wear out or malfunction. This enables maintenance teams to fix issues pre-emptively, minimising production interruptions [3, 4].

Key Technologies: Machine learning algorithms (often supervised or semi-supervised) using historical sensor data to identify patterns preceding failure.

Value Proposition: Reduced downtime, lower maintenance costs, fewer catastrophic breakdowns.

Regulatory Considerations

- If an AI system merely recommends maintenance schedules, it may be considered limited risk under the EU AI Act.
- If it can autonomously shut down critical equipment upon detecting anomalies, the potential safety implications might push it into the high-risk category, requiring rigorous testing and post-market monitoring.

Supply Chain Optimisation

Manufacturing success depends heavily on efficient supply chains. AI-enabled tools can forecast demand, track shipments, adjust inventory levels in real time, and even factor in external variables such as weather patterns or political events that could disrupt supplies.

Key Technologies

Predictive Analytics: Using historical sales data and macroeconomic indicators for demand forecasting.

Routing Algorithms: Real-time route planning to account for traffic, fuel prices, or unexpected border delays.

Value Proposition: Lower inventory costs, quicker response to market shifts, reduced lead times.

Regulatory Considerations

- Typically **limited risk** because of its indirect effect on public safety.
- Transparency obligations (Article 52) might apply if AI-driven insights directly impact contractual or consumer-facing decisions, such as product availability or pricing.

Quality Control and Inspection

One of the most popular AI applications is **computer vision** for automated quality control. Instead of relying solely on human inspectors, AI-driven cameras and algorithms can identify minute defects or irregularities at a speed and consistency unattainable by manual checks [1, 5].

Key Technologies

Convolutional Neural Networks (CNNs) for image recognition.

Classical Machine Vision with rule-based logic for simpler defect detection tasks.

Value Proposition: Higher accuracy in defect identification, reduced scrap rates, consistent product standards, and lower labour costs.

Regulatory Considerations

- If the AI's decisions only flag defects for human review, risk is fairly low.
- For safety–critical components (e.g. aircraft parts or medical implants), mistakes could endanger lives, potentially putting the system in the **high-risk** bracket.

Digital Twins and Simulation

Digital twins are virtual replicas of physical assets or systems that mirror real-time conditions. By applying AI-driven simulations, manufacturers can test new process parameters, anticipate failures, or evaluate potential design improvements without interrupting actual production [1].

Key Technologies

Large-scale data integration from sensors, enterprise resource planning (ERP) systems, etc.

Modelling and Simulation frameworks often enhanced with machine learning to approximate complex, nonlinear behaviours.

Value Proposition: Minimises trial-and-error in physical production, supports rapid prototyping, and fosters continuous improvement.

Regulatory Considerations

- Typically **limited risk** because the AI is used for decision support rather than direct control.
- If a digital twin directly commands an autonomous process (e.g. adjusting the speed of a conveyor belt in real time), it may elevate risk if malfunctions could cause accidents.

Human–Robot Collaboration

More advanced than standard robotics, **collaborative robots** (cobots) share workspace with human operators [1, 3]. Machine learning algorithms interpret sensor data (including force sensors, vision systems) to detect the proximity or motions of people, enabling safe cooperation.

Value Proposition: Humans handle complex, dexterous, or creative tasks; robots handle repetitive, heavy, or precise ones.

Regulatory Considerations

- High risk is likely if collision or operational failures endanger workers.
- The EU AI Act requires robust human oversight features, emergency stop mechanisms, and thorough risk assessments.

Cutting-edge use-cases increasingly rely on secure, sovereign data spaces such as GAIA-X and the automotive Catena-X network [6, 7]. These federations implement the Data Governance Act principles introduced in Chap. 1 and will soon underpin the Digital Product Passport (DPP) scheme proposed under the EU Sustainable Products Initiative [8].

For manufacturers, the implication is twofold:

Compliance teams must evaluate whether AI models ingesting shared production data inherit joint-controller obligations under both the DPP and the AI Act's provider/deployer split. Contract clauses must balance IP protection with mandatory interoperability.

Early participation in such data spaces can therefore unlock supply chain visibility and accelerate AI-driven optimisation, while simultaneously satisfying traceability duties demanded by Annex IV [9].

4.4 The Intersection of AI Applications and Risk Classification

As detailed in Chap. 2, the EU AI Act enforces a four-tier classification: prohibited, high risk, limited risk, and minimal risk. For manufacturing, the major distinction is often between limited risk and high risk. Minimal-risk systems, such as a basic AI scheduling assistant, pose few compliance hurdles beyond general best practices. Prohibited systems (such as subliminal manipulation or social scoring) rarely arise in typical industrial contexts, although one must still be vigilant about potential misuse (e.g. employee tracking verging into ethically dubious territory).

When evaluating an AI application's risk level, manufacturers should consider:

1. **Potential Harm**: Could incorrect decisions physically harm people or property?
2. **Context**: Is the AI used for safety–critical applications (e.g. controlling a crane in real time)?
3. **Autonomy**: Does the AI merely suggest actions, or does it autonomously execute them without immediate human intervention?
4. **Impact on Fundamental Rights**: Could the system inadvertently discriminate or breach privacy?

Many advanced AI deployments in manufacturing do not solely fall under limited or minimal categories because they can have a direct, tangible impact on human safety, leading them to be labelled high risk. This classification then triggers obligations such as data governance (Article 10), conformity assessments (Article 43), and post-market monitoring (Article 61) [9].

4.5 Transforming the Engineering Lifecycle

AI's influence goes beyond the production phase. Increasingly, it shapes the entire engineering lifecycle, from conceptual design to maintenance and recycling. This expanded role has direct bearing on how manufacturers comply with the EU AI Act.

AI-Aided Design and Prototyping

Computer-aided design (CAD) systems have long assisted engineers in blueprinting products, but AI now augments that process through generative design. Such AI might propose structural shapes optimised for minimal weight or maximum strength. While these design suggestions typically do not pose direct physical risks themselves, errors could produce flawed prototypes if engineers rely too heavily on the AI's results without human review.

In most cases, these remain limited-risk applications though if they feed directly into automated manufacturing lines, oversight mechanisms become crucial. A flawed generative design that is automatically approved for production could compromise structural integrity, especially in industries like aerospace or automotive engineering.

Digital Thread and Lifecycle Management

Modern product lifecycle management (PLM) systems create a "digital thread," a continuous record of each component from concept to disposal. AI can parse these records to identify inefficiencies, predict supply chain bottlenecks, or anticipate product failures in the field.

Risk Considerations: Minimal if the AI provides decision support. However, if the system automatically triggers changes to production processes or product designs, the stakes rise.

Ethical Considerations

Data privacy and intellectual property concerns abound, especially if the PLM integrates partner or customer data. Managing these aspects ties back to the Act's data governance and transparency requirements, particularly in cross-border collaborations (see Chap. 3).

4.6 Impact on Workers and Organisational Structures

Beyond the technical dimension, AI in manufacturing reshapes how human workers interact with machinery and data. This has direct implications for compliance under the EU AI Act, as well as for broader legal and ethical considerations.

Skills and Training

AI-driven equipment frequently demands a workforce with new skill sets, such as:

Data Analysis: Operators who can interpret machine learning outputs and performance dashboards.

Maintenance Expertise: Technicians skilled in diagnosing sensor or algorithmic errors, not just mechanical faults.

Digital Tool Proficiency: Engineers comfortable with augmented reality, digital twins, or advanced simulation platforms.

If employees cannot effectively oversee or override AI systems, manufacturers may inadvertently breach the Act's human oversight (Article 14) obligations, risking classification of systems as high risk without sufficient safety nets.

Role of Human Oversight

While AI can automate decisions, human-centric design remains pivotal. The EU AI Act demands that employees using high-risk AI systems have the tools and knowledge to detect malfunctions or anomalies and the authority to intervene. This might involve:

- Emergency shutdown features for robots.
- Dashboards that visualise real-time sensor data and AI inferences in an understandable format.

- Protocols for employees to raise concerns or report anomalies without fear of reprisal.

Such measures not only comply with legal norms but also foster trust in AI-driven processes. Workers who trust the systems they operate are more likely to collaborate proactively with these tools, offering feedback that further refines AI performance.

Socio-Economic Shifts

A deeper consideration is the possible displacement of jobs through AI-driven automation, or the redefinition of roles in ways that demand continuous upskilling. While the EU AI Act does not directly legislate on labour transitions, one must bear in mind that the Act's ethical and transparency elements reflect societal expectations that workers be treated fairly and informed of AI's functions. Employers implementing advanced AI without proper planning for workforce impacts risk reputational damage and strained labour relations.

4.7 Ethical and Environmental Dimensions

The EU AI Act is built on a foundation of maintaining trust, safeguarding rights, and reinforcing ethical development. As AI becomes more powerful, it amplifies both opportunities for efficiency and risks of misuse or oversight, including ecological impact.

Energy Efficiency and Sustainability

Factory-scale machine learning models can be computationally intensive. High energy usage not only drives operational costs but also undermines sustainability targets, a growing concern for governments and consumers alike.

Eco-optimised AI: Manufacturers increasingly adopt more energy-efficient hardware and algorithms, aligning with corporate sustainability goals.

Lifecycle Footprint: AI can reduce waste by improving product quality and predictive maintenance, thus indirectly promoting greener manufacturing.

Adherence to the EU AI Act (e.g. robust data governance, accurate sensors, real-time monitoring) can reinforce sustainability by ensuring efficient resource usage and minimising machine failures or rework.

Fairness and Non-discrimination

While consumer-facing industries face scrutiny over AI bias in services like lending or recruitment, manufacturers are not immune. Bias can arise in algorithms that evaluate worker performance or allocate tasks, potentially disadvantaging certain groups. Under the Act, any AI that significantly impacts individual rights could be high risk or at least subject to thorough audits. Though less common in a purely manufacturing environment, certain HR or recruitment tools for factory staff incorporate AI and thus require attention to data representativeness and fairness.

4.8 Navigating Compliance in Practice

For manufacturers leveraging AI in engineering contexts, bridging innovation and compliance can be challenging. Below are some practical steps to ensure alignment with the EU AI Act whilst harnessing AI's transformative power:

- **Identify Critical Use Cases**

Inventory all AI systems, prioritising those with direct human safety implications (robotics, safety monitoring) or significant impacts on product reliability (quality inspection for high-stakes components).

- **Categorise by Risk**

Apply the EU AI Act's risk classification criteria (see Chap. 2) to decide whether a system is high, limited, or minimal risk. This classification influences subsequent obligations for data governance, transparency, and oversight.

- **Develop Clear Human Oversight Protocols**

For each AI-driven process, define how operators can monitor decisions, override automated actions, and escalate issues. Ensure they are trained appropriately and that user interfaces are intuitive.

- **Integrate Testing and Documentation**

Maintain detailed records of AI model training, validation sets, and real-world testing. For high-risk AI, expect possible audits or third-party conformity assessments (Article 43).

- **Monitor Performance Continuously**

Post-market monitoring (Article 61) requires collecting data on how the system performs in real conditions and addressing emergent safety or bias issues. Closed-loop feedback from factory floors ensures ongoing compliance and fosters iterative improvements.

- **Embed Ethical Guidelines**

In line with the Act's broader ethical objectives, consider fairness, data privacy, and worker impacts. This cultural shift can reduce risk of non-compliance and strengthen your brand's reputation.

When undertaken systematically, these measures not only conform to regulatory demands but also enrich the overall manufacturing process. By championing safe, transparent AI applications, organisations are likely to discover new efficiencies, reduce downtime, and future-proof their offerings against evolving market conditions.

4.9 Case Vignettes: Illustrative Examples

To bring these ideas into clearer focus, consider two brief vignettes from the manufacturing world:

Welding Robot Integration

A medium-sized European automotive supplier invests in advanced welding robots powered by machine learning. Sensors detect the thickness and material composition of metal pieces, adjusting welding parameters on the fly. Because an error can compromise structural integrity and endanger workers nearby, these robots fall under high risk. The company must:

- Conduct a risk assessment enumerating possible failure points (sensor misreadings, algorithmic drift).
- Establish human oversight protocols: an operator can halt or override the robot if weld quality dips below certain thresholds.
- Document performance thoroughly, especially as the algorithm refines its parameters over time.

Warehouse Inventory System

Meanwhile, a British retailer adopts an AI-based system that predicts stock levels across multiple warehouses based in Europe, automatically reordering supplies from manufacturers. If the AI's forecasts err, the result is limited to overstocking or shortfalls an economic inconvenience but not a safety concern. Under the EU AI Act, the system would thus be limited risk. The primary requirement is transparency about AI usage to relevant personnel, ensuring they know how decisions are reached and that human operators can intervene if needed.

4.10 Conclusion

The march of AI through manufacturing and engineering stands as one of the defining trends in today's industrial world, offering potential for immense gains in efficiency, product quality, and competitiveness. From predictive maintenance to robotic collaboration, the technology can streamline operations and open doors to groundbreaking design possibilities. Yet this progress also brings new challenges, particularly regarding safety, oversight, and ethical considerations, areas where the EU AI Act plays a pivotal regulatory role.

Key themes have emerged:

- **Comprehensive AI Deployment**: AI extends from the shop floor to the boardroom, implicating every phase of the product lifecycle.
- **Risk Differentiation**: Classifying AI systems according to their potential harm is essential, as the EU AI Act mandates different obligations for high-risk versus limited-risk systems.
- **Human Oversight**: In manufacturing contexts, ensuring operators can intervene remains crucial to compliance and fosters worker trust in AI technologies.
- **Documentation and Monitoring**: Thorough record-keeping, robust testing, and continuous tracking of AI performance are indispensable for meeting legislative requirements and ensuring reliability.

Ethical and Environmental Dimensions

- Manufacturers cannot overlook broader societal impacts of AI usage, from labour considerations to sustainability.

As AI continues to evolve, so too will the legislative frameworks meant to guide its development. In adapting to the EU AI Act's stipulations, manufacturers are not only safeguarding themselves against regulatory risk but also positioning their businesses at the forefront of responsible innovation. In the next chapter, Chap. 5: Compliance Requirements, we move from an application-centric discussion to a focus on the specific legal obligations that manufacturers must meet under the Act covering critical areas like conformity assessments, data governance, and post-market duties in far greater detail.

Key Takeaways

I. AI is reshaping virtually every facet of manufacturing, from automated welding and assembly to predictive maintenance and supply chain forecasting.
II. The EU AI Act's classification means not all manufacturing AI is subject to the same obligations. Systems influencing human safety or critical outcomes are likely high risk and thus more heavily regulated.
III. AI extends from initial design (e.g. generative design algorithms) to maintenance and recycling, creating a continuous chain of responsibilities.
IV. Manufacturers must train their workforce to supervise AI systems effectively, ensuring compliance with requirements for human oversight and transparency.
V. Aligning with best practices on fairness, data management, and environmental sustainability benefits both compliance and corporate reputation.
VI. Through rigorous testing, documentation, and cross-functional collaboration, manufacturers can fully leverage AI's benefits while adhering to the Act's provisions.

Moving forward, a more granular look at compliance requirements will enable manufacturers to transform these conceptual insights into operational action plans.

References

1. Siemens (2024) Artificial intelligence in industry. https://www.siemens.com/global/en/products/automation/topic-areas/artificial-intelligence-in-industry.html
2. World Economic Forum (2024) How we can unleash the power of AI in manufacturing. https://www.weforum.org/stories/2024/01/how-we-can-unleash-the-power-of-ai-in-manufacturing/
3. IBM (2024) How AI is being used in manufacturing. https://www.ibm.com/think/topics/ai-in-manufacturing
4. Nordcloud (2024) 10 examples of AI in manufacturing to inspire your smart factory. https://nordcloud.com/blog/10-examples-of-ai-in-manufacturing-to-inspire-your-smart-factory/
5. Built In (2024) 13 AI in manufacturing examples to know. https://builtin.com/artificial-intelligence/ai-manufacturing-robots-automation
6. GAIA-X Association (2025) About GAIA-X. https://gaia-x.eu/about/
7. Catena-X Consortium (2025) Catena-X. https://catena-x.net
8. European Commission (2025) EU digital product passport. https://data.europa.eu/en/news-events/news/eus-digital-product-passport-advancing-transparency-and-sustainability
9. European Union (2024) EU AI act, annex III high risk list. https://artificialintelligenceact.eu/annex/3/

Chapter 5
Compliance Requirements

Abstract This chapter translates the Act's legal text into an engineer-friendly blueprint. It details risk management duties (Art. 9), data governance and bias mitigation rules (Art. 10), transparency obligations (Arts. 13 and 52), human oversight mandates (Art. 14), technical robustness standards (Art. 15), conformity assessment pathways and CE marking (Arts. 11, 17, 43, 47), and post-market monitoring (Art. 61). Summaries, flowcharts, and compliance checklists enable readers to audit existing systems or design new ones that meet the regulation's letter and spirit. Penalties and enforcement mechanics (Arts. 71–73 and 99) are also outlined to emphasise the cost of non-conformity.

5.1 Introduction

Building on the foundation laid in previous chapters, we now move into the core compliance obligations mandated by the EU Artificial Intelligence (AI) Act [1, 2]. Whether an AI system is categorised as high risk, limited risk, or minimal risk (see Chap. 2), certain legal requirements must be met to ensure safe and transparent operation. These requirements are not mere bureaucratic hurdles; they reflect an overarching ethos of accountability, data governance, and human-centric oversight, all central to the EU's vision of responsible AI innovation.

This chapter offers a structured breakdown of the Act's key compliance measures, helping manufacturers and engineers understand precisely what is expected of them. We begin by surveying the horizontal obligations that apply across risk tiers (e.g. risk management systems, data governance, and transparency) before drilling down into requirements that scale with risk. We then examine additional, more stringent responsibilities for high-risk systems, including human oversight mechanisms, conformity assessments, and post-market monitoring. Throughout, we provide examples relevant to manufacturing contexts, emphasising how these obligations can be integrated into standard organisational processes. By the end of the chapter, readers should have a clear grasp of the essential steps needed to align their AI deployments with EU legal standards.

© The Author(s), under exclusive license to Springer Nature Switzerland AG 2025
M. Kauffman, *Navigating the EU AI Act: A Practical Guide for Global Manufacturing and Engineering Leaders*, https://doi.org/10.1007/978-3-031-99613-9_5

5.2 Overview of Risk-Based Compliance Pathways

The EU AI Act adopts a risk-based approach to compliance. In broad terms, the higher the risk an AI system poses to individual rights or public safety, the more rigorous the compliance obligations become. While prohibited (unacceptable) AI systems are outright banned, the remaining tiers—high risk, limited risk, and minimal risk—face escalating requirements. Before diving into each specific obligation, it helps to have a top-level view of the Act's compliance structure:

1. **Minimal-Risk AI**
   - No binding obligations beyond existing horizontal EU law; voluntary codes of conduct are encouraged.
   - Adherence to general principles recommended but not mandated.
2. **Limited-Risk AI (Article 52)**
   - Must comply with transparency obligations (e.g. informing users they are interacting with an AI).
   - May need basic user instructions or disclaimers, especially if AI systems produce synthetic or automated outputs.
3. **High-Risk AI (Articles 6–29)**
   - Must implement a documented risk management system (Article 9) and undergo conformity assessment before market entry (Article 43).
   - Subject to strict rules on data governance, human oversight, documentation, accuracy, robustness, cybersecurity, quality management systems (Article 17) and post-market monitoring (Article 61).
4. **Unacceptable-Risk AI (Prohibited AI Article 5)**
   - Strictly prohibited. Examples include real-time biometric surveillance for law-enforcement or AI that manipulates vulnerable groups.

Because manufacturing often involves safety–critical processes or machinery that can directly affect human well-being, many AI systems in factories and engineering settings fall under the high-risk category. Consequently, a large portion of this chapter focuses on the compliance steps relevant to high-risk AI. Nonetheless, manufacturers should also remain mindful of the limited-risk category, particularly where AI interacts with staff or generates user-facing insights.

Non-EU manufacturers that act as 'providers' are directly responsible for compliance and, where required, must appoint an authorised representative established in the European Union (Articles 16 and 54) [3, 4]. Importers and distributors established in the EU share specific verification and traceability duties.

5.3 Risk Management Data Governance Obligations (Articles 9 and 10)

Data forms the backbone of any AI system. In line with the EU's broader emphasis on data protection and reliability, Articles 9 and 10 of the AI Act outline governance measures specifically for high-risk applications, although the principles are valuable across the board.

High-risk AI providers must establish, implement, document, and maintain a continuous, iterative risk management system covering design, testing, deployment and decommissioning phases.

Quality, Accuracy, and Representativeness

High-risk AI systems must be trained, validated, and tested on datasets that accurately reflect the range of operating conditions and scenarios the system will face. This requirement aims to minimise bias or inaccurate decision-making that could jeopardise safety or lead to discriminatory outcomes.

- **Example**: A robotic welding arm that adapts to varying steel thicknesses must be trained on data from multiple grades of steel, not just a single type. Similarly, if different temperature ranges are typical in factories across Europe, relevant sensor data from each range should be included in the training process.

Traceability of Data Sources

Manufacturers must be able to document the origin of the data used for development (Article 12). This includes noting whether the data was generated in-house, sourced from a vendor, or scraped from public datasets. Traceability is crucial for verifying compliance, especially if a problem emerges post-deployment and regulators need to ascertain whether the data itself was flawed or incomplete.

Data Management Practices

The AI Act encourages (and, for high-risk systems, effectively mandates) robust data lifecycle management:

- **Data Minimisation**: Collect only as much data as is strictly necessary for the AI's intended purpose.
- **Version Control**: Keep track of dataset versions and training processes, since changes in data composition can alter AI performance.
- **Security Protocols**: Enforce data access controls, encrypt sensitive data, and maintain secure storage facilities.

Article 10's requirement for "relevant, representative, free-of-errors" data dovetails with ISO/IEC 5259-1:2024 (analytics/ML data-quality metrics) and ISO 8000-8 (conceptual, syntactic, and pragmatic dimensions of data quality) [5, 6]. Embedding those metrics in data-pipeline gates provides objective evidence for conformity files and aligns with the Big Data fairness guidance.

Given the EU's prior experience with the General Data Protection Regulation (GDPR), manufacturers might already have some data protection structures in place, though the AI Act is not limited solely to personal data. Even sensor or operational data should be subject to validated governance rules to ensure compliance and reliability.

5.4 Transparency Requirements (Articles 13 and 52)

For all but the most trivial (minimal-risk) AI systems, the EU AI Act imposes transparency obligations. While these requirements are especially pronounced for high-risk applications, limited-risk systems also face certain demands.

Disclosure of AI's Nature

When users, be they factory operators, end consumers, or business partners, interact with an AI system, they must be informed that they are dealing with AI rather than a human or a purely mechanical process. This rule preserves user autonomy, ensuring people can make informed decisions about how much they trust or rely on AI outputs.

- **Manufacturing Context**: An AI chatbot assisting technicians in troubleshooting machine faults must clarify that it is AI-driven. Alternatively, if a scheduling system autonomously assigns shift patterns, employees should be made aware that an algorithm is making the recommendations.

Clear Instructions and Labelling

Beyond simply acknowledging AI usage, manufacturers deploying high-risk systems must supply user manuals or instructions detailing the system's capabilities, limitations, and conditions for safe operation.

- **High-Risk Example**: An AI-driven forklift should come with operational guides explaining under what conditions the AI can be overridden, possible error rates, and recommended safety checks.
- **Limited-Risk Example**: A basic AI recommendation engine for reorder levels might only require a short disclaimer in the software interface.

Regulators accept concise artefacts that expose dataset provenance and model limitations. Datasheets for Datasets, Data Cards, and Model Cards give engineering teams lightweight templates that already satisfy most of Annex IV §2–§4 items and can be auto-generated from machine learning platforms [7, 8].

Identifying Synthetic Outputs

If an AI system produces synthetic text, images, or other media that might be confused with genuine human output, the Act demands labelling to prevent deception. While this aspect is more prominent in consumer-facing contexts (e.g. chatbots), it can also matter in manufacturing if generated images or instructions might mislead operators.

5.5 Human Oversight (Article 14)

One of the cornerstones of the EU AI Act is preserving the **human-in-the-loop** principle. Even if an AI system has advanced decision-making capabilities, a human operator must be able to supervise, interpret, and, when necessary, override those decisions. Operators should be trained, and training records retained, to demonstrate competence. Avoiding automation bias is critical; organisational policies should empower employees to question AI outputs. This requirement applies most strongly to high-risk AI.

Defining the Scope of Oversight

The Act outlines that human oversight mechanisms should be designed and integrated such that operators can:

- **Monitor AI Decisions**: View real-time information on what the AI is doing or predicting.
- **Intervene if Needed**: Interrupt or halt the AI's operations, especially in safety–critical scenarios.
- **Interpret Outputs**: Access meaningful explanations of how the AI reached its decision (where feasible).
- **Illustration**: A safety officer in a steel mill might have a dashboard showing how an AI-based system is controlling blast furnace temperatures. If the system goes beyond a safe operating threshold, the officer must be able to step in or shut the system down.

Training of Personnel

Human oversight is only effective if operators understand how the AI system works, what its limitations are, and how to detect anomalies. The Act thus implies that training programmes should be instituted, covering both technical aspects (interface navigation, interpreting AI-driven alerts) and safety protocols (when to override the AI).

Best Practice: Document all training sessions, ensuring traceability, and a clear demonstration of compliance if regulators ever request proof that staff are competent to supervise the system.

Avoiding Automation Bias

Automation bias can occur when human operators overly rely on AI outputs, assuming the system cannot be wrong. Effective oversight includes organisational policies that encourage critical thinking and support employees who question or counter AI recommendations when they observe inconsistencies. Reinforcing a culture of accountability helps maintain alignment with the Act's spirit of human-centric AI.

5.6 Accuracy, Robustness, and Cybersecurity (Article 15)

High-risk AI systems must not only be transparent; they also require reliable, resilient engineering, capable of maintaining performance in real-world conditions. Article 15 addresses three central pillars: accuracy, robustness, and cybersecurity.

Accuracy

To ensure user safety and trust, the Act stipulates that AI systems need consistent, validated accuracy in the tasks they perform. If an AI's error rate rises above a certain threshold, it might jeopardise safety or cause discriminatory outcomes.

- **Example**: A quality control AI classifying defective parts must maintain a high detection rate with low false positives. If the system's accuracy is insufficient, it could either reject too many parts (economic loss) or miss dangerous defects (safety concerns).

Manufacturers are advised to define performance benchmarks early in the design phase, validate them under realistic test conditions, and continue tracking them after deployment.

Robustness

Real-world industrial settings can be unpredictable environments which might vary in temperature, humidity, or lighting; sensor data can degrade; or anomalous materials might enter the production line. High-risk AI solutions should be robust to such variations, ensuring safe operation even under less-than-ideal conditions.

- **Practically**: Incorporate error-handling logic, redundant sensors, or fail-safe procedures so the system does not produce catastrophic outcomes if data inputs go awry.

Cybersecurity

AI systems, especially those connected to the internet or integrated into complex networks, can be targets of hacking or tampering. A compromised AI in a factory could trigger sabotage scenarios, from halting operations to creating defective goods deliberately.

- **Requirements**: The EU AI Act expects secure development practices (e.g. secure coding, intrusion detection, encryption of data in transit and at rest). For high-risk systems, manufacturers must outline how vulnerabilities are assessed and mitigated.
- **Implementation**: Regular penetration tests, incident response plans, and swift patch deployment are essential to maintain compliance.

5.7 Conformity Assessment and CE Marking (Articles 11, 17, 43 and 47)

A conformity assessment is the process by which manufacturers prove that a high-risk AI system meets the Act's technical requirements before it can be placed on the EU market. Depending on the specific application or relevant directives, this assessment might be done through self-assessment or require a notified body.

- Self-assessment may be possible where harmonised standards fully apply; otherwise, a notified body must review the quality management system and technical documentation.
- Technical documentation must follow Annex IV and be kept for 10 years [9, 10].
- Upon positive assessment, the provider issues an EU Declaration of Conformity and affixes the CE mark.
- Non-EU providers must ensure their authorised representative can supply this documentation on request.

Self-Assessment Versus Third-Party Evaluation

Self-Assessment: In some cases, the manufacturer can carry out an internal review, checking all compliance criteria [11]. If the system is not overly complex or the legislation permits, self-assessment may suffice.

Third-Party Assessment: For more critical or intricate high-risk systems, the Act mandates external auditing by a notified body designated by EU authorities. This body examines the technical documentation, performance tests, and governance measures, issuing a certificate if satisfied.

Technical Documentation (Article 11)

Regardless of who conducts the assessment, the manufacturer must supply detailed technical documentation, covering:

- System architecture, algorithms used, and data workflows.
- A thorough risk analysis pinpointing how hazards have been mitigated.
- Records of test results, audits, or simulations used to validate safety.
- Maintenance and update policies to ensure ongoing compliance after the system is deployed.

Tip: Align the structure of your technical documentation with any existing product compliance documentation (e.g. under the Machinery Directive), reducing duplication and confusion.

CE Marking and Registration

Similar to other EU product regulations, once a high-risk AI system achieves compliance, it may bear a **CE marking** (if part of a product requiring CE marking under EU harmonisation legislation). Additionally, certain systems must be registered in an EU database, making them traceable should safety incidents arise. This central registry fosters transparency, allowing authorities and end-users to verify a product's compliance status.

5.8 Post-Market Monitoring (Article 61)

Compliance does not end at the time of deployment or sale. The EU AI Act imposes obligations to continuously monitor the performance of high-risk AI, collecting data on errors, near misses, or adverse events. The rationale is that AI can evolve, degrade, or encounter unforeseen circumstances over time.

Ongoing Data Collection

Manufacturers should establish processes to gather real-world operational data such as sensor logs, operator feedback, and error reports, and compare these operational results to the expected performance baselines. Significant deviations could signal the need for re-training, software updates, or more profound design changes.

Incident Reporting

In line with a broader trend towards proactive risk management, if an AI system malfunctions with potentially dangerous consequences, the manufacturer must swiftly inform the relevant national authority. Delays in reporting can magnify harm and draw heavier scrutiny or penalties.

Model Retraining and Updates

Some AI applications rely on adaptive or continuous learning. If the system automatically retrains on live data, manufacturers must ensure each update meets the same safety and transparency standards. A new model version that inadvertently introduces bias or lowers safety margins could result in compliance breaches. Maintaining version control logs and reapplying crucial validation steps is therefore essential.

5.9 Penalties and Enforcement (Articles 71–73 and 99)

Much like the General Data Protection Regulation (GDPR), the EU AI Act wields substantial enforcement powers, including:

- **Administrative Fines**: For severe violations, fines could go up to millions of euros or a percentage of worldwide turnover, whichever is higher.

  Up to €35 million or 7% of worldwide annual turnover for prohibited practices.
  Up to €15 million or 3% for other obligations.
  Up to €7.5 million or 1% for supplying false or misleading information.

- **Market Removal**: The EU can demand that non-compliant high-risk AI be withdrawn or recalled from the market.
- **Reputational Damage**: Public scrutiny or blacklisting can deter potential partners and damage brand credibility.

Although exact penalty structures can vary by Member State, the overarching message is unambiguous: non-compliance can be costly financially and reputationally.

5.10 Integrating Compliance into Organisational Processes

To navigate these obligations without stifling innovation, successful manufacturers treat EU AI Act compliance as an ongoing organisational function rather than a peripheral add-on. Below are some strategies to get your organisation started:

Continuous Learning and Staff Training

Train employees on new or updated obligations, especially those relevant to their roles (human oversight, incident reporting, data management). Regular refreshers ensure institutional knowledge remains robust, and new hires get up to speed quickly.

Create an AI Inventory and Risk Register

Establish a living document identifying each AI system, its risk classification, and compliance status. Regularly update it when systems evolve or new technologies are introduced.

Establish Cross-Functional Committees

Involve legal teams, data scientists, operational managers, and frontline workers in oversight committees. Such collaboration helps interpret EU AI Act requirements in practical terms and ensures that no aspect, be it data governance, user training, or cybersecurity is overlooked.

Adopt a Lifecycle Approach (Compliance by Design!)

Integrate compliance checks at every stage: concept, design, prototype, testing, deployment, and maintenance. This is often referred to as "compliance by design."

Leverage Industry Standards

Many areas of the EU AI Act align with established international standards (ISO, IEC). Aligning with these standards can streamline conformity assessments, reduce rework, and demonstrate good faith in meeting or exceeding regulatory expectations.

5.11 Conclusion

For manufacturers, the EU AI Act's compliance requirements may appear wide-ranging—encompassing data, transparency, oversight, security, and ongoing monitoring. Yet, these measures ultimately reinforce best practices for safe and trustworthy AI. Meeting the Act's obligations is not merely a box-ticking exercise; it is an investment in the long-term reliability and market viability of AI-driven products.

Key compliance areas highlighted in this chapter include:

- **Data Governance**: Ensuring datasets are accurate, representative, and well-documented (Article 10).
- **Transparency**: Informing users about AI's nature, especially in limited- or high-risk systems, and disclosing synthetic outputs (Article 52).
- **Human Oversight**: Designing AI systems with user-friendly interfaces and override capabilities (Article 14).
- **Accuracy, Robustness, and Cybersecurity**: Maintaining rigorous test procedures and resilience measures (Article 15).
- **Conformity Assessment**: Potentially involving a notified body for high-risk AI, backed by comprehensive technical documentation (Article 43).
- **Post-Market Monitoring**: Ongoing vigilance to detect and correct real-world performance issues (Article 61).

By weaving these requirements seamlessly into engineering processes and quality management structures, manufacturers can ensure they reap the economic and operational gains of AI while remaining within legal bounds. The next phase—Chap. 6: Impact on Innovation—will explore how complying with these obligations can serve not as a brake on technological advancement but as a catalyst for safer, more user-focused AI innovations in manufacturing and engineering.

Key Takeaways

I. Higher-risk AI incurs more stringent compliance measures, whereas limited-risk systems primarily need transparency disclosures.
II. Scrupulous data management underpins robust, bias-free AI, requiring accurate, representative training datasets and clear traceability.
III. Clear labelling of AI nature and user-facing disclosures are mandatory for limited- and high-risk systems.
IV. Operators must have the authority and skill to monitor and intervene in AI decisions, reducing dependence on automated judgements alone.
V. Meticulous records of design, testing, risk mitigation, and version control ensure readiness for audits and user inquiries.
VI. The AI Act enforces an ongoing duty to track performance, handle incidents, and update systems responsibly.
VII. Successful compliance entails weaving these obligations into organisational processes, from R&D to maintenance and beyond.

References

1. European Union (2024) Regulation (EU) 2024/1689, Arts. 8–18, 43, 47, 61 and 99. https://eur-lex.europa.eu/eli/reg/2024/1689/oj/eng
2. European Commission (2021) Proposal for a regulation on AI
3. WilmerHale (2024) Obligations for deployers, providers, importers and distributors of high-risk AI systems in the EU AI Act. https://www.wilmerhale.com/en/insights/blogs/wilmerhale-privacy-and-cybersecurity-law/20240826-obligations-for-deployers-providers-importers-and-distributors-of-high-risk-ai-systems
4. A&O Shearman (2024) Zooming in on AI #10: EU AI act—what are the obligations for 'high-risk AI systems'?. https://www.aoshearman.com/insights/ao-shearman-on-tech/zooming-in-on-ai-10-eu-ai-act-what-are-the-obligations-for-high-risk-ai-systems
5. ISO/IEC 5259-1:2024 (2024) AI—data quality for analytics and ML. https://bsol.bsigroup.com/Bibliographic/BibliographicInfoData/000000000030424807
6. ISO 8000-8:2015 (2015) Data quality—Part 8. https://bsol.bsigroup.com/Bibliographic/BibliographicInfoData/000000000030262729
7. Gebru T et al (2021) Datasheets for datasets. https://doi.org/10.1145/3458723
8. Google PAIR (2024) Data cards playbook. https://sites.research.google/datacardsplaybook/
9. Securiti (2024) Article 11: technical documentation|EU AI act. https://securiti.ai/eu-ai-act/article-11/
10. European Union (2024) Annex IV: technical documentation referred to in Article 11(1)|EU AI act. https://artificialintelligenceact.eu/annex/4/
11. OneTrust (2024) Conformity assessments under the proposed EU AI act: a step-by-step guide. https://www.onetrust.com/resources/conformity-assessments-under-the-proposed-eu-ai-act-a-step-by-step-guide/

Chapter 6
Impact on Innovation

Abstract This chapter addresses the perennial concern that regulation stifles creativity. Using historical analogies (e.g. GDPR) and fresh empirical evidence, we show how the EU AI Act can stimulate innovation through better data hygiene, user-centric design, and safer experimentation. Separate sections explore the distinct opportunities and constraints faced by SMEs versus multinationals, the competitive advantages of "compliance by design," and the Act's influence on research funding and academic curricula. The chapter concludes with strategic recommendations for turning regulatory compliance into an R&D accelerator.

6.1 Introduction

From robotic arms on the factory floor to advanced algorithms guiding supply chains, AI is transforming manufacturing and engineering at an unprecedented pace [1]. Because the EU AI Act applies extraterritorially to providers wherever they are established, global manufacturers must now ask whether this regulation stifles or stimulates innovation [2]. Early fears that new rules would hamper cutting-edge AI development have gradually given way to recognition that well-crafted regulation can, in fact, foster a more robust and trustworthy environment for innovation. In this chapter, we examine how the EU AI Act influences technological progress, product design, market competition, and broader research trajectories in manufacturing and engineering.

We begin by highlighting the delicate balance that policymakers have tried to achieve, ensuring that AI continues to flourish while safeguarding society against unmanageable risks. We then delve into the specific ways that EU AI Act compliance can spur innovative thinking, for example, by motivating stronger data governance, encouraging user-focused design, and opening new avenues of market differentiation. Furthermore, we discuss the perspective of small- and medium-sized enterprises (SMEs), which often worry about bearing the brunt of compliance costs yet may reap unique advantages by adapting early. We conclude with practical insights on

© The Author(s), under exclusive license to Springer Nature Switzerland AG 2025

M. Kauffman, *Navigating the EU AI Act: A Practical Guide for Global Manufacturing and Engineering Leaders*, https://doi.org/10.1007/978-3-031-99613-9_6

how to align regulatory obligations with creative exploration, turning the EU AI Act into a catalyst for responsible innovation.

6.2 The Delicate Balance of Regulation

Laws and regulations fundamentally aim to protect societal values such as safety, privacy, fairness, and economic stability. Historically, certain rules have been seen as impediments to business. For instance, when environmental standards or worker safety measures were first introduced, industries resisted, warning of spiralling costs, and diminished competitiveness. Over time, though, many companies discovered that meeting these new requirements frequently led to efficiency gains, brand enhancement, and market opportunities.

With AI, the stakes are particularly high. Rapid technological progress can bring about significant benefits such as greater efficiency, personalised products, real-time analytics, and more. Yet the same technology can be misused or produce unintended harm, from dangerous robotic malfunctions to discriminatory decision-making algorithms. The EU AI Act aims to ensure the advantages of AI are not overshadowed by hazards that could undermine public trust. By tying obligations to an AI system's risk level, it aspires to encourage rather than suppress beneficial AI [3].

Lessons from the GDPR

An illuminating parallel is the General Data Protection Regulation (GDPR), enacted in 2018. Early predictions suggested that strict data privacy requirements might thwart digital innovation and burden European businesses. In reality, while companies did incur compliance costs, many also saw improved data practices, heightened consumer trust, and, in some cases, reduced data management inefficiencies. Tech giants and small ventures alike discovered that strong data protection could be reframed as a selling point, demonstrating accountability to clients and investors [1]. Over time, GDPR became a global reference for data privacy.

Likewise, the EU AI Act could prove to be a blueprint for responsible AI worldwide, especially because of the "Brussels Effect" (discussed in Chap. 1). By adhering to well-defined safety and transparency rules from the outset, organisations may find themselves better positioned to meet evolving international standards. Indeed, some multinational enterprises already align product designs with EU regulations by default, understanding that early compliance paves the way for easier global expansion.

Addressing Complex Risks

Manufacturers routinely deal with safety rules, from product-specific directives (e.g. Machinery Directive) to environmental impact requirements. AI governance is, in many respects, the next logical step. Instead of retarding development, compliance can force engineers to integrate best practices at inception, like robust data governance, thorough testing, and risk assessment. This sense of "compliance by design" minimises haphazard or unsafe experimentation, leading to more secure, ethically grounded innovation.

6.3 How Compliance Can Foster Innovation

For many businesses, especially in manufacturing and engineering, adjusting to new regulations necessitates rethinking product design, supply chain management, and workforce training. While this reorganisation can present short-term difficulties, it can also unlock fresh perspectives and creative problem-solving. Participation in sectoral European data spaces (for example, Catena-X for automotive or Manufacturing-X) could transform compliance cost into data-as-a-service revenue [4]. By contributing anonymised production data, firms gain preferential access to shared AI models and unlock new offerings such as predictive-quality benchmarks, a clear case where regulation catalyses innovation.

Data-Driven Excellence

As noted in Chap. 5, the EU AI Act demands rigorous data governance (Article 10) for high-risk AI systems. This rule compels organisations to scrutinise their datasets for accuracy, completeness, and bias. In turn, data professionals must develop advanced methods of cleaning, labelling, and validating machine data. While such processes can be costly, they often catalyse the development of high-quality, standardised data repositories. With better data, AI models become more reliable and can be repurposed for other innovative applications in predictive maintenance, yield improvement, or new product lines.

Example: A British aerospace component manufacturer invests in advanced data pipelines to comply with the AI Act's requirements for accurate, representative training data. This same data infrastructure then enables the company to explore machine learning for material innovation developing new alloys or composites by analysing production variables in unprecedented detail. By embedding data excellence at the heart of compliance, the firm finds itself poised for cutting-edge research, not just fulfilling regulatory box-ticking.

User-Centric and Ethical Design

Another crucial facet of compliance is ensuring transparency (Article 52) and human oversight (Article 14). Far from being burdensome, these obligations can prompt manufacturers to adopt user-centric design principles. For instance, if a robotic system demands an intuitive interface so operators can readily understand and override AI decisions, the engineering team might put more effort into user experience (UX) research, simplified dashboards, and real-time alerts. This results in not only a safer system but also one that workers feel more confident adopting, thereby boosting productivity.

Additionally, focusing on fairness and non-discrimination can lead to new solutions or services [2]. As an example, an AI used to allocate tasks on assembly lines might evolve from a simplistic algorithm to one that responsibly balances workload, skill development, and even well-being metrics among employees. By framing employees as crucial stakeholders, manufacturers deepen trust and loyalty—a factor that can enhance retention in sectors where skilled labour is pivotal.

Leveraging Post-Market Monitoring

The EU AI Act's emphasis on post-market monitoring (Article 61) encourages businesses to adopt a continuous improvement mindset. Monitoring an AI solution's real-world performance feeds back into iterative updates, refining model accuracy and reliability. This iterative loop can reveal novel use cases or performance enhancements. For example, if a predictive maintenance solution deployed in multiple factories uncovers novel failure modes through post-market analytics, the manufacturer can capitalise on these findings to create advanced diagnostic solutions. Thus, compliance fosters a structure where constant learning, product improvement, and innovative add-ons become ingrained habits.

6.4 SMEs, Knowledge Gaps, and the Innovation Imperative

While large corporations often have the resources, legal, financial, and technical to navigate new regulations, SMEs can be more vulnerable to compliance pressures. Nevertheless, these firms can also discover unique advantages, provided they approach the EU AI Act strategically.

Potential Barriers for SMEs

- **Limited Budgets**: SMEs may struggle to fund comprehensive conformity assessments, third-party audits, or full-time compliance officers.
- **Specialist Skills**: Hiring data scientists or AI ethics specialists can be challenging, especially for smaller firms outside major tech hubs.
- **Complex Supply Chains**: SMEs typically partner with larger clients who impose stringent compliance conditions. This can lead to complex negotiations or require adopting external standards rapidly.

Despite these hurdles, SMEs that embrace compliance early can differentiate themselves in a competitive market. Large corporations seeking to de-risk their supply chains often prefer suppliers or partners that confidently meet or exceed EU regulatory benchmarks [2].

Turning Compliance into Opportunity

- **Niche Innovations**: Smaller companies can specialise in tackling high-risk AI challenges where they can become experts, e.g. a startup focusing on ethical data annotation or an SME offering advanced testing frameworks for AI-driven machinery.

- **Investor Appeal**: Venture capitalists and institutional investors increasingly value governance and ethics [5]. Demonstrating compliance can signal professionalism and reduce risk, making SMEs more attractive for funding.
- **Collaboration and Partnerships**: By forming alliances with research institutions or joining consortia, SMEs share the burden of compliance while gaining access to cutting-edge knowledge. They can co-develop new solutions that meet EU standards from inception, opening routes to broader markets.

6.5 Large Manufacturers and Global Competitiveness

For multinational conglomerates, the EU AI Act represents both a substantial compliance project and a gateway to shaping international standards. Companies leading in high-tech engineering or automotive manufacturing frequently see regulation not as a barrier but as part of a corporate social responsibility narrative.

Scaling Compliance for Global Consistency

Large manufacturers with sprawling global footprints often unify their policies under the strictest regulatory regime for practical efficiency. For instance, if a German automotive parts manufacturer also has plants in Mexico and China, it might adopt the EU's rigorous approach to AI governance across all sites. This ensures the same safety, transparency, and ethical standards globally, simplifying the supply chain and centralising training programmes.

By doing so, the firm not only remains competitive in the EU but also anticipates emergent rules in other jurisdictions. When governments or trade associations in Asia or the Americas move towards more stringent AI regulations, the manufacturer is already well-prepared, capitalising on its existing compliance structure as a market differentiator.

Leading Through R&D

Large manufacturers also benefit from robust research and development (R&D) budgets. While the EU AI Act requires thorough risk assessments and documentation, these processes can spur advanced R&D aimed at building AI systems that are inherently safe, explainable, and resilient. Investing in explainable AI (XAI), for example, is a rising trend among corporations seeking to demystify complex models, thereby meeting regulatory expectations of transparency and interpretability.

Indeed, some corporations now view compliance labs as innovation hubs. Instead of grudgingly creating a separate compliance function, they integrate compliance-minded experts into engineering teams to propose design improvements and next-gen prototypes. This synergy fosters solutions that surpass minimal regulatory thresholds and elevate brand leadership.

6.6 Market Differentiation and Branding

In an era of heightened consumer and partner awareness, being known for compliance and responsible AI can enhance a manufacturer's standing. Here are specific branding benefits:

- **Customer Trust**: Whether your customers are industrial buyers, public sector entities, or end consumers, demonstrating alignment with the EU AI Act's ethical and safety standards can instil confidence.
- **Partner Collaboration**: Large original equipment manufacturers (OEMs) frequently vet suppliers for compliance before integrating their parts or modules. Achieving EU regulatory certification or third-party audits can set a supplier apart.
- **Thought Leadership**: Companies that openly discuss their methods for data governance, AI fairness, and worker safety can become sought-after voices in industry forums. White papers detailing compliance best practices and innovation successes help position these businesses as thought leaders.

A leading factory automation firm, for instance, may publish case studies on how it integrated robust human oversight protocols into collaborative robots. Potential buyers see these robots not just as meeting a regulatory requirement but as exemplars of advanced engineering, reliability, and social responsibility.

6.7 Overcoming Perceived Constraints

While the EU AI Act undoubtedly imposes new processes, it needs not stifle creativity. Below are strategies for preserving an innovative culture even in a highly regulated environment:

Build Cross-Functional Collaboration

Compliance teams must not work in isolation. By bringing together engineers, data scientists, legal advisers, and frontline workers, novel ideas flourish. Such collaboration often yields simpler, more elegant solutions that exceed compliance minimums.

Adopt Agile Methodologies

Agile development emphasises iterative testing and quick feedback loops—perfect for reconciling fast-moving AI projects with compliance demands. Instead of final compliance checks near product launch, agile teams integrate mini-checks throughout development.

Encourage Pilot Projects

Firms can create "sandbox" environments where new AI features are tested under realistic conditions but with minimal real-world risk. In these controlled environments, compliance experts can evaluate the system's alignment with the EU AI Act before it goes into full-scale production. This approach unleashes experimentation while containing liabilities.

Leverage External Expertise

Startups and SMEs, in particular, may lack dedicated compliance or AI ethics staff. Partnering with academic institutions or technology consultancies can fill these knowledge gaps. Joint research or pilot programmes also provide an external viewpoint, often spurring fresh insights that internal teams might miss.

6.8 The EU AI Act's Influence on Research and Education

Beyond direct industrial impacts, the AI Act is already shaping how universities, vocational schools, and research institutes approach AI. Curricula increasingly blend technical proficiencies (machine learning, systems engineering) with modules on ethics, law, and policy. Graduates enter the workforce aware of regulatory expectations and adept at building AI solutions that prioritise safety and transparency from the outset.

Example: A European engineering school might offer a course called "AI Risk Management in Manufacturing," teaching students how to design and assess AI-driven machinery. By collaborating with local factories, students gain hands-on experience, and manufacturers glean fresh perspectives on compliance and innovation. Graduates from such programmes become valuable hires for organisations needing professionals who already understand the complexities of the EU AI Act.

Similarly, research funding bodies, both in the EU and beyond, now factor compliance aspects into grants for AI projects in robotics, digital twins, or generative design. Award criteria may include demonstration of bias mitigation, robust data collection practices, or potential for real-world safety improvements. Thus, the Act drives a portion of AI research towards socially beneficial ends, rather than purely profit-driven or speculative ventures.

6.9 Looking to the Future: Harmonisation and Competitive Positioning

With the global race to harness AI intensifying, the EU AI Act's risk-based approach could become a template for other regions. As standards converge internationally, manufacturers that have long embraced compliance may find it easier to adapt to new markets or bilateral agreements.

For instance, if the United States eventually introduces a federal AI law mirroring the EU's commitments to ethical and safe design, companies already aligned with the AI Act would be a step ahead. This potential for regulatory harmonisation reduces friction for globally oriented enterprises, enabling them to focus on refining AI products rather than juggling disparate legal frameworks.

Open Innovation Ecosystems

Some experts anticipate that the Act will spur open innovation ecosystems where multiple stakeholders (manufacturers, AI vendors, universities, regulators) collaborate to ensure compliance and share best practices [4]. This might involve:

- Common data standards: enabling safer model training across different factories.
- Shared testbeds: neutral environments to pilot high-risk AI systems and gather evidence for conformity assessments.
- Coordinated incident reporting: making it easier to learn from near-misses across an entire sector, not just one company.

Such ecosystems could result in a virtuous cycle: regulation encourages safer AI design, safer AI design earns public trust, trust supports broader adoption, and broader adoption fuels further innovation. Rather than a zero-sum trade-off between regulation and creativity, manufacturers can establish synergy for growth.

6.10 Conclusion

The EU AI Act undoubtedly adds a new dimension of regulatory compliance to the manufacturing and engineering fields. Yet, it would be a mistake to see it merely as an obstacle to innovation. By enforcing robust data governance, user-centric design, and post-market monitoring, the Act can push organisations to adopt higher-quality, safer, and more ethically grounded AI solutions. This shift can lead to stronger consumer and investor confidence, as well as internal efficiencies through better documentation and more reliable operations.

From small, specialised start-ups to global corporations with extensive supply chains, forward-thinking manufacturers are already discovering ways to weave compliance into the fabric of their innovation processes. By doing so, they reduce legal risk and carve out fresh opportunities for differentiation, streamlined data practices, and cross-functional creativity. As other jurisdictions move towards regulating

AI, companies with a head start on EU compliance are likely to remain competitive and adapt more fluidly.

In essence, the EU AI Act functions not as a barrier but as a boundary, guiding AI development towards outcomes that are technologically pioneering yet socially responsible. Market forecasts support this trajectory with estimations that the AI manufacturing sector is set to grow to over $20 billion by 2028, yet only a small fraction of companies have realised meaningful ROI from AI to date [6]. Regulation that enforces data readiness, transparence and oversight could serve as the structural foundation required to unlock this untapped potential. Up next, Chap. 7: Steps to Compliance offers pragmatic guidance on taking your organisation from theoretical knowledge to tangible action plans, ensuring that innovation proceeds safely, ethically, and profitably under the EU's regulatory framework.

Key Takeaways

I. The EU AI Act safeguards public values while supporting AI-driven progress, echoing lessons from the GDPR.
II. Compliance pushes organisations to adopt better data governance (Article 10), user-centric interfaces (Articles 13 and 52) and iterative monitoring (Article 61), ultimately producing more reliable and creative AI solutions.
III. Both small firms and large manufacturers can leverage compliance as a competitive edge, whether through niche specialisation, enhanced brand image, or streamlined global expansion.
IV. Transparency and fairness requirements (Articles 14 and 52) compel businesses to design AI that is comprehensible and respects user autonomy, often spurring product improvements.
V. The Act's demand for consistent testing and documentation (Annex IV; Articles 17 and 43) fosters a culture of continuous improvement and advanced research in explainable AI.
VI. Adapting early to EU standards positions manufacturers advantageously for a global regulatory environment that is likely to converge on similar principles.

With an understanding of how regulation can be harnessed to invigorate, rather than restrain, innovation, we now move towards the practical steps required for organisations to adhere to the Act, guidance that unfolds in Chap. 7.

References

1. Deloitte NL (2024) The EU artificial intelligence act: a comprehensive deep dive. https://www.deloitte.com/nl/en/services/legal/perspectives/the-EU-Artificial-Intelligence-Act-deep-dive.html
2. Holistic AI (2024) EU AI act now in effect: a practical guide for global enterprises. https://www.holisticai.com/blog/eu-ai-act-in-effect-guide-for-global-enterprises
3. European Commission (2025) Regulatory framework for AI. https://digital-strategy.ec.europa.eu/en/policies/regulatory-framework-ai

4. International Data Spaces Association (2025) The dataspace protocol. https://www.sintef.no/... dataspace-protocol.pdf
5. EY (2024) The EU AI act: what it means for your business. https://www.ey.com/en_ch/insights/forensic-integrity-services/the-eu-ai-act-what-it-means-for-your-business
6. World Economic Forum (2024) How we can unleash the power of AI in manufacturing. https://www.weforum.org/stories/2024/01/how-we-can-unleash-the-power-of-ai-in-manufacturing/

Chapter 7
Steps to Compliance

Abstract Offering a practical roadmap, this chapter guides organisations through seven sequential actions: compiling an AI inventory; classifying risk; producing technical documentation and a quality management system; completing conformity assessment and EU database registration; instituting human oversight training; establishing post-market monitoring; and embedding continuous improvement loops. A running case study of a Spanish robotics supplier illustrates how each step unfolds in real time, while decision matrices help readers adapt the roadmap to firm size and resource constraints.

7.1 Introduction

Successfully meeting the requirements of the EU Artificial Intelligence (AI) Act involves more than merely reviewing a legal checklist [1]; it demands a coherent, strategic approach that aligns your organisation's technical, operational, and legal processes. Because the Act applies even to non-EU providers whose AI outputs are used in the EU, global manufacturers must coordinate risk management (Article 9), data governance (Article 10), human oversight (Article 14), technical documentation (Article 11 and Annex IV), conformity assessment (Article 43), and post-market monitoring (Article 61) within a single compliance programme.

This chapter distils the path to compliance into a series of tangible steps, each building upon insights from the previous chapters. We begin by outlining how to conduct an AI inventory, ensuring that no AI system flies under the radar. We then move through the process of risk classification, emphasising the significance of correctly identifying which systems are high risk or limited risk. From there, we examine technical documentation and conformity assessment procedures, demystifying what needs to be compiled and when an external notified body might be required. Next, we discuss human oversight and training measures, essential to bridging any gap between technology and its day-to-day users. Finally, we address post-market monitoring, the ongoing vigilance that keeps AI deployments safe and

© The Author(s), under exclusive license to Springer Nature Switzerland AG 2025
M. Kauffman, *Navigating the EU AI Act: A Practical Guide for Global Manufacturing and Engineering Leaders*, https://doi.org/10.1007/978-3-031-99613-9_7

legally compliant after their initial launch. By chapter's end, you will have an actionable framework to guide your organisation's compliance journey.

7.2 Step 1: Conduct an AI Inventory and Risk Register

Why an AI Inventory Matters

The first and most foundational step to compliance is developing a comprehensive inventory of your AI systems. In many manufacturing or engineering organisations, AI might be used in siloed projects or embedded in existing software, making it easy to overlook certain tools. However, the EU AI Act's broad definition of AI (see Chap. 2) means any software that infers, predicts, or makes decisions qualifies [2]. Missing even a seemingly minor AI module could leave you unknowingly non-compliant.

Practical Tips for Building Your Inventory

Cross-Departmental Survey

Interview managers, engineers, and IT staff across departments. Ask them to identify all software or hardware solutions that employ AI or advanced analytics. Proactively explore areas like quality control, supply chain tools, and research projects.

Examine Supply Chain Integration

Investigate whether AI is embedded in machines or components you purchase from external suppliers. If those products influence safety or produce outputs relevant to EU consumers, you may share compliance obligations.

Document Key Parameters

For each identified AI system, log its name, function, type of AI (machine learning, rule-based system, etc.), data sources, user groups, and deployment scope. This record ensures transparency and forms the baseline for further assessment.

Establish Ownership

Assign a clear owner or "system champion" for each AI solution. This individual will manage compliance tasks, like documentation and risk assessments, and serve as a point of contact for queries.

Crafting a Central Register

Many organisations opt for a central register or database, accessible to compliance officers, data scientists, and relevant engineers. This register not only lists the AI systems but also their compliance status, version histories, provider/importer/authorised-representative so downstream actors can see role allocations (Articles

16–24) and associated risk level [3]. Whenever you introduce, update, or decommission an AI tool, the register should be updated, reflecting a living compliance structure.

7.3 Step 2: Assess Risk Classification (Articles 6 & Annex III)

Relevance of Risk Tiers

Once you have an exhaustive list of AI systems, the next step is determining which risk category each belongs in: unacceptable risk (banned), high risk, limited risk, or minimal risk. This classification guides the depth and extent of your compliance measures (see Chap. 2).

- **Unacceptable risk**: Systems that manipulate users subliminally or contravene fundamental rights are prohibited.
- **High risk**: AI applications influencing safety or critical rights, including many scenarios in manufacturing (e.g. robotics, equipment control).
- **Limited risk**: AI that interacts with humans or automates non-critical decisions, subject primarily to transparency rules.
- **Minimal risk**: AI with negligible potential for harm, requiring no specific compliance obligations beyond general good practices.

Practical Criteria for Determining Risk

Potential for Physical Harm: Does the AI control or significantly influence machinery with potential to injure workers or produce defective goods that reach consumers?

Influence on Individuals' Rights: Could AI decisions affect employment, privacy, or non-discrimination in ways that fundamentally alter users' rights?

Level of Automation: Does the AI autonomously execute actions, or does it merely offer suggestions to humans who retain the final say?

Regulated Product Context: Is the AI part of a product already under EU harmonisation legislation, such as medical devices, automotive equipment, or certain categories of machinery?

By cross-referencing these queries with the official categories and **Annex III** of the Act, you can ascertain which AI systems merit in-depth requirements for data governance, oversight, and audits [4].

Finalising Your Risk Register

Consolidate each system's risk classification into your AI register. Clearly mark high-risk AI, as these will demand the most rigorous processes. Flag any safety-component

AI falling under Annex I product legislation (e.g. Machinery Regulation), because these are automatically high-risk. Note justifications for your classification decisions; if an inspector or notified body questions your assessment, you will need documented reasoning.

7.4 Step 3: Develop Technical Documentation and Governance Frameworks

Role of Technical Documentation

For high-risk AI systems, the EU AI Act mandates extensive documentation that covers design principles, data usage, system architecture, testing protocols, and risk mitigation strategies. Even for limited-risk AI, maintaining some level of documentation is advisable, as it clarifies how transparency and human oversight are managed.

Components of a comprehensive documentation set include:

System Description

Provide an overview of the AI's intended purpose, operational scope, and integration points. Highlight essential functionalities and user interfaces, noting their relationship to risk factors (e.g. controlling robotic arms in hazardous settings).

Algorithmic Details

Summarise the machine learning model(s) or rule-based logic in use, emphasising how data flows from input to decision output. For proprietary algorithms, you need not reveal trade secrets but should describe enough to demonstrate compliance (e.g. training methods, performance benchmarks).

Data Governance and Management

Under Article 10, detail the data sources, quality checks, cleansing procedures, and version control [4]. If any personal data is processed, cross-reference GDPR compliance measures.

Risk Assessment and Mitigation

Outline known failure modes and potential harms, including safety, bias, and cybersecurity issues. Identify mitigations (redundant sensors, anomaly detection, or fallback procedures). Use formal risk management frameworks if possible, such as ISO 31000 or similar.

Testing and Validation Reports

Include results from simulations, real-world pilots, or stress tests. Demonstrate how you have validated accuracy, robustness, and safety performance under typical and edge-case conditions.

Datasheet/Data Card/Model Card Bundles

Each high-risk AI solution should include three artefacts in its dossier:

- Datasheet—provenance, sampling, cleaning, and known gaps of every training dataset [5].
- Model Card—architecture, intended use, performance slices and ethical considerations [6].
- Data Card—condensed, audience-friendly overview for deployers/operators [7].

These lightweight documents satisfy Annex IV traceability yet remain comprehensible to non-experts.

Human Oversight Plan

Clarify who is authorised to supervise or override the system, along with the interface or dashboard they will use (Article 14). Document training regimens for these supervisors.

Maintenance and Updates

Provide procedures for post-market monitoring, including how model updates or retraining events are tracked, tested, and validated to avoid unanticipated risk increases.

Governance Framework Implementation

While preparing technical documentation, it is prudent to institute an internal governance body or steering group responsible for AI oversight. This group reviews design decisions, ensures version control, and audits compliance readiness. Engineers, legal advisors, quality managers, and data scientists should all participate in this body.

7.5 Step 4: Conformity Assessment and Database Registration

Understanding Conformity Assessment

For many high-risk AI systems, the EU AI Act requires a conformity assessment before you can place the product on the EU market (Article 43). This step verifies that your system adheres to all applicable requirements. Depending on the system's nature, the law may stipulate either a self-assessment or a third-party (notified body) assessment.

Self-Assessment: You conduct your own compliance checks, culminating in a formal declaration of conformity. Typically permitted for less complex or lower-stakes high-risk uses.

Third-Party Assessment: A notified body performs an external review of your technical documentation, risk analyses, and test results [8, 9]. Often mandated for more sensitive or complex high-risk systems.

Preparing for a Third-Party Audit

1. **Finalise Documentation**

Before contacting a notified body, ensure that your documentation set is complete and coherent. Missing or inconsistent data can delay approvals.

2. **Select a Notified Body**

EU Member States accredit certain organisations to handle conformity assessments in specific sectors (e.g. machinery, medical devices). Choose one that has experience relevant to your AI's domain.

3. **Engage with Openness**

During the assessment, the notified body may ask for clarifications, additional tests, or site visits. Cooperate proactively—this shortens timelines and fosters trust.

4. **Address Non-Conformities**

If the assessor highlights shortcomings (e.g. incomplete data logging or insufficient oversight mechanisms), treat them as learning opportunities. Rectify them thoroughly, updating both your system and documentation.

CE Marking and Registration

If your AI system is integrated into a product subject to EU harmonisation legislation, you may place the CE marking on the product once certification or a positive assessment is concluded. In addition, providers (or their authorised representative) must register the high-risk AI in the EU database before market entry (Article 51).

7.6 Step 5: Enforce Human Oversight and Staff Training

Training Personnel for Oversight Duties

Even the most rigorous documentation and technical safeguards will fail if front-line operators do not understand how to supervise and intervene. Article 14 of the EU AI Act emphasises the "human-in-the-loop" principle, granting humans the power to override AI decisions in safety–critical scenarios.

Identify Training Needs: Ascertain which staff roles require in-depth knowledge (maintenance engineers, shift supervisors, quality inspectors, etc.).

Design Training Modules: Cover key topics like how the AI processes data, interpretive dashboards, handling alerts, and emergency override procedures.

Hands-On Drills: Simulate error conditions or anomalies in a controlled environment so workers can practise responding.

Documenting and Verifying Competency

Keep thorough records of who has received training, the materials or modules used, and the outcomes of any tests or drills. This information may be crucial if an incident occurs and an inspector wants proof that adequate oversight measures were in place.

Encourage employees to report anomalies and voice concerns about the AI's performance. Offer a clear communication channel—perhaps via an internal app or hotline—so staff can swiftly highlight near misses or suspicious behaviour. This culture of vigilance is key to preventing small issues from escalating into dangerous or non-compliant situations.

7.7 Step 6: Establish Post-Market Monitoring

Why Continuous Monitoring Matters

After a system is deployed, real-world conditions can deviate from the test environment. Sensor degradation, unexpected workloads, or environmental shifts can affect AI decisions. The EU AI Act (Article 61) compels manufacturers to gather performance data and watch for potential hazards or emerging biases.

Setting Up Monitoring Processes

1. **Data Collection Infrastructure**: Equip the deployed system to log operational parameters, user interventions, and any error states or collisions (in case of robotics).
2. **Regular Performance Reviews**: Analyse logs for deviations from baseline metrics. If accuracy or reliability declines, investigate root causes and initiate corrective measures.
3. **Feedback Loops**: Maintain channels for user feedback, both from operators and, if applicable, end consumers.

Handling Incidents and Updates

Should a serious malfunction or near miss occur, immediate reporting to relevant authorities may be required, depending on the severity. If the AI adapts over time (online learning), introduce a "model version control" system. Each iteration or retraining event must undergo a mini risk assessment to ensure compliance remains intact.

7.8 Step 7: Engage in Continuous Improvement

Compliance is not a one-off exercise. Manufacturing processes evolve, market conditions shift, and AI technology advances rapidly. By baking iterative refinement into your compliance cycle, you can stay abreast of shifting standards and capitalise on new insights.

Internal Audits and Drills

Regularly schedule internal audits that mimic an external conformity assessment. Test your documentation for completeness, review incident logs, and re-run risk analyses. Such drills foster preparedness for any official inspection and reduce last-minute scrambling.

Track Regulatory Updates

Keep an eye on evolving EU AI Act guidance, amendments, or supplementary rulings by the European Commission or Member State authorities. Harmonised standards or "common specifications" may be introduced, refining what compliance means for certain high-risk applications.

Embrace the Learning Mindset

Each compliance step, from the AI inventory to post-market monitoring, generates data. Use it not only to stay legally sound but also to discover new opportunities for innovation, as highlighted in Chap. 6. For instance, feedback from a conformity assessment might spur creative design refinements that enhance safety and operational efficiency.

7.9 Case Study: A Roadmap in Action

To visualise these steps, consider the example of a fictitious Spanish manufacturer, **AutoWeld SA**, introducing an AI-driven robotic welding system targeted for car assembly lines in Germany and France:

AI Inventory: AutoWeld identifies the welding robot's machine learning submodule as a high-risk AI, given its direct impact on product and worker safety.

Risk Classification: A formal risk analysis confirms the system must be treated as high risk due to potential harm if it malfunctions.

Documentation: The engineering team compiles architecture diagrams, sensor calibrations, training datasets, and algorithmic parameters into a single, consistent technical file.

Conformity Assessment: AutoWeld engages a notified body specialising in industrial automation. The resulting audit highlights some improvements needed in cybersecurity. After addressing these gaps, AutoWeld obtains a positive assessment.

Staff Training: Operators in the client's facility receive specific instructions on reading the robot's diagnostic dashboard and performing emergency stops. Supervisors log all training sessions, verifying comprehension.

Post-Market Monitoring: Deployed robots send performance logs to AutoWeld's central server. Any anomalies or extended downtime events trigger immediate internal review.

Continuous Improvement: AutoWeld analyses data from multiple factories, refining the welding algorithm to handle new material types. They re-run partial conformity checks to ensure compliance stands for updated versions.

Following these steps meticulously, AutoWeld enters the EU market with an AI product that not only meets the Act's obligations but also fosters confidence amongst clients seeking advanced, reliable robotics solutions.

7.10 Conclusion

Compliance with the EU AI Act is a complex but navigable endeavour if tackled systematically. By following these seven key steps: (1) conducting an AI inventory, (2) assessing risk classification, (3) developing technical documentation and governance, (4) completing conformity assessments, (5) enforcing human oversight and training, (6) implementing post-market monitoring, and (7) embracing continuous improvement; manufacturers can transition from merely understanding the legislation to fully integrating it within their operational frameworks.

Not only do these measures reduce the risk of non-compliance and related penalties, but they also provide a foundation upon which safer, more trustworthy AI products can flourish. The end result is a win–win scenario: the EU's goal of promoting responsible AI is met, and businesses gain robust systems that align with industry best practices, reduce liability, and earn market credibility. Equipped with this roadmap, you can approach the rest of the EU AI Act's lifecycle, particularly collaboration with EU partners (Chap. 8) and the strategic use of technology for compliance (Chap. 9), with clarity and confidence.

Key Takeaways

I. Compliance efforts should be broken down into clear steps, from initial inventory to ongoing monitoring, ensuring no important detail is missed.
II. Pinpointing a system's risk level is pivotal. High-risk AI requires more extensive documentation, testing, and oversight.
III. Comprehensive records covering design, data governance, and risk mitigation strategies are non-negotiable for high-risk AI.

IV. Staff empowerment is fundamental. Operators must know how to monitor AI operations and intervene if needed, backed by well-designed user interfaces and guidelines.
V. Depending on the AI's complexity, self-assessment or third-party audits may be necessary. Preparation is key to smooth, efficient certification.
VI. Performance tracking and incident reporting are ongoing duties, integral to maintaining compliance and safeguarding end-users.
VII. Regular audits, culture-building, and iterative learning ensure that compliance becomes a driver of ongoing innovation and risk management excellence.

References

1. ISACA (2024) Understanding the EU AI act (white paper). https://www.isaca.org/resources/white-papers/2024/understanding-the-eu-ai-act
2. ArtificialIntelligenceAct.eu (2024) EU AI act compliance checker. https://artificialintelligenceact.eu/assessment/eu-ai-act-compliance-checker/
3. PwC (2024) AI act compliance tool. https://www.pwc.com/cz/en/sluzby/technologie-a-data/ai-act/ai-compliance-tool.html
4. NAVEX (2024) EU AI act compliance requirements. https://www.navex.com/en-us/solutions/regulations/eu-ai-act/
5. Gebru T et al (2021) Datasheets for datasets. https://doi.org/10.1145/3458723
6. Google PAIR (2024) Data cards playbook. https://sites.research.google/datacardsplaybook/
7. Mitchell M et al (2018) Model cards for model reporting. https://arxiv.org/abs/1810.03993
8. OneTrust (2024) Conformity assessments under the proposed EU AI act: step-by-step guide. https://www.onetrust.com/resources/conformity-assessments-under-the-proposed-eu-ai-act-a-step-by-step-guide/
9. Holistic AI (2024) Conformity assessments in the EU AI act: what you need to know. https://www.holisticai.com/blog/conformity-assessments-in-the-eu-ai-act

Chapter 8
Collaborating with EU Partners

Abstract Compliance rarely happens in isolation. This chapter explains how to map complex supply chains, craft flow-down clauses, and conduct supplier audits so that every tier meets EU requirements. It also covers constructive engagement with national authorities and notified bodies, the value of cross-border R&D consortia, and contractual mechanisms that allocate liability, data-sharing duties, and incident-reporting responsibilities. Communication strategies, internal and external around the guidance, fostering transparency and trust across stakeholder groups.

8.1 Introduction

Achieving compliance with the EU AI Act [1] is rarely a solitary pursuit. For most manufacturers and engineering firms, partnerships, supply chains, and customer relationships stretch across multiple borders, especially within Europe. Whether you are a non-EU company seeking to export AI-integrated products into European markets or an EU-based organisation reliant on global suppliers, collaboration is indispensable. Building robust alliances, sharing compliance intelligence, and fostering mutual trust can ease the operational burdens of meeting the Act's detailed requirements.

This chapter focuses on collaborative strategies that enable smoother compliance journeys. We begin by examining supply chain responsibilities, including how manufacturers can align their global networks to meet EU risk classifications and technical documentation standards. Next, we look at cooperative engagements with regulators, including the role of national supervisory authorities and notified bodies. Then, we explore cross-border research and development (R&D) collaborations and how joining forces can accelerate both innovation and compliance. We will also investigate how to formalise partnerships through legal contracts that assign roles for AI governance among suppliers, distributors, and clients. By the end of the chapter, you will appreciate the manifold ways collaboration can transform what might seem like an isolated regulatory burden into a collective effort that benefits all stakeholders.

© The Author(s), under exclusive license to Springer Nature Switzerland AG 2025

M. Kauffman, *Navigating the EU AI Act: A Practical Guide for Global Manufacturing and Engineering Leaders*, https://doi.org/10.1007/978-3-031-99613-9_8

8.2 Why Collaboration Is Essential

A Complex Ecosystem of Partners

Manufacturing is a network of interdependent relationships. Even a seemingly simple AI-driven product could have multiple layers of external inputs:

- **Component Suppliers**: Providing sensors, circuit boards, or software libraries that power an AI system.
- **Software Developers**: Creating core AI algorithms, either as in-house teams or subcontractors.
- **Distributors**: Ensuring finished products reach various EU markets, often requiring region-specific adaptations or labelling.
- **Service Providers**: Offering maintenance, updates, or compliance consultancy.

Each contributor has a role in maintaining the safety and reliability of the final AI solution, making it impossible for a single entity to handle the entire compliance process. The EU AI Act recognises these interdependencies by assigning obligations to "providers," "importers," and "distributors," among others (see Chap. 3).

Shared Responsibility Under Articles 16–24, Collective Benefits

When done well, collaboration simplifies compliance. Suppliers that maintain up-to-date technical documentation can pass these records to manufacturers for inclusion in broader conformity assessments. Manufacturers that work alongside distributors can better tailor user manuals, ensuring transparency in local languages. This synergy reduces redundancy, eases administrative loads, and clarifies accountability in the event of audits or incidents.

The Act assigns distinct duties to providers (often the manufacturer), importers, distributors and, where the provider is outside the Union, an authorised representative (Article 54) [2]. Importers must verify CE marking, technical documentation and EU database registration before placing high-risk AI on the market; distributors must ensure labelling and storage conditions do not compromise compliance.

Beyond compliance efficiency, robust collaboration can also boost innovation, as knowledge-sharing and co-creation can produce safer, more reliable AI systems (see Chap. 6). If all parties in the supply chain are aligned on ethical data practices, risk classification, and user-centric design, the final product is more likely to succeed, building consumer trust and competitive advantage.

8.3 Supply Chain Collaboration

Mapping the Chain

The first step to collaborative compliance is supply chain mapping. Many organisations underestimate how complex their chains have become, with Tier 1 suppliers relying on Tier 2 and Tier 3 partners, and so forth. By charting all links, from raw materials to final assembly, manufacturers can identify exactly where AI is introduced, who is responsible for its development, and what data is being used.

Example: A French automotive firm might source an AI-based sensor array from a German supplier, which, in turn, licenses certain machine learning libraries from a Swedish tech company. Each link introduces potential compliance obligations regarding data usage, algorithmic reliability, and transparency. Without clear visibility, the automotive firm risks gaps in its technical documentation and conformity assessments.

Contractual Obligations and Flow-Down Clauses

To ensure each party meets its responsibilities, manufacturers commonly incorporate flow-down clauses into their supply contracts. These clauses specify that suppliers must:

1. Follow the EU AI Act (where relevant to their subsystem).
2. Add a requirement that non-EU suppliers appoint an authorised representative established in the EU when they act as providers of high-risk AI subsystems (Article 54), ensuring a single contact point for market-surveillance authorities.
3. Provide comprehensive documentation on data governance, algorithmic design, and testing.
4. Audit checklists should cover Article 10 data-governance evidence, Article 17 quality management system, and 10-year record-keeping (Article 18).
5. Cooperate with external audits if a notified body or regulatory authority requests additional evidence.

Such contractual clauses clarify accountability [3]. If a high-risk AI component malfunctions, the manufacturer can trace the documentation trail back to the supplier, verifying whether due diligence was performed. This approach also sets expectations early, so suppliers build compliance into their processes rather than scrambling post-production.

Supplier Qualification and Audits

Supplier qualification is another collaborative tool. Before entering into a relationship, a manufacturer can conduct a compliance audit, perhaps modelled on ISO or other international standards to ascertain a supplier's readiness for EU AI Act obligations. Where deficiencies arise (e.g. insufficient data logging, inadequate cybersecurity), the supplier is given time to rectify them, with follow-up checks ensuring improvement.

In some cases, especially for critical high-risk AI modules (like robotic control systems or safety sensors), the manufacturer might even request that the supplier obtains third-party conformity assessment for their subsystem, reducing the complexity of the manufacturer's overall product certification. While this does impose an additional burden on suppliers, it can also create new business opportunities for them if they gain a reputation for delivering EU-compliant AI components.

8.4 Working with National Authorities and Notified Bodies

Role of National Supervisory Authorities

Each EU Member State designates national authorities to oversee AI compliance [4]. These agencies liaise with businesses, handle inquiries, and enforce the EU AI Act at a national level. Although direct communication with them may seem daunting, early and proactive engagement can:

- **Clarify Ambiguities**: If your engineers are unsure how certain articles of the Act apply to a novel AI technology, you can seek informal guidance.
- **Accelerate Approvals**: In some instances, especially for new or experimental AI applications, open dialogue with authorities fosters understanding and shortens the path to conformity [4].
- **Prevent Missteps**: Regulators appreciate being looped in on emerging issues, from data usage to borderline risk classifications, rather than discovering them during an unannounced inspection.

Harmonised Standards and Technical Specifications

In certain areas, harmonised standards or EU-endorsed "common specifications" may be referenced, delineating best practices for AI design and testing. Engaging with regulators who help shape these standards ensures your organisation stays ahead of shifting benchmarks. This is especially relevant for manufacturers in fast-evolving fields like collaborative robotics or additive manufacturing.

For high-risk AI systems, a notified body (see Chap. 5) might be mandatory to perform conformity assessments. Rather than treating these third-party evaluations as mere box-ticking exercises, savvy businesses cultivate collaborative partnerships with their chosen notified bodies:

Pre-assessment Consultation: Before official auditing begins, share your preliminary documentation. Request feedback on potential gaps to reduce friction during the formal review.

Iterative Approach: If the AI system will undergo multiple updates or expansions, schedule periodic check-ins. The notified body can confirm continuing compliance as your technology evolves.

Cross-industry Insights: Notified bodies often assess numerous companies and can share anonymised best practices (within their confidentiality boundaries) that might guide your own compliance strategy.

Such collaborative approaches expedite certifications, cut rework costs, and can even refine your product's design based on the notified body's operational experience.

8.5 Cross-Border R&D Alliances

Joint Ventures and Consortia

R&D collaborations frequently extend across national borders, involving universities, private labs, start-ups, and established manufacturers. Under the EU AI Act, such alliances can be a two-edged sword: they introduce diverse expertise that fosters innovation but also multiply potential compliance pitfalls if roles and data protocols are not clearly delineated.

Example: A Dutch maritime engineering firm teams up with a Greek research institute to develop an AI-based collision avoidance system for autonomous ships. Although the institute primarily handles early-stage algorithm design, the Dutch partner is responsible for the final product's conformity assessment and user documentation. Should the institute fail to track data usage or prove model robustness properly, the entire project could be compromised.

To pre-empt such hazards, partners in R&D consortia should collectively map responsibilities early. Often, they create a joint compliance committee to coordinate data governance, risk classifications, and project milestones relevant to AI Act requirements.

Funding Programmes and EU Initiatives

Various EU-funded programmes (e.g. Horizon Europe) encourage cross-border projects in AI, awarding grants that prioritise not only scientific excellence but also ethical and legal alignment. Collaborating within these frameworks can yield several advantages:

Regulatory Guidance: Funded projects often receive support or direct mentorship from EU agencies well-versed in legal compliance.

Networking: Partnerships formed under these programmes deepen relationships with academia, start-ups, and potential end-users.

Showcasing Compliance Leadership: Being part of an EU-endorsed project signals to clients and competitors alike that your work is at the forefront of responsible AI development.

8.6 Collaborative Contracting for AI Governance

Defining Roles and Responsibilities

One crucial but sometimes overlooked aspect of collaboration is structuring robust legal agreements. While a standard supply contract covers price, delivery terms, and warranties, dealing with AI requires additional clauses that detail:

1. **Risk Assessments**: Which party undertakes them, and who reviews results?
2. **Data Protection**: How will data be shared, stored, and protected, particularly if it includes personal or sensitive information that intersects with the GDPR?
3. **Technical Documentation**: Which side is responsible for drafting or updating the documents needed for a final product's conformity assessment?
4. **Oversight and Intervention**: If the AI system is operational at the partner's facility, who ensures compliance with the EU AI Act's human oversight mandates?
5. **Incident Reporting**: Protocols for notifying each other and relevant authorities if anomalies or safety–critical incidents arise. Contracts should mirror Article 62 by specifying a 15-day deadline for partners to report serious incidents to the provider, importer, and competent authority.

Allocating Liability and Indemnification

Allocating liability is essential to avoid disputes if something goes wrong. Contracts can include:

- **Liability Sharing**: If a defect arises in a subsystem provided by one party, define whether that party bears partial or full responsibility for non-compliance penalties.
- **Indemnification Clauses**: Stipulating that if one partner's negligence leads to regulatory fines or lawsuits, they must compensate the other.

Although such discussions can be legally complex, clarity here prevents misunderstandings and fosters a more transparent and trusting partnership. Both parties understand that meeting EU AI standards is in their mutual interest.

Service-Level Agreements (SLAs)

For AI-based solutions with continuous learning or updates, partners often sign service-level agreements detailing ongoing obligations:

- **Update Frequency**: How often the AI model is retrained or patched for security.
- **Performance Metrics**: Targets for accuracy, availability, or anomaly detection.
- **Reporting and Review Mechanisms**: How results are shared and how often performance reviews occur, ensuring the system stays within compliance thresholds over its lifecycle.

An SLA can embed compliance reviews as part of the performance metrics, for instance requiring biannual audits or data usage checks [5]. This approach merges business objectives and legal obligations into a single operational framework.

8.7 Building Trust Through Transparency and Communication

Collaboration is not restricted to external partners; it also includes synergy among internal departments, from engineering and legal to sales and marketing. A robust internal communication strategy ensures each department understands the significance of the EU AI Act, the need for accurate documentation, and the potential consequences of misalignment. Regular cross-functional meetings can:

- Update stakeholders on new regulatory developments.
- Address emerging risks or anomalies.
- Gather feedback from sales teams, who may have direct insights into client compliance demands.

This holistic approach creates a culture of compliance, where everyone recognises AI governance as part of daily operations rather than an afterthought.

External Transparency with Customers

End customers, whether businesses or individual consumers, are also partners in AI compliance. The EU AI Act's transparency requirements (Articles 13 and 52) dictate that users must be informed when interacting with AI, understand any limitations and know-how to seek help or override decisions if needed. Proactive communication helps:

- **Manage Expectations**: Users grasp that the system is AI-driven, not foolproof, and that human oversight remains vital.
- **Empower Safe Usage**: By providing thorough user manuals and disclaimers, manufacturers encourage responsible engagement, reducing risk of misuse.
- **Reinforce Brand Reputation**: Being open about how you handle data, fairness, and safety fosters trust, which can translate into long-term loyalty.

Shared Learning Portals

In manufacturing ecosystems with multiple partners, some forward-thinking organisations create shared learning portals or digital platforms. These might host best-practice guidelines, templates for risk assessments, frequently asked questions on the EU AI Act, or relevant case studies. Such portals accelerate collective understanding, saving each partner from reinventing compliance protocols in isolation. In essence, the group "crowdsources" knowledge and updates the portal as the regulatory environment evolves.

8.8 Overcoming Collaboration Barriers

Despite the apparent benefits of cooperative strategies, barriers do exist. For instance:

- **Competitive Secrecy**: Partners might hesitate to share detailed technical data or algorithmic designs, worried about losing intellectual property advantages. Carefully crafted non-disclosure agreements and clear data-sharing protocols can mitigate these fears.
- **Jurisdictional Confusion**: Cross-border deals can involve varying legal frameworks. For instance, a supplier in the United States might find EU compliance demands unfamiliar. Invest in targeted legal advice or designate an EU-based representative who can guide them.
- **Resource Constraints**: Smaller businesses may struggle to fund audits or joint R&D. In such cases, explore EU grants or cost-sharing models, ensuring that resource limitations do not sabotage compliance initiatives.

8.9 Case Example: A Multi-Partner Approach

To illustrate how collaboration can shape compliance, consider **AeroTech**, a fictional multinational consortium producing AI-driven drone systems for industrial inspections across Europe:

Supplier Network

AeroTech sources flight control software from a Swedish developer and sensors from a Belgian electronics firm. Both suppliers sign contracts pledging adherence to data governance norms and providing transparent documentation about their modules.

Collaborative Workshops

Every quarter, the three partners hold virtual workshops to align on the EU AI Act's updates, share testing results, and refine risk assessments. They collectively decide how to handle newly discovered edge cases (e.g. extreme weather flight).

Conformity Assessment Plan

Recognising the drone's potential to operate near humans, all parties agree it is a high-risk system. Together, they compile technical documentation. The Belgian firm addresses sensor calibration data, the Swedish developer provides algorithmic performance metrics, and AeroTech integrates everything into a cohesive package for the notified body.

Regulator Engagement

AeroTech proactively reaches out to a French national authority for guidance, as the initial market launch is in France. Authorities provide clarifications on transparent labelling for drone operators and potential limitations for flight in public spaces.

Shared Post-Market Monitoring

Once deployed, the drone logs flight data to a cloud platform accessible by all partners. If an anomaly occurs, each partner analyses its domain (sensors, flight control software, or integration layer) to identify root causes and propose fixes.

Through systematic collaboration, AeroTech meets the Act's standards more rapidly and cost-effectively than if each firm worked in silos. Their success also demonstrates to prospective customers and regulators that this consortium takes AI governance seriously, reinforcing trust and brand credibility.

8.10 Conclusion

Collaboration emerges as the linchpin for meeting the EU AI Act's multifaceted requirements. A single organisation, no matter how resourceful, rarely has total control over every component or subsystem in a modern AI-driven manufacturing product. By mapping supply chains, coordinating with notified bodies, engaging regulators proactively, and sharing knowledge across partners, compliance ceases to be an isolated struggle and becomes a collective endeavour. When partners fail to cooperate, penalties for inaccurate or incomplete information can reach €7.5 million or 1% of global turnover (Article 99 §3).

Simultaneously, collaborative frameworks foster deeper innovation. Through open channels of communication, partners exchange domain insights, refine risk classifications together and converge on best practices for data handling and system oversight. Contracts and legal agreements formalise these relationships, addressing liability, data management, and incident reporting in ways that build mutual confidence. In short, well-structured partnerships unlock both regulatory efficiency and the innovative potential that arises when diverse expertise converges around a common goal.

As you refine your collaborative strategies, the next step, Chap. 9: Leveraging Technology for Compliance, highlights the tools and platforms that can streamline many of these shared tasks, from automated risk assessments to centralised compliance dashboards. By combining strong partnerships with technological enablers, manufacturers can develop an even more integrated and agile response to the EU AI Act, ensuring that AI products are not just lawful but truly best in class.

Key Takeaways

I. AI manufacturing relies on global supply chains where each partner—supplier, importer, distributor—carries a share of compliance responsibility.

II. Effective collaboration reduces redundancy, clarifies accountability, and accelerates risk management, leading to safer, more innovative AI solutions.

III. Flow-down clauses, supplier audits, and shared risk assessments are practical tactics for ensuring compliance alignment across tiers.

IV. Proactive dialogue with national authorities and notified bodies fosters transparency, speeds up certifications, and clarifies evolving standards.
V. Cross-border projects and EU-funded initiatives benefit from collaborative compliance committees that weave the Act's principles into pioneering AI research.
VI. Contracts and service-level agreements must address AI-specific concerns such as data governance, oversight responsibilities, and liability arrangements to prevent confusion and disputes.
VII. Cultivating a culture of transparency, both internally and externally, supports ongoing vigilance, encourages feedback on anomalies, and cements trust.

Next, we explore how technological solutions, including auditing tools, model management platforms, and compliance dashboards can fortify these collaborative efforts, making it simpler and faster to remain in alignment with EU AI standards.

References

1. European Union (2024) Regulation (EU) 2024/1689 (EU AI Act Arts 23–24). https://eur-lex.europa.eu/eli/reg/2024/1689/oj/eng
2. Taylor Wessing (2024) Obligations on importers and distributors. https://www.taylorwessing.com/-/media/taylor-wessing/files/uk/2024/eu-ai-act-obligations/2409_obligations_on_importers_and_distributors.pdf
3. Baker Donelson (2024) Who's who under the EU AI act: spotlight on key actors. https://www.bakerdonelson.com/whos-who-under-the-eu-ai-act-spotlight-on-key-actors
4. Steptoe (2024) EU AI act decoded—issue 7: obligations of distributors. https://www.steptoe.com/a/web/ik8A63DVoevXhwuzn92U5t/eu-ai-act-decoded-issue-7-obligations-distributors-final.pdf
5. Latham and Watkins (2024) EU AI Act obligations for deployers of high-risk AI systems. https://www.lw.com/en/insights/eu-ai-act-obligations-for-deployers-of-high-risk-ai-systems

Chapter 9
Leveraging Technology for Compliance

Abstract This chapter surveys the emerging tech stack that can make EU AI Act adherence faster and less error-prone. Topics include AI auditing software for bias and robustness testing, model management platforms for traceability and version control, real-time compliance dashboards, and cybersecurity plus data governance tools aligned with Articles 10 and 15. Integration advice, change management tactics, cost considerations, and a worked example of a German automotive supplier's "compliance ecosystem" equip readers to choose and deploy the right solutions.

9.1 Introduction

For many manufacturers, the compliance journey under the EU Artificial Intelligence (AI) Act may feel daunting [1]. Meeting obligations for risk classification, technical documentation, human oversight, and post-market monitoring often demands significant staff time, detailed record-keeping, and extensive cross-functional coordination. Fortunately, technology itself can become an ally in this process. A growing ecosystem of tools and platforms, from AI auditing software to compliance dashboards helps automate tasks, detect risks, and streamline collaboration [2]. The Act's technical requirements such as event logging (Article 12), data governance (Article 10), quality management (Article 17) and 10-year record retention (Article 18), all lend themselves to digitalisation.

In this chapter, we explore how organisations can leverage technology to meet EU AI Act requirements more efficiently. We begin by examining AI auditing tools, which systematically evaluate algorithmic fairness, robustness, and accuracy, offering transparent performance metrics. Next, we delve into the role of model management platforms that track data provenance, model versions, and testing results, making it easier to maintain the technical documentation demanded by the legislation. We then turn to compliance dashboards, which serve as real-time command centres for monitoring risk classification, incident reporting, and post-market analytics. Finally, we highlight cybersecurity and data governance solutions

© The Author(s), under exclusive license to Springer Nature Switzerland AG 2025
M. Kauffman, *Navigating the EU AI Act: A Practical Guide for Global Manufacturing and Engineering Leaders*, https://doi.org/10.1007/978-3-031-99613-9_9

that help fortify AI systems against unauthorised manipulation—a crucial aspect of the Act's requirements for high-risk AI.

By the end, you will have a clearer sense of which types of technological solutions might best suit your organisation's specific compliance landscape, along with the strategic considerations for integrating these tools into everyday workflows. Properly deployed, these platforms do more than reduce regulatory burdens; they also help embed best practices in quality, safety, and innovation at the core of your AI-driven operations.

9.2 Why Digital Tools Matter

Complexity of Regulatory Obligations

The EU AI Act's multi-layered requirements from data governance (Article 10) to post-market monitoring (Article 61) can be intricate. Without structured tools, teams risk missing subtleties in risk classifications or losing track of training datasets during model updates [3]. Manual approaches to record-keeping or auditing could work in small-scale projects, but they are far less feasible when your AI is woven into multiple production lines or global supply chains.

Compliance platforms, dashboards, and automated checks help cut human error and reduce administrative overhead. By codifying risk assessment steps into software, or by building AI-driven anomaly detection modules for post-market monitoring, organisations can standardise processes, lower costs, and free staff to tackle higher-level tasks like exploring new engineering solutions or refining ethical guidelines.

Tools that facilitate compliance do more than just meet the letter of the law. They embed a culture of *data excellence*, *iterative testing*, and *transparent governance* [2]. When properly integrated, these tools yield benefits that extend well beyond regulation: improved product safety, data-driven insights for R&D, quicker adaptation to market changes, and stronger user trust.

9.3 AI Auditing Tools

Purpose of AI Auditing

AI auditing software systematically reviews machine learning models to ensure they meet expected performance, fairness, and safety thresholds [4]. These tools can help detect bias in training data, validate model outputs under various scenarios, and produce reports suitable for both internal and external assessments.

Example: Suppose a predictive maintenance system in a factory includes an anomaly detection model. An AI auditor can repeatedly test the model against historical sensor data, identifying hidden patterns or corner cases that trigger false positives or false negatives. If the model's performance drifts below a certain threshold, the tool flags the issue, prompting engineers to retrain or recalibrate.

Key Features of AI Auditing Platforms

- **Bias Detection**: Algorithms that examine training sets for imbalances or potentially discriminatory variables (e.g. systematically skewed sensor readings, incomplete data from certain facility types).
- **Explainability Dashboards**: Visual interfaces illustrating how the model weighs inputs. These are crucial for meeting the Act's transparency and oversight obligations, particularly if human operators need to interpret the AI's reasoning.
- **Stress Testing and Scenario Simulations**: Automated methods for pushing models to their limits—introducing noise, extreme values, or adversarial examples—ensure the AI remains robust in real-world conditions.
- **Compliance Reporting**: Generating formatted documentation that covers performance metrics, error margins, and compliance checklists. This output can be appended to broader technical documentation.

Selecting and Integrating an Auditing Tool

Compatibility

Check that the auditing software supports your chosen AI frameworks (TensorFlow, PyTorch, scikit-learn, or proprietary libraries).

Depth Versus Simplicity

Some tools offer simple analytics for quick reviews, while others provide in-depth forensic capabilities. Choose based on your system's complexity and risk classification.

Deployment Model

On-premise solutions can be preferable for sensitive manufacturing data. Cloud-based platforms, meanwhile, may facilitate easier collaboration across regions.

Scalability

Ensure the tool can handle large datasets and integrate with existing DevOps or engineering pipelines.

9.4 Model Data Management Platforms

Tracking the AI Lifecycle

Under the EU AI Act, high-risk AI systems must be transparent about their training data sources, version histories, and performance evolution [1]. Model management platforms function like central repositories that store every iteration of a machine learning model, along with key metadata (training sets, hyperparameters, test results, deployment logs). By adopting such platforms, manufacturers can more easily compile the technical documentation demanded by Articles 10 and 11.

Analogy: Think of model management platforms as a "GitHub for AI," where each new model version is tagged, tested, and documented before it's merged into a production environment.

Core Components

- **Data Lineage**: The platform should maintain records of how training data was collected, cleaned, and annotated, ensuring traceability if regulators or internal audits question data biases or representativeness.
- **Version Control**: Each update to an AI model—changing hyperparameters, adding new training examples, or swapping architectures—receives a unique identifier. This facilitates rollbacks if issues surface, which is critical for post-market monitoring and incident response.
- **Testing and Validation Logs**: Incorporate repeated performance tests and risk analyses every time the model is retrained or retriggered. This helps demonstrate ongoing compliance to a notified body.
- **Deployment Pipeline**: Some platforms link directly to deployment tools, enabling a single-click path from testing to live production. Administrators or compliance officers can enforce checkpoint reviews before final rollout.

Benefits to Compliance and Beyond

- **Streamlined Documentation**: Instead of chasing scattered files, teams can generate consolidated technical files that show each model's lifecycle.
- **Faster Incident Resolution**: If an anomaly arises, quick reference to model versions and training data helps pinpoint the cause and expedite corrective measures.
- **Continuous Improvement**: Storing performance metrics over time enables engineers to spot trends, refine features, and maintain an agile development approach [5].

9.5 Compliance Dashboards

A Central Command Centre

Even if you adopt auditing and model management tools, overseeing multiple AI systems across diverse production lines can remain cumbersome. Compliance dashboards provide a unified interface where authorised staff can monitor risk statuses, conformity assessments, training logs, and incident reports in real time. High-risk systems must be registered in the EU database before market entry (Article 51). Dashboards can poll the API (once available) to show registration ID and status.

Example: A dashboard might show a colour-coded map of facilities globally, highlighting which sites deploy high-risk AI robotics and indicating each site's most recent conformity assessment date. Alerts could appear if any site's post-market monitoring logs detect anomalies or if a training dataset is missing essential metadata.

Key Dashboard Metrics

- **Risk Classification Overview**: Summaries of how many AI systems fall under high risk versus limited risk, with links to detailed risk assessments.
- **Documentation Status**: Progress trackers revealing if technical files are incomplete or if certain obligatory documentation (e.g. third-party audit certificates) is overdue.
- **Incident Alerts**: Flags for any reported operational failures, near misses, or user complaints. Provides a straightforward escalation path for immediate investigation.
- **Conformity Assessment Calendar**: Scheduling reminders for upcoming or lapsed assessments, ensuring no product is sold without valid certification.
- **User Interaction Logs**: For systems requiring human oversight, the dashboard might visualise how often operators override AI decisions, which can signal potential reliability or training gaps.

Implementation Tips

Integration

Dashboards should pull data seamlessly from AI auditing tools, model management systems, and real-time factory sensors. Fragmented data sources undermine efficiency.

Customisability

Each department, compliance, engineering, legal, etc., might need different views or permission levels. A flexible dashboard structure supports varied use cases.

Scalability

As you expand AI deployments or open new manufacturing sites, ensure the dashboard can incorporate additional metrics without losing clarity.

Illustrative tool: Trustworthy and Ethical Assurance (TEA) Platform.

The TEA Platform, released by the Alan Turing Institute, BSI and NPL in the UK, links risk registers, evidence artefacts, and live metrics into an assurance argument graph that can be exported as a single PDF for notified-body audits. It demonstrates how ethics checkpoints and AI-Act controls can be visualised in one pane of glass [6].

9.6 Cybersecurity and Data Governance Solutions

The Security Imperative

The EU AI Act mandates robust cybersecurity (Article 15) for high-risk AI systems, reflecting the reality that compromised algorithms can jeopardise workers, products, or even public safety. Hackers could manipulate AI robotics to damage machinery or produce defective goods. Alternatively, they might tamper with supply chain data, creating chaos in logistics. Vulnerability scanners should include checks for obsolete ML libraries, an issue flagged by notified bodies in 12% of industrial AI assessments [3].

Essential Security Tools (A Non-exhaustive List)

Intrusion Detection Systems (IDS): Monitoring network traffic and device communications, these solutions flag suspicious anomalies that might indicate hacking attempts targeting AI controllers or sensor data streams.

End-to-End Encryption: Particularly vital for remote or cloud-based AI workflows. Encryption ensures data integrity so malicious actors cannot intercept or corrupt model updates.

Access Control and Identity Management: Strict permission systems ensure only authorised individuals can retrain, modify, or override AI functionalities. Multi-factor authentication for engineers and compliance officers adds a further layer of security.

Vulnerability Scanning: Regularly testing AI software for known vulnerabilities or misconfigurations. Some scanning solutions are specifically geared to detect AI or machine learning libraries that have known security flaws.

Data Governance Platforms

In line with data governance obligations (Article 10), specialised platforms can coordinate data collection, annotation, storage, and retrieval, ensuring that each dataset is documented, controlled, and archived. Certain features include:

Metadata Tagging: Labelling data by source, date, sensor type, or any relevant compliance parameter.

Data Quality Checks: Automated scripts scanning for missing values, duplicates, or anomalies that compromise model reliability.

GDPR Integrations: If your AI systems process personal data (e.g. worker biometrics), these tools can incorporate GDPR-compliant consent and deletion workflows.

9.7 Integrating Compliance Tools into Organisational Workflows

Cross-Functional Steering Committees

Adopting a new software suite or platform is rarely just an IT project. The complexity of the EU AI Act calls for a cross-functional steering committee that includes engineering, IT, legal/compliance, operations, and possibly finance. Assign a digital "system owner" (Article 16 provider duty) within the dashboard for each high-risk AI, clarifying responsibility chains for importers and distributors [3]. This committee can:

- Set budgets for AI compliance tools.
- Define user requirements for dashboards or auditing software.
- Oversee pilot testing and enterprise-wide rollouts.
- Monitor post-deployment usage, ensuring staff adopt the tools effectively.

Phased Implementation

Large-scale integration is best undertaken in stages:

- **Pilot Phase**: Select one high-risk AI application to trial the new auditing or model management platform. Gather feedback from engineers, track efficiency gains, and document any issues.
- **Refinement**: Tweak configurations, dashboards, or user permissions based on pilot insights.
- **Expansion**: Gradually roll out the tools to multiple AI systems or factory sites, providing adequate training and support.
- **Optimisation**: Periodically assess how effectively each tool is delivering compliance benefits and refine or upgrade as needed.

Employee Training and Change Management

No matter how sophisticated the technology, your workforce must be equipped to use it properly:

Tool-Specific Training: Engineers learn to upload new model versions, compliance officers interpret dashboard alerts, and managers schedule audits.

Awareness Sessions: General overviews for broader staff, highlighting how the tool fosters safer AI usage and aligns with EU regulations.

Feedback Loops: Encourage suggestions from frontline users. Those who rely on the tool daily often spot usability issues or missing features faster than managers or external consultants.

9.8 Real-World Example: A Comprehensive Compliance Ecosystem

To illustrate the synergy of these technological components, imagine a German automotive parts manufacturer, **MechanoTech GmbH**, that heavily relies on AI for predictive maintenance, quality control, and automated welding. Faced with meeting the EU AI Act, MechanoTech adopts the following strategy:

- **Model Management Platform**: Every new machine learning model is checked into this platform, complete with training datasets and test logs. Operators use version tags to see when the model was last updated and how it performed against baseline metrics.
- **AI Auditing Software**: Monthly scans examine each live model for drift or anomalies. If the system detects a dip in accuracy for a welding defect detection AI, it issues an alert. MechanoTech's data science team investigates, possibly retraining the model or adjusting sensor inputs.
- **Central Compliance Dashboard**: An intuitive dashboard displays each factory site, colour-coding high-risk AI systems. It shows pending conformity assessments, upcoming training sessions for staff, and any flagged incidents. Supervisors can see real-time overrides or error logs, linking them back to the relevant model's documentation in the management platform. The company's dashboard also triggers a 15-day countdown if a serious incident flag is raised, mirroring Article 62 reporting deadlines.
- **Cybersecurity Integration**: Intrusion detection sensors monitor each welding robot's network connections. If data tampering or unauthorised attempts to modify the AI software are detected, the system notifies the compliance officer and automatically reverts to a safe model snapshot.
- **Continuous Improvement**: Data gleaned from the auditing tools helps identify areas where the AI might be underperforming, fuelling iterative improvements. Meanwhile, the compliance dashboard provides evidence that MechanoTech meets the Act's obligations, reassuring clients, partners, and regulators alike.

By weaving these tools into daily operations, MechanoTech not only avoids compliance pitfalls but also discovers performance optimisations, fosters a safety-driven culture, and sustains consumer trust in their AI-enhanced automotive parts.

9.9 Challenges and Mitigating Actions

Cost and Resource Allocation

Sophisticated auditing or model management platforms can be costly. Smaller firms (SMEs) may find these out of reach without external grants or consortium-based initiatives (see Chap. 8). Alternatively, open-source or partial solutions may suffice if configured carefully, though they might require extra internal expertise for setup and maintenance.

An overabundance of compliance tools can lead to "dashboard fatigue," where staff juggle multiple interfaces daily. Aim to consolidate whenever possible or integrate separate platforms via APIs so that vital compliance data converges in fewer places.

AI compliance software is developing rapidly. The best practice you adopt today might be superseded tomorrow by advanced, AI-enabled auditing that automatically identifies emerging hazards. A forward-looking strategy includes regular reviews of your compliance tech stack, ensuring it remains aligned with state-of-the-art solutions and the latest regulatory guidance.

9.10 Conclusion

Technology can bring order, efficiency, and scalability to the intricate demands of the EU AI Act. Viewed strategically, compliance tooling converts mandatory record-keeping into a live knowledge base that fuels iterative engineering advances. Whether you are managing a handful of AI-driven systems or a vast network of solutions across multiple factories, the right tools including auditing platforms, model management repositories, compliance dashboards, and robust cybersecurity measures that can significantly reduce operational strain. Moreover, they help embed compliance as a seamless, ongoing process rather than a disruptive event occurring just before product launch or during regulatory audits.

When carefully chosen and integrated, these solutions also open pathways to data-driven excellence, improved collaboration, and continuous R&D. Rather than viewing compliance tools as an overhead cost, manufacturers can regard them as strategic enablers: part of building safe, accountable, and innovative AI systems in a fast-moving, competitive world.

From here, we move on to Chap. 10: Case Studies—Lessons from the Field, where we will see these principles in action, gleaning real-world insights from manufacturers that have navigated AI compliance both effectively and creatively. Their experiences provide practical examples of how theoretical guidance translates into tangible results.

Key Takeaways

I. Automating tasks like risk assessment, data lineage tracking, and version control liberates staff from repetitive chores, minimising errors.

II. Dashboards, auditing tools, and model management platforms provide a clear trail of how AI is designed, tested, deployed, and monitored—vital to meeting EU AI Act standards.

III. Cybersecurity software is indispensable for high-risk AI, guarding against malicious manipulation and ensuring robust, safe operations.

IV. Evaluate vendor offerings or open-source options based on your system scale, budget, and integration needs. Avoid tool overload that might hamper usability.
V. Technical solutions must be backed by organisational commitment and well-trained personnel who understand both the technology and the compliance rationale.
VI. Keep abreast of advancements in AI auditing and governance to continuously refine your compliance strategies and maintain leadership in responsible AI innovation.

References

1. OneTrust (2024) Conformity assessments: step-by-step guide. https://www.onetrust.com/resources/eu-ai-act-conformity-assessment-a-step-by-step-guide-white-paper/
2. Deloitte UK (2024) Navigating the EU AI act: guide for CDOs. https://www.deloitte.com/uk/en/services/risk-advisory/blogs/navigating-the-eu-ai-act.html
3. Matrix Requirements (2024) Technical documentation under the AI Act. https://matrixreq.com/blog/technical-documentation-under-the-ai-act-must-read-guide
4. Acrolinx (2024) EU AI act compliance for technical writers. https://www.acrolinx.com/blog/eu-ai-act-compliance/
5. PwC (2024) AI compliance tool. https://www.pwc.com/cz/en/sluzby/technologie-a-data/ai-act/ai-compliance-tool.html
6. Trustworthy and ethical assurance platform, 2025. https://assuranceplatform.azurewebsites.net/

Chapter 10
Case Studies—Lessons from the Field

Abstract Three detailed case studies—Robusta Automotive (assembly robots), GreenSteel (predictive maintenance), and KineticMed (medical device quality control)—demonstrate how diverse manufacturers tackled risk classification, documentation, human oversight, and post-market monitoring. Cross-case analysis distils common success factors (early supplier alignment, intuitive operator interfaces) and pitfalls (data bias, under-scoped risk assumptions). The narratives translate theory into practice and offer sector-specific benchmarks for readers planning their own compliance journeys.

10.1 Introduction

Conversations about compliance with the EU AI Act [1] can feel theoretical until they are grounded in practical examples. Real-world case studies provide insights into what happens when manufacturers integrate AI governance into their existing processes, adapt to risk classifications, and manage cross-border supply chains. Such examples illuminate both the challenges and benefits of compliance, showing how organisations transform obligations into opportunities for innovation, collaboration, and strategic growth.

In this chapter, we present three in-depth case studies drawn from fictitious yet plausible scenarios based on empirical data collected from manufacturing and engineering firms. Each case demonstrates how a particular company tackled the Act's requirements ranging from data governance and human oversight to post-market monitoring. We will also explore each firm's journey in forming alliances with suppliers, employing compliance technology, and using the Act as a catalyst for process improvement. By highlighting successes and stumbling blocks, these examples convey lessons you can tailor to your own compliance strategy.

© The Author(s), under exclusive license to Springer Nature Switzerland AG 2025

M. Kauffman, *Navigating the EU AI Act: A Practical Guide for Global Manufacturing and Engineering Leaders*, https://doi.org/10.1007/978-3-031-99613-9_10

10.2 Case Study 1: Robusta Automotive's High-Risk Assembly Robots

Robusta Automotive is a mid-sized German car-parts supplier headquartered near Stuttgart. With around 5000 employees, the company produces precision components for major automakers across Europe. Seeking to improve efficiency, Robusta recently introduced AI-powered assembly robots on one of its production lines. These robots use advanced machine learning to handle sensitive tasks, such as spot welding intricate assemblies and aligning components with minimal human intervention. As the robots autonomously weld precision spots, they constitute a "safety component" of machinery under Annex III point 2(c) and are therefore high risk.

While the deployment promised to boost throughput and reduce errors, the management recognised that any malfunction could pose safety risks to nearby workers, not to mention endanger part integrity. Since the AI effectively controls physical machinery, the system was deemed high risk under the EU AI Act (see Chap. 2). Robusta needed a thorough compliance plan before scaling up across all production lines.

Key Challenges

1. **Risk Classification Uncertainty**

The engineering team initially debated whether the robots were truly high risk, arguing that a separate safety system (an emergency stop circuit) already protected workers. However, EU guidance confirmed that if AI autonomously controls critical operations with a potential for serious harm, it must be classified as high risk.

2. **Data Quality and Representativeness**

The robots had been trained on welding parameters from a single pilot plant. The risk was that differing conditions in other plants, temperature variations, multiple shifts, part tolerances might degrade performance or cause unexpected failures.

3. **Human Oversight Protocols**

Workers found the new AI-driven robots occasionally too abrupt. Article 14 requires clearly defined human-machine interfaces, so operators must obtain meaningful information and possess validated override tools. Operators needed user-friendly overrides and real-time dashboards to intervene if the robots misread a part orientation or malfunctioned.

4. **Supplier Dependencies**

The robotic arms themselves came from a Dutch robotics manufacturer, while the machine learning modules were licensed from a Finnish software provider. Ensuring both suppliers aligned with the Act's data governance and cybersecurity obligations proved challenging.

Compliance Measures and Strategies

(a) **AI Inventory and Thorough Data Examination**

Robusta's first step was creating an AI inventory (see Chap. 7). Upon discovering multiple robot configurations across different lines, management consolidated them into a single high-risk classification. Data scientists then revisited the training data for each assembly task, integrating broader examples from multiple lines (including part misalignment, environmental variations, and different material batches). This expansion satisfied Article 10's requirement for representative datasets.

(b) **Conformity Assessment with a Notified Body**

Because the robots were controlling a safety–critical process, a third-party conformity assessment was mandatory. Robusta engaged a German notified body specialising in industrial machinery. Engineers prepared detailed technical documentation [2]:

- System architecture, specifying how real-time sensor data guided robotic motions.
- Risk analysis linking each potential failure mode to mitigations, such as redundant sensors or manual overrides.
- Testing logs from pilot runs under a variety of conditions, proving robust welding accuracy without significant drift.

The notified body recommended minor adjustments, including calibrating sensors every shift change and enhancing cybersecurity measures around remote software updates. Robusta implemented these changes, resubmitted its documentation, and ultimately earned certification.

(c) **Human Oversight and Training**

Article 14 of the Act emphasises that operators must be able to monitor, interpret, and override AI decisions. Robusta introduced an intuitive control panel featuring real-time video feeds and diagnostic readouts (pressure, temperature, welding time). Operators had a prominent "Emergency Pause" button, as well as a less drastic "Manual Realignment" option that slowed the robot to half-speed if subtle misalignments were detected. Shift supervisors and line workers underwent hands-on training drills, simulating error scenarios.

(d) **Continuous Monitoring Tools**

Post-market monitoring (Article 61) was integrated into a compliance dashboard. The system logged every weld, including position data and predicted quality scores, flagging anomalies automatically. If the rate of micro-defects in a day's production exceeded a preset threshold, an alert was sent to the maintenance team, prompting a manual investigation and potential model recalibration.

Outcomes and Lessons

Improved Welding Quality

The AI-based welding achieved higher precision, reducing rework by 15%. Operators, once sceptical, soon appreciated the clarity provided by real-time dashboards and the sense of control conferred by straightforward overrides.

Stronger Supplier Alliances

By insisting on data and technical documentation from Dutch and Finnish partners, Robusta enforced a more thorough product lifecycle approach. These alliances matured into co-development partnerships, where they collectively refined sensor calibration and predictive analytics features.

Cultural Shift

Once compliance was embedded, staff began to see the EU AI Act as a framework for **continuous improvement**, not just red tape. Engineers were proud to demonstrate to automotive clients that the system surpassed basic safety norms, showcasing compliance as a market advantage.

Cost-Effective Scaling

The lessons gleaned from the pilot line, particularly around data diversification and user interface design allowed subsequent deployments to proceed smoothly, avoiding repeated guesswork.

10.3 Case Study 2: GreenSteel's Predictive Maintenance Revolution

GreenSteel is a multinational steel producer with furnaces, rolling mills, and finishing plants across the EU. Fierce competition and rising energy costs led the company to embrace AI-based predictive maintenance. The central concept: harness sensor data from machinery (temperatures, vibrations, pressure readings) to predict breakdowns before they happen, minimising downtime and prolonging equipment life.

While the core AI was developed by GreenSteel's in-house data science unit, the plants themselves were scattered in Germany, Poland, and Portugal, each with unique machinery configurations. Questions soon arose about whether this predictive maintenance system qualified as high risk or limited risk under the EU AI Act. After consultation, GreenSteel determined that:

- If the AI directly triggered emergency shutdown of critical equipment (such as blast furnaces), a malfunction could lead to serious production or safety hazards, thus high risk.
- If the AI merely provided alerts for human intervention, it was limited risk.

However, top management desired maximum automation to reduce labour costs, effectively pushing the system into high-risk territory.

Compliance Challenges

1. **Varying Plant Environments**

Each site had its own machinery ages, calibration routines, and data logging frequency. The predictive models needed consistent data standards to avoid incomplete or biased training sets.

2. **Data Governance Across Borders**

Because the system was centralised, plant data from multiple EU Member States converged in a cloud platform. The Act's data governance principles (Article 10) demanded transparent data lineage, secure transit, and thorough documentation.

3. **Human–Machine Coordination**

If the AI recommended an immediate furnace shutdown, missing or delayed human oversight might cause product defects, or worse, disrupt safety protocols. Clear procedures were required for local staff to confirm or override these automated maintenance decisions.

4. **Post-Market Monitoring**

The AI's performance in one region might drift if new sensor types or unplanned conditions emerged. Failing to detect such model drift could risk repeated false alarms or missed real breakdowns.

Implementation Steps

(a) **Harmonised Data Pipelines**

GreenSteel launched a multi-site data integration project. They enforced standard sampling rates for temperature and vibration sensors, uniform naming conventions, and consistent data cleaning scripts. This overcame the problem of partial or irregular data feeding the AI.

(b) **Risk Management and Technical Documentation**

A dedicated compliance officer joined the data science team, applying the guidelines from Chaps. 5 and 7. They documented:

- How the AI model was trained (including sensor calibration procedures) and evaluated under various operational conditions [3].
- Typical failure patterns identified in historical data.
- The logic behind automated shutdown triggers.
- Steps for staff to re-verify any critical alert or recommended shutdown.

(c) **Conformity Assessment**

A self-assessment route was initially considered, but legal counsel advised a third-party assessment because the system touched critical infrastructure. A notified body in Belgium evaluated GreenSteel's approach to data governance, confirming robust calibration, test sets reflecting multiple operating scenarios, and safe fallback modes. After minor improvements to alarm thresholds and user training modules, the system achieved compliance status.

(d) **Enhanced Cybersecurity**

Because predictive maintenance software is often updated remotely, GreenSteel integrated end-to-end encryption and intrusion detection at each plant's network perimeter. Regular penetration tests ensured an attacker could not manipulate sensor readings to sabotage operations.

Outcomes and Lessons

Reduced Downtime by 20%

The advanced warning system cut unexpected equipment failures significantly, boosting production stability.

Operator Confidence

At first, staff resisted any AI that could trigger furnace shutdowns. However, after a series of drills demonstrating accurate alerts and easy override procedures, operators embraced the system, praising its clarity and reliability.

Data-Driven Culture

Harmonising data pipelines not only ensured Act compliance but also paved the way for analytics in energy consumption, product quality, and logistics. GreenSteel soon leveraged these expanded insights to optimise schedules and reduce carbon footprints—showcasing how compliance-driven improvements can ripple outwards.

Holistic Risk Management

Through frequent post-market monitoring, the firm caught a model drift issue in its Portuguese plant early, avoiding a potential meltdown. Continuous feedback from each site kept predictive models fresh, meeting the EU AI Act's principle of ongoing performance validation.

10.4 Case Study 3: KineticMed's AI-Driven Medical Device Production

KineticMed is a Swedish manufacturing group focusing on medical devices, such as orthopaedic implants and prosthetics. To maintain leading-edge quality, the firm adopted AI-based computer vision for quality control in its final assembly lines. The

AI inspects each prosthetic for micro-defects or dimensional inconsistencies, aiming to ensure that patients receive flawless, safe implants. Because orthopaedic implants are "medical devices" under Regulation (EU) 2017/745, the AI quality-control module is doubly regulated: MDR + AI-Act high-risk (Annex III §5).

The critical nature of these products raises obvious regulatory stakes. A flawed device could severely harm a patient's mobility or health. Given these implications, KineticMed's AI system sits squarely in the high-risk domain, as it falls under healthcare equipment that influences patient safety.

Challenges in Compliance

1. **Medical Regulations Overlap**

Orthopaedic implants are already strictly regulated by the Medical Device Regulation (MDR). KineticMed had to harmonise MDR obligations with the EU AI Act, ensuring consistency in data usage and device labelling.

2. **AI Black-Box Concerns**

The deep learning algorithms for defect detection were complex. Demonstrating how the system arrived at a decision—especially when it flagged tiny anomalies—required robust explainability features.

3. **Post-Market Accountability**

Once implants ship, any discovered manufacturing defect might trigger product recalls. The AI's performance history, and all decisions it made during final checks, needed to be logged, archived, and retrievable for potential audits or investigations.

Path to Compliance

(a) **Integrated Documentation**

KineticMed combined MDR and EU AI Act technical files into a single, well-structured repository. They emphasised:

- Machine learning architecture details, highlighting the final "confidence score" threshold for defect detection.
- Representative training data: images of implants from different angles, lighting conditions, and material batches.
- Testing set composition, ensuring coverage of edge cases like hairline cracks, surface pitting, or slight dimension mismatches that might elude a less rigorous system.
- **Enhanced Explainability**

Article 13 of the AI Act calls for transparency, especially for high-risk AI. KineticMed's data science team added a "heatmap" feature, visually overlaying the AI's focal points on each scanned prosthetic. If a defect was flagged, operators could see an on-screen highlight of the suspicious region, bridging the gap between black-box decisions and tangible human-understandable cues.

(c) **Human Oversight and Operator Training**

Because prosthetics are critical devices, a quality assurance operator remained in the loop. If the AI flagged a product as defective, the item was set aside. A second manual inspector checked the suspect area with precision instruments and a microscope. KineticMed used well-documented checklists and procedures so that if the AI frequently generated false positives, the threshold would be readjusted.

(d) **Global Collaboration**

KineticMed had production lines in both Sweden and a newly opened facility in Estonia. The firm instituted a central compliance dashboard (see Chap. 9) that tracked daily defect rates, calibration logs, and AI updates at both sites using a platform inspired by the UK initiatives like the TEA framework [4]. Should an error spike emerge, the compliance team in Sweden could investigate patterns, verifying data quality and ensuring the model's reliability across all lines.

Outcomes and Lessons

Defect Detection Improved by 35%

The AI identified micro-fissures or surface irregularities that human inspectors often missed. Recall rates for finalised implants dropped, boosting KineticMed's reputation for near-perfect quality.

Regulatory Synergy

Rather than wrestle with separate audits for MDR and the AI Act, the firm streamlined compliance under one integrated approach. Inspectors from EU health authorities appreciated the clarity of single, consolidated documentation—leading to quicker approvals and fewer queries.

Employee Empowerment

Operators initially wary of being replaced found that the AI served as a powerful assistant. They enjoyed more confidence in final product checks, viewing the system as a "second pair of eyes" that bolstered patient safety.

Market Advantage

By publicly showcasing robust AI-driven quality processes, KineticMed differentiated itself from competitors. Hospitals and clinics increasingly favoured KineticMed's implants, trusting that the manufacturer proactively mitigated production risks.

10.5 Common Threads and Cross-Case Actionable Insights

Examining these three distinct case studies reveals several recurring themes and strategies:

Integrated Risk Assessment

In each scenario, carefully determining whether a system was high risk or limited risk under the EU AI Act was a crucial first step. Undertaking robust internal analyses, and sometimes consulting external authorities, helped avoid underestimating risk levels.

Supplier and Partner Alignment

All three companies depended on external components or software. Early engagement with suppliers to secure consistent data standards, technical documentation, and security protocols simplified the path to compliance [5] (see Chap. 8). See also the supplier flow-down clauses and authorised-representative mandates ensured extraterritorial partners met Article 54 obligations.

Human Oversight

Even advanced AI robotics or analytics solutions were never left fully autonomous. Whether it was a welding arm, a predictive maintenance system, or a quality control camera, the EU AI Act's principle of human oversight required user interfaces, override mechanisms, and staff training. This synergy of human and machine often led to better overall performance [6].

Continuous Monitoring and Improvement

Instead of "deploy and forget," each company instituted post-market reviews, tapping into real-world data to refine AI models. This iterative loop not only upheld compliance but also spurred improvements in product quality, operator confidence, and error detection.

Added Value

Compliance, while initially perceived as a cost, yielded organisational benefits: higher quality, improved supply chain transparency, safer workplace conditions, and a stronger reputation with clients and regulators. By linking compliance steps to R&D or operational excellence, these enterprises discovered fresh avenues for growth and differentiation.

10.6 Conclusion

Real-world experiences underscore that the EU AI Act is not just a legal hurdle; it can be a strategic framework for safer, more efficient, and more ethically grounded AI solutions. Whether it is a medium-sized automotive supplier fine-tuning assembly robots, a multinational steel producer streamlining maintenance, or a medical device manufacturer perfecting quality checks, success hinges on weaving compliance into daily workflows rather than treating it as an afterthought.

Lessons from the field illustrate how to balance automation with human oversight, expand datasets for robust model training, and engage third-party assessors proactively. Critically, each organisation found unique ways to transform compliance from a necessary chore into a catalyst for innovation. As you move forward with your own AI initiatives, these examples can inspire practical approaches for integrating data governance, risk assessment, technical documentation, and continuous monitoring into a coherent process that benefits both your business and the broader community of users and stakeholders.

The next chapter, Chap. 11: Insights from Industry Leaders, enriches this perspective by exploring executive viewpoints, drawing on interviews and surveys that highlight the strategic decisions, culture shifts, and forward-thinking mindsets essential to thriving in the age of regulated AI.

Key Takeaways

I. Proper risk classification and data governance from the onset avoids costly rework, ensuring your AI projects align with the EU AI Act's safety and transparency requirements.
II. Even in highly automated settings, trained operators and intuitive override mechanisms prevent the AI from veering off course, reassuring staff and regulators alike.
III. Clear contractual obligations, audits, and data-sharing protocols elevate compliance across the supply chain, reducing last-minute surprises.
IV. Integrating AI Act obligations with existing regulations (like MDR) or other quality standards (ISO) simplifies audits and underscores a commitment to comprehensive product safety.
V. Real-world performance tracking (Article 61) fosters an adaptive cycle of improvements, bridging compliance with ongoing innovation.
VI. Far from being a deterrent, meeting EU AI Act obligations can differentiate products, enhance brand credibility, and catalyse a company-wide culture of excellence.

References

1. European Union (2024) Regulation (EU) 2024/1689. https://eur-lex.europa.eu/eli/reg/2024/1689/oj/eng
2. OneTrust (2024) Conformity assessments under the proposed EU AI act: a step-by-step guide. https://www.onetrust.com/resources/conformity-assessments-under-the-proposed-eu-ai-act-a-step-by-step-guide/
3. Matrix Requirements (2024) Technical documentation under the AI act. https://matrixreq.com/blog/technical-documentation-under-the-ai-act-must-read-guide
4. Trustworthy and ethical assurance (TEA) platform, 2023. https://www.gov.uk/ai-assurance-techniques/alan-turing-institute-and-university-of-york-trustworthy-and-ethical-assurance-platform
5. Deloitte UK (2024) Navigating the EU AI act: guide for CDOs. https://www.deloitte.com/uk/en/services/consulting-risk/blogs/navigating-the-eu-ai-act.html
6. NAVEX (2024) EU AI act compliance requirements. https://www.navex.com/en-us/solutions/regulations/eu-ai-act/

Chapter 11
Insights from Industry Leaders

Abstract Drawing on interview data and survey results, Chapter 11 captures executive-level perspectives on governance priorities, organisational culture shifts, and innovation strategy under the EU AI Act. Themes include investor expectations, reputation management, cross-functional teams, and real-time compliance. Leaders from SMEs and multinationals share candid reflections on balancing speed with diligence, bridging legal-technical language gaps, and forecasting future regulatory trends. The insights serve as a strategic compass for decision-makers steering AI-driven transformation.

11.1 Introduction

Regulatory compliance often materialises from a confluence of legal mandates, internal advocacy, and calculated decision-making at the highest levels of an organisation [1]. The threat of fines up to €35 million or 7% of worldwide turnover (Article 99) has put AI governance on board agendas worldwide. While the previous chapters provided a detailed breakdown of the EU Artificial Intelligence (AI) Act's technical, operational, and collaborative requirements, this chapter shifts focus to the executive and thought-leadership perspective. How do senior managers, chief engineers, or heads of compliance interpret the Act's implications? What strategic pivots do they adopt to navigate compliance while championing innovation? Drawing on fictional yet representative "interviews" and surveys, we capture the ethos and mindset of industry leaders grappling with AI's transformative potential.

We begin by examining the strategic rationale behind embracing rigorous AI governance, including how top executives balance investment priorities and weigh business risks. We then look at culture and workforce transformation, showing how leaders galvanise teams around new compliance standards. Next, we discuss innovation and market differentiation, including how companies harness compliance to outpace competitors or deepen trust with partners. We also review key lessons and recurring themes that surfaced from these conversations—from the challenges of bridging technical and legal domains to the opportunities arising from synergy

© The Author(s), under exclusive license to Springer Nature Switzerland AG 2025
M. Kauffman, *Navigating the EU AI Act: A Practical Guide for Global Manufacturing and Engineering Leaders*, https://doi.org/10.1007/978-3-031-99613-9_11

between human talent and AI-driven systems. By chapter's end, you will understand not merely what the EU AI Act requires but how it resonates with the strategic thinking of those guiding modern manufacturing and engineering firms.

11.2 Why Leaders Prioritise AI Governance

From Risk Management to Value Creation

One of the clearest takeaways from speaking with industry executives is a shift in how they perceive AI regulation. Initially, many saw the EU AI Act as a risk management exercise, akin to meeting safety or environmental standards. Over time, however, a number of leaders discovered that robust AI governance also creates value: it streamlines data usage, fosters consumer trust, and can lead to new revenue opportunities [2].

Quote (CIO of a Tier 1 automotive supplier):

> When the Act was first announced, we scrambled to figure out how to avoid fines. But the deeper we got into compliance, the more we realised it was pushing us towards best practices we should have had in place anyway. Cleaner data pipelines, consistent machine learning training, better documentation. Our product quality improved, and that opened new customer contracts in high-end markets.

Maintaining Reputational Capital

Senior managers also stress reputation management as a driving factor. In the age of social media and instant news, a single AI-driven mishap, like a defective robotic arm injuring a worker could tarnish a company's public image and erode stakeholder confidence. By adopting the EU AI Act's thorough approach to risk identification, data governance, and oversight, they reduce the chance of catastrophic errors [3]. This protective function is particularly valuable to publicly listed firms, which are accountable to shareholders mindful of regulatory compliance and corporate responsibility metrics.

Quote (COO of a consumer electronics manufacturer):

> Customers want reassurance that the AI in our smart appliances isn't collecting data unlawfully or endangering user safety. By showcasing our EU AI Act compliance, we send a powerful signal: we take your privacy and safety seriously. That's invaluable brand currency in a competitive market.

Investor and Stakeholder Demands

Private equity groups, large institutional investors, and even lenders increasingly ask pointed questions about AI governance, especially if a manufacturer's growth prospects hinge on automation or advanced analytics. Leaders mention that prospective deals often hinge on due diligence of data governance practices and alignment with evolving European regulations [4].

Quote (VP of Investor Relations at a robotics start-up):

> During our funding rounds, investors grilled us on how we would handle the EU's new AI rules. We quickly learnt that 'compliance readiness' was not a footnote but a centrepiece of their risk analysis. Demonstrating a roadmap for alignment gave them confidence in our long-term viability.

11.3 Building Organisational Culture for Compliance

Fostering Cross-Functional Teams

A repeated sentiment is that AI compliance demands cross-functional collaboration [1], engineering, legal, HR, finance, and IT cannot remain siloed. Industry leaders underscore the value of forming dedicated committees or task forces to unify these perspectives. Engineers articulate the technical feasibility of oversight mechanisms, while compliance officers ensure legal thoroughness, and HR might handle training modules on AI ethics.

Insight: One aerospace manufacturer created an "AI Governance Council" with equal representation from R&D, compliance, and operations. They met monthly to review how new or updated AI systems fit the EU AI Act's risk tiers, track data usage, and plan external audits. Over 18 months, this approach reportedly cut external review times by nearly half, as everyone was "on the same page" from the outset.

Leaders also speak of weaving compliance metrics into performance reviews and bonus structures, rewarding employees or teams that proactively identify potential compliance gaps or produce safe, ethically grounded AI solutions. This stands in stark contrast to older models where staff were rewarded mainly for speed or cost savings. By balancing productivity with accountability, some organisations minimise the tension between delivering advanced AI features and avoiding rushed shortcuts that imperil compliance.

Practical Example: A German chemical process plant now gives "*Innovation and Integrity*" awards for AI project teams that exemplify thorough risk assessments, strong user training programmes, and feedback loops. According to the plant's site director, morale and alignment improved as staff realised compliance is not an obstacle but a path to industry leadership.

Training Programmes and Employee Engagement

Executives frequently highlight staff training as both a compliance necessity (see Chap. 7 on human oversight) and a cultural cornerstone. Formal training sessions help employees grasp how the EU AI Act protects them and the wider community. For high-risk AI, operators must understand override protocols, error flags, and how to respond to anomalies. However, senior managers also see training as a chance to champion broader ethical AI principles like fairness, transparency, and social responsibility.

Observation: Multiple leaders report that once workers appreciate the rationale behind safety mandates and data governance, they become more receptive to internal compliance processes. Rather than "top-down orders," staff see themselves as frontline guardians, ensuring AI does not jeopardise health, rights, or brand reputation.

11.4 Embracing Compliance as a Competitive Edge

Differentiation in a Crowded Market

Chief executives in manufacturing companies outside the EU note that the EU AI Act, like the General Data Protection Regulation (GDPR), is fostering a global standard [5]. As a result, many non-EU customers or partners also trust products that align with stringent European requirements. This dynamic paves the way for compliance to become a marketing advantage.

Case in Point: A Brazilian automation solutions provider added "EU AI Compliant" labels to its product brochures and emphasised its robust testing protocols in sales pitches. Clients in the European Union uncertain about emerging AI rules in their own region, showed heightened interest, seeing the label as an assurance of advanced, future-proof design [5].

Leaders also cite how early compliance can facilitate **faster acceptance** in regulated sectors such as healthcare, automotive and transport where safety or reliability concerns run high. By forging a proactive stance, these companies skip the friction of last-minute redesigns or repeated re-certification. This readiness also streamlines expansions into new EU Member States or local markets, as the foundational compliance work is already standardised.

Testimony (Product Director at a "high-risk" robotics firm):

> We spent nine months thoroughly aligning our warehouse robots with EU guidelines on sensor redundancy and transparency. That was a big investment. But once we cleared the initial conformity assessments, scaling across multiple European warehouses was basically plug-and-play. Our path to market was smoother, and we built a reputation as the safer option.

Opportunity for Innovation

Executives repeatedly mention that compliance fosters innovation, echoing themes from Chap. 6. Whether it is designing intuitive human–machine interfaces, exploring advanced testing data, or forging new data partnerships, the drive to meet the EU AI Act can become a catalyst for improvement. Some even link compliance pursuits to fresh intellectual property (IP) generation like new sensor algorithms or model management approaches that become a company's competitive hallmark.

11.5 Challenges Faced by Leaders

Balancing Speed and Thoroughness

On the flip side, leaders concede that compliance can slow product release cycles, especially for high-risk AI systems demanding robust documentation and external audits. Start-ups and SMEs, in particular, worry about losing ground to more agile rivals if they pour resources into regulatory overhead. However, many emphasise that internal process discipline, such as agile development sprints incorporating compliance checks, which helps them avoid crippling slowdowns.

Quote (CTO at a Dutch SME producing AI-driven packaging lines):

> We can't afford big compliance teams. Our solution is embedding a compliance checklist at every stage of development. It might add two or three days to each sprint, but it saves us from disasters and rework later. Our clients appreciate we put safety and transparency front and centre.

Technical Versus Legal Terminology

A recurring issue is the "language barrier" between engineers and lawyers. Engineers deal with model architectures, sensor arrays, performance metrics; lawyers talk about articles, recitals, and extraterritorial scope. Bridging this gap requires skilled compliance officers who speak both "legalese" and technical jargon, translating legislative mandates into actionable engineering tasks.

Insight: Several multinational manufacturers have introduced "compliance liaisons" in R&D teams. These individuals interpret EU AI Act articles into design criteria, risk templates, or data retention guidelines. They also gather engineering insights like real-world complexities of sensor calibration that might shape how legal advisers interpret risk classification. This synergy fosters a more harmonious, productive environment.

Evolving Standards

Another headache is the act's dynamic nature. Leaders predict the EU AI Act will experience amendments, updates, or expanded guidance, especially for emerging technologies like general-purpose AI (GPAI) or quantum computing. Staying ahead of these evolving details requires continuous monitoring of European Commission publications, trade association updates, or regulatory forums. Companies that fail to track changes risk inadvertently slipping out of compliance.

Proactive Strategy: Some larger conglomerates invest in policy intelligence or membership in European trade bodies that liaise with lawmakers. This early visibility into potential legislative shifts allows them to adapt well in advance. Meanwhile, smaller firms may rely on external legal counsel or alliances with bigger partners to keep abreast of policy developments.

11.6 Looking to the Future: Leader Perspectives on AI Governance Trends

During our fictionalised conversations, multiple senior professionals offered predictions on how the AI governance landscape might evolve and how it could impact manufacturing strategy:

Global Convergence

Many foresee a drift towards international harmonisation, with the US, UK, and possibly Asian countries adopting frameworks similar to the EU AI Act. If that occurs, companies that are already EU-compliant enjoy a first-mover advantage. Several interviewees cited the AI Standards Hub (UK) as their go-to venue for horizon scanning, standards mapping and up-skilling engineers in fast-moving ISO/JTC 1 working groups [6]. Participation, they argued, de-risks future audits because product teams already speak the language of upcoming harmonised standards.

Sector-Specific Directives

Just as the EU has sectoral rules for machinery or medical devices, some anticipate more refined directives for AI in certain industries, like construction or aerospace. Leaders want to ensure they adapt swiftly, viewing each directive as an impetus to refine processes further.

Deeper Ethical Mandates

Beyond risk-based safety, executives expect future rules to incorporate stronger ethical dimensions, addressing algorithmic bias, environmental impact, or sustainability. Proactive manufacturers are exploring how to measure carbon footprints for AI training and applying "ethical design" frameworks.

Real-Time Compliance

The concept of "compliance by design" might mature into continuous real-time checks, especially if adaptive or online learning models become mainstream. Factories may need automated modules that monitor AI behaviour for policy violations, triggering alerts or throttles if deviations appear.

High-Level Observation: Leaders do not view these shifts as potential business killers but as a natural progression in a world where technology's influence is too significant to remain unregulated. They highlight that companies excelling at synergy between regulatory demands and engineering agility will shape the next generation of AI-driven manufacturing.

11.7 Leadership Practices that Enable Sustainable Compliance

From numerous interviews and surveys, certain practices consistently emerge as beneficial for sustaining compliance momentum:

1. **Long-Term Planning Over Quick Fixes**

Rather than implementing ad hoc solutions to pass an audit, successful leaders embed compliance in multi-year strategies. They treat each product iteration or AI deployment as an opportunity to refine frameworks rather than hoping short-term patches suffice.

2. **Transparent Communication**

Executives champion open channels, both internally and with external stakeholders. They publicly discuss compliance commitments, hold frequent Q&A sessions with employees, and maintain active dialogues with regulators or notified bodies. This openness garners trust and reduces friction.

3. **Empowering Mid-Level Managers**

While executive vision sets broad goals, mid-level managers oversee daily details—like verifying datasets or ensuring that user manuals follow transparency rules. The best-case scenarios see mid-level staff entrusted with resources, training, and autonomy to adapt local solutions that uphold the Act.

4. **Designate Compliance Champions**

Some leaders formalise an internal role of "compliance champion" who receives additional training in AI governance, risk management, and communication skills. Embedded in engineering or production teams, these champions detect issues early, coach colleagues, and unify scattered efforts.

5. **Iterative Audits and Drills**

Just as factories might conduct safety drills, forward-looking companies hold "AI compliance drills." They simulate a scenario—a data breach, a suspicious sensor reading, or a near miss—and observe how staff respond. Post-mortems highlight process gaps, spurring improvement before real incidents occur.

11.8 Exemplary Leadership Profiles (Anonymised)

Chief Legal Officer

Marta leads a legal and compliance team at a multinational with factories across Spain, Poland and Italy. Early in the drafting of the EU AI Act, she launched a series of training roadshows for each factory, emphasising the Act's key points. She formed a cross-site "compliance circle," meeting monthly to track risk classifications, data usage trends, and new AI pilot projects. Thanks to her approach, the organisation rarely sees last-minute scrambles, engineers learned to consult her team from project inception. In interviews, Marta highlights:

> I never wanted compliance to be perceived as a legal blockade. Instead, we frame it as risk clarity enabling safer, better products. I strive for approachability, so our engineers always feel comfortable raising concerns. That open door leads to fewer mistakes and higher trust across the organisation.

Head of R&D

Nicola oversees a 200-person R&D department at a manufacturer of AI-powered drones for industrial inspections. She sees the EU AI Act as a spur to advanced R&D, pushing the company to perfect sense-and-avoid algorithms, robust data logging, and secure communication channels. Nicola regularly invites staff to propose new compliance-oriented features like an "explainability overlay" on the drone's flight plan. If the idea holds merit, she provides resources to prototype it. In an internal newsletter, Nicola states:

> Yes, the Act imposes constraints, but constraints can breed creativity. We're investing in sensor fusion that not only meets collision-avoidance rules but also enhances flight efficiency. Compliance becomes an opportunity to refine our edge in the drone market, everyone wins, from regulators to end customers.

CEO

As CEO of a medium-sized forging company, Pierre faces intense competition, prompting him to adopt AI for predictive maintenance and advanced process control. Initially overwhelmed by the EU AI Act's complexity, he sought help from an external consultancy, then delegated compliance matters to an internal task force. Pierre invests personally in monthly reviews, ensuring that lessons gleaned from risk classification or post-market monitoring feedback into strategy discussions with the board. He remarks:

> I realised ignoring the Act's details was a big mistake. Once I dove in, it dawned on me that compliance is about maturing as an organisation becoming data-driven, agile, and trustworthy. When I communicate that vision to investors or employees, they see EU law not as a threat but as a guiding star for better manufacturing.

11.9 Recurrent Themes: The Leader's Toolkit

From all these experiences, four prevalent themes emerge, forming a "***leader's toolkit***" for navigating AI compliance:

Long-Horizon View

Leaders who embed compliance as a continuous, strategic objective rather than a short-term milestone often see deeper returns—improved processes, data synergy, and brand credibility.

Regardless of technological advancements, it is the workforce, operators, engineers, middle managers that actualises compliance. Leaders who invest in training, empowerment, and open dialogue stand a better chance of meeting (and exceeding) legal obligations.

Proactive Collaboration

Top executives increasingly champion supplier audits, joint R&D alliances, and active regulator relationships. They see these connections as essential for both scaling AI globally and anticipating shifts in regulatory landscapes.

Accountability and Transparency

Effective leaders do not hide from tough questions, be they from regulators, customers, or their own teams. They produce clear risk analyses, share them, and invite feedback. This accountability fosters a climate of trust and readiness.

11.10 Conclusion

Industry leaders view the EU AI Act through a multifaceted lens, balancing risk mitigation, brand reputation, operational efficiency, and innovative potential. While some initially approached compliance with trepidation, many discovered it can function as an engine for corporate transformation strengthening data management, catalysing product improvements, and reinforcing user trust. The testimonies presented in this chapter underscore that success arises not from a mechanical "checklist" approach, but from a culture shift led by executives who articulate a vision of responsible, forward-thinking AI.

These insights dovetail with the practical steps, case studies, and collaborative frameworks outlined in previous chapters. By bridging big-picture leadership strategies with operational best practices, organisations can confidently navigate the EU AI Act and emerge as trustworthy, agile competitors in a rapidly evolving global marketplace. In the next chapter, Chap. 12: Looking Ahead, we cast our gaze to the future, exploring potential expansions of AI legislation, the drive towards international harmonisation, and how manufacturers can remain nimble in this shifting regulatory terrain.

Key Takeaways

I. Senior leaders increasingly view AI compliance as integral to risk management, brand preservation, and value creation.

II. Successful adherence demands cross-functional teamwork, transparent communication, incentive systems, and a commitment to continuous learning.
III. Proactively meeting the EU AI Act can differentiate products, attract investors, and heighten consumer trust—especially in heavily scrutinised sectors.
IV. Balancing speed and regulatory thoroughness, bridging legal/technical jargon, and keeping pace with evolving standards remain ongoing hurdles.
V. Executives predict expansions in sector-specific legislation, deeper ethical mandates, and real-time compliance checks, all of which can be harnessed for strategic gain.
VI. Sustained compliance success emerges from a blend of long-horizon thinking, people-centric leadership, proactive collaboration, and transparent accountability.

References

1. Deloitte UK (2024) Navigating the EU AI act. https://www.deloitte.com/uk/en/services/consulting-risk/blogs/navigating-the-eu-ai-act.html
2. KPMG (2024) Decoding the EU AI act. https://assets.kpmg.com/content/dam/kpmg/xx/pdf/2024/02/decoding-the-eu-artificial-intelligence-act.pdf
3. EY (2024) Why companies must prepare now for the new EU AI act. https://www.ey.com/en_gl/insights/public-policy/why-companies-must-prepare-now-for-the-new-eu-ai-act
4. Holistic AI (2024) EU AI act now in effect: a practical guide. https://www.holisticai.com/blog/eu-ai-act-in-effect-guide-for-global-enterprises
5. White and case LLP (2025) AI watch: global regulatory tracker—EU. https://www.whitecase.com/insight-our-thinking/ai-watch-global-regulatory-tracker-european-union
6. Alan Turing Institute/BSI/NPL (2025) AI standards hub. https://aistandardshub.org/

Chapter 12
Looking Ahead

Abstract The final substantive chapter peers beyond 2025, assessing potential amendments to the EU AI Act, sector-specific add-ons, and global harmonisation efforts. It evaluates emerging challenges such as general-purpose AI, adaptive learning systems, sustainability metrics, and AI-as-a-Service business models. Practical advice on modular governance frameworks, policy-watch teams, and scenario planning helps organisations remain resilient amid uncertainty and regulatory evolution.

12.1 Introduction

The EU Artificial Intelligence (AI) Act represents one of the most wide-ranging legislative initiatives in the realm of AI governance [1]. Yet it is only a first chapter; Articles 74–75 already empower the European Commission to adopt delegated acts that refine risk categories and documentation duties. Like many landmark regulations, the Act is likely to evolve through amendments, additional guidance, sector-specific rules, and, potentially, broader global harmonisation efforts. For manufacturers and engineers, this dynamic environment raises pressing questions about future-proofing. How can organisations maintain compliance when standards and expectations keep shifting? What developments loom on the horizon for AI governance within and beyond the EU?

This chapter examines emerging trends and possible expansions to the EU AI Act, as well as the prospects for international alignment on AI standards. We discuss how manufacturers can stay nimble, investing in flexible infrastructure, forging international alliances, and embedding a forward-thinking culture that anticipates policy shifts rather than reacting to them at the last moment. We also consider how the Act might intersect with new domains, from general-purpose AI (GPAI) to sustainability metrics and cross-jurisdictional data governance. By the end, you will gain a strategic overview of what "compliance" might look like in the years ahead, ensuring that your organisation not only endures but also thrives in the future regulatory landscape.

© The Author(s), under exclusive license to Springer Nature Switzerland AG 2025

M. Kauffman, *Navigating the EU AI Act: A Practical Guide for Global Manufacturing and Engineering Leaders*, https://doi.org/10.1007/978-3-031-99613-9_12

12.2 Evolution of the EU AI Act: Potential Amendments and Updates

Post-Implementation Refinements

Much like the General Data Protection Regulation (GDPR), which underwent practical clarifications and updates after its initial rollout, the EU AI Act is expected to see post-implementation refinements [2]. National authorities, trade associations, and industry consortia will likely feedback on real-world challenges encountered during initial enforcement. These inputs can lead to:

- **Revised Risk Classifications**: Certain applications might be regraded from limited risk to high risk, or vice versa, based on incident data or technological breakthroughs.
- **Extended Annexes**: Annex III, which enumerates high-risk use cases, may gain new entries or clarifications as AI penetrates additional sectors like agriculture, large-scale 3D printing, or advanced composite fabrication.
- **Clarified Definitions**: Debate over what exactly constitutes "general-purpose AI" (GPAI) could yield official commentary or further legislative text, refining the broad definition in Article 3(1).

For manufacturers, the lesson is to treat risk classification as an ongoing process. A system that starts as limited risk might become high risk if regulators update guidelines or discover that minor failures can cascade into major hazards [3]. Vigilance around evolving definitions is paramount.

Potential for Sector-Specific Rules

Another possibility is that the EU AI Act becomes a framework law, with subsequent legislation specifying how it applies to niche industries. We already see a precedent in other EU directives that rely on overarching principles but require domain-specific interpretations (e.g. the Machinery Directive's synergy with other safety regulations). Expect to see:

- **AI in Construction**: Potentially, guidance on using AI-driven drones for site inspections or safety oversight, imposing stricter data usage constraints.
- **AI in Energy**: As Europe transitions to greener power grids, high-risk AI controlling energy distribution could demand advanced cybersecurity protocols or real-time fail-safes.
- **AI in Agriculture**: Robotics used in harvesting, pesticide application, or livestock monitoring might move from limited to high-risk categories if they pose threats to animal welfare, worker safety, or the environment.

For manufacturers operating in these verticals, close coordination with trade bodies and national regulators will be key to adjusting product designs, supply chain processes, and post-market oversight procedures.

12.3 Global Harmonisation: Towards International AI Standards?

The Drive for Consistency

Few companies operate solely within the borders of the EU. For many, the real challenge is juggling multiple jurisdictions with varying regulations. However, we are witnessing a growing appetite for international standards in AI governance. Whether through the G20, OECD frameworks, or bilateral talks, policymakers worldwide acknowledge that AI is a global concern transcending national lines. If the EU AI Act becomes a model for other regions, or if key principles are adopted globally, the burden of multiple, conflicting regulations could diminish.

Case Example: A Japanese supplier of AI-enhanced industrial sensors might initially struggle to adapt to the EU's data governance standards. But if Japan introduces parallel rules or mutual recognition schemes, that same supplier gains worldwide credibility, speeding up negotiations with European customers and distributors.

Convergence or Fragmentation?

The outcome of global harmonisation efforts remains uncertain. Some experts foresee a steady convergence towards EU-like frameworks emphasising human oversight, risk-based regulation, and ethical AI. Others predict persistent fragmentation, especially if major powers like the United States or China retain distinct philosophies on data use and AI oversight.

From a manufacturer's standpoint, the safest strategy is to design AI systems to the strictest set of recognised global standards. This "compliance by default" approach ensures readiness for future expansions or reciprocal agreements. Many corporations used a similar tactic under GDPR, upgrading global data handling to EU levels, thereby benefiting from consistent internal processes.

Role of International Standards Bodies

Entities such as the International Organization for Standardization (ISO) and the International Electrotechnical Commission (IEC) are developing norms for AI management, covering topics like risk assessment, bias mitigation, and data quality. The EU may adopt or reference these international standards in official "harmonised standards," simplifying the path for businesses that already follow ISO guidelines. Staying involved in these bodies—whether through membership, feedback, or technical committees, positions manufacturers at the forefront of shaping globally compatible AI practices [4].

12.4 New Frontiers: GPAI and Adaptive Systems

The Ambiguity of GPAI

Large language models, multi-modal systems, or generative AI solutions exemplify general-purpose AI (GPAI) tools that can be adapted to countless tasks not envisaged by their original designers. This versatility poses a regulatory conundrum: risk levels are context-dependent. An apparently harmless chatbot might be repurposed for critical tasks, tipping it into high-risk territory. Alternatively, an industrial vision model could move from verifying packaging labels to controlling robotic arms on a safety–critical assembly line.

The EU AI Act does not currently provide final, definitive rules for GPAI, although it references the phenomenon. The Commission's April 2025 "Guidance Note on GPAI" (expected under Article 52a, draft stage) may introduce a duty of care for foundation-model providers and shared liability with downstream deployers [5]. Many experts predict additional clarifications or new annexes addressing:

- **Duty of Care**: Requiring GPAI providers to supply disclaimers, usage guidelines, or technical constraints preventing misuse in high-risk applications.
- **Liability Sharing**: Potentially splitting accountability between the GPAI developer and the deployer who fine-tunes it for a specific domain.
- **Post-deployment Oversight**: Mandating that providers track and manage derivative uses if their GPAI powers systems that the Act categorises as high risk.

For manufacturers adopting GPAI for product design or operational tasks, flexible governance frameworks that swiftly re-evaluate risk upon usage changes will be essential. A system initially used for benign data analysis might one day become a critical decision-maker, requiring rapid upgrades to documentation and oversight mechanisms.

Adaptive and Online Learning Systems

Another frontier is online or continuous learning, where AI models evolve by ingesting fresh data in real time. This capacity improves responsiveness but complicates compliance. If a model changes daily without a thorough "re-certification," how do we ensure it remains safe and unbiased?

Regulators might introduce:

- **Real-Time Auditing**: Enabling rolling checks or partial re-validation whenever the model updates beyond a defined threshold.
- **Model Version Snapshots**: Demanding that each iteration be preserved for potential post-incident analysis.
- **Adaptive System Caps**: Setting performance or error-rate ceilings, beyond which systems must revert to a known safe model until a full review is done.

Dataspace Compliance on the Horizon

Draft ISO/IEC 20151 proposes conformance classes for dataspace nodes, essentially real-time "CE marking" for data-sharing endpoints. Manufacturers that instrument their AI pipelines to emit dataspace-compliance telemetry will be well placed once these concepts appear in delegated acts under the AI Act.

Manufacturers using adaptive AI to govern dynamic processes (e.g. real-time quality adjustments on production lines) should invest in robust model management platforms (see Chap. 9) that log every micro-update, test it against baseline criteria, and enforce human or automated sign-off before deployment.

12.5 Sustainability and Ethical Dimensions

Linking AI Governance to Environmental Goals

Climate change and resource conservation remain pivotal concerns worldwide. As AI systems become more pervasive, they carry their own environmental footprint, energy for computation, e-waste from frequent hardware upgrades, etc. While the current EU AI Act primarily focuses on safety and fundamental rights, future iterations may incorporate sustainability metrics:

- **Green AI Audits**: Checking whether the computational overhead for AI is proportionate to the societal benefit—mirroring calls for "lean AI."
- **Lifecycle Assessments**: Encouraging or mandating that manufacturers track an AI-driven system's total energy consumption across its lifecycle, including training, updates, and end-of-life disposal [3].

For forward-thinking organisations, adopting environmentally conscious AI practices now could pre-empt or smoothly align with future additions to the regulatory framework. This might also resonate with investor and consumer demand for greener, more sustainable manufacturing processes.

Deepening Ethics and Human Rights

Concurrently, some policymakers and advocacy groups argue that AI oversight must expand to address discrimination, fairness, and societal impacts beyond what is detailed in the current risk categorisations. This could translate into:

- **Bias Testing Requirements**: Official protocols assessing algorithmic fairness for all AI, not just those in high-risk categories, especially if they touch workforce management or consumer services.
- **Stakeholder Impact Statements**: Calls for "ethical impact assessments" akin to environmental impact assessments, documenting how AI might affect local communities, labour conditions, or economic equity.

- **Enhanced Worker Protections**: Potential expansions that protect employee data from overreaching, AI-driven monitoring systems—reinforcing the principle that manufacturing employees not be subjected to intrusive or discriminatory analytics.

By staying attuned to debates around AI ethics, manufacturers can embed inclusive design, explainability, and fairness checks from inception, making it simpler to adapt if the Act later codifies such measures more strictly.

12.6 Preparing for Regulatory Change: Organisational Strategies

Flexible Governance Frameworks

As the Act evolves, rigid or static compliance structures risk obsolescence. Many leading companies now champion modular governance frameworks:

Scalable Conformity Workflows

Instead of a single fixed checklist, they maintain layered procedures that can expand to incorporate new articles or annexes.

Cross-Functional Watch Teams

Small, agile groups keep tabs on policy developments, integrate new rules, and pilot revised processes.

Living Documentation

Digital documentation that adapts to new references or updated classification guidelines, rather than static PDF files that are rarely revisited.

Future-Focused Partnerships

Collaboration remains a powerful hedge against shifting regulation (see Chap. 8). Through trade associations, standardisation committees, or research consortia, manufacturers can glean early insights into potential amendments and shape them with real-world technical feedback. This approach also helps create informal alliances that share best practices, reducing the cost and time of repeated compliance expansions.

Practical Tip: Join or form dedicated AI compliance working groups within your industry. These groups can track legislative proposals, coordinate responses to public consultations, and unify sectoral stances on pressing topics like adaptive AI, data ethics, or new domain expansions.

Workforce and Cultural Readiness

All the regulatory foresight in the world will not matter if employees remain ill-prepared. Some best practices include:

- **Ongoing Training**: Instead of one-off compliance modules, schedule quarterly refreshers that incorporate new guidelines, real-life case studies, or emergent best practices.
- **Multi-Skilled Teams**: Encourage engineers to learn compliance fundamentals and compliance officers to understand basic AI functionalities. This cross-pollination shortens reaction times when new rules emerge.
- **Innovation Tolerance**: Create an environment where staff feel empowered to propose safer or more transparent AI solutions, even if that means rethinking established processes. Reward creative alignment with prospective regulations, bridging compliance with forward-looking R&D.

12.7 Shifting Business Models and AI-As-A-Service

Rise of Service-Based Approaches

Another developing trend is the transformation of AI components from in-house modules to external "AI-as-a-Service" (AIaaS) solutions. Manufacturers may rely on cloud platforms or specialised vendors to handle training, inference, or analytics. This approach can reduce internal overheads but also complicates compliance because:

Shared Liability

The AIaaS provider manages certain aspects of data governance or model updates, but the manufacturer remains liable under EU law if the system is deployed in a high-risk scenario.

Cross-Border Data Flows

If providers store or process data in non-EU jurisdictions, additional clarity might be needed regarding GDPR compliance and extraterritorial aspects of the AI Act.

Real-Time Transparency

Manufacturers must ensure that the AIaaS platform logs relevant performance metrics, anomalies, and version changes, enabling consistent post-market monitoring.

Platform Ecosystems and Certification

We may see the emergence of certified AI platforms pre-approved to meet certain baseline EU compliance standards. Partnering with such platforms could accelerate time-to-market for new AI-driven devices. However, leaders emphasise that platform certification alone does not negate a manufacturer's duty to verify final usage aligns with the correct risk classification. Nonetheless, forging relationships with these "trusted AI platforms" might offer a smoother compliance path for organisations lacking deep AI or legal expertise in-house.

12.8 Anticipating Cross-Jurisdictional Conflict

Differing Legal Philosophies

Despite harmonisation efforts, friction can arise when AI deployments intersect with laws that follow contrasting philosophies, for example, regions emphasising national security or open innovation over data privacy and robust oversight. Manufacturers must keep watch for:

Conflicting Data Requirements

One country might require indefinite data retention, while the EU imposes minimisation or strict usage limitations [6].

Regulatory Overlaps

A single AI product could be scrutinised by multiple authorities under slightly different criteria. This is particularly common in automotive or healthcare sectors with diverse national safety codes.

Trade Barriers

If certain AI solutions meet EU standards but clash with foreign rules, manufacturers face the dilemma of running parallel compliance tracks or ceding certain markets.

Advanced planning, legal counsel with cross-border experience, and possibly product design that toggles different compliance modes can help mitigate such complexities.

The Role of Mutual Recognition

One positive development is the concept of mutual recognition agreements, bilateral deals recognising each other's conformity assessments or certifications. Although primarily used in simpler product categories, there is a chance the EU may negotiate such arrangements for AI if partner countries adopt comparable frameworks. Firms that lead compliance in the EU might then expand globally with fewer re-certification burdens, reinforcing the argument that early, thorough compliance is an investment in worldwide market agility [7].

12.9 Conclusion

The EU AI Act is far from static. Evolving definitions, new annexes, sector-specific expansions, ethical mandates, and environmental considerations are all on the horizon, reflecting the ongoing metamorphosis of AI technologies and societal expectations. Globally, the appetite for harmonisation or reciprocal recognition may grow, but consistent frameworks are not guaranteed. In essence, the future of AI governance stands at a crossroads, merging localised legal approaches with the potential for truly international norms.

For manufacturers and engineers, the key to thriving in this environment lies in adaptability: building flexible compliance structures, fostering cross-functional expertise, and continuously monitoring policy trends. By intertwining compliance with innovation, businesses can transform regulatory uncertainties into opportunities for ethical leadership, process excellence, and competitive differentiation. Regulatory foresight, whether in data governance, general-purpose AI, or sustainability, allows organisations to remain a step ahead, shaping rather than merely reacting to tomorrow's AI landscape.

In the final chapter, Chap. 13: The Ethical Imperative, we return to the core values underpinning responsible AI innovation. Bridging legal obligations with ethical considerations, we explore how manufacturers can embed fairness, transparency, and respect for human dignity into the DNA of their AI systems, ensuring that compliance is not just about following rules but about building trust and integrity in a technology-driven future.

Key Takeaways

I. The EU AI Act may expand or refine its scope, especially around sector-specific rules, updated risk classifications, and advanced AI domains like adaptive learning.
II. Harmonisation efforts are ongoing, yet manufacturers should prepare for partial fragmentation, designing to the strictest consistent standards to reduce future rework.
III. As general-purpose and continuously learning AI spread, new obligations—like real-time auditing or snapshot-based oversight—are likely to emerge.
IV. Future AI regulations could integrate sustainability metrics, ethical impact statements, or bias prevention measures more deeply.
V. Building scalable, modular governance frameworks, tracking policy developments, and investing in cross-functional training can help companies pivot swiftly when standards shift.
VI. Adopting AIaaS or platform ecosystems demands clear liability boundaries and consistent logging to maintain robust compliance.
VII. Collaborations with trade groups, notified bodies, and standard-setting organisations allow manufacturers to anticipate changes, shape policy, and share the load of compliance updates.

References

1. European Commission (2025) Guidelines on prohibited AI practices. https://digital-strategy.ec.europa.eu/en/library/commission-publishes-guidelines-prohibited-artificial-intelligence-ai-practices-defined-ai-act
2. WilmerHale (2024) Standardisation for compliance in the EU AI act. https://www.wilmerhale.com/en/insights/blogs/wilmerhale-privacy-and-cybersecurity-law/20241204-standardization-for-compliance-in-the-european-unions-ai-act

3. Future of Life Institute (2024) Implementing the European AI act. https://futureoflife.org/project/eu-ai-act/
4. ISO/IEC CD 20151 (Draft), Dataspace concepts and characteristics. https://www.iso.org/standard/86589.html
5. Pinsent Masons (2025) EU AI act rules on GPAI models under DeepSeek review. https://www.pinsentmasons.com/out-law/analysis/eu-ai-act-gpai-deepseek-review
6. Jones Day (2025) EU AI act—first rules take effect on prohibited AI systems. https://www.jonesday.com/en/insights/2025/02/eu-ai-act-first-rules-take-effect-on-prohibited-ai-systems#:~:text=The%20Development%3A%20The%20AI%20Act's,became%20applicable%20the%20same%20day
7. Biometric Update (2025) EU AI act's impact on businesses inspires simplification efforts. https://www.biometricupdate.com/202503/eu-ai-acts-impact-on-businesses-inspires-simplification-efforts#:~:text=The%20European%20Union's%20AI%20Act,adaptation%20challenges%20and%20compliance%20costs

Chapter 13
The Ethical Imperative

Abstract Concluding the book, this chapter links legal compliance to a broader ethical mandate. It revisits core values embedded in the EU AI Act, such as human dignity, fairness, transparency, and accountability, and translates them into actionable design principles for manufacturing and engineering contexts. Topics span ethical-by-design methodologies, bias audits, worker empowerment safeguards, and sustainability. The chapter argues that ethical AI not only mitigates risk but also bolsters brand trust, attracts talent, and fuels long-term innovation, positioning responsible manufacturers at the forefront of the AI revolution.

13.1 Introduction

Amid the technical discussions and legal requirements of the EU Artificial Intelligence (AI) Act, it is easy to lose sight of the Act's deeper purpose: promoting AI that respects human dignity, upholds fundamental rights, and fosters public trust. Legal provisions on data governance, human oversight, and risk classification all serve as stepping stones towards a more profound goal, establishing *ethical* principles for AI development and usage. For manufacturers and engineers, fulfilling the EU AI Act is not merely about avoiding penalties or securing market access. It is about ensuring that every stage of AI's lifecycle from data collection to deployment, honours ethical values such as fairness, transparency, and accountability [1].

In this final chapter, we examine the ethical dimension of AI compliance. We begin by detailing how ethical principles underpin the EU AI Act's legislative framework, particularly regarding safety, equity, and human rights. We then illustrate what "ethical design" looks like in manufacturing contexts, exploring real steps businesses can take to embed moral considerations alongside technical solutions. We also discuss how building a culture of ethical AI goes well beyond ticking boxes, instead requiring a shared vision that resonates across the organisation. Finally, we address how ethical compliance can bolster reputation, facilitate long-term resilience, and foster a deeper sense of purpose in AI-driven industries. By embracing the "ethical imperative," manufacturers can transform compliance into a guiding compass, ensuring

© The Author(s), under exclusive license to Springer Nature Switzerland AG 2025

M. Kauffman, *Navigating the EU AI Act: A Practical Guide for Global Manufacturing and Engineering Leaders*, https://doi.org/10.1007/978-3-031-99613-9_13

AI innovations benefit people and society without compromising on responsibility or integrity.

13.2 Why Ethics Matters in AI Governance

Bridging Legal and Moral Obligations

The EU AI Act is grounded in the EU's broader legal commitments, protecting human rights, privacy, non-discrimination, and democracy. However, laws often provide only minimum thresholds, leaving room for interpretation and moral discretion. "Ethics" fills that space by urging organisations to ask why certain AI designs or data uses may be right or wrong, not simply whether they are lawful.

Illustration: A manufacturing firm might develop an AI to monitor worker efficiency in real time. While technically permissible under certain labour and privacy rules, the firm might question whether such pervasive surveillance could erode trust or infringe upon workers' sense of dignity. Ethics invites the deeper reflection, are we respecting personal autonomy or simply aiming to maximise short-term output?

The EU's Human-Centric Approach

Since the European Commission's earliest communications on AI, an overarching theme has been human-centric development [2]. This means AI should enhance human capabilities rather than supplant them entirely, respect privacy and equality, and remain understandable and controllable by humans. That ethos extends to manufacturing, where an "ethical AI" mindset ensures workers and end-users remain at the forefront of design considerations, rather than seeing them as expendable components in an automated process.

13.3 Key Ethical Principles Underpinning the EU AI Act

Although the Act does not explicitly designate a separate "ethical code," its articles reflect foundational moral principles:

Respect for Human Dignity

Provisions banning subliminal manipulation or social scoring (Article 5) protect individual autonomy, ensuring AI does not degrade humans into mere subjects of algorithmic manipulation.

Fairness and Non-discrimination

Data governance (Article 10) and bias mitigation requirements underscore the principle that AI systems should not disadvantage individuals based on gender, ethnicity, or other protected attributes.

Autonomy and Oversight

The emphasis on human oversight (Article 14) preserves the user's ability to monitor, understand, and override AI decisions, upholding the moral right to self-determination.

Transparency

Requirements to inform users they are dealing with AI systems (Article 52) and to provide clarity about AI outputs reflect the moral duty to avoid deception and grant individuals informed choice.

Accountability

Through conformity assessments and post-market monitoring, the Act aligns with the concept that developers, deployers, and providers bear responsibility for AI's impacts.

These principles guide technical requirements, yet they also hold broader ethical weight—challenging manufacturers to reflect on how their AI innovations can safeguard human welfare beyond regulatory minimums.

13.4 Designing AI Systems with Ethical Foundations

Ethical-by-Design Methodology

To translate ethics from abstract ideals into concrete engineering actions, some businesses adopt ethical-by-design methodologies. This approach parallels "privacy-by-design" under the GDPR, emphasising the incorporation of moral safeguards from the earliest design stages.

Typical Steps

Ethical Impact Assessment

- Before coding begins, map potential ethical risks—data misuse, worker autonomy infringement, and hidden biases [3].
- **Stakeholder Interviews**: Gather insights from employees, end-users, consumer advocacy groups, or trade unions to understand real-world concerns and moral priorities.
- **Multi-Disciplinary Design Teams**: Involve ethicists or social scientists alongside engineers and compliance officers, ensuring a holistic approach to product development.
- **Iterative Prototyping**: Build small-scale versions of the AI system, then evaluate them for potential harms or negative externalities, readjusting design choices accordingly.

Fair and Diverse Datasets

Bias in manufacturing AI often stems from unrepresentative training data—sensor readings or product samples skewed to a single set of conditions. Ethically, if the AI inadvertently discriminates against certain workforce demographics or material types, it could perpetuate unfairness or safety gaps.

Best Practices

- **Inclusive Data Gathering**: Ensure the data reflects various factory settings, climates, or operational conditions.
- **Bias Audits**: Employ AI auditing tools (see Chap. 9) to detect patterns indicating potential discrimination or erroneous assumptions about worker behaviour.
- **Ongoing Revision**: Update training sets as new data streams in, especially when expanding operations to new plants or markets.

Meaningful Human Oversight

While the Act mandates oversight (Article 14), an ethical perspective pushes deeper [4]: How do we equip operators with genuine authority and comprehension over AI-driven machinery, rather than merely fulfilling a legal formality?

Intuitive Interfaces

Dashboards that visualise AI decisions in human-readable terms, so staff are not mystified by "black-box" outputs.

Operator Consultation

Involve floor workers or line managers in shaping how oversight is implemented, ensuring they find it practical and empowering.

Empowerment Versus Policing

Use oversight tools to assist, not micromanage, employees. Ethically minded design fosters collaboration between humans and AI, rather than imposing top-down surveillance.

13.5 Beyond Compliance: Building a Culture of Ethical AI

Leadership Vision and Values

Establishing an ethical AI culture starts at the top. If senior executives frame ethics purely as a risk management or public relations tactic, staff will mirror that instrumental attitude. Conversely, leaders who articulate a genuine commitment to human welfare, integrity, and sustainability can ignite a shared sense of purpose. This vision must be reinforced through resource allocation and consistent communication.

Practical Example: A British automotive tier one manufacturer's CEO explicitly ties AI governance metrics to corporate values, emphasising safety, worker empowerment, and community trust. Internal newsletters showcase how thoughtful oversight protects both the workforce and the final product's end-users (passengers). By repeatedly connecting moral imperatives to day-to-day tasks, the CEO positions ethics as a core business driver.

Ethical Committees and Training

Many organisations now appoint AI ethics committees, cross-functional bodies that review AI projects for moral alignment, beyond just legal compliance. Tasks may include:

- Reviewing new AI proposals or expansions for ethical red flags.
- Receiving and addressing staff concerns about algorithmic decisions that could be unfair or intrusive.
- Proposing ethical guidelines, codes of conduct, or reward structures for employees who champion moral considerations in AI design.

Similarly, ongoing training beyond technical instructions, familiarises staff with ethical principles, prompting them to consider how their decisions or code changes can affect human rights or environmental sustainability.

Transparency with Internal and External Stakeholders

Ethics thrives in an environment of open dialogue. Internally, staff must feel safe questioning whether an AI system might exploit user data or push workers beyond reasonable limits. Externally, consumers, regulators, and suppliers appreciate honest explanations of how AI-driven processes function and what data they collect. Communicating proactively can prevent suspicions or controversies that undermine brand reputation.

Case Insight: A Swiss medical device manufacturer conducts quarterly ethical briefings, inviting employees from all levels and representatives from local hospitals to discuss emerging AI modifications. By showcasing prototypes and collecting real feedback, the firm refines both ethical approaches and user-friendly designs, solidifying trust in its AI-based quality checks.

13.6 Aligning Ethics with Business Outcomes

Reputation and Market Trust

Manufacturers that prioritise ethical AI not only avert reputational crises, they often see tangible commercial rewards. Clients, especially major OEMs or public bodies, are increasingly selective, preferring partners who can prove robust data protection, fair algorithms, and worker respect.

Quote (Procurement Manager at a French automotive group):

> *When short-listing robotics suppliers, we considered not just features and price but also how responsibly they handled AI. If a vendor could demonstrate an ethical approach and compliance readiness, that gave them an edge. Our brand depends on trustworthy supply chains, after all.*

Sustainable Workforce Engagement

A factory reliant on AI can strain worker morale if staff fear job displacement or intrusive monitoring. Ethically designed systems, by contrast, emphasise collaboration: AI tackles repetitive or dangerous tasks while humans handle complex, value-adding roles. By highlighting how AI preserves worker health or allows upskilling, manufacturers foster a supportive workforce rather than adversarial relationships.

Practical Example: A Spanish electronics plant introduced cobots (collaborative robots) that handle heavy lifting. Workers initially worried about losing roles. Management undertook an ethical "cobot integration" plan, ensuring transparent dialogues, retraining for advanced maintenance tasks, and clarifying the human-in-the-loop structure. The result was higher job satisfaction and fewer musculoskeletal injuries [5].

Ethical lapses can be costly. AI-driven crises, be it widespread product defects or discriminatory workforce allocation, can spark legal battles, mass recalls, or public outcry. Manufacturers who consistently embed ethics enjoy risk mitigation that extends beyond EU territory, strengthening resilience in uncertain global markets.

13.7 The Ethical Imperative in Specific Manufacturing Scenarios

Worker Monitoring Versus Worker Empowerment

One contentious area is using AI for **worker performance tracking**. A system might measure task completion times or detect safety lapses. Ethically, there is a thin line between useful productivity insights and invasive micromanagement.

Fair Data Practices

Only gather data genuinely needed to ensure safety and efficiency. Avoid capturing excessively personal metrics.

Transparency and Consent

Inform workers about what data is collected and why, clarifying it will not be misused or cause unfair disciplinary action.

Positive Interventions

Instead of punishing slight performance dips, AI triggers supportive interventions—like scheduling additional training or rotating tasks to reduce strain.

Automated Quality Control in High-Stakes Products

Quality control AI for products like medical implants, aerospace components, or heavy machinery must weigh the risk of false positives (discarding safe items) against false negatives (approving flawed, potentially dangerous products). Ethically:

- **Acceptable Risk Thresholds**: Determine tolerances aligned with real-world safety [4]. If an item is erroneously rejected, it costs money. If an item is wrongly approved, it jeopardises lives. Ethics dictates erring on the side of caution without becoming unreasonably conservative.
- **Operator Confirmation**: Provide a second line of manual checking for borderline rejections, balancing trust in AI's speed with the moral imperative to protect users.
- **Stakeholder Communication**: Inform end-users or regulatory bodies about the QA system's design. This fosters trust in the manufacturing process's rigour.

Algorithmic Scheduling and Shift Allocation

AI-based scheduling can reduce logistical inefficiencies by assigning shifts or tasks automatically. Yet if poorly designed, these systems may inadvertently discriminate or ignore worker well-being (e.g. overworking certain employees or failing to accommodate personal constraints).

Ethical Interventions

Fairness Checks

Before deploying, simulate different scenarios. Does the AI routinely assign less favourable shifts to older or less tech-savvy staff, for instance?

Feedback Mechanisms

Create channels for workers to appeal scheduling decisions, encouraging a dialogue between human managers and algorithmic outputs.

Work-Life Balance

Implement caps or constraints that reflect agreed labour standards, preserving rest days and not overloading any individual unnecessarily.

13.8 Implementing an Ethical Roadmap

Step 1: Establish Ethical Guidelines

Begin with drafting a clear set of **ethical guidelines** tailored to the company's operational realities. Incorporate references to:

- Human-centric design principles (e.g. user oversight, data minimisation).
- Worker dignity and rights (limiting intrusive data collection).
- Product safety thresholds (engineering fail-safes, robust testing).
- Inclusivity and fairness (avoiding bias against specific user demographics).

Senior leaders should endorse these guidelines publicly, signposting that the company values ethical AI from boardroom to factory floor.

Step 2: Align Existing Processes with Principles

Next, conduct an **ethical gap analysis**. Identify where current AI initiatives might clash with the newly established guidelines. For instance, a predictive maintenance tool might glean worker data without explicit consent, or a supply chain optimiser might inadvertently penalise smaller suppliers.

Action: Introduce updated standard operating procedures (SOPs) that bake in consent checks, equality safeguards, or additional risk thresholds for safety–critical tasks.

Step 3: Foster Awareness and Skills

Invest in **training** that covers not just the "how" of compliance but the "why" of ethical reflection:

- Regular workshops or e-learning modules on ethical AI fundamentals, case studies, and moral dilemmas.
- Simulation exercises: walking staff through hypothetical scenarios like an AI glitch that pressures staff to skip break times or compromises product safety and brainstorming ethical resolutions.

This continuous learning environment keeps ethics top-of-mind, rather than an afterthought.

Step 4: Deploy Oversight Tools and Committees

Where possible, automate **ethical checks**. This might mean integrating bias detection modules into AI auditing pipelines, setting up real-time compliance dashboards, or scheduling recurring model reviews by an ethics panel. Build ethical sign-offs into engineering sprints or product rollouts, akin to how software teams conduct code reviews. When the AI or data usage crosses certain risk thresholds, an ethics committee might hold a "go/no-go" meeting to evaluate potential harm or infringement.

13.9 Ethics as a Catalyst for Future-Ready Innovation

Enhancing R&D and Competitiveness

Meeting ethical requirements can spark creative problem-solving. Engineers grappling with transparency mandates devise more interpretable models, which can then inspire novel functionalities or user interfaces that also enhance overall product performance. Data scientists refining bias mitigation strategies can unearth fresh insights about operational variability, leading to improved robustness. In short, ethics-driven design frequently **pushes boundaries** in ways that purely technical or commercial motivations might not.

Attracting Talent and Partners

Millennial and Generation Z professionals increasingly prioritise employers whose mission aligns with societal values. By weaving ethical AI into the company's identity, manufacturers become more appealing to top-tier engineers, data scientists, and project managers who seek meaningful careers. Likewise, partners, suppliers, and customers evaluating potential collaborations will often favour businesses known for ethical stewardship, reducing friction in contract negotiations.

13.10 Potential Pitfalls and Limitations

Ethical "Greenwashing" or "AI-Washing"

Some organisations may publicly champion ethical AI while making only superficial changes. If discovered, this contradiction damages credibility more than not mentioning ethics at all. True ethical practice must be integrated systemically, not merely used as marketing spin.

Embedding ethics thoroughly can be resource-intensive, requiring multidisciplinary committees, training budgets, or advanced auditing tools [6]. Smaller manufacturers or those operating on tight margins may struggle to justify upfront costs. Nevertheless, incremental, iterative steps like simpler bias checks or pilot-scale ethical reviews can still yield meaningful progress.

Global Variances in Ethical Norms

Manufacturers with multinational footprints face ethical dilemmas if local norms conflict with the EU's emphasis on data privacy, transparency, or worker rights. For example, certain jurisdictions might permit or even encourage extensive AI-based worker surveillance. Maintaining consistent ethical standards across all operations may spark tensions, but it ultimately fosters a stable, principled reputation.

13.11 Conclusion

The EU AI Act, through its rigorous focus on risk-based controls, transparency, and oversight, reminds us that AI technology wields enormous influence over people's lives. For manufacturers and engineers, ethical compliance with these principles is not a peripheral concern or a mere legal box to tick. It is a profound imperative, guiding decisions about how we build, deploy, and manage AI-driven machinery and processes.

By anchoring AI systems in a culture of responsibility, from inclusive data gathering and user-friendly interfaces to robust oversight and fairness checks, organisations can transcend compliance into genuine ethical leadership. These efforts bolster brand reputation, strengthen workforce morale, and reduce liability risks. More importantly, they ensure that innovation advances in harmony with societal needs and human dignity.

In a world where technology evolves at breakneck speed, ethics provides a stable compass. It prompts the critical questions: Are we respecting the people who design, use, or are affected by our AI? Does our pursuit of efficiency or profit overshadow our commitment to safeguarding human values and safety? Through the EU AI Act and beyond, embracing the "ethical imperative" answers these questions with clarity and conviction. The result is AI that fosters better products, more sustainable business practices, and, above all, enduring trust from employees, consumers, regulators, and global communities alike.

Key Takeaways (the Final One!!!)

I. The EU AI Act's rules embody fundamental ethical values protecting human dignity, fairness, autonomy, and accountability which go beyond mere technical obligations.
II. Ethical design involves impact assessments, diverse datasets, meaningful oversight, and balancing organisational objectives with moral responsibilities.
III. Embedding ethics in AI requires visible leadership commitment, cross-functional dialogue, and training that equips staff to see compliance as a positive force.
IV. Organisations that champion ethical AI often enjoy stronger brand trust, better workforce alignment, fewer legal entanglements, and a competitive advantage in increasingly conscientious markets.
V. Ethics, like compliance, evolves alongside technology. Regular reviews, stakeholder engagement, and honest reflection ensure AI remains aligned with shifting societal expectations.
VI. A manufacturer that treats ethics as an integral dimension of AI architecture is better poised to weather crises, adapt to new regulations, and attract the next generation of socially aware talent.

References

1. TrustPath (2024) Understanding transparency obligations in the EU AI act. https://www.trustpath.ai/blog/what-do-transparency-obligations-under-the-eu-ai-act-mean-for-ai-systems#:~:text=of%20AI%20systems.-,Understanding%20Transparency%20Obligations%20in%20the%20EU%20AI%20Act,the%20rationale%20behind%20their%20outputs
2. European Commission (2024) Artificial intelligence—Q&As. https://ec.europa.eu/commission/presscorner/detail/en/qanda_21_1683
3. DataGuard (2024) The EU AI act: what are the obligations for providers?. https://www.dataguard.com/blog/the-eu-ai-act-and-obligations-for-providers/#:~:text=Providers%20must%20prepare%20and%20maintain,that%20need%20to%20be%20regulated
4. IAPP (2024) EU AI act shines light on human oversight needs. https://iapp.org/news/a/eu-ai-act-shines-light-on-human-oversight-needs
5. World Economic Forum (2024) How we can unleash the power of AI in manufacturing. https://www.weforum.org/stories/2024/01/how-we-can-unleash-the-power-of-ai-in-manufacturing/
6. Farrer and Co. (2024) Safety first: the EU GPSR. https://www.farrer.co.uk/news-and-insights/safety-first-the-eus-general-product-safety-regulation-is-now-in-force

Glossary

Below is a list of terms frequently used throughout the book, defined in the context of AI and the EU AI Act 2024.

Term	Concise definition (EU-AI-act context)	Key chapter(s)
Adaptive (online learning) AI	A model that continues to update its parameters after deployment, often in real time, requiring extra version control and monitoring	12, 9
Algorithm	A finite sequence of instructions an AI system follows to transform inputs into outputs such as predictions or decisions	2
Annex III	Schedule to the EU AI act listing use cases automatically deemed *high risk* (e.g. safety components, employment decisions)	2, 5
Artificial intelligence (AI)	Software that, for given human-defined objectives, can generate outputs—including predictions or decisions—that influence physical or virtual environments (Art. 3§1)	1, 2
Authorised representative	EU-based natural or legal person responsible for compliance liaison on behalf of a non-EU provider of high-risk AI or GPAI	3, 5, 7
Bias	Systematic error causing unfair or discriminatory outcomes due to data, model or deployment factors	2, 13
Brussels effect	Global tendency for firms outside the EU to adopt EU regulatory standards (e.g. GDPR, EU AI act) to retain market access	Intro, 3
CE marking	Symbol affixed to indicate conformity of a high-risk AI system (or its host product) with EU legislation after successful assessment	5, 7

(continued)

© The Editor(s) (if applicable) and The Author(s), under exclusive license to Springer Nature Switzerland AG 2025

M. Kauffman, *Navigating the EU AI Act: A Practical Guide for Global Manufacturing and Engineering Leaders*, https://doi.org/10.1007/978-3-031-99613-9

(continued)

Term	Concise definition (EU-AI-act context)	Key chapter(s)
Compliance by design	Engineering practice of embedding regulatory and ethical requirements into every lifecycle phase rather than bolting them on later	5, 7
Conformity assessment	Procedure (self-assessment or third-party) verifying that a high-risk AI system meets all Act requirements before market placement	5, 7
Cobot	Collaborative robot designed to operate safely alongside humans;may trigger *high-risk* classification when AI-controlled	4, 10
Cybersecurity (Art. 15)	Measures ensuring AI systems remain resilient to intentional or accidental manipulation throughout their lifecycle	5, 9
Data governance	Organisational policies ensuring data quality, integrity, representativeness, and lawful use across the AI lifecycle (Art. 10)	5, 7
Dataset datasheet/model card	Lightweight, standardised metadata artefacts documenting provenance, assumptions and limitations of training data or models,helpful for compliance with Annex IV	5, 7, 9
Deployer	Any user (e.g. factory operator) exercising control over an AI system under their authority; distinct from provider	3, 5
Digital twin	Virtual replica of a physical process or product used for simulation, optimisation, and monitoring, often powered by AI	4, 10
Distributor	Entity—neither provider nor importer—that makes an AI system available on the EU market	3, 5
EU database for high-risk AI	Public registry (maintained by the European Commission) in which providers must register high-risk systems before market placement	5, 7
EU declaration of conformity	Legally binding document in which the provider attests that the AI system meets all relevant provisions (Art. 47)	5
General-purpose AI (GPAI)	Broad, versatile AI model that can be adapted to multiple downstream tasks; subject to obligation outlined in Article 54	2, 12
High-risk AI	Systems that may significantly affect safety or fundamental rights; subject to strict obligations (Annex III + Art. 6)	2, 5
Human oversight	Design feature and organisational process ensuring humans can monitor, interpret, and override AI decisions (Art. 14)	5, 7, 13

(continued)

(continued)

Term	Concise definition (EU-AI-act context)	Key chapter(s)
Importer	EU-based entity placing a non-EU manufacturer's AI system on the EU market, assuming compliance duties (Art. 23)	3, 5
Limited-risk AI	AI subject mainly to transparency duties (e.g. chatbot disclosure) but not to the full high-risk regime	2, 5
Machine learning (ML)	Subfield of AI in which models learn patterns from data rather than being hard-coded; most current industrial AI is ML-based	2, 4
Minimal-risk AI	Applications with negligible potential harm; not subject to additional obligations under the Act	2
Model inventory	Organisational register listing each AI system, its purpose, risk tier, ownership, and compliance status; foundational Step for internal compliance see Chap. 7	7
Notified body	Independent organisation designated by an EU Member State to perform third-party conformity assessments of high-risk AI	5, 7
Operator (in manufacturing)	Human interacting with or supervising AI-enabled machinery; typically the "deployer" for Act purposes	4, 5
Post-market monitoring	Structured collection and analysis of operational data to ensure ongoing compliance and prompt corrective action (Art. 61)	5, 7, 9
Predictive maintenance	Data-driven technique using AI to forecast equipment failures and schedule maintenance, often *high risk* if automatic shutdowns occur	4, 10
Provider	Entity (often the manufacturer) that develops or places an AI system on the EU market under its own name or trademark (Art. 16)	3, 5
Quality management system (QMS)	Documented organisational framework (Art. 17) covering processes, resources, and responsibilities to ensure continuous AI compliance	5, 7
Risk management system	Continuous, documented process for identifying, analysing, and mitigating risks across the AI lifecycle (Art. 9)	5, 7
Synthetic data	Artificially generated data used to augment or replace real data, often to improve privacy or balance training sets	5, 12
Technical documentation	Evidence set (Annex IV) demonstrating an AI system's compliance must be kept for 10 years post-market	5, 7
Transparency obligations	Duties to inform users that they interact with AI, disclose synthetic content, and provide clear instructions (Arts. 13 and 52)	5

(continued)

(continued)

Term	Concise definition (EU-AI-act context)	Key chapter(s)
Unacceptable-risk AI	AI practices banned outright (e.g. manipulative subliminal techniques, untargeted social scoring, unlawful biometric ID)	2, 5

GPSR Compliance

The European Union's (EU) General Product Safety Regulation (GPSR) is a set of rules that requires consumer products to be safe and our obligations to ensure this.

If you have any concerns about our products, you can contact us on ProductSafety@springernature.com

In case Publisher is established outside the EU, the EU authorized representative is:

Springer Nature Customer Service Center GmbH
Europaplatz 3
69115 Heidelberg, Germany

Batch number: 10105467

Printed by Printforce, the Netherlands

Lecture Notes in Electrical Engineering

Volume 983

Series Editors

Leopoldo Angrisani, Department of Electrical and Information Technologies Engineering, University of Napoli Federico II, Naples, Italy
Marco Arteaga, Departament de Control y Robótica, Universidad Nacional Autónoma de México, Coyoacán, Mexico
Bijaya Ketan Panigrahi, Department of Electrical Engineering, Indian Institute of Technology Delhi, New Delhi, Delhi, India
Samarjit Chakraborty, Fakultät für Elektrotechnik und Informationstechnik, TU München, Munich, Germany
Jiming Chen, Zhejiang University, Hangzhou, Zhejiang, China
Shanben Chen, Materials Science and Engineering, Shanghai Jiao Tong University, Shanghai, China
Tan Kay Chen, Department of Electrical and Computer Engineering, National University of Singapore, Singapore, Singapore
Rüdiger Dillmann, Humanoids and Intelligent Systems Laboratory, Karlsruhe Institute for Technology, Karlsruhe, Germany
Haibin Duan, Beijing University of Aeronautics and Astronautics, Beijing, China
Gianluigi Ferrari, Università di Parma, Parma, Italy
Manuel Ferre, Centre for Automation and Robotics CAR (UPM-CSIC), Universidad Politécnica de Madrid, Madrid, Spain
Sandra Hirche, Department of Electrical Engineering and Information Science, Technische Universität München, Munich, Germany
Faryar Jabbari, Department of Mechanical and Aerospace Engineering, University of California, Irvine, CA, USA
Limin Jia, State Key Laboratory of Rail Traffic Control and Safety, Beijing Jiaotong University, Beijing, China
Janusz Kacprzyk, Systems Research Institute, Polish Academy of Sciences, Warsaw, Poland
Alaa Khamis, German University in Egypt El Tagamoa El Khames, New Cairo City, Egypt
Torsten Kroeger, Stanford University, Stanford, CA, USA
Yong Li, Hunan University, Changsha, Hunan, China
Qilian Liang, Department of Electrical Engineering, University of Texas at Arlington, Arlington, TX, USA
Ferran Martín, Departament d'Enginyeria Electrònica, Universitat Autònoma de Barcelona, Bellaterra, Barcelona, Spain
Tan Cher Ming, College of Engineering, Nanyang Technological University, Singapore, Singapore
Wolfgang Minker, Institute of Information Technology, University of Ulm, Ulm, Germany
Pradeep Misra, Department of Electrical Engineering, Wright State University, Dayton, OH, USA
Sebastian Möller, Quality and Usability Laboratory, TU Berlin, Berlin, Germany
Subhas Mukhopadhyay, School of Engineering and Advanced Technology, Massey University, Palmerston North, Manawatu-Wanganui, New Zealand
Cun-Zheng Ning, Department of Electrical Engineering, Arizona State University, Tempe, AZ, USA
Toyoaki Nishida, Graduate School of Informatics, Kyoto University, Kyoto, Japan
Luca Oneto, Department of Informatics, BioEngineering, Robotics and Systems Engineering, University of Genova, Genova, Genova, Italy
Federica Pascucci, Dipartimento di Ingegneria, Università degli Studi "Roma Tre", Rome, Italy
Yong Qin, State Key Laboratory of Rail Traffic Control and Safety, Beijing Jiaotong University, Beijing, China
Gan Woon Seng, School of Electrical and Electronic Engineering, Nanyang Technological University, Singapore, Singapore
Joachim Speidel, Institute of Telecommunications, Universität Stuttgart, Stuttgart, Germany
Germano Veiga, Campus da FEUP, INESC Porto, Porto, Portugal
Haitao Wu, Academy of Opto-electronics, Chinese Academy of Sciences, Beijing, China
Walter Zamboni, DIEM—Università degli studi di Salerno, Fisciano, Salerno, Italy
Junjie James Zhang, Charlotte, NC, USA

The book series *Lecture Notes in Electrical Engineering* (LNEE) publishes the latest developments in Electrical Engineering—quickly, informally and in high quality. While original research reported in proceedings and monographs has traditionally formed the core of LNEE, we also encourage authors to submit books devoted to supporting student education and professional training in the various fields and applications areas of electrical engineering. The series cover classical and emerging topics concerning:

- Communication Engineering, Information Theory and Networks
- Electronics Engineering and Microelectronics
- Signal, Image and Speech Processing
- Wireless and Mobile Communication
- Circuits and Systems
- Energy Systems, Power Electronics and Electrical Machines
- Electro-optical Engineering
- Instrumentation Engineering
- Avionics Engineering
- Control Systems
- Internet-of-Things and Cybersecurity
- Biomedical Devices, MEMS and NEMS

For general information about this book series, comments or suggestions, please contact leontina.dicecco@springer.com.

To submit a proposal or request further information, please contact the Publishing Editor in your country:

China

Jasmine Dou, Editor (jasmine.dou@springer.com)

India, Japan, Rest of Asia

Swati Meherishi, Editorial Director (Swati.Meherishi@springer.com)

Southeast Asia, Australia, New Zealand

Ramesh Nath Premnath, Editor (ramesh.premnath@springernature.com)

USA, Canada

Michael Luby, Senior Editor (michael.luby@springer.com)

All other Countries

Leontina Di Cecco, Senior Editor (leontina.dicecco@springer.com)

**** This series is indexed by EI Compendex and Scopus databases. ****

Dae-Ki Kang · Rayner Alfred ·
Zamhar Iswandono Bin Awang Ismail ·
Aslina Baharum · Vinesh Thiruchelvam
Editors

Proceedings of the 9th International Conference on Computational Science and Technology

ICCST 2022, 27–28 August, Johor Bahru, Malaysia

Springer

Editors
Dae-Ki Kang
Dongseo University
Busan, Korea (Republic of)

Zamhar Iswandono Bin Awang Ismail
Higher Colleges of Technology
Abu Dhabi, United Arab Emirates

Vinesh Thiruchelvam
Asia Pacific University of Technology and Innovation
Kuala Lumpur, Malaysia

Rayner Alfred
Creative Advanced Machine Intelligence Research Centre, Faculty of Computing and Informatics
Universiti Malaysia Sabah
Kota Kinabalu, Sabah, Malaysia

Aslina Baharum
Faculty of Computer and Mathematical Sciences
Universiti Teknologi MARA (UiTM)
Melaka, Malaysia

ISSN 1876-1100 ISSN 1876-1119 (electronic)
Lecture Notes in Electrical Engineering
ISBN 978-981-19-8405-1 ISBN 978-981-19-8406-8 (eBook)
https://doi.org/10.1007/978-981-19-8406-8

© The Editor(s) (if applicable) and The Author(s), under exclusive license to Springer Nature Singapore Pte Ltd. 2023
This work is subject to copyright. All rights are solely and exclusively licensed by the Publisher, whether the whole or part of the material is concerned, specifically the rights of translation, reprinting, reuse of illustrations, recitation, broadcasting, reproduction on microfilms or in any other physical way, and transmission or information storage and retrieval, electronic adaptation, computer software, or by similar or dissimilar methodology now known or hereafter developed.
The use of general descriptive names, registered names, trademarks, service marks, etc. in this publication does not imply, even in the absence of a specific statement, that such names are exempt from the relevant protective laws and regulations and therefore free for general use.
The publisher, the authors, and the editors are safe to assume that the advice and information in this book are believed to be true and accurate at the date of publication. Neither the publisher nor the authors or the editors give a warranty, expressed or implied, with respect to the material contained herein or for any errors or omissions that may have been made. The publisher remains neutral with regard to jurisdictional claims in published maps and institutional affiliations.

This Springer imprint is published by the registered company Springer Nature Singapore Pte Ltd.
The registered company address is: 152 Beach Road, #21-01/04 Gateway East, Singapore 189721, Singapore

Preface

Computational science and technology is a rapidly growing multi- and interdisciplinary field that uses advanced computing and data analysis to understand and solve complex problems. The absolute size of many challenges in computational science and technology demands the use of supercomputing, parallel processing, sophisticated algorithms, and advanced system software and architecture. The ICCST2022 conference provides a unique forum to exchange innovative research ideas, recent results, and share experiences among researchers and practitioners in the field of advanced computational science and technology. Building on the previous eight conferences that include Regional Conference on Computational Science and Technology (RCSST 2007) and a series of International Conference on Computational Science and Technology (ICCST2014, ICCST2016, ICCST2017, ICCST2018, ICCST2019, ICCST2020, ICCST2021), the Ninth International Conference on Computational Science and Technology (ICCST2022) program offers practitioners and researchers from academia and industry the possibility to share computational techniques and solutions in this area, to identify new issues, and to shape future directions for research, as well as to enable industrial users to apply leading-edge large-scale high-performance computational methods. This volume presents a theory and practice of ongoing research in computational science and technology. The focus of this volume is on a broad range of methodological approaches and empirical references points including artificial intelligence, mathematical computational, knowledge discovery and machine learning, smart technologies, augmented and virtual reality, system applications and architecture, and final soft computing and analysis. The carefully selected contributions to this volume were initially accepted for oral presentation during the Ninth International Conference on Computational Science and Technology (ICCST22) held on August 27–28, 2022, virtually, from Johor. The level of contributions corresponds to that of advanced scientific works, although several of them could be addressed also to non-expert readers. The volume brings together 52 chapters. The volume brings together 52 chapters divided into six main parts that include *Machine Learning and Knowledge Discovery*, *Information Systems and Technology*, *Networking and IoT*, *Computer Vision and Virtual Reality*, *Smart Technologies*, and *Mathematics and Modelling*. In concluding, we would also like to express our deep gratitude and appreciation to all the program committee members, panel reviewers, organizing committees, and volunteers for your efforts to make this conference a successful event. It is worth emphasizing that much theoretical and empirical work remains to be done. It is encouraging to find that more research on computational science and technology is still required. We sincerely hope the readers will find this book interesting, useful, and informative, and it will give then a valuable inspiration for original and innovative research.

Busan, Korea (Republic of) — Dae-Ki Kang
Kota Kinabalu, Malaysia — Rayner Alfred
Abu Dhabi, United Arab Emirates — Zamhar Iswandono Bin Awang Ismail
Melaka, Malaysia — Aslina Baharum
Kuala Lumpur, Malaysia — Vinesh Thiruchelvam

Keynote Speakers

Prof. Dr. Amir Akramin Bin Shafir, Director, Research Management Centre, International Islamic University Malaysia, Malaysia.

Prof. Dr. Chan Chee Seng, Department of Artificial Intelligence, Faculty of Computer Science and Information Technology, Malaya University, Malaysia.

Assoc. Prof. Dr. Yuto Lim, School of Information Science, Security and Networks Area, Center for Trustworthy IoT Infrastructure, Japan Advanced Institute of Science and Technology, Japan.

Mohd Helmy Abd Wahab, Advanced Telecommunication Research Center (ATRC), Department of Computer Engineering, Faculty of Electrical and Electronic Engineering, Universiti Tun Hussein Onn Malaysia, Malaysia.

Contents

The Equity of Pedagogical Approaches in E-Learning Initiatives in Higher Education Institutions in United Arab Emirates

Lina Daouk, Nafeth Hashlamoun, Mai Ezzat, Zamhar Iswandono, and Rayner Alfred

Abstract The slow adoption of e-learning in Higher education institutions (HEIs) in the United Arab Emirates (UAE) has been attributed to many Information Technology (IT) success factors or factors related to the management of IT. While many researchers were able to identify legitimate factors that have impact on the adoption of e-learning, the impact of the cultural factors has not been thoroughly studied. The HEIs in the UAE have implemented e-learning solutions developed in Western countries. When these e-learning solutions are used without any cultural pedagogical consideration, they are creating inequitable learning conditions for UAE students. This case study aimed to investigate the role of the western-oriented culture e-learning pedagogy on the adoption of e-learning among UAE female students. Data was collected qualitatively from interviews with colleagues who are using western-oriented cultural pedagogies in their e-learning classes. A pattern matching technique was used to analyse the collected data. The cultural constructs of the Unified Theory of Acceptance and Use of Technology (UTAUT) were used as a lens to guide this study. The findings revealed the impact of the western-oriented pedagogy of Blackboard Learn (BBLearn) on the slow adoption of e-learning in the researchers' department. This study confirmed the importance of the local UAE culture on the adoption of e-learning and calls for further studies in that domain.

Keywords Technology enhanced learning · Technology adoption models · E-learning first section

L. Daouk (✉) · N. Hashlamoun · M. Ezzat · Z. Iswandono
Higher Colleges of Technology, Abu Dhabi, United Arab Emirates
e-mail: ldaouk@hct.ac.ae

Z. Iswandono
e-mail: zismail@hct.ac.ae

R. Alfred
Creative Advanced Machine Intelligence Research Centre, Faculty of Computing and Informatics, Universiti Malaysia Sabah, Kota Kinabalu, Sabah, Malaysia
e-mail: ralfred@ums.edu.my

© The Author(s), under exclusive license to Springer Nature Singapore Pte Ltd. 2023
D.-K. Kang et al. (eds.), *Proceedings of the 9th International Conference on Computational Science and Technology*, Lecture Notes in Electrical Engineering 983,
https://doi.org/10.1007/978-981-19-8406-8_1

1 Introduction

Over the last two decades, Higher Education institutions in the UAE have spent millions of dollars on e-learning initiatives in an attempt to leverage their benefits [36]. Noesgaard and Ørngreen listed 19 distinct criteria to investigate the effectiveness of e-learning initiatives [25]. Learning outcomes, students' performance, adoption, and users' satisfaction are among the most commonly used criteria. The rate of adoption as described by the UTAUT is among the highest used criterion to judge the effectiveness of e-learning initiatives [1, 8, 22, 23, 39]. The rate of adoption of e-learning among UAE students is still slow when compared to the large investments and the number of e-learning courses in the UAE HEIs [13, 36]. This is confirmed by the existence of only one e-learning university offering accredited e-learning courses out of the total 76 licensed universities in the UAE by the ministry of higher education ("CAA," n.d.). The executives of the UAE HEIs are raising legitimate concerns about the reasons for this slow adoption of e-learning initiatives.

Many researchers have used information technology (IT) success factors to investigate the adoption of e-learning in HEIs [2, 16, 32, 38]. While these studies have identified some legitimate factors that impact the adoption of e-learning in HEIs, we found that the cultural factors have been neglected [9]. The current pedagogy in e-learning initiatives in HEIs in THE UAE is founded on western cultural values, language, orientations, and epistemology. This western-oriented culture e-learning pedagogy creates inequitable learning conditions [6] for many UAE female students at my department. This paper attempts to study the impact of this western-oriented culture e-learning pedagogy on the adoption of e-learning among UAE female students at the institution of the Higher education institutions (HEIs) using a case study research methodology.

Section 2 provides the detail of the literature review or related works. Section 3 presents the proposed approach, Sect. 4 presents the results and discussion and finally Sect. 5 concludes this paper.

2 Related Works

This literature review aims to provide an overview of the impact of culture on the adoption of e-learning in higher education institutions. It is organized as follows: The cultural constructs of the UTAUT are first presented and discussed. This is followed by a discussion about the impact of culture on the adoption of e-learning in HEIs. This is followed by an investigation about the cultural pedagogy of BBLearn. The effect of THE UAE culture on learning pedagogies is then examined after that. The literature review concludes by identifying the gaps about the effects of culture on the adoption of e-learning in HEIs in the UAE.

2.1 UTAUT Cultural Constructs

Based on a full review of the users' technology adoption literature, Venkatesh, Morris, Davis, and Davis proposed a Unified Theory of Acceptance and Use of Technology (UTAUT) that integrated elements from eight theoretical technology acceptance models [35]. UTAUT included four constructs that were empirically proven to have a significant role in determining user acceptance and usage behavior: performance expectancy (PE), effort expectancy (EE), social influence (SI), and facilitating conditions (FC) [35]. While the attributes of the PE, EE, and FC constructs are all based on IT success factors theory [12] and theory of planned behavior [3], we notice that the attributes of the SI are based on the cultural aspects of users. The attributes of the SI construct are based on the individual's internalization of the reference group's subjective cultures. The PE, EE, and FC were used by many scholars to study the adoption rate of e-learning in HEIs. For example, PE, EE, and FC constructs of UTAUT were used to develop a theoretical framework for evaluating instructors' adoption of e-learning learning management systems [3]. In a similar study, Lee empirically tested the same three constructs based on data collected from 363 learners of a web-based learning program to investigate the slow adoption of e-learning in HEIs [19]. Lee concluded that the satisfaction attributes in these three constructs had the most significant effect on the rate of adoption. While the literature includes many studies that used these three constructs to investigate e-learning adoption in HEIs, we found that fewer studies have used the attributes of the SI construct to investigate its impact on the adoption of e-learning in HEIs [5, 15, 18, 31].

2.2 Adoption of E-Learning and Culture

Prior studies have focused on the adoption of e-learning from an information system user satisfaction point of view [4, 37]. However, the role of culture in the adoption of e-learning has been overlooked [27]. Pan et al. stated that eastern students and teachers suffer sever disadvantages when instructional techno-logical interventions are superimposed on their existing traditional pedagogical models. The authors recommended a set of pedagogical strategies to address the eastern students' problems [27]. Capatîna analysed the effect of culture on e-learning. The author concluded that the strategic development and design of e-learning should be based on the cultural needs of learners [9]. The effect of culture on e-learning courses has also been studied and recommended that changes should be adopted to suite the culture of the targeted learners [29]. On the other hand, Topor and Dinu studied the impact of e-learning on cultural evolution [33]. The author argued that the cultural value transfer in e-learning is exercising a direct influence on learners' culture. The authors concluded that a positive contribution is achieved in human capital development through cultural awareness and improving social competence. Whether adjustment should be done to e-learning to suit the receiving learners' culture or the learners' culture should be changed to suit the e-learning culture, it is clear from the discussion above that culture has an impact on the adoption of e-learning and that should be studied further.

2.3 Blackboard Learn Cultural Pedagogy

Keller, Lindh, Hrastinski, Casanovas, and Fernandez stated that instructors in HEI are using e-learning tools for administrative purposes rather than pedagogical purposes [17]. Lack of knowledge about the e-learning pedagogies was identified as the main reason for this administrative usage only. There is a growing demand in the UAE for the adoption of e-learning systems that are developed in western counties [24]. While these e-learning softwares provide flexibility for teachers to use them according to their pedagogical preferences, we find that most teachers in the UAE are using them for administrative purposes [17].

BBLearn is the most popular used e-learning system in HEI and in the UAE [7, 21]. However, the usage in these HEIs is mainly limited for administrative purposes [24]. BBLearn is developed by the company Blackboard Incorporation which is based in Washington DC ("Blackboard—Education Technology and Services," n.d.-a). The administrative part of BBLearn is based on the Sharable Content Object Reference Model (SCORM) standard. SCORM is a standard that defines a specific way for developing both the learning management systems and the training contents ("SCORM Explained," n.d.). While this plug and play functionality of BBLearn provides a powerful tool for creating and sharing contents across organizations, it creates a pedagogical cultural bias problem in the UAE HEIs. This cultural bias problem is due to the fact that SCROM is based on the western e-learning cultural pedagogies and the UAE HEIs are adopting it at the administrative level without applying any pedagogical changes. As stated by Bennett, the equity pedagogy dimension is often overlooked in multicultural education [6]. We can conclude that BBLearn when used only for administrative purposes in the UAE HEIs is creating an equity pedagogy cultural bias problem.

2.4 The UAE Culture and Learning Pedagogies

The UAE culture has its roots in the Arabic-Islamic culture [30]. The UAE population is 9.5 million consisting of 25% Emirati nationals and 75% expatriates. Culturally, the UAE nationals have a stronghold of Muslim-Arab values. The UAE is at the upfront edge of all the Arab countries in modernization and globalization. Studying the pressure between the Arabic-Islamic rooted culture and the push to modernization, Emirati society has achieved a balance between commercialism and traditionalism. The UAE has adopted and used many of the commercial realities of the global culture while sustaining their traditional values yet in a progressive way. It was concluded that the UAE will continue this trend of seeking a balance between the benefits of modern, progressive, and innovative approaches with a very real and sincere focus on the cultural heritage, religious foundation, and communal relationship that made the UAE one of the most intriguing social, economic, and political success stories in the middle east region. Richardson studied the possible influence of the Arabic-Islamic culture on the reflective practices pedagogical approach that was proposed for a new degree at the Higher education institutions (HEIs) [28]. After studying the UAE cultural traits and the supervised teaching practices in the UAE, the author concluded that the reflection western pedagogical learning approach as conceived and implemented in western cultures is not the best pedagogical learning approach for teachers in the UAE institutions. This is one of a very few studies that investigate the western pedagogies cultural relevance when applied in the UAE settings. We can conclude that the UAE culture has its origin in the Arabic-Islamic culture and it will continue to adopt new commercial innovations and open up on global culture.

2.5 *A Brief Description of the Gaps*

Bennett stated that there are limited studies on culturally relevant pedagogy with Latinos and Asian and that this dimension is often overlooked [6]. The situation with cultural equity of e-learning pedagogies is not better. Particularly, the role of culture in the adoption of e-learning has been overlooked [9, 27, 33]. With very limited studies on culture and its role in the adoption of e-learning in the UAE, and with the western-oriented culture inherited in BBLearn, the issue of ascertaining the impact of this western-oriented culture e-learning pedagogy on the adoption of e-learning among the UAE female students at my college merits special attention.

3 Theoretical Framework

The social influence construct of the UTAUT determinants of intention and usage provides a theoretical lens that was used to direct this case study. Based on the previous work of [17] who studied the impact of national cultural on e-learning implementation, teaching and learning western cultural attributes were selected as part of the social influence construct to guide the development of this case study. The selected attributes are presented in Figs. 1 and 2.

3.1 *Epistemology and Ontology*

Epistemology and ontology are used interchangeably [14]. However, as a researcher we understand that epistemology is about the way we know things while ontology is about what things are. This research is a qualitative case study where out ontology is

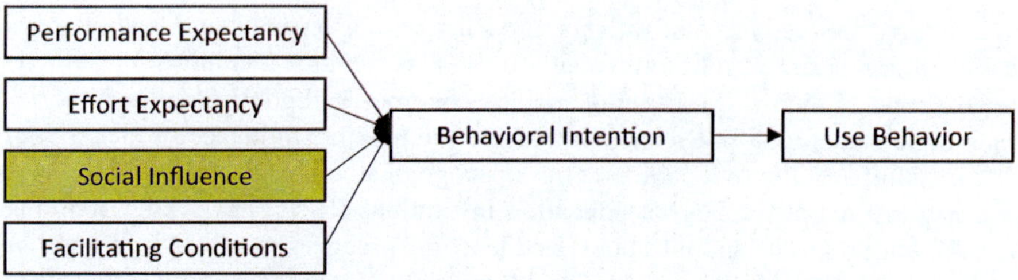

Fig. 1 Unified Theory of Acceptance and Use of Technology (UTAUT)

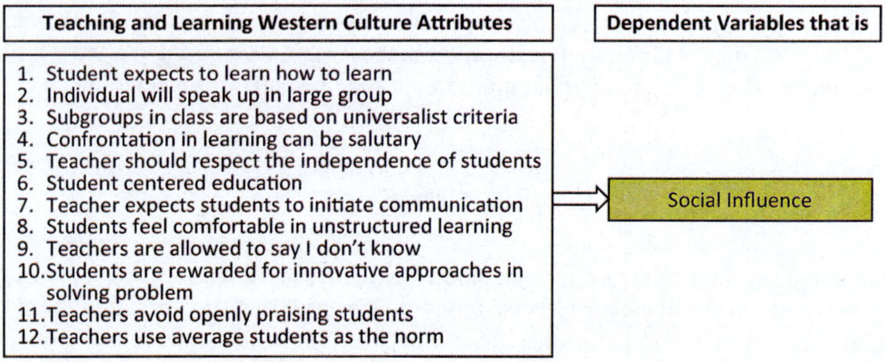

Fig. 2 Teaching and learning western culture attributes that are part of the social influence construct

to seek answers to the research questions by analyzing participants' answers and conducting a literature review. This paper will continue to have an epistemology stance that is based on postpositivist philosophy. Postpositivist accepts that theories, background, knowledge, and values of the researchers might influence what is observed [41]. Qualitative research will allow us to understand and increase our knowledge about social science endeavors. This research follows the postpositivist paradigm as described by Phillips and Burbules where educational research studies should not be solely based on unchallengeable rock solid foundations, but rather should allow space for human conjectures.

4 Research Methodology

The research methodology used in this study is a case study. A case study research is considered as one of the most challenging research methodologies that can examine complex issues and contemporary real-life situations [40]. Case studies allow researchers to examine cases where large samples of participants may not be easily available, and this is exactly the case of this research [11, 40].

Table 1 Summary of the sample demographics

Demo	Age	Gender	Total experience	BBL experience	Job title	Dept	Edu
P1	57	Female	20	12	Lecturer	CIS Business Solution	MS
P2	38	Female	9	5	Assistant Professor	CIS E-Commerce	PhD
P3	62	Female	23	15	Assistant Professor	CIS Business Solution	PhD

4.1 Data Collection

Data was collected from open-ended questions conducted in face to face interviews. Interviews are listed among the most important source for data in case studies [26]. Leedy and Ormrod stated that the most common types of interview questions are the open-ended and semi-structured ones. Three of my colleagues from my CIS department who are currently involved in developing courses using BBLearn agreed to participate in this research [20]. The demographic profiles of the three participants referred to as P1, P2, and P3 are summarized in Table 1.

The design of the 9 interview questions of this research were of open-ended and semi-structured types that allowed us to question participants about facts and their opinions and to follow up on certain answers. All interviews were completed in a maximum one hour. All data collected were text generated from the immediate transcription of the participants' answers. The 9 interview questions related to the western oriented culture e-learning pedagogy are listed below.

- **Question 1**: What do you believe are the reasons behind the slow adoption of e-learning among the female students of our department/college?
- **Question 2**: How do you describe your e-learning pedagogy from a cultural point of view? (western oriented pedagogy or eastern one). Please elaborate.
- **Question 3**: What are the pedagogical cultural challenges encountered by female students in your e-learning courses?
- **Question 4**: Using Hofstede's 4-D model of cultural differences among societies, how do you describe the cultural e-learning pedagogy of the Blackboard Learn in our college?
- **Question 5**: Given the five attributes of the individualism as listed by Hofstede, elaborate on the existence of these attributes in the teacher-student relationship in your e-learning classes using BBLearn?
- **Question 6**: Given the six selected attributes of the small power distance culture as listed by Hofstede, discuss the existence of these attributes in the teacher-student relationship in your e-learning classes using BBLearn?

- **Question 7**: Given the five selected attributes of the weak uncertain avoidance culture as listed by Hofstede, discuss the existence of these attributes in the teacher-student relationship in your e-learning classes using BBLearn?
- **Question 8**: Given the five selected attributes of the feminism culture as listed by Hofstede, discuss the existence of these attributes in the teacher-student relationship in your e-learning classes using BBLearn?
- **Question 9**: What are the aspects of the UAE culture that you think should be incorporated into the e-learning pedagogy to improve the adoption of e-learning among female students at our college?

4.2 Quality and Trustworthiness of the Study or Findings

In order to improve the quality and trustworthiness of this study, we made sure that the selected colleagues for the interviews came from different cultural background. While we understand that triangulation of evidences by collecting data from other sources like websites, archival documents, policies, and artifacts will improve the trustworthiness of this study, the time frame of this study will not permit to conduct such triangulation. However, the selection of colleagues from different cultural backgrounds combined with the literature review and the pattern matching technique used in this study should be enough to bring the quality and trustworthiness of this research to an acceptable level.

4.3 Ethical Issues and Limitations

Data collected in this research will be used only for the sole purpose of this research and will not be shared with anyone. No personal information will be identified. We will keep the digital data for three years and then it will be deleted. This research is under the exempt research type as described in [10] since the risk to harm the participants is not greater than the risk they encountered during their normal working activities.

5 Data Analysis Technique

A pattern matching logic was used to analyse the collected data [34]. Pattern matching is considered as the core procedure of theory-testing in case studies. Sufficient condition proposition was used in the pattern matching. It states that an outcome Y is always present when condition X is present. The outcome Y in this case study is the "increased rate of e-learning adoption" and the condition X is "the existence of teaching and learning western culture attributes". This proposition is tested by first

Table 2 Top 10 used words in the answers of the first 4 questions

Word	Length	Count	Weight %	Similar words
Culture	7	5	8.77	Culture, culture
Learning	8	4	7.02	Learn, learning
Pedagogy	8	3	5.26	Pedagogy
Western	7	3	5.26	Western
Bblearn	7	2	3.51	Bblearn
Oriented	7	2	3.51	Oriented
Pedagogical	11	2	3.51	Pedagogical
Use	3	2	3.51	Use, used
Adoption	8	1	1.75	Adoption
Among	5	1	1.75	Among

selecting a case in which outcome Y is absent and use the case study to search for X. In this study, Y is absent since we don't have an increased rate of e-learning adoption among the female students in our college. The expected pattern is: X is absent which means there is a lack of teaching and learning western culture attributes. Pattern matching in this study consists of checking what the attributes of X are in the observed pattern. The collected data was first organized by grouping the answers for the same question from the three participants together. In the second step, NVivo was used to discover the pattern matching. This was achieved by using text query and word matching queries available in NVivo. In the last step, we compared the discovered patterns from the collected data with the predicated ones from the selected cultural attributes in our theoretical model.

6 Findings

This section reports the findings of the interviews. The first four questions covered the participants' stand on the impact of culture on e-learning. Questions 4–8 asked participants to comment on the cultural e-learning pedagogies of BBLearn using the teaching and learning western cultural attributes. The last question was about the aspects of the UAE culture that should be incorporated.

All participants answered the first four questions with full details. In these answers, culture was mentioned as one of the main reasons for the slow adoption. The participants' answers to the first four questions were coded in a node based on cultural relevance. A text search query was applied on the created node and it revealed that both words "culture" and "western" are among the highest repeated words with 5 and 3 counts respectively. Table 2 summarizes the frequency of the first 10 words used in these answers.

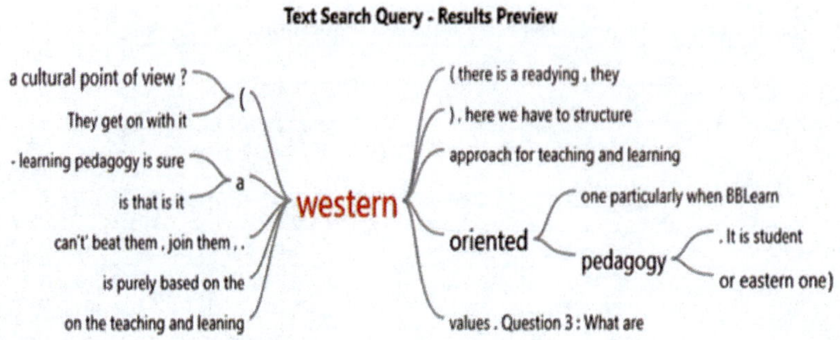

Fig. 3 The text search results on the word western and its relationship with the other words used in the answers of the first 4 questions

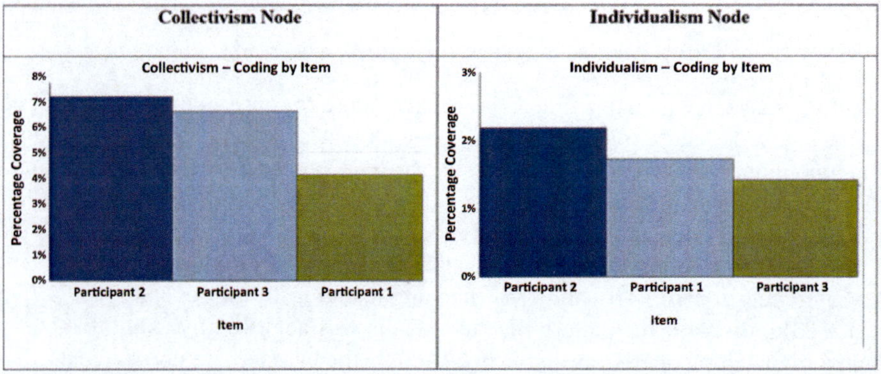

Fig. 4 Participants' answers to question 5 coded in NVivo based on collectivism versus individualism references. *Note* Higher reference rates in collectivism node versus the individualism node

The three participants confirmed that their e-learning pedagogy is a culturally western oriented one. The western e-learning pedagogy of BBLearn was mentioned by the three participants as one of the main challenges encountered by the female students at our college. A text search on the word "western" in the answers of the first 4 questions, after applying cultural relevance coding, supports the above finding as per Fig. 3. We can see that the three participants agree and confirm that the western-oriented pedagogy of BBLearn is one of the main reasons behind the slow adoption of e-learning in our department.

Questions 5–8 aimed to investigate the existence of the condition X which is "the existence of teaching and learning western culture attributes". All Participants answered these questions with full details. Answers to question 5 revealed that students prefer to learn from their teachers the content of the courses and not to learn how to learn. Students are not comfortable to speak in large groups, and the three participants confirmed that subgroups are formed based on social and personnel criteria

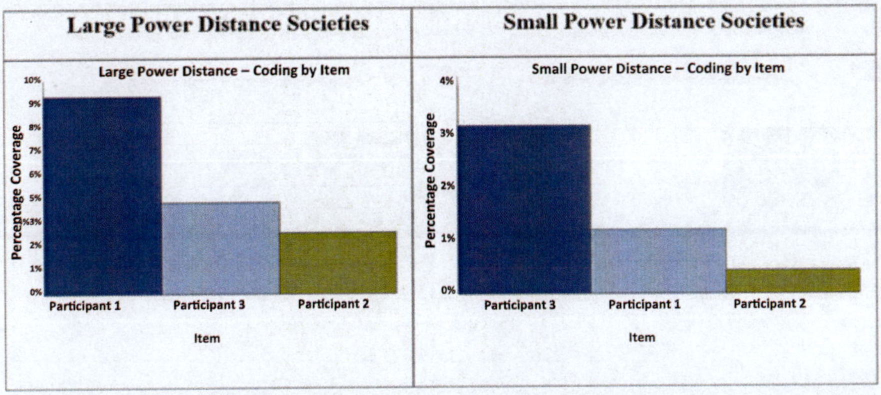

Fig. 5 Participants' answers to question 6 coded in NVivo based on large power distance societies versus small ones. *Note* Higher reference rates in large power node versus the small power node

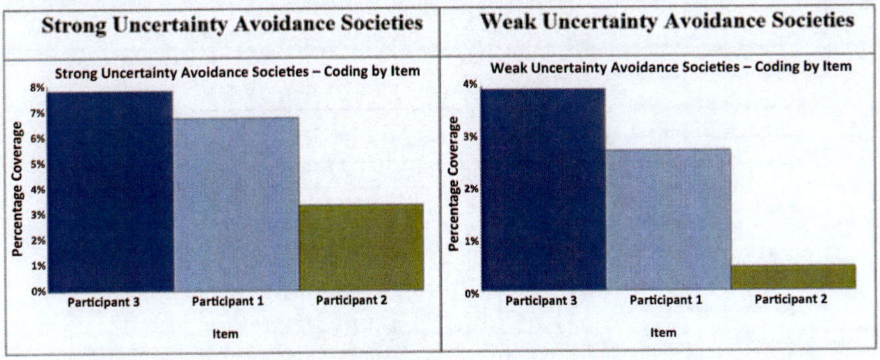

Fig. 6 Participants' answers to question 7 coded in NVivo based on strong uncertainty avoidance societies versus weak ones. *Note* Higher reference rates in strong uncertainty avoidance node versus the weak uncertainty node

and not universalist ones. Students don't enjoy confrontation during their learning and they look for certificates more than looking for competences (Please see Fig. 4).

Answers to questions 6, 7 and 8 were consistent with what was discovered in question 5. In particular, the three participants revealed that their students don't have the attributes of small power distance culture. While the participants did not elaborate on what attributes the students have, there is an agreement between the three participants that while teachers prefer student-centred education, students to find their own way, unstructured learning, innovative approach in solving problems, and other feminism attributes, the students' teaching and learning preferences are different. From the answers to these questions, it is clear that teachers prefer to use teaching and learning western culture attributes. However, students may not be comfortable to learn in such pedagogy. The three participants confirmed that some

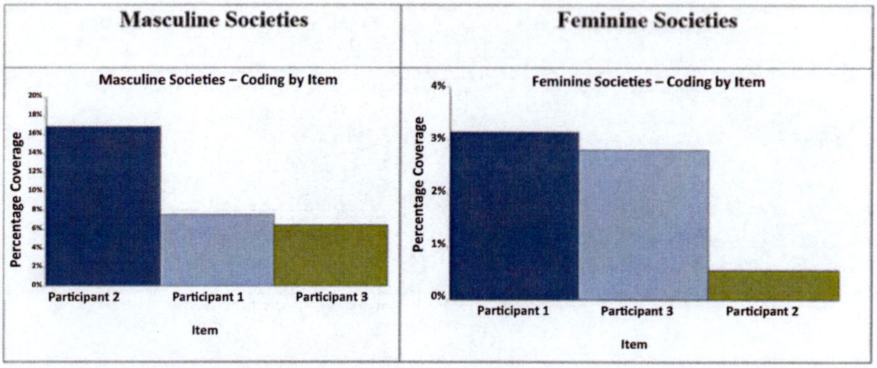

Fig. 7 Participants' answers to question 8 coded in NVivo based on feminine societies versus masculine societies. *Note* Higher reference rates in masculine node versus the feminine node

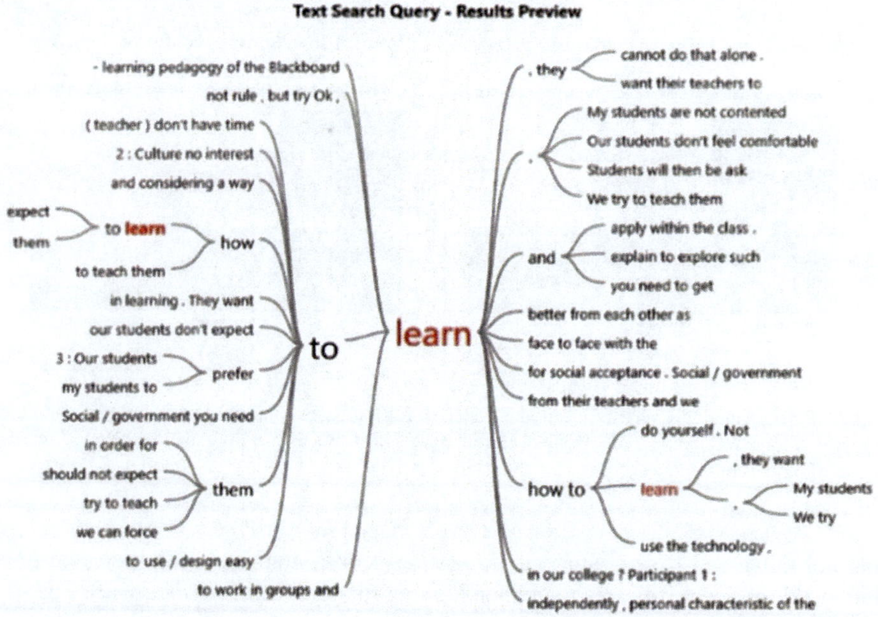

Fig. 8 The text search results on the word learn and its relationship with the other words used in all answers

adjustments are needed in order to motivate their students to learn (Please see Figs. 5, 6 and 7).

Participants' answers were coded in NVivo. Nodes were created based on the following cultural constructs: Individualism versus collectivism, large power distance societies versus small distance societies, strong uncertainty avoidance societies versus weak uncertainty avoidance societies, and masculine versus feminine societies.

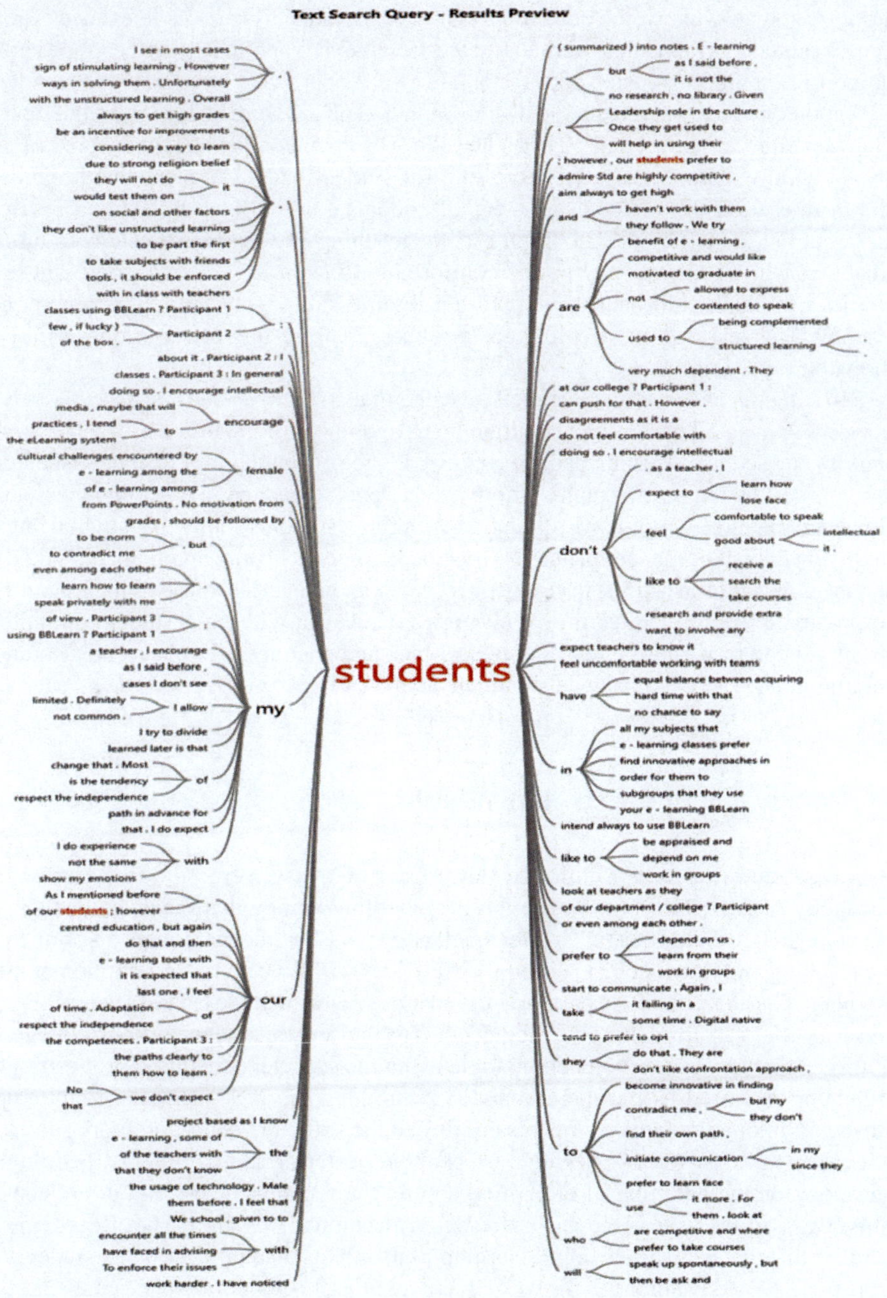

Fig. 9 The text search results on the word student and its relationship with the other words used in all answers

Participants' answers were coded after that in each node based on its relevance. The figures included graphs that were extracted from NVivo and they provide empirical support of the findings listed above.

Another two text search queries were conducted on the words Learn and Students. The word-tree for Learn had 20 branches and 9 of them contain negative words like "No", "Don't", and "Not". The word-tree for student had 77 branches with many contrasting words like "However", "but", "students don't". Most of the branches in these queries state that teachers prefer to use teaching and learning western culture attributes while students may not be comfortable to learn in such pedagogy. These queries provided additional support to the findings above on the nonexistence of teaching and learning western culture attributes. The results of these queries are shown in Figs. 8 and 9.

Participants' answers to question 9 revealed that collectivist cultural aspects such as working in small groups and creating them randomly are suggested by the participants to be used in e-learning pedagogies. Avoidance of small power distance aspects is suggested by the participants to motivate students to engage in e-leaning classes. However, the three participants agree that with time, students should be taught using small power distance teaching and learning pedagogies. The three participants stated that the UAE cultural aspects in strong uncertainty avoidance and masculinism should be first included in the e-learning pedagogies. However, with time, students should be introduced to western e-learning pedagogies and encouraged to learn in such an environment. Overall, a balance should be achieved between the two.

7 Conclusion and Recommendations

This case study research confirmed the impact of the western culture e-learning pedagogy of BBLearn on the adoption of e-learning among the female students in our college. It revealed that this western pedagogy if used without any consideration to the teaching and learning local cultural differences will slow down the adoption of e-learning. The analysis of the participants' answers confirmed the lack of the teaching and learning western cultural attributes among the female students in our college. In this case study a pattern matching logic using the sufficient condition proposition argument was used as the main analytic technique. The lack of western cultural attributes among the female students confirmed the sufficient condition proposition. It revealed the existence of a negative effect of the western-oriented culture e-learning pedagogy on the adoption of eLearning among the female students at our college. This negative effect answered the first research question of this study. The effect of the cultural differences while using e-learning platforms is also noticed when students from the UAE study abroad in the western universities. For example, some challenges listed stated that e-learners from the UAE face when enrolled in Western Universities. One of them is the irrelevance of Western universities' curricula to socio-cultural values of UAE. Based on the findings, collectivist cultural activities and avoidance of small power distance attributes were among the aspects of the UAE culture that participants had suggested to be incorporated into their e-learning pedagogies to improve e-learning adoption among female students in my college. This answered the second research question of this study. In conclusion, this case study research study explored and investigated the role of the western-oriented culture e-learning pedagogy on the adoption of e-learning among female students in our college. Further research is needed to extend the findings of this study to other departments in my college and to other UAE tertiary institutions.

References

1. Abbad MM, Morris D, De Nahlik C (2009) Looking under the bonnet: factors affecting student adoption of e-learning systems in Jordan. Int Rev Res Open Distrib Learn 10(2)
2. Abdullah F, Ward R (2016) Developing a general extended technology acceptance model for e-learning (getamel) by analysing commonly used external factors. Comput Hum Behav 56:238–256
3. Ajezen I (1991) The theory of planned behavior, organizational behavior and human decision processes
4. Arbaugh JB (2002) Managing the on-line classroom: a study of technological and behavioral characteristics of web-based MBA courses. J High Technol Manag Res 13(2):203–223
5. Arenas-Gaitán J, Ramírez-Correa PE, Rondán-Cataluña FJ (2011) Cross cultural analysis of the use and perceptions of web based learning systems. Comput Educ 57(2):1762–1774
6. Bennett C (2001) Genres of research in multicultural education. Rev Educ Res 71(2):171–217
7. Bradford P, Porciello M, Balkon N, Backus D (2007) The blackboard learning system: the be all and end all in educational instruction? J Educ Technol Syst 35(3):301–314
8. Buche MW, Davis LR, Vician C (2012) Does technology acceptance affect e-learning in a non-technology-intensive course? J Inf Syst Educ 23(1):41
9. Capatîna G (2015) The implications of culture in e-learning. In: The international scientific conference eLearning and software for education, vol 3. "Carol I" National Defence University, p 95
10. Cozby PC (1977) Methods in behavioral research
11. Creswell JW (1966) Research design: qualitative, quantitative, and mixed methods approaches
12. DeLone W, McLean E (2003) Model of information systems success: a ten years' update. J Manage
13. Elango R, Gudep VK, Selvam M (2008) Quality of e-learning: an analysis based on e-learners' perception of e-learning. Electron J e-Learn 6(1):29–41
14. Goodson L, Phillimore J (2004) Qualitative research in tourism: ontologies, epistemologies and methodologies
15. Graham CR, Woodfield W, Harrison JB (2013) A framework for institutional adoption and implementation of blended learning in higher education. Internet High Educ 18:4–14
16. Kassim ES, Jailani SFAK, Hairuddin H, Zamzuri NH (2012) Information system acceptance and user satisfaction: the mediating role of trust. Procedia Soc Behav Sci 57:412–418
17. Keller C, Lindh J, Hrastinski S, Casanovas I, Fernandez G (2009) The impact of national culture on e-learning implementation: a comparative study of an Argentinean and a Swedish University. Educ Media Int 46(1):67–80
18. King E, Boyatt R (2015) Exploring factors that influence adoption of e-learning within higher education. Br J Educ Technol 46(6):1272–1280
19. Lee MC (2010) Explaining and predicting users' continuance intention toward e-learning: an extension of the expectation-confirmation model. Comput Educ 54(2):506–516
20. Leedy PD (1974) Practical research: planning and design

21. Machado M, Tao E (2007) Blackboard vs. moodle: comparing user experience of learning management systems. In: 2007 37th annual frontiers in education conference—global engineering: knowledge without borders, opportunities without passports, pp S4J-7–S4J-12. https://doi.org/10.1109/FIE.2007.4417910
22. Mirza AA, Al-Abdulkareem M (2011) Models of e-learning adopted in the Middle East. Appl Comput Inform 9(2):83–93
23. Moon JW, Kim YG (2001) Extending the TAM for a world-wide-web context. Inf Manag 38(4):217–230
24. Mouakket S, Bettayeb AM (2015) Investigating the factors influencing continuance usage intention of learning management systems by university instructors: the blackboard system case. Int J Web Inf Syst 11:491–509
25. Noesgaard SS, Ørngreen R (2015) The effectiveness of e-learning: an explorative and integrative review of the definitions, methodologies and factors that promote e-learning effectiveness. Electron J e-Learn 13(4):277–289
26. Padgett DK (2004) The qualitative research experience
27. Pan CC, Tsai MH, Tsai PY, Tao Y, Cornell R (2003) Technology's impact: symbiotic or asymbiotic impact on differing cultures? Educ Media Int 40(3–4):319–330
28. Richardson PM (2004) Possible influences of Arabic-Islamic culture on the reflective practices proposed for an education degree at the higher colleges of technology in the United Arab Emirates. Int J Educ Dev 24:429–436
29. Rogers PC, Graham CR, Mayes CT (2007) Cultural competence and instructional design: exploration research into the delivery of online instruction cross-culturally. Educ Technol Res Dev 55(2):197–217
30. Stone LA (2002) The Islamic crescent: Islam, culture and globalization. Innov Eur J Soc Sci Res 15:121–131
31. Tapanes MA, Smith GG, White JA (2009) Cultural diversity in online learning: a study of the perceived effects of dissonance in levels of individualism/collectivism and tolerance of ambiguity. Internet High Educ 12(1):26–34
32. Tarhini A, Hone K, Liu X, Tarhini T (2017) Examining the moderating effect of individual-level cultural values on users' acceptance of e-learning in developing countries: a structural equation modeling of an extended technology acceptance model. Interact Learn Environ 25(3):306–328
33. Topor S, Dinu MS (2014) Culture and e-learning-global challenges and perspectives. In: The international scientific conference eLearning and software for education, vol 1. "Carol I" National Defence University, p 497
34. Trochim WM (1989) Outcome pattern matching and program theory. Eval Program Plann 12:355–366
35. Venkatesh V, Morris MG, Davis GB, Davis FD (2003) User acceptance of information technology: toward a unified view. MIS Q 425–478
36. Vrazalic L, Macgregor RC, Behl D, Fitzgerald J (2009) E-learning barriers in the United Arab Emirates: preliminary results from an empirical investigation
37. Wang LCC, Bagaka's JG (2002) Understanding the dimensions of self-exploration in web-based learning environments. J Res Technol Educ 34(3):364–373
38. Wang YS, Wang HY, Shee DY (2007) Measuring e-learning systems success in an organizational context: scale development and validation. Comput Hum Behav 23(4):1792–1808
39. Wu J, Tsai RJ, Chen CC, Wu Y (2006) An integrative model to predict the continuance use of electronic learning systems: hints for teaching. Int J e-Learn 5(2)
40. Yin RK (1984) Case study research: design and methods
41. Young B (2001) Postpositivism and educational research. Phillips DC, Burbules NC (2000) Postpositivism and educational research. Rowman and Littlefield Inc., Lanham, MD. ISBN: 0-8476-9122-5 $15.95. Int J Educ Manag 15:206

Malay Fake News Classification Using a Deep Learning Approach

Wafa Athilah Fikriah Binti Nordin, Rayner Alfred, Chin Pei Yee, Siti Hasnah Tanalol, Raymond Victor Loudin, and Zamhar Iswandono

Abstract In today's world, most people spend their time interacting online through social media platforms. They regularly consume news from online media sources such as social media and online newspapers. People also tend to spread the information received without confirming its truth. As the spread of fake news online grows, the public and researchers have become more interested in identifying fake news. The content of fake news that has manipulative information has negative effects on individuals and society. Most existing research investigates different kinds of features from news articles to develop fake news detection. The existing research is not capable of determining fake news for the Malay language due to lake of resources for Malay Language. The goal of this paper is to evaluate the performance of the proposed Bidirectional RNN deep learning approach to classify fake Malay news by varying the dropout rate of the proposed RNN model. In this work, four different dropout's values are used to evaluate the performance of the RNN models, which are 0.1 (Model $M1$), 0.3 (Model $M2$), 0.5 (Model $M3$) and 0.8 (Model $M4$). Based on the results obtained, the smaller is the dropout rate the less number of epochs is required to train the RNN model as it is observed that Model $M1$ required more num-

W. A. F. B. Nordin · R. Alfred (✉) · C. P. Yee · S. H. Tanalol
Creative Advanced Machine Intelligence Research Centre, Faculty of Computing and Informatics, Universiti Malaysia Sabah, Kota Kinabalu, Sabah, Malaysia
e-mail: ralfred@ums.edu.my

W. A. F. B. Nordin
e-mail: bi18110022@student.ums.edu.my

C. P. Yee
e-mail: peiyee@ums.edu.my

S. H. Tanalol
e-mail: hasnah@ums.edu.my

R. V. Loudin
Infinity Enterprise, Sabah, Malaysia
e-mail: raymondv@infinityenterprise.my

Z. Iswandono
Higher Colleges of Technology, Abu Dhabi, United Arab Emirates
e-mail: zismail@hct.ac.ae

© The Author(s), under exclusive license to Springer Nature Singapore Pte Ltd. 2023
D.-K. Kang et al. (eds.), *Proceedings of the 9th International Conference on Computational Science and Technology*, Lecture Notes in Electrical Engineering 983,
https://doi.org/10.1007/978-981-19-8406-8_2

ber of epochs to train the RNN model. However, the best accuracy can be obtained when the dropout rate was 0.3 with 90.1% accuracy in Model $M2$. Having higher percentage of dropout rate will not produce a trained model that can perform with high accuracy value. In conclusion, maintaining the percentage of dropout to be 0.3 and below enables the LSTM to produce good values of accuracies. The accuracy of the forecasted result is highly influenced by the length of the text.

Keywords Malay fake news · Detection and classification · Deep learning approach

1 Introduction

Nowadays, most people spend their time interacting online through social media platforms [21]. They tend to consume news from media sources such as social media feeds and online news rather than traditional media outlets such as printed newspapers, radio, and television. The reason for this trend is that online media is less costly and easier to access news anywhere at any time. It is also easier for people to share, comment, and talk about the news or any topic with friends and other readers. For instance, 62% of U.S. adults consume news on social media in 2022 [21]. According to a study published by the Malaysian Communications and Multimedia Commission (MCMC) in 2020, more than 90% of Malaysians use the internet for social purposes [14]. Even though there are benefits of social media, the quality of the news received is much lower than the traditional news media. Since social media is less expensive to produce news and much faster to disseminate such information, a large amount of fake news was produced and rapidly propagated with different purposes such as financial and political gain.

People tend to spread the information they received without confirming its truth. This is where the issue of fake news comes from. The manipulative information propagated will lead to a negative impact on individuals and society. The credibility of the news is doubted because fake news is intentionally created to manipulate audiences. The sentiment of fake news could be negative or positive [7]. This will bring a significant influence on the way society views and responds to real news. Therefore, it is important to generate methods that automatically detect fake news to resolve the serious impact caused by fake news. Detecting fake news on social media is very important yet it is also technically a challenging task. In order to combat the growth of fake news, there are various fact-checking websites deployed to the public to help them determine the content of the news. For example, PolitiFact, Snopes, FactCheck.org and many more in different countries [8]. Malaysia's government also implemented a variety of procedures, including a new law and a fact-checking platform which is called `Sebenarnya.my`. These fact-checking websites are helpful in combating fake news, but they need expert analysis in order to detect fake news. As the spread of fake news online grows, the public and researchers have become more interested in identifying fake news. Most existing research is using the machine learning method for fake news detection. However, they are not capable of detecting fake news in the Malay language. Neural network models have recently outperformed traditional machine learning methods due to their superior ability to extract features.

Hence, the aim of the paper is to develop a fake news detection system that applies a deep learning algorithm that is able to classify fake and non-fake Malay news. For that reason, the objectives for this paper include designing a suitable framework for Malay fake news detection using a deep learning-based algorithm, implementing and optimizing the proposed deep learning-based to detect fake news in the Malay language and finally evaluating the performance of the proposed approach. Section 2 provides the detail of the literature review or related works. Section 3 presents the proposed approach, Sect. 4 presents the results and discussion and finally Sect. 5 concludes this paper.

2 Related Works

There are different features used for fake news detection. It includes the content of an article, public responses, and the source of the article [22]. A significant number of research within the field of machine learning have been conducted to identify fake news. An automatic detection of fake content in online news has also been proposed in [16]. They first introduced two novel datasets covering seven different domains collected from crowdsourced and the web. They also conducted comparative analyses on the identification of linguistic features differences in fake news and real news content. They built a fake news detector using a linear SVM classifier. Then, they conducted several evaluations using five-fold cross validation based on the linguistic features. The obtained result indicated that their model achieved accuracy up to 76%. Reference [20] utilized the dataset aforementioned to evaluate their two proposed deep learning models. The first model has several layers, including the FastText embedding layer, a Bi-GRU encoding layer, Word level attention, and a multi-layer perceptron (MLP). The second model embedding layer is based on the Embedding for Language Model (Elmo) and the MLP Network, which is the same as applied in Model 1. Four sets of experiments were conducted which include the Multi-Domain Analysis, Cross-Domain Analyses, Multi-Domain Training and Domain-wise Testing, and Domain-wise Training and Domain-wise Testing. The results showed that the two proposed methods significantly outperformed the existing models. The results of the experiments have also shown that the type of domain played an important role in the identification of fake news. This study can be further improved by utilizing the BERT and XLNet embedding based method and using WordNet to retrieve connections between words which may influence the fake news detection.

Reference [17] surveyed different features for fake news detection in which, the features were implemented and evaluated for effectiveness in several supervised learning classifiers to differentiate fake from real stories on the dataset. The supervised learning classifiers included were k-Nearest Neighbors (KNN), Naive Bayes

(NB), Random Forests (RF), Support Vector Machine with RBF kernel (SVM), and XGBoost (XGB). The best results were obtained from the RF and XGB classifiers with 85% and 86% respectively. This research can be further explored by looking at the Named-Entity Recognition (NER) [3] in a larger corpus and applying deep learning approach to get a better performance. A deep learning model was also introduced that incorporates three features of fake news, which are text, response, and source [19]. The model was called CSI that consists of three modules Capture, Score, and Integrate. The first module uses a Recurrent Neural Network to capture a given article's temporal pattern of user activity, and the second module learns the source feature based on user behaviour. Then, with the third module, the two modules are combined to perform the classification tasks. They showed an experimental study of real-world data from Twitter and Weibo. The results indicated that CSI achieves greater accuracy than existing models, with 89.2% on Twitter and 95.3% on Weibo.

Convolutional Neural Network (CNN) was also used with images contained in the article to perform the classification [24]. Their model contains two major branches: text branch and image branch. The text branch utilizes two features, textual explicit and text latent. The imaging branch is also made of two features, visual explicit and visual latent. The findings indicated that it truly works better using images. This study proposed to further their investigation in other languages news data and other features such as social network and user behaviour.

Reference [18] developed several deep learning models to detect and classify fake news into the predefined fine-grained categories. The model was developed based on the Convolutional Neural Network (CNN) and Bi-directional Long Short-Term Memory (Bi-LSTM) networks. For the final classification, the representations gathered from CNN and Bi-LSTM were combined and evaluated in the Multi-layer Perceptron Model (MLP). The experiments were performed on a benchmark dataset and showed impressive results that outperformed the existing state of the art with an overall accuracy of 44.87%. This study highlighted the problem in requiring more labelled datasets to train models and the history of fake statements of a user to better understand the pattern of user behaviours.

Various models and classifiers have been studied for fake news detection [10]. The author first introduced his dataset which was collected, mostly from Kaggle. He utilized word representation such as n-gram and Term Frequency—Inverse Document Frequency (TF-IDF). The classifier models that were used are Naive Bayes (NB), Decision Tree, Random Forest, K-Nearest Neighbor, Convolutional Neural Network (CNN), and Long Short-Term Memory (LSTM). A combination of CNN and LSTM was also developed which obtained the highest accuracy of 97.3% compared to other models. This study can be further explored by looking at the performance of the classifiers using different features to detect fake news. Reference [1] suggested a model that can accurately identify the stance of the headline and news articles in order to identify fake news. The Fake News Challenge dataset was used for the experiment. The data was pre-processed using NLTK and the word vector representation has experimented with two techniques which were Bag of Words (BOW) and Term Frequency—Inverse Document Frequency (TF-IDF). This study utilized several neural network architectures such as Dense Neural Network (DNN),

Convolutional Neural Network (CNN), and Recurrent Neural Network (RNN). The best performing model from the proposed solution was the TF-IDF Vectors with DNN. The result obtained up to 94.31% accuracy. The study suggested extending the work by performing analysis on different dataset from social media and in other languages. They also planned to further investigate the propagation of the news that will be predicted faster using simple techniques.

Reference [15] presented a geometric deep learning-based automated fake news detection model. Their research focused on expanding conventional convolutional neural networks to graphs, which allows for the integration of large datasets including content, user profile and behaviour, social graph, and news dissemination. The experiments showed that the structure of social networks and how they were disseminated, were important factors in accurately identifying fake news (92.7% ROC AUC). They then observed that, after only a few hours of propagation, fake news can be reliably detected at an early point. The aging of the model was tested separately in time for training and test results. Authors also furthered their study by introducing adversarial techniques to improve the graph neural network for making decisions. A framework was also introduced that utilized the Long Short-Term Memory (LSTM) and Gated Recurrent Unit (GRU) for classification purposes [23]. They also used Natural Language Toolkit (NLTK) for data pre-processing, such as spell-checking [5], stemming [13] and Part of Speech (POS) tagger [4]. They used FastText as a word embedding for their dataset. They experimented on a dataset consisting of news articles. The result obtained for LSTM is accuracy of 94.3%, while GRU has 91.9% accuracy. The study suggested that hosting a server with an interactive user interface that will automatically scrape news and keep the algorithm training by itself. Kresnakova et al. presented the use of various deep learning models used to detect fake news from the text [12]. They used a dataset from the Kaggle competition which consisted of 20,386 articles from the political news area. The pre-processing involves the Gensim implementation of word2vec for the word embedding. The four classification models used to train the data comprised feed-forward neural networks, Convolutional Neural Networks (CNN) with one convolutional layer, Convolutional Neural Networks (CNN) with more convolutional layers, and Long Short-Term Memory (LSTM). The result obtained proves that most of the models gained consistent performance however convolution and LSTM models proved to be the most effective. This research can be further extended by exploring more on the training data in different languages within the various areas and more features to have a better result for the detection of fake news.

Reference [11] proposed four similar neural networks where each model was trained using different text vectors of news articles. N-gram vectors with news title were inserted to the first model, N-gram vectors with news content to the second model, sequence vectors with news title to the third model, and sequence vectors with news content to the fourth model. Data was collected from an open-source news dataset and pre-processed before being fed into the proposed model. The authors developed Keras neural network models with some dense layers and it was trained using the TensorFlow framework to perform classification tasks. The result showed that the model could achieve up to 90.3% accuracy. From the results observed, models trained with news content were performing better than the models trained with news titles. Further research can be done in other languages such as Malay and Chinese. A combination of different technique algorithms may enhance more fake news detection. A comparison study of fake news detection using machine learning and deep learning techniques was also conducted by [2]. They used two different datasets of news articles from open source that focus on political news type. There were five classification models selected to compare the results obtained using several datasets. The models used were Naive Bayes (NB), Support Vector Machine (SVM), Neural Network Keras, Neural Network TensorFlow, and Long Short-Term Memory (LSTM). They implemented Doc2vec to each model as a word embedding. The results obtained shown that deep learning models like LSTM, NN-Keras, and NN-TensorFlow outperformed traditional machine learning models like NB and SVM. LSTM classifier provided the best results among the other deep learning with an average accuracy of 94.21%. This study suggested that more investigations should be made on datasets of other languages such as Malay and Arabic also considering the removal of stop words [6] to enhance the fake news detection.

3 Methods

The proposed framework of fake news detection using the deep learning method is illustrated in Fig. 1 shown. The first step of the framework is data collection. The dataset will be collected from open source and freely available online. This study will be using a dataset that was obtained from the prior works called Malaya Fake News Preprocessed Dataframe [9]. As stated in the study, the dataset contains news mostly related to politics and entertainment. This dataset will be used because it provides a dataset in Malay language which is crucially needed in this work since the objective of this work is to formulate a framework to classify news in Malay language. The dataset will then be split into train and test sets to train and evaluate the proposed RNN model. Since this work focuses on the textual data of news articles, cleaning and transforming textual data into appropriate format that can be interpreted by a computer is needed when building automated machine learning models on text data. The NLP technique that will be used for the data cleaning is tokenization. Tokenization is the separation of text or sentence into a smaller unit called tokens. Text to sequence transformation process will also be used to transform data into list of sequences and the data will be padded in order to have the same length for all data. The next step is to train the RNN model. The RNN model will be used as it is powerful for modelling sequence data such as text. RNN architecture basically consists of three layers which are input, hidden, and output layers, as shown in Fig. 2. RNNs can have several hidden recurrent units which have memory gates. Long Short-Term Memory (LSTM) is a unit of RNN. LSTM will be added to the architecture since it is capable of learning long-term dependencies by remembering information for long periods of time. A dropout regularization which is a technique used to avoid overfitting when training neural networks will also be added together

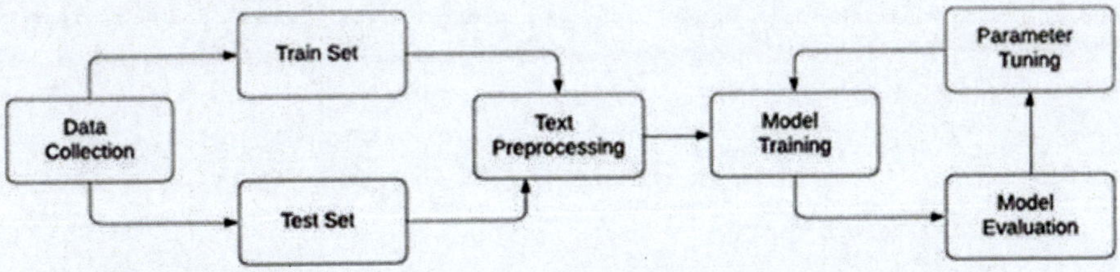

Fig. 1 Proposed framework of Malay fake news detection system

Fig. 2 RNN unit

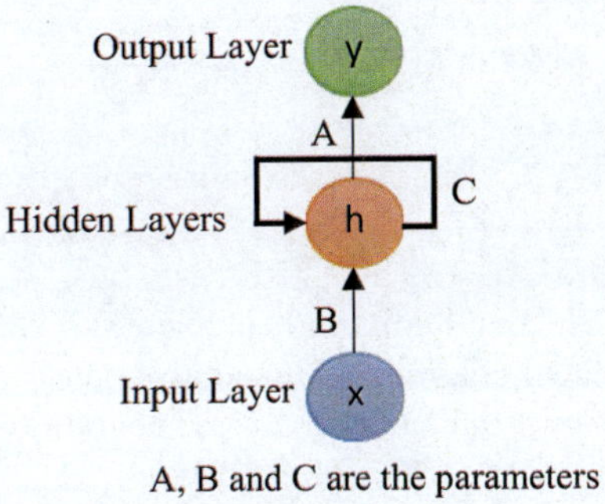

with LSTM. The dense layer will be used as the output layer. Then, all the layers in RNN will be compiled together. After that the model will go through the process of model evaluation using the test data. During this process, the model will also be fine-tuned with hyperparameters (e.g., dropout percentage) to make sure the model is able to produce the best accuracy result.

3.1 Datasets

The dataset that will be used in this work is a collection of news retrieved from different websites [9]. Some of the news articles in the Malaya dataset cannot be correctly classified by their proposed model, thus they were treated as outliers and filtered out. The result of their project had produced a filtered Malay fake news dataset which was being used in this work. The dataset has been cleaned up where all texts have been converted to lowercase, tokenized, and rejoined for more sophisticated dataset used for training model purposes. Each of the news articles has been labeled as real (1) and fake (0). Therefore, this work uses this dataset as an input for training the proposed deep learning model. The dataset contains an imbalanced set of data which real news (1) has 25,030 sample news while fake news (0) has 12,562 sample news.

The dataset is split into train and test sets before the these texts are sent to the pre-processing stage. The dataset consists of two columns, news and label columns.

```
#Normalizing the data with lower case, get rid of extra spaces, and url links.
def normalize(data):
        normalized = []
        for i in data:
                i = i.lower()
                # get rid of urls
                i = re.sub('https?://\S+|www\.\S+', '', i)
                # get rid of non-words and extra spaces
                i = re.sub('\\W', ' ', i)
                i = re.sub('\n', '', i)
                i = re.sub(' +', ' ', i)
                i = re.sub('^ ', '', i)
                i = re.sub(' $', '', i)
                normalized.append(i)
        return normalized
X_train = normalize(X_train)
X_test = normalize(X_test)
```

Fig. 3 Normalizing data using regular expression

The news column is defined as features that will be used to train the model and the label column is defined as targets for the model training which consist of 0 as fake news and 1 as real news. The train set consist of 80% of the dataset while the test set consist of 20% of the dataset.

3.2 Text Pre-processing

The dataset will then be pre-processed in the next stage. This procedure makes use of Python's Natural Language Toolkit (NLTK) package, regex, and scikit-learn library. The data is normalized using the regular expression library. The train data and test data are normalized to lower case. The data also cleaned by getting rid of URL links that exist in the data as well as extra spaces and non-words as shown in Fig. 3. Figure 4 shows the tokenization, text to sequences, and pad process. Tokenization process splits the data into a list of tokens. After that, the data is transformed to numerical data using the text to sequence function. The maximum vocabulary is initialized to 1000 vocabularies from the data. The transformed data then being padded with 500 maximum lengths to have the same lengths for each article.

3.3 Building the RNN Model

Figure 5 outlines the code snippet of building the RNN model using sequential structures. The first layer of the model consists of embedding layer with maximum vocabulary that has been initialized as 1000 and has 128 number of neurons. A Bidirectional layer is also applied to the model so that the model processes both the forward and backward direction of the data. It consists of 128 number of neurons.

```
max_vocab = 1000
tokenizer = Tokenizer(num_words=max_vocab)
tokenizer.fit_on_texts(X_train)

#Convert text to vectors since the classifier only takes numerical data.
#Tokenize the text into vectors
X_train = tokenizer.texts_to_sequences(X_train)
X_test = tokenizer.texts_to_sequences(X_test)

#Apply padding to have the same length for each article
maxlen = 500
X_train = tf.keras.preprocessing.sequence.pad_sequences(X_train,
            padding='post', maxlen=maxlen)
X_test = tf.keras.preprocessing.sequence.pad_sequences(X_test,
            padding='post', maxlen=maxlen)
```

Fig. 4 Tokenization, text to sequences, and pad process

```
#Building the RNN
model = tf.keras.sequential([
        tf.keras.layers.Embedding(max_vocab, 128),
        tf.keras.layers.Bidirectional(tf.keras.layers.LSTM(128)),
        tf.keras.layers.Dense(64, activation='relu'),
        tf.keras.layers.Dropout(0.1),
        tf.keras.layers.Dense(1, activation='sigmoid')
])
```

Fig. 5 Code snippet of building the RNN model

```
Model: "sequential"
_________________________________________________________________
Layer (type)                    Output Shape          Param #
=================================================================
embedding (Embedding)           (None, None, 128)     128000
bidirectional (Bidirectional)   (None, 256)           263168
dense (Dense)                   (None, 64)            16448
dropout (Dropout)               (None, 64)            0
dense_1 (Dense)                 (None, 1)             65
=================================================================
Total params: 407,681
Trainable params: 407,681
Non-trainable params: 0
_________________________________________________________________
```

Fig. 6 RNN model summary

The first dense layer has a unit of 64 neurons with ReLU activation function. A dropout layer also added, and the rate are varied with different value to find the best model with highest accuracy. In this work, four different dropout's values are used which are 0.1 (Model 1), 0.3 (Model 2), 0.5 (Model 3) and 0.8 (Model 4). The last layer has the second dense layer with 1 unit and Sigmoid activation function which return values between 0 and 1. Figure 6 shows the summary of the model created.

```
#The earlyStopping function is used to stop the training when the is no
#Improvement in the loss for the two consecutive epochs.
early_stop = tf.keras.callbacks.EarlyStopping(monitor='val_loss',
                    patience=2, restore_best_weights=True)

model.compile(loss='binary_crossentropy',
                    optimizer='adam',
                    metrics=['accuracy'])

history = model.fit(X_train, y_train, epochs=100, validation_split=0.1,
                    batch_size=64, shuffle=True, callbacks=[early_stop])
```

Fig. 7 Training the RNN model

```
#Evaluate the testing set.
model.evaluate(X_test, y_test)
```

Fig. 8 Evaluating the RNN model

3.4 Training the RNN Model

The model training is performed using the Google Colaborator programming tools by implementing Tensorflow, Keras, and scikit-learn libraries. Based on Fig. 7, the model is compiled with binary cross entropy, Adam optimizer, and accuracy metrics for training. The binary cross entropy loss function is employed in this study because this project is learning binary classification problems. A callback function called EarlyStopping is used to stop the training when there is no improvement in the validation loss value for two consecutive epochs. The model is fit with 100 epochs and 64 batch size with validation data split to 10% from the train set.

3.5 Evaluating the RNN Model

The next step involves the evaluation of the model using the test data set. The metrics calculation also used to evaluate the model. The metrics used are accuracy, precision, recall, and F1-score. The confusion matrix calculation is also defined to summarize the performance of the trained model. Figure 8 outlines the code to perform the evaluation of the RNN model and Fig. 9 outlines how the model can be used to perform prediction for the test dataset. A threshold of 0.5 is set to make sure the predicted values of data return as 1 for more than 0.5 value and return as 0 if less than 0.5.

```
pred = model.predict(X_test)
binary_predictions = []
for I in pred:
        if I >= 0.5:
                #Real and print 1
                binary_predictions.append(1)
        else:
                #Fake and print 0
                binary_predictions.append(0)
```

Fig. 9 Prediction using the RNN model

Table 1 Evaluation results of the trained RNN models

Model	Dropout rate	Accuracy	Precision	Recall	F1-score
1	0.1	88.6	90.7	92.3	91.5
2	0.3	90.1	93.9	91.6	92.8
3	0.5	89.4	91.5	92.7	92.1
4	0.8	88.1	92.9	89.9	91.3

4 Results and Discussion

Table 1 shows the performance result obtained from the model evaluation. The experiment conducted on 4 similar RNN model with four different dropout rate which are 0.1, 0.3, 0.5, and 0.8. The first model with dropout rate 0.1 shows the performance evaluation result with the accuracy of 88.6%, precision of 90.7%, recall of 92.3%, and F1-score of 91.5%. The second model with dropout rate 0.3 shows the performance evaluation result with the accuracy of 90.1%, precision of 93.9%, recall of 91.6%, and F1-score of 92.8%. The third model with dropout rate 0.5 shows the performance evaluation result with the accuracy of 89.4%, precision of 91.5%, recall of 92.7%, and F1-score of 92.1%. The fourth model with dropout rate 0.8 shows the performance evaluation result with the accuracy of 88.1%, precision of 92.9%, recall of 89.9%, and F1-score of 91.3%. Overall, Model 2 shows the highest result which has the dropout percentage of 0.3.

We can further see the training and validation loss of Model $M1$ to see whether the model is able to fit the data given dropout rate of 0.1. Figure 10 shows the training and validation accuracy of Model $M1$ throughout the training over 100 epochs. The values of the training accuracy is stabilized when the number of epoch reaches 20. Model $M1$ stops training at 40 epochs as there is no further improvement in validation loss after 30 epochs. Figure 11 shows the result obtained for Model $M2$. The model stops training at 7 epochs as there is no further improvement in validation loss after 6 epochs. Figure 12 shows the result obtained for the model loss and accuracy in Model $M3$. The model stops training at 7 epochs as there is no further improvement in validation loss after 6 epochs. Figure 13 shows the result obtained for the model

Table 2 Confusion matrices for $M1$, $M2$, $M3$, $M4$

Model		Predicted labels	
$M1$	Actual labels	Fake	Real
	Fake	2057	474
	Real	382	4606
$M2$	Actual labels	Fake	Real
	Fake	2001	307
	Real	438	4773
$M3$	Actual labels	Fake	Real
	Fake	2074	433
	Real	365	4647
$M4$	Actual labels	Fake	Real
	Fake	1907	362
	Real	532	4718

loss and accuracy in Model $M4$. The model stops training at 3 epochs as there is no further improvement in validation loss after 2 epochs.

The smaller is the dropout rate the less number of epochs required to train the RNN model, as seen in Figs. 12 and 13. Model $M1$ required more number of epochs to train the RNN model as seen in Fig. 10. However, the best accuracy can be obtained when the dropout rate was 0.3 with 90.1% accuracy in Model $M2$. Having higher percentage of dropout rate will not produce a trained model that can perform with high accuracy value as seen in Fig. 13.

The comparison of the confusion matrices for all models can be referred from Table 2. Based on the results tabulated in Table 2, for Model $M1$, 2057 test data that were correctly predicted as fake news and 4606 test data that were correctly predicted as real news. Model $M1$ also has incorrectly predicted the test data where 474 of fake news were predicted as real news and 382 real news were predicted as fake news. Hence, the accuracy for Model $M1$ was 88.6%. In Model $M2$, 2001 test data were correctly predicted as fake news and 4773 test data were correctly predicted as real news. However, Model $M2$ also has incorrectly predicted the test data where 307 of fake news were predicted as real news and 438 real news were predicted as fake news, with accuracy of 90.1%. In contrast, In Model $M3$, 2074 test data were correctly predicted as fake news and 4647 test data were correctly predicted as real news. At the same time also, Model $M3$ also has incorrectly predicted the test data where 433 of fake news were predicted as real news and 365 real news were predicted as fake news, with accuracy of 89.4%. Finally, in Model $M4$, 2001 test data correctly were predicted as fake news and 4773 test data were correctly predicted as real news and incorrectly predicted the test data where 307 of fake news were predicted as real news and 438 real news were predicted as fake news, with accuracy of 88.1%.

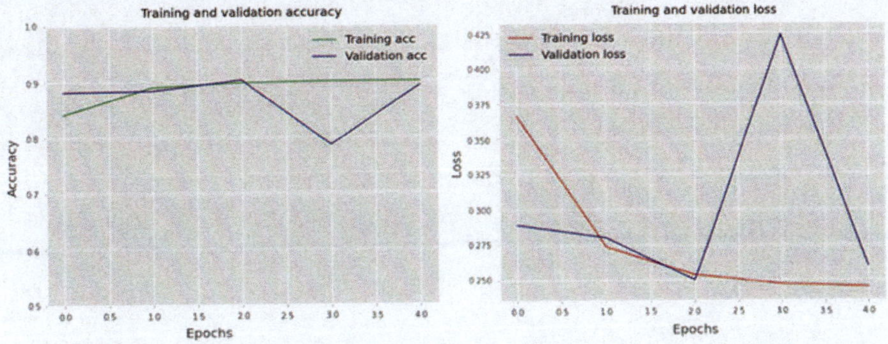

Fig. 10 Validation of accuracy and loss for the trained Model 1

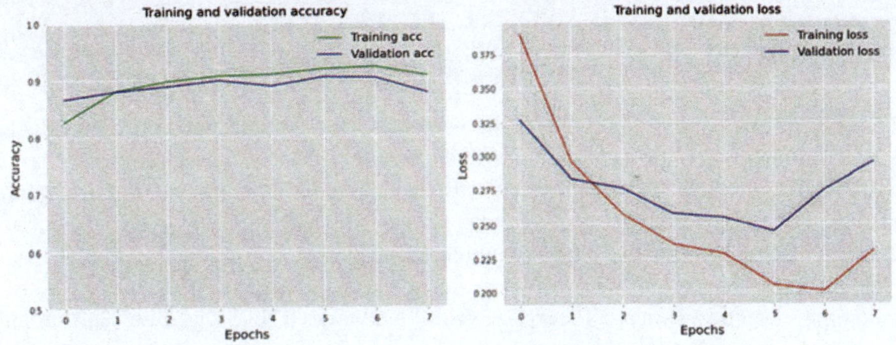

Fig. 11 Validation of accuracy and loss for the trained Model 2

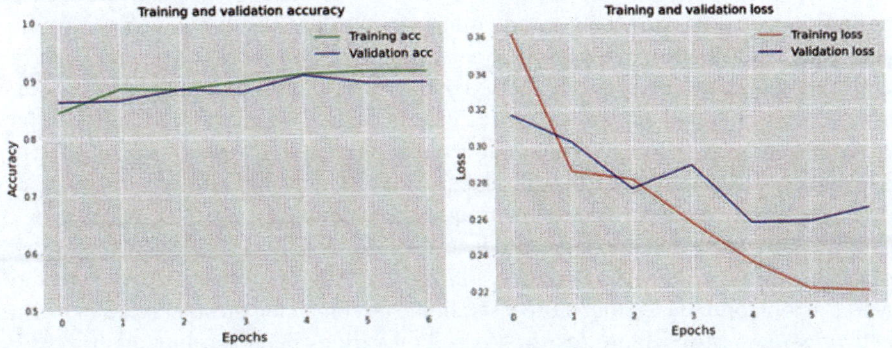

Fig. 12 Validation of accuracy and loss for the trained Model 3

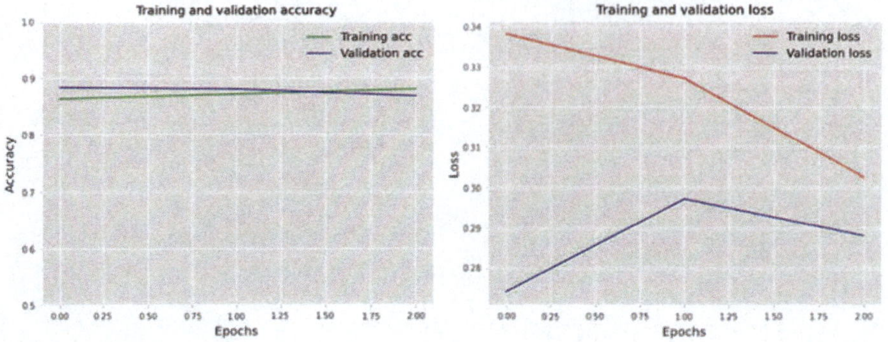

Fig. 13 Validation of accuracy and loss for the trained Model 4

5 Conclusion

The goal of this paper is to evaluate the performance of the proposed Bidirectional RNN deep learning approach to classify fake Malay news. Currently, there are many works that have proposed various methods for fake news detection. However, most of the studies only focused on detecting fake news for English language because a significant number of the datasets which are available online created mainly for English language works. Thus, this paper proposes a suitable framework for Malay fake news detection using a Recurrent Neural Network (RNN) algorithm. Different dropout rates was also used to optimize the proposed framework for obtaining the best possible results of the accuracy. Based on the results obtained, the RNN Model $M1$ with dropout rate 0.3 can detect fake news based on the dataset used with the highest performance evaluation result among other models. The result obtained shows the performance evaluation result with the accuracy of 90.1%, precision of 93.9%, recall of 91.6%, and F1-Score of 92.8%. Based on these findings, the smaller is the dropout rate the less number of epochs is required to train the RNN model as it is observed that Model $M1$ required more number of epochs to train the RNN model. Having higher percentage of dropout rate will not produce a trained model that can perform with high accuracy value. In conclusion, maintaining the percentage of dropout to be 0.3 and below enables the RNN model to produce good values of accuracies. The accuracy of the forecasted result was also highly influenced by the length of the text. For future works, it is recommended that more tests should be conducted on other types of neural networks. As a result, more research based on various deep learning algorithms is recommended for future work. Furthermore, the current dataset was limited to news mostly related to politics and entertainment. A more comprehensive and relevant labelled dataset containing news data in the Malay language from many domains such as health, sport, and more should be established. Fake news detection should not be limited to text articles detection; instead, other elements such as photos and videos should be explored. This is significant since it will increase the performance of the model. However, labelled data that can be used for model training in fake news detection is not easily available. As a result, research into the use of unsupervised machine learning algorithms for detecting fake news should be done.

References

1. Abedalla A, Al-Sadi A, Abdullah M (2019) A closer look at fake news detection: a deep learning perspective. In: Proceedings of the 2019 3rd international conference on advances in artificial intelligence, ICAAI 2019. Association for Computing Machinery, New York, NY, pp 24–28. https://doi.org/10.1145/3369114.3369149
2. Alameri SA, Mohd M (2021) Comparison of fake news detection using machine learning and deep learning techniques. In: 2021 3rd international cyber resilience conference (CRC), pp 1–6. https://doi.org/10.1109/CRC50527.2021.9392458
3. Alfred R, Leong LC, On CK, Anthony P, Fun TS, Razali MNB, Hijazi MHA (2013) A rule-based named-entity recognition for Malay articles. In: Motoda H, Wu Z, Cao L, Zaiane O, Yao M, Wang W (eds) Advanced data mining and applications. Springer Berlin Heidelberg, Berlin, Heidelberg, pp 288–299
4. Alfred R, Mujat A, Obit JH (2013) A ruled-based part of speech (RPOS) tagger for Malay text articles. In: Selamat A, Nguyen NT, Haron H (eds) Intelligent information and database systems. Springer Berlin Heidelberg, Berlin, Heidelberg, pp 50–59
5. Basri SB, Alfred R, On CK (2012) Automatic spell checker for Malay blog. In: 2012 IEEE international conference on control system, computing and engineering, pp 506–510. https://doi.org/10.1109/ICCSCE.2012.6487198
6. Chekima K, Alfred R (2016) An automatic construction of Malay stop words based on aggregation method. In: Berry MW, Mohamed AH, Yap BW (eds) Soft computing in data science. Springer Singapore, Singapore, pp 180–189
7. Chekima K, Alfred R (2018) Sentiment analysis of Malay social media text. In: Alfred R, Iida H, Ibrahim AAA, Lim Y (eds) Computational science and technology. Springer Singapore, Singapore, pp 205–219
8. Chowdhary KR (2020) Natural language processing. Springer India, New Delhi, pp 603–649. https://doi.org/10.1007/978-81-322-3972-7_19
9. Huseinzol05. Huseinzol05/Malaya: natural language toolkit for Bahasa Malaysia. https://malaya.readthedocs.io/. https://github.com/huseinzol05/malaya
10. Kaliyar RK (2018) Fake news detection using a deep neural network. In: 2018 4th international conference on computing communication and automation (ICCCA), pp 1–7. https://doi.org/10.1109/CCAA.2018.8777343
11. Kong SH, Tan LM, Gan KH, Samsudin NH (2020) Fake news detection using deep learning. In: 2020 IEEE 10th symposium on computer applications and industrial electronics (ISCAIE), pp 102–107. https://doi.org/10.1109/ISCAIE47305.2020.9108841
12. Krešňáková VM, Sarnovský M, Butka P (2019) Deep learning methods for fake news detection. In: 2019 IEEE 19th international symposium on computational intelligence and informatics and 7th IEEE international conference on recent achievements in mechatronics, automation, computer sciences and robotics (CINTI-MACRo), pp 000,143–000,148. https://doi.org/10.1109/CINTI-MACRo49179.2019.9105317
13. Leong LC, Basri S, Alfred R (2012) Enhancing Malay stemming algorithm with background knowledge. In: Anthony P, Ishizuka M, Lukose D (eds) PRICAI 2012: trends in artificial intelligence. Springer Berlin Heidelberg, Berlin, Heidelberg, pp 753–758
14. MCMC annual reports | Malaysian communications and multimedia (2020). https://www.mcmc.gov.my/en/resources/publications/annual-reports
15. Monti F, Frasca F, Eynard D, Mannion D, Bronstein MM (2019) Fake news detection on social media using geometric deep learning. arXiv arXiv:1902.06673

16. Pérez-Rosas V, Kleinberg B, Lefevre A, Mihalcea R (2018) Automatic detection of fake news. In: Proceedings of the 27th international conference on computational linguistics. Association for Computational Linguistics, Santa Fe, New Mexico, USA, pp 3391–3401. https://aclanthology.org/C18-1287
17. Reis JCS, Correia A, Murai F, Veloso A, Benevenuto F (2019) Supervised learning for fake news detection. IEEE Intell Syst 34(2):76–81. https://doi.org/10.1109/MIS.2019.2899143
18. Roy A, Basak K, Ekbal A, Bhattacharyya P (2019) A deep ensemble framework for fake news detection and multi-class classification of short political statements. In: Proceedings of the 16th international conference on natural language processing. NLP Association of India, International Institute of Information Technology, Hyderabad, India, pp 9–17. https://aclanthology.org/2019.icon-1.2
19. Ruchansky N, Seo S, Liu Y (2017) CSI: a hybrid deep model for fake news detection. In: Proceedings of the 2017 ACM on conference on information and knowledge management, CIKM'17. Association for Computing Machinery, New York, NY, USA, pp 797–806. https://doi.org/10.1145/3132847.3132877
20. Saikh T, De A, Ekbal A, Bhattacharyya P (2019) A deep learning approach for automatic detection of fake news. In: Proceedings of the 16th international conference on natural language processing. NLP Association of India, International Institute of Information Technology, Hyderabad, India, pp 230–238. https://aclanthology.org/2019.icon-1.27
21. Shearer E, Mitchell A (2022) News use across social media platforms in 2020. https://www.pewresearch.org/journalism/2021/01/12/news-use-across-social-media-platforms-in-2020/
22. Shu K, Sliva A, Wang S, Tang J, Liu H (2017) Fake news detection on social media: a data mining perspective. SIGKDD Explor Newsl 19(1):22–36. https://doi.org/10.1145/3137597.3137600
23. Verma A, Mittal V, Dawn S (2019) Find: fake information and news detections using deep learning. In: 2019 twelfth international conference on contemporary computing (IC3), pp 1–7. https://doi.org/10.1109/IC3.2019.8844892
24. Yang Y, Zheng L, Zhang J, Cui Q, Li Z, Yu PS (2018) TI-CNN: convolutional neural networks for fake news detection. arXiv arXiv:1806.00749

Iterative Method for Solving Nonlinear Fredholm Integral Equations Using Quarter-Sweep Newton-PKSOR Method

L. H. Ali, J. Sulaiman, and A. Saudi

Abstract This paper studies the combination of the quarter-sweep approach with the Newton-PKSOR iteration to solve the nonlinear Fredholm integral equations of the second kind. We call them the quarter-sweep Newton-PKSOR (QSNPKSOR) iterative method. For this purpose, the discretization process is conducted using a first-order quadrature scheme with the quarter-sweep approach to form quarter-sweep approximation equation, which leads to a nonlinear system. Then, we apply Newton's method to convert it into a linear form. Following that, we solve the generated linear system using PKSOR iterative method. The proposed method is motivated by the complexity reduction technique to gain an approximate solution to the nonlinear Fredholm integral equation. Thus, to validate the findings, we compared the results with full- and half-sweep Newton-PKSOR methods on several numerical examples. The comparison of total computing operations and percentage reduction involved in these iterative methods are also included. The numerical results show that the quarter-sweep NPKSOR method has significantly reduced the iteration time required to solve the problem but poses less accuracy compared to the standard form.

Keywords Nonlinear Fredholm integral equations · Newton-PKSOR · Quarter-sweep approach · Half-sweep approach · Quadrature scheme

L. H. Ali (✉) · J. Sulaiman
Faculty of Science and Natural Resources, Universiti Malaysia Sabah, Kota Kinabalu, 88400 Sabah, Malaysia
e-mail: labiyana15@gmail.com

J. Sulaiman
e-mail: jumat@ums.edu.my

A. Saudi
Faculty of Computing and Informatics, Universiti Malaysia Sabah, Kota Kinabalu, 88400 Sabah, Malaysia
e-mail: azali@ums.edu.my

© The Author(s), under exclusive license to Springer Nature Singapore Pte Ltd. 2023

D.-K. Kang et al. (eds.), *Proceedings of the 9th International Conference on Computational Science and Technology*, Lecture Notes in Electrical Engineering 983,
https://doi.org/10.1007/978-981-19-8406-8_3

1 Introduction

The term integral equation was first coined by du Bois-Reymond in 1888 [1]. It is known to appear naturally in physics, chemistry, biology, and engineering problems [2]. An enormous application of integral equations has been mentioned by many researchers [3–5] for instance geophysics, electricity and magnetism, elasticity, radiation, heat transfer, quantum mechanics, optimization, mathematical economics, diffraction problems, and many more. Usually, solving nonlinear integral equations yield a considerable amount of difficulties compared to linear integral equations. However, great attention has been paid to solving nonlinear integral equations due to their important application in various research fields. In this paper, we will restrict our discussion to one of the class of nonlinear integral equations which is the nonlinear Fredholm integral equations. The solution of nonlinear Fredholm integral equations has been a subject of interest to many researchers. Many methods have been developed to solve nonlinear Fredholm integral equations. In recent studies, Amiri et al. [6] developed a numerical method based on cosine-trigonometric basis functions, Maturi [7] presented the successive approximation method, Najafi [8] introduced the Nyström-quasilinearization method, Rashidinia et al. [9] applied non-polynomial spline functions, and Ramadan and Osheba [10] studied a new hybrid function based on a combination of Bernstein and improved block-pulse functions to solve nonlinear Fredholm integral equations. In Parand et al. [11], developed the solution of linear and nonlinear Fredholm integral equations using least square-support vector regression. Besides that, Asif et al. [12] proposed the numerical solution of linear and nonlinear integral equations using the Legendre multi-wavelets collocation method and Karamollahi et al. [13] presented the approximate solution of nonlinear Fredholm integral equations using a class of Hermite interpolation polynomials. However, those methods are in their standard form and the complexity reduction techniques has not been implemented. Thus, the focus of this study is to see how effective the complexity reduction techniques in solving nonlinear Fredholm integral equations.

The complexity of the method has been subject to the discussion of many researchers, especially in solving problems numerically. Many innovations have been applied in order to speed up the iteration time of the methods using half- and quarter-sweep approaches in solving the scientific problems. Half-sweep approach was introduced by Abdullah [14] to solve Poisson equation. It then has been applied in solving 2D Helmholtz equations [15], two-dimensional Helmholtz equations [16] and 2D free space wave propagation [17]. The concept of half-sweep is to take only a half of total node points from the solution domain to be considered in the iteration while the remaining node points are computed after the convergence criterion is satisfied. Following the introduction of half-sweep approach, quarter-sweep approach then was invented by Othman and Abdullah [18] to solve the two-dimensional Poisson equation using Modified Explicit Group (MEG). The concept of quarter-sweep is an extension of the half-sweep approach to form a faster solver compared to the half-sweep approach. The concept of quarter-sweep is to take only a quarter of the total node points from the solution domain compared to just half of the total node points in half-sweep approach. It has been applied to solve numerous scientific problems efficiently compared to the standard method and method with half-sweep approach. For instance, steady state problems [19], linear complementarily problem [20], and Poisson image blending problem [21].

Motivated by the concept of complexity reduction techniques, this study combines the quarter-sweep approach with the Newton-PKSOR iterative method to solve nonlinear Fredholm integral equations. The form method is called the quarter-sweep Newton-PKSOR iterative method. The concept of half- and quarter-sweep is to apply the complexity reduction techniques to the iterative method to speed up the computational time. Together with the development of iterative methods in scientific studies, these concepts have been applied with iterative methods in solving many important equations and problems. Some of them are quarter-sweep preconditioned Gauss-Seidel [22] quarter-sweep Newton-Gauss-Seidel [23] quarter-sweep SOR [24, 25] quarter-sweep KSOR [26], quarter-sweep preconditioned relaxation method [27], and quarter-sweep modified Accelerated Overrelaxation method [28]. It was shown in the stated literature that the implementation of half- and quarter-sweep approaches has reduced the computational complexity of the proposed method, resulting in faster computational time in obtaining the approximate solution of the problems compared to the full-sweep approach.

Recently, the study in [29] have shown that the Newton-PKSOR to be efficient in solving the nonlinear Fredholm integral equations by lowering the number of iterations and computational time compared to the Newton-GS and Newton-KSOR methods. Following that, the study in [30] have proposed the half-sweep Newton-PKSOR method which shown that the half-sweep Newton-PKSOR can compete the standard Newton-PKSOR in solving the nonlinear Fredholm integral equations. By taking the quarter-sweep approach to an advantage, this study combines the approach with the Newton-PKSOR to form quarter-sweep Newton-PKSOR method. As witnessed by the literature, the quarter-sweep method for solving nonlinear Fredholm integral equations has not been studied. Moreover, the PKSOR method introduced by Constantinescu et al. [31] had not been combined with the quarter-sweep approach. Thus, this study aims to improve the efficiency of Newton-PKSOR by implementing the method with the quarter-sweep approach to solve nonlinear Fredholm integral equations. To demonstrate the efficiency of quarter-sweep Newton-PKSOR (QSNPKSOR), we compare the results to the full- and half-sweep Newton-PKSOR represent by FSNPKSOR and HSNPKSOR, respectively.

Here is the outline of this study: in Sect. 2, we discuss the discretization of nonlinear Fredholm integral equations using quadrature rule to generate a nonlinear system. Then, we discuss the formulation and algorithm of the proposed method to solve the nonlinear Fredholm integral equations. In Sect. 3, we present several numerical examples and discuss the results of this study. Finally, this study ends with brief conclusions and suggestions in Sect. 4.

2 Methodology

The topic of this study is the nonlinear Fredholm integral equations of the second kind, which are defined by Ebrahimi and Rahidinia [32]

$$x(s) = y(s) + \int_a^b K(s, t, x(t))\mathrm{d}t, \quad x \in [a, b], \tag{1}$$

where $K(s, t, x(t))$ is known as the kernel of the integral equations which continuous on interval $[a, b]$, $y(s)$ is known function and $x(s)$ is the unknown function to be determined. As indicated before, this paper study the QSNPKSOR iterative method to solve Eq. (1). For this purpose, discretization process is conducted to discretize the nonlinear Fredholm integral equation using quadrature scheme. The discussion will cover full-, half-, and quarter-sweep approaches. To discretize the integral term of Eq. (1), we need to consider the solution domain $[a, b]$ be divided into n equidistance subintervals which represent by $s_0, s_1, s_2, \ldots, s_n$ into $a = s_0 < s_1 < s_2 < \cdots < s_{n-1} < s_n = b$. The distance between two node points is presented by h, where $h = (b - a)/n$. So now we can classified the grid points into type ●, ○, and □ as illustrated in Fig. 1 [24].

Figure 1 shows the distribution of grid points on interval $[a, b]$ for half-sweep and quarter-sweep methods. The half-sweep method involves of two-types of node points, which is ● and ○. Meanwhile, the quarter-sweep iteration involves several different types of node points classified as ●, ○, and □. The node points of type ● are computed using iterative methods while the remaining node points of type ○ and

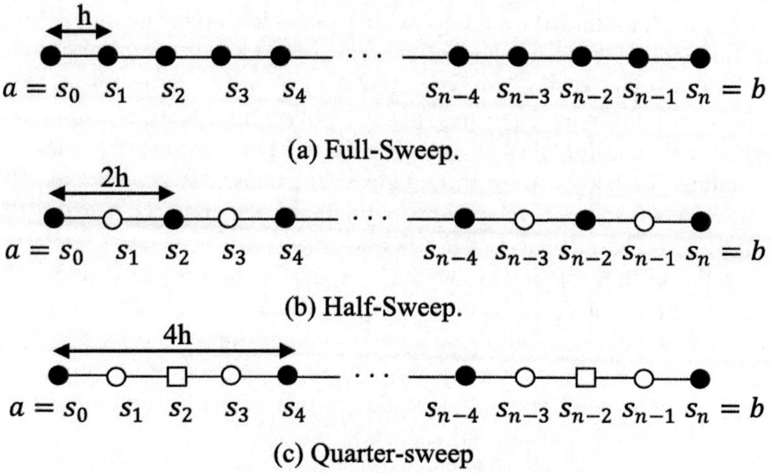

Fig. 1 Distribution of grid points on the solution domain using full-, half- and quarter-sweep approach

□ are computed using direct method [26]. The distance between two node points of type ● is represented by a step size, h. The quarter-sweep method has 4 h step size which bigger than the full- and half-sweep iteration. Based on Fig. 1, it is clear that the implementation of half- and quarter-sweep with iterative method will only consider half and quarter from all the interior node points to be considered in the iteration respectively.

Following that, we will discuss the discretization process on Eq. (1) using the first-order quadrature rule to generate a system of nonlinear equations. The general equation of quadrature scheme is given by Muthuvalu and Sulaiman [33]

$$\int_a^b y(s)\mathrm{d}t = \sum_{j=0}^{n} A_j y(s_j) + \epsilon_n(y), \tag{2}$$

where $s_j, (j = 0, p, 2p, \ldots, n)$ are abscissas of the partition points of the integration interval on the interval $[a, b]$, A_j are the numerical coefficients and $\epsilon_n(y)$ is the truncation error. The coefficient of A_j on Eq. (2) based on the Trapezoidal rule can be defined as follows

$$A_j = \begin{cases} \frac{1}{2}ph, & j = 0, n \\ ph, & otherwise \end{cases}, \tag{3}$$

where the value of p corresponds to 1, 2, and 4 for full-, half-, and quarter-sweep cases, respectively. In general, the application of the quadrature scheme towards the integral term of Eq. (1) will form the following equation

$$x_i - \sum_{j=0,p,2p,\ldots,n}^{n} A_j k(s_i, t_j, x_j) = y_i, \quad i = 0, p, 2p, \ldots, n \tag{4}$$

Then, the expansion of Eq. (4) based on Eq. (3) will obtained the nonlinear approximation equation

$$\begin{aligned} x_i &- \frac{1}{2}phk(s_i, t_0, x_0) - phk(s_i, t_p, x_p) - phk(s_i, t_{2p}, x_{2p}) \\ &- \cdots - phk(s_i, t_{n-p}, x_{n-p}) - \frac{1}{2}phk(s_i, t_n, x_n) = y_i \end{aligned} \tag{5}$$

The nonlinear function of Eq. (5) can be defined as

$$\begin{aligned} &f_i(x_0, x_p, x_{2p} \ldots, x_n) = x_i - \frac{1}{2}phk(s_i, t_0, x_0) - phk(s_i, t_p, x_p) \\ &- phk(s_i, t_{2p}, x_{2p}) - \cdots - phk(s_i, t_{n-p}, x_{n-p}) \\ &- \frac{1}{2}phk(s_i, t_n, x_n) - y_i \end{aligned} \tag{6}$$

where $i = 0, p, 2p, \ldots, n$. Flowing that, we can obtain the following system of nonlinear equations

$$f_i\left(x_0, x_p, x_{2p} \ldots, x_n\right) = 0 \tag{7}$$

Now, we perform the linearization process using Newton's method to convert Eq. (7) to be a linear system as follows [34]

$$J\left(\underline{x}^{(k)}\right)\Delta\underline{x}^{(k)} = -f_i\left(\underline{x}^{(k)}\right) \tag{8}$$

where

$$J\left(\underline{x}^{(k)}\right) = \begin{bmatrix} \frac{df_0}{dx_0} & \frac{df_0}{dx_p} & \frac{df_0}{dx_{2p}} & \cdots & \frac{df_0}{dx_n} \\ \frac{df_p}{dx_0} & \frac{df_p}{dx_p} & \frac{df_p}{dx_{2p}} & \cdots & \frac{df_p}{dx_n} \\ \frac{df_{2p}}{dx_0} & \frac{df_{2p}}{dx_p} & \frac{df_{2p}}{dx_{2p}} & \cdots & \frac{df_{2p}}{dx_n} \\ \vdots & \vdots & \vdots & \ddots & \vdots \\ \frac{df_n}{dx_0} & \frac{df_n}{dx_p} & \frac{df_n}{dx_{2p}} & \cdots & \frac{df_n}{dx_n} \end{bmatrix}_{\left(\frac{n}{p}+1\right)\times\left(\frac{n}{p}+1\right)}$$

$$\Delta\underline{x}^{(k)} = \left[\Delta x_0 \; \Delta x_p \; \Delta x_{2p} \; \cdots \; \Delta x_n\right]^{\mathrm{T}}$$

and the solution vector is determined by the following expressions

$$\underline{x}_i^{(k+1)} = \underline{x}_i^{(k)} + \Delta\underline{x}_i \tag{9}$$

The generated linear system in Eq. (8) also can be written in a general form as

$$\sum_{j-1}^{n} a_{ij}x_j = b_i, \quad i = 0, p, 2p, \ldots n. \tag{10}$$

Since we have formed a linear system, it can be solved using iterative method. In this study, we consider the PKSOR iterative method to solve the corresponding linear system. Consider the following general formulation of the PKSOR iteration [31]

$$x_i^{[k+1]} = \frac{1}{\left(1+\omega_2^*\right)}\left(x_i^{(k)} + \frac{\omega_1^*}{a_{ii}}\left(b_i - \sum_{j=1}^{i-1} a_{ij}x_j^{(k+1)} - \sum_{j=i+1}^{n} a_{ij}x_j^{(n)}\right)\right) \tag{11}$$

where $\omega_1^* \in \mathbb{R}$, $\omega_2^* \in \mathbb{R} - [-2, 0]$ and $k = 0, 1, 2, \ldots$ represents the iteration. The linear system can also represent in the matrix form as follows

$$A\underline{x} = \underline{f} \tag{12}$$

where

$$A = \begin{bmatrix} A_{0,0} & A_{0,p} & A_{0,2p} & \cdots & A_{0,n} \\ A_{p,0} & A_{p,p} & A_{p,2p} & \cdots & A_{p,n} \\ A_{2p,0} & A_{2p,p} & A_{2p,2p} & \cdots & A_{2p,n} \\ \vdots & \vdots & \vdots & \ddots & \vdots \\ A_{n,0} & A_{n,p} & A_{n,2p} & \cdots & A_{n,n} \end{bmatrix}_{\left(\frac{n}{p}+1\right)\times\left(\frac{n}{p}+1\right)},$$

$$\underline{x} = \left[\, x_0 \; x_p \; x_{2p} \; \cdots \; x_n \,\right]^{\mathrm{T}},$$

$$\underline{f} = \left[\, f_0 \; f_p \; f_{2p} \; \cdots \; f_n \,\right]^{\mathrm{T}},$$

Thus, the formulation of PKSOR iteration can also defined in the matrix from as follows

$$x^{[k+1]} = \frac{1}{\left[(1+\omega_2^*)D - \omega_1^* L\right]}\left(D + \omega_1^* U\right)x^{[k]} + \omega_1^* b. \tag{13}$$

The coefficient matrix A is based on the decomposition of $A = D - L - U$, where $A \leftarrow A - dD$ and $\frac{\omega_2^* - \omega_1^*}{\omega_2^*}$, meanwhile D, L and U are diagonal matrix, strictly lower triangular matrix, and strictly upper triangular matrix, respectively. From the coefficient matrix A, with replacing $p = 4$ it can clearly show the application of the quarter-sweep approach will reduce the size of the matrix by a quarter if compared to the full-sweep and half-sweep methods. More precisely, we illustrate the comparison of the FSNPKSOR, HSNPKSOR, and QSNPKSOR in the context of computational complexity in Table 1. Table 1 shows the comparison of total arithmetic operations involved in implementing FSNPKSOR, HSNPKSOR, and QSNPKSOR per iteration. Based on Fig. 1 and Algorithm 1, it can be concluded that the total arithmetic operations of HSNSOR and QSNSOR per iteration have been reduced by approximately 1/2 and 3/4 of FSNPKSOR, respectively.

Algorithm 1: Family of Newton-PKSOR iteration towards nonlinear Fredholm integral equations

(i) Let $q = 0$ $\underline{x}^{(0)} = q$, $k = 0$, and $\varepsilon = 10^{-10}$.
(ii) Set the optimum values of ω_1^* and ω_2^*.

Table 1 Comparison of total arithmetic operation involves for implementing FSNPKSOR, HSNPKSOR, and QSNPKSOR per iteration

Methods	Arithmetic operation	
	$+$ *or* $-$	$\times$ *or* $\div$
FSNPKSOR	$(n+3)(n+1)$	$(n+4)(n+1)$
HSNPKSOR	$\left(\frac{n}{2}+3\right)\left(\frac{n}{2}+1\right)$	$\left(\frac{n}{2}+4\right)\left(\frac{n}{2}+1\right)$
QSNPKSOR	$\left(\frac{n}{4}+3\right)\left(\frac{n}{4}+1\right)$	$\left(\frac{n}{4}+4\right)\left(\frac{n}{4}+1\right)$

(iii) Set $\Delta\underline{x}^{(0)} = 0$ and compute

(a) Matrix $A = J\left(\underline{x}^{(k)}\right)$,
(b) and $\underline{b} = -G\left(\underline{x}^{(k)}\right)$.

(iv) Calculate the current value, $\Delta\underline{x}^{(k+1)}$

(a) For $i = 0, p, 2p, \ldots, n$ compute Eq. (13) iteratively.
(b) Run the convergence test, $\left|\Delta x_i^{(k+1)} - \Delta x_i^{(k)}\right| \leq \varepsilon$. If satisfied, proceed to step vi, otherwise repeat step iv(a).

(v) Calculate the current value using Eq. (9).
(vi) Test the convergence with condition, $\left|G\left(\underline{x}^{(k+1)}\right)\right| \leq \varepsilon$. If satisfied, proceed to step vii, otherwise repeat step iii.
(vii) Compute the node points type □ and the node points type ○ by second order Lagrange interpolation.
(viii) Display the results.

3 Numerical Experiments

In this section, we present three numerical examples to illustrate the efficiency of the proposed methods as follows.

Example 1 [35] For the first example, we consider the following nonlinear Fredholm integral equation

$$x(s) = 1 - \frac{5}{12}s + \int_0^1 st[x(t)]^2 \mathrm{d}t, \tag{14}$$

where the exact solution is given by $x(s) = 1 + \frac{1}{3}s$.

Example 2 [36] Then, we consider the problem of nonlinear Fredholm integral equation as follows

$$\begin{aligned} x(s) = \cos(s) + \frac{5}{2} - s + \frac{1}{2}\cos(1)\sin(1) \\ + 2\sin(1) + \int_0^1 \left(s - x(t)^2\right)\mathrm{d}t, \end{aligned} \tag{15}$$

with the exact solution given by $x(s) = \cos(s) + 1$.

Example 3 [2] Lastly, we consider the following nonlinear Fredholm integral equation

$$x(s) = 1 - \frac{1}{3}s + \int_0^1 st^2x^3(t)\mathrm{d}t, \tag{16}$$

where the exact solution is $x(s) = 1$.

The experimental tests for this study were conducted on five sizes of grid points, $m = 256, 512, 1024, 2045$, and 4096, so that the total grid points can be divided into half- and quarter-sweep cases. Based on Algorithm 1, the experimental tests are conducted using Borland C++. To illustrate the efficiency of QSNPKSOR, we compared the results with the FSNPKSOR and HSNPKSOR methods. The data for this study were obtained by running the programme to obtain the optimum values of the ω_1^* and ω_2^* for NPKSOR methods. However, since we are implementing the half- and quarter-sweep approach, we used the stored values from the FSNPKSOR for HSNPKSOR and QSNPKSOR. Therefore, for QSNPKSOR, we only need to find the optimum values for grid point 256 while the remaining grid points are computed from the recorded optimum values of HSNPKSOR method. After all the optimum parameters are recorded, we run the programme to get the number of iterations (k), computational time (time) measured in seconds, and maximum absolute error (MAE) for FSNPKSOR, HSNPKSOR, and QSNPKSOR. The numerical results of this study are recorded in Tables 2, 3, 4 and 5.

The results obtained clearly show that the implementation of HSNPKSOR and QSNPKSOR has reduced the computational time for all tested examples. Also, based on the technique explained in previous paragraph, the implementation of HSNPKSOR and QSNPKSOR methods are easy to be implemented by using the restored values of ω_1^* and ω_2^* from FSNPKSOR method. The computational time for HSNPKSOR and QSNPKSOR have reduced by approximately 70–76% and 91–94%, respectively compared with the standard FSNPKSOR. In terms of iteration number, QSNPKSOR recorded the almost same number of iterations compared to FSNPKSOR and HSNPKSOR but it recorded lower number of iterations for grid size 256. Also, in terms of maximum absolute error, QSNPKSOR possess less accurate results than FSNPKSOR and HSNPKSOR due to the implementation of direct method. However, as the grid sizes increase, this method's accuracy is also increased, resulting in more accurate approximate solutions than small grid sizes. Referring to Tables 2, 3 and 4, it can be seen that the pattern of the results for QSNPKSOR is identical to the FSNPKSOR and HSNPKSOR. For instance, the results for QSNPKSOR ($m = 1024$) is identical to the HSNPKSOR ($m = 512$) and FSNPKSOR ($m = 256$). This is due to the implementation of half-and quarter-sweep approaches where the results are dragged forward to be more efficient.

Table 2 Numerical results of FSNPKSOR, HSNPKSOR, and QSNPKSOR for Example 1

	Methods	Grid size				
		256	512	1024	2048	4096
k	FSNPKSOR	115.0 ($\omega_1^* = -4.0769$) ($\omega_2^* = -4.0731$)	115.0 ($\omega_1^* = -4.0723$) ($\omega_2^* = -4.0685$)	114.0 ($\omega_1^* = -4.0700$) ($\omega_2^* = -4.0662$)	109.0 ($\omega_1^* = -4.0563$) ($\omega_2^* = -4.0525$)	107.0 ($\omega_1^* = -4.0563$) ($\omega_2^* = -4.0525$)
	HSNPKSOR	108.0 ($\omega_1^* = -4.2143$) ($\omega_2^* = -4.2103$)	115.0 ($\omega_1^* = -4.0769$) ($\omega_2^* = -4.0731$)	115.0 ($\omega_1^* = -4.0723$) ($\omega_2^* = 4.0685$)	114.0 ($\omega_1^* = -4.0700$) ($\omega_2^* = -4.0662$)	109.0 ($\omega_1^* = -4.0563$) ($\omega_2^* = -4.0525$)
	QSNPKSOR	108.0 ($\omega_1^* = -4.2385$) ($\omega_2^* = -4.2344$)	108.0 ($\omega_1^* = -4.2143$) ($\omega_2^* = -4.2103$)	115.0 ($\omega_1^* = -4.0769$) ($\omega_2^* = -4.0731$)	115.0 ($\omega_1^* = -4.0723$) ($\omega_2^* = -4.0685$)	114.0 ($\omega_1^* = -4.0700$) ($\omega_2^* = -4.0662$)
Time	FSNPKSOR	0.17	0.61	2.37	9.00	35.34
	HSNPKSOR	0.04	0.17	0.62	2.40	9.02
	QSNPKSOR	0.01	0.05	0.17	0.63	2.39
MAE	FSNPKSOR	1.27165E−05	3.17897E−06	7.94747E−07	1.98698E−07	4.96621E−08
	HSNPKSOR	5.08758E−05	1.27165E−05	3.17897E−06	7.94747E−07	1.98698E−07
	QSNPKSOR	2.03662E−04	5.08758E−05	1.27165E−05	3.17897E−06	7.94747E−07

Table 3 Numerical results of FSNPKSOR, HSNPKSOR, and QSNPKSOR for Example 2

	Methods	Grid size				
		256	512	1024	2048	4096
k	FSNPKSOR	459.0 ($\omega_1^* =$ 0.8812) ($\omega_2^* =$ 1.1518)	492.0 ($\omega_1^* =$ 0.7433) ($\omega_2^* =$ 0.9557)	497.0 ($\omega_1^* =$ 0.7511) ($\omega_2^* =$ 0.9209)	497.0 ($\omega_1^* =$ 0.7481) ($\omega_2^* =$ 0.9175)	498.0 ($\omega_1^* =$ 0.7491) ($\omega_2^* =$ 0.9182)
	HSNPKSOR	458.0 ($\omega_1^* =$ 0.9276) ($\omega_2^* =$ 1.1736)	459.0 ($\omega_1^* =$ 0.8812) ($\omega_2^* =$ 1.1518)	492.0 ($\omega_1^* =$ 0.7433) ($\omega_2^* =$ 0.9557)	497.0 ($\omega_1^* =$ 0.7511) ($\omega_2^* =$ 0.9209)	497.0 ($\omega_1^* =$ 0.7481) ($\omega_2^* =$ 0.9175)
	QSNPKSOR	440.0 ($\omega_1^* =$ 1.0441) ($\omega_2^* =$ 1.3573)	458.0 ($\omega_1^* =$ 0.9276) ($\omega_2^* =$ 1.1736)	459.0 ($\omega_1^* =$ 0.8812) ($\omega_2^* =$ 1.1518)	492.0 ($\omega_1^* =$ 0.7433) ($\omega_2^* =$ 0.9557)	497.0 ($\omega_1^* =$ 0.7511) ($\omega_2^* =$ 0.9209)
Time	FSNPKSOR	0.74	2.70	10.68	42.49	172.08
	HSNPKSOR	0.22	0.65	2.74	10.67	42.55
	QSNPKSOR	0.06	0.18	0.67	2.71	10.71
MAE	FSNPKSOR	7.03927E−07	1.76056E−07	4.40509E−08	1.10567E−08	2.80573E−09
	HSNPKSOR	2.84035E−06	7.06879E−07	1.76231E−07	4.40657E−08	1.10583E−08
	QSNPKSOR	1.14609E−05	2.84035E−06	7.06879E−07	1.76231E−07	4.40657E−08

Table 4 Numerical results of FSNPKSOR, HSNPKSOR, and QSNPKSOR for Example 3

	Methods	Grid size				
		256	512	1024	2048	4096
k	FSNPKSOR	85.0 ($\omega_1^* = -$ 4.9366) ($\omega_2^* = -$ 4.9355)	85.0 ($\omega_1^* = -$ 4.9279) ($\omega_2^* = -$ 4.9268)	85.0 ($\omega_1^* = -$ 4.9115) ($\omega_2^* = -$ 4.9104)	85.0 ($\omega_1^* = -$ 4.9088) ($\omega_2^* = -$ 4.9077)	85.0 ($\omega_1^* = -$ 4.9066) ($\omega_2^* = -$ 4.9055)
	HSNPKSOR	84.0 ($\omega_1^* = -$ 5.1901) ($\omega_2^* = -$ 5.1889)	85.0 ($\omega_1^* = -$ 4.9366) ($\omega_2^* = -$ 4.9355)	85.0 ($\omega_1^* = -$ 4.9279) ($\omega_2^* = -$ 4.9268)	85.0 ($\omega_1^* = -$ 4.9115) ($\omega_2^* = -$ 4.9104)	85.0 ($\omega_1^* = -$ 4.9088) ($\omega_2^* = -$ 4.9077)
	QSNPKSOR	79.0 ($\omega_1^* = -$ 5.3406) ($\omega_2^* = -$ 5.3051)	84.0 ($\omega_1^* = -$ 5.1901) ($\omega_2^* = -$ 5.1889)	85.0 ($\omega_1^* = -$ 4.9366) ($\omega_2^* = -$ 4.9355)	85.0 ($\omega_1^* = -$ 4.9279) ($\omega_2^* = -$ 4.9268)	85.0 ($\omega_1^* = -$ 4.9115) ($\omega_2^* = -$ 4.9104)
Time	FSNPKSOR	0.14	0.52	2.04	8.03	31.83
	HSNPKSOR	0.04	0.15	0.52	2.03	7.95
	QSNPKSOR	0.01	0.04	0.14	0.51	2.01
MAE	FSNPKSOR	1.01732E−05	2.54317E−06	6.35783E−07	1.58943E−07	3.97335E−08
	HSNPKSOR	4.07015E−05	1.01732E−05	2.54317E−06	6.35783E−07	1.58943E−07
	QSNPKSOR	1.62944E−04	4.07015E−05	1.01732E−05	2.54317E−06	6.35783E−07

Table 5 Percentage reduction of computational time for HSNPKSOR and QSNPKSOR compared with FSNPKSOR

Methods	Computational time (%)		
	Example 1	Example 2	Example 3
HSNPKSOR	72.13–76.47	70.27–75.93	71.43–75.02
QSNPKSOR	91.80–94.12	91.89–93.78	92.31–93.69

4 Conclusions

This study has presented the combination of the quarter-sweep method with Newton-PKSOR, called QSNPKSOR to solve the nonlinear Fredholm integral equations. Based on the results, the QSNPKSOR recorded relatively faster computational time compared to the FSNSOR and HSNPKSOR. This is due to the complexity reduction technique, which reduces the total arithmetic operations involved in QSNPKSOR. This study also found the iteration number of FSNPKSOR, HSNPKSOR, and QSNPKSOR approaching the same number as the grid size becomes large. The QSNPKSOR method also possess a relatively less accurate maximum absolute error than FSNPKSOR and HSNPKSOR due to the implementation of the direct method to compute the remaining node points in the large solution domain. However, the accuracy of QSNPKSOR is increased as the grid sizes increase, resulting in more accurate approximate solutions. To improve the accuracy of HSNPKSOR and QSNPKSOR, we suggest further research regarding this issue to solve the nonlinear integral equations.

References

1. Kress R (2014) Linear integral equations, 3rd edn. Springer, New York, NY
2. Wazwaz A-M (2015) A first course in integral equations, 2nd edn. World Scintific Publishing Co. Pvt. Ltd.
3. Maglevanny II, Karyakina TI (2020) Numerical solution of Urysohn type nonlinear second kind integral equations by successive quadratures using embedded Dormand and Prince scheme 5(4). Comput Res Model 12(2):275–300
4. Kaur M, Singh I (2020) Hermite wavelets collocation method for solving nonlinear Fredholm integral equations. J Phys Sci Eng Technol 12(2):82–88
5. Almousa M (2020) Adomian decomposition method with modified bernstein polynomials for solving nonlinear Fredholm and Volterra integral equations. Math Stat 8:278–285
6. Amiri S, Hajipour S, Beleanu D (2020) On accurate solution of the Fredholm integral equations of the second kind. Appl Numer Math 150:478–490
7. Maturi DA (2019) The successive approximation method for solving nonlinear Fredholm integral equation of the second kind using Maple. Adv Pure Math 9:832–843
8. Najafi E (2020) Nyström-quasilinearization method and smoothing transformation for the numerical solution of nonlinear weakly singular Fredholm integral equations. J Comput Appl Math 368:1–13
9. Rashidinia J, Maleknejad K, Jalilian H (2019) Convergence analysis of non-polynomial spline functions for the Fredholm integral equation. Int J Comput Math 1–15

10. Ramadan M, Osheba HS (2020) A new hybrid orthonormal Bernstein and improved block-pulse functions method for solving mathematical physics and engineering problems. Alex Eng J 59:3643–3652
11. Parand K, Aghaei AA, Jani A, Ghodsi A (2021) A new approach to the numerical solution of Fredholm integral equations using least squares-support vector regression. Math Comput Simul 180:114–128
12. Asif M, Khan I, Haider N, Al-Mdallal Q (2020) Legendre multi-wavelets collocation method for numerical solution of linear and nonlinear integral equations. Alex Eng J 59:5099–5109
13. Karamollahi N, Heydari M, Loghmani GB (2021) Approximate solution of nonlinear Fredholm integral equations of the second kind using a class of Hermite interpolation polynomials. Math Comput Simul 187:414–432
14. Abdullah AR (1991) The four point-Explicit Decoupled Group (EDG) method: a fast poisson solver. Int J Comput Math 38:61–70
15. Akhir MKM, Othman M, Sulaiman J, Majid ZA, Suleiman M (2011) The four point-EDGMSOR iterative method for solution of 2d Helmholtz equations. Commun Comput Inf Sci 253CCIS(Part 3):218–227
16. Akhir MKM, Othman M, Sulaiman J, Majid ZA, Suleiman M (2011) Half-sweep modified successive OverRelaxation for solving two-dimensional helmholtz equations. Aust J Basic Appl Sci 5(12):3033–3039
17. Hasan MK, Othman M, Abbas Z, Sulaiman J, Ahmad F (2007) Parallel solution of high speed low order FDTD on 2D free space wave propagation. In: Lecture notes in computer science (including subseries lecture notes in artificial intelligence and lecture notes in bioinformatics), vol 4706 LNCS(Part 2), pp 13–24
18. Othman M, Abdullah AR (2000) An efficient four points modified explicit group Poisson solver. Int J Comput Math 76:203–217
19. Hoe NY, Hasan MK (2013) Investigation of steady state problems via quarter-sweep schemes. Sains Malaysiana 42(6):837–844
20. Koh WS, Sulaiman J, Mail R (2010) Quarter-sweep projected modified gauss-seidel algorithm applied to linear complementarity problem. Am J Appl Sci 7(6):790–794
21. Eng JH, Saudi A, Sulaiman J (2019) Implementation of quarter-sweep approach in poisson image blending problem. In: Lecture notes in electrical engineering, vol 481, pp 127–136
22. Sunarto A, Sulaiman J, Lung JCV (2021) Numerical solution of the time-fractional diffusion equations via quarter-sweep preconditioned Gauss-Seidel method. Int J Appl Math 34(1):111–125
23. Lung JCV, Sulaiman J (2020) On quarter-sweep finite difference scheme for one-dimensional porous medium equations. Int J Appl Math 33(3):394–450
24. Suardi MF, Radzuan NZFM, Sulaiman J (2019) Performance of quarter-sweep SOR iteration with cubic B-spline scheme for solving two-point boundary value problems. J Eng Appl Sci 14(3):693–700
25. Chew JVL, Aruchunan E, Sunarto A, Sulaiman J (2022) Quarter-sweep successive over-relaxation approximation to the solution of porous medium equations. IAENG Int J Appl Math 52(2)
26. Muhiddin FA, Sulaiman J, Sunarto A (2020) Numerical evaluation of quarter-sweep KSOR method to solve time-fractional parabolic equations. Int J Eng Trends Technol 63–69
27. Sunarto A, Agarwal P, Sulaiman J, Lung JCV, Momani S (2021) Quarter-sweep preconditioned relaxation method, algorithm and efficiency analysis for fractional mathematical equation. Fractal Fract 5(98):1–14
28. Dahalan AA, Saudi A, Sulaiman J (2022) Pathfinding for mobile robot navigation by exerting the quarter-sweep modified accelerated overrelaxation (QSMAOR) iterative approach via the Laplacian operator. Model Simul Eng 9388146
29. Ali LH, Sulaiman J, Saudi A (2022) Newton-PKSOR with quadrature scheme in solving nonlinear Fredholm integral equations. Math Stat 10(1):246–250
30. Ali LH, Sulaiman J, Saudi A, Ming XM (2022) Numerical solution of nonlinear Fredholm integral equations using half-sweep newton-PKSOR iteration. Math Stat 10(4):868–874

31. Constantinescu R, Poenaru RC, Pop F, Popescu PG (2019) A new version of KSOR method with lower number of iterations and lower spectral radius. Soft Comput 23(22):11729–11736
32. Ebrahimi N, Rahidinia J (2015) Collocation method for linear and nonlinear Fredholm and Volterra integral equations. Appl Math Comput 270:156–164
33. Muthuvalu MS, Sulaiman J (2011) Numerical solution of second kind linear Fredholm integral equations using QSGS iterative method with high-order Newton-Cotes quadrature schemes. Malays J Math Sci 5(1):85–100
34. Chew JVL, Sulaiman J (2016) On quarter-sweep finite difference scheme for one-dimensional porous medium equations. Glob J Eng Technol Rev 1(1):9–16
35. Maleknejad K, Nediasl K (2011) Application of Sinc-collocation method for solving a class of nonlinear Fredholm integral equations. Comput Math Appl 62:3292–3303
36. Mustafa M, Al-Zubaidy KA (2011) Use of Bernstein polynomial in numerical solution of nonlinear Fredholm integral equation. Eng Technol J 29(1):110–115

Analysis on the Performance of Healthcare Industry in Malaysia with TOPSIS Model

Weng Hoe Lam, Weng Siew Lam, Kah Fai Liew, and Mohd Azam Bin Din

Abstract The COVID-19 pandemic was characterized as a global pandemic by the WHO in March 2020. The COVID-19 pandemic had major effects, short- and long-term impacts on the healthcare company and health market. Hence, healthcare companies should analyze their financial performance so that they can sustain themselves in this critical period. This paper aims to propose a research framework for assessing and ranking the healthcare companies' financial performance in Malaysia with Technique for Order Preference by Similarity to Ideal Solution (TOPSIS) model using the financial ratios such as dividend yield, current ratio, earnings per share, profit margin, debt to equity ratio, price earnings ratio and return on equity. The outcomes of this paper indicate HAIO obtains the best performance, followed by AHEALTH, DKSH, IHH, BJCORP and lastly ADVENTA. This analysis is significant as it assists to assess the healthcare companies' financial performance with the proposed research framework based on TOPSIS model.

Keywords TOPSIS model · Ideal solution · Financial ratio · Ranking

W. H. Lam · W. S. Lam (✉) · K. F. Liew · M. A. B. Din
Department of Physical and Mathematical Science, Faculty of Science, Universiti Tunku Abdul Rahman, Kampar Campus, Jalan Universiti, Bandar Barat, 31900 Kampar, Perak, Malaysia
e-mail: lamws@utar.edu.my

W. H. Lam
e-mail: whlam@utar.edu.my

K. F. Liew
e-mail: liewkf@utar.edu.my

M. A. B. Din
e-mail: azam@utar.edu.my

© The Author(s), under exclusive license to Springer Nature Singapore Pte Ltd. 2023

D.-K. Kang et al. (eds.), *Proceedings of the 9th International Conference on Computational Science and Technology*, Lecture Notes in Electrical Engineering 983,
https://doi.org/10.1007/978-981-19-8406-8_4

1 Introduction

Nowadays, healthcare companies play an extremely crucial role in providing healthcare services to the community. The COVID-19 pandemic has impacted the healthcare system; hence it is also caused global panic [1]. According to recent studies, healthcare companies have experienced much higher anxiety, stress, and burnout during this difficult period [2–6]. Since the healthcare companies have been affected by the COVID-19 pandemic, therefore a comprehensive analysis should be performed on studying the financial performance (FP) of the companies in the healthcare sector. The FP of companies can be used to determine the financial health and financial status of a company. Based on previous studies, the researchers have adopted the financial ratios to investigate the FP of companies [7–14].

TOPSIS model helps to deal with the multi-criteria decision making (MCDM) problems [15]. MCDM problem involves identifying the best solution by considering multiple decision criteria [16–19]. Decision making models have been applied to solve MCDM problems in various fields [20–27]. TOPSIS model helps the decision-makers to do the comparison and ranking of a set of decision alternatives. TOPSIS model is employed to determine the alternatives' ranking according to the longest proximity from the negative ideal solution (NIS) and the shortest proximity from the positive ideal solution (PIS).

The applications of the TOPSIS model are found in the field of predictive handoff for heterogeneous networks [28], dam site selection using Geographic Information Systems [29], solar farm site selection [30], decision support system in top management [31], decision support system [32], prediction of diabetes based on general characteristics of humans [33], public blockchain evaluation [34], prioritizing watersheds [35], smart city recommendations [36] and medical clinic for disease diagnosis [37].

The past studies on FP have been done in banks [38, 39], construction companies [40, 41], technology companies [42, 43] and portfolio selection [44–56]. According to previous studies, there is no study has been done comprehensively on the investigation of FP of healthcare companies in Malaysia using TOPSIS model. Hence, this paper aims to evaluate the FP of healthcare companies in Malaysia with the proposed research framework based on TOPSIS model. The remainder of this manuscript is organized as follows. Section 2 presents the data and methodology of this study. The empirical results of this paper are illustrated in the next section. Lastly, the conclusion is made in the final part of the article.

Table 1 Proposed research framework

Level	
Objective	Evaluation of the FP of healthcare companies
Decision criteria (financial variables)	Current ratio (CR), dividend yield (DY), earnings per share (EPS), profit margin (PM), return on equity (ROE), debt to equity ratio (DER), price earnings ratio (PER)
Decision alternatives (healthcare companies)	ADVENTA, AHEALTH, BJCORP, DKSH, HAIO, IHH

2 Data and Methodology

2.1 Data

In this paper, the data consists of healthcare companies which are listed in Malaysian stock market. The data are gathered between the years 2017 and 2019. The healthcare companies such as ADVENTA, AHEALTH, BJCORP, DKSH, HAIO and IHH are investigated in terms of FP. The proposed research framework is demonstrated in Table 1.

The financial ratios, for example, CR, DY, EPS, PM, ROE, DER and PER are utilized to assess the FP of decision alternatives. In this study, the financial variables, for instance, DER and PER are needed to be kept to a minimum. In contrast, CR, DY, EPS, PM and ROE are the financial variables that should be optimized [57, 58].

2.2 TOPSIS

TOPSIS model helps the decision-makers to do the comparison and ranking of a set of decision alternatives. TOPSIS model is employed to rank the alternative based on the longest proximity from the NIS and the shortest proximity from the PIS. The TOPSIS model comprises the steps that are indicated below [59, 60]:

Step 1: Draw the performance matrix $((x_{ij})_{m\times n})$.

Draw a performance matrix $(x_{ij})_{m\times n}$ which comprised of n criteria and m alternatives, where x_{ij} is the score of the companies according to the financial variables.

$$(x_{ij})_{m\times n} = \begin{bmatrix} x_{11} & x_{12} & \dots & x_{1n} \\ x_{21} & x_{22} & \dots & x_{2n} \\ \cdot & & & \cdot \\ \cdot & & & \cdot \\ \cdot & & & \cdot \\ x_{m1} & x_{m2} & \dots & x_{mn} \end{bmatrix} \quad (1)$$

Step 2: Establish the normalized decision matrix

$$r_{ij} = \frac{x_{ij}}{\sqrt{\sum_{i=1}^{m} x_{ij}^2}}, \ i = 1, 2, \ldots, m, \ j = 1, 2, \ldots, n \tag{2}$$

$$\mathrm{R} = (r_{ij})_{m\times n} \tag{3}$$

Step 3: Establish the weighted normalized decision matrix (T)

$$\mathrm{T} = \begin{bmatrix} w_1 r_{11} & w_2 r_{12} & \ldots & w_n r_{1n} \\ w_1 r_{21} & w_2 r_{22} & \ldots & w_n r_{2n} \\ \cdot & & & \cdot \\ \cdot & & & \cdot \\ \cdot & & & \cdot \\ w_1 r_{m1} & w_2 r_{m2} & \ldots & w_n r_{mn} \end{bmatrix} \tag{4}$$

where $\sum_{j=1}^{n} w_j = 1$. The financial ratios have the same importance to measure the performance of the companies [61, 62].

Step 4: Determine the PIS (A_b) solution and NIS (A_w) solution

$$\begin{aligned} A_b &= \{\langle \max(t_{ij} | i = 1, 2, \ldots, m) | j \in J_+\rangle, \langle \min(t_{ij} | i = 1, 2, \ldots, m) | j \in J_-\rangle\} \\ &\equiv \{t_{bj} | j = 1, 2, \ldots, n\}, \end{aligned} \tag{5}$$

$$\begin{aligned} A_w &= \{\langle \min(t_{ij} | i = 1, 2, \ldots, m) | j \in J_+\rangle, \langle \max(t_{ij} | i = 1, 2, \ldots, m) | j \in J_-\rangle\} \\ &\equiv \{t_{wj} | j = 1, 2, \ldots, n\}. \end{aligned} \tag{6}$$

Step 5: Measure proximity of the alternatives from PIS (d_{ib}) and NIS (d_{iw})

$$d_{ib} = \sqrt{\sum_{j=1}^{n} \left(t_{ij} - t_{bj}\right)^2}, \ i = 1, 2, \ldots, m \tag{7}$$

$$d_{iw} = \sqrt{\sum_{j=1}^{n} \left(t_{ij} - t_{wj}\right)^2}, \ i = 1, 2, \ldots, m \tag{8}$$

Step 6: Measure the relative closeness from the ideal solution (s_{iw})

$$s_{iw} = \frac{d_{iw}}{d_{ib} + d_{iw}}, \ 0 \le s_{iw} \le 1, \ i = 1, 2, \ldots, m \tag{9}$$

According to the alternatives' performance, each of the decision alternatives will obtain a respective s_{iw}.

Step 7: Rank the decision alternatives.

The decision alternatives' ranking is determined according to the s_{iw}. The company with the biggest value of s_{iw} is treated as the best performing company in terms of FP.

3 Empirical Results

Table 2 demonstrates the normalized decision matrix.

Table 3 demonstrates the weighted normalized decision matrix.

Table 4 displays the NIS and PIS for each financial ratio.

The proximity of the healthcare companies from PIS (d_{ib}) and NIS (d_{iw}) is tabulated in Table 5.

Table 2 Normalized decision matrix

Name of company	CR	DY	EPS	PM	ROE	DER	PER
ADVENTA	0.3478	0.4416	−0.0162	0.2310	−0.1658	0.2113	0.9345
AHEALTH	0.4736	0.4323	0.4935	0.4369	0.5373	0.1175	0.0564
BJCORP	0.1821	0.2127	−0.0405	0.0133	0.0111	0.3327	−0.0097
DKSH	0.2489	0.3994	0.6892	0.0420	0.3040	0.8909	0.0441
HAIO	0.6815	0.6400	0.4953	0.8226	0.7637	0.0650	0.0705
IHH	0.3084	0.0611	0.1852	0.2779	0.0897	0.1814	0.3412

Table 3 Weighted normalized decision matrix

Name of company	CR	DY	EPS	PM	ROE	DER	PER
ADVENTA	0.0497	0.0631	−0.0023	0.0330	−0.0237	0.0302	0.1335
AHEALTH	0.0677	0.0618	0.0705	0.0624	0.0768	0.0168	0.0081
BJCORP	0.0260	0.0304	−0.0058	0.0019	0.0016	0.0475	−0.0014
DKSH	0.0356	0.0571	0.0985	0.0060	0.0434	0.1273	0.0063
HAIO	0.0974	0.0914	0.0708	0.1175	0.1091	0.0093	0.0101
IHH	0.0441	0.0087	0.0265	0.0397	0.0128	0.0259	0.0487

Table 4 Negative ideal (A_w) and positive ideal (A_b) solutions

	CR	DY	EPS	PM	ROE	DER	PER
A_w	0.0260	0.0087	−0.0058	0.0019	−0.0237	0.1273	0.1335
A_b	0.0974	0.0914	0.0985	0.1175	0.1091	0.0093	−0.0014

Table 5 The proximity from the PIS (d_{ib}) and NIS (d_{iw})

Name of company	d_{ib}	d_{iw}
ADVENTA	0.0566	0.0139
AHEALTH	0.0068	0.0521
BJCORP	0.0461	0.0257
DKSH	0.0357	0.0340
HAIO	0.0009	0.0779
IHH	0.0330	0.0216

The proximity of the healthcare companies with NIS and PIS is displayed in Figs. 1 and 2, respectively.

As can be seen from Fig. 1, the separation distance of HAIO (0.0779) from the NIS is the longest. Therefore, HAIO is able to achieve a good FP compared to other healthcare companies. The second farthest separation distance is achieved by AHEALTH (0.0521), followed by DKSH (0.0340), BJCORP (0.0257), IHH (0.0216) and finally ADVENTA (0.0139). As a result, ADVENTA is the closest to the NIS, thus ADVENTA did not perform well in terms of FP in the study period.

According to Fig. 2, HAIO (0.0009) has a very close separation distance to the PIS, this implies that HAIO has achieved a very good performance in most of the financial ratios relative to other healthcare companies. Based on this result, HAIO can be served as a good benchmark for other companies for further improvement. The

Fig. 1 The proximity from NIS

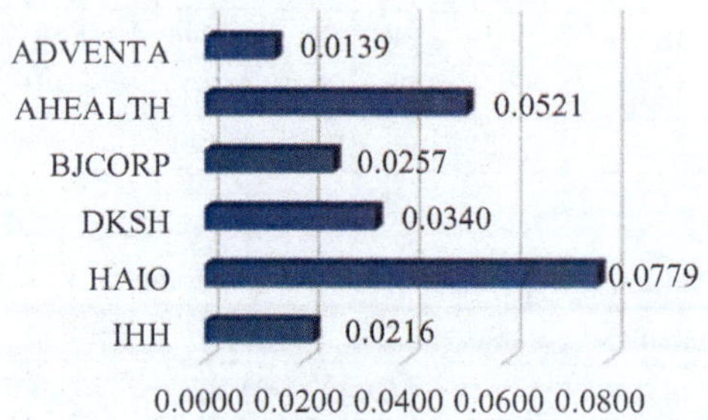

Fig. 2 The proximity from PIS

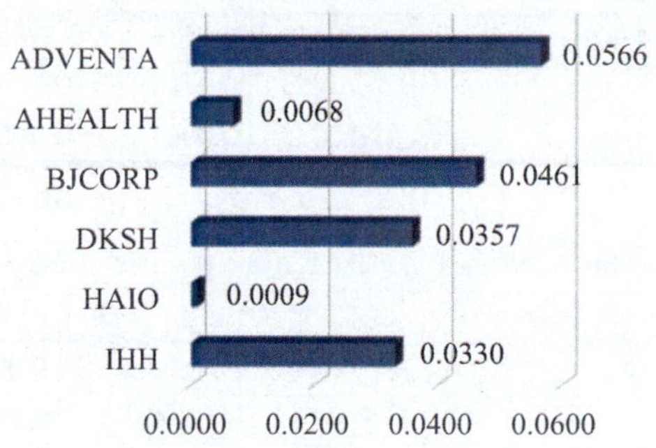

Table 6 FP of companies

Name of company	s_{iw}	Ranking
ADVENTA	0.1973	6
AHEALTH	0.8849	2
BJCORP	0.3577	5
DKSH	0.4876	3
HAIO	0.9886	1
IHH	0.3956	4

separation distance to the PIS for AHEALTH is 0.0068, which is the second shortest distance to the PIS. The next one is obtained by IHH with the separation distance of 0.0330, followed by DKSH (0.0357), BJCORP (0.0461) and lastly ADVENTA (0.0566). Hence, it is believed that ADVENTA achieves the worst FP with the farthest proximity from the PIS and the nearest proximity to the NIS.

Table 6 tabulates the relative closeness as well as the ranking of the healthcare companies.

Figure 3 demonstrates the relative closeness of the healthcare companies.

As presented in Table 6 and Fig. 3, since HAIO obtains the biggest value of s_{iw} (0.9886), therefore HAIO obtains the first place among the healthcare companies. This indicates that HAIO has outperformed the other healthcare companies. In other words, in terms of FP, HAIO is identified as the best healthcare company.

The s_{iw} for AHEALTH, DKSH, IHH, BJCORP and ADVENTA is 0.8849, 0.4876, 0.3956, 0.3577 and 0.1973, respectively. Therefore, the second until sixth ranking is achieved by AHEALTH, DKSH, IHH, BJCORP and ADVENTA, respectively. In this study, a well-performed company can be served as a benchmark for the companies to enhance their FP by minimizing the proximity to the PIS and maximizing the proximity from the NIS. Based on the results, TOPSIS model is robust to measure the FP of companies in the healthcare sector based on important financial ratios.

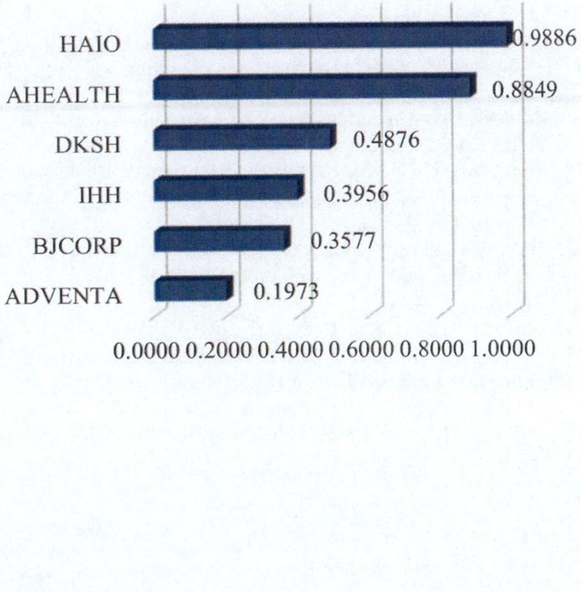

Fig. 3 Relative closeness of the healthcare companies

4 Conclusion

The paper aims to analyze the FP of healthcare companies based on TOPSIS model. The major outcomes of this paper indicate that HAIO has outperformed the other healthcare companies in Malaysia. Hence, HAIO is known as the ideal healthcare company in terms of FP among the studied companies using TOPSIS model. The second ranking is obtained by AHEALTH, followed by DKSH, IHH, BJCORP and lastly ADVENTA. This investigation is significant as it aims to evaluate the FP of healthcare companies using TOPSIS model.

Acknowledgements The study is supported by Universiti Tunku Abdul Rahman (UTAR), Malaysia.

References

1. Algunmeeyn A, El-Dahiyat F, Altakhineh MM, Azab M (2020) Understanding the factors influencing healthcare providers' burnout during the outbreak of COVID-19 in Jordanian hospitals. J Pharm Policy Pract 13(1):1–8
2. Chemali Z, Ezzeddine FL, Gelaye B, Dossett ML, Salameh J, Bizri M, Dubale B, Fricchione G (2019) Burnout among healthcare providers in the complex environment of the Middle East: a systematic review. BMC Public Health 19(1):1–21
3. Ayati N, Saiyarsarai P, Nikfar S (2020) Short- and long-term impacts of COVID-19 on the pharmaceutical sector. DARU J Pharm Sci 28(2):799–805
4. Chan HF, Brumpton M, Macintyre A, Arapoc J, Savage DA, Skali A, Stadelmann D, Torgler B (2020) How confidence in health care systems affects mobility and compliance during the COVID-19 pandemic. PLoS ONE 15(10):e0240644
5. Chen S, Lai Y, Tsay S (2020) Nursing perspectives on the impacts of COVID-19. J Nurs Res 28(3):85–89
6. Sim MR (2020) The COVID-19 pandemic: major risks to healthcare and other workers on the front line. Occup Environ Med 77(5):281–282
7. Price JE, Haddock MD, Brock HR (1993) College accounting, 10th edn. Macmillan/McGraw-Hill, New York
8. Akguc O (2010) Financial statement analysis, 13th edn. Arayis Publication, Istanbul
9. Khrawish HA (2011) Determinants of commercial banks performance: evidence from Jordan. Int Res J Financ Econ 81(1):148–159
10. Liew KF, Lam WS, Lam WH (2017) An empirical evaluation on the efficiency of the companies in Malaysia with data envelopment analysis model. Adv Sci Lett 23(9):8264–8267
11. Lam WS, Liew KF, Lam WH (2018) An optimal control on the efficiency of technology companies in Malaysia with data envelopment analysis model. J Telecommun Electron Comput Eng 10(1):107–111
12. Lam WH, Lam WS, Liew KF (2017) Improvement on the efficiency of technology companies in Malaysia with data envelopment analysis model. In: Lecture notes in computer science, vol 10645, pp 19–30
13. Liew KF, Lam WS, Lam WH (2016) Financial analysis on the company performance in Malaysia with multi-criteria decision-making model. Syst Sci Appl Math 1(1):1–7
14. Lam WS, Liew KF, Lam WH (2018) Investigation on the efficiency of financial companies in Malaysia with data envelopment analysis model. J Phys Conf Ser 995(1):012021
15. Hwang CL, Yoon K (1981) Multi-attribute decision making. Springer, Germany

16. Lam WS, Leong WB, Lam WH (2015) Selection of mobile network operator based on multi-criteria decision-making model using analytic hierarchy process. Math Stat J 1(1):12–18
17. Lam WS, Chen JW, Lam WH (2016) An empirical study on the selection of fast-food restaurants among the undergraduates with AHP model. Am J Inf Sci Comput Eng 2(3):15–21
18. Lam WS, Lee WK, Lam WH (2015) Multi-criteria decision making in job selection problem using analytic hierarchy process model. Math Stat J 1(2):3–7
19. Lam WS, Lam WH, Jaaman SH, Liew KF (2021) Performance evaluation of construction companies using integrated entropy-fuzzy VIKOR model. Entropy 23(3):320–335
20. Lam WS, Lam WH, Liew KF, Chen JW (2018) An empirical study on the preference of fast food restaurants in Malaysia with AHP-TOPSIS model. J Eng Appl Sci 13:3226–3231
21. Lam WS, Bakar MA, Lam WH, Liew KF (2017) Multi-criteria decision making in the selection of mobile network operators with AHP-TOPSIS model. J Eng Appl Sci 12:6382–6386
22. Lam WS, Lam WH, Bakar M, Sek XY (2018) An empirical evaluation on the preference of mobile broadband in Malaysia with analytic hierarchy process model. Adv Sci Lett 24(11):8573–8577
23. Lam WS, Bishan RS, Lam WH (2017) An empirical study on the mold machine-tool selection in semiconductor industry with analytic hierarchy process model. Adv Sci Lett 23(9):8286–8289
24. Lam WH, Din MA, Lam WS, Chen JW (2018) Evaluation on the performance of suppliers in Malaysia with TOPSIS model. J Fundam Appl Sci 10(6S):406–415
25. Lam WH, Lam WS, Liew KF, Wong SC (2018) Data driven decision analysis on the selection of course programmes with AHP-TOPSIS model. Int J Supply Chain Manag 7(4):202–208
26. Lam WS, Bishan RS, Lam WH, Liew KF (2018) An empirical evaluation on the preference of supermarkets with analytic hierarchy process model. J Phys Conf Ser 012037
27. Lam WS, Bakar MA, Lam WH, Chen JW, Ma HL (2018) Evaluation on the preference of coffee shops among undergraduate students with analytic hierarchy process model. Int J Supply Chain Manag 7(4):209–215
28. Kabiri S, Kalbkhani H, Lotfollahzadeh T, Shayesteh MG, Solouk V (2016) Technique for Order of Preference by Similarity to Ideal Solution based predictive handoff for heterogeneous networks. IET Commun 10(13):1682–1690
29. Jozaghi A, Alizadeh B, Hatami M, Flood I, Khorrami M, Khodaei N, Ghasemi TE (2018) A comparative study of the AHP and TOPSIS techniques for dam site selection using GIS: a case study of Sistan and Baluchestan Province, Iran. Geosciences 8(12):494–516
30. Nazari MA, Aslani A, Ghasempour R (2018) Analysis of solar farm site selection based on TOPSIS approach. Int J Soc Ecol Sustain Dev (IJSESD) 9(1):12–25
31. Rahim R, Siahaan APU, Wijaya RF, Hantono H, Aswan N, Thamrin S, Sari DAP, Agustina S, Santosa RB, Muttaqin WM, Sujito S (2018) Technique for Order of Preference by Similarity to Ideal Solution (TOPSIS) method for decision support system in top management. Int J Eng Technol 8(2):290–293
32. Rahim R, Supiyandi S, Siahaan APU, Listyorini T, Utomo AP, Triyanto WA, Irawan Y, Aisyah S, Khairani M, Sundari S, Khairunnisa K (2018) TOPSIS method application for decision support system in internal control for selecting best employees. J Phys Conf Ser 1028(1):012052. IOP Publishing
33. Zulqarnain M, Dayan F, Saeed M (2018) TOPSIS analysis for the prediction of diabetes based on general characteristics of humans. Int J Pharm Sci Res 9(7):2932–2939
34. Tang H, Shi Y, Dong P (2019) Public blockchain evaluation using entropy and TOPSIS. Expert Syst Appl 117:204–210
35. Meshram SG, Alvandi E, Meshram C, Kahya E, Al-Quraishi AMF (2020) Application of SAW and TOPSIS in prioritizing watersheds. Water Resour Manage 34(2):715–732
36. Nasution N, Bhawika GW, Wanto A, Ginantra NLWSR, Afriliansyah T (2020) Smart city recommendations using the TOPSIS method. IOP Conf Ser Mater Sci Eng 846(1):012028. IOP Publishing
37. Zulqarnain RM, Abdal S, Ali B, Ali L, Dayan F, Ahamad MI, Zafar Z (2020) Selection of medical clinic for disease diagnosis by using TOPSIS method. Int J Pharm Sci Rev Res 61(1):22–27

38. Lam WS, Chen JW, Lam WH (2017) Data driven decision analysis in bank financial management with goal programming model. In: Lecture notes in computer science, vol 10645, pp 681–689
39. Chen JW, Lam WS, Lam WH (2019) Optimization and comparison of bank financial management in Malaysia with goal programming model. J Adv Res Dyn Control Syst 11(12S):39–46
40. Lam WS, Liew KF, Lam WH (2019) Investigation on the performance of construction companies in Malaysia with entropy-TOPSIS model. Earth Environ Sci 385:012006
41. Lam WS, Liew KF, Lam WH (2021) Performance evaluation of construction companies in Malaysia with entropy-VIKOR model. Eng J 25(1):297–305
42. Lam WH, Yeoh HB, Lam WS, Chen JW (2018) Analysis on the performance of technology companies with Z score model. Bull Electr Eng Inform 7(4):633–639
43. Lam WH, Yeoh HB, Lam WS, Lai AJX (2020) Evaluation on the financial performance of technology companies in Malaysia with Zmijewski model. Malays J Math Sci 14(S), 81–92
44. Lam WS, Jaaman SH, Ismail H (2015) The impact of human behaviour towards portfolio selection in Malaysia. Procedia Soc Behav Sci 172:674–678
45. Lam WS, Jaaman SH, Ismail H (2014) Index tracking modelling in portfolio optimization with mixed integer linear programming. J Appl Sci Agric 9(18):47–50
46. Lam WS, Jaaman SH, Ismail H (2015) The impact of different economic scenarios towards portfolio selection in enhanced index tracking problem. Adv Sci Lett 21(5):1285–1288
47. Lam WS, Jaaman SH, Ismail H (2015) An empirical comparison of different optimization models in enhanced index tracking problem. Adv Sci Lett 21(5):1278–1281
48. Lam WS, Lam WH (2016) Strategic decision making in portfolio management with goal programming model. Am J Oper Manage Inf Syst 1(1):34–38
49. Lam WS, Jaaman SH, Ismail H (2014) Comparison between two-stage regression model and variance model in portfolio optimization. J Appl Sci Agric 9(18):36–40
50. Lam WS, Lam WH (2015) Portfolio optimization for index tracking problem with mixed-integer programming model. J Sci Res Dev 2(10):5–8
51. Lam WS, Jaaman SH, Ismail H (2015) An empirical study on the characteristics of high-risk aversion behavior in portfolio decision making using regression model. Adv Environ Biol 9(7):17–20
52. Lam WS, Jaaman SH, Ismail H (2015) Investigation on relationship between human behavior and portfolio selection problem in Malaysia using decision making model. Adv Environ Biol 9(7):6–10
53. Lam WS, Jaaman SH, Lam WH (2019) An enhanced Mean-Gini extended model in portfolio optimization with different level of risk aversion. ASM Sci J 12(6):41–46
54. Lam WS, Jaaman SH, Lam WH (2021) Higher moment portfolio optimization with polynomial goal programming approach. J Phys Conf Ser 1988:012001
55. Lam WS, Jaaman SH, Lam WH (2019) Mathematical modelling of risk in portfolio optimization with Mean-Gini approach. J Phys Conf Ser 1212:012031
56. Lam WS, Lam WH, Jaaman SH (2021) Portfolio optimization with a mean-absolute deviation-entropy multi-objective model. Entropy 23(10):1266
57. Lam WS, Lam WH, Liew KF (2018) Data analysis on the performance of technology sector in Malaysia with entropy-TOPSIS model. In: Communications in computer and information science, vol 886, pp 194–203
58. Lai AJX, Lam WH, Lam WS (2018) Optimization on the credit risk of companies in Malaysia with data envelopment analysis model. Int J Eng Technol 7:13–16
59. Roszkowska E (2011) Multi-criteria decision-making models by applying the TOPSIS method to crisp and interval data. Multi Criteria Decis Mak Univ Econ Katowice 6(1):200–230
60. Bakirci F, Shiraz SE, Sattary A (2014) Financial performance analysis of iron, steel metal industry sector companies in the Borsa Istanbul: DEA super efficiency and TOPSIS methods. Ege Acad Rev 14(1):9–19

61. Lam WH, Lam WS, Liew KF (2019) Performance analysis on telecommunication companies in Malaysia with TOPSIS model. Indones J Electr Eng Comput Sci 13(2):744–751
62. Lam WS, Liew KF, Lam WH (2017) Evaluation on the financial performance of the Malaysian banks with TOPSIS model. Am J Serv Sci Manage 4(2):11–16

Prosthetic Parts for Protecting Mobile Robots

Kamal Hussein, Pablo Otero Roth, Alicia Triviño Cabrera, Georgios Tsaramirsis, Izzat Aldarraji, Saman Esfandiari, Rayner Alfred, and Zamhar Iswandono

Abstract Mobile robotics are currently designed to be worked in difficult and complicated conditions of applications such as disasters and dangerous areas in industries. Such types of robots need to protect themselves from damage against unexpected effects. In this paper we propose five different prosthetic parts, which can be installed on the mobile robots to increase survival chances. These protection mechanisms were inspired by how some animals protect themselves in nature. The five proposed prosthetic parts are named as umbrella, four-bar mechanism, retractable roof, turtle shell, and box shell. A visual design of all five methods is included in this paper. The analysis of all the five shells is based on the ABS Plastic as the material of

K. Hussein · P. O. Roth · A. T. Cabrera
University of Malaga, 29071 Málaga, Spain
e-mail: 0619813174@uma.es

P. O. Roth
e-mail: Pablo.otero@uma.es

A. T. Cabrera
e-mail: atc@uma.es

G. Tsaramirsis
Higher Colleges of Technology Women's Campus, Abu Dhabi, United Arab Emirates

I. Aldarraji
Automated Manufacturing Department, University of Baghdad, Baghdad, Iraq
e-mail: Izzat.a@kecbu.uobaghdad.edu.iq

S. Esfandiari
Independent Consultant, Westhafen Tower, Westhafenplatz 1, Frankfurt am Main, Germany

R. Alfred (✉)
Creative Advanced Machine Intelligence Research Centre, Faculty of Computing and Informatics, Universiti Malaysia Sabah, Kota Kinabalu, Sabah, Malaysia
e-mail: ralfred@ums.edu.my

Z. Iswandono
Higher Colleges of Technology, Abu Dhabi, United Arab Emirates
e-mail: zismail@hct.ac.ae

© The Author(s), under exclusive license to Springer Nature Singapore Pte Ltd. 2023
D.-K. Kang et al. (eds.), *Proceedings of the 9th International Conference on Computational Science and Technology*, Lecture Notes in Electrical Engineering 983,
https://doi.org/10.1007/978-981-19-8406-8_5

the shell, mainly due to its low density and adequate strength, and 1000 N distributive external static force. The results showed the ability of equipping the introduced shells into the mobile robot platform without needing mechanical modifications.

Keywords Shells · Mobile robots · Adaptive mechanisms · Robot's self-preservation · Prosthetic parts

1 Introduction

Robotics is one of the pillars of Industry 4.0 and its importance is expected to increase over the coming years [1]. Researchers in the field of robotics tend to focus more on the robot's artificial intelligence and maneuverability, neglecting its self-preservation. This is contradicting a common application of robotics which is their use in the fields which are dangerous for humans to work. In such scenarios, it makes sense to develop technologies that can increase the robot's survival chances. In nature, some animals, such as snails or turtles, carry with them a shell that they can use as a protection mechanism [2]. If the animal senses danger, it will hide in the shell. Examples of danger can be the presence of a predator, a certain weather condition, such as heat or rain, or fall from a height. These are some example conditions, in which the animal is going to hide inside its shell [3]. The self-preservation may vary from simple tasks like protecting the robot on a rainy day [4] to a complex one such as marching on planet Mars [5]. For instance, a robot might be used for probing and sampling an area which has a rainy climate. The regular robots must find a place to cover every time it starts to rain; in some critical situations in which the robot fails to find a proper cover, the whole project may be subject to failure or an undesired long delay. Furthermore, in an area which has been destroyed by the earthquake, due to the risks of aftershock or the collapses of the remaining ruins, it might be dangerous to do explorations by humans. Therefore, robots can play a major role in saving lots of lives [6, 7]; however, robots are subject to such jeopardies as well. Moreover, in the explorations in space, such as marching on mars which are necessarily being done by robots, predicting a mechanism to secure the robot in the cases of falling from a high cliff or any unexpected situations is vital [8]. Finally, such necessity can be stated for various fields such as: inspector robots in refineries [9]; autonomous industrial robots; or in any environment which might be dangerous for the robots to operate in [10, 11]. Therefore, due to the dramatic increase of robot's applications in our life, predicting a mechanism to protect them in unexpected dangerous situations is inevitable.

Although self-preservation has been offered by different methods, they are mainly about their artificial intelligence of the robots. It varies from simple self-charging to complex self-balancing methods. For instance, [12], offers a system to resupply a self-contained robot. Developers in [13] suggest a method for dynamically balancing the robot by using a power system in which "a dynamically balancing robot including an electrically activated component that is employed when the robot is brought to a stable shutdown configuration". Likewise, an algorithm was used to lower the damage of NAO robots when they fall. In this method, whenever the robot loses its stability and concludes that falling is unavoidable, it takes its hands in front of its body to lower the total damage [14].

In this research, we take a different approach. Being inspired by nature, we are migrating the idea of protection shells to mobile robots. This paper proposes five mechanical methods for preservation of various kinds of robots and evaluates their reaction under a constant disturbed force of 1000 N.

2 Animals with Shells

Turtles use their shell for protection; however, it may be interesting to know that some species of turtle, especially the sea turtle, are not able to hide in their shell. Although there are some other kinds of animals, such as: sea urchins; crustaceans; armadillos; cockles; and snails, the mechanism of the American box turtle is the aim of this research. This mechanism is activated when the turtle feels a probable danger, mainly being sensed by its eyes. In the case of robots, several sensors will enable the robot to be fully aware of the possible dangers that may come towards it. For example, by using the ultrasonic sensors on different sides of the robot, it will be able to detect a falling object over or being shot towards it. Same sensor when used at its bottom, will help the robot to notice if it has fallen [15]. Additionally, a temperature sensor will alert the robot of a fire nearby [16]; by contrast, a humid sensor will alert it of rain or waterfalls [17]. In all these cases, by hiding in its shell, the robot will be able to significantly lower the damage to its body. This hiding might be either temporary or it may totally deactivate the robot until the rescue team arrives. Lastly, as a mechanical evaluation of this mechanism, the turtle hides its head inside the shell during a complex spinal bend motion of its neck. As an example, for adoption of such motion on the robots, the head of the robot, which is mainly a camera or a communication device, can hide inside the shell by a simple four-bar linkage mechanism.

3 Proposed Protection Mechanisms

In this research we propose the use of five different protection mechanisms that can be applied to different types of robots in different situations. In general, this is a simple moderate cost solution. The main cost of the prosthetic part is about its material which mainly depends on the mission of the robot and the probable condition it may work in. Such materials may vary from fiberglass for a humid environment to complex strong alloys for environments with the risk of fall of heavy objects.

Determining the exact situation, a robot may face is impossible. For example, it may fall into water or be caught in a fire, it may get captured by a storm or even fall from a height; therefore, in this research, we focus on the proof of concept and propose five prosthetic parts for protecting robots. The main advantage of all the offered methods is their ability to be installed on the current robots. No mechanical modification needs to be done on the currently operating robots and these shells can be implemented on any currently existing robots. The analysis of all the methods is based on the ABS Plastic as the material of the shell, mainly due to its low density and adequate strength, and 1000 N distributive external static force. Even if we could not cover all cases, it is the belief of the authors that the proposed approach may inspire other researchers to modify the proposed prosthetics for the protection of their robots.

3.1 Umbrella

Maybe the most convenient and accessible prosthetic for the robots would be an Umbrella shape protection part. Due to the simplicity of the design, the low cost of manufacturing, and its suitability for implementation on any kind of robots, it might be the first choice for reaching the purpose of this research. Following this, as shown in Fig. 1, is a simple suggestion for the umbrella part, subject to change depending on the type of the robot. The arc form for the shape has been chosen as it would let the materials falling on it to slide down (Figs. 2, 3, 4; Table 1).

As potentially expected, the maximum stress of 6E6 is on the joint point of the shell to the robot. Since it is much lower than the Elastic modulus of 2E9, ABS plastic is relatively suitable for this case.

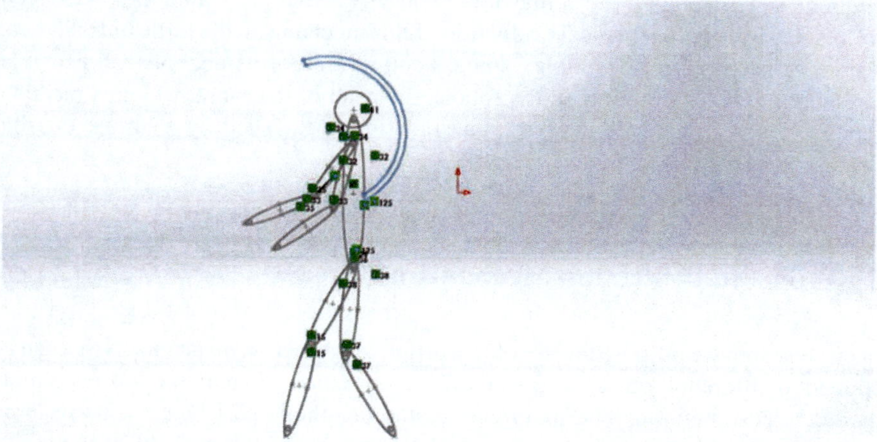

Fig. 1 Proposed umbrella part

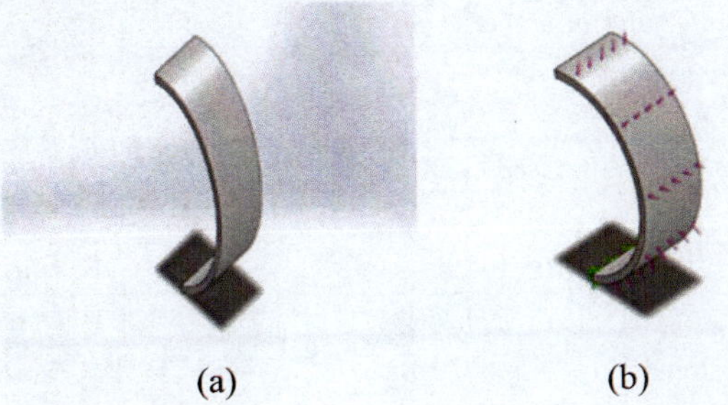

Fig. 2 Umbrella part: **a** 3D model, **b** force diagram

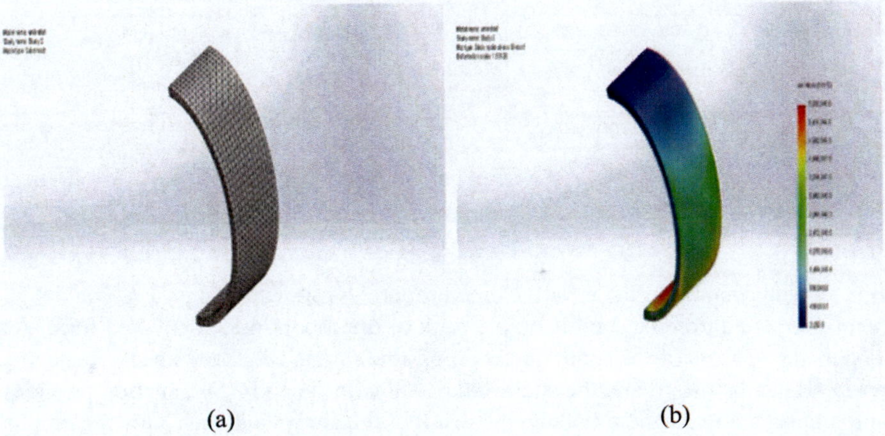

Fig. 3 Umbrella part: **a** mesh nodes, **b** stress analysis

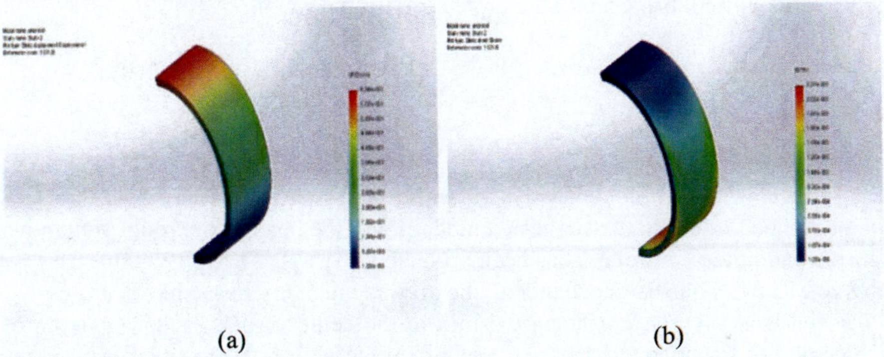

Fig. 4 Umbrella part: **a** displacement, **b** strain analysis

Table 1 Features of umbrella part

Material properties		Reaction forces	Analysis result	
Mass	10.834 kg			
Volume	0.0106216 m^3	X = 630.105	Mesh size	21.9851 mm
Weight	106.174 N		Min stress	2050.51 N/m^2
Type	ABS, plastic	Y = 156.305	Max stress	2050.51 N/m^2
Tensile strength	3e + 007 N/m^2		Min displacement	0 mm
Elastic modulus	2e + 009 N/m^2	Z = − 1.15618	Max displacement	62.4806 mm
Poisson's ratio	0.394		Max strain	0.00221664
Mass density	1020 kg/m^3	Resultant = 649.203		
Shear modulus	3.189e + 008 N/m^2			

3.2 *Four-Bar*

This model is mainly suitable for the biped robots. As shown in Figs. 5, 6, 7 and 8, in the normal condition the shell is on the back of the robot, in the detracted form. As soon as the sensors of the robot detect a possible threat, the robot kneels, bends its head and hands inward and the shell retracts all over its body. Such action provides a broad protection for the robot and the result of the analysis in the following graphs proves the efficiency of this method (Table 2).

Although the shell may need to be replaced after the crush due to the extreme deformation, it will bring a reliable protection for the robot.

3.3 *Retractable Roof*

The retractable roof becomes activated any time a risk of falling object threatens the quadruped robot. In the suggested model, only the head of the robot containing sensors and micro controllers has been protected (Figs. 9, 10, 11, 12; Table 3).

Same as the 'Four-bar mechanism', the displacement might seem to be excessive. However, it is slightly more than 0.5 mm in the center of the shell which can be neglected in comparison to the relatively high price of the camera and sensors at the top of the robot.

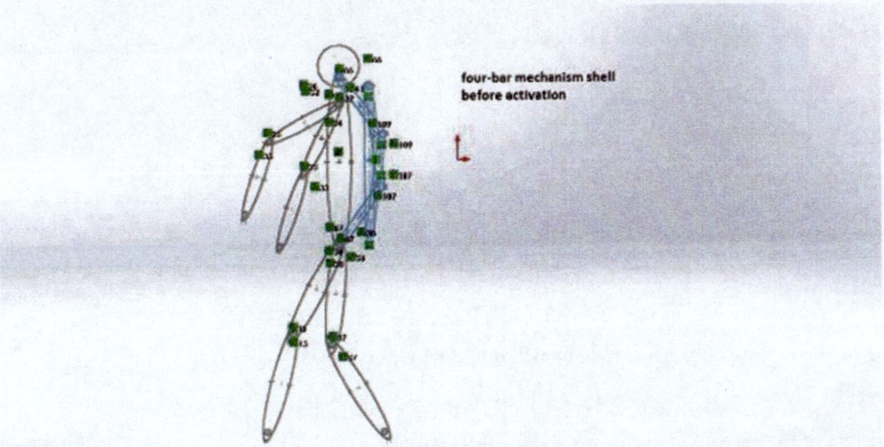

(a)

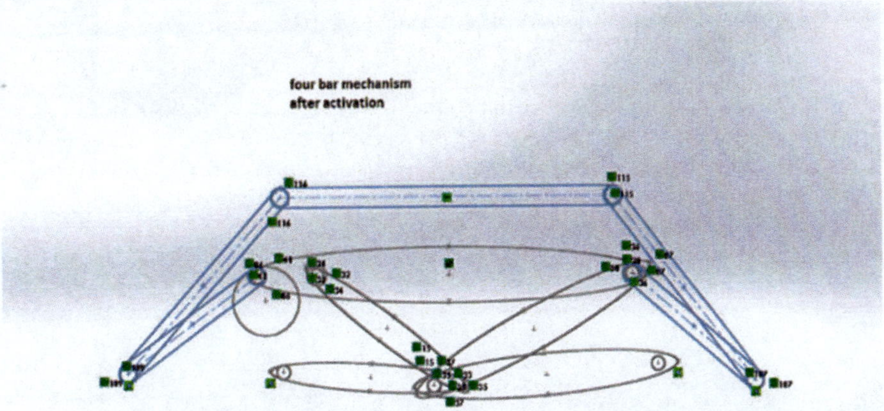

(b)

Fig. 5 Proposed four-bar part, **a** before activation, **b** after activation

3.4 Turtle

Between varieties of the defending mechanisms, hiding inside a rough shell in turtles and snails is by far the most suitable solution in this case. Several reasons such as simplicity, low cost and inclusivity explain the suitability of this mechanism. Firstly, on one hand, such mechanisms shouldn't lower the maneuverability of a robot; on the other hand, it should be simple to be designed. By using this shell, no extra mechanism needs to be added to the current robots and if it be properly fitted to the

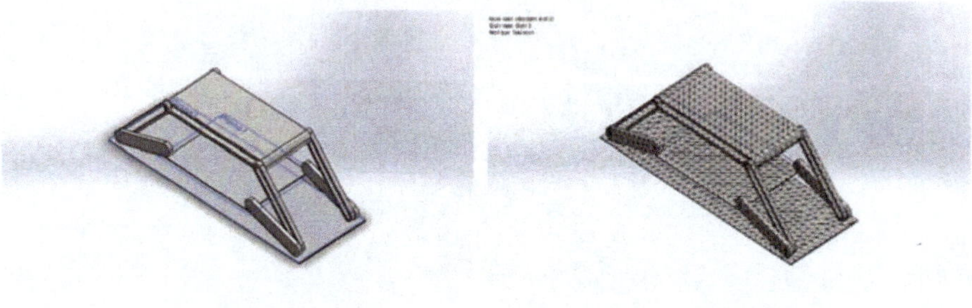

(a) (b)

Fig. 6 Four-bar mechanism, **a** 3D model, **b** mesh nodes analysis

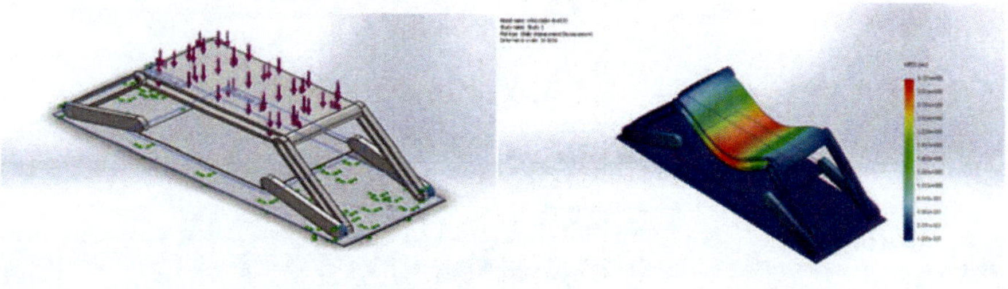

(a) (b)

Fig. 7 Four-bar mechanism, **a** force diagram, **b** stress analysis

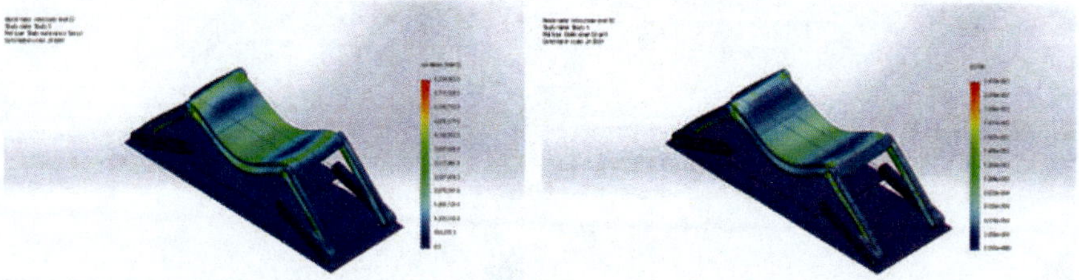

(a) (b)

Fig. 8 Four-bar mechanism, **a** displacement, **b** strain analysis

main body, it won't reduce their maneuverability. Therefore, the main advantage of this idea is that it can be implemented to the already existing robots and the only difference will be a modification to their algorithm. Secondly, the total cost of this mechanism should be low enough to be considered for mass production.

A shell like the one of the turtle's was the main idea of this research. Such a shell provides full protection for the robot. For the shell to be adaptive for already existing

Table 2 Features of four-bar mechanism

Material properties		Reaction forces	Analysis result	
Mass	10.834 kg			
Volume	0.0106216 m^3	X = 0.0441222	Mesh size	22.2881 mm
Weight	106.174 N		Min stress	0 N/m^2
Type	ABS, plastic	Y = 999.86	Max stress	6.2349e + 006 N/m^2
Tensile strength	3e + 007 N/m^2		Min displacement	0 mm
Elastic modulus	2e + 009 N/m^2	Z = −0.175363	Max displacement	3.33731 mm
Poisson's ratio	0.394		Max strain	0.00240988
Mass density	1020 kg/m^3	Resultant = 999.86		
Shear modulus	3.189e + 008 N/m^2			

robots, some appropriate slots need to be made through it to provide enough space for the head and legs of the robot to hide in (Fig. 13).

3.5 Box Shell

A Box shell allows the robot to hide in is a different approach for this purpose. This method is suitable for fixed robots, such as industrial robots. The industrial robots are continuously subject to the environmental dangers such as the leakage or breakage of the water pipes or the particles of wood cutting or metal welding. Therefore, considering a shell beside the robot for the crucial times provides a safe space for it.

A membrane may ensure that no fluid enters the shell and, moreover, a pump can be put inside the shell to evacuate any liquid which may enter the shell (Fig. 14).

4 Conclusions and Future Work

In this paper, we propose the use of prosthetic parts for protecting mobile robots. These protecting parts can vary, from parts that partly cover the robots to complete shells, where the robot can hide inside, protecting the robot from various hazardous conditions. To promote the idea of using protection parts and to prove the concept, we introduced five different protection parts that can be applied to different kinds of robots and offer protection against various hazards. These are the umbrella, the four-bar, the retractable roof, the turtle shell, and the box shell. The mechanical designs

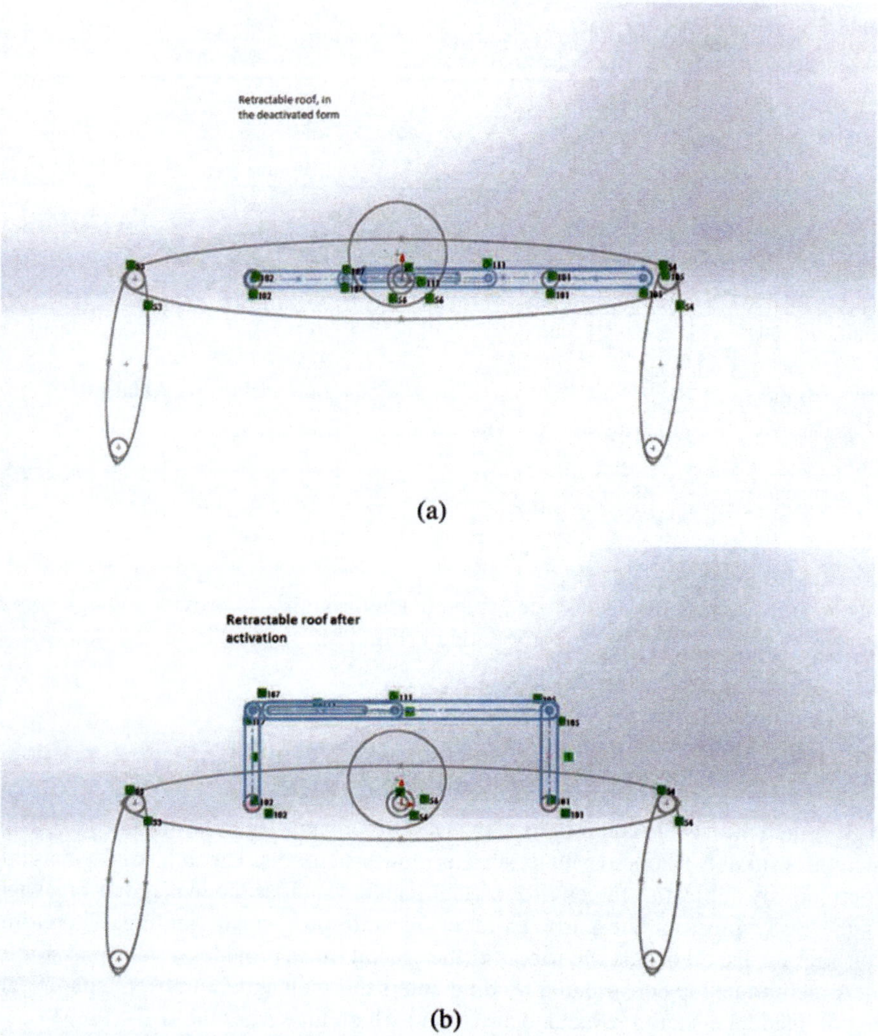

Fig. 9 Retractable roof, **a** deactivated form, **b** after activation

were also included as part of this research. The main limitation of this work is that the proposed prosthetic parts do not offer any "smart" features as it does not include any sensors or any other type of electronics.

As future work, we plan to work more in the box shell design so it can house mobile robots, sensors, and other sensitive equipment. In the future version, we will add sensors that will allow it to open and close based on the environmental conditions. Even if the methods proposed in this paper cannot protect robots from any possible threat, they can be used as inspiration for other researchers or they can be used as parts of more complex integrated solutions.

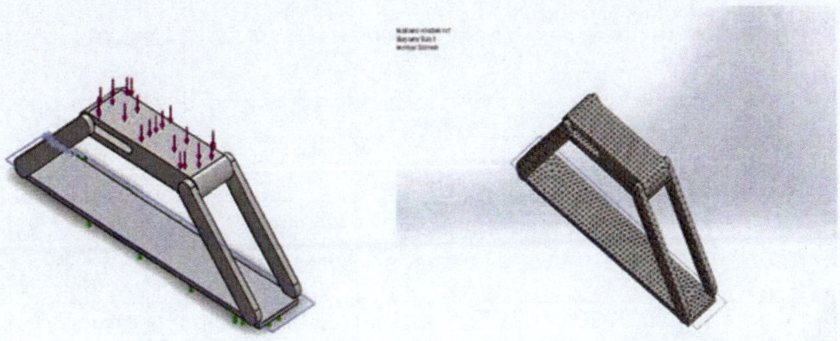

(a) (b)

Fig. 10 Retractable roof, **a** force diagram, **b** mesh nodes

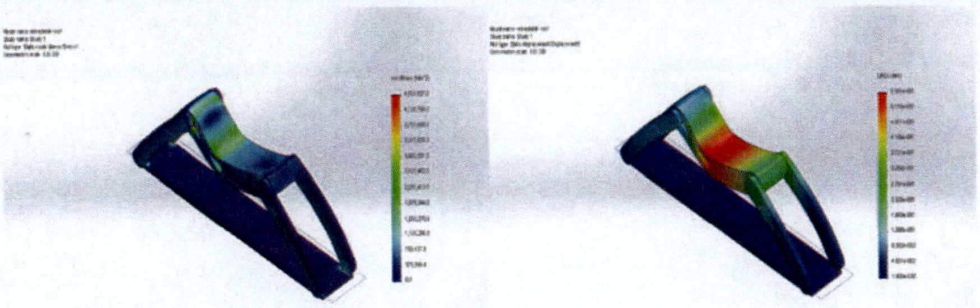

(a) (b)

Fig. 11 Retractable roof, **a** stress analysis **b** displacement

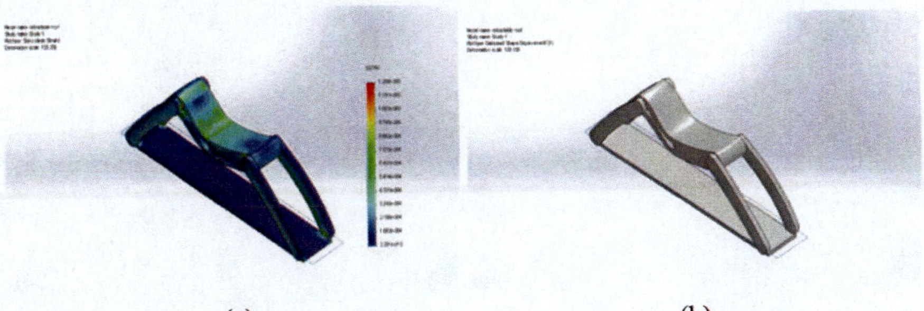

(a) (b)

Fig. 12 Retractable roof, **a** strain, **b** 3D model

Table 3 Features of retractable roof

Material properties		Reaction forces	Analysis result	
Mass	10.834 kg			
Volume	0.0106216 m^3	X = − 0.0810661	Mesh size	14.2013 mm
Weight	106.174 N		Min stress	0.000579966 N/m^2
Type	ABS, plastic	Y = 999.881	Max stress	4.50083e + 006 N/m^2
Tensile strength	3e + 007 N/m^2		Min displacement	0 mm
Elastic modulus	2e + 009 N/m^2	Z = 0.049305	Max displacement	0.55811 mm
Poisson's ratio	0.394		Max strain	0.00129934
Mass density	1020 kg/m^3	Resultant = 999.881		
Shear modulus	3.189e + 008 N/m^2			

(a)

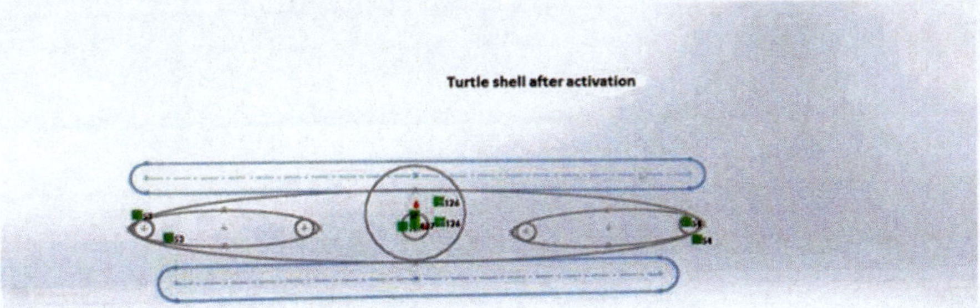

(b)

Fig. 13 Turtle shell, **a** before activation, **b** after activation

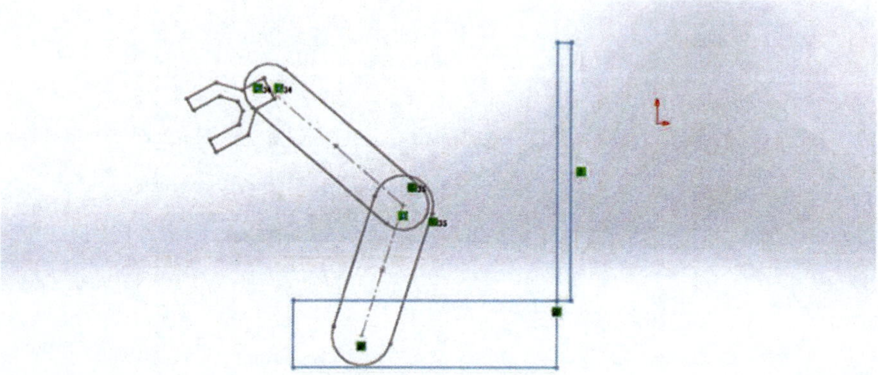

(a)

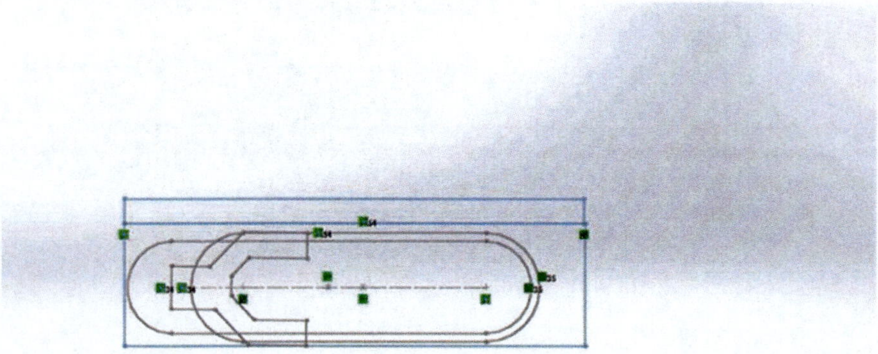

(b)

Fig. 14 Box shell, **a** side view, **b** top view

References

1. Tsaramirsis G, Kantaros A, Al-Darraji I, Piromalis D, Apostolopoulos C, Pavlopoulou A, Alrammal M, Ismail Z, Buhari SM, Stojmenovic M, Tamimi H, Randhawa P, Patel A, Khan FQ (2022) A modern approach towards an industry 4.0 model: from driving technologies to management. J Sens 2022:18 pages. Article ID 5023011. https://doi.org/10.1155/2022/5023011
2. Paci P, Mancini C, Nuseibeh B (2022) The case for animal privacy in the design of technologically supported environments. Front Vet Sci 8:784794. https://doi.org/10.3389/fvets.2021.784794
3. Brookes JI, Rochette R (2007) Mechanism of a plastic phenotypic response: predator-induced

shell thickening in the intertidal gastropod *Littorina obtusata*. J Evol Biol 20(3):1015–1027. https://doi.org/10.1111/j.1420-9101.2007.01299.x. PMID: 17465912

4. Wei C, Wang P, Ma Y, Li R (2022) Innovative design of a healthy and environmentally friendly intelligent cleaning robot in a smart city public environment. J Environ Public Health 2022:1–12. https://doi.org/10.1155/2022/1602125
5. Tunstel E, Maimone M, Trebi-Ollennu A, Yen J, Petras R, Willson R (2005) Mars exploration rover mobility and robotic arm operational performance, vol 2, pp 1807–1814. https://doi.org/10.1109/ICSMC.2005.1571410
6. Cai X, Ning H, Dhelim S, Zhou R, Zhang T, Xu Y, Wan Y (2021) Robot and its living space: a roadmap for robot development based on the view of living space. Digit Commun Netw 7(4):505–517. ISSN 2352-8648. https://doi.org/10.1016/j.dcan.2020.12.001
7. Smids J, Nyholm S, Berkers H (2020) Robots in the workplace: a threat to—or opportunity for—meaningful work? Philos Technol 33:503–522. https://doi.org/10.1007/s13347-019-00377-4
8. Dutta V, Zielińska T (2021) Cybersecurity of robotic systems: leading challenges and robotic system design methodology. Electronics 10:2850. https://doi.org/10.3390/electronics10222850
9. Li H, Li R, Zhang J, Zhang P (2020) Development of a pipeline inspection robot for the standard oil pipeline of China National Petroleum Corporation. Appl Sci 10:2853. https://doi.org/10.3390/app10082853
10. Wong C, Yang E, Yan X-T, Gu D (2018) Autonomous robots for harsh environments: a holistic overview of current solutions and ongoing challenges. Syst Sci Control Eng 6(1):213–219. https://doi.org/10.1080/21642583.2018.1477634
11. Trevelyan JP, Kang SC, Hamel WR (2008) Robotics in hazardous applications. In: Siciliano B, Khatib O (eds) Springer handbook of robotics. Springer, Berlin, Heidelberg. https://doi.org/10.1007/978-3-540-30301-5_49
12. Colens A (1999) Patent No. US5869910, 9 Feb 1999
13. Blackwell T, Casner D, Nelson B, Wiley S (2011) Patent No. US 8,041,456 B1, 18 Oct 2011
14. Ruiz-del-Solar J, Palma-Amestoy R, Marchant R, Parra-Tsunekawa I, Zegers P (2009) Learning to fall: designing low damage fall sequences for humanoid soccer robots. Robot Auton Syst 57(8):796–807. ISSN 0921-8890. https://doi.org/10.1016/j.robot.2009.03.011
15. Al-Darraji I, Derbali M, Jerbi H, Qudus Khan F, Jan S et al (2021) A technical framework for selection of autonomous UAV navigation technologies and sensors. Comput Mater Contin 68(2):2771–2790
16. Khan F, Xu Z, Sun J, Khan FM, Ahmed A, Zhao Y (2022) Recent advances in sensors for fire detection. Sensors 22:3310. https://doi.org/10.3390/s22093310
17. Munawar HS, Hammad AWA, Waller ST (2022) Remote sensing methods for flood prediction: a review. Sensors 22:960. https://doi.org/10.3390/s22030960

Secret Key Design Using an Algebraic Procedure (KAP) for Encrypted Energy Internet-of-Things (EIoT) Contents

Abbas M. Al-Ghaili, Hairoladenan Kasim, Naif M. Al-Hada, Zainuddin Hassan, Ridha Omar, Marini Othman, and Ibraheem Shayea

Abstract This study designs a secret Key using an Algebraic Procedure (KAP). KAP has the capability to act in a different manner dependent on the plaintext input. Two different secret key length approaches are presented in this work. Increase the unpredictability and privacy of plaintext by using such a scheme is the main objective of this design presented in this paper. A sequence of algebraic procedures has been utilized in order to construct a particular coefficient by which the link between plaintext size and key length is customizable. This design of secret key enhances the privacy of sensitive data notably utilized with smart Internet-of-Things (IoT) based

A. M. Al-Ghaili (✉)
Institute of Informatics and Computing in Energy (IICE), Universiti Tenaga Nasional (UNITEN), 43000 Kajang, Selangor, Malaysia
e-mail: abbas@uniten.edu.my

A. M. Al-Ghaili · H. Kasim · Z. Hassan · R. Omar
College of Computing and Informatics (CCI), UNITEN, 43000 Kajang, Selangor, Malaysia
e-mail: hairol@uniten.edu.my

Z. Hassan
e-mail: zainuddin@uniten.edu.my

R. Omar
e-mail: ridha@uniten.edu.my

N. M. Al-Hada
Shandong Key Laboratory of Biophysics, Institute of Biophysics, Dezhou University, Dezhou 253023, China
e-mail: naifalhada@yahoo.com

M. Othman
INTI International University (NIU), Nilai, Malaysia
e-mail: marini.othman@newinti.edu.my

I. Shayea
Faculty of Electrical and Electronics Engineering, Istanbul Technical University, Istanbul, Turkey
e-mail: ibr.shayea@gmail.com

A. M. Al-Ghaili
Department of Computing, College of Computing & Informatics (CCI), UNITEN, Kajang, Selangor 43000, Malaysia

© The Author(s), under exclusive license to Springer Nature Singapore Pte Ltd. 2023

D.-K. Kang et al. (eds.), *Proceedings of the 9th International Conference on Computational Science and Technology*, Lecture Notes in Electrical Engineering 983,
https://doi.org/10.1007/978-981-19-8406-8_6

applications. Calculations on how long it takes to produce and encrypt a single secret key have shown that the suggested architecture is efficient. Additionally, the findings have demonstrated that when a directly proportional connection between user entries to size-in-bits, the relationship between user entries and secret key algebraic functions is reversely proportional thus higher unpredictability of plaintext and secret key may be attained.

Keywords Algebraic procedure · Quick Response Code · Energy Internet-of-Things · Secret key

1 Introduction

Various Internet-of-Things (IoT) apps have recently started adopting the method of Quick Response Code (QR-Code) to let numerous smart facilities and activities be done where the computation time is short. The number of investigations concentrating on the utilization of QR-Code is steadily rising [1–5]. As a result, safeguarding data from unwanted access is an ongoing concern. This work focuses on creating an alternative encryption technique for plaintext and secret key to enhance the unpredictability of plaintext behavior.

In order to construct a successful and precise system, IoT technology depends on a few additional key elements [6, 7]. One of these is the encryption scheme through which the smart system is resilient. The verification technique is nonetheless another crucial instrument for a strong smart application. Usually, the encryption strategy has to be followed by a rigorous verification mechanism [8]. Therefore, any the encryption strategy is very vital for smart and IoT applications since it influences the overall smart system including the verification process as well [9–13].

Physical and cloning attacks against IoT devices are possible because of their tiny size, cheap cost, and restricted resources and that is a very important procedure to be taken [14, 15]. Device authentication and secret-key generation may be done using physical unclonable functions, which are comparatively recent security primitives for IoT devices [16, 17]. In [18], the author devises an efficient method for generating secret codes. The coding we employ is called a polar code. Polarization and suitable code building design will demonstrate that the leaking of helper data regarding the secret key may be minimized. It has been shown that the suggested method significantly improves the dependability of the system.

In order to produce symmetric cryptographic keys, the physical properties of the vehicle-to-vehicle communication link are exploited to provide a lightweight solution known as physical-layer security (PLS) [19, 20]. By using empirical parameter choices, current algorithm designs do not achieve optimal key-generation performance. The paper presented in [21] provides a unique channel response quantization approach to generate a key for PLS in V2V communications. Allows the algorithm to deal with non-reciprocity in V2V channel answers at valid entities.

It was proposed in [22] that an arbitrary number of multi-antenna LNs (LN) exist in mesh topology with an eavesdropper that has an arbitrary number of antennas. Selected antennas on each node send out pilot signals. A weighted combination of the estimated channel information is broadcast successively by each node in the second phase. The eavesdropper can't get the channel information needed to generate the group key, but the other LNs can. By encoding and quantizing the predicted channels, each node may then construct a group key. For quantization, it is relying on well-established methods. Detailed instructions for determining the antennas and coefficients used in the broadcast phase for group-key generation are given.

A two-way randomness method has been used to come up with a secret key (SKG) [23, 24]. The legitimate transceivers send each other their own random signal over a reciprocal wireless channel. Then, the multiplication of the transmitted and received signal is used to generate secret keys. Wireless channel based, one-way randomness, and two-way randomness are three ways to do SKG. The theoretical SKG rates (SKGRs) of each method are calculated and compared. As a result, Scheme-1bit and Scheme-3bit, which are both practical SKG schemes based on two-way randomness, have been designed and simulated in full. Scheme-1bit is usually used in situations where the signal-to-noise ratio (SNR) is low [25].

We will review a number of related works that concern the issue of creating and generating secret keys. In [23], a two-way randomness-based secret key production mechanism is suggested. Mutual transmission of random signals over wireless channels generates a shared randomness that is used to produce secret keys for legal transceivers. The theoretical secret key generation rates of the three secret key generation approaches, wirelessly channel based, one-way randomness, and two-way randomness, are developed and compared in quasi-static channels. Another research study has mentioned that due to the fact that as more than a destination might receive information, the multicast service type poses a security risk. Multicast is a significant service type in terrestrial as well as space communication networks. By using a dynamic satellite as a relay node, it is possible to accomplish end-to-end point-to-point distribution of a quantum key across a great distance [26–28]. Because of this, figuring out how to give secret keys for distant multicast services and how to properly disperse secret keys between the data packet and several destination nodes for lengthy secure multicast services becomes an extremely important challenge. In this article, a point-to-multipoint space relay mechanism is suggested, and a quantum satellite network that is meant to fulfill point-to-multipoint key distribution is constructed. The quantum satellite network has a unique node topology. The approach of auxiliary topology is used to dynamically represent the weighted information of the connection, and then a mobile relaying structure is built. In conclusion, this work suggests a space-based relay algorithm for packet solutions to find the solution of lengthy point-to-multipoint key switch as well as enhance the success rate of packet requests. The presented algorithm will route as well as allocate key money to enhance the packet requests [29]. However, in [30], a secret key production based on communication networks is susceptible to malicious activities in environments with wireless channels because of the openness and invariance of the public pilot. This is especially true for man-in-the-middle attacks [31–34], in which the threat actor transparently behaves to manipulate channel metrics and deduce the produced codes. So, to defend against such an assault, a vibrant private flyer can be devised. The private flyer and the secret key have been obtained from the qualities of the wireless channels, and neither has been shared with any third parties. The unpredictability of wireless channels may be improved by using singular value decomposition methods. The private flyer has the capability to decrypt as well as to verify the connections, which has the potential to minimize the amount of wireless information. According to the findings of the research and modeling, the suggested SKG scheme has the ability to prevent active and potential man-in-the-middle threats and actions [21, 27, 35, 36].

The complexity of guessing the spread spectrum steganographic system's secret key is a significant factor in the system's overall level of security [37]. Within the context of the known-message attack, many estimation techniques were suggested as a means of estimating the secret key. Whenever the number of monitored data is insufficient, however, the estimate accuracy rates of the various available estimation methods are not up to par. At the moment, developing more efficient estimators remains a tough and open subject. In this research, an estimator for estimating the secret key is offered that is developed on the basis of equivalent keys. This estimation may be used for the classic and the secure SS steganographic techniques. The overlap of a single spherical shell and a hypercone is referred to as an equivalent area, which is formed by keys that are equal to one another. After running a Monte Carlo simulation, it is concluded that it is possible to calculate the secret key by putting up the comparable keys that have been equally taken from the equivalent area. Consequently, the basic probability chooses comparable keys from completely random matrices by taking use of the pairings of embedded signals and the texts that are encoded in them. For the purpose of determining how accurate the suggested predictor is, a conceptual investigation is carried out on it. The findings in the study validate the conceptual analysis and demonstrate that the suggested estimator is beneficial of estimating that are already in use. In addition, the study demonstrates the vulnerability of the more secure SS steganographic approaches.

Establishing cryptographic keys amongst the various gadgets in a smart home is necessary in order to ensure the safety of the communications carried via home area networks. In spite of the fact that several key agreement methods have been suggested, it is impossible to put any of them into direct practice for the simple reason in certain situations, smart objects do not have access to an internet trustworthy third party. In addition, the components of the smart house may come from a variety of sources depending on the particular circumstances. As a consequence of this, it is difficult to safely generate the transmission keys by making use of the secrets that have been stored by various manufacturers. There are several data encryption strategies that are produced without the use of an internet trustworthy third party or stored secrets. Yet, several issues with these plans to consider can be found. Within that piece, it is expanded upon previous schemes and developed for a data encryption scheme for home automation. The main goal was to make it such that the gadgets in the house could independently and safely determine the communications key. In particular, our plan is devised in such a way that it takes advantage of the private incompatibility issue. A proof that the proposed method is secure, as well as a prototype implementation of the technique will be given. In addition to this, a performance analysis on the presented system has been done and in relation to the results of other works in the same field, it has been evaluated. The analysis has demonstrated that the presented and suggested data encryption method has the potential to resolve the issues that have been raised by similar studies; nevertheless, it does so at the expense of certain multiple communication overconsumption [38].

To produce a good level of protection, the creation of keys for the physical layer security relies on exploiting unexpected characteristics of communication networks. This kind of security demands a high level of tolerance between more than a user aiming to establish symmetric keys. To reduce the impact of device biometric interference, the research presented here explored the feasibility of improving channel reciprocity in time-division duplex communications networks by employing a loop-back scheme that makes use of various frequency bands. Through conceptual model as well as real observations, it has been shown that the system is resistant to passive and active eavesdropping as well as attacks using a man-in-the-middle [39]. The next step is to build a mechanism for the production of an encryption key. The effectiveness of the suggested approach to the production of secret keys is after that tested using both simulation studies and hands-on experimentation. Results have shown that the suggested method can successfully minimize non-reciprocity and surpasses the traditional time-division duplex scheme in both the key dispute rate and the key encryption rate. This is shown by the fact that the presented algorithm surpasses the conventional time-division duplex strategy [40].

In order for encryption solutions to function properly, keys are an absolutely necessary component. It is known for some ways to manufacture encryption techniques, whether they be hardware or software. In this work, a rapid technology method for generating a private key is presented and discussed. The suggested technique takes into consideration as data the customer's list of three answers pairings and the customer's email-ID, which is one of a kind for each end consumer. The first thing that is done is the generation of a seeding, which involves mixing a triplet with a singular parameter. The question-and-answer pair is first represented as three separate points, then a harmonic set building notion is used to complete the creation procedure. The triangle that is generated with this design, alongside customer's email address, is used to establish a variable that is one of a kind. In the subsequent stage, a one-of-a-kind key is produced with the assistance of cyclic unusual architecture with the seed being used as an input. This method of generating private keys is put through its paces in terms of accuracy, unpredictability, disparity, data, dependability, and resistance to a number of different security risks [41].

This work proposes an unexpected secret key utilizing an algebraic process (KAP) to encrypt QR-Code data for use with IoT-based smart applications.

This paper is organized as follows: Sect. 2 introduces a schematic representation about the conception of KAP. In Sect. 3, the detailed steps of the proposed KAP design are discussed and explained supported with graphs for more clarification. Results and discussion are provided in Sect. 4. Conclusion has been drawn in Sect. 5.

2 The Proposed Secret Key Design Using an Algebraic Procedure (KAP) Schematic Representation

The suggested KAP is important to encrypt information provided by user. The method of encryption approach deals with the way the gathered information is being shielded. KAP is meant to maintain information protected from being modified in an unauthorized way. The goal of this research is to create the secret key required to decrypt KAP plaintext. A generalized visual portrayal of the primary stages of KAP and their connection to the verification tool is shown in Fig. 1 showing the proposed KAP flowchart.

In Fig. 1, the conception of IoT environment is connected to KAP for EIoT applications and systems where the sensitive data will be processed and preserved secure after it has been secured via encrypting its secret key using KAP. Once original data from the user has been received, the KAP algorithm will start to extract certain contents. Then, the process of secret key design and generation could start. After this process and once the secret key is generated, a related QR code is generated also where the security scheme is being more complicated. By then, the verification of QR code and secret code contents could be carried out and its authentication can be checked. The verification procedure is at the receiver party. Before the verification procedure, the QR code normally will be sent to the user (receiver) party. The verification can be performed by the IoT application to make sure the QR code and the secret key are valid.

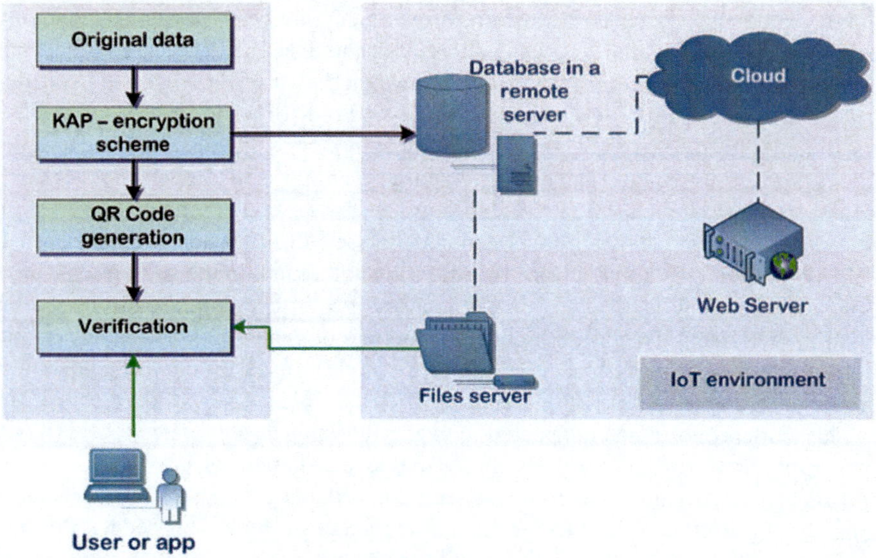

Fig. 1 KAP flowchart—a connection to IoT environment: a graphical representation

When the secret key has been generated, a valid copy will be transferred to the Cloud but before the secret key needs to be stored in a database connected to the Cloud as can be seen in Fig. 1. A copy of the secret key will be stored in a "Files server" from which the verification procedure will restore the original data related to the secret key and QR code. If the data stored in "Files server" is identical to data owned (inserted) by the user or by the application, it means it is authentic, valid, and source-original.

3 The Proposed KAP Design

Here, we'll go through the basics of the KAP design and how it may be put to use with an EIoT application verification tool.

3.1 KAP Inputs Design

A user's whole data set is encrypted when it is first received by KAP. As seen in Fig. 2, information is made up of a variety of values, including plain text, numeric data, and combinations of both.

In Fig. 2, categories of data are 'string', 'number', and 'string or number' are allocated to the variables: 'name', 'mobile', 'address' fields, respectively.

However, to add a robust layer of encryption to KAP, additional new fields have been developed to incorporate new forms of information and data types. The most

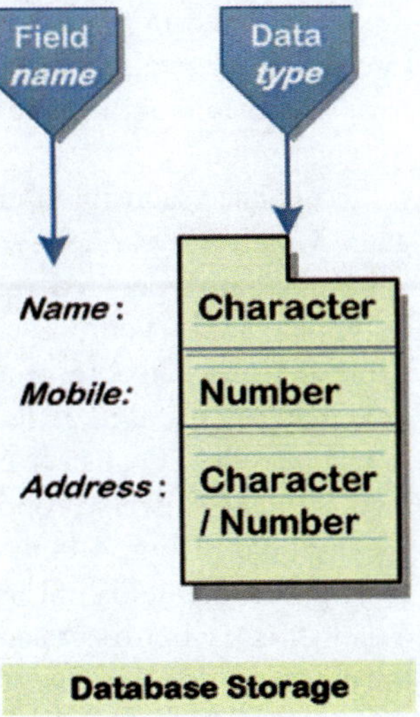

Fig. 2 Description of the data storage for the user

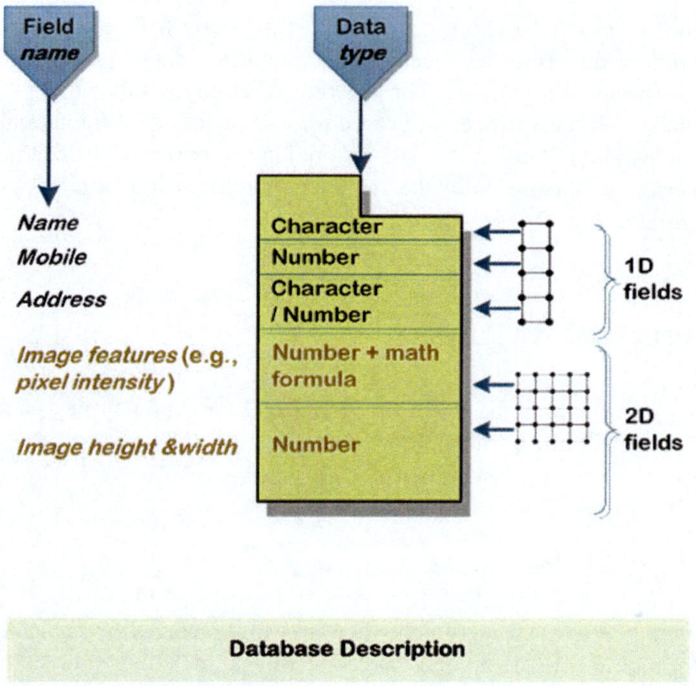

Fig. 3 Extended description of the data storage for the user

critical data to be protected are the characteristics and information about a person's face, fingerprints, the answer to a security question, and their unique user ID, that is shown in Fig. 3.

As observed in Fig. 3, new data types e.g., image characteristics, image pixels' intensities are evaluated for encryption. Additionally, two-dimension data (image rows and columns), instead of one-dimension numbers as in Fig. 2, are considered for encryption.

To enhance security of encryption technique, additional matter has been investigated which is a variable with an attribute that is competent to routinely adjust its value and data type. Generally, the user has varied activities. The user may be asked to provide an answer to a security question in this way. This method is quite helpful to provide alternative responses with varying values providing amounts by which the field may be increased or decreased in instance the answer is a number. As projected, in instance the SQ answer evolves, for instance, from textual to numerical, its data type could be shifting proportionally. This suggested strategy is depicted in Fig. 4.

The field in Fig. 4 is dynamic and constantly exchanging in quantity and data type, unlike the fields in Figs. 2 and 3 whose values and data types are fixed. When a danger is trying to get access to or alter data, it is difficult to predict the value of the data.

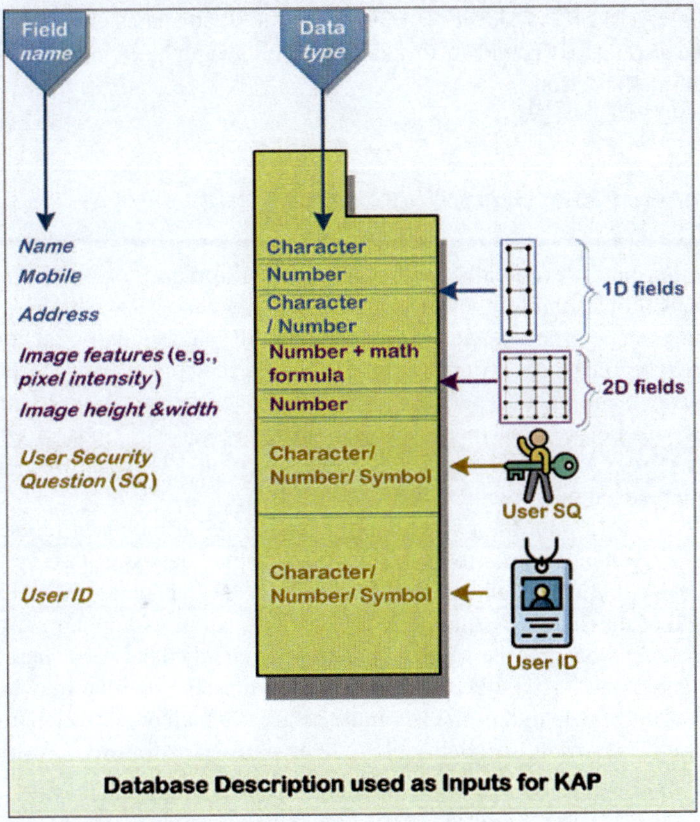

Fig. 4 A design of KAP's database description

Inputs for KAP will include the following: facial image features (FIF), fingerprint feature (FPF), image size (ImgS), user SQ (USQ), and user ID (as shown in Fig. 4). (UID). A description to inputs of KAP can be shown in Table 1.

Table 1 An explanation to KAP related variables

KAP input feature/field	Data explanation			
	Type	Array size	Status	
			Data type	Field value
Face image feature (FIF)	Numerical/string	2D	Altered	Variable
Finger-print feature (FPF)	Numerical/string	2D	Altered	Variable
Image size (ImgS)	Numerical	2D	Fixed	Variable
USQ	Numerical/string/symbol	1	Altered	Variable
UID	Numerical/string	1	Fixed	Constant

In Table 1, it is observed, the 'FIF' includes these attributes: 'numerical and or string', because status related to the data may be swapping, sometimes it can be 'numerical' and of 'string'.

3.2 Proposed KAP Secret-Key Length Design

These five attributes, eventually, will be merged in a variety of matrices utilizing various ways. For example, the facial image will be changed to different sorts of images using image processing and analysis methods; subsequently, specific values and features are retrieved. After then, these values are mixed with the other data before they are encrypted. To perform the KAP, a succession of in-series encryption procedures will be done on these variables utilizing various kinds of encryption algorithms. The KAP output is the QR-Code, encrypted UID, SQ response. The practical methods are comprehensively explained.

2^{i} and $2^{i} + 2^{(i-1)}$ are the two techniques that KAP has chosen to use for key length (i − 1). When it comes to key size and encrypted data size, the *i* is supposed to be flexible. A hypothetical change in the 'field value' or 'data type' might result in the length of the 2nd key being increased by $2^{(i-1)}$. For example, in Table 1, the SQ increment may occur if the value is increased or changed between characters and numbers. The $2^{i} + 2^{(i-1)}$ is the maximum size of a secret key that may be used in this design. Thus, the length of the key must be greater than or equivalent to the size of the dataset. As a result, inequality (1) may be phrased as follows:

$$2^{i} \leq key\,length \geq 2^{i} + 2^{i-1} \tag{1}$$

Therefore, the following two formulas (Eqs. 2 and 3) are used for the secret key length:

$$2^{i} \leq key_1\,length \tag{2}$$

$$key_2\,length \geq 2^{i} + 2^{i-1} \tag{3}$$

3.3 KAP Input (Plaintext)—Math-Based Procedure

The amount of user data that is encrypted is not inversely proportional to the length of the key. To put it another way, the encryption system will only take a certain amount of information (i.e., plaintext) from this information, and not the whole user's information. As previously indicated, the amount of plaintext being encrypted is liable to vary if the type data varies. KAP treats this data as plaintext input because of its changeable length, which results in two unique bit/byte sizes. In the design of the key length, consideration has been given to both min and max input sizes. This criterion is a necessary and is provided in inequality (4) and (5):

$$u_{data_{\min}} < 2^i \tag{4}$$

$$u_{data_{\max}} < 2^i + 2^{i-1} \tag{5}$$

while,

- $u_{data_{\min}}$ denote the min length of user's input information.
- $u_{data_{\max}}$ denote the max length of user's input information.

Thus, in order to determine the relationship between the key length and data size with account of the adjustable variable (i) firstly, inequalities (2) and (4) are used to establish inequality (6); secondly, inequalities (3) and (5) are used to obtain inequality (7), as follows:

The following are the inequalities (2) and (4) that are used to find out the relation between the key size and data size when taking into consideration the adjustable variable I first, inequality (2) and inequality (4) will be used to produce the formulation as in inequality (6); second, inequality (3) and inequality (5) will be used to produce the formulation as in inequality (7), as follows:

$$u_{data_{\min}} < 2^i \leq key_1\, length \tag{6}$$

$$u_{data_{\max}} < 2^i + 2^{i-1} \leq key_2\, length \tag{7}$$

Thus, inequalities (6) and (7) are going to generate the following expression:

$$size\, of\ data < adjustable\, variable \leq key\, length \tag{8}$$

3.4 KAP Secret-Key—Scheme and Block Diagram

The proposed scheme of KAP is graphically represented as depicted in Fig. 5.

4 Results and Discussion

Obtained results related to the plaintext time complexity and computation time can be found in this section of this paper. Then, it assesses the performance of the proposed design (KAP). The results obtained can be different from the other works in terms of

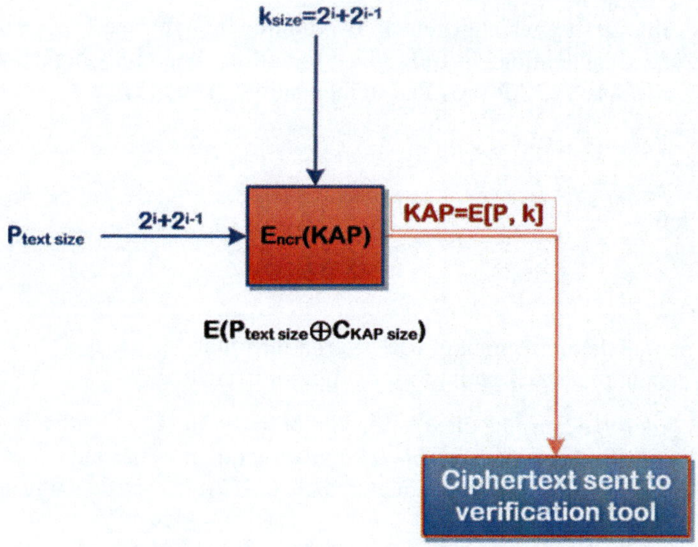

Fig. 5 KAP scheme—a block diagram

time complexity and computation time. Besides, there are algebraic operations and formulas applied to the data being processed. Another factor that is different from the other is that, the data processed depends on the data entered by the users so if the user has entered different two forms of data the produced key will be different in terms of its length and computation time.

4.1 Time Complexity

The suggested secret key design's running time has been described using Big-O-Notation. The collecting of plaintext inputs and the production of secret key values via the application of associated equations are the two basic methods executed, as previously described. In most cases, O(n) is determined by the inputs and outputs provided by the user.

As with FIF, there are a defined size of picture e.g., 640 × 480. In this situation, O(n) relies on values being taken from that picture. A set number of rows' and columns' items, say: k1 elements, is fixed. An array A FIF [k1] is devoted to storing k entries. As a result, the following is how O(n) is defined in this context and represented in Eqs. (9)–(13):

$$O(n)_{A_{FIF}[k1]} = O(k1) \tag{9}$$

Likewise, O(n) for FPF, ImgS, USQ, and UID adopts the similar technique while these entries are only placed to an array. Therefore:

$$O(n)_{A_{FPF}[k2]} = O(k2) \tag{10}$$

$$O(n)_{\mathrm{A_{ImgS}}[k3]} = O(k3) \tag{11}$$

$$O(n)_{\mathrm{A_{USQ}}[k4]} = O(k4) \tag{12}$$

$$O(n)_{\mathrm{A_{UID}}[k5]} = O(k5) \tag{13}$$

Finally, Big-O-Notation for n for KAP can be a totalization formulation of Eqs. (9)–(13); thus: time complexity for KAP can be defined as in Eq. (14):

$$O(n)_{\mathrm{KAP}} = O(k) \tag{14}$$

A time complexity number that equates to O(k), seems fair if it is set side by side with different techniques that concern the time complexity parameter.

The results obtained have depended on the applying of a number of mathematical and algebraic formulas and formulas.

If the user has typed numerical data, the encrypted form will be different from that textual form that has the same length (in terms of bytes-based size). The two forms (textual and numerical data) will be different in term of length, secret key length, secret key contents, even though they are the same length. The reason is that they are different of types.

4.2 *Time Based Efficiency*

The following is an explanation of the process used to collect the samples used in this comparison: In this study, a variety of tests were conducted using various samples of entries. There were seven experiments in which a varied number of samples have been taken into account. In accordance with the suggested architecture, the plaintext user entries have been supplied to the KAP inputs, which are denoted by the letters $P_{text_{size}}$ and k_{size}. For the first test that has been implemented, a total of 10 samples are evaluated, with each sample producing a total of 10 secret keys. It is possible to record the calculation time for each sample, while the computation time given in Fig. 6 is the mean computing time for all samples of the linked test. Similar procedures are followed for the other tests, each with a varied number of samples. As for the approaches utilized to generate the same number of words, either for certificate or for password approach will be used.

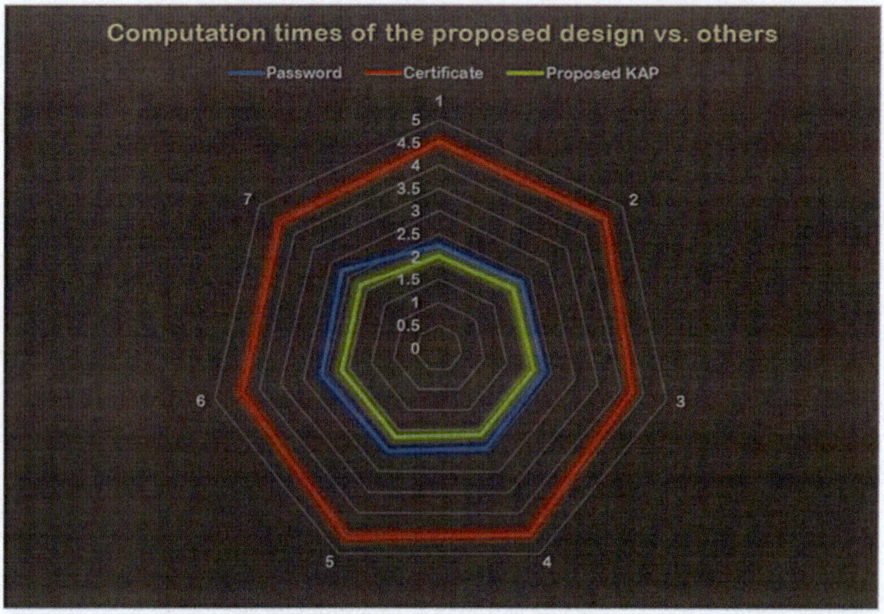

Fig. 6 Mean calculation times related to KAP against a number of competitive research works

According to the amount of processing time necessary to generate one secret key, the suggested design plan for secret key design is as follows: It has been determined that the proposed KAP outperforms existing methods such as the Password and Certificate processes in terms of calculation times for the secret key and textual processing. It is possible to observe the final result in Fig. 6.

For the sake of this comparison, the following is an explanation of the sampling procedure: It was thought of doing several tests on various samples of the submissions. Various numbers of samples were used in seven distinct experiments, each with a slightly different outcome. The KAP inputs $P_{text_{size}}$ and k_{size} have been fed plaintext user entries as part of the suggested design. For the first test, a total of ten samples are used, and ten secret keys are generated correspondingly. Each sample's computing time (CT) is documented. The CT is the mean CT for all tests in the relevant test. In the same way, CT has been performed for other tests with varied quantity of tests. Other approaches mentioned in Fig. 6 have carried the same quantity of words.

In term of computation time, in the event that the user inputs numerical data, the encrypted form will be distinct from the textual form, despite the fact that both forms have the same length (in terms of bytes-based size). Even if they both have the same length, the two types of data (textual and numerical data) will have distinct secret key lengths and contents of secret key even though they both have the same length. The reason for this is because they come in a variety of forms.

On the right-hand side of Fig. 6, there have been seven tests; they have been analyzed. The CT for each sample is calculated in this manner, and the meant CT for whole samples of each test is produced in this manner as well. According to the illustration, the relevant outcome is provided in Fig. 6.

As can be seen, the proposed design takes less time than the Certificate and Password processes, with the exception of two tests in the case of the Password method. The reason for this is because as contrasted to the Certificate technique, the length of the secret key is considered to be the maximum length that is considered acceptable.

5 Conclusion

This study proposes a dynamic architecture for a secret key depending on plaintexts provided by the users of the application. To accomplish this design, several algebraic procedures are formulated and suggested. The main component in developing it is to retain a random and unexpected performance of encryption algorithm (e.g., KAP) considering the key size and the manner the values are created. A verification means based on a QR tag could compare the security of such IoT applications.

Acknowledgements This work has been fully funded by Universiti Tenaga Nasional (UNITEN). Grant Codes are: (J510050002/2021048) and (J510050002/2021156). This research is also supported by the TNB SEED FUND (Code: U-TL-RD-21).

References

1. Liu Z, Choo KKR, Grossschadl J (2018) Securing edge devices in the post-quantum internet of things using lattice-based cryptography. IEEE Commun Mag 56(2):158–162. https://doi.org/10.1109/MCOM.2018.1700330
2. Neisse R, Baldini G, Steri G, Ahmad A, Fourneret E, Legeard B (2017) Improving internet of things device certification with policy-based management. In: 2017 global internet of things summit (GIoTS), 6–9 June 2017, pp 1–6. https://doi.org/10.1109/GIOTS.2017.8016273
3. Rane S, Dubey A, Parida T (2017) Design of IoT based intelligent parking system using image processing algorithms. In: 2017 international conference on computing methodologies and communication (ICCMC), pp 1049–1053. https://doi.org/10.1109/ICCMC.2017.8282631
4. Xiao-Long W, Chun-Fu W, Guo-Dong L, Qing-Xie C (2017) A robot navigation method based on RFID and QR code in the warehouse. In: 2017 Chinese automation congress (CAC), 20–22 Oct 2017, pp 7837–7840. https://doi.org/10.1109/CAC.2017.8244199
5. Ghaffari M, Ghadiri N, Manshaei MH, Lahijani MS (2017) P^{4}QS: a peer-to-peer privacy preserving query service for location-based mobile applications. IEEE Trans Veh Technol 66(10):9458–9469. https://doi.org/10.1109/TVT.2017.2703631
6. Kirkham T, Armstrong D, Djemame K, Jiang M (2014) Risk driven smart home resource management using cloud services. Future Gener Comput Syst 38:13–22. https://doi.org/10.1016/j.future.2013.08.006
7. Chen YH, Tsai MJ, Fu LC, Chen CH, Wu CL, Zeng YC (2015) Monitoring elder's living activity using ambient and body sensor network in smart home. In: 2015 IEEE international conference on systems, man, and cybernetics, pp 2962–2967. https://doi.org/10.1109/SMC.2015.515

8. Zhang Y (2018) Test and verification of AES used for image encryption. 3D Res 9(1):1–27
9. Thammarat C, Kurutach W (2019) A lightweight and secure NFC-base mobile payment protocol ensuring fair exchange based on a hybrid encryption algorithm with formal verification. Int J Commun Syst 32(12):e3991
10. Thammarat C, Kurutach W (2018) A secure fair exchange for SMS-based mobile payment protocols based on symmetric encryption algorithms with formal verification. Wireless Commun Mob Comput 2018
11. Wu D, Gan Q, Wang X (2018) Verifiable public key encryption with keyword search based on homomorphic encryption in multi-user setting. IEEE Access 6:42445–42453
12. Chen L, Chen Z (2019) Practical, dynamic and efficient integrity verification for symmetric searchable encryption. In: International conference on cryptology and network security. Springer, pp 163–183
13. Zhen L, Ma K (2021) Encryption and verification scheme of source IPv6 address between internet domains. In: 2021 2nd international conference on computer communication and network security (CCNS). IEEE, pp 126–133
14. Zhang J, Shen C, Su H, Arafin MT, Qu G (2021) Voltage over-scaling-based lightweight authentication for IoT security. IEEE Trans Comput 71(2):323–336
15. Badhib A, Alshehri S, Cherif A (2021) A robust device-to-device continuous authentication protocol for the internet of things. IEEE Access 9:124768–124792
16. Mall P, Amin R, Das AK, Leung MT, Choo K-KR (2022) PUF-based authentication and key agreement protocols for IoT, WSNs and smart grids: a comprehensive survey. IEEE Internet Things J
17. Deebak B et al (2022) Lightweight blockchain based remote mutual authentication for AI-empowered IoT sustainable computing systems. IEEE Internet Things J
18. Chen B, Willems FMJ (2019) Secret key generation over biased physical unclonable functions with polar codes. IEEE Internet Things J 6(1):435–445. https://doi.org/10.1109/JIOT.2018.2864594
19. Khan AR et al (2022) DSRC technology in vehicle-to-vehicle (V2V) and vehicle-to-infrastructure (V2I) IoT system for intelligent transportation system (ITS): a review. In: Recent trends in mechatronics towards industry 4.0, pp 97–106
20. Mun H, Seo M, Lee DH (2021) Secure privacy-preserving V2V communication in 5G-V2X supporting network slicing. IEEE Trans Intell Transp Syst
21. Bottarelli M, Karadimas P, Epiphaniou G, Ismail DKB, Maple C (2021) Adaptive and optimum secret key establishment for secure vehicular communications. IEEE Trans Veh Technol 70(3):2310–2321. https://doi.org/10.1109/TVT.2021.3056638
22. Thai CDT, Lee J, Prakash J, Quek TQS (2019) Secret group-key generation at physical layer for multi-antenna mesh topology. IEEE Trans Inf Forensics Secur 14(1):18–33. https://doi.org/10.1109/TIFS.2018.2837661
23. Zhang S, Jin L, Lou Y, Zhong Z (2018) Secret key generation based on two-way randomness for TDD-SISO system. China Commun 15(7):202–216. https://doi.org/10.1109/CC.2018.8424614
24. Peechara RR, Sucharita V (2021) A chaos theory inspired, asynchronous two-way encryption mechanism for cloud computing. PeerJ Comput Sci 7:e628
25. Jin L, Zhang S, Lou Y, Xu X, Zhong Z (2019) Secret key generation with cross multiplication of two-way random signals. IEEE Access 7:113065–113080
26. Mao Y, Zeng P, Chen TY (2021) Recent advances on quantum key distribution overcoming the linear secret key capacity bound. Adv Quant Technol 4(1):2000084
27. Wang H, Zhao Y, Tornatore M, Yu X, Zhang J (2021) Dynamic secret-key provisioning in quantum-secured passive optical networks (PONs). Opt Express 29(2):1578–1596
28. Liao Q, Liu H, Zhu L, Guo Y (2021) Quantum secret sharing using discretely modulated coherent states. Phys Rev A 103(3):032410
29. He X et al (2022) Routing and secret key assignment for secure multicast services in quantum satellite networks. J Opt Commun Netw 14(4):190–203. https://doi.org/10.1364/JOCN.445621

30. Huang Y, Jin L, Wei H, Zhong Z, Zhang S (2018) Fast secret key generation based on dynamic private pilot from static wireless channels. China Commun 15(11):171–183. https://doi.org/10.1109/CC.2018.8543098
31. Hayashi M, Vázquez-Castro Á (2020) Physical layer security protocol for Poisson channels for passive man-in-the-middle attack. IEEE Trans Inf Forensics Secur 15:2295–2305
32. Devi OR, Vallabhaneni SP, Hussain MA, Kumar TK (2020) Security analysis on remote authentication against man-in-the-middle attack on secure socket layer. IOP Conf Ser Mater Sci Eng 981(2):022015. IOP Publishing
33. Stepień K, Poniszewska-Marańda A (2019) Security measures in the vehicular ad-hoc networks—man in the middle attack. In: International conference on mobile web and intelligent information systems. Springer, pp 136–147
34. Al-Shareeda MA, Anbar M, Manickam S, Hasbullah IH (2020) Review of prevention schemes for man-in-the-middle (MITM) attack in vehicular ad hoc networks. Int J Eng Manage Res 10
35. Lu X, Lei J, Shi Y, Li W (2021) Intelligent reflecting surface assisted secret key generation. IEEE Signal Process Lett 28:1036–1040
36. Jagadeesh H, Joshi R, Rao M (2021) Group secret-key generation using algebraic rings in wireless networks. IEEE Trans Veh Technol 70(2):1538–1553
37. You J, Wang YG, Zhu G, Wu L, Zhang H, Kwong S (2022) Estimating the secret key of spread spectrum watermarking based on equivalent keys. IEEE Trans Multimed 1. https://doi.org/10.1109/TMM.2022.3147379
38. Zhang Y, Zhao H, Xiang Y, Huang X, Chen X (2019) A key agreement scheme for smart homes using the secret mismatch problem. IEEE Internet Things J 6(6):10251–10260. https://doi.org/10.1109/JIOT.2019.2936884
39. Mishra P, Kumar S, Garg U, Pilli ES, Joshi R (2021) Security perspectives of various IoT cloud platforms: a review & case study. In: 2021 international conference on computing, communication, and intelligent systems (ICCCIS). IEEE, pp 727–731
40. Peng L, Li G, Zhang J, Woods R, Liu M, Hu A (2019) An investigation of using loop-back mechanism for channel reciprocity enhancement in secret key generation. IEEE Trans Mob Comput 18(3):507–519. https://doi.org/10.1109/TMC.2018.2842215
41. Pandey F, Dash P, Samanta D, Sarma M (2022) Efficient and provably secure intelligent geometrical method of secret key generation for cryptographic applications. Comput Electr Eng 101:107947. https://doi.org/10.1016/j.compeleceng.2022.107947

Overview on Online Examination Proctoring Using Machine Learning

Ain Najiha Mohamad Nor, Kauthar Mohd Daud, Noramalina binti Mohd Hatta, and Nurul Izrin Md Saleh

Abstract Pandemic COVID19 has brought changes to the education sector, explicitly teaching and assessment. Online proctoring during the examination has been applied for distance learners. However, the malpractice during the online examination is one of the most challenging aspects of academic integrity due to various cheating behaviors. To curb these cheating behaviors, different online proctoring integration services have been developed and used in various academic institutions to maintain the sanctity of the examination. However, most of the proctoring integration services are paid basis. This review discusses online proctoring integration services and highlights different proctor attributes provided by each proctoring integration service. Next, this paper reviews recent work implementing machine learning in online proctoring. In addition, a generalized framework using ML is developed to highlight the potential and promise of using ML in assisting educators and learners with online distance learning.

Keywords Machine learning · Online examination · Proctoring · Optimization algorithms

1 Introduction

With the existence of the internet, online gaming such as PlayerUnknown's Battlegrounds (PUBG) and Mobile Legends, online video-sharing platforms such as

A. N. M. Nor · K. M. Daud (✉)
Fakulti Teknologi dan Sains Maklumat, Centre for Artificial Intelligence, Universiti Kebangsaan Malaysia, Bangi, Selangor, Malaysia
e-mail: kauthar.md@ukm.edu.my

N. M. Hatta
Pusat Genius@Pintar Negara Universiti Kebangsaan Malaysia, Bangi, Selangor, Malaysia
e-mail: noramalina@ukm.edu.my

N. I. M. Saleh
Department of Data Science, Universiti Malaysia Kelantan, Pengkalan Chepa, Kelantan, Malaysia
e-mail: nurulizrin.ms@umk.edu.my

© The Author(s), under exclusive license to Springer Nature Singapore Pte Ltd. 2023
D.-K. Kang et al. (eds.), *Proceedings of the 9th International Conference on Computational Science and Technology*, Lecture Notes in Electrical Engineering 983,
https://doi.org/10.1007/978-981-19-8406-8_7

YouTube and Spotify, and others have emerged and gained popularity, particularly among Generation Y and beyond. One remarkable transformation that has garnered popularity is e-learning. The e-learning platforms such as Google Classroom and Massive Open Online Courses (MOOCs) like Coursera and EdX have optimized learning environments. In Malaysia, the Ministry of Education has rebranded the digital learning platform to "Digital Educational Learning Initiative Malaysia" (DELIMa), including Google Classroom, Apple Teacher Learning Center, and Microsoft O365 services for 10,000 schools, 370,000 teachers, and 2.5 million students. Online education has been widely implemented worldwide as it allows learners to be anywhere, they are.

However, in early 2020, the world was hit by the rapid spread of SARS-CoV-2. This pandemic affected all sectors, including education, economy, and tourism. The spreading of this virus has transformed human life whereby working from home and social distancing to name a few, have been imposed to prevent this virus's spreading [1]. The transformation of traditional face-to-face learning to online education has occurred worldwide, including in China, Indonesia, and Malaysia [2–4]. The impact of the deadly coronavirus has caused the mandatory closure of academic institutions. Therefore, online distance learning has been implemented to curb the spreading of this virus [3].

Although many efforts have been focused on creating, developing, and applying the best online learning platforms such as Moodle, Google Classroom, and others, there is less effort focused on e-invigilation. Before the pandemic, the examinations were conducted physically, whereby the presence of invigilators was mandatory during the assessment process in a defined venue. During the pandemic, such assessment has been shifted to online proctoring or online invigilation, whereby three online proctoring systems are being practiced: (1) automated proctoring using machine learning or artificial intelligence, (2) safe-exam browser, and (3) live proctoring session by hired invigilators [5]. However, malpractices during examinations, such as plagiarism and cheating, still occur [6].

In this paper, we focused on the related online proctoring using machine learning in Sect. 2; the proctoring integration services will be reviewed in Sect. 3; the proposed generalized framework of automated proctoring using machine learning is discussed in Sect. 4. Finally, the conclusion and closing annotations and proposed possible future work are concluded in Sect. 5.

2 Online Proctoring

The application of technology in learning and teaching has been on the rise for the past few years with the growth in online distance learning (ODL) offered by universities and academic institutions. These offerings are targeted in the local area and globally due to their flexibility and distance. Nevertheless, the flexibility and distance between learners and educators may affect the integrity of the online assessment, whereby the misbehaviour of dishonesty, cheating, and others are persisted.

Several surveys and studies have been conducted to identify students' perceptions of online assessment. Most students indicated that cheating in an online environment is more manageable than in traditional face-to-face classes [7]. Moreover, many students acknowledged that cheating during examinations is a substantial issue in online examinations [6, 8]. In addition, students are four times more likely to conduct malpractices in online examinations than in the traditional classroom.

Online proctoring has been implemented ever since online distance learning was introduced and offered to ODL learners in universities and institutions. Several online proctoring platforms, Examity, MeritTrac, ProctorU, ExamOnline, SpeedExam, and others, have been developed to monitor the candidates from conducting dishonesty during the exam session. Figure 1 illustrates the basic workflow of online proctoring.

As shown in Fig. 1, online proctoring involves the monitoring activities of candidates during the exam session. The candidates need to authenticate their identity using student ID and image, whereby the candidates need to turn on their web camera during the session. Then, they will sit for the exam while being monitored by the invigilator or proctor. Table 1 shows the three types of online proctoring systems that have been identified [9].

The live proctoring occurs in a real-time situation with a human proctor that monitors the candidates with audio–video live streaming. For this type of proctoring, the examination is scheduled priorly. Thus, this type of proctoring is inconvenient for part-time students as they need to suit their timetable and working hours. Furthermore, it requires a high-end device compatible with the proctoring, and the ratio of one human proctor to candidates is 1:32. Thus, it needs to hire more human proctors

Fig. 1 The basic workflow of online proctoring

Table 1 Types of online proctoring system

Type of online proctoring	Real-time proctoring	Presence of a human proctor	Video recorded	Time flexibility	Date flexibility	Generated feedback	Number of candidates
Live	✓	✓	×	×	×	Real-time	Low
Recorded	×	✓	✓	✓	✓	Provided upon request	Medium
Automated	×	×	✓	✓	✓	Not provided	High

for a more significant number of candidates. Next, the recorded proctoring involves recording candidates' video images and activities logs during the exam. The human proctor will later analyze the recorded video to identify any misconduct during the examination. Meanwhile, automated proctoring involves using artificial intelligence (AI) algorithms to replace the human proctor. In this online proctoring, different key events of possible fraud or cheating will be recognized and identified by AI algorithms and only be reviewed by the human proctor, if possible, misconduct occurs [1, 9].

In conclusion, the three types of online proctoring require a good and uninterrupted internet connection and the availability of a web camera on a candidate's devices. Nevertheless, a lockdown has been introduced to ensure the examination's sanctity. The term "computer or browser lockdown" prevents the candidates from using any other applications on the same device and eliminates the user-computing processes [10].

3 Review of Proctoring Integration Services

The examination proctoring can be considered as one of the tedious works since it involves the process of monitoring students throughout the examination period. The process will become costly if the number of exam candidates increases; thus, more supervisors must be monitored. There are three main reasons why students tend to cheat in an examination. First, they fear failure, especially in studies [11]. Next, they are eager to take some risks due to poor inspections during exams. Other than that, include no fear of getting caught cheating. The less effective action taken by institutions can lead to this factor.

Examining proctoring has also advanced as the world has stepped into a more technological and less artificial situation. Recently, with the urge of COVID19 that has brought to more contactless activities, several approaches to exam proctoring have been made. One of the approaches is the automation of exam proctoring, where AI is applied to monitor students' activities.

Many researchers have researched students' misconduct in examinations and proposed several ways to tackle the solutions throughout the year—some papers utilized artificial intelligence and computer-based strategies in online examinations [12]. In 2021, Saba and the team suggested that an eye-tracking mechanism be installed to evaluate whether students are focusing or watching around. Other than that, monitoring exams using a webcam can detect any unusual activities. The image captured is subtracted from the reference image based on the same research. It is considered cheating if the image outcome is more than the threshold value.

Aini et al. implemented background subtraction to perceive any strange motions during the examination by placing a camera to capture an image or video. The image processing will be displayed on the surveillance laptop, and a camera frame will be divided according to the exam candidates' seating place [13]. Any movements of the participants are detected if it is beyond the frame divided [13]. Thus, this could help the supervisor in proctoring the examination as it is impossible to keep an eye on at the same time.

Apart from utilizing face and camera features, there is another obstacle in conducting online examinations: eye contact and movements [12]. Based on Javed and Aslam [14], eye movements in online exams can be extracted using a biometric algorithm. Javed and Saba et al. stated that the sample of the eye at a normal state could be compared with the current eye movements to decide whether there is any cheating or otherwise. An algorithm is used to detect the characteristics of eye motion, from corner to edge, that is obtained from facial images of the candidates. They also added that machine learning systems could be integrated to use the eye regions and non-eye templates for further eye recognition [14].

Another crucial part of conducting an online examination is user authentication. User authentication is critical to ensure that the person who is supposed to sit for the exam is the valid person instead of a third party [15]. Karim and Shukur listed several systems that aid the process of user authentication. The systems include ProctorU, ProctorCam, Webassessor BioSig-ID, and SecureXam. Some of the systems required payment to activate. These systems mainly protect online exam sessions from online threats such as hacking and third parties' involvement [16].

For ProctorCam and ProctorU, Karim and Shukur said both systems require a person's accurate time proctor throughout the examination session. Based on this, there is a low risk of having online threats during the period. Other systems, for instance, Webassessor, acquire exam candidates for facial recognition before sitting for the exam [15]. This can prevent any third-party candidate from sitting the exam. Table 2 shows the comparisons between five proctoring systems that have been installed in institutions around the world. The comparisons were based on proctoring features such as proctor existence, connection requirements, browser blocking, and candidates' authentication options.

4 Machine Learning Implementation in Online Proctoring

ML has been a great help in different areas such as cancer detection, sentiment analysis [17], computer vision, email filtering, and others, as it learns patterns from data and gradually improves the prediction outcomes. ML has been used in online proctoring to identify the malpractices pattern in eye-tracking, face-recognition, gesture-detection, and others by detecting different features. In everyday practices of online proctoring, face images through built-in cameras have been used as a dataset to discover cheating behavior among candidates. Face recognition has been studied extensively in the literature using image processing algorithms. However, many improvements need to be made due to noise in images and lighting conditions that affect the variations in human facial appearances, scales, poses, and others. These problems hindered the accuracy and effectiveness of detecting malpractices during online proctoring.

Table 2 The comparisons of five proctoring systems in proctor features

Proctor attributes	ProctorU	ProctorCam	Webassessor BioSig-ID	SecureXam	Webassessor
Require live proctoring	Yes	Yes	No	Yes	Yes
Connections requirements	Yes	Yes	Yes	Yes	Yes
Proctor attributes	ProctorU	ProctorCam	Webassessor BioSig-ID	SecureXam	Webassessor
Flexible exam time	Yes	Yes	No	No	No
Data security	Yes	Yes	Yes	Yes	Yes
Interaction between students and proctor live	Yes	Yes	No	Yes	No
Exam question released at a specific time	Yes	Yes	Yes	Yes	Yes
Proctor can monitor online	Yes	Yes	No	Yes	No
Able to block proctor from seeing students	No	No	Yes	No	Yes
Recording available	No	Yes	Yes	Yes	Yes
Pause/cancel exam	No	Yes	No	No	No
Automatic proctoring	No	Yes	Yes	No	Yes
Audio recording	No	Yes	No	Yes	No
Browser lock	No	Yes	Yes	Yes	Yes
Authentication options	Yes	Yes	Yes	Yes	Yes
Web camera obligatory	Yes	Yes	Yes	Yes	Yes
Proctor logging report	No	Yes	Yes	Yes	Yes
Recording copy save	Yes	Yes	Yes	Yes	Yes
Customize features based on the institution	No	Yes	Yes	Yes	Yes
Browser options characteristics					
Availability on win/mac	Yes	Yes	Yes	Yes	Yes
Plug in browsers	No	No	No	No	No
Page navigation freeze (forward/back)	No	No	Yes	No	Yes

(continued)

Table 2 (continued)

Proctor attributes	ProctorU	ProctorCam	Webassessor BioSig-ID	SecureXam	Webassessor
Block printing option	No	No	Yes	No	Yes
Avoid copy paste from outside browsers	No	No	Yes	No	Yes
Stop other running apps	No	No	Yes	No	Yes
Authentication options					
User authentication	Yes	Yes	Yes	Yes	Yes
Username requirement	Yes	Yes	Yes	Yes	Yes
Password requirements	Yes	Yes	Yes	Yes	Yes
Students' ID requirements	Yes	Yes	Yes	Yes	Yes
Pointing device for signature	No	No	Yes	No	Yes
Facial recognition	No	No	Yes	No	No
Voice recognition	No	No	Yes	No	No
Thumbprint recognition	No	No	Yes	Yes	No
Eye movement recognition	Yes	Yes	No	Yes	No
Keystroke rhythms	No	No	Yes	No	Yes

An incremental training on deep learning face recognition was proposed by Asep and Bandung [18] to improve face recognition accuracy during the examination period. The study gradually updates the training set of the user's face by acquiring from lecture sessions. Moreover, the study proposed image size reduction for face detection and lower resolution images produced higher accuracy than high-resolution images. The study also shows that the proposed incremental training only affects the training time, yet the accuracy rate is similar to one batch training (Fig. 2).

An automatic method was developed in a study by comparing the pixels of captured still images with the pixels of reference images. The activity is considered cheating if the difference exceeds the threshold value [19]. In another study, a face-based method using Discrete Cosine Transformation (DCT), Kernel Discriminant Analysis (KDA), and Radial Basis Function Network (RBFN) has been proposed for security features in online proctoring in distinguishing different individuals. The study found that the proposed method could identify the face in a shorter time with a lesser error rate than the other methods [20].

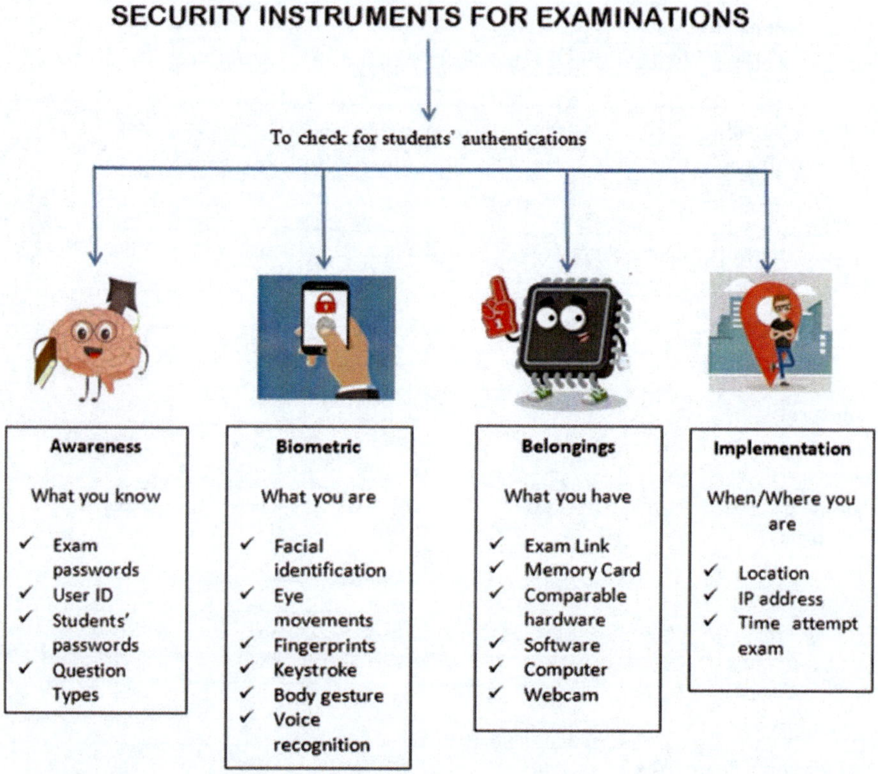

Fig. 2 Security instruments frameworks for online examinations

Meanwhile, a recurrent neural network has been applied to the anomaly detection algorithm to identify the cheating scenarios in the exam. According to the research, an anomaly detection algorithm is used to identify any potential misconducted behaviors of students by studying the students' typing patterns, learning processes, and others. In this study, three scores (quizzes, midterm exam, and final exam) are used as test cases to identify the correlation with cheating in exams. The study found that the proposed method achieved higher accuracy in detecting cheating cases than other anomaly detection algorithms, including naïve Bayes, local outlier factor (LOF), robust covariance, and isolation forest algorithm. Indirectly, the recurrent neural network optimized the anomaly detection algorithms.

Another study examined different head postures and hand movements to identify cheating-related abnormal behavior. The study used Tensorflow-PoseNet and Multiple Pose Estimation to extract the video camera's input footage, classify the images, and achieve 94 and 97% accuracy for abnormal head pose and paper trading. Meanwhile, Nishchal et al. proposed combining mood and gesture interpretation to discover the cheating activities of exam candidates [11]. They used an open pose tool analysis and convolutional neural network (CNN) to detect any exam deception activities. A total of four classes are assimilated with movements such as looking straight, turning backward, bending down, and stretching arms towards the back. According to their study, the model accuracy is 77.8% accurate.

Students' activities have also been researched by categorizing the students' activities during the exam using a deep learning approach. The proposed model combined characteristics of deep AlexNet and SqueezeNet for features extraction. The extracted features are selectively optimized using an atom search optimization algorithm. The optimized features are then passed to different classifiers in which the Fine KNN classifier produces higher accuracy of 93.88% compared to other classifier algorithms. In another study, a voice recognition tool was used to determine and validate student identity. The online oral assessment tool, Real Talk, was integrated with the Deep Speaker algorithm to facilitate educators and academic institutions in oral assessment. The audio data of students' discussions were used as a sample for identifying and verifying the identity of examiners.

A fusion-based live online authentication and biometric user authentication using a hybrid LCNN-Salp swarm algorithm (LCNN-SSO) has also been developed in which this study uses a modified wolf optimization algorithm to extract the optimal features from face images, fingerprints, and keystrokes. The optimal features are fused using the optimal feature fusion (OFF) algorithm, which is then used to classify normal and abnormalities using LCNN-SSO. Their research achieved higher accuracy, precision, recall, F-measure, sensitivity, and specificity compared with different classifier algorithms, including SVM, KNN, LSTM, and DBN. The proposed algorithm (LCNN-SSO) achieved a lower false-positive rate (FPR) than the others.

Most of the developed systems fully implement machine learning or artificial intelligence concepts. Combining different identifiers or features in these systems improves traditional proctoring as it incorporates multi-modal authentication. The multi-modal authentication outperformed the unimodal process in terms of accuracy and security as it includes different attributes or features, including keystrokes, iris monitoring, face, gaze detection, and others. Based on the previous works, the proposed framework is to integrate machine learning and optimization for analyzing and identifying optimal multi-features.

Figure 3 illustrates the proposed integrative schemes in the online proctoring domain. In Fig. 3, the reconciliation of different features that could be used as input for feature selection can improve the accuracy of detecting whether a particular student is conducting an abnormal or normal behaviour during the exam. Additionally, the optimization algorithm is included in selecting the optimal features, which machine learning will analyse and classify.

5 Conclusion

Due to COVID-19, the education sectors have implemented various proctoring systems to help proctors identify students' abnormal behavior during online examinations. The abnormal behavior includes cheating, substitute students, and others.

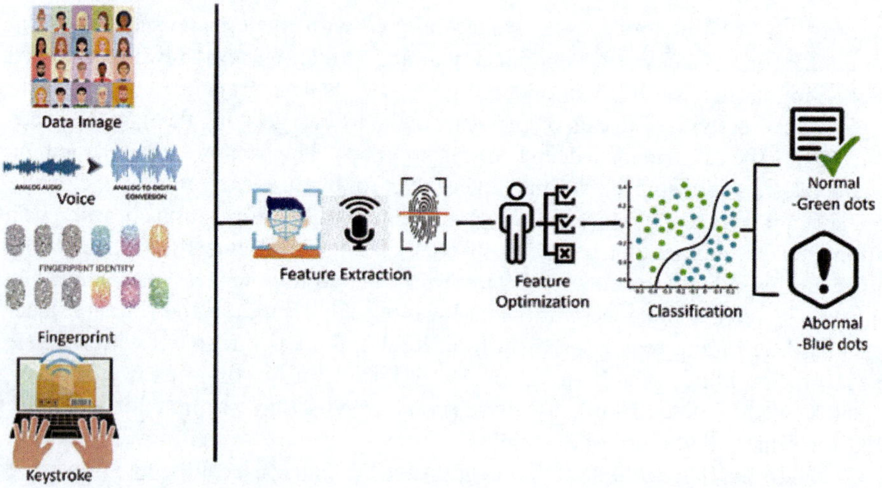

Fig. 3 Proposed integrative machine learning and optimization for optimal multi-features

Different online proctoring software has been used to identify abnormal behavior to mitigate student cheating. In this review, we started with the description of online proctoring and its type. Furthermore, we also review different proctoring integration services available in the market. The reviewed proctoring integration services include ProctorU, ProctorCam, Webassessor BioSig-ID, SecureXam, and Webassessor. However, due to the high implementation cost of these services, thus, machine learning and optimization algorithms were proposed. Eventually, it improves the performance in terms of cost. However, most of the proposed integrated machine learning and optimization algorithms only focused on one feature. Consequently, the fusion of different features, including face image, keystrokes, fingerprints, and voice, may improve the accuracy of detecting abnormalities. Regardless of the different strategies for detecting abnormal behaviors, combining machine learning and optimization algorithms may optimize the feature selection and thus improve the accuracy.

References

1. Manoharan S, Ye X (2020) On upholding academic integrity in online examinations. In: 2020 IEEE conference on e-learning, e-management and e-services, IC3e 2020, pp 33–37. https://doi.org/10.1109/IC3e50159.2020.9288468
2. Widiyatmoko A (2021) The effectiveness of google classroom as a tool to support online science learning: a literature review. J Phys Conf Ser 1918:052069. https://doi.org/10.1088/1742-6596/1918/5/052069
3. Razami HH, Ibrahim R (2021) Distance education during COVID-19 pandemic: the perceptions and preference of university students in Malaysia towards online learning. Int J Adv Comput Sci Appl 12:118–126. https://doi.org/10.14569/IJACSA.2021.0120416

4. Lei M, Medwell J (2021) Impact of the COVID-19 pandemic on student teachers: how the shift to online collaborative learning affects student teachers' learning and future teaching in a Chinese context. Asia Pac Educ Rev 22:169–179. https://doi.org/10.1007/s12564-021-09686-w
5. Haiyan C, Irwin K (2020) Education technology for online learning in times of crisis. In: IEEE TALE 2020—an international conference on engineering, technology, and education, pp 758–763
6. Dendir S, Maxwell RS (2020) Cheating in online courses: evidence from online proctoring. Comput Hum Behav Reports 2:100033. https://doi.org/10.1016/j.chbr.2020.100033
7. D'Souza KA, Siegfeldt DV (2017) A conceptual framework for detecting cheating in online and take-home exams. Decis Sci J Innov Educ 15:370–391. https://doi.org/10.1111/dsji.12140
8. Cleophas C, Hoennige C, Meisel F, Meyer P (2021) Who's cheating? Mining patterns of collusion from text and events in online exams. SSRN Electron J 1–19. https://doi.org/10.2139/ssrn.3824821
9. Hussein MJ, Yusuf J, Deb AS et al (2020) An evaluation of online proctoring tools. Open Prax 12:509. https://doi.org/10.5944/openpraxis.12.4.1113
10. Alessio HM, Malay N, Maurer K et al (2017) Examining the effect of proctoring on online test scores. Online Learn J. https://doi.org/10.24059/olj.v21i1.885
11. Nishchal J, Reddy S, Navya PN (2020) Automated cheating detection in exams using posture and emotion analysis. In: 2020 IEEE international conference on electronics, computing and communication technologies (CONECCT), July 2020. IEEE, pp 1–6
12. Saba T, Rehman A, Jamail NSM, Marie-Sainte SL, Raza M, Sharif M (2021) Categorizing the students' activities for automated exam proctoring using proposed deep L2-GraftNet CNN network and ASO based feature selection approach. IEEE Access 9:47639–47656
13. Aini YK, Saleh A, Darwito HA (2020) Implementation of background subtraction method to detect movement of examination participants. In: 2020 international electronics symposium (IES), Sept 2020. IEEE, pp 494–500
14. Javed A, Aslam Z (2013) An intelligent alarm based visual eye tracking algorithm for cheating free examination system. Int J Intell Syst Appl 5(10):86
15. Karim NA, Shukur Z (2016) Using preferences as user identification in the online examination. Dynamics 19(20):21
16. Karim NA, Shukur Z (2015) Review of user authentication methods in online examination. National University of Malaysia, Selangor, Malaysia. Asian J Inf Technol 14(5):166–175
17. Im TL, San PW, On CK, Alfred R, Anthony P (2013) Analysing market sentiment in financial news using lexical approach. In: 2013 IEEE conference on open systems (ICOS), pp 145–149
18. Ganidisastra AH, Bandung Y (2021) An incremental training on deep learning face recognition for m-learning online exam proctoring. In: 2021 IEEE asia pacific conference on wireless and mobile (APWiMob), pp 213–219
19. Jalali K, Noorbehbahani F (2017) An automatic method for cheating detection in online exams by processing the students webcam images. In: 3rd conference on electrical and computer engineering technology (E-Tech 2017), Tehran, Iran
20. Irfan CMA, Nomura S, Ouzzane K, Fukumura Y (2009) Face-based access control and invigilation tool for e-learning systems. In: Proceedings—2009 international conference on biometrics and Kansei engineering, ICBAKE 2009, pp 40–44. https://doi.org/10.1109/ICBAKE.2009.43

Facial Micro-expressions Analysis: Its Databases, Feature Extraction, and Classification Methods

Wee How Khoh, Alvin Fang Chuen Lim, Ying Han Pang, and Hui Yen Yap

Abstract Facial micro-expressions (ME) are a subtle expression that appears within a shorter duration compared to facial macro-expressions. This expression is a type of universal non-verbal communication that appears on someone's face unconsciously when he or she is attempting to hide their true emotions. It reveals a person's true emotional state, which is vital for both mental health diagnosis and security purposes. Because micro-expressions are typically low in intensity and short in duration, humans and/or machine learning models can be specifically trained to recognize some expressions, but the obtained performance is usually consistently poor. The research on facial micro-expression recognition is relatively new in the field of computer vision. This paper presents an overview of facial micro-expressions and their taxonomy. Each posed and spontaneous database, as well as feature representation and classification techniques, are examined and expanded upon. In addition, we conclude the overview by pointing out a few flaws in micro-expression recognition methods and making some recommendations for future research directions that could aid researchers in making progress in this field.

Keywords Facial expressions · Micro-expressions · Recognition · Biometrics

W. H. Khoh (✉) · A. F. C. Lim · Y. H. Pang · H. Y. Yap
Faculty of Information Science and Technology, Multimedia University, 75450 Melaka, Malaysia
e-mail: whkhoh@mmu.edu.my

A. F. C. Lim
e-mail: alvinlim.fangchuen@gmail.com

Y. H. Pang
e-mail: yhpang@mmu.edu.my

H. Y. Yap
e-mail: hyyap@mmu.edu.my

© The Author(s), under exclusive license to Springer Nature Singapore Pte Ltd. 2023

D.-K. Kang et al. (eds.), *Proceedings of the 9th International Conference on Computational Science and Technology*, Lecture Notes in Electrical Engineering 983,
https://doi.org/10.1007/978-981-19-8406-8_8

1 Introduction

Researchers have been paying a lot of attention to automatic facial micro-expressions (ME) recognition in recent years because of its potential applications in fields like psychology and neuroscience. Micro-expressions are small muscle movements that occur beneath the skin of the face. It usually shows up unconsciously on the face when a person is attempting to hide his or her true emotions. Haggard and Isaacs [1] first discovered it when scanning recorded psychotherapy films to search for nonverbal cues in communication. It was also noticed by Ekman and Friesen [2] when they were reviewing a psychiatric patient's videotaped interview, who was attempting to mask her genuine emotions by pretending that she was no longer suicidal. When the interview tape was examined in slow motion, a transient expression of sadness on the patient's face was discovered, followed by an instant fake smile every time she answered a question. These instant expressions reveal a person's true feelings and emotions, which is vital and critical in mental health diagnosis, lie detection, and forensics purposes.

In this paper, a brief overview of recent studies in the field of micro-expressions is provided. The following is how the paper is structured: The taxonomy of micro-expressions is first described. Section 2 examines several publicly accessible databases. Handcrafted features and learning-based features are two of the common feature representations in micro-expressions, as described in Sect. 3. Section 4 discusses some of the most recent classification techniques in micro-expression recognition. Finally, Sect. 5 brings the overview to a conclusion.

1.1 Taxonomy of Micro-expressions

Ekman and Friesen [2] created and recognized a taxonomy of six (6) basic emotions in the 1970s. These emotions include happiness, sadness, fear, surprise, disgust, and anger. Each emotion's expression is uniquely exhibited and can be identified through automated computer vision algorithms. It was then extended into seven (7) types of emotions that are commonly captured in facial micro-expressions [3]. Besides the six emotions that were proposed by Ekman et. al. previously, a "contempt" emotion is added. Figure 1 shows the seven basic emotions for micro-expressions.

In addition, three key moments in the extraction of micro-expressions are used to identify the appearance of the expressions, which are onset, apex, and offset. The onset is when the expression first appears, the apex is when the expression reaches its peak, and the offset is when the expression begins to fade out [4]. The apex frames are the most important since they contain the most expressive emotions. Furthermore, when compared to the onset and offset frames, the apex frames have the largest magnitude features. Hence, most of the micro-expression recognition research were conducted using apex frames.

Facial micro-expressions are generally divided into three categories based on how they are elicited : stimulated expressions, neutralized expressions, and masked

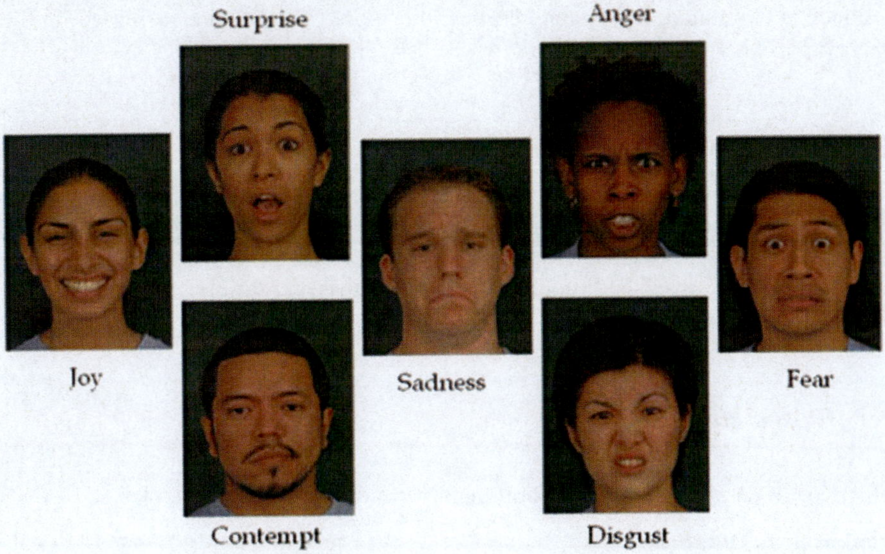

Fig. 1 The basic emotions for micro-expressions [4]

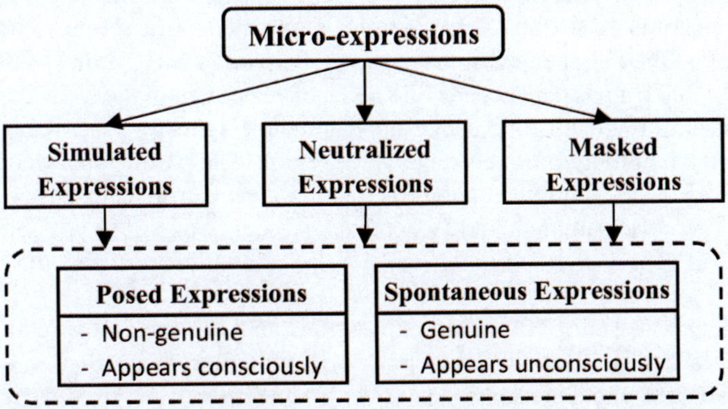

Fig. 2 Taxonomy of micro-expressions

expressions [5]. The simulated expressions occur if an emotional expression is not accompanied by a genuine expression that occurs for a brief moment, neutralized expressions occur when genuine expressions are suppressed to remain blank expressions, and masked expressions occur when genuine expressions are completely disguised by a falsified expression [6]. As shown in Fig. 2, all of these expressions can be further characterized as posed or spontaneous. Spontaneous expressions could be explained when a genuine expression appears unconsciously by a person while posed expressions are non-genuine expressions that appear consciously. For instance, a simulated expression can be either posed or spontaneous based on the ways it appears on a person's face (consciously/unconsciously).

2 Micro-expressions Databases

This section describes the publicly available and well-known micro-expression databases. As aforementioned, there are two types of databases based on the content of the expression: posed and spontaneous type databases. A brief description of each database is elaborated.

2.1 Posed Databases

This type of database consists of non-genuine micro-expression samples.

Polikovsky's Database

Polikovsky's database is one of the earliest publicly available posed micro-expressions databases that was produced by Polikovsky et al. [7]. There are a total of 13 samples in this database, which were gathered from 6 Asian and 4 Caucasian university students. The Point Grey Pictures Company's Grasshopper camera was used to capture the facial expressions of the participants with a setting of 640 × 480 resolution at 200 frames per second (FPS). The environment of the facial expressions recording is bright with uniform background and three lights for shadow cancellation to maximize the number of pixels on the face region. The participants were asked to perform seven basic emotions (happiness, sadness, fear, surprise, disgust, anger, and contempt) for a brief moment and switch back to neutral as fast as possible.

USF-HD

The USF-UD consists of 100 micro-expressions and 181 macro-expressions. This database was collected at the University of South Florida by Shreve et al. [6]. The JVC-HD100 and Panasonic AG-HM40 camcorder were used to record the expressions of the participants with a resolution of 720 × 1280 at 30 FPS. The recording environment is under normal lighting conditions. Each participant was given some films that contain both micro-expressions and micro-expressions. Then, the participants were then instructed to imitate the expressions as closely as possible.

Iranian Kinect Face Database

The IKFDB database, developed by Mousavi and Mirinezhad [8], which contains both the macro-expressions and micro-expressions that were collected from 40 participants. The recording sessions took place in a lab-controlled environment that had several lights installed for good illumination to capture the expressions in a better condition with fewer noises. A stool was placed 1 m away from the sensor for the participants to sit. The Kinect version 2 motion sensor camera was used to capture the emotions of participants at the rate of 30 FPS. During the enrolment, the participants were requested to execute seven basic facial expressions while shifting their heads 45° in four directions (right, left, top, and bottom). This database contains raw and cropped samples which result in a total of four different types of samples. The resolution for the raw samples is 1920 × 1080 pixels and 512 × 424 pixels for both micro-expressions and micro-expressions, respectively. It was then cropped into a resolution of 256 × 400 pixels and 128 × 200 pixels to reduce its size.

2.2 *Spontaneous Databases*

This type of database consists of genuine micro-expressions samples.

Canal-9

The Canal-9 database contains 43 h of spontaneous expression samples from 70 political debates that were broadcasted in Switzerland collected by Vinciarelli et al. [9]. The Canal 9 TV station was used to capture the expressions with a resolution of 720 × 576 pixels. The total number of participants is 190 (25 women and 165 men). The political debates were conducted at a studio with good lighting conditions. There was a moderator who asked the participants some political questions and they were required to answer with a yes or no.

SMIC

The SMIC database, collected by Li et al. [10] contains a total of 164 micro-expressions samples. All the expression samples were captured from 7 Caucasians and 8 Asians. The participants were asked to watch a movie and remain poker face during the recording. A camera was attached at the top of the monitor used by the participants to watch the movie clip. The expressions of participants were monitored during the recording sessions and a penalty of answering a long questionnaire was given to those who failed to conceal their true emotions. There are two types of samples collected in this database. The differences between these types of samples are the resolution, the number of frames, and the type of camera used for recordings. Both samples were recorded in a resolution of 640 × 480 pixels with different FPS settings. The first samples were recorded with a high-speed camera at 100 FPS and the second samples were recorded with a normal-speed near-infrared camera at 25 FPS.

CASME

The CASME is the first database published by the Chinese Academy of Sciences by Yan et al. [11]. It contains a total of 195 samples that were recorded with two resolution settings: 640 × 480 pixels and 1280 × 720 pixels, at 60 FPS from 35 participants (13 females and 22 males). All participants were requested to always keep a neutral face during the enrollment process in order for the micro-expressions to emerge unconsciously on their faces. To achieve accurate labeling for each emotion, each micro-expression was given an action unit. As a result, this database was successfully collected with a total of 8 emotion classes, including amusement, sadness, disgust, fear, surprise, repression, contempt, and tension.

CASME II

The CASME II [12] is an upgraded version of CASME that was recorded at a faster frame rate of 200 FPS while maintaining the same resolution as the prior database (CASME). It contains 247 samples from 26 participants with 5 emotion classes, including happiness, disgust, surprise, repression, and tension. There is an additional class named 'Others' to label all the unidentified emotions.

CAS(ME)^2

CAS(ME)^2 is the latest database produced by the Chinese Academy of Sciences by Qu et al. [13]. The Logitech Pro C920 camera was used to capture the facial expression of the participants. All samples were captured in 640 × 480 pixels resolution at 30 FPS. It has a lower quality in terms of frame rates as compared with the previous version (CASME and CASME II). Each participant was told to remain neutral while watching 9 emotional videos. The participants were then instructed to evaluate the recorded film of their own expressions and complete a self-report of their feelings for expression labeling purposes.

SAMM

The SAMM contains 159 spontaneous micro-expressions samples collected by Davison et al. [14]. All the expression samples were collected from 32 participants of different ethnicity. The Basler Ace acA2000-340km camera was used to capture the expressions with a resolution of 2040 × 1088 pixels at 20 FPS. Each participant was exposed to 7 different stimuli to induce their emotions. There was an observer to monitor the participants in different rooms and a monetary reward was given to those who hid their emotions the best.

3 Features Extraction

This section provides some of the existing micro-expressions research. In general, there are two types of features in these works: handcrafted and learning-based features.

3.1 Handcrafted Features

The handcrafted features are features that are extracted manually using conventional machine learning algorithms. Adegun and Vadapalli [15] employed the Local Binary Pattern (LBP) to extract apex frame features from micro-expression image sequences whereas the Local Binary Pattern from Three Orthogonal Planes (LBP-TOP) was used to extract spatiotemporal features. The values of the center pixel and neighboring pixels were compared to generate values for each feature. Duan et al. [16] developed an eyeME framework that used features from the eye's region instead of the whole face for micro-expressions recognition. In the work, the eyes region was firstly separated from the images using the rule of "Three court five eyes". Then, LBP-TOP was applied to the eye region to extract spatial features from the intra-frame local binary mode and temporal features from the inter-frame sequential mode. The downside of the LBP-TOP is that it only extracts feature from the three planes, which consist of motion and appearance features, and ignores other essential information such as simultaneous motion features in both horizontal and vertical planes. To address this limitation, Esmaeili et al. [17] used Local Binary Pattern from Six Intersection Planes (LBP-SIP) to extract more features from micro-expressions. The LBP-SIP able to extract more informative features from the concatenated histogram of six temporal planes which intersect in the central pixel. The temporal planes consist of two temporal planes from LBP-TOP and four planes in horizontal and vertical directions of facial muscle motion. Esmaeili and Shahdi [18] used Cubic-LBP to concatenate the LBP code on fifteen planes for micro-expressions recognition. It was introduced as an alternative to alleviate the limitation of LBP-TOP. Cubic-LBP able to extract subtle facial muscle movements from short ME image sequences, containing 6 vertical planes, 6 horizontal planes, and 3 diagonal planes. The vertical planes show the appearance changes and display motion changes whereas the horizontal planes show the changes of one row from frames. On the other hand, the diagonal planes show the appearance and motion changes.

3.2 Learning-Based Features

The learning-based features are features that are automatically learned and extracted using deep learning methods. Liong et al. [19] proposed a Shallow Triple Stream Three-dimensional Convolutional Neural Network (STSTNet) to extract and learn the features of micro-expressions by using the onset and apex frames. Three-dimensional convolutional layers are used in the network and an input cube which containing horizontal optical flow, vertical optical flow, and optical strain is used to learn a compact yet discriminative feature representation. Peng et al. [20] proposed an end-to-end trainable two-stream network named Dual Temporal Scale Convolutional Neural Network (DTSCNN) for micro-expressions features extraction. The streams in the STSCNN consist of a shallow network which have three-dimensional convolutional and pooling layers. A slow fusion model mechanism is also employed in the networks to process the ME sequences internally in order to avoid overfitting issues. The purpose of the dual streams in the networks is to adapt to the inconsistent frame rates of ME samples. To obtain higher-level features in shallow networks, the proposed network is fed with optical-flow sequences for additional data. Furthermore, Thuseethan et al. [21] proposed a hybrid framework based on a combination of three-dimensional and two-dimensional deep convolutional neural network-based architectures, denoted as HDCNN. The spatiotemporal features of ME are extracted using the HDCNN and a fusion mechanism is added to detect the intensity changes from the videos to enhance classification efficiency. Li et al. [22] proposed Three-dimensional flow-based Convolutional Neural Networks to extract features from micro-expressions video samples. The proposed networks consist of 12 layers and each layer has 3 data stream sub-networks that receive input from one tube channel. Several three-dimensional convolutional layers are used to extract features from micro-expressions samples and the extracted features are flattened and concatenated before passing to the classification layer.

4 Classification

This section presents some state-of-the-art classification techniques that are commonly employed in classifying micro-expressions in recent years.

4.1 Support Vector Machine (SVM)

The SVM [23] is the most widely used technique for micro-expression recognition. It is a supervised binary classification algorithm that analyses data for classification and regression analysis. Takalkar et al. [24] proposed a method where manually extracted features and learned features were combined. In the proposed method, the Local Binary Pattern-Three Orthogonal Planes (LBP-TOP) is used to combine spatial and time analysis to encapsulate regional face movement for handcrafted features whereas the convolutional neural network (CNN) was utilized for learning the features of micro-expressions. The SVM classifier is then trained with the combined features from LBP-TOP and CNN. The reported empirical results showed that the proposed method was able to achieve an accuracy of 80% on the CASME database.

Pan et al. [25] proposed a three-layer structure model for micro-expressions recognition and introduced the Hierarchical Support Vector Machine (H-SVM) to mitigate the influence of imbalance micro-expressions samples distribution that typically happened in multi-classification. In a multi-classification problem, samples are categorized into a binary class (*i*th class as positive while the rest as negative). For example, a "surprise" expression is labeled as positive while the rest is labeled as negative. The SVM on the first layer is only used to train to classify the expression of "surprise". The samples will then be sent to the subsequent SVM layer to classify the remaining classes as illustrated in Fig. 3. The highest accuracy rate obtained from the experiments was 64.63% on SMIC, CASME II, and SAMM databases. Although improvement has been reported with H-SVM, the hierarchy could be significantly increased as the class increases.

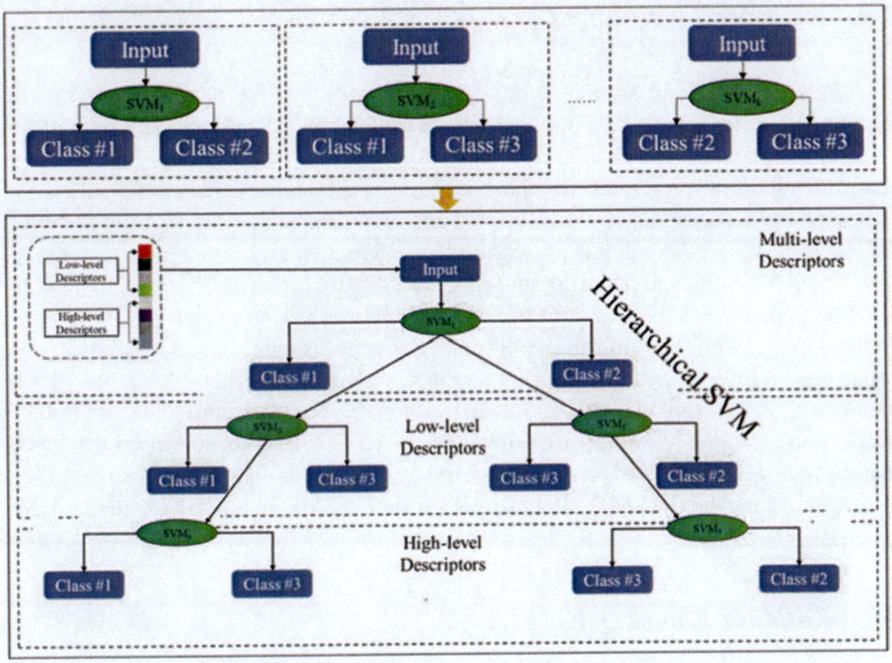

Fig. 3 Hierarchical SVM proposed by Pan et al. [26]

4.2 *Extreme Learning Machine (ELM)*

The ELM is a learning algorithm for the single hidden layer feedforward neural networks (SLFN) that has the capability to alleviate the limitations of traditional feed-forward neural networks since it does not require any parameter tuning to increase learning speed [26]. Adegun and Vadapalli [15] used LBP for apex frame features and LBP-TOP for spatiotemporal features extraction from the micro-expressions videos samples that were divided into image sequences. Both SVM and ELM were used in this work. The experiment conducted using the ELM achieves an accuracy of 97.65 and 96.26% on the SVM. In the works of Guo et al. [27], the Centralized Binary Pattern from Three Orthogonal Panels (CBP-TOP) features extraction method is used on the image sequence to obtain the spatiotemporal features of the micro-expressions. The authors conducted the experiments on the CASMA II database with the ELM classifier and obtained a promising result of 82% accuracy.

4.3 K-Nearest Neighbors (k-NN)

The *k*-NN [28] is a non-parametric classification algorithm that can make a prediction based on the neighboring data points. Happy and Routray [29] used Fuzzy Histogram of Oriented Optical Flow (FHOFO) features for micro-expressions recognition, three nearest neighbors were used for the *k*-NN to classify the expressions. Guo et al. [30] used the LBP-TOP for micro-expressions features extraction and the value of the spatial distance between each feature was calculated and classified using the *k*-NN. The KNN achieved a similar classification rate on both FHOFO and LBP-TOP features which were 65.83% and 65.99%, respectively. Kang et al. [31] proposed a micro-expression recognition method using the multimodal fusion approach. The combined architecture of convolutional neural network (CNN) and long short-term memory (LSTM) were used for feature extraction. Both temperature data and micro-expressions samples were fed into the CNN + LSTM network to obtain the fused feature vector. Then the *k*-NN is used to classify the features and obtained an accuracy rate of 63.2% on the CASME database, 67.1% on CASME II, and 64.3% on SAMM.

4.4 Random Forest (RF)

The RF [32, 33] is a supervised learning algorithm that is used for classification and regression. This algorithm operates by constructing a multitude of decision trees. Zhang et al. [34] proposed a feature fusion method where the features extracted from optical flow orientation histogram and LBP-TOP were concatenated to generate local statistical features. The experiments were conducted on the CASME II database and obtained an accuracy of 62.5% using the RF classifier on the fused micro-expressions features. Ibrahim et al. [35] utilized CNN to extract micro-expression features. The VGG-19 model architecture was adopted for feature extraction, and the random forest was used to classify the features. The proposed method achieved an accuracy of 99% on the SMIC-Cropped database.

4.5 Sparse Representation Classifier (SRC)

The SRC [36, 37] is a technique that uses sparse representation and learned redundant dictionaries for classification. Hao [38] proposed a combination of the 2D Gabor Filter and Sparse Representation for micro-expressions recognition (2DGSR). The increasing discrimination power of 2D Gabor filters is able to enhance the robustness of micro-expressions sample variations for better feature extraction. The extracted features of the 2D Gabor filter were classified using the SRC. An accuracy rate of 64.88% was obtained on the CASME II database using the proposed method.

4.6 Advanced Genetic Algorithm (AGA)

Liu et al. [39] proposed Advanced Genetic Algorithm (AGA) that utilized several existing features descriptor and classifiers for micro-expressions recognition. The AGA is a search heuristic algorithm based on natural evolution that stimulates the features extracted for replication, crossover, and mutation. The Bi-Weighted Oriented Optical Flow (Bi-WOOF), Optical Flow Features from Apex frame Network (OFF-ApexNet), and Shallow Triple Stream Three-dimensional CNN (STSTNet) are used for feature extraction. Next, the AGA is applied for feature integration. As for the classification, the SVM, RF, *k*-NN and Discriminant Analysis (DA) are adopted for classifying the integrated features. An accuracy rate of 85% was obtained on the proposed method which was relatively higher than most state-of-art methods.

4.7 Spatio-Temporal Recurrent Convolutional Networks (STRCN)

Xia et al. [40] proposed a spatio-temporal recurrent convolution network for micro-expression recognition, coined STRCN. The recurrent convolutional layers in STRCN are used to extract visual features from micro-expressions samples whereas the classificatory layer is used to categorize the types of the expressions. There are two types of STRCN: the appearance-based namely STRCN-A and the geometric-based namely STRCN-G. The STRCN-A is used to preserve the features of the facial regions by catenating the frames to transform the micro-expressions sequence into a matrix. On the other hand, The STRCN-G is used to extract geometric information from facial muscle movements by computing the optical flow field on the matrix values of onset and offset frames. The proposed method was conducted on three datasets (SMIC, CASME, SAMM) and obtained an accuracy of 83.6% on STRCN-A, as well as 82.7% on STRCN-G.

4.8 Graph Convolutional Networks (GCN)

Lo et al. [41] proposed graph convolutional neural networks that utilized relation modeling technique for micro-expressions recognition. The three-dimensional CNN in the network is used to extract action unit features from micro-expressions samples and the dependency laying between the action unit nodes is identified by GCN before the classification layer. The proposed method obtained an accuracy of 58.8% using the k-fold validation testing protocol on the CASME II database. Additionally, Xie et al. [42] proposed an Action Units assisted Graph Attention convolutional network (AU-GACN) that combines the emotion category label and action units for micro-expressions recognition. The AU-GACN is able to outperform the GCN and achieves promising accuracy of 71.2% on CASME II, SAMM, and SMIC databases. Tang et al. [43] proposed a dual stochastic GCN that fuses spatio-temporal features. The proposed model improved feature learning by extracting both spatial and temporal features. The model was first trained using the macro-expression databases. Then, transfer learning was performed using the trained model on micro-expressions samples. The proposed method obtained an accuracy of 72.7% on MMEW and 75.0% on SAMM databases.

5 Conclusion

Unlike other state-of-the-art studies, this research provided an overview of micro-expressions by introducing accessible micro-expression databases and categorizing their methodologies based on common techniques in feature extraction and classification approaches. In summary, most state-of-the-art facial methods are assessed by using the laboratory collected samples but have not been tested in a real-world environment. Due to this reason, the advancement in micro-expressions recognition is practical for real-world applications. Real-world experiments are always the ultimate target since the proposed methods or models are practical and useful to solve real-world problems. Each existing method can be evaluated more accurately and subsequently improved by conducting experiments in real-world environments. As for the dataset improvement, real-world environment samples from different ethnicities should be collected and analyzed to help the research forward faster. However, it is not an easy task to collect such datasets due to privacy protection. Permission is required from multiple disciplines and an all-inclusive strategy is required to address the privacy concerns of the public.

With the growth and maturity of CNN, specifically in recognition capabilities, the trend of micro-expression is shifting from low-level handcraft features to high-level approaches. In that, sophisticated pre-processing techniques could be abolished as CNN could extract its own features, and learn and classify them automatically, making the micro-expression recognition easier as even.

Acknowledgements This research is supported by the Multimedia University internal research fund, IR Fund, under grant No. MMUI/210110.

References

1. Haggard EA, Isaacs KS (1966) Micromomentary facial expressions as indicators of ego mechanisms in psychotherapy. In: Methods of research in psychotherapy, pp 154–165
2. Ekman P, Friesen WV (1969) Nonverbal leakage and clues to deception. Psychiatry 32:88–106
3. Matsumoto D, Keltner D, Shiota MN, Frank MG, O'Sullivan M (2008) What's in a face? Facial expressions as signals of discrete emotions. In: Lewis M, Haviland JM, Feldman Barret L (eds) Handbook of emotions. Guilford Press, New York, pp 211–234

4. Danescu R, Borza D, Itu R (2018) Detecting micro-expressions in real time using high-speed video sequences. In: Intelligent video surveillance
5. Ekman P (2001) Telling lies: clues to deceit in the marketplace politics and marriage. W.W. Norton and Company
6. Shreve M, Godavarthy S, Goldgof D, Sarkar S (2011) Macro- and micro-expression spotting in long videos using spatio-temporal strain. In: 2011 IEEE international conference on automatic face & gesture recognition (FG), pp 51–56
7. Polikovsky S, Kameda Y, Ohta Y (2009) Facial micro-expressions recognition using high speed camera and 3D-gradient descriptor. In: 3rd international conference on imaging for crime detection and prevention (ICDP 2009), pp 1–6
8. Mousavi SMH, Mirinezhad SY (2021) Iranian kinect face database (IKFDB): a color-depth based face database collected by kinect v.2 sensors. SN Appl Sci 1–17
9. Vinciarelli A, Dielmann A, Favre S, Salamin H (2009) Canal9: a database of political debates for analysis of social interactions. In: International conference on affective computing and intelligent interaction and workshops, pp 1–4
10. Li X, Pfister T, Huang X, Zhao G, Pietikäinen M (2013) A spontaneous micro-expression database: inducement, collection and baseline. In: International conference and workshops on automatic face and gesture recognition (FG), pp 1–6
11. Yan WJ, Wu Q, Liu Y, Wang SJ, Fu X (2013) CASME database: a dataset of spontaneous micro-expressions collected from neutralized faces. In: 2013 10th IEEE international conference and workshops on automatic face and gesture recognition (FG), pp 1–7
12. Yan WJ, Li X, Wang SJ, Zhao G, Liu YJ, Chen YH, Fu X (2014) CASME II: an improved spontaneous micro-expression database and the baseline evaluation. PLoS ONE 9(1)
13. Qu F, Wang SJ, Yan WJ, Li H, Wu S, Fu X (2017) CAS(ME)^2: a database for spontaneous macro-expression and micro-expression spotting and recognition. IEEE Trans Affect Comput 424–436
14. Davison AK, Lansley C, Costen N, Tan K, Yap MH (2016) SAMM: a spontaneous micro-facial movement dataset. IEEE Trans Affect Comput 116–129
15. Adegun IP, Vadapalli HB (2020) Facial micro-expression recognition: a machine learning approach. Sci Afr 8
16. Duan X, Daia Q, Wang X, Wang Y, Hua Z (2016) Recognizing spontaneous micro-expression from eye region. Neurocomputing 27–36
17. Esmaeili V, Feghhi MM, Shahd SO (2021) Automatic micro-expression apex frame spotting using local binary pattern from six intersection planes. CoRR
18. Esmaeili V, Shahdi SO (2020) Automatic micro-expression apex spotting using cubic-LBP. Multimed Tools Appl 79:20221–20239
19. Liong ST, Gan YS, See J, Khor HQ, Huang YC (2019) Shallow triple stream three-dimensional CNN (STSTNet) for micro-expression recognition. In: International conference on automatic face and gesture recognition
20. Peng M, Wang C, Chen T, Liu G, Fu X (2017) Dual temporal scale convolutional neural network for micro-expression recognition. Front Psychol
21. Thuseethan S, Rajasegarar S, Yearwood J (2019) Detecting micro-expression intensity changes from videos based on hybrid deep CNN. In: Yang Q, Zhou ZH, Gong Z, Zhang ML, Huang SJ (eds) Advances in knowledge discovery and data mining. PAKDD 2019. Lecture notes in computer science, vol 11441. Springer, Cham
22. Li J, Wang Y, See J, Liu W (2019) Micro-expression recognition based on 3D flow convolutional neural network. Pattern Anal Appl 1331–1339
23. Suykens J, Vandewalle J (1999) Least squares support vector machine classifiers. Neural Process Lett 293–300
24. Takalkar MA, Xu M, Chaczko Z (2020) Manifold feature integration for micro-expression recognition. Multimed Syst 26(5):535–551
25. Pan H, Xie L, Wang Z, Lv Z, Li J (2020) Hierarchical support vector machine for facial micro-expression recognition. Multimed Tools Appl 79(3):1–15

26. Huang GB, Zhu QY, Siew CK (2004) Extreme learning machine: a new learning scheme of feedforward neural networks. In: 2004 IEEE international joint conference on neural networks (IEEE Cat. No. 04CH37541), vol 2, pp 985–990
27. Guo Y, Xue C, Wang Y, Yu M (2015) Micro-expression recognition based on CBP-TOP feature with ELM. Optik 126:4446–4451
28. Fix E, Hodges JL (1951) Discriminatory analysis: nonparametric discrimination: consistency properties
29. Happy SL, Routray A (2019) Fuzzy histogram of optical flow orientations for micro-expression recognition. IEEE Trans Affect Comput 10:394–406
30. Guo Y, Tian Y, Gao X, Zhang X (2014) Micro-expression recognition based on local binary patterns from three orthogonal planes and nearest neighbor method. In: 2014 international joint conference on neural networks (IJCNN), pp 3473–3479
31. Kang J, Chen XY, Liu QY, Jin SH, Yang CH, Hu C (2021) Research on a microexpression recognition technology based on multimodal fusion. Complexity 2021
32. Ho TK (1995) Random decision forests. In: Proceedings of the 3rd international conference on document analysis and recognition, pp 278–282
33. Cover T, Hart P (1967) Nearest neighbor pattern classification. IEEE Trans Inf Theory 21–27
34. Zhang S, Feng B, Chen Z, Huang X (2017) Micro-expression recognition by aggregating local spatio-temporal patterns. In: International conference on multimedia modeling, pp 638–648
35. Ibrahim AF, Ristiawanto SP, Setianingsih C, Irawan B (2021) Micro-expression recognition using VGG19 convolutional neural network architecture and random forest. In: 2021 4th international symposium on agents, multi-agent systems and robotics (ISAMSR), pp 150–156
36. Donoho D, Elad M (2003) Optimally sparse representation in general (nonorthogonal) dictionaries via L1 minimization. Proc Natl Acad Sci 2197–2202
37. Plenge E, Klei SS, Niessen WJ, Meijering E (2015) Multiple sparse representations classification. PLoS ONE
38. Hao Z (2017) Micro-expression recognition based on 2D gabor filter and sparse representation. J Phys Conf Ser 787
39. Liu KH, Jin QS, Xu HC, Gan YS, Liong ST (2021) Micro-expression recognition using advanced genetic algorithm. Signal Process Image Commun 93
40. Xia Z, Hong X, Gao X, Feng X, Zhao G (2020) Spatiotemporal recurrent convolutional networks for recognizing spontaneous micro-expressions. IEEE Trans Multimed 22
41. Lo L, Xie HX, Shuai HH, Cheng WH (2020) MER-GCN: micro-expression recognition based on relation modeling with graph convolutional networks. In: 2020 IEEE conference on multimedia information processing and retrieval, pp 79–84
42. Xie HX, Lo L, Shuai HH, Cheng WH (2020) AU-assisted graph attention convolutional network for micro-expression recognition. In: Proceedings of the 28th ACM international conference, pp 2871–2880
43. Tang H, Chai L, Lu W (2022) Transferring dual stochastic graph convolutional network for facial micro-expression recognition

Development of 'SiPantik' for the Teachers' Test Items Production

Zyaliah Maya Ramdan, Aslina Baharum, and Christina Andin

Abstract The fulfillment of the intended learning outcomes depends on producing test items that adhere to the principles outlined in Principle 9-Points. In addition, the necessity of follow-up activities to optimize students' abilities because of the effective test item measures cannot be underestimated. However, there is currently no system that offers teachers systematic direction for producing test items that adhere to the principles of item construction and item evaluation based on the principle 9-Points. Consequently, the primary purpose of this study is to establish a systematic guidance system that can generate test items that adhere to the principle 9-Points. System Development Life Cycle adapted Waterfall method used for guiding the development of the Systematic Guidance System, namely SiPantik. With SiPantik, it is certain that teachers will be able to generate effective test items efficiently and effectively. Effective test items can measure student achievement and enable teachers to arrange suitable follow-up actions based on student capacities. Finally, compliance with aspects of Pupil Assessment Management as required in Malaysian Education Quality Standards Wave 2 (SKPMg2) can be realized.

Keywords Test item · Principle 9-Points · Systematic guidance system · Effective

---

Z. M. Ramdan
Sabah State Education Department, Ministry of Education Malaysia, Kota Kinabalu, Malaysia
e-mail: zyaliah_maya_mp21@iluv.ums.edu.my

A. Baharum (✉)
Faculty of Computer and Mathematical Sciences, Universiti Teknologi MARA (UiTM), Jasin Campus, Melaka, Malaysia
e-mail: aslinabaharum@uitm.edu.my

C. Andin
Faculty of Psychology and Education, Universiti Malaysia Sabah, Kota Kinabalu, Malaysia
e-mail: christina@ums.edu.my

© The Author(s), under exclusive license to Springer Nature Singapore Pte Ltd. 2023

D.-K. Kang et al. (eds.), *Proceedings of the 9th International Conference on Computational Science and Technology*, Lecture Notes in Electrical Engineering 983,
https://doi.org/10.1007/978-981-19-8406-8_9

1 Introduction

Test items are crucial to assessment success. Therefore, it becomes a necessity to produce effective test items. The production of test items using computer technology has been a concern for the past four decades [1, 2]. Automated Item Generation (AIG) is believed to be efficient and effective in test development [3, 4]. Two groups comprise the AIG model: the special model and the framework model [3]. The AIG framework concept provides a foundation for algorithmically producing items. Previously, the AIG framework [3] was separated into two techniques: automated [4–6] and semi-automatic [7, 8]. The schemas and concepts provided by the two frameworks represent their distinction. Special Model AIG also uses unique development as an alternative to the basic framework [9, 10]. Within the item generating infrastructure, this model differs from the framework model in structure. This model is used to manufacture products based on certain materials and processes. This approach solves problems that cannot be solved with the AIG framework, or the data set provided. In this study, the item production focuses more on the item construction and item scoring guidelines based on the principle 9-Points (P9P) described by [11]. The Waterfall model was adapted to develop a system that can facilitate teachers to produce items that meet the characteristics found in P9P. The main goal of this study was to come up with a set of guidelines for making test items. By combining the process of item production and the development of this system, the test items produced by the teacher are certainly effective test items that meet the procedures required.

2 Literature Review

2.1 The Concept of Producing Test Items

Teachers must apply the principles and techniques of formulating and preparing items so that the objective and subjective tests are of good quality [12]. From [11] shows that the production of test items needs to go through two processes: the item construction process and the item evaluation process. The production of practical test items should adhere to systematic guidelines. The study of [13, 14] show similarities in the step of producing items, i.e., by emphasizing planning before item construction and then making an item review. This indicates that the process of producing items that follow the guidelines will produce quality test items. There are guidelines and P9P in the [11, 15], pages 79–83, as a guide to the test item production process.

Item Construction Guidelines. Figure 1 shows the workflow diagram for the item construction procedure. In item construction, the creation of a Table of Test Specification (ToTs) [16] will be followed by a Table of Item Specification (ToIS). After ensuring that the context, construct, and level of difficulty of the ToIS items are aligned with the ToTS, the developer will proceed to write the items. Before beginning to write an item, the developer must first determine whether the item's type is objective or subjective, then determine the type of evidence (facts, concepts, generalizations, judging, or rational abilities), and then design context-related stimuli to measure constructs (text, tables, or graphs). Next, the developer writes the item's stimuli, verbs, and appropriate instructions to encourage students to respond as intended. Finally, the developer writes the item's answer choices or scoring guidelines. The workflow chart of the item construction process does not clearly show the use of ToIS during item construction. Still, in the Malaysian Examinations Board release letter in 2019, attached is an Item Construction Specification Form (also known as ToIS), as shown in Fig. 2. This indicates that the constructed item must comply with the specifications in the form.

Item Evaluation Based on the Principle 9-Point. The item evaluation process is evaluating an item using the nine aspects found in P9P, as in Fig. 3. An item evaluation process must be implemented to validate test items before assessing students. [13] stated that it is important to evaluate the item to reach the desired level. Supported by [16], the item evaluation stage serves to reject or improve items and check for technical errors. Under the requirements of the [11], the item evaluation process ends

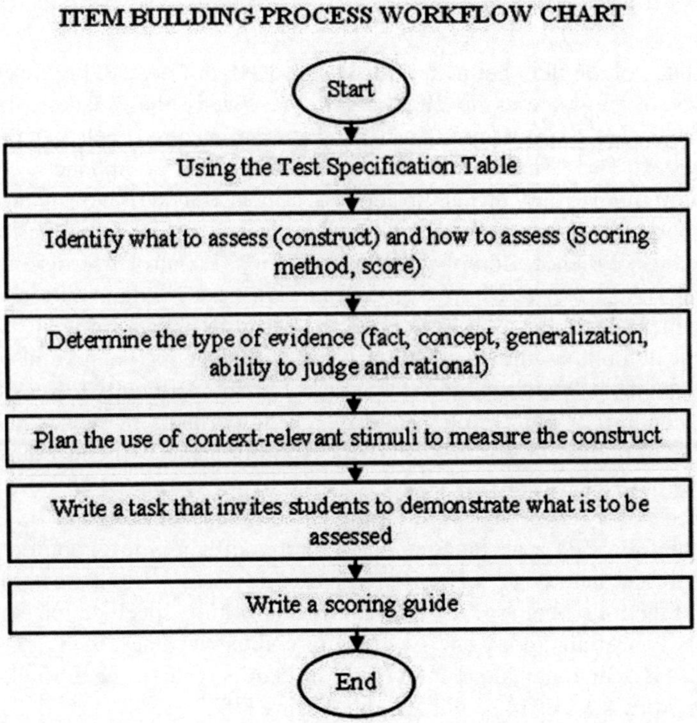

Fig. 1 Item construction process workflow chart

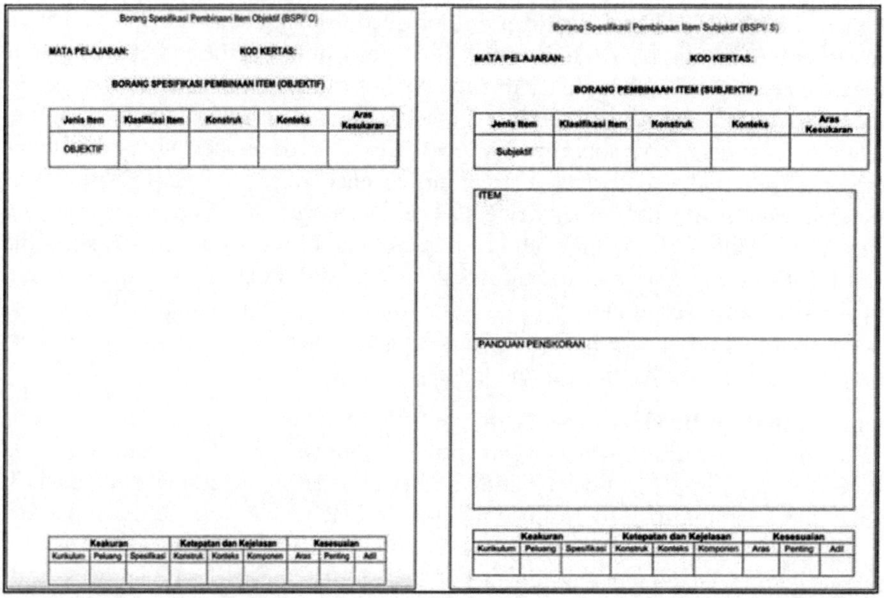
Borang Spesifikasi Pembinaan Item Objektif (BSPI/ O)

MATA PELAJARAN: KOD KERTAS:

BORANG SPESIFIKASI PEMBINAAN ITEM (OBJEKTIF)

Jenis Item	Klasifikasi Item	Konstruk	Konteks	Aras Kesukaran
OBJEKTIF				

Keakuran			Ketepatan dan Kejelasan			Kesesuaian		
Kurikulum	Peluang	Spesifikasi	Konstruk	Konteks	Komponen	Aras	Penting	Adil

Borang Spesifikasi Pembinaan Item Subjektif (BSPI/ S)

MATA PELAJARAN: KOD KERTAS:

BORANG PEMBINAAN ITEM (SUBJEKTIF)

Jenis Item	Klasifikasi Item	Konstruk	Konteks	Aras Kesukaran
Subjektif				

ITEM

PANDUAN PENSKORAN

Keakuran			Ketepatan dan Kejelasan			Kesesuaian		
Kurikulum	Peluang	Spesifikasi	Konstruk	Konteks	Komponen	Aras	Penting	Adil

Fig. 2 Item construction specification form

with the status of the item being accepted, improved, or rejected. Figure 3 shows a process flow of the process of evaluating an item. Briefly, the P9P consists of three domains: consistency, accuracy or clarity, and appropriateness. Each domain consists of three aspects. The Domain of Conformity is Curriculum Compliance - complying with the curriculum in the Curriculum and Assessment Standards Document (DSKP); Accept Opportunity - adequate equipment, teacher attendance, teacher expertise, actual student experience; Comply with Specifications - control in terms of relevance and coverage. Accuracy or Clarity is Accurate Construct - assignments in items can indicate evidence that can assess the expected construct; Exact Context - according to the context found in the curriculum; Clear Components - items must be clear and understandable. The Suitability Domain is Suitable Difficulty Level - fits items to a low, medium, or high level; Appropriately Important - the ideas and content assessed are important in learning; and Fair Appropriateness - items do not benefit certain categories such as race, gender, economic status, and place of residence. Items are accepted if they meet all P9Ps. At the same time, items need to be improved if one of the following aspects: component clarity, difficulty level appropriateness, essential appropriateness, or fair appropriateness are not met. But if the item does not meet one of the following aspects: curriculum conformity, specification conformity, opportunity conformity, construct accuracy or context accuracy, then items need to be rejected [15]. In conclusion, every time an item is ready to be drafted, the item evaluation process needs to be implemented using P9P.

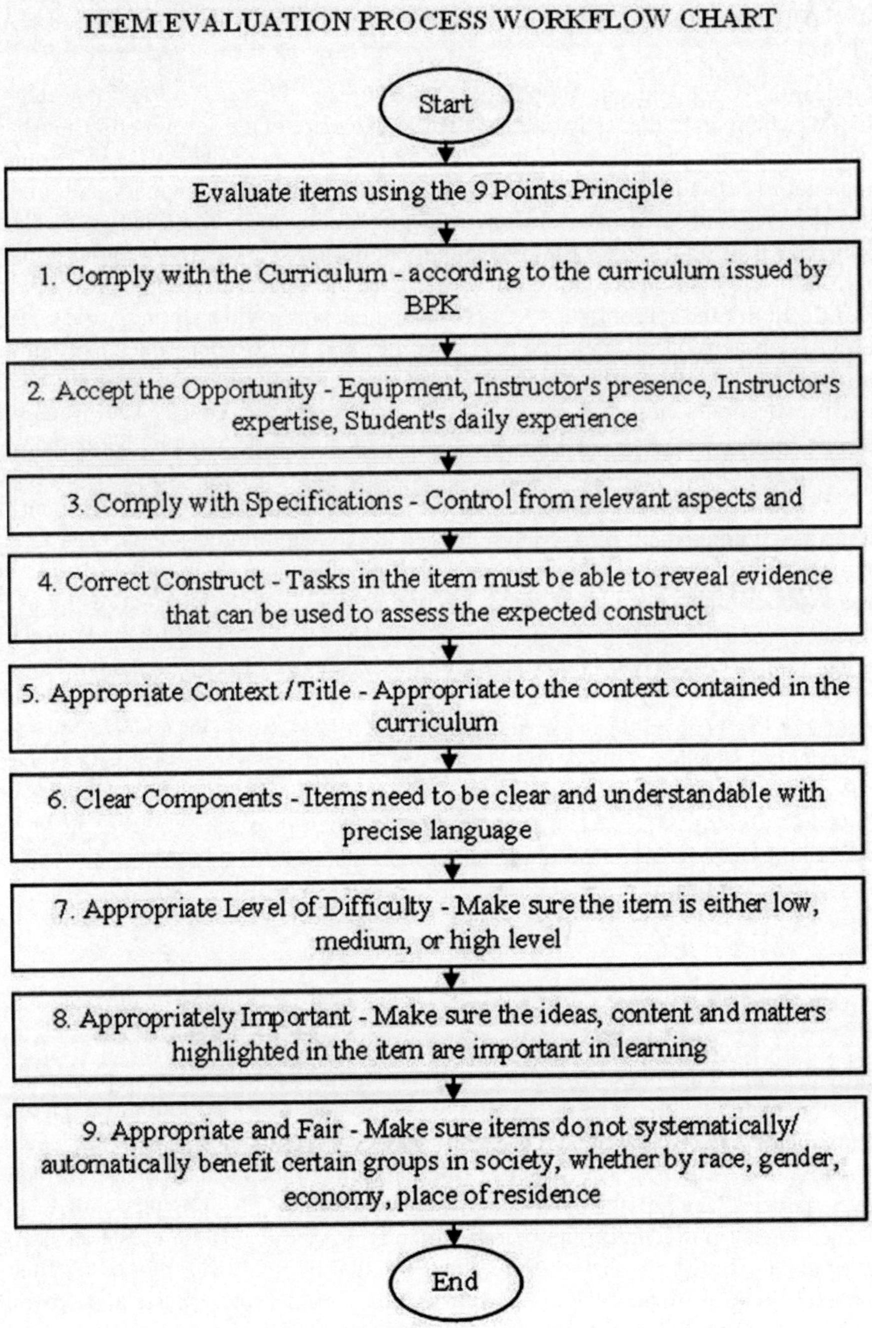

Fig. 3 Item evaluation process workflow chart

2.2 System Development Life Cycle

Application systems are developed according to specific needs to facilitate daily tasks. The findings of the study indicate that the existence of the developed system has made examination management more efficient [17], saving teachers time in learning preparation [18] and facilitated teachers to check answers and assign marks automatically [19]. This indicates that each system developed aims to facilitate daily work. The study of [20] stated that the system development life cycle (SDLC) is a form of organized and systematic methodology in describing the system development process. The selection of the appropriate system development phase will have an impact on the developed system [20]. Reference [21] identifies two SDLC approaches: predictive and adaptive. Well-understood, low-risk projects utilize a predictive strategy, while poorly understood, high-risk initiatives use an adaptive approach. The Waterfall Model is predictive, while the Iterative, Increment, and Agile Models are adaptable. The models differ in approach, but the purpose is to construct a system. Therefore, it can be stated that the Waterfall Model can be applied to system development if all needs are known and fully comprehended at the beginning of the project. Even more so, the Waterfall Model focuses on the development process based on the initial challenges that have been recognized. In contrast, the Agile Model focuses on user feedback throughout the system development process [22]. As a result, the Waterfall Paradigm is an easy-to-use and effective model [23–25]. In 1970 Winston W. Royce proposed the Waterfall model that is now regarded as the "traditional life cycle" because it has been utilized for many years field [26–28]. Plan, analyze, design, construct, test, evaluate, and document are the seven-phase in establishing a system according to the modified Waterfall model studied by [29, 30]. But there are just five phases in the study of [23, 31] that are included in their research. However, it shows that planning, analyzing, designing, and creating are the most important processes in system development.

3 Methodology

The Waterfall model phases were adapted in combination with the item production standards from [11] to make the SiPantik development process more thorough and comprehensive. This agrees with prior studies that explain the Waterfall model can be adjusted [23, 28–30]. The Waterfall-SiPantik model includes the phase of planning, analyzing, designing, constructing, and testing. In the planning phase, there are two activities; (a) to identify the item production process and (b) to create a system development strategy by listing the system's goals. In Analyzing phase, the functional requirements of the designed system are considered. Prior research and existing practice are examined during this phase to gather specific information. In the third phase, all requirements derived from the analysis phase are transformed into a design for system development, which consists of five activities: (a) flowcharts design, (b) process flow diagrams design, (c) entity-relationship diagrams design, (d) functional hierarchy diagrams design, and (e) user interface design. All system designs will be transformed into actual systems in the construction phase. The SiPantik construction process comprises two primary activities: (a) interface construction and (b) code construction utilizing VBA Microsoft Excel. Also, integrating the required guidelines and P9P by [11]. In this testing phase, the purpose is to guarantee the system runs smoothly. Three main activities in the final phase; (a) developing the manual handbook, (b) evaluation by the expert, and (c) enhancements after expert recommendations.

Flowchart. Flowcharts employ graphic symbols to depict algorithm instructions, as depicted in Fig. 4 Developer Flow Diagram and Assessor Flow Diagram. As a developer, after entering the password, ToTS need to be planned by choosing the right context, construct, and difficulty level for each item to be created. The next step is for the developer using the ToIS to select an item classification based on the construct's appropriateness. Developers then begin to write stimulus components, task terms or instructions, and answer options. Once the item is drafted, the item needs to go through a review process. The assessor enters a password and selects the items to be reviewed for evaluation. All aspects of the P9P will be used to evaluate each item. An item's status is either approved, improved, or rejected at the end of the evaluation process.

Functional Hierarchy Diagram. The Functional Hierarchy diagram illustrates system navigation visually. It also shows how the system works as a whole and makes sure that development is consistent and meets user needs. SiPantik consists of three functionalities. First: Home; intended for user selection. Second: the developer; responsible for planning the ToTS and constructing items in ToIS; and third:

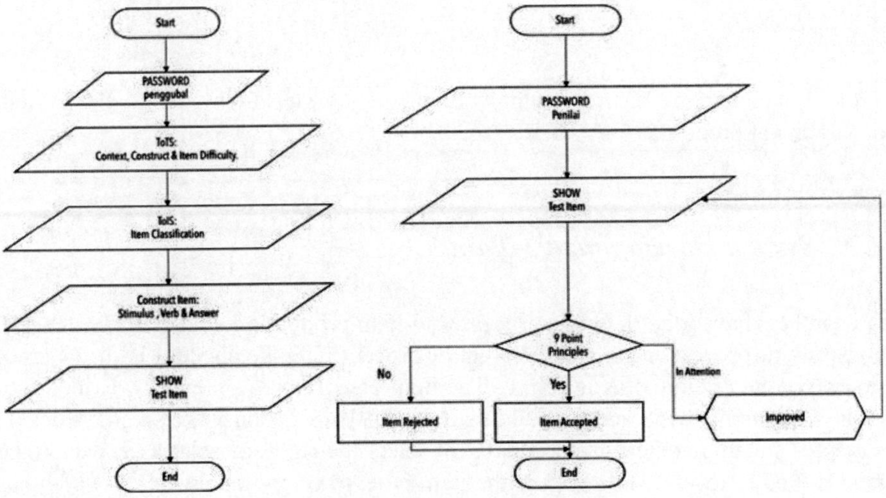

Fig. 4 Developer flow diagram and assessor flow diagram

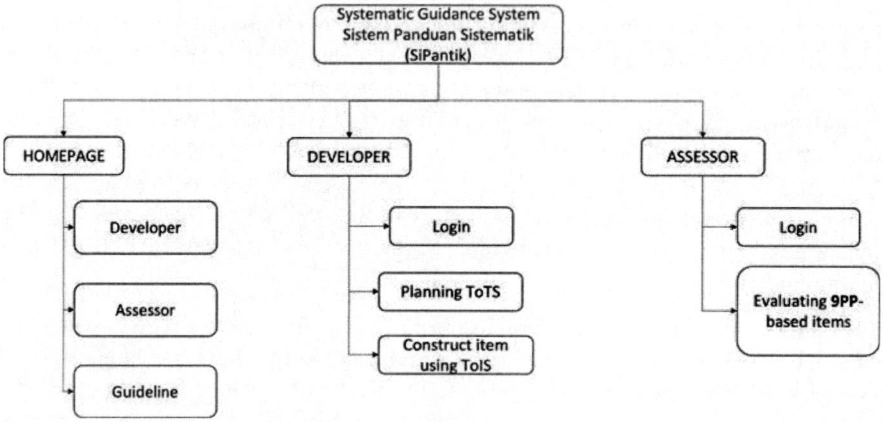

Fig. 5 Functional diagram hierarchy of SiPantik

the assessor; responsible for evaluating P9P-based items. This function is shown in Fig. 5.

4 Finding and Discussion

This section will detail the system that has been built to meet the Malaysian Examinations Board's recommendations.

4.1 Result

Table 1 summarizes the results of adopting the Waterfall model to the system development process.

4.2 System Development SiPantik

Researchers have identified two processes in item production [11], namely the item construction process and the item evaluation process. The sketch of the item construction process in detail is as in Fig. 6. The item construction process is divided into two; (a) planning ToTS and (b) building items on ToIS. It consists of eight steps; (1) choose a suitable context from DSKP, (2) select the right construct, (3) decide the item difficulty level, (4) determine the item type, (5) copy the construct which has been selected from ToTS, (6) copy the context which has been chosen from ToTS,

Table 1 Summary of system development process

Step	Activities	Result
Planning	Recognizing item production process	Item construction process
		Item evaluation process
	List of system goals	Designing ToTS
		Using ToIS build test items
		Evaluate test items based on P9P
Analyzing	Item construction process flowchart	Figure 6
Designing	Flowcharts	Figure 4
	Functional hierarchy diagrams	Figure 5
Constructing	Homepage, login, guideline and notes page	Figure 7
	ToTS	Figure 8
	ToIS	Figure 9
	9PP	Figure 10
Testing	Developing the manual handbook	https://anyflip.com/afrlr/mbtu/
	Evaluation by the expert	–

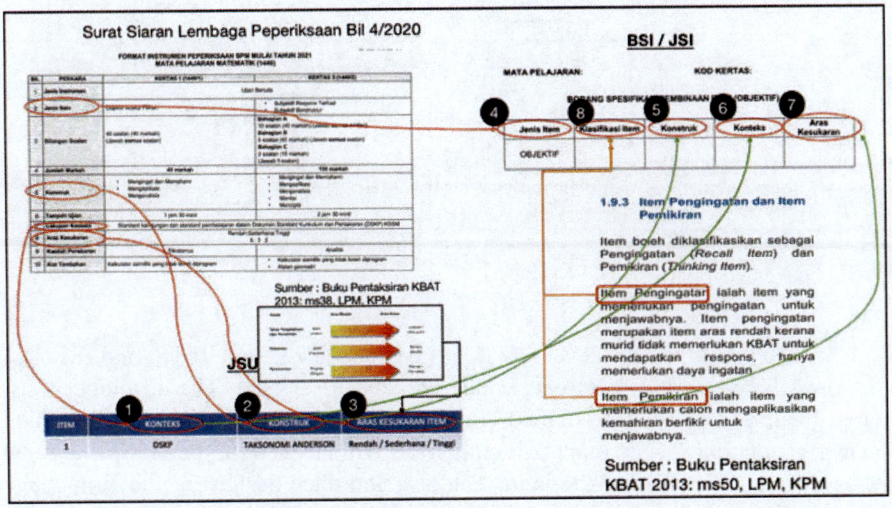

Fig. 6 Item construction process flow

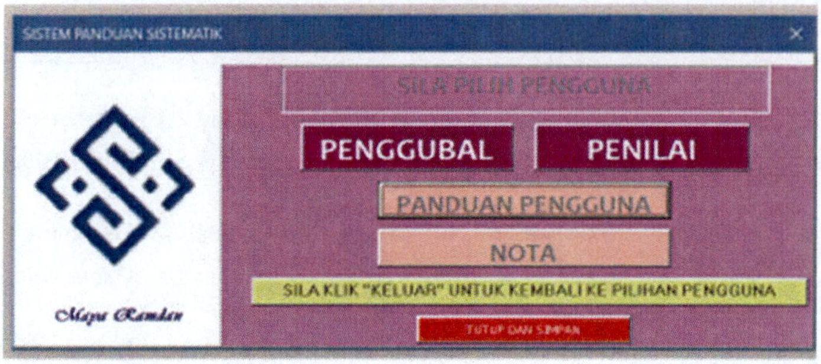

Fig. 7 Homepage user interface

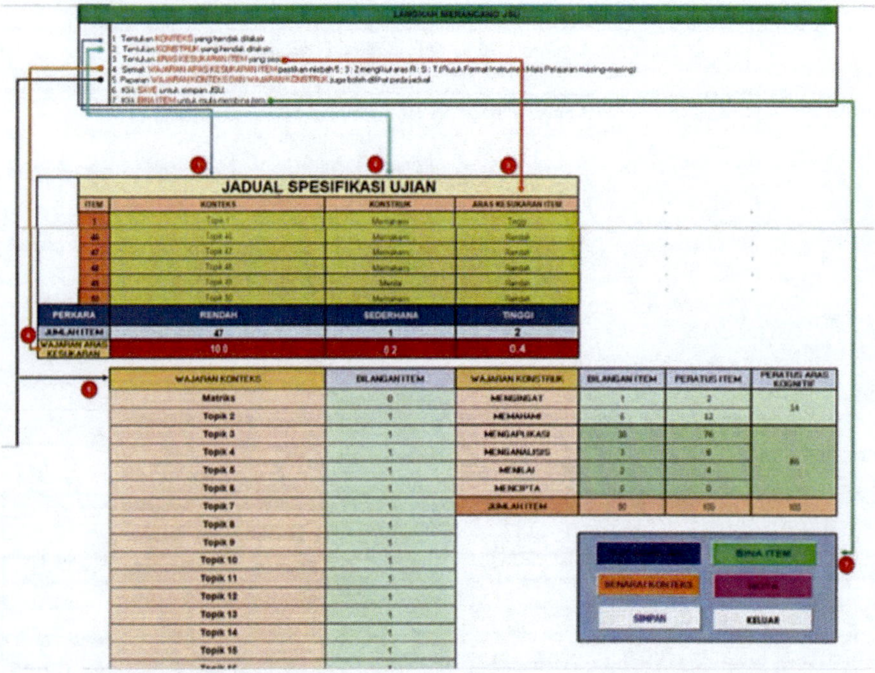

Fig. 8 Table of test specification (developer)

(7) copy the item difficulty level which has been decided from ToTS, and (8) classify items based on the constructs which have been selected. The developer must comply with every aspect contained in the [32, 33] when planning the ToTS. Then the aspects that have been planned in the ToTS will be used on the ToIS as a guide to the developer in writing the item. The user interface design for the Homepage shown in Fig. 7 displays a button for the user to key in the password field and refer to the guideline and notes before starting to use the system. In ToTS, after entering

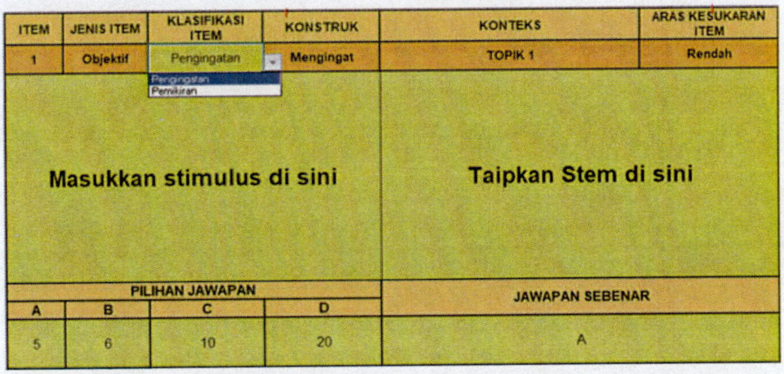

Fig. 9 Table of item specification (developer)

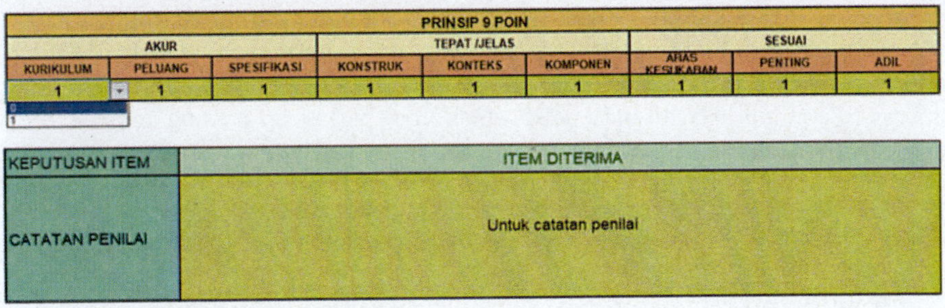

Fig. 10 Assessor form

the password, the developer will begin to design the ToTS, as shown in Fig. 8, by choosing the item’s construct, context, and difficulty level. Then, the weight of the item difficulty level will be checked to suit the format of [32]. In addition, this system is also used to balance the number of contexts and constructs according to suitability. Next, the developer will write the items in the ToIS, as shown in Fig. 9, according to the item components: stimulus, stem, answer choice and answer key. The last step of the system development is to make an item evaluation using the P9P in the evaluation form shown in Fig. 10. At this site, the assessor needs to evaluate professionally one item at a time based on each aspect found in P9P. The assessor will give a score of 1 if the item conforms to the aspect and a score of 0 if the item does not conform to the aspect. The evaluation process will end with the item being accepted, rejected, or improved. The evaluator should also write a brief note as a reference to the developer on the items that need to be improved.

4.3 Discussion

This study's significant findings guide the production of structured and systematic items. Thus, this manual can be automatically adapted to the system that has been developed to speed up the production process. In addition to the rapid production of producible items, test items also acquire all the P9P characteristics. Student progress can be measured against learning objectives effectively with P9P-compliant test items. As a result, learning activities aligned with the course's goals can be arranged effectively. According to the findings [3], expert manual validation was recommended for items with quality or validity issues. Item evaluation using the 9-point approach is simplified and applied using SiPantik in response to a recommendation [3]. Individuals who use SiPantik should also be able to improve their skills in making test items. At the same time, because verbs are used so often in constructs, they will be good at giving classroom tests. This study has some limitations that should be considered. The weakness of this study is that to make this item, and the user must still come up with the idea of writing it. An idea is a thought or suggestion about something that could be achieved. When writing an item, ideas do not come up on their own. The process of coming up with these ideas should be sparked by reading and attending conferences or seminars. SiPantik's effectiveness on teachers who create summative assessment items could be studied in further research.

5 Conclusion

In conclusion, using SiPantik, this study describes the design and development of a P9P-based guideline item structure. Teacher-friendly tools that meet the requirements of the P9P have been developed because of this effort. Because of this adaptation of the Waterfall model, this guidance system has successfully provided test items. Future works will continue to evaluate the developed SiPantik in terms of efficiency and effectiveness in producing test items among teachers. The result from this study will help educators improve the quality of current test items and provide follow-up learning activities that are more specific to the context, construct, and ability level.

References

1. Fremer J, Anastasio EJ (1969) Computer-assisted item writing—1 (spelling items) 1. J Educ Meas 6(2):69–74
2. Anastasio EJ (1969) Computer-Assisted Item Writing—II (Sentence Completion Items)
3. Prasetyo SE, Adji TB, Hidayah I (2020) Automated item generation: model and development technique. In: 2020 7th international conference on information technology, computer, and electrical engineering (ICITACEE). IEEE, pp 64–69
4. Rafatbakhsh E, Ahmadi A, Moloodi A, Mehrpour S (2021) Development and validation of an automatic item generation system for English idioms. Educ Meas Issues Pract 40(2):49–59

5. Liu M, Calvo RA, Aditomo A, Pizzato LA (2012) Using Wikipedia and conceptual graph structures to generate questions for academic writing support. IEEE Trans Learn Technol 5(3):251–263
6. Shin E (2021) Automated item generation by combining the non-template and template-based approaches to generate reading inference test items
7. Gierl MJ, Lai H, Turner SR (2012) Using automatic item generation to create multiple-choice test items. Med Educ 46(8):757–765
8. Prasetyanto AAB, Adji TB, Hidayah I (2020) Automatic question generator system conceptual model for mathematic and geometry parallel question replication. J Phys: Conf Ser 1577(1):012023
9. Liu M, Rus V, Liu L (2017) Automatic Chinese multiple choice question generation using mixed similarity strategy. IEEE Trans Learn Technol 11(2):193–202
10. von Davier M (2018) Automated item generation with recurrent neural networks. Psychometrika 83(4):847–857
11. Lembaga Peperiksaan Malaysia (2019) Surat Siaran Lembaga Peperiksaan Bil 1/2019 Pemerkasaan Pentaksiran Tingkatan 3(PT3) Di Sekolah Mulai Tahun 2020
12. Izard J (2005) Overview of test construction. KN Ross. Quantitative research methods in educational planning. UNESCO International Institute for Educational Planning, Paris
13. Adnan M, Nawi NM, Abdullah MFNL, Ahmad CNC, Arifin NS (2018) Pembinaan item Kemahiran Berfikir Aras Tinggi Matematik Tingkatan Satu untuk topik Pecahan. Jurnal Pendidikan Sains Dan Matematik Malaysia 8(1):46–54
14. Amoako FQI. Attitude of senior high school teachers toward test construction: developing and validating a standardised instrument
15. Lembaga Peperiksaan Malaysia (2013) Pentaksiran Kemahiran Berfikir Aras Tinggi
16. Adom D, Mensah JA, Dake DA (2020) Test, measurement, and evaluation: understanding and use of the concepts in education. Int J Eval Res Educ 9(1):109–119
17. Othman J, Rosly WNSWM, Samsudin N, Abdullah SSS, Mazeni NA, Hamat M, Zahari MF, Hasan S, Daud N, Kassim AM (2020) Penyata peperiksaan komprehensif (ComprESs): ke arah pengurusan peperiksaan yang efisien. Jurnal Inovasi Malaysia 3(2)
18. Sutisna MR, Yuniarti Y, Windayana H, Hendriyana H (2021) Design of a Web-Based Digital Learning Resource Center to assist online learning with mathematics content in primary schools. J Phys: Conf Ser 1987(1):012005
19. Marifati IS, Mulyanto JD, Kesuma C, Ubaidillah U, Suripah S (2021) Computer-based test application to support e-learning in pandemic situation. PIKSEL: Penelitian Ilmu Komputer Sistem Embedded and Logic 9(2):185–196
20. Silitonga PD, Purba DER (2021) Implementasi System Development Life Cycle Pada Rancang Bangun Sistem Pendaftaran Pasien Berbasis Web. Jurnal Sistem Informasi Kaputama (JSIK) 5(2):196–203
21. Ding SX, Jeinsch T, Frank PM, Ding EL (2000) A unified approach to the optimization of fault detection systems. Int J Adapt Control Signal Process 14(7):725–745
22. Fagarasan C, Popa O, Pisla A, Cristea C (2021) Agile, waterfall and iterative approach in information technology projects. IOP Conf Ser: Mater Sci Eng 1169(1):012025
23. Rahayu TK (2020) Application report process of Islamic school based on Pesantren boarding using waterfall model. J Phys: Conf Ser 1569(2):022025
24. Nasir FM, Baharuddin FW, Nizar FM, Hazli N (2020) Pembangunan Sistem Pengurusan Program Secara Dalam Talian. Multidiscip Appl Res Innov 1(1):21–29
25. Anam K, Asyhar B, Saddhono K, Setyawan BW (2021) E-SIP: website-based scheduling information system to increase the effectivity of lecturer's performance and learning process. Ingénierie des Systèmes d'Information 26(3):265–273
26. Santoso MA, Fiati R, Nindyasari R (2021) The system development life cycle model implementation on information system of performance reporting IT asset case study: PT Kereta Api Indonesia (Persero). Matrix: Jurnal Manajemen Teknologi dan Informatika 11(2):95–106
27. Munassar NMA, Govardhan A (2010) A comparison between five models of software engineering. Int J Comput Sci Issues (IJCSI) 7(5):94

28. Krishna ST, Sreekanth S, Perumal K, Kumar Reddy KR (2012) Explore 10 different types of software development process models. Int J Comput Sci Inf Technol 3(4):4580–4584
29. Taufik D (2020) PDCA cycle method implementation in industries: a systematic. IJIEM (Indones J Ind Eng Manag) 1:157–166
30. Abdul Hamid NHA, Mustopa NN (2020) PRESIMS: requirements and design for preschool management system using usability theory. ESTEEM Acad J 16:51–64
31. Saputro H, Wahyudin I, Iskandar A (2020) Development of a web-based online library information system at the national university using the waterfall method. Jurnal Mantik 4(1):300–310
32. Lembaga Peperiksaan Malaysia (2020) Surat Siaran Lembaga Peperiksaan Bil. 4 Tahun 2020 Format Instrumen Peperiksaan Sijil Pelajaran Malaysia (SPM) Dan Ujian Pencapaian Bahasa Antarabangsa (UPBA) Mulai Tahun 2021 Bagi Mata Pelajaran Akademik Kurikulum Standard Sekolah Menengah (KSSM)
33. Bahagian Pembangunan Kurikulum (2018) KSSM Matematik Tingkatan 4 dan 5

Face Illumination Reduction Using MADPIP Restoration Approach to Biometric Patient Authentication System

Rayner Henry Pailus, Rayner Alfred, Joe Henry Obit, Aswami Fadillah bin Mohd Ariffin, Mohammad Zaharudin bin Ahmad Darus, and Nazri Ahmad Zamani

Abstract Despite of the rapid development of computer vision technology and the widely advancement of digital image processing systems used in many fields, some uncontrollable factors often exist during the process of image acquisition, resulting in various image defects. In particular, under poor illumination conditions, such as indoors, night-time, or cloudy days, the light reflected from the object surface may be weak; consequently, the image quality of such a low-light image may be seriously degraded due to colour distortions and noise. Reducing illumination by various techniques such as edges extraction for example the High Pass Filter is only good in performing one of the image restoration tasks, which is sharpening adjoining areas with slight difference in the brightness. However, in order to produce a complete restoration for a very low level of illuminance image, restoration process has to have a balance process of blurring and sharpening processes. The Low-Pass Filter is one of blurring filter methods that can be used to blur images. In order to find the best hybrid of image restoration algorithm, several combinations of blurring and sharpening processes have to be conducted. For that reason, this paper proposes a new method called the Multiple Adaptive Derivatives for Passive Image Processing (MADPIP) which is a module that performs a series of synergy algorithm combinations in restoring illuminated images by reducing illumination noise and revealing the facial outlines and facial features. In this work, the MADPIP global and local features extraction algorithms were investigated. It was concluded that the Zero Padding

R. H. Pailus · R. Alfred (✉) · J. H. Obit
Creative Advanced Machine Intelligence Research Centre, Faculty of Computing and Informatics, Universiti Malaysia Sabah, Kota Kinabalu, Sabah, Malaysia
e-mail: ralfred@ums.edu.my

J. H. Obit
e-mail: joehenry@ums.edu.my

A. F. M. Ariffin · M. Z. A. Darus · N. A. Zamani
CyberSecurity Malaysia, Cyberjaya, Selangor Darul Ehsan, Malaysia
e-mail: asmawi@cybersecurity.my

M. Z. A. Darus
e-mail: zaharudin@cybersecurity.my

N. A. Zamani
e-mail: nazri.az@cybersecurity.my

© The Author(s), under exclusive license to Springer Nature Singapore Pte Ltd. 2023
D.-K. Kang et al. (eds.), *Proceedings of the 9th International Conference on Computational Science and Technology*, Lecture Notes in Electrical Engineering 983,
https://doi.org/10.1007/978-981-19-8406-8_10

+ Laplacian Masked and Border Replication + Low Pass Filter (ZPL + BRLPF) combination has the best score for the MSE assessment and thus best algorithm for global feature extraction. On the other hand, for local features extraction, the 2L derivative with 3 colour Logarithmic and HPF-Unsharp Masking with 8 scaling factors has produced the best result in retrieving the local features extraction.

Keywords Face illumination reduction · Image restoration · Biometric patient authentication · Global features · Local features

1 Introduction

Some uncontrollable factors often exist during the process of image acquisition, resulting in various image defects and thus affecting the quality of face recognition or face classification tasks. In particular, under poor illumination conditions, such as indoors, night-time, or cloudy days, the light reflected from the object surface may be weak; consequently, the image quality of such a low-light image may be seriously degraded due to colour distortions and noise [46]. As a result, one may require to have a robust algorithm that is able to acquire image under Very Low-Light (VLL), Low-Light (LL) and Mild Illuminance (MI) environments in which these environments only allow one to capture dark face images or half dark face images and face images with various mild disoriented lighting respectively.

The robustness and adaptive capabilities of the CNN based methods are insufficient to meet the requirements of practical applications because most current trend of low-light image enhancement methods apply corrective adjustment during specific testing environment only [4]. As a result, if the same method is used in different environment with different degree of brightness, the CNN based methods must be re-trained again. The ideal method should be able to adaptively adjust to different application conditions and different types of low-light images. Thus, the ideal method should be able to deal with illumination without dependent solely on training and testing data processes and less preprocessing is required to improve the source in the initial stage [4].

One of the passive illumination reduction approaches is to reduce illumination by applying kernel convolution operation towards raw digital images or visual images. Raw images were taken from ordinary color visual digital camera or also known as visual spectra. Those images taken using visual spectra are very dependent on luminance or lighting or brightness of the surrounding environment. Global features of face image are features obtained from the entire face image, whereas local features describe the textures of facial components like eyebrows, eyes, nose, top and bottom lips, cheek, chin and forehead.

Since the beginning of face recognition research started, many state-of-the-art algorithms are not robust enough to deal with datasets with large range of illumination weights differences or very low light illumination [13, 29]. One of the illuminated

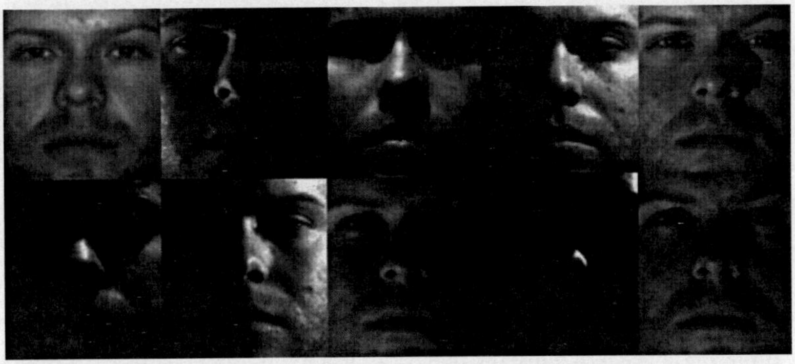

Fig. 1 Sample of images with invariant illumination of extended Yale B dataset

invariant datasets which has 64 classes of illumination is the Extended Yale 4 dataset which has the highest number of illumination classification [39].

As seen in Fig. 1, most of these images do not look like they are from the same person or from the same class due to illumination. Therefore, before these images can be used for feature extraction for the purpose of face recognition, those face images need to go through several processes of restoration to reveal those images face main structures, edges, and features by eliminating unnecessary noise. Face structures and edges are the facial outlines and facial features are the facial components, like eyes, nose, and mouth.

The aim of this paper is to propose and investigate a robust method that combines multiple adaptive algorithms which is able to reduce illumination of acquired images using the MADPIP approach through restoration and enhancement processes to produce derivatives with the lowest value for the Mean Square Error (MSE) [31], Euclidean distance results (ED) [24] or to produce derivatives with the highest value for the Structural Similarity Index (SSIM) [34], Peak Signal-to-Noise Ratio (PSNR) [1]. The proposed Multiple Adaptive Derivatives for Passive Image Processing (MADPIP) is a module that performs a series of synergy algorithm combinations in restoring illuminated images by reducing illumination noise and revealing the facial outlines and facial features. In addition to that, the proposed method also is able to improve the prediction tasks of Support Vector Machines (SVM) [25] and K-Fold-Cross Validation selection with Random Forest (CV) results [18] and getting the highest accuracy of Mean Successful Recognition Rate (MSRR). Section 2 provides the detail of the literature review or related works. Section 3 presents the proposed approach, Sect. 4 presents the results and discussion and finally Sect. 5 concludes this paper.

2 Related Works

For face recognition with illumination variation, two main categories of approaches have been introduced: (i) those methods that attempt to model the appearance variation caused by illumination, and (ii) those methods that attempt to build an invariant representation of the face. The CNN based model is the best example of modelling the appearance variation caused by illumination [40]. Modelling illumination solution is one of the most tedious processes because multiple images are required to cumulate all the possible changes of illumination, one missing illumination model can create a huge error of misrepresentation recognition. On the other hand, the Retinex illumination model is an example of the second method that build an illumination-invariant representation of the face which allows a single training image to be used for each person [26]. Retinex algorithms [22] separate reflectance, from the illumination, and this can enhance the image details. The Retinex illumination model explains the observed brightness at each point on the face in terms of the intrinsic reflectivity at that point, together with the magnitude and angle of the incident illumination. Given this model, it is commonly assumed that the low spatial frequency components of the face image represent the illumination information while the high spatial frequency components represent the intrinsic face reflectivity, that should be recovered for recognition.

2.1 Low Light Illumination Invariant Method

Solving low-light illumination has gained much attention from the research groups of pattern recognition and machine learning since the early 1990s [15]. Face recognition is a difficult task due to the issues with low-light illumination variance which affects the identification rate [15]. In the beginning, Principal Component Analysis (PCA) [36] was used to minimise low light illumination but eigen-features failed to classify between low light images and high illumination image of the same class. A new framework for face recognition and feature extraction is proposed with kernel fisher discriminant analysis and fisher Linear Discriminant Analysis (LDA) [45], however instead of minimising illumination, LDA was used to connect low light illuminated image with high illumination images and found out that more of the same low-light illumination of different class being misclassification.

The illumination varies while capturing the face photographs, even when in outdoor environment while creating person's identity in real-time, the pixel representation varies from low-light illumination to high illumination and may come in a split seconds without consciously realizing their changes [44]. Face Recognition in low lighting conditions using Fisherface [5] method and CLAHE [28] was proposed for face recognition using Rahman et al. [33]. CLAHE alone was not able to reduce illumination in fact created additional noise, as a result Fisherface's eigenfaces are still affected by illumination and for 1300 test images only 57.07% recognition accuracy for AR, Yale and ORL datasets [33].

Illumination is one of the main challenges of face detection. Poor illumination may cause dark images with no face image to be detected. For a non-robust face recognition system, it will stop here. Convolution Neural Network (CNN) or (ConvNets) or CNN-based model [40] defines an exceptionally powerful class of one-to-many models. CNN-based models achieving state-of-the-art results in classification, semantic segmentation, and action recognition tasks, amongst others. Nonetheless, CNN-based models have their limitations, and they have fundamental drawbacks and sometimes it's quite easy to fool a network [15]. When face objects are hidden to a certain extent by illumination or coloured background or only allow to display different figure of the face, the human visual system still able to find significant pieces of information and identify what they have seen but not CNN based [15]. Creating a CNN-based model which has the ability to recognize objects at the same level as humans has been proven to be difficult. Regardless of where the object is present in the image a well-trained CNN-based model cannot identify the object present in the image.

2.2 *Passive Illumination Reduction Approaches*

To understand passive illumination reduction on human face is best described in Johann Heinrich Lambert, book titled "Photometria". Human skin is a Lambertian surface which displays Lambertian reflectance. Lambertian reflectance is the property that defines an ideal "matte" or diffusely reflecting surface [38].

Retinex algorithms able to capture the low and high frequency in the Low-Light images beneath the dark image by performing Retinex algorithm through the mix of High Pass Filter and Low Pass Filer. These methods not only increase the contrast and brightness of an image but also able to extract the three colours of an image without noise. Each of passive illumination reduction filters has its own specialty, such as Gaussian convolutional kernel does not have the ability to preserve edges, consequently, they may lead to halo phenomena in some regions with sharp boundaries or cause the whole image to be too bright [7].

Face image passive normalization approach includes all types of image filtration and enhancement techniques such as High Pass Filter (HPF) [11], Low Pass Filter (LPF), Logarithms, Laplacian and other image filtration operation algorithms are useless without the right combination of filters [10]. Mustafa et al. has mentioned that face image normalization has two approaches [30]. The first method is image restoration, where this is the process of retrieving image formation model using filtering approaches and the second method is the image enhancement, where criteria are used to increase the perceived or the apparent quality of the image approaches by applying methods such as HE and Gama Correction (GC) approaches. When face image is restored globally, face outline is improved, however, facial components like eyes, mouth and nose are not revealed totally because of local features are heavily illuminated.

Histogram equalization (HE) is the most common method used for image enhancement when images have illumination variations [41]. Even for images under controlled illumination, HE improves the recognition results by flattening the histogram of pixel intensities of the images. However, HE has many restrictions on brightness management [8] and contrast distribution always deteriorate from over-improvement which lead to impractical results. Shan et al. proposed Gamma correction (GC) alone to enhance image [35] but Rahman et al. believed GC method should normalization illumination along with HE [32].

2.3 Active Illumination Reduction Approaches

Face recognition is used in applications like access control and surveillance but without able to adapt to low light causing face recognition labelled as a failure biometric system [23]. Variations in ambient lighting produces significant degradation in face recognition performance [43]. Thermal infrared (thermal IR) has been used in facial recognition systems with some success against the ambient illumination. Near infrared (NIR) has the potential to overcome the problems associated with visible [27] and thermal Infrared face recognition which is more robust against illumination variations and face detection. NIR is useful for face detection as the bright eye effect [17] allows the eyes to be localized and skin reflectance properties at just above and below 1.4 μm which highlights the face regions clearly, however under different type of illumination, such as Natural illumination, Artificial illumination and Infrared illumination, NIR will display face image differently.

Thermography is the detection of the heat that emits from a person's body and face [12]. Just like visible camera, thermal also a non-contact heat captures whereby a good LWIR able to capture a clear pattern of human's heat especially from a face because it was not covered unlike our body that is covered with cloths [12]. Thermography can be taken "on-the-fly" to maintain identification of a person especially when there is no presence of light, for example, during emergency evacuations most of the time in the situation of total darkness, visual camera will not capture any image and only infrared camera can. Below are some of algorithm which were introduced pertaining on reducing illumination using active illumination reduction approaches.

Socolinsky et al. presented a study on the face recognition performance using a PCA algorithm for both visible and thermal images [37]. The most interesting conclusion of their study is that face recognition using thermal images degrades more sharply than with visible images when training and test images are chosen from different sessions. By improving Socolinsky's method Arya et al. using hyperspectral imaging system to minimize several limitations arise in the existing and classical face recognition system because the skin spectra resulting with three-dimensional cubic dataset illustrates the unique features for an individual [3].

Krišto et al. initiated Infrared cameras capture thermal patterns from the blood vessels pattern under the skin. These vessels transport warm blood all over the body [20]. This active illumination approach was not only can improve the illumination challenges but also provides the advantages of capturing unique facial thermogram of a person which is independent from the surrounding ambient lights. Knowing this because not only that each person's head and facial components are unique but also having the advantages of being unique by using the algorithm which compare fingerprints.

A face recognition under cross-spectral and intra-spectral of distinct nature was also introduced by Kakkirala et al. [42]. Their studies have tried the performance of various infrared against visual images in the same environment as well as in different environment and distance of images captured (VIS, SWIR, LWIR). They found out that LWIR is the most consistent at distance of 6–7 m at any illumination variation throughout the day and night. In a natural environment NIR and SWIR can produce thermal details at the most distance of 3–4 m at night and at the distance of 6–7 m during the day. It shows here that NIR and SWIR are still depending on the presence of light.

3 Methods

In a basic face recognition image processing, it starts with face detection by using the Multiple Adaptive Derivatives (MAD) algorithm for face detection (MFD) (see Fig. 2). However, if there is no face detected in the image, a non-robust face recognition will stop there. Unlike to the state-of-the-art method, Convolutional Neural Network (CNN) which immediately performs convolutional method without performing normalization towards the source image. In other words, "Garbage in Garbage out" will occur when using the traditional CNN method. CNN based methods have no mechanism that makes it smart enough to determine whether to proceed or continue with the training or not. In other words, CNN method is too dependent on human supervision. CNN methods have no mechanism to recover images from low light illuminated image [16, 29]. Based on the recent studies, CNN based alone is incapable to solving low-light problems but has to borrow a very old method such as Retinex theory. Retinex theory was developed and presented by Land and McCann [21] explained the basic principles governing the process of image formation or scene perception of an image model approaches to reduce the illumination distortion. This work emphasizes on face recognition robustness, meaning just because the picture is showing a total dark image does not mean that there is no face image in there. Without lighting or illuminance, no face image can be seen at all and if there is only a partial illuminance, it means that only a half part of the face image can be detected. Next, if any of facial components are not seen in the image, that image will proceed to the restoration process.

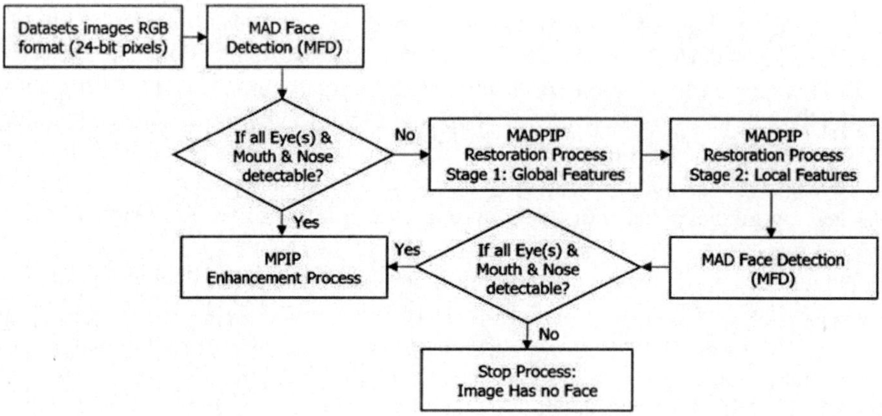

Fig. 2 Multiple Adaptive Derivatives for Passive Image Processing (MADPIP) framework

3.1 Multiple Adaptive Derivatives for Passive Image Processing (MADPIP)—Global and Local Features

MADPIP Global Features Restoration The Multiple Adaptive Derivatives for Passive Image Processing (MADPIP) is designed to retrieve global features by using Laplacian Mask combined with Low Pass Filter (LPF) [19] and processed with Zero Padding and Border Replication to produce the base filter or a foundation image in Stage 1. This combination is known as $2L$. Border Replica (BR) padding and Zero Padding (ZP) refined the connection between 3×3 matrix pixels. The unique of Low Light RBG image is that it allows Laplacian filter to find all the edge outside or within the face for every 3 by 3 kernel and it improves the texture by using Zero padding, which is a technique that allows one to preserve the original texture input size and BR increases the edges sizes that it may be and when these kernels find the original face texture.

MADPIP Local Features Restoration In Stage 2, the local restoration will be performed by using Logarithm and sharpen further when added with High Pass Filter (HPF) [19] derivative. If all facial components are seen, then it proceeds with the enhancement process by applying the Adaptive Gamma Adjustment (AGA), Outline Texture and Features (OTF) and Homogenous Mapping Indexes (HMI) algorithms.

For the local features restoration, MADPIP proposes the multiple configuration combinations of 2L derivative with three colour Logarithmic and the HPF-Unsharp Masking. The dynamic range of an image can be compressed by replacing each pixel value with its logarithm [9]. This is also know as $3LH$. The combination between the three colour Logarithmic and the HPF-Unsharp Masking process is labelled as LH. For algorithm labelled as LHprocessing $2L$ derivative, the result will be labelled as $3LH$. The logarithmic operator is an exponential process which mapping function is a logarithmic curve [6]. In other words, each pixel value is replaced with its logarithm value. Most implementations take either the natural logarithm or the base 10 logarithm [2]. However, the basis does not influence the shape of the logarithmic curve, only the scale of the output values which are scaled for display on an 8-bit system. Hence, the basis does not influence the degree of compression of the dynamic range.

The $3LH$ process starts when image from dataset comes in, processed using the $2L$, the image is converted to arrays of pixels in the form of float, in which pixels i and j are processed up until to the size of the image. Pixel's separation is performed according to colour and normalization separately but according to global min and max. Logarithmic process is performed and later these three 8-bit arrays are assembled. Once these 8-bit logarithmic layers assembled back to 24-bit image, an extraction process of HPF Unsharp Masking has to be performed before merged it back to 24-bit 3 colour Logarithm layers image which labelled as $3LH$.

3.2 Dataset

In this work, the extended Yale B dataset [14] will be used. The extended Yale B dataset has 10 subjects with 64 illumination conditions. Dataset has all the classes of illumination challenges and total image is 640. For each class in this dataset there is an image which is a neutral image with no illumination challenges nor incorrect facial component sizes due to eyes closed or facial expression. The proposed MADPIP mission is to convert all images of the same person/class, which has 64 types of illumination classifications into a single type of illumination. Figure 3 depicts some of the images that will be used to evaluate the performance of the proposed MADPIP in this work.

3.3 Experimental Setup

Global Features Retrieval The experimental setup for the global features retrieval in this work is divided into 2 parts. The first part of the experiment focuses on finding the best Laplacian configuration that is able to reduce the most illumination noise by getting the lowest MSE and Euclidean Distance and the highest SSIM and PSNR. Table 1 tabulates 13 Laplacian's algorithms with various settings that will be assessed in this works. Based on the results shown in Table 4, LM5 was found to be the best Laplacian configuration. As a result, LM5 will be used in the second part of the experiment.

After finding the best Laplacian configuration (e.g., LM5, see Table 4) that is able to reduce the most illumination noise, the second part of the experiment setup is designed to evaluate the multiple combinations of the best Laplacian configuration (e.g., LM5) and Low Pass Filter (LPF) padded with zero padding and border replication ($2L$ Variations). This experiment will calculate the total average for Mean (M),

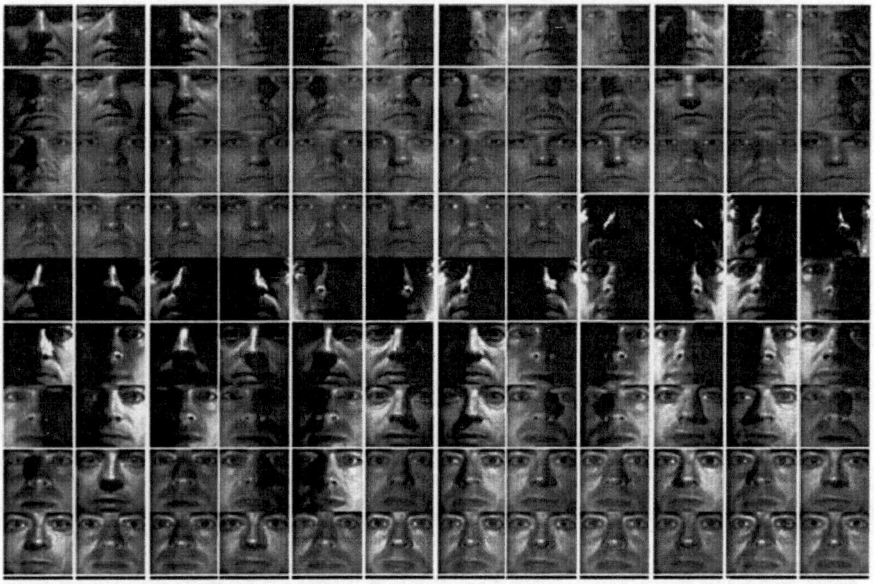

Fig. 3 Extended Yale B dataset

Table 1 13 Laplacian's algorithms

Laplacian configurations	Abbreviation	Function
Laplacian mask (+ c) without scale	LM + CWOS	Edge detection
Laplacian mask (+ c) with scale	LM + CWS	Edge detection
Laplacian mask (− c) without scale	LM − CWOS	Edge detection
Laplacian mask (− c) with scale	LM − CWS	Edge detection
Laplacian mask plus 1	LM1	Edge detection and sharpening
Laplacian mask plus 2	LM2	Edge detection and sharpening
Laplacian mask plus 3	LM3	Edge detection and sharpening
Laplacian mask plus 4	LM4	Edge detection and sharpening
Laplacian mask plus 5	LM5	Edge detection and sharpening
Laplacian mask plus 7	LM7	Edge detection and sharpening
Laplacian mask plus 8	LM8	Edge detection and sharpening
Laplacian mask plus 9	LM9	Edge detection and sharpening

Table 2 The multiple combinations of the best Laplacian configuration and low pass filter (LPF) padded with zero padding and border replication ($2L$ variations)

No	Kernel configurations	Abbreviation	Remarks
1	Laplacian (basic) filter	LB	
2	Best Laplacian masked	L	Best Laplacian masked obtained from Table 1
3	Low pass filter	LPF	
4	Zero padding + Laplacian masked	ZPL	Best Laplacian masked with zero padding
5	Border replication + Laplacian masked	BRL	Best Laplacian masked with border replication
6	Zero padding + Laplacian masked and zero padding + low pass filter	ZPL + ZPLPF	Best Laplacian masked with zero padding and low pass filter with zero padding
7	Border replication + Laplacian masked and border replication + low pass filter	BRL + BRLPF	Best Laplacian masked with border replication and low pass filter with border replication
8	Zero padding + Laplacian masked and border replication + low pass filter	ZPL + BRLPF	Best Laplacian masked with zero padding and low pass filter with border replication
9	Border replication + Laplacian masked and zero padding + low pass filter	BRL + ZPLPF	Best Laplacian masked with border replication and low pass filter with zero padding

Standard Deviation (SD) and Coefficient of Variation (CV) of Mean Squared Error (MSE), Structural Similarity Index (SSIM), Peak Signal-to-Noise Ratio (PSNR), between two images and Euclidean Distance (ED) for all proposed combinations in this work. The Structural Similarity Index (SSIM) is a perceptual metric that quantifies image quality degradation caused by processing such as data compression or by losses in data transmission. The PSNR block computes the peak signal-to-noise ratio, in decibels, between two images. This ratio is used as a quality measurement between the original and a compressed image. These 2L variations include the following settings as shown in Table 2.

Local Features Retrieval The experimental setup for the local features retrieval in this work is designed to assess the performance of several combinations of $3LH$ algorithms with several different scaling factors. After every image processed with the proposed $3LH$ algorithms in performing Local Restoration process, assessment tools MSE, SSIM, PSNR and Euclidean Distance (ED) will be used to evaluate the proposed $3LH$.

Table 3 Various combination of $3LH$ algorithms

No	Algorithms	Abbreviation
1	2L derivative	2L
2	Logarithmic and the HPF-unsharp masking process	LH
3	Logarithmic image processing	LIP
4	Generalized logarithmic image processing	GLIP
5	Logarithmic-HE	LHE
6	Symmetric generalized logarithmic image processing	SGLIP
7	2L derivative with 3 colour logarithmic and HPF-unsharp masking with 4 scaling factors	3LH4
8	2L derivative with 3 colour logarithmic and HPF-unsharp masking with 5 scaling factors	3LH5
9	2L derivative with 3 colour logarithmic and HPF-unsharp masking with 6 scaling factors	3LH6
10	2L derivative with 3 colour logarithmic and HPF-unsharp masking with 7 scaling factors	3LH7
11	2L derivative with 3 colour logarithmic and HPF-unsharp masking with 8 scaling factors	3LH8
12	2L derivative with 3 colour logarithmic and HPF-unsharp masking with 9 scaling factors	3LH9
13	2L derivative with 3 colour logarithmic and HPF-unsharp masking with 10 scaling factors	3LH10

The performance of the proposed $3LH$ algorithms will also be compared to existing conventional or published algorithms that perform similar function such as Logarithmic Image Processing (LIP) [9], Generalized Logarithmic Image Processing (GLIP) [9], Logarithmic-HE (LHE) [6] and Symmetric Generalized Logarithmic Image Processing) (SGLIP) [9]. In this experiment, there are 13 algorithms as shown in Table 3. This include the four existing conventional or published algorithms that perform similar function with our proposed $3LH$, such as LIP, GLIP, LHE and SGLIP. The main component of the proposed $3LH$ such as $2L$ and LH and the other 7 types of $3LH$ possible configuration combination scaling factor from 4 to 10. The various combination of $3LH$ algorithms are tabulated in Table 3.

Table 4 Assessment for the 13 Laplacian's algorithms based on MSE, SSIM, PNSR and ED

No	Algos	MSE			SSIM			PNSR			ED		
		M	SD	CV	M	SD	CV	M	SD	CV	M	SD	CV
1	None	10,865	7837	0.7	0.20	0.10	0.50	12	11	0.94	3537 K	816 K	0.2
2	LM + CWOS	10,463	5816	0.6	0.17	0.15	0.88	12.46	9	0.72	3415 K	766 K	0.2
3	LM + CWS	10,346	5354	0.5	0.16	0.14	0.87	11.47	7.96	0.69	3408 K	693 K	0.2
4	LM – CWOS	9516	5078	0.5	0.16	0.16	0.99	14.28	10.25	0.72	3360 K	652 K	0.2
5	LM – CWS	9248	6017	0.7	0.17	0.16	0.96	14.37	10.09	0.70	3421 K	752 K	0.2
6	LM1	10,043	6194	0.6	0.17	0.17	1.02	14.92	10.58	0.71	3351 K	650 K	0.2
7	LM2	10,086	5656	0.6	0.17	0.17	1.01	15.79	8.23	0.52	3320 K	649 K	0.2
8	LM3	9938	5251	0.5	0.18	0.17	0.95	15.8	10.19	0.64	3370 K	651 K	0.2
9	LM4	9846	5968	0.6	0.18	0.18	0.97	15.98	10.68	0.67	3358 K	652 K	0.2
10	LM5	**9690**	**4898**	**0.5**	**0.19**	**0.16**	**0.85**	**16.0**	**2.5**	**0.78**	**3343 K**	**689 K**	**0.2**
11	LM6	9839	4439	0.5	0.19	0.18	0.94	15.76	11.29	0.72	3399 K	648 K	0.2
12	LM7	9881	4592	0.5	0.19	0.15	0.79	15.36	10.33	0.67	3382 K	643 K	0.2
13	LM8	9907	4760	0.5	0.19	0.15	0.76	15.12	12.93	0.86	3364 K	640 K	0.2
14	LM9	9926	4841	0.5	0.19	0.12	0.64	15.02	11.03	0.73	3351 K	637 K	0.2

4 Results and Discussion

4.1 Global Features Retrieval

Table 4 tabulates the results of the 13 Laplacian's algorithms that were assessed based on the MSE, SSIM, PNSR and ED values. LM5 was found to produce the lowest MSE and Euclidean Distance and the highest SSIM and PSNR. Thus, it will be used as the basic Laplacian algorithm to be combined with other algorithms.

Table 5 tabulates the Assessment for the best combination of 2*L* (Laplacian and LPF and Zero padding and Border Replica Padding) based on MSE, SSIM, PNSR and ED. The Zero Padding + Laplacian Masked and Border Replication + Low Pass Filter (ZPL + BRLPF) combination has the best score for the MSE assessment.

When images are getting brighter and clearer, both mean SSIM and PSNR values are increasing as well, and when coefficient of variation values are getting smaller, this indicates that more and more images of the same classes starting to look similar.

In this assessment, all four assessment tools are used. This is because all four assessment instruments are important tools to understand the level of problem that illumination can cause. MSE is able to the get average differences of all pixel compared on every pixel location of an image, SSIM evaluates on the pixel similarity based on small windows, where this evaluations is good in comparing the main structure of the image, the more similarity the higher the SSIM value, PSNR evaluate the ratio between the maximum possible power of an image and the power of corrupting noise that affects the quality of its representation and Euclidean Distance (ED)

Table 5 Assessment for the best combination of $2L$ (Laplacian (LM5) and LPF and zero padding and border replica padding) based on MSE, SSIM, PNSR and ED

No	Algos	MSE			SSIM			PNSR			ED		
		M	SD	CV	M	SD	CV	M	SD	CV	M	SD	CV
1	None	10,865	7837	0.7	0.2	0.1	0.5	12	11	0.942	3537 K	816 K	0.2
2	LB	10,463	5816	0.6	0.2	0.1	0.6	12	9	0.724	3415 K	766 K	0.2
3	L	9690	4898	0.5	0.2	0.2	0.8	16	13	0.781	3343 K	689 K	0.2
4	LPF	10,103	5872	0.6	0.2	0.1	0.5	16	12	0.73	3419 K	751 K	0.2
5	ZPL	9221	4783	0.5	0.2	0.1	0.5	17	13	0.77	3357 K	649 K	0.2
6	BRL	9212	5863	0.6	0.2	0.1	0.6	17	14	0.76	3317 K	647 K	0.2
7	ZPL + ZPLPF	9242	5312	0.6	0.2	0.1	0.6	17	12	0.61	3316 K	646 K	0.2
8	BRL + BRLPF	9025	5038	0.6	0.2	0.1	0.7	17	12	0.72	3337 K	649 K	0.2
9	ZPL + BRLPF	**8601**	**4783**	**0.6**	**0.2**	**0.1**	**0.5**	**17**	**13**	**0.69**	**3356 K**	**649 K**	**0.2**
10	BRL + ZPLPF	8932	4993	0.6	0.2	0.2	0.7	16	11	0.7	3347 K	650 K	0.2

comparing the sum of distance differences between pixels of two images, where the lesser distance between these pixels the better it is. Thus, it can be concluded that the Zero Padding + Laplacian Masked and Border Replication + Low Pass Filter (ZPL + BRLPF) combination has the best score for the MSE assessment and thus best algorithm for global feature extraction.

4.2 Local Features Retrieval

Table 6 shows that an algorithm with 8.0 scaling factors, was able to restore 63 illuminated images better than other algorithms that used Logarithmic concepts as it produced the lowest value of MSE.

$3LH$ is the combination algorithm of $2L$ and LH. This experiment demonstrated that with $2L$ and LH, either one will not be able to achieve the best performance similar to the $3LH$ with 8.0 scaling factor algorithm. Table 6 shows that the MSE values for $2L$ and LH are greater when compared to the MSE value for the proposed $3LH8$. For local features extraction, the 2L derivative with 3 colour Logarithmic and HPF-Unsharp Masking with 8 scaling factors has produced the best result in retrieving the local features extraction.

Table 6 Performance assessment of various combination of $3LH$ algorithms based on MSE, SSIM, PNSR and ED

No	Algos	MSE			SSIM			PNSR			ED		
		M	SD	CV	M	SD	CV	M	SD	CV	M	SD	CV
1	None	10,865	7837	0.71	0.15	0.08	0.533	12.01	11.31	0.94	3537 K	816 K	0.213
2	2L	8410	4731	0.563	0.24	0.17	0.708	17.95	11.81	0.66	3312 K	645 K	0.195
3	LH	7972	4580	0.575	0.21	0.19	0.881	17.73	13.55	0.76	3337 K	647 K	0.194
4	LIP	8456	6821	0.807	0.17	0.15	0.882	12.46	9	0.72	3415 K	766 K	0.224
5	GLIP	8390	6941	0.827	0.19	0.18	0.947	16.03	12.52	0.78	3403 K	689 K	0.202
6	LHE	8023	5903	0.736	0.19	0.17	0.895	15.82	11.54	0.73	3419 K	751 K	0.22
7	SGLIP	8002	5783	0.723	0.2	0.19	0.95	17.23	13.2	0.77	3357 K	649 K	0.193
8	3LH4	7901	4363	0.552	0.21	0.18	0.857	17.23	13.89	0.76	3317 K	647 K	0.195
9	3LH5	7802	4312	0.553	0.21	0.17	0.81	17.23	11.67	0.61	3316 K	646 K	0.195
10	3LH6	7825	4338	0.554	0.21	0.19	0.905	17.23	12.32	0.72	3337 K	649 K	0.195
11	3LH7	7733	4323	0.559	0.25	0.19	0.76	17.23	13.13	0.68	3356 K	649 K	0.194
12	3LH8	**7725**	**3038**	**0.393**	**0.26**	**0.2**	**0.692**	**17.81**	**11.4**	**0.64**	**3257 K**	**636 K**	**0.2**
13	3LH9	7728	3328	0.431	0.23	0.18	0.783	17.23	12.4	0.65	3317 K	647 K	0.195
14	3LH10	7729	3318	0.429	0.25	0.2	0.8	17.18	13.69	0.72	3317 K	648 K	0.195

5 Conclusion

The work presented in this paper was aimed to assess the multiple combination of kernels Laplacian (e.g., LM5) and Low Pass Filter (LPF) padded with zero padding and border replication. In this work, the total average for Mean, Standard Deviation and Coefficient of Variation (CV) of MSE, SSIM, PSNR and ED for every combinations were computed. It was concluded that the Zero Padding + Laplacian Masked and Border Replication + Low Pass Filter (ZPL + BRLPF) combination has the best score for the MSE assessment and thus best algorithm for global feature extraction. On the other hand, for local features extraction, the 2L derivative with 3 colour Logarithmic and HPF-Unsharp Masking with 8 scaling factors has produced the best result in retrieving the local features extraction. This experiment demonstrated that with $2L$ and LH, either one will not be able to achieve the best performance similar to the $3LH$ with 8.0 scaling factor algorithm.

Acknowledgements This work was supported in part by research grants from Universiti Malaysia Sabah (SDN0057).

References

1. Ahmed AS (2018) Comparative study among Sobel, Prewitt and canny edge detection operators used in image processing. J Theor Appl Inf Technol 96(19):6517–6525
2. Ali AM, Entesar A (2019) Some new properties of logarithms
3. Arya S, Pratap N, Bhatia K (2015) Future of face recognition: a review. Procedia Comput Sci 58:578–585

4. Batchuluun G, Koo JH, Kim YH, Park KR (2021) Image region prediction from thermal videos based on image prediction generative adversarial network. Mathematics 9(9):1053
5. Belhumeur PN, Kriegman DJ (1998) What is the set of images of an object under all possible illumination conditions? Int J Comput Vis 28(3):245–260
6. Chaudhury S, Raw S, Biswas A, Gautam A (2015) An integrated approach of logarithmic transformation and histogram equalization for image enhancement. In: Proceedings of fourth international conference on soft computing for problem solving. Springer, pp 59–70
7. Chaves J (2008) Introduction to nonimaging optics. CRC Press
8. Chen L, Zhang L, Hu Y, Li M, Zhang H (2003) Head pose estimation using fisher manifold learning. In: 2003 IEEE international SOI conference. Proceedings (Cat. No. 03CH37443). IEEE, pp 203–207
9. Deng G (2016) The symmetric generalized lip model and its application in dynamic range enhancement. J Math Imaging Vis 55(3):253–265
10. Deswal S, Gupta S, Bhushan B (2015) A survey of various bilateral filtering techniques. Int J Signal Process Image Process Pattern Recognit 8(3):105–120
11. Dogra A, Bhalla P (2014) Image sharpening by Gaussian and Butterworth high pass filter. Biomed Pharmacol J 7(2):707–713
12. Du Q, Xu H, Ma Y, Huang J, Fan F (2018) Fusing infrared and visible images of different resolutions via total variation model. Sensors 18(11):3827
13. El Mahouachi D, Akhloufi MA (2022) Deep adaptive convolutional neural network for near infrared and thermal face recognition. In: Infrared technology and applications XLVIII, vol 12107. SPIE, pp 417–428
14. Georghiades A, Belhumeur P (1998) Illumination cone models for faces recognition under variable lighting and pose. IEEE Trans Pattern Anal Mach Intell 23:6
15. Gu J, Wang Z, Kuen J, Ma L, Shahroudy A, Shuai B, Liu T, Wang X, Wang G, Cai J et al (2018) Recent advances in convolutional neural networks. Pattern Recognit 77:354–377
16. Heo J, Kong SG, Abidi BR, Abidi MA (2004) Fusion of visual and thermal signatures with eyeglass removal for robust face recognition. In: 2004 conference on computer vision and pattern recognition workshop. IEEE, p 122
17. Huang CT, Wang Z, Kuo CCJ (2016) Visible-light and near-infrared face recognition at a distance. J Vis Commun Image Represent 41:140–153
18. Jung Y, Hu J (2015) A K-fold averaging cross-validation procedure. J Nonparametr Stat 27(2):167–179
19. Khan M, Alam M, Masud M, Amin A (2016) Importance of high order high pass and low pass filter. World Appl Sci J 34(9):1261–1268
20. Krišto M, Ivasic-Kos M (2018) An overview of thermal face recognition methods. In: 2018 41st international convention on information and communication technology, electronics and microelectronics (MIPRO). IEEE, pp 1098–1103
21. Land EH, McCann JJ (1971) Lightness and retinex theory. J Opt Soc Am 61(1):1–11
22. Lee CH, Shih JL, Lien CC, Han CC (2013) Adaptive multiscale retinex for image contrast enhancement. In: 2013 international conference on signal-image technology & internet-based systems. IEEE, pp 43–50
23. Lee H, Park SH, Yoo JH, Jung SH, Huh JH (2020) Face recognition at a distance for a stand-alone access control system. Sensors 20(3):785
24. Liberti L, Lavor C, Maculan N, Mucherino A (2014) Euclidean distance geometry and applications. SIAM Rev 56(1):3–69
25. Linge S, Langtangen HP (2020) Loops and branching. In: Programming for computations-python. Springer, pp 59–77
26. Liu X, Liu C, Lan H, Xie L (2020) Dehaze enhancement algorithm based on retinex theory for aerial images combined with dark channel. Open Access Libr J 7(4):1–12
27. Ma J, Chen C, Li C, Huang J (2016) Infrared and visible image fusion via gradient transfer and total variation minimization. Inf Fus 31:100–109
28. Ma J, Fan X, Yang SX, Zhang X, Zhu X (2018) Contrast limited adaptive histogram equalization-based fusion in YIQ and HSI color spaces for underwater image enhancement. Int J Pattern Recognit Artif Intell 32(07):18540186

29. Mahouachi D, Akhloufi MA (2021) Adaptive deep convolutional neural network for thermal face recognition. In: Thermosense: thermal infrared applications XLIII, vol 11743. SPIE, pp 15–22
30. Mustafa WA, Yazid H, Abdul-Nasir AS, Jaafar M (2017) An illumination normalization on face images: a comparative. Computer 35(1):1–9
31. Niyogi S, Freeman WT (1996) Example-based head tracking. In: Proceedings of the second international conference on automatic face and gesture recognition. IEEE, pp 374–378
32. Rahman S, Rahman MM, Abdullah-Al-Wadud M, Al-Quaderi GD, Shoyaib M (2016) An adaptive gamma correction for image enhancement. EURASIP J Image Video Process 2016(1):1–13
33. Rahman MF, Sthevanie F, Ramadhani KN (2020) Face recognition in low lighting conditions using fisherface method and CLAHE techniques. In: 2020 8th international conference on information and communication technology (ICoICT). IEEE, pp 1–6
34. Setiadi DRIM (2021) PSNR vs SSIM: imperceptibility quality assessment for image steganography. Multimed Tools Appl 80(6):8423–8444
35. Shan S, Gao W, Cao B, Zhao D (2003) Illumination normalization for robust face recognition against varying lighting conditions. In: 2003 IEEE international SOI conference. Proceedings (Cat. No. 03CH37443). IEEE, pp 157–164
36. Sirovich L, Kirby M (1987) Low-dimensional procedure for the characterization of human faces. J Opt Soc Am A 4(3):519–524
37. Socolinsky DA, Selinger A (2004) Thermal face recognition in an operational scenario. In: Proceedings of the 2004 IEEE computer society conference on computer vision and pattern recognition, 2004. CVPR 2004, vol 2. IEEE, p II
38. Sun B, Edgar MP, Bowman R, Vittert LE, Welsh S, Bowman A, Padgett MJ (2013) 3D computational imaging with single-pixel detectors. Science 340(6134):844–847
39. The extended Yale face database B. http://vision.ucsd.edu/~leekc/ExtYaleDatabase/ExtYaleB.html
40. Valueva MV, Nagornov N, Lyakhov PA, Valuev GV, Chervyakov NI (2020) Application of the residue number system to reduce hardware costs of the convolutional neural network implementation. Math Comput Simul 177:232–243
41. Wang Z, Tao J (2006) A fast implementation of adaptive histogram equalization. In: 2006 8th international conference on signal processing, vol 2. IEEE
42. Weidlich VA (2021) Thermal infrared face recognition. Cureus 13(3)
43. Wu Z, Fuller N, Theriault D, Betke M (2014) A thermal infrared video benchmark for visual analysis. In: Proceedings of the IEEE conference on computer vision and pattern recognition workshops, pp 201–208
44. Xiang M, Liu Y, Liao T, Zhu X, Yang C, Liu W, Shi H (2021) The 3rd grand challenge of lightweight 106-point facial landmark localization on masked faces. In: 2021 IEEE international conference on multimedia & expo workshops (ICMEW). IEEE, pp 1–6
45. Yang J, Frangi AF, Yang JY, Zhang D, Jin Z (2005) KPCA plus LDA: a complete kernel Fisher discriminant framework for feature extraction and recognition. IEEE Trans Pattern Anal Mach Intell 27(2):230–244
46. Yang KF, Zhang XS, Li YJ (2019) A biological vision inspired framework for image enhancement in poor visibility conditions. IEEE Trans Image Process 29:1493–1506

Smart Cattle: Cattle Live Weight Estimation Based on a Deep Learning Approach

Nur Lydia Jane Binti Mohd Jaini, Rayner Alfred, Januarius Gobilik, Joe Henry Obit, Florence Sia Fui, Samry Mohd Shamrie Sainin, Raymond Victor Loudin, and Zamhar Iswandono

Abstract Smart Cattle is a suite of technologies for boosting the quality and quantity of production in dairy farms. Measuring cattle's live weight is essential as the captured data could be used to monitor cattle's health and nutrition management. However, weighting cattle and consistently monitoring its weights of is time consuming and require intensive labour. Profitability of beef production system depends highly on the annual live weight gain, stocking rate, feed quality, feed quantity, and the balance between these factors. The lighter cattle causes farmers to lose profit. In feedlot, where the feed can be systematically rationed, this problem can be mitigated with feeding correction. The limitation of this approach is to monitor the live-weight and growth rate of the cattle, as the farmers need to weigh the animals at consistent interval such as once in a month during the production period. This task is laborious, time consuming

N. L. J. B. M. Jaini · R. Alfred (✉) · J. H. Obit · F. S. Fui · S. M. S. Sainin
Creative Advanced Machine Intelligence Research Centre, Faculty of Computing and Informatics, Universiti Malaysia Sabah, Kota Kinabalu, Sabah, Malaysia
e-mail: ralfred@ums.edu.my

N. L. J. B. M. Jaini
e-mail: bi18110154@student.ums.edu.my

J. H. Obit
e-mail: joehenry@ums.edu.my

F. S. Fui
e-mail: florence.sfs@ums.edu.my

S. M. S. Sainin
e-mail: shamrie@ums.edu.my

J. Gobilik
Faculty of Sustainable Agriculture, Universiti Malaysia Sabah, Sabah, Malaysia
e-mail: jgobilik@ums.edu.my

R. V. Loudin
Infinity Enterprise, Sabah, Malaysia
e-mail: raymondv@infinityenterprise.my

Z. Iswandono
Higher Colleges of Technology, Abu Dhabi, United Arab Emirates
e-mail: zismail@hct.ac.ae

© The Author(s), under exclusive license to Springer Nature Singapore Pte Ltd. 2023
D.-K. Kang et al. (eds.), *Proceedings of the 9th International Conference on Computational Science and Technology*, Lecture Notes in Electrical Engineering 983,
https://doi.org/10.1007/978-981-19-8406-8_11

and costly. Hence, the aim of this work is to develop a live weight estimation system that will be able to perform a physical weight estimation of a cattle using a deep learning approach. The objectives of this work are to design a deep learning model, specifically a Convolutional Neural Network (CNN) for cattle weighting estimation system based on features extracted from the images, to implement an optimized deep learning model for cattle physical weight estimation based on high resolution images and finally to evaluate the performance of the proposed cattle weight estimation system using a deep learning model. This approach can be used to replace traditional methods of weighing using electronic scale and direct observation and measurement can be made based on the cattle images. In this work, the impact of dropout rates on the performance of convolutional neural networks will also be studied for image classification. Based on the results obtained, the proposed CNN model with dropout regularization produced higher performance accuracy of 54.0%.

Keywords Cattle weight estimation · Live weight · Deep learning approach · Convolutional neural network · Smart farming

1 Introduction

Profitability of beef production system will depend on the annual live weight gain, stocking rate, feed quality, feed quantity, and the balance between these factors [5, 18]. In tropical region, cattle feeding on pasture would grow only 0.2–0.8 kg/day or sometimes, 1 kg/day, but only for a short period [18]. A better growth rate or 1.12–1.33 kg/day can be achieved when the cattle are fed with supplement (feed of higher quality), which is usually attained more consistently once the cattle are reared in feedlot [5]. Cattle reared in feedlot, however, are not growing uniformly. The lighter cattle cause farmers to lose profit. In feedlot, where the feed can be systematically rationed, this problem can be mitigated with feeding correction. The limitation of this approach is to monitor the live-weight and growth rate of the cattle, the farmers need to weigh the animals at consistent interval, such as once in a month during the production period [?]. Moreover, monitoring and maintaining cattle live weight is important for farmers to make decision regarding their health, nutrition, management and if it is in a healthy environment hence determine whether the cattle are suitable to buy or sell to market.

There are several methods used to weigh cattle either by measuring its body length and heart girth or by measuring with a scale. However, measuring the exact live body weight can be challenging when trying to lead them to scales as it could be laborious, time consuming and costly where there may be some farmers that could not afford electric weighing scale as it is expensive, and the measurement process can induce stress for both animal and the farmers when the process is done manually. Hence, in the present research, to help cattle farmers in tropical regions to improve the productivity of their farm, the monitoring and data analysis could be replaced with the use of computer vision and machine learning approach.

Several previous works conducted by researchers deals with the use of machine learning approach and it has shown to be very effective to estimate the weight of a cattle with computer vision involved. Deep learning techniques that are slowly emerging and are shown to be able to provide better results. However, most studies involved the use of depth images where extra equipment is needed to acquire those images where the use of 2D images can also provide more accurate results. Thus, this work will investigate the effectiveness of applying deep learning approach to estimate cattle's weight using 2D images. This work also investigates the effects of optimizing the parameter settings of the CNN model on the performance accuracy. This work will overcome the challenges faced by farmers of manually weighing the cattle by using the optimized CNN model. This is due to the fact that the convolutional neural network (CNN) model that uses a deep learning approach has been found to be very effective in images classification [9, 10].

A CNN is a neural network with the fundamental architecture comprising a sequence of three-layer types: convolution, pooling (or subsampling) and fully connected where the model coefficients being used for feature extraction and fully connected layer being used for classification that determined in the training process [9, 20]. In addition to that, CNN also has functionalities that can find itself important through training and reduce the need for time consuming manual engineering [9, 10].

Hence, the aim of this work is to develop a live weight estimation framework that will be able to perform a live weight estimation of a cattle using a deep learning approach. The objectives of this research include formulating the framework of the proposed Convolutional Neural Network (CNN) for the cattle's live weight estimation, implementing and optimizing the proposed CNN approach and finally evaluating the performances of the proposed CNN approach in terms of prediction accuracy. Section 2 provides the detail of the literature review or related works. Section 3 presents the proposed approach, Sect. 4 presents the results and discussion and finally Sect. 5 concludes this paper.

2 Related Works

Various techniques for predicting animal weight using 2D and 3D vision-based approaches have been proposed and validated. While there have been several machine learning-based studies, the number of deep learning-based techniques in the sector is still very small. Some deep learning algorithms are discussed in this section.

In existing visual approaches, four main cattle body areas are often used for identification: muzzle, face, back and trunk. Robust features were extracted from cattle muzzle print photos for cattle identification using the Weber Local Descriptor [8]. In another work, Cai and Li presented a facial representation model of cattle based on local binary pattern texture features [7]. The work in [14], suggested a method for identifying cattle based on extracted cattle facial traits.

Recently, deep learning approaches with more powerful feature extraction and image representation capabilities have been widely used in the fields of visual classification and recognition [21, 23]. A CNN method was proposed for cow identification based on trunks features extracted from the raw images. It was further suggested that the CNN-based method can be developed for identifying individual cattle using a primary muzzle point picture pattern [14]. Zin et al. trained a CNN model to recognize individual cattle using the back photos of the cattle [25]. This method omitted data from the head and legs, which contain useful identification information such as contour and texture traits. In [21], a CNN network was fine-tuned with side-view shots of dairy cows and was utilized to identify animals. Previous image frame-based techniques, on the other hand, did not consider behavior such as walking posture.

In animal weighting, most studies presented multiple linear regression approaches by using the body weight of the animal as dependent variable and extra traits as independent variables to estimate cattle's weight. Ashwini used Stepwise Multiple Regression Analysis for the crossbred cattle to predict body weight involving combination of heart girth, height at wither and chest depth [3]. The data are grouped by sex wise female and male groups and female group was further divided based on the age of the cattles. Results have shown that the live weight of female cattle can be estimated using the multiple regression model based on the combination of heart girth, wither height and age information [3]. A regression model that predicts body weight based on body and on body condition scores of beef cows has also been introduced [16]. It was shown that the combination of body condition scores with body measurement resulting in an increase of accuracy compared to using heart girth measurement alone [16].

Another work that used Stepwise method used to find which parameters are important for measurements as there are many variables obtained. The hip width with dorsum area shows confidence level of 95% and correlation and coefficient of 0.71 and it is a recommended measurement for estimating body weight [24].

Computer vision techniques have been used to automatically measure morphological traits in cattle in which, image processing techniques were used along with GLCM methods to learn 2D images of cattle [17]. The GLCM method was useful for pattern recognition where the method was able to eliminate noise from the backdrop of other elements from the data. Weight calculations were calculated based on pixel, area, and perimeter measurements [17].

For 3-dimensional (3D) vision, Computer vision system and Image Processing toolboxes in Matlab were used to process raw 3D images of cow [22]. The raw images captured includes information related to the cow body surface, frames of selection box, image noise and weighing scale. The six steps that were involved in the process were the segmentation of cow body, removal image noise, transformation, interpolation, anatomical landmark identification, and lastly morphological trait quantification. Then, a multiple linear regression model is used to model the hip width, DIM, age, parity, and actual body weight in order to estimate the live weight of the cattle [15, 22].

Recent studies have shown that machine learning techniques can be used to accurately predict weight [11, 19]. For instance, the Random Forest Regression method was found to be able to produce better prediction of cattle's body weight via various body measurements taken manually with tape compared to linear regression model [11, 19].

In [24], a Bagging algorithm for cattle's body weight prediction method has been introduced based on the body measurements of dorsal area extracted from 2-D images. `LabelMe` software was used to segment the image by cutting the original image in RGB color space and mask it. Then, the Correlation-based Feature Selection (CFS) was used to select relevant features for cattle's live body weight estimation.

In addition to that, studies conducted using Artificial Neural Network (ANN) have shown better results compared to traditional linear regression model [13]. In [6], a Barqi breed sheep weight prediction method has been developed by combining ANN with regression model. The side photo of sheep and age and gender were taken as input to train the machine learning model. While in [12], a method that combined ANN and linear regression was used to estimate pig weight using image processing where top view images were used to perform the weight prediction.

3 Methods

3.1 Dataset Collection

The dataset contains cattle images and its actual weight that are obtained from GitHub dataset and mixed with data obtained from FPL, Universiti Malaysia Sabah. The dataset contains 473 of cattle images with weights as label. In this work, all cattles are categorized into 5 classes (e.g., Lighter, Light, Average, Heavy, Heavier) according to their actual weights. The actual weights of the cattles are discretized into 5 partitions with classes label according to the discretized weights [1, 2] as follows,

- Lighter (160–219 kg)
- Light (220–241 kg)
- Average (242–264 kg)
- Heavy (265–297 kg)
- Heavier (298 kg and above kg).

The images are manually cropped to filter out other objects to focus more on the cattle's body. Then splitting the whole dataset into three sets, 80% training set, 10% validation set, and 10% test for evaluation process. Figures 1, 2, 3, 4 and 5 depicts the ten samples of cattles images from each category of the train dataset. For the training dataset, 360 cattle images will be used for training the CNN model, 54 cattle images will be used for validating the CNN model and finally, 59 cattle images will be used for testing the CNN model. All Images are labelled as `[Cattle's weight]_[file number]`. For instance, in Fig. 1, the first photo of cattle labeled with 181_6.jpg represents cattle ID 6 weighting 181 kg.

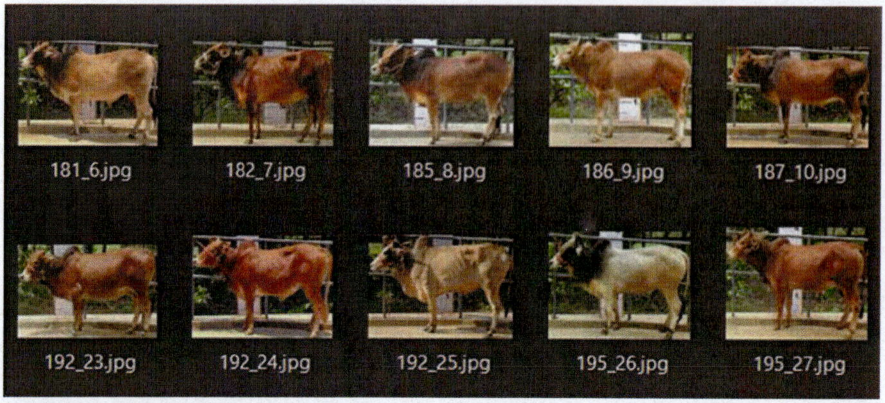

Fig. 1 Lighter (0–219 kg) cattles

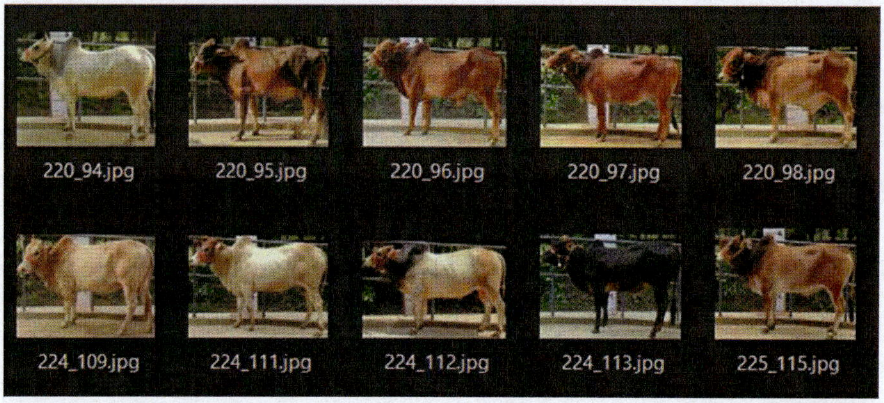

Fig. 2 (220–241 kg) cattles

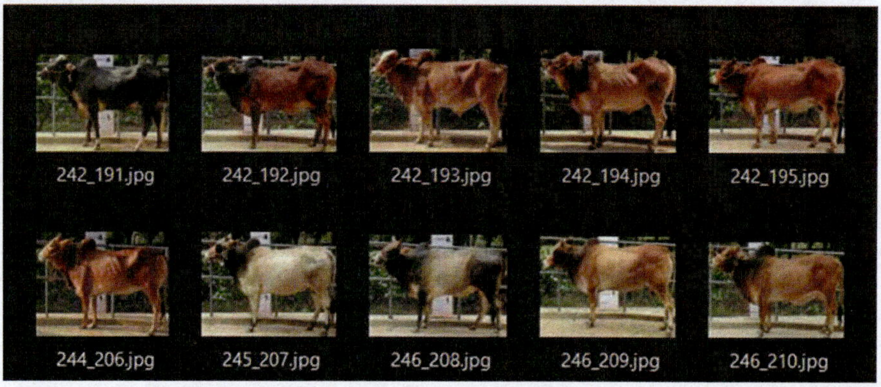

Fig. 3 Average (242–264 kg) cattles

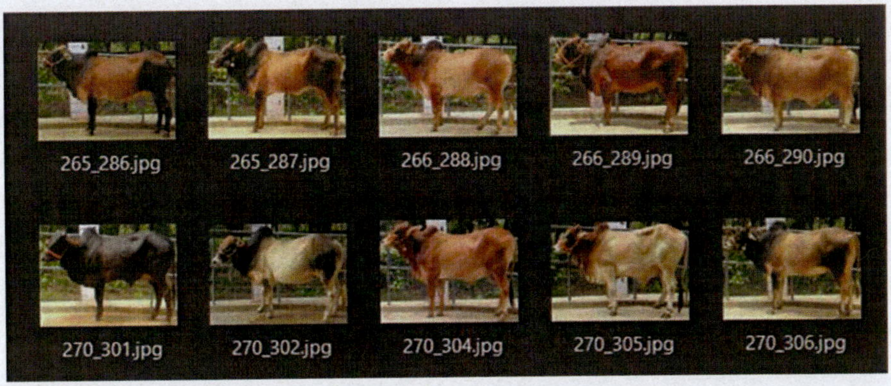

Fig. 4 Heavy (265–297 kg) cattles

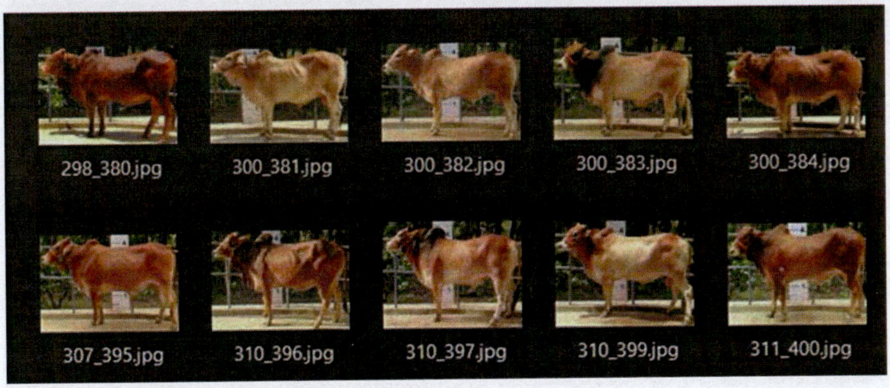

Fig. 5 Heavier (298 kg and above kg) cattles

3.2 Designing the CNN Model

The initial values for the CNN parameters are set based on the input of the CNN model. The input of the CNN has the shape of (batch size, height, width, depth). The first component represents the batch size of the image. The second component represents the image dimensions which are the height, width and the RGB image of depth of 3. The input of the CNN model has the following values for all the components:

- Height = 128
- Width = 128
- Size of Feature Map = 128 × 128 × 3
- Kernel Size = 3 × 3
- Batch Size = 32
- Filter Size = 32

- Epochs = 100
- Dropout Rate = 0.2 (For CNN model with dropout setting)
- Activation= ReLU and SoftMax activation.

Since this work deals with image classification, SoftMax activation is applied in the fully connected layer with dense of 5 units indicates the five class.

A three layer of convolution layer are used with the first layer to define image shape and other two layers as the hidden layers. As shown in Fig. 6, the proposed CNN model consists of three layers of Convolutional Layers followed by three layers of Max-Pooling Layers, and one layer of Fully Connected Layers with dropout in all layers for regularization and the total of trainable parameters are 220,293.

Dropout Regularization Since there is a large number of parameters in the deep neural network, the model produced could be overfitting, requiring the use of regularization techniques. To prevent overfitting, dropout regularization technique will be applied with 0.5 in every layer [4]. In this study, the impact of dropout regularization on the performance of convolutional neural networks will be studied for image classification, specifically for cattle weight classification. Figure 6 depicts the comparison between CNN model with and without dropout regularization.

4 Results and Discussion

Figures 8 and 9 showed the performance comparison between CNN methods with and without dropout regularization. With each method uses three layers with filters of 32, size kernel of images size and 3 for RGB input and dense of 32 and 5 units. It can be seen that the accuracy is higher when dropout regularization is used with 54% accuracy. Figure 7 illustrates the confusion matrix for the CNN model with dropout regularization having only 32 out of 59 cattle images that were correctly classified.

The limitation of this work is that the appropriate number of partitions for the discretized cattle's weight was not fully investigated as lower number or higher number of partitions may increase the accuracy of the prediction of the proposed CNN model. Another limitation that can be observed here is that the optimized hyperparameter setting was not investigated. As a result, the optimum performance of the proposal CNN model was not investigated. Both Figs. 8 and 9 illustrates that the precision and recall values for categories Heavier and Lighter are higher compared to the other categories. Thus, the range of the discretized cattle's weight also has a great influence on the performance of the CNN model.

Figure 10 shows the model accuracy and loss for the CNN model without dropout regularization and it shows some kind of overfitting as you can see that the loss value is greater than the CNN model with dropout regularization as shown in Fig. 11.

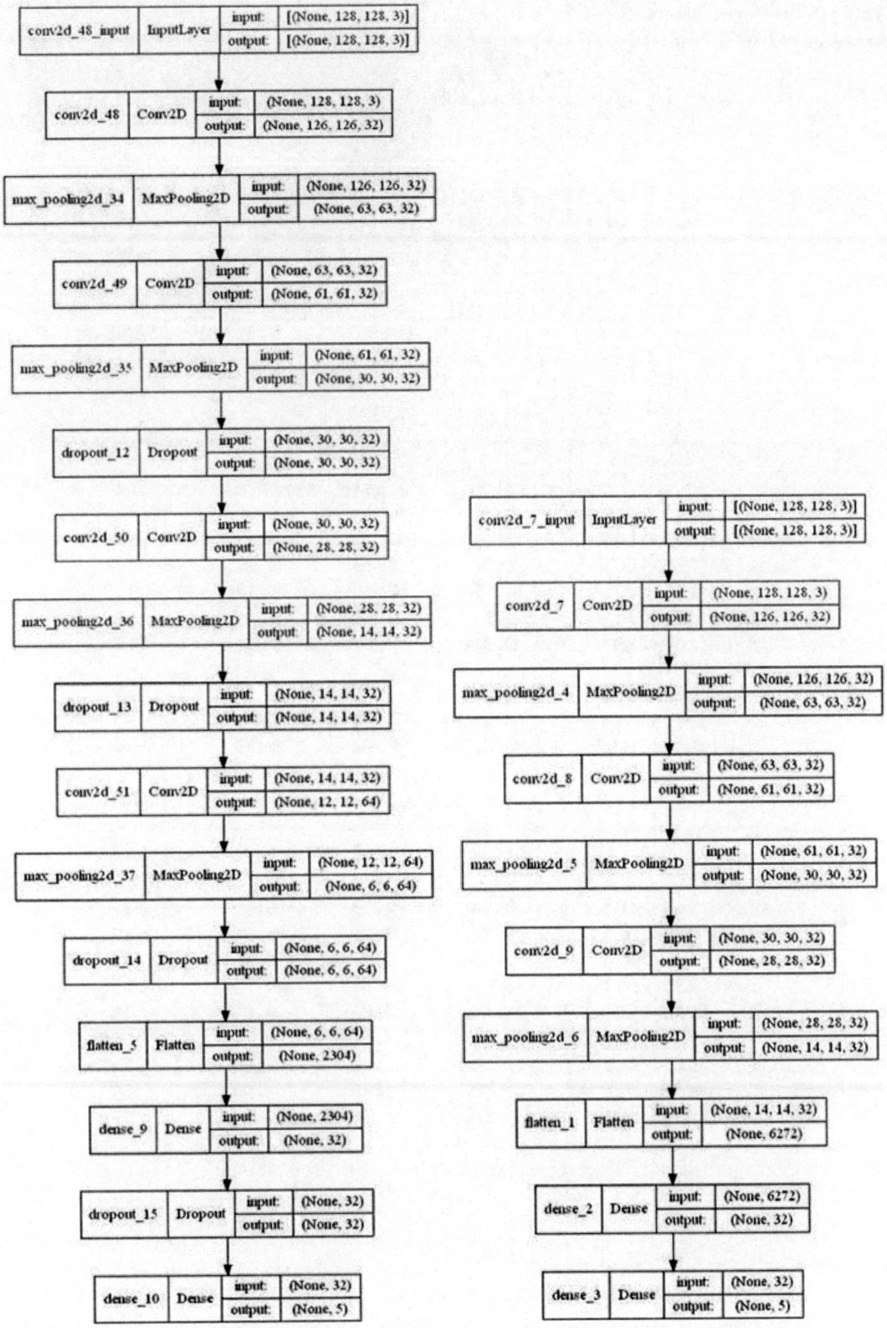

Fig. 6 CNN architecture with (left diagram) and without (right diagram) dropout

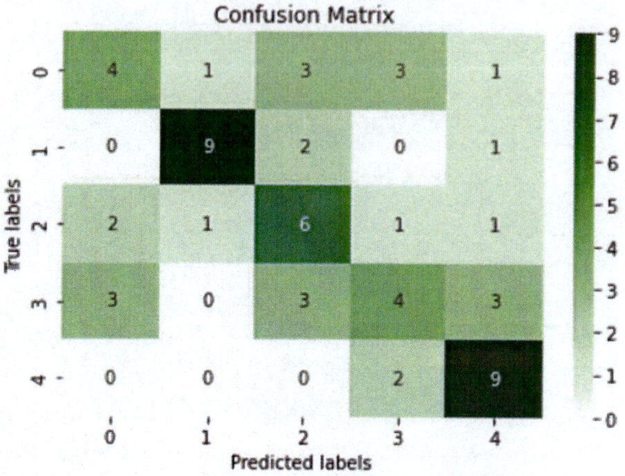

Fig. 7 Confusion matrix for the results of the CNN model

	precision	recall	f1-score	support
Average(242-264)	0.22	0.33	0.27	12
Heavier(298-960)	1.00	0.83	0.91	12
Heavy(265-297)	0.15	0.18	0.17	11
Light(220-241)	0.15	0.15	0.15	13
Lighter(160-219)	0.60	0.27	0.37	11
accuracy			0.36	59
macro avg	0.43	0.36	0.37	59
weighted avg	0.42	0.36	0.37	59

Fig. 8 Precision, recall and F1-score of the CNN model without dropout regularization

	precision	recall	f1-score	support
Average(242-264)	0.44	0.33	0.38	12
Heavier(298-960)	0.82	0.75	0.78	12
Heavy(265-297)	0.43	0.55	0.48	11
Light(220-241)	0.40	0.31	0.35	13
Lighter(160-219)	0.60	0.82	0.69	11
accuracy			0.54	59
macro avg	0.54	0.55	0.54	59
weighted avg	0.54	0.54	0.53	59

Fig. 9 Precision, recall and F1-score of the CNN model with dropout regularization

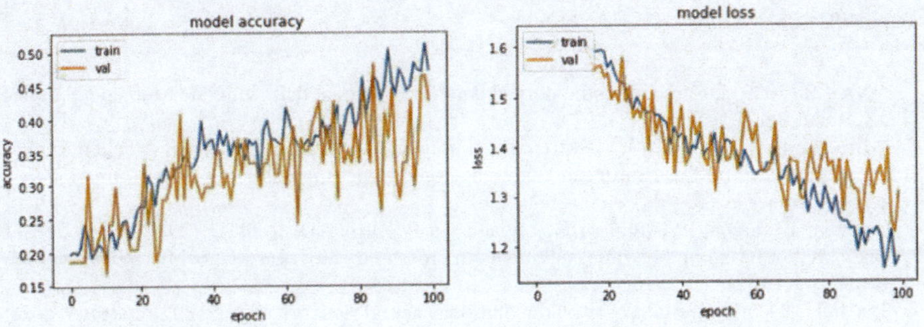

Fig. 10 Model accuracy and loss for the CNN model without dropout regularization

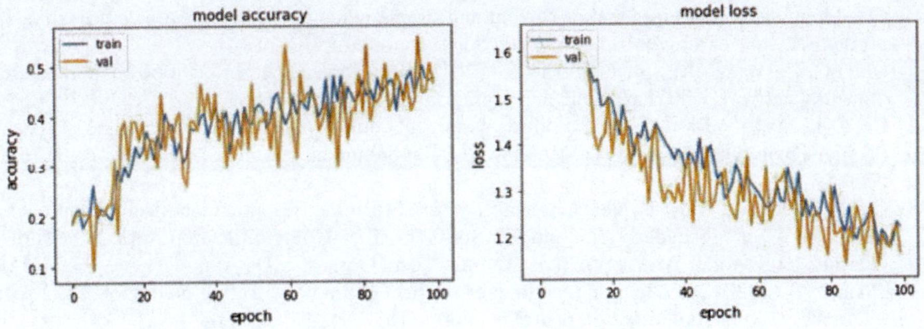

Fig. 11 Model accuracy and loss for the CNN model with dropout regularization

5 Conclusion

The work presented in this paper was aimed to develop a live weight estimation approach that is able to perform a physical live weight estimation of a cattle using a deep learning approach. Based on the findings obtained in this work, the dropout regularization affected the performance of the CNN model. For future works, one may investigate the appropriate number of partitions for the cattle's weight in order to improve the prediction capability of the proposed CNN model. Another future work that can be explored is the investigation to find the optimum performance accuracy of the proposed CNN model by optimizing the values of the hyperparameter settings of the CNN model.

Acknowledgements This work was supported in part by research grants from Universiti Malaysia Sabah (DFK2002).

References

1. Alfred R (2009) Discretization numerical data for relational data with one-to-many relations. J Comput Sci 5(7):519
2. Alfred R, Kazakov D (2007) Discretization numbers for multiple-instances problem in relational database. In: East European conference on advances in databases and information systems. Springer, pp 55–65
3. Ashwini JP, Sanjay P, Amipara G, Lunagariya P, Parmar D, Rank D (2019) Prediction of body weight based on body measurements in crossbred cattle. Int J Curr Microbiol Appl Sci 8(03):1597–1611
4. Bačanin Džakula N et al (2019) Convolutional neural network layers and architectures. In: Sinteza 2019-international scientific conference on information technology and data related research. Singidunum University, pp 445–451
5. Barbero R, Malheiros E, Araújo T, Nave R, Mulliniks J, Berchielli T, Ruggieri A, Reis R (2015) Combining Marandu grass grazing height and supplementation level to optimize growth and productivity of yearling bulls. Anim Feed Sci Technol 209:110–118
6. Bhatt C, Hassanien AE, Shah NA, Thik J (2018) Barqi breed sheep weight estimation based on neural network with regression. arXiv preprint arXiv:1807.10568
7. Cai C, Li J (2013) Cattle face recognition using local binary pattern descriptor. In: 2013 Asia-Pacific signal and information processing association annual summit and conference. IEEE, pp 1–4
8. Gaber T, Tharwat A, Hassanien AE, Snasel V (2016) Biometric cattle identification approach based on Weber's local descriptor and AdaBoost classifier. Comput Electron Agric 122:55–66
9. Gjergji M, de Moraes Weber V, Silva LOC, da Costa Gomes R, De Araújo TLAC, Pistori H, Alvarez M (2020) Deep learning techniques for beef cattle body weight prediction. In: 2020 international joint conference on neural networks (IJCNN). IEEE, pp 1–8
10. Hijazi S, Kumar R, Rowen C et al (2015) Using convolutional neural networks for image recognition, vol 9. Cadence Design Systems Inc., San Jose, CA, USA
11. Huma ZE, Iqbal F (2019) Predicting the body weight of Balochi sheep using a machine learning approach. Turk J Vet Anim Sci 43(4):500–506
12. Kaewtapee C, Rakangtong C, Bunchasak C (2019) Pig weight estimation using image processing and artificial neural networks. J Adv Agric Technol 6(4)
13. Khorshidi-Jalali M, Mohammadabadi M, Koshkooieh AE, Barazandeh A, Babenko O (2019) Comparison of artificial neural network and regression models for prediction of body weight in Raini Cashmere goat
14. Kumar S, Pandey A, Satwik KSR, Kumar S, Singh SK, Singh AK, Mohan A (2018) Deep learning framework for recognition of cattle using muzzle point image pattern. Measurement 116:1–17
15. Miller GA, Hyslop JJ, Barclay D, Edwards A, Thomson W, Duthie CA (2019) Using 3D imaging and machine learning to predict liveweight and carcass characteristics of live finishing beef cattle. Front Sustain Food Syst 3:30
16. Nilchuen P, Yaigate T, Sumon W (2021) Body measurements of beef cows by using mobile phone application and prediction of body weight with regression model. Songklanakarin J Sci Technol 43(6)
17. Nugraha U, Wahyu AP (2018) Weight measurement and identification of cow type using computer vision method. Int J Eng Technol 7(4.34):291–294
18. Poppi DP, Quigley SP, Silva TACCD, McLennan SR (2018) Challenges of beef cattle production from tropical pastures. Rev Bras Zootecn 47
19. Ruchay A, Kolpakov V, Kalschikov V, Dzhulamanov K, Dorofeev K (2021) Predicting the body weight of Hereford cows using machine learning. IOP Conf Ser Earth Environ Sci 624:012056. IOP Publishing
20. Santos L, Santos FN, Oliveira PM, Shinde P (2019) Deep learning applications in agriculture: a short review. In: Iberian robotics conference. Springer, pp 139–151

21. Shen W, Hu H, Dai B, Wei X, Sun J, Jiang L, Sun Y (2020) Individual identification of dairy cows based on convolutional neural networks. Multimed Tools Appl 79(21):14711–14724
22. Song X, Bokkers E, van der Tol P, Koerkamp PG, Van Mourik S (2018) Automated body weight prediction of dairy cows using 3-dimensional vision. J Dairy Sci 101(5):4448–4459
23. Tompson JJ, Jain A, LeCun Y, Bregler C (2014) Joint training of a convolutional network and a graphical model for human pose estimation. In: Advances in neural information processing systems, vol 27
24. Weber VADM, Weber FDL, Gomes RDC, Oliveira Junior ADS, Menezes GV, Abreu UGPD, Belete NADS, Pistori H (2020) Prediction of Girolando cattle weight by means of body measurements extracted from images. Rev Bras Zootecn 49
25. Zin TT, Phyo CN, Tin P, Hama H, Kobayashi I (2018) Image technology based cow identification system using deep learning. In: Proceedings of the international multiconference of engineers and computer scientists, vol 1, pp 236–247

The Design of a New IT Governance Framework to Optimize the Utilization of Information Technology for MSME

Yusuf Wibisono and Yanti Rosmunie

Abstract The presence of industry 4.0 brings various conveniences in operational and marketing processes for companies, including micro, small and medium enterprises (MSMEs). All companies are required to be able to utilize information technology in order to win the competition. Based on 2019 data, the number of MSMEs in Indonesia is 99.99% of the total number of companies and contributes to a GDP of 61.1%. With many MSMEs in Indonesia, their existence also opens up wider job opportunities with more equitable economic benefits for the community. To win the competition, MSMEs must optimize the use of information technology. To use information technology to align with business goals, MSMEs need IT Governance. Implementation of the IT Governance Framework generally runs in large companies. Constraints for small companies are usually related to the cost and complexity of the existing IT Governance Framework. This paper aims to develop an IT Governance Framework suitable for MSMEs and can be applied easily and efficiently. The initial stage is to develop a framework based on a literature review of the existing frameworks, namely COBIT, ITIL, CMMI, TOGAF, and relevant research. The proposed framework consists of six main components: Architectural Readiness, Technology Adoption, Service Management, Continuous Improvement, Leader Engagement, and Network Capability. Subsequently, improvements were made to the proposed framework by conducting quantitative and qualitative research on MSMEs to produce a better framework. As a validation stage, the enhanced framework will then be discussed with experts and the response from the MSMEs. Thus, it is hoped that the proposed framework can help MSMEs effectively utilize information technology that supports and is in line with MSME's business values.

Keywords IT Governance · Framework · MSME · Indonesia

Y. Wibisono (✉)
Mulia University, Balikpapan 76114, Kalimantan Timur, Indonesia
e-mail: wibisono@universitasmulia.ac.id

Y. Rosmunie
Universiti Malaysia Sarawak, 94300 Kota Samarahan, Sarawak, Malaysia
e-mail: byanti@unimas.my

© The Author(s), under exclusive license to Springer Nature Singapore Pte Ltd. 2023
D.-K. Kang et al. (eds.), *Proceedings of the 9th International Conference on Computational Science and Technology*, Lecture Notes in Electrical Engineering 983,
https://doi.org/10.1007/978-981-19-8406-8_12

1 Introduction

Micro, Small and Medium Enterprises (MSMEs) are businesses whose number of employees, sales, or loan size is below a certain threshold, and each country may have different criteria. The meaning of Micro in MSME, according to [1, 2], is a company with a maximum initial capital of up to IDR 1 billion (USD 67,252) or a maximum total annual sales of IDR 2 billion (USD 134,504), which usually has limited employees and technology.

Based on data from the Indonesian Central Statistics Agency, the growth of MSMEs in Indonesia is relatively rapid. In 2012, the number of MSMEs reached 55,206,444. In the same year, the number of big businesses was 4952 companies. So, the number of MSMEs is 99.99% of all companies. In 2018, the number of MSMEs was 64,194,057 companies, and the number of large businesses was 5460. The percentage of MSMEs to all companies is also 99.99%. For six years, the growth of MSMEs has reached 16.28%, and large companies have reached 10.26%.

The growth of MSMEs is also driven by various facilities provided by Industry 4.0 to introduce and market products to the public through technology and social media. Industry 4.0 also provokes new startups that start small businesses and develop into big businesses. Most of these MSMEs have implemented information technology as a natural part of supporting their business. MSMEs must do this if they do not want to lose out on the tight competition based on information technology. To optimize the use of Information Technology, MSMEs need IT Governance.

Well-known IT governance, such as ITIL and COBIT, is very complex and requires significant financial outlay, especially in the form of consulting fees [3]. Hence, they are adopted mainly by large organizations with significant IT budgets. Due to the complexity of frameworks, such as ITIL and COBIT, small businesses have barely reached the conditions for adopting such frameworks. Some modules can help, but they are complex to implement in small businesses. Meanwhile, the high cost and complexity of handling large amounts of digital information can be intimidating for small businesses [4].

In Indonesia, MSMEs generally only use technology as competitors do. The government hopes that MSMEs will be present in the digital world and support business through the effective use of IT [5]. Based on [6], only about 13% or only about 8.3 million of the 64.2 million MSME entrepreneurs in Indonesia utilize digital technology, especially e-commerce platforms, even though the number of active smartphone users in Indonesia is estimated to be around 42% or more than 100 million people in 2018.

Companies in micro, small and medium scale have different characteristics from large companies. These differences can be seen from the revenue value, asset value, number of employees, and influence on the global economy. MSMEs have a significant impact from an economic and social perspective. However, they generally have a low quality of information regarding access to credit markets [7].

From the various characteristics of MSMEs, applying IT Governance Framework standards such as COBIT, ITIL, CMMI, and TOGAF, is not easy due to the complexity and cost of consulting in its implementation. On the other hand, each existing framework generally focuses on a specific area, such as ITIL as service management, TOGAF, which focuses on enterprise architecture, and COBIT, which focuses on enterprise governance. With this difference in focus, ideally, companies need to implement multiple frameworks to be complete. Of course, this is not efficient for MSMEs.

This condition is why the importance of an IT Governance framework that is easy to apply to MSMEs. A framework that focuses on reducing implementation complexity and using an approach that fits the pattern and style of MSMEs. In today's era, MSMEs may not implement overall IT management as is generally done by large companies. However, some are instead directly using technology to support their latest business operations, according to Industry 4.0. This study aims to develop an IT Governance framework that can be used by MSME companies more efficiently.

2 Literature Review

In this study, it is necessary first to describe some of the underlying theoretical bases. It will present the theory IT Governance, including the definition of the IT Governance Framework, the importance of IT Governance, the condition of IT Governance in Indonesia, and the impact of IT Governance. The following section will explain what MSME is, the condition of MSME in Indonesia, the types of MSME, and the importance. Furthermore, it will explain the adoption of IT Governance in the company by presenting four frameworks, namely COBIT, ITIL, CMMI, and TOGAF.

2.1 IT Governance Framework

IT Governance Framework is a framework that defines ways and methods as organizational guidelines in implementing, managing, and monitoring IT Governance in an organization [8]. The IT Governance Framework helps provide a roadmap and evaluate the performance and effectiveness of IT governance. Several existing IT Governance Frameworks include COBIT, ITIL, CMMI, and TOGAF.

COBIT. COBIT is a framework for information technology governance for the entire enterprise model [9]. This information technology includes all information technology and processing to achieve company goals. Governance and management are clearly distinguished in COBIT, as they have different activities, different organizational structures and serve different purposes.

The research was conducted to analyze the capability of the IT governance level in Indonesia Air Asia using the COBIT 5 framework [10], which aims to provide recommendations for effective and efficient IT management and solutions. The results of the IT governance assessment at AirAsia Indonesia using the COBIT 5 framework produce an average score on the degree of capability from the whole process, indicating that most of the governance activities in the assessed domains have been monitored adequately. To raise the higher level, companies need to start designing innovations according to the company mission, such as vending machines ticket innovation and chatbots can make it easier for customers to order tickets and improve service quality by using information technology. Another study on the implementation of COBIT was conducted to explore the implementation of the COBIT framework to measure IT risk control of internet financial companies in China [11] and the implementation of the COBIT framework to define and maintain information technology infrastructure in the West Nusa Tenggara Regional General Hospital [12].

ITIL. ITIL is a best practice framework for managing IT services in organizations [13]. ITIL is the leading IT service management framework used in the world today. ITIL as a concept began in the 1980s, and many companies adopted it in the 1990s to help manage their IT infrastructure. ITIL has developed into a leading framework for IT services that help increase the competitiveness of enterprises.

There is a considerable amount of research and application of ITIL in organizations. Among these is research on implementing ITIL process management to mitigate operating risk on cloud architecture infrastructure for the banking and financial services industry [14, 15]. Another study was also carried out to build a cognitive model by adopting ITIL to improve IT services for companies in India [16]. Meanwhile, a study was also conducted to find forms of improvement in the System Change Management section based on the ITIL version 3 framework in making recommendations based on the values expected by management and evaluating the Continuous Service Improvement process [17]. Another study was also conducted [18] to explore how ITIL 4 can provide a framework for managing security.

CMMI. CMMI is a process and behavior model that assists organizations in process improvement. CMMI also supports productive behavior to reduce risk in software, product, and service development [19]. The Software Engineering Institute develops the CMMI as a process improvement tool for a project, division, or organization. CMMI helps improve performance and supports businesses in developing consistently better products and services. CMMI can also help create structures to encourage productive and efficient organizational behavior.

The following are some studies and applications of CMMI carried out by other researchers. A study in Sudan was conducted to provide a solution by encouraging SMEs to implement a software model in their software development process using CMMI [20]. Other research focuses on conducting a study to build an Intelligent Electric Meter Software Quality Evaluation Model based on the CMMI framework [21]. Meanwhile, a study was also conducted to look at the application of CMMI in Mass Spectrometer Software Project Development, which showed that the application of CMMI can efficiently increase the productivity and level of visual management, thus leading to broad application prospects [22].

TOGAF. TOGAF is an enterprise architecture framework developed in 1995 by The Open Group [23]. This framework contains a set of rules for organizational development designed to help companies achieve development goals quickly and cost-effectively. TOGAF contains guidelines for creating, interpreting, analyzing, and using the different elements of a single strategy. TOGAF also offers solutions with clear ROI, solid communication, and good perspective.

TOGAF is a framework for enterprise architecture and has been extensively researched and implemented in various projects. A study was conducted to see the application of TOGAF's ADM (Architecture Development Method) methodology to develop Enterprise Architecture for healthcare services by identifying target architectures, looking for gaps, and using gaps as recommendations to solve problems [24]. The TOGAF ADM Framework is also applied to design an IT master plan based on the TOGAF ADM framework in PDAMs by implementing the Preliminary Requirement Management, Architecture Vision, Business Architecture, Information System Architecture, Technology Architecture, to Opportunity and Solution [25]. A study was also conducted to examine the application of TOGAF to build an IS/IT framework for forest fire disaster mitigation by utilizing the Architectural Vision, Business Architecture, Information System Architecture, and Technology Architecture [26].

The latest developments for each IT Governance Framework are as follows (Table 1).

Table 1 IT governance framework development

Framework	Version (Release)	Development
COBIT	COBIT 2019 (2018)	COBIT 2019 has six governance principles. There was a change in terminology, and the number of governance support processes became 40 from the previous 37. Governance principles were added, and management performance is based on the CMMI performance management scheme instead of ISO/IEC 33000. COBIT 2019 covers new technologies and business trends in IT, which can integrate with company standards, guidelines and regulations [27]
ITIL	ITIL 4 (2019)	The updates focused on customers, value flows, and digital transformation, including artificial intelligence and cloud computing. In ITIL 4, service dimensions are not only about IT. Conducted value co-creation and service value chain development. Focus on practice, not just on the process [28]
CMMI	CMMI 2.0 (2018)	Focus on improving the ability to develop quality products and solutions and provide quality service offerings that meet market and customer needs. Also, improve supply chain management and reduce risk [29]
TOGAF	TOGAF 10 (2022)	The focus is strengthened on agile environments and digital transformation. The documentation is extended to assist organizations in implementing the framework and adapting it to business needs. TOGAF 10 is also easier to customize according to business needs [30]

2.2 Micro, Small, Medium Enterprises (MSME)

Each country has its definition for MSMEs. In Indonesia, regulations regarding MSMEs are stated in [1], which is then regulated in more detail in [2], providing the definition based on business capital as follows: Micro Business if the business capital is a maximum of IDR 1 billion (USD 67,252), Small Business if the business capital is more than IDR 1 billion (USD 67,252) and a maximum of IDR 5 billion (USD 336,261), and Medium Business if the business capital is more than IDR 5 billion (USD 336,261) to IDR 10 billion (USD 672,522). If the company has carried out activities before this regulation is implemented, then the following criteria will be applied based on annual sales results: Micro Business if annual sales are a maximum of IDR 2 billion (USD 134,504), Small Business if the annual sales are more than IDR 2 billion (USD 134,504) to a maximum of IDR 15 billion (USD 1,008,783), and Medium Business if annual sales are more than IDR 15 billion (USD 1,008,783) up to a maximum of IDR 50 billion (USD 3,362,610).

Based on the size, MSME businesses are divided into Micro, Small, and Medium Enterprises. By type of business, the Indonesian Central Statistics Agency divides MSMEs simply into two types, namely agricultural and non-agricultural businesses. Meanwhile, MSME practitioners are trying to divide the types of MSMEs in Indonesia. MSME businesses divide into 7 business types (Putra, 2018), namely: Culinary, Fashion, Technology, Cosmetics, Automotive, Souvenir, and Agribusiness. Meanwhile, the MSMEs also divide into 12 business types (rumah.com), namely: Culinary, Fashion, Educational, Automotive, Agribusiness, Tour & Travel, Creative Product, Internet Technology, Beauty, Event Organizer, Cleaning Service, and Children's Needs.

According to [31], the number of MSMEs in 2018–2019 was 64.2 million, consisting of 63,350,222 micro-companies, 783,132 small companies, and 60,702 medium companies. Meanwhile, the number of large companies is 5500. So the percentage of MSME is 99.9914%, and large companies are 0.0086%. The absorption capacity of MSME workers is 117 million workers or 97% of the labour absorption capacity. Meanwhile, the contribution of MSMEs to the national economy (GDP) is 61.1%, and large businesses contribute the remaining 38.9%.

2.3 Related Research

A study was conducted [32] to examine the alignment of digital strategies among 43 SMEs in the UK, Ireland, Italy, and Spain operating in five industrial sectors. The approach is divided into five levels: Passive Acceptance, Connection, Immersion, Fusion, and Transformation. Analysis was conducted based on interviews and questionnaires to classify each of the 43 companies into one of five levels of technology use. The result of this study indicate that Passive Acceptance is 9.3%, Connection is 30.23%, Immersion is 44.19%, Fusion is 9.30%, and Transformation is 6.98%.

A study on improving the performance of MSMEs through digital platforms was also carried out [33]. This study examines the effect of digital platform capabilities and network capabilities on the financial performance of SMEs. This study also examines how exploitation and exploration orientations moderate this relationship. Based on an analysis of 230 Swedish SMEs, the results show that digital platform capabilities positively affect SME performance through network capabilities. The results show that SMEs can improve their performance through digital platform capabilities by aligning these capabilities with their orientation.

A study in Spain [34] shows that SMEs must recognize the importance of defining innovation strategies in applying embedded technologies, advanced robotics, cyber-physical systems, and IoT. It is necessary to do the right strategy before implementing the technology. The research also shows evidence of the positive influence of Industry 4.0 on SME innovation, which positively affects company growth and product quality.

Another study conducted in Italy [35] showed that many SMEs assume that any technology investment will automatically generate a positive impact on performance. In many cases, companies underestimate the costs and difficulties of introducing new technical solutions in an organized system and lack the resources to support organizational innovation. This study concludes that companies need to plan well in implementing each new technology.

From some of these studies, it can be summarized that it is essential for companies to prepare themselves well to face future challenges. The new technology brought by Industry 4.0 does provide many benefits. However, companies need to assess readiness and prepare the best strategy for adopting technology. In this position, proper IT governance is needed for MSME so that technology implementation can be in line with organizational goals.

3 Method

This research begins with a literature review by examining various relevant sources and continues with a combination of qualitative and quantitative research. The proposal for the framework at the initial stage will be carried out by combining studies of the existing framework with relevant research. Furthermore, the proposed framework will be refined in several stages (see Fig. 1).

This research model is designed in six stages: Literature Review, Proposal of the Draft Framework, Quantitative and Qualitative Research, Revising Framework, Validating Framework, and Finalizing Framework.

- Literature Review: focus on studies on the existing framework and conduct studies on previous studies relevant to this research.
- Proposal of the draft framework: propose a framework based on studies of the existing framework and previous research.

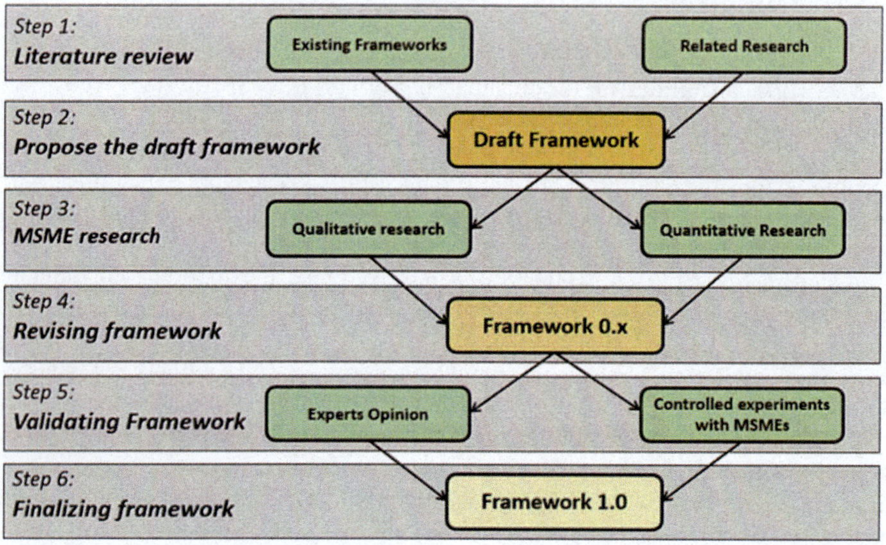

Fig. 1 The research process

- Quantitative and Qualitative Research: conduct a survey to support quantitative research on MSMEs to obtain valid data on the characteristics of MSMEs in Balikpapan. Meanwhile, the interview and observation will be conducted to support the qualitative research to obtain information on managing MSMEs in Balikpapan.
- Revising framework: based on the research results on MSME, further revisions are made as needed based on the data and information obtained. In this phase, gradual improvements are made to the proposed framework. The proposed framework is named 0.x, which means 0.1, 0.2, or 0.3, to anticipate versions of changes. This change was made to meet the needs based on the research results.
- Validating framework: the revised framework then consulted with the experts and conducted controlled experiments with MSMEs.
- Finalizing framework: The framework is finalized based on expert opinion and the result of the controlled experiment and will be named 1.0.

4 Research Framework

The research is in the early stages of designing a framework based on a literature review and combined with a study of related research. The framework proposed in the initial stage consists of six main components: Architecture Readiness, Technology Adoption, Service Management, Continual Improvement, Leader Involvement, and Network Capability (See Fig. 2).

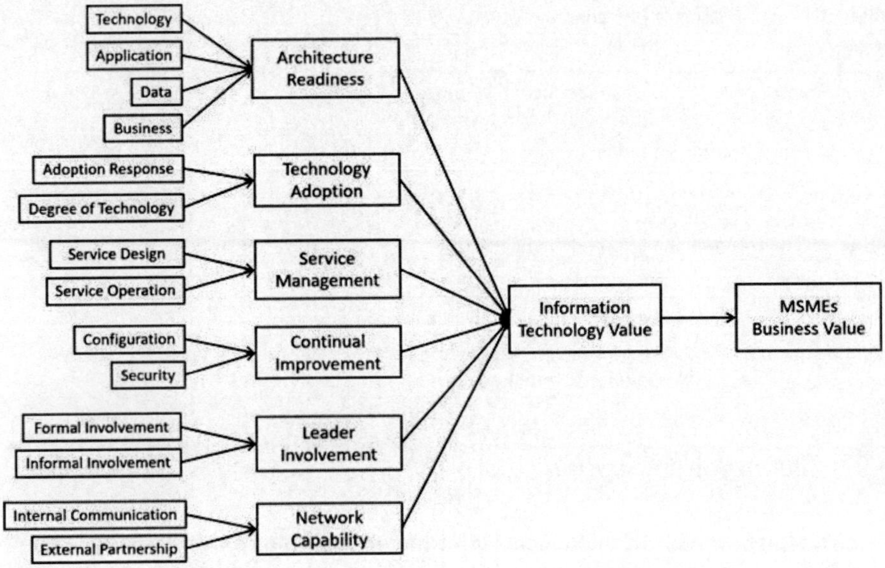

Fig. 2 The proposed framework

4.1 Architecture Readiness

This part identifies the architectural readiness of MSMEs. Although the information technology architecture is generally simple or very simple, in principle, MSMEs must still have a good perception of the architecture used. This element of architecture readiness is based on [36], who researched the implementation of TOGAF for MSME, which divided the architecture into four groups, namely technology architecture, application architecture, data architecture, and business architecture. The following Table 2 explains the architecture readiness.

Table 2 The components of architecture readiness

Components	Descriptions
Technology architecture	Focus on how to design the technology model within the company
Application architecture	Describes how applications are designed within the company and the relationship between them
Data architecture	Describes the design of data and the relationship between databases
Business architecture	Describes the business model, business strategy, and business functionality

Table 3 The components of technology adoption

Components	Descriptions
Adoption response	Describes how the company's level of acceptance or response to the presence of new technology
Degree of technology	Indicates the level of sophistication of the technology used

Table 4 The components of service management

Components	Descriptions
Service design	See how companies design their IT services
Service operation	See how the company provides IT services to support the company's operational activities

4.2 Technology Adoption

In looking at how MSME implements information technology, it is essential to know how the company's level of adoption of technology is. With the rapid development of technology, the winners are those who can choose the best technology to support their business activities. The principle in Technology Adoption is based CMMI concept in maturity level and also referred to [32], who researched digital strategy alignment in SME, and divided the company's adoption readiness into five levels: passive acceptance, connection, immersion, fusion, and transformation. The following Table 3 gives the explanation of Technology Adoption.

4.3 Service Management

Service Management is developed from the ITIL Framework. There are five components in ITIL, namely service strategy, service design, service transition, service operation, and continual service improvement [13]. This research tries to simplify and only takes two components: service design and service operation. The following Table 4 gives the explanation of Service Management.

4.4 Continual Improvement

One of the critical things in the company is maintaining business continuity [37]. Concerning this in the implementation of information technology and the provision of services to support its business operations, it is essential to see how the company makes improvement efforts to improve the quality and aspects of technology security.

Table 5 The Components of Continual Improvement

Components	Descriptions
Configuration	See how the company configures technology to achieve the best performance and improve it over time
Security	See how the company has made security for configurations and services in order to ensure continuity and business improvement

Table 6 The Components of Leader Involvement

Components	Descriptions
Formal involvement	Leadership involvement is shown formally in the IT organizational structure
Informal involvement	Leadership is involved but not clearly stated in the formal organizational structure

This continual improvement was developed from the principles in COBIT and ITIL. The following Table 5 gives the explanation of Continual Improvement.

4.5 Leader Involvement

The success of implementing information technology also depends on how the leadership is involved with it. The level of leadership involvement will affect the speed of IT improvement because, in general, it will be related to IT investment. The following Table 6 gives the explanation of Continual Improvement.

4.6 Network Capability

The success of organizational performance is undoubtedly influenced by how communication and relationships run. Obstacles in relationships within the organization or relationships with external parties will be a barrier for the company to be able to develop better. Network Capability is developed from [33], focusing on how MSMEs compete in digital platform capabilities, network capabilities, and ambidexterity. The following Table 7 gives the explanation of Continual Improvement.

5 Discussion

Based on the variables that have been studied previously, the complete outline of the proposed framework can be seen in Fig. 2.

Table 7 The components of network capability

Components	Descriptions
Internal communication	Focus on how communication within the organization goes well
External partnership	Focus on how the organization builds partnerships with external parties in order to support operations and business development

Based on Fig. 2, all proposed framework components can be seen. First, the MSME must have an idea of the desired architecture. This architecture can describe the technology, applications, data, and business. This architectural description may be simple but straightforward enough.

After having a picture of the desired architecture, the next step is determining the Technology Adoption to support the applied architecture. At this point, the ability to adapt can be measured from two factors: the level of depth of the technology used and how the company responds to the technology.

The combination of the chosen architecture and technology will become an IT service in the company. Service Management is formed from two factors, namely the service model that is designed and how the service is carried out to support the company's activities.

At this point, the company must also prepare efforts for business continuity by conducting Continual Improvement. This improvement strategy is implemented through the proper configuration and how the company makes security efforts for its overall information technology.

The next aspect is how the leadership is involved in IT decision-making. This Leader's Involvement can be formal because there is an IT organization, and the top leadership has the highest role in decision making. The involvement of leaders can also be informal; that is, they do not have an IT organization specifically, but decisions in IT always involve top management.

The last aspect is related to Network Capability, namely the company's ability to optimize communication within the team and external parties. IT implementation is closely related to communication patterns between users and the IT team and how to use outside partners to support the services provided.

6 Conclusion

MSME has obstacles, challenges, and opportunities to implement information technology to support its business activities. The number of MSMEs in a country is vast compared to the number of large companies. However, the economic impact of MSMEs in the country is still not optimal financially and is still relatively small when viewed from the comparison of the number of MSMEs with large companies. One way to increase the economic role of MSMEs is to optimize the use of information technology to support their business activities. One way to optimize the use of information technology is to apply the appropriate IT Governance Framework. Therefore, it is essential to conduct and support research to develop an ideal framework for MSMEs.

Acknowledgements This research paper is supported and funded by Faculty of Computer Science and Information Technology, University Malaysia Sarawak, Malaysia and Faculty of Computer Science, Mulia University, Indonesia.

References

1. Kementerian Sekretariat Negara RI (2020) Undang-Undang Cipta Kerja. Indonesia (2020).
2. Pemerintah Indonesia (2021) Peraturan Pemerintah Republik Indonesia Nomor 07 Tahun 2021 tentang Kemudahan, Pelindungan, dan Pemberdayaan Koperasi dan Usaha Mikro, Kecil, dan Menengah. Indonesia
3. He MQY, Peiris KDA, Myers M (2019) IT governance strategies for SMEs in the fourth industrial revolution. In: Proceeding of 23rd Pacific Asia conference inference system security ICT platform 4th industrial revolution, PACIS 2019
4. Ramaswamy M (2019) Southern U.: information technology strategies for small businesses. Issues Inf Syst 20:216–225
5. Masduki T (2022) Teten Kecewa Banyak UMKM Masih Belum Go Digital, Kenapa?, https://www.cnbcindonesia.com/tech/20200830201547-37-183127/teten-kecewa-banyak-umkm-masih-belum-go-digital-kenapa. Last accessed 03 Feb 2022
6. Kominfo (2022) Tingkatkan produktivitas dan Nilai Tambah UMKM melalui Teknologi Digital. https://www.kominfo.go.id/content/detail/30276/tingkatkan-produktivitas-dan-nilai-tambah-umkm-melalui-teknologi-digital/0/berita. Last accessed 02 Mar 2022
7. Dinutistomo IA, Lubis AW (2021) MSME lending and bank efficiency: Evidence from Indonesia. Banks Bank Syst 16:93–103. https://doi.org/10.21511/bbs.16(3).2021.09
8. Techopedia (2022) IT governance framework. https://www.techopedia.com/definition/30607/it-governance-framework#:~:text=IT governance framework is a, and processes within an organization. Last accessed 04 Jan 2022
9. ISACA (2018) Governance and management objectives
10. Pratiwi AN, Suharjito, Sukmandhani AA (2020) Analysis of capability level in dealing with IT business transformation competition using cobit framework 5 (case study at Airasia Indonesia). In: Proceeding of 2020 international conference on information management and technology ICIMTech 2020, vol 5, pp 609–614. https://doi.org/10.1109/ICIMTech50083.2020.9211116.
11. Yubo H (2020) IT risk control for internet finance based on COBIT. In: Proceedings—2020 international conference on big data artificial intelligence and software engineering ICBASE 2020, pp 275–278. https://doi.org/10.1109/ICBASE51474.2020.00064
12. Agitha N, Anjarwani SE, Azizah MI, Yunus IR, Witjaksono RW (2020) Implementation of COBIT 4.1 to define and maintain infrastructure of information technology at regional public hospital in West Nusa Tenggara. In: 2020 International conference on advancement in data science e-learning information systems ICADEIS 2020, pp 5–9. https://doi.org/10.1109/ICADEIS49811.2020.9277015.
13. Buehring S (2018) ITIL: what is ITIL®? http://www.itsmfi.org/forums/posts.aspx?group=155312&topic=1131678&page=1&hhSearchTerms=&#post_1131678%5Cnhttp://files/2549/posts.html
14. Mahalle A, Yong J, Tao X (2020) ITIL process management to mitigate operations risk in cloud architecture infrastructure for banking and financial services industry. Web Intell 18:239–248. https://doi.org/10.3233/WEB-200444

15. Ming C, On CK, Rayner A, Guan TT, Patricia A (2018) The determinant factors affecting cloud computing adoption by small and medium enterprises (SMEs) in Sabah, Malaysia. J Telecommun Electron Comput Eng 10:83–88
16. Dayal R, Vijayakumar V, Kushwaha RC, Kumar A, Ambeth Kumar VD, Kumar A (2020) A cognitive model for adopting ITIL framework to improve IT services in Indian IT industries. J Intell Fuzzy Syst 39:8091–8102. https://doi.org/10.3233/JIFS-189131
17. Arisenta R, Suharjito, Sukmandhani AA (2020) Evaluation model of success change management in banking institution based on ITIL V3 (case study). In: Proceeding of 2020 international conference information management technology ICIMTech 2020, vol 3, pp 470–475. https://doi.org/10.1109/ICIMTech50083.2020.9211191
18. Berger D, Shashidhar N, Varol C (2020) Using ITIL 4 in security management. In: 8th International symposium digital forensics security ISDFS 2020, pp 1–6. https://doi.org/10.1109/ISDFS49300.2020.9116257
19. White SK (2022) What is CMMI? A model for optimizing development processes. https://www.cio.com/article/2437864/process-improvement-capability-maturity-model-integration-cmmi-definition-and-solutions.html. Last accessed 03 Jan 2022
20. Mahmoud Z, Solyman A, Elhag AAM (2019) Harmonized software quality improvement models for Sudanese SME based on CMMI. In: Proceedings international conference computer control electrical electronics engineering 2019, ICCCEEE 2019. https://doi.org/10.1109/ICCCEEE46830.2019.9071148
21. Li W, Wang Y, Hou H, Zhao X (2020) Software quality evaluation model of smart electricity meters based on CMMI. In: 2020 5th international conference computer and communication systems ICCCS 2020, pp 157–161. https://doi.org/10.1109/ICCCS49078.2020.9118592
22. Wang W, Lu X, Jia R, Li F (2018) Development of mass spectrometer software project based on CMMI. In: 2017 3rd IEEE international conference computer communication ICCC 2017, pp 2508–2511. https://doi.org/10.1109/CompComm.2017.8322987
23. White SK (2022) What is TOGAF? An enterprise architecture methodology for business. https://www.cio.com/article/3251707/what-is-togaf-an-enterprise-architecture-methodology-for-business.html. Last accessed 10 Mar 2022
24. Girsang AS, Abimanyu A (2021) Development of an enterprise architecture for healthcare using TOGAF ADM. Emerg Sci J 5:305–321. https://doi.org/10.28991/esj-2021-01278
25. Sardjono W, Vijayanto RM (2021) Designing of IT master plan based on TOGAF ADM framework in the regional water utility company. In: IOP conference series earth environmental science, vol 729. https://doi.org/10.1088/1755-1315/729/1/012016
26. Dores A, Fitrianah D, Meilina P (2019) IS/IT framework for disaster mitigation on forest fires using TOGAF. In: 5th International conference information management ICIM 2019, pp 252–256. https://doi.org/10.1109/INFOMAN.2019.8714663
27. Harisaiprasad K (2022) COBIT 2019 and COBIT 5 comparison title. https://www.isaca.org/resources/news-and-trends/industry-news/2020/cobit-2019-and-cobit-5-comparison. Last accessed 13 June 2022
28. ITIL Training: what's changed with the ITIL 4 update? https://www.itiltraining.com/uk/ITIL4-whats-new. Last accessed 15 June 2022
29. ISACA: CMMI institute expands CMMI V2.0 to include services and supplier management. https://www.isaca.org/en/why-isaca/about-us/newsroom/press-releases/2018/cmmi-institute-expands-cmmi-v2-to-include-services-and-supplier-management. Last accessed 15 June 2022
30. White SK (2022) What's new in TOGAF 10? https://www.cio.com/article/400374/whats-new-in-togaf-10.html%0A. Last accessed 11 June 2022
31. Kementerian Koperasi dan Usaha Kecil dan Menengah: data UMKM. https://kemenkopukm.go.id/data-umkm/?Z5t48x8N2JyP34BJEKormtKjqZCBJZsacc8No7AzrN6pq9ohsl. Last accessed 18 Feb 2022
32. Canhoto AI, Quinton S, Pera R, Molinillo S, Simkin L (2021) Digital strategy aligning in SMEs: a dynamic capabilities perspective. J Strateg Inf Syst 30. https://doi.org/10.1016/j.jsis.2021.101682

33. Cenamor J, Parida V, Wincent J (2019) How entrepreneurial SMEs compete through digital platforms: the roles of digital platform capability, network capability and ambidexterity. J Bus Res 100:196–206. https://doi.org/10.1016/j.jbusres.2019.03.035
34. Somohano-Rodríguez FM, Madrid-Guijarro A, López-Fernández JM (2020) Does industry 4.0 really matter for SME innovation? J Small Bus Manag 00:1–28. https://doi.org/10.1080/00472778.2020.1780728
35. Cimini C, Boffelli A, Lagorio A, Kalchschmidt M, Pinto R (2021) How do industry 4.0 technologies influence organisational change? An empirical analysis of Italian SMEs. J Manuf Technol Manag 32:695–721. https://doi.org/10.1108/JMTM-04-2019-0135
36. Arman AA, Ruhman G, Hurriyati R (2018) Design of EA development guideline for small enterprises based on TOGAF 9.1. Adv Intell Syst Comput 657:210–219. https://doi.org/10.1007/978-3-319-67223-6_20
37. Kato M, Charoenrat T (2018) Business continuity management of small and medium sized enterprises: evidence from Thailand. Int J Disaster Risk Reduct 27:577–587. https://doi.org/10.1016/j.ijdrr.2017.10.002

A Safe Box Safety System with Facial Authentication for Multiple Users

Min Ning Ooi, Kim Gaik Tay, and Chang Choon Chew

Abstract A safe box with a security system protects personal and valuable items from being stolen, tampered or damaged. The most used security mechanism is using keys and cards, or the use of passwords, personal identification numbers (PINS), and fingerprints for personal identification. Face recognition is one of the most effective security confirmation techniques for a safe box management system. This project proposed a comprehensive face-authentication-based security system that includes software, hardware, and the Internet of Things (IoT). In the software system, the OpenCV built-in library and face recognition algorithm written in Python were implemented. A total of 150 images (i.e., 50 images per subject from 3 individuals) were registered in the database for training and testing the developed face recognition system. The control and working of the hardware are performed by a Raspberry Pi, to which a Pi camera, an infrared sensor, and a solenoid lock are connected. The IoT enables the communication between the Raspberry Pi and Telegram cloud, allowing real-time monitoring of the safe box and alerting the owner by capturing images of approaches. The developed system shows an accuracy ranging from 0.79 to 0.91, the precision range of 0.571–1, the recall range of 0.64–0.88, and the F1-score ranging from 0.6 to 0.81.

Keywords Face authentication · Safe box · Raspberry Pi · Internet of Things (IoT)

1 Introduction

With the advancement of technology, most facilities have incorporated innovative practices or methods into their routine operations to make daily life easier. The success in the technologies has also led to the illegal misuse of technology for

M. N. Ooi · K. G. Tay (✉) · C. C. Chew
Universiti Tun Hussein Onn Malaysia, 86400 Parir Raja, Batu Pahat, Johor, Malaysia
e-mail: tay@uthm.edu.my

C. C. Chew
e-mail: chewcc@uthm.edu.my

© The Author(s), under exclusive license to Springer Nature Singapore Pte Ltd. 2023

D.-K. Kang et al. (eds.), *Proceedings of the 9th International Conference on Computational Science and Technology*, Lecture Notes in Electrical Engineering 983,
https://doi.org/10.1007/978-981-19-8406-8_13

burglaries and thefts [1]. These issues are a challenge, especially with the advancement in crime methods. Therefore, it is essential to have a reliable safe technology to protect one's assets and privacy, especially for application in a household (i.e., to ensure the safekeeping of important documents, cash and cheques, and belongings). A fireproof and waterproof safe box could further protect one's valuables from floods, fires, tsunamis, earthquakes, or accidents [2]. Furthermore, the safe system is often designed to save the users' convenience at home by limiting their number of walk-in visits to the office or facilities (e.g., a bank for cash). Lastly, the safe box's weight makes it less likely to be stolen, protecting the valuables stored within it.

Safe boxes have been used to protect assets for thousands of years. Traditional safe box security methods include the use of keys and locks, passwords, and personal identification numbers (PINs). Each of these conventional methods has its limitations, wherein a non-owner can duplicate one's key or destroy its lock, compromising the security of the system. Even though password and PIN authentication provide higher security means using complex passwords, this method often leads to confusion and has not provided sufficient flexibility [3].

To prevent duplication and counterfeiting, biometric authentication technology can be implemented in a safe box safety system. Fingerprint identification is the earliest biometric technology because each person's fingerprint is unique. [4]. However, this method can be overly sensitive to missing fingerprints caused by injury, sweat, or dirt adhering to the skin. Facial recognition technology is an alternative means of biometric systems. Facial features recognition offers advantages over other biometric features in terms of accuracy, simplicity, and convenience [5], especially the fact that the system can perceive one's facial characteristics without requiring its users to interact with the system.

Face recognition has been applied in smart home security [6–8], or control appliances in [9] and attendance systems [10] using various face detection and recognition algorithm as summarized in Table 3.

This project aims to develop a face authentication system for the safe box application using a Raspberry Pi board. The algorithm is developed in Python based on the OpenCV and its built-in libraries for the recognition of the face correctly and reliably. This system allows multiple registered users to access the safe box while preventing unauthorized unlocking. One of the most notable advantages of this system is the use of IoT. When the infrared sensor detects an object approaching the box, the Pi camera would take a picture and send it to the owners to alert them of the possible intrusion.

2 Materials and Methods

The project design is divided into three major phases: enrollment, authentication, and IoT. This development process is followed by the performance evaluation of the system. The overall project design flowchart is shown in Fig. 1.

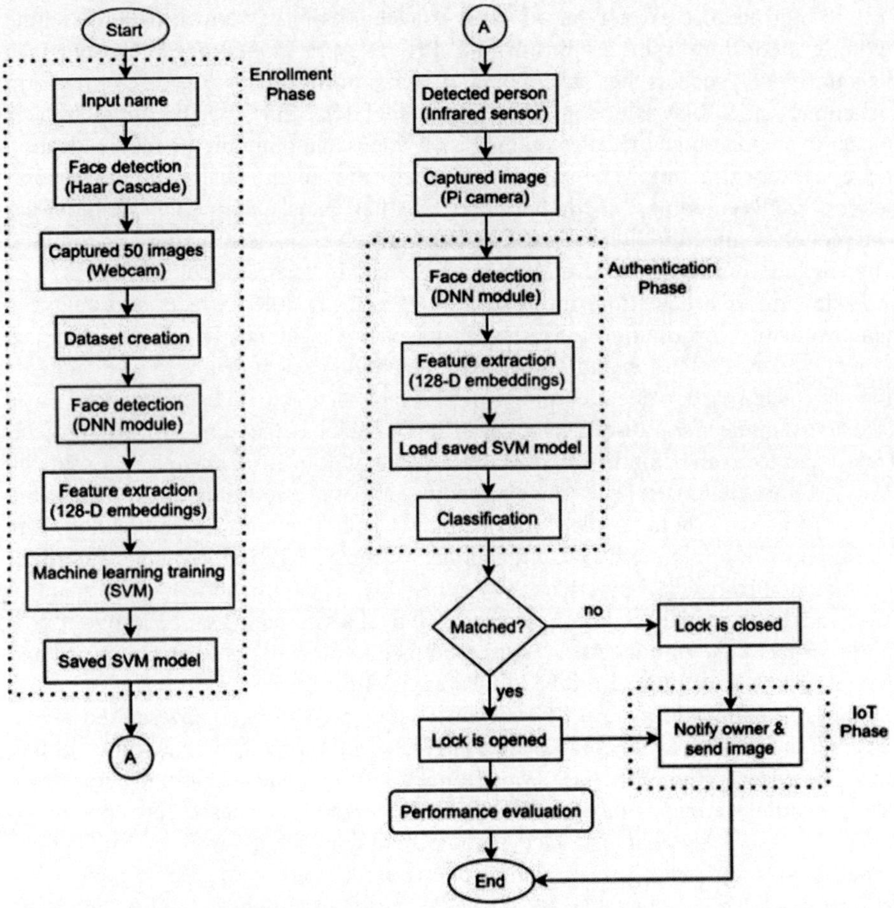

Fig. 1 Overall project design flowchart

2.1 *Enrollment Phase*

The enrollment phase consists of creating and training three registered users in the database. It begins storing the user's personal information, such as their names, in the system. After the users' names were added, the Haar cascade algorithm which was saved in "`haarcascade_frontalface_default.xml`" file was loaded to determine whether the image contains a human face. An image was captured and stored locally if the face's location was pinpointed. The face image was associated with the registered individuals one at a time. After fifty images were captured for each user, these images were stored in a dataset.

OpenCV itself cannot perform face recognition [11]. Thus, deep learning in conjunction with OpenCV was used to build this authentication system. Deep learning was implemented in the past for many processes, such as object recognition

[12, 13] and feature extraction, which is conventionally performed using machine learning algorithms such as that in [14, 15]. The pre-trained deep learning face detector model, such as the Caffe model, was first utilized with OpenCV to perform fast and accurate face detection. The Caffe model determines whether human faces appear in a wide range of scales in images or videos and pinpoint the faces' location to prepare for extraction. When utilizing the Caffe model with OpenCV's deep neural network (DNN) module, two files namely ".prototxt" and ".caffemodel" are required. The model architecture, which is the layers themselves, is described by ".prototxt", whereas the weights for the even layers are defined in ".caffemodel" [16, 17].

Following face detection, the feature extraction module extracts and analyzes human identity information [18]. The FaceNet model comes from the OpenFace project, which is a Python and Torch implementation of deep learning face recognition [18]. The Pytorch deep learning model, i.e., "openface_nn4.small2.v1.t7", was used to compute the 128-dimensional (128-D) facial embeddings in this project. This "openface_nn4.small2.v1.t7" has 93% recognition in Labeled Faces in the Wild (LFW) benchmark [19]. The embedding process includes the input data to the network followed by the triplet loss function to calculate the face embedding. The neural network computes 128-D embeddings for each face before using the triplet loss function to tweak the network's weights [11]. Thus, the network can quantify faces and deliver highly accurate performance by discriminating embeddings appropriate for face recognition. As a result, each face's embeddings were computed and the output was saved in the "embeddings.pickle" file.

These face embeddings were then trained using SVM to recognize an individual [11]. Scikit-learn's implementation of SVM is the common machine learning model. The Scikit-learn (LabelEncoder) was initialized to encode the user's name from the embeddings file into labels followed by the SVM model used for recognizing faces. The label encoder and face recognizer model were saved as pickle files. Label encoder and face recognizer models were named as "le.pickle" and "recognizer.pickle" respectively. Here, "le.pickle" is an output label encoder containing each user's name labels, whereas "recognizer.pickle" is an SVM-based output model that recognizes faces. The face recognizer model uses this output label encoder to differentiate between the registered users and unknown individuals. The authentication phase makes use of these SVM-based output models for the identification problem.

2.2 Authentication Phase

Before proceeding to the authentication phase, the infrared sensor and Raspberry Pi camera were connected to detect an individual's presence and capture the high-resolution image of that individual. After the image was captured, the authentication process began by loading the DNN module and 128-D embeddings to perform face detection and feature extraction. The operation of these two processes is as described in the preceding phase. After generating the 128-D vector, the result was compared to

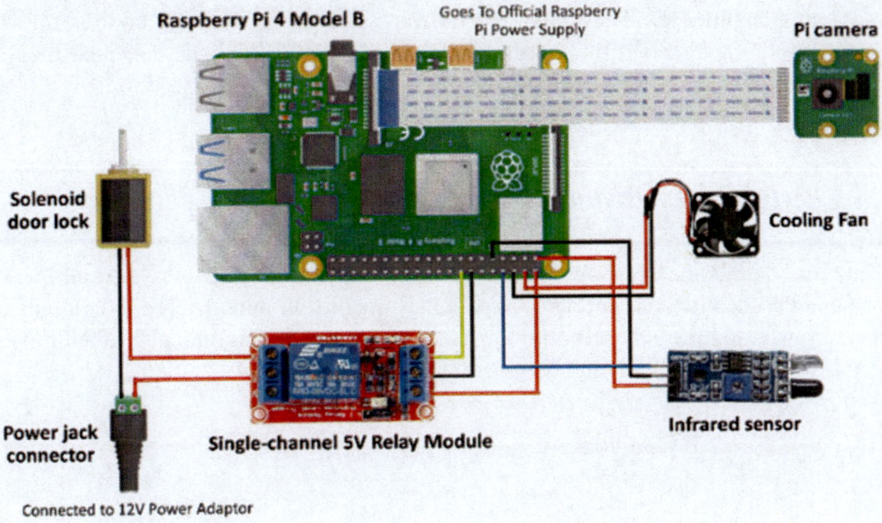

Fig. 2 Schematic diagram of safe box safety system with face authentication

the previously saved SVM model. The SVM model predicts the individual's identity by taking the highest probability. Suppose the computed probability exceeds the minimum confidence level of 0.8, in this case, the associated name would be displayed along with the probability for the face. In contrast, when the calculated probability is less than the required confidence level, the word "unknown" appears above the rectangle box for the face.

Following the authentication phase, a decision would be taken by the Raspberry Pi, which is connected to a Pi camera, a solenoid door lock, an infrared sensor, and a single-channel relay module, as shown in Fig. 2. This solenoid lock is activated to grant access to the user when the individual's face matches the dataset. In contrast, the lock remains closed if the individual's data does not match the dataset.

2.3 *Internet of Things (IoT) Phase*

In the IoT phase, whether the individual's face matches the dataset or not, the owner would be notified via IoT. The notification is sent along with the individual's image captured by the Pi camera, indicating that this individual is attempting to open the safe box. Even if the owner is not nearby, this IoT alerts them that someone is approaching the safe box.

IoT was applied in Telegram through its bot API. Firstly, a user's bot was created through BotFather to obtain a token to access HTTP APIs for the bot's access and control. Then, a private channel was set up and the user's bot was added as a channel administrator. Later, a "Hello" message was forwarded from the channel to IDBot to obtain a channel ID. The token and ID were inserted in Python's code to allow access for the face recognition system to send photos to the channel taken by the Pi camera.

2.4 Performance Evaluation Phase

Since this face authentication system has been carried out four classes, i.e., three registered users and an unknown group (consisting of four unregistered individuals), the confusion matrix is based on multi-class machine learning models. In addition, accuracy, precision, recall, and F1-score based on Eqs. (1)–(4)

$$Accuracy = \frac{TP + TN}{TP + TN + FP + FN} \tag{1}$$

$$\text{Precision} = \frac{TP}{TP + FP} \tag{2}$$

$$\text{Recal} = \frac{TP}{TP + FN} \tag{3}$$

$$\text{F1 - score} = \frac{2TP}{2TP + FP + FN} \tag{4}$$

where *TP*, *FP*, *TN*, *FN* represent true positive, false positive, true negative, and false negative rate, respectively were used to evaluate the performance of the developed face authentication system.

3 Results and Discussion

Figure 3 shows the testing performance of the system on the first registered user, "MinHan" from the live streaming of the videos from a different angle. The system has successfully recognized the person as "MinHan" with a high probability of 88.41% in Fig. 3a. In Fig. 3b, the user was misclassified as another user, "MinYan", with a probability of 89.01%, and as "unknown" with a probability of 72.22% in Fig. 3c. Of the 25 times testing attempts, this user was recognized as "MinHan", and misclassified as "MinYan", and "unknown" for 16, 2, and 7 times respectively.

Figure 4 illustrates the results of the second registered user, "MinYan", using the developed face authentication system. The results showed that the user is recognized as "MinYan" with a probability of 95.94% in Fig. 4a, "MinNing" with a probability of 82.59% in Fig. 4b, and "unknown" with a probability of 69.41% in Fig. 4c. Of the 25 attempts, the user was recognized as "MinYan", "MinNing", and "unknown" 19, 4, and 2 times, respectively. Figure 5 presents the results for the third registered

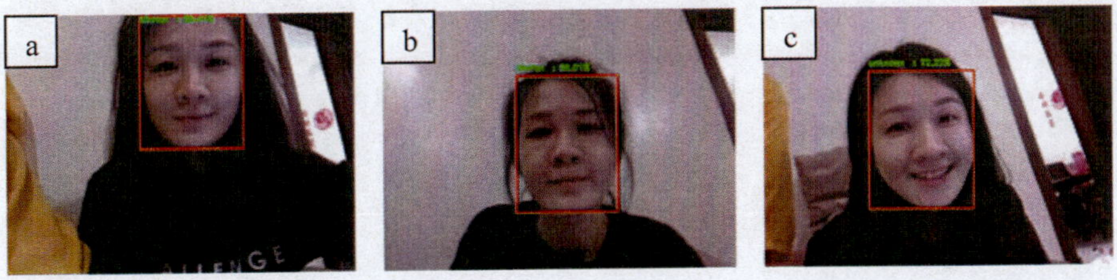

Fig. 3 Face authentication result for the first registered user, "MinHan", being recognized as **a** "MinHan" **b** "MinYan" **c** "unknown"

user, "MinNing. The system recognized this user as "MinNing" with a probability of 96.04% in Fig. 5a and "unknown" with a probability of 77.79% in Fig. 5b. Through the 25 trials, the user was recognized as "MinNing" 22 times and "unknown" for 3 times. Figure 6 shows the first unknown user tested with the face authentication system with different authentication results. The results recognized this user as "unknown" with a probability of 67.80% in Fig. 6a and "MinYan" with a probability of 84.90% in Fig. 6b. The testing was run for 7 times and the results of being recognized as "unknown" and "MinYan" were 4 times and 3 times, respectively.

Figure 7 illustrates the second unknown user tested with the face authentication system. The system identified the individual as "unknown" with a probability of 61.73% in Fig. 7a and "MinNing" with a probability of 95.40% in Fig. 7b. The testing was conducted 6 times and the results of being recognized as "unknown"

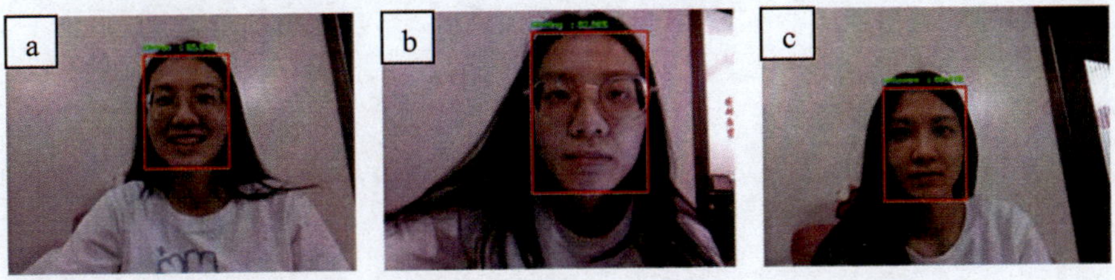

Fig. 4 Face authentication result for the second registered user, "MinYan", being recognized as **a** "MinYan" **b** "MinNing" **c** "unknown"

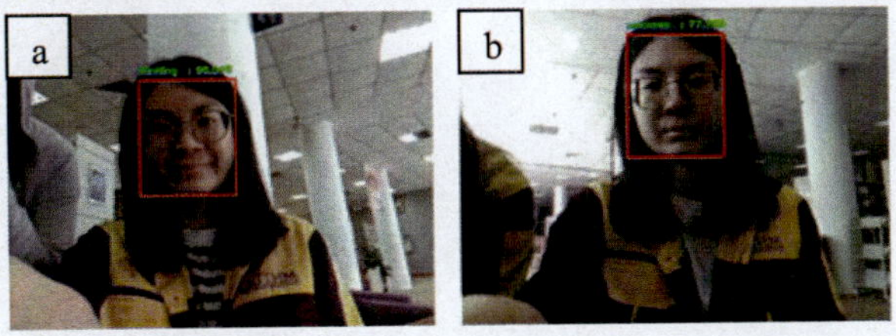

Fig. 5 Face authentication result for the third registered user, "MinNing", being recognized as **a** "MinNing" **b** "unknown"

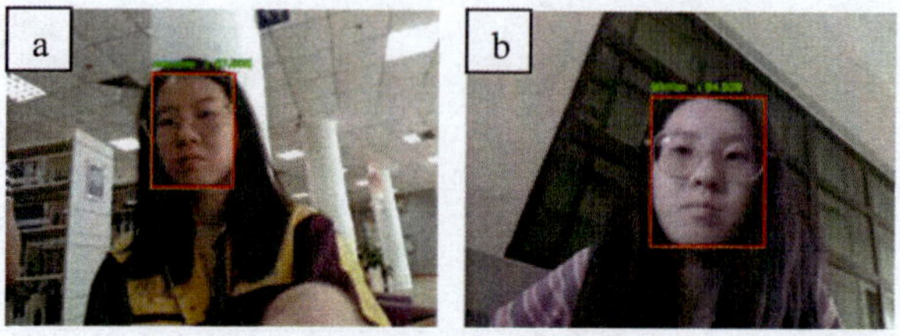

Fig. 6 Face authentication result for the first unknown individual being recognized as **a** "unknown" **b** "MinYan"

and "MinNing" were both 3 times, respectively. Figure 8 depicts the third unknown user recruited for system testing. The system indicates that this user is recognized as "unknown" with a probability of 76.99% in Fig. 8a and "MinYan" with a probability of 80.80% in Fig. 8b. The test was repeated 6 times and the results of being recognized as "unknown" and "MinYan" were both 3 times, respectively. Figure 9 shows that the face authentication system tested the fourth unknown user. The subject was identified as "unknown" with a probability of 58.26%. This testing was conducted 6 times and all results returned as "unknown".

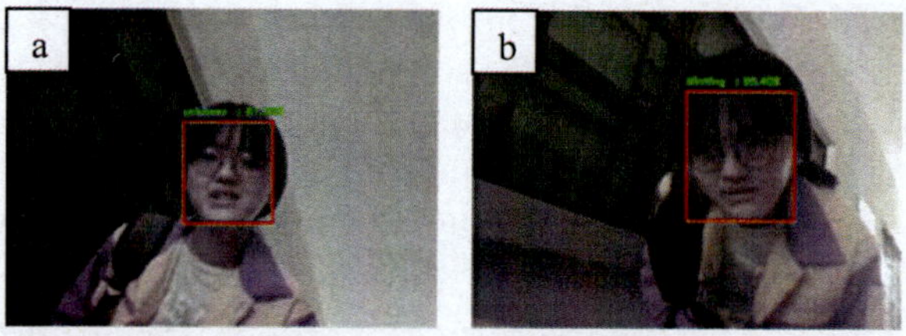

Fig. 7 Face authentication result for the second unknown individual being recognized as **a** "unknown" **b** "MinNing"

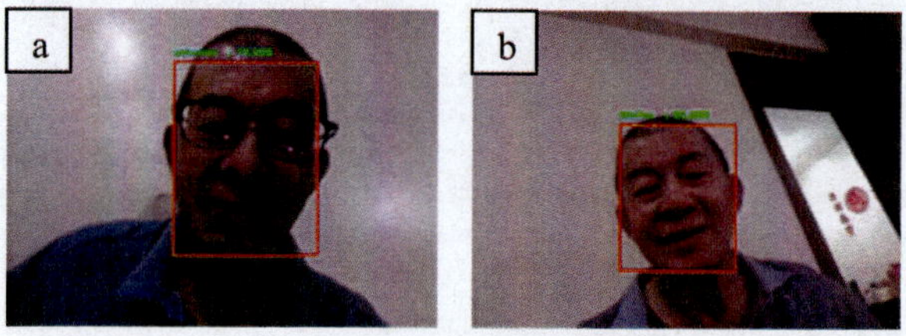

Fig. 8 Face authentication result for the third unknown individual being recognized as **a** "unknown" **b** "MinYan"

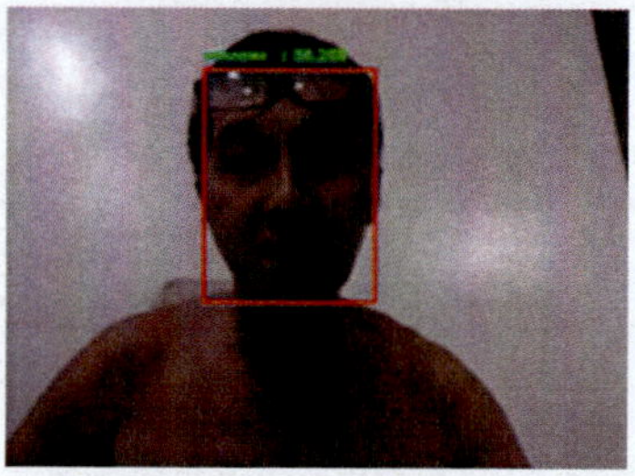

Fig. 9 Face authentication result for the fourth unknown individual being recognized as "unknown"

Table 1 Confusion matrix for multi-class classification

Confusion matrix for multi-class classification		True class			
		"MinHan"	"MinYan"	"MinNing"	"Unknown"
Predicted class	"MinHan"	16	0	0	0
	"MinYan"	2	19	0	6
	"MinNing"	0	4	22	3
	"Unknown"	7	2	3	16

Table 1 shows the confusion matrix for the multi-class classification of three registered users and the unknown group.

The confusion matrix in binary classification was used to measure the most common performances such as accuracy, precision, recall, and F1-score for each class (i.e., user) as given in Table 2. The classification accuracy for the three registered users is approximately 1, indicating that the face authentication system functions with reasonable accuracy when recognizing registered users. In contrast, the accuracy obtained is relatively lower for the unknown group. This is due to the misidentification of an unknown individual as one of the registered users. Since the testing images are obtained from different angles, backgrounds and illuminations, the performance for each class is random and unpredictable. Nonetheless, the good results in Table 2 validates the robustness of the developed system. It must be emphasized here that the performance of each class shown in Table 2 can further be improved with the installation of a fixed imaging system.

Previous study [7] has similar methodology with us, they obtained 97% classification accuracy with 40 images per subject for training. Their studies used three subjects, but it is unclear number of test image used per subject to obtain accuracy and what is the recognition rate for each subject. Our recognition accuracy for each class is shown in Table 3 ranging from 79 to 91% using 25 images per class. It can be concluded that our study is comparable to previous study [7] and other techniques as summarized in Table 3. Also it is comparable to the embedding model that we used, "openface_nn4.small2.v1.t7" which has 93% recognition in LFW benchmark. Even though our recognition accuracy is a bit lower, but the accuracy may increase if we increased number of test images.

Table 2 Summary of the performance evaluation for each class

Evaluation class	Accuracy	Precision	Recall	F1-score
"MinHan"	0.91	1	0.64	0.78
"MinYan"	0.86	0.704	0.76	0.73
"MinNing"	0.9	0.759	0.88	0.81
"Unknown"	0.79	0.571	0.64	0.60

Table 3 Comparison of recognition accuracy with literature

No.	Studies	Methods	Accuracy (%)
1	[6]	Histogram of Oriented Gradients HOG and SVM classifier	92.68
2	[7]	FaceNet algorithm, DN and, triplet loss function to detect face, HOG features and SVM classifier	97
3	[8]	Siamese NN, FaceNet	87.5
4	[9]	Haar Cascade classifier, ConvNet, Fisher Face algorithm and support vector machine (SVM) classifier	92.86
5	[10]	Haar Cascade algorithm to detect faces, LBP features and Euclidean distance to classify face	77
6	Our study	Haar cascade algorithm to detect face in training dataset, Caffe model, OpenCV's deep neural network (DNN) module to detect face in testing stage and SVM classifier	79–91

4 Conclusion

A safe box safety system with face recognition for multiple users was developed by combining software, hardware, and IoT systems. In the software development, the OpenCV–Python was used to perform face recognition between the three registered users and four unknown individuals. The software is incorporated into the hardware system, which includes a Raspberry Pi, a Pi camera, a solenoid lock, and an infrared sensor to simulate the safe's security mechanism. The real-time face authentication monitoring of the safe box is supported by the internet facilities that allow communication between the Raspberry Pi and Telegram Cloud. In this work, the safe box access would be granted is for 10 s when the confidence level of the captured image is more than a probability of 0.8 or vice versa. The testing of the system showed an accuracy ranging between 0.79 and 0.91, a precision range of 0.571 and 1, the recall ranging from 0.64 to 0.88, and the obtained F1-score is between 0.6 and 0.81. Based on the relatively good accuracy, we can conclude that the developed face recognition system is relatively reliable. However, improvement in the system's accuracy, precision, recall, and F1-score can be enhanced by increasing number of the training and testing images captured using a fixed camera. Future works also include the use of other feature extraction methods and classifiers to validate and verify our proposed method. Lastly, the existing Python code can be added with these classifiers before calculating the average probability given by the different classifiers to increase the accuracy and precision of recognizing the registered users correctly.

Acknowledgements This research was made possible by funding from Fundamental Research Grant Scheme number FRGS/1/2020/STG06/UTHM/03/7 provided by the Ministry of Higher Education, Malaysia.

References

1. Yedulapuram S et al (2020) Automatic door lock system by face recognition. IOP Conf Ser Mater Sci Eng 981(3):1–6
2. Ltd, Advantages of using a safe box information technology essay. UK Essays. https://www.ukessays.com/essays/information-technology/advantages-of-using-a-safe-box-information-technology-essay.php?vref=1. Accessed 30 July 2022
3. Zhu Z, Cheng Y (2020) Application of attitude tracking algorithm for face recognition based on OpenCV in the intelligent door lock. Comput Commun 154:390–397
4. Heenaye MMNM, Subramanian RK (2015) A study of dorsal vein pattern for biometric security. Univ Mauritius Res J 15(1):17–19
5. Zhao X, Wei C (2017) A real-time face recognition system based on the improved LBPH algorithm. In: 2017 IEEE 2nd international conference on signal and image processing (ICSIP), pp 72–76
6. Ganjoo R, Purohit A (2021) Anti-spoofing door lock using face recognition and blink detection. In: 2021 6th international conference on inventive computation technologies (ICICT), pp 1090–1095
7. Rameswari R, Kumar SN, Aananth MA, Deepak C (2021) Automated access control system using face recognition. Mater Today Proc 45:1251–1256
8. Saxena N, Varshney D (2021) Smart home security solutions using facial authentication and speaker recognition through artificial neural networks. Int J Cogn Comput Eng 2:154–164
9. Jaihar J et al (2020) Smart home automation using machine learning algorithms. In: 2020 international conference for emerging technology (INCET), pp 1–3
10. Chinimilli BT, Anjali T, Kotturi A, Kaipi VR (2020) Face recognition based attendance system using Haar cascade and local binary pattern histogram algorithm. In: 2020 4th international conference on trends in electronics and informatics (ICOEI), pp 701–704
11. Rosebrock A. OpenCV face recognition. PyImage search. https://www.pyimagesearch.com/2018/09/24/opencv-face-recognition/. Accessed 30 July 2022
12. Lim ST et al (2021) Breast cancer diagnosis system using hybrid support vector machine-artificial neural network. Int J Electr Comput Eng (IJECE) 11(4):3059–3069
13. Chin SW et al (2021) Dorsal hand vein authentication system using artificial neural network. Indones J Electr Eng Comput Sci 21(3):1837–1846
14. Tay KG, Choy YY, Huong A (2018) Forecasting electricity consumption by multiple linear regression. Int J Eng Technol 7(4):3515–3520
15. Huong A, Ngu X (2014) The application of extended Modified Lambert Beer model for measurement of blood carboxyhemoglobin and oxyhemoglobin saturation. J Innov Opt Health Sci 7(3):1450026(1)–1450026(7)
16. Rosebrock A. Face detection with OpenCV and deep learning. PyImageSearch. https://www.pyimagesearch.com/2018/02/26/face-detection-with-opencv-and-deep-learning/. Accessed 30 July 2022
17. Vish V. Face detection with OpenCV and deep learning—Vinu Vish. Medium. https://medium.com/@vinuvish/face-detection-with-opencv-and-deep-learning-90bff9028fa8. Accessed 30 July 2022
18. Thabet AB, Amor NB (2015) Enhanced smart doorbell system based on face recognition. In: 2015 16th international conference on sciences and techniques of automatic control and computer engineering (STA), pp 373–377
19. Model and Accuracies. https://cmusatyalab.github.io/openface/models-and-accuracies/

Depression Detection Based on Features of Depressive Behaviour Through Social Media Analytic: A Systematic Literature Review

Nur Amira Mat Ripah, Aliza Abdul Latif, Zaihisma Che Cob, Sulfeeza Mohd Drus, Rina Md Anwar, and Husni Mohd Radzi

Abstract People are becoming more conscious of the importance of mental health as time goes on. Thus, the detection of mental diseases is becoming a significant concern. Due to the multifaceted nature of each mental problem, many psychiatrists have had difficulty diagnosing mental illness in a patient, making it challenging to provide proper therapy before it is too late. However, because social media has become so ingrained in people's daily lives, it has created an environment where more information about a patient's mental illness is potentially available. This research was carried out as a Systematic Literature Review (SLR), a method of locating, evaluating, and interpreting publicly available materials to answer a set of research questions. The purpose of this study is to answer questions about text-based depression detection based on people who depressive behavior might have shown in their social media postings. The findings reveal that specific aspects of how these people use social media can help diagnose depression early on. This SLR discovered that the chosen social media data is basically the country's leading social site. However, some of the papers indirectly mentioned their challenges during the process. The main challenges highlighted are regarding the ethical issues of the data available. Furthermore, it is also shown that various machine learning algorithms are used,

N. A. Mat Ripah (✉) · A. Abdul Latif · Z. Che Cob · S. Mohd Drus · R. Md Anwar · H. Mohd Radzi
Department of Informatics Technology, Universiti Tenaga Nasional, Kajang, Selangor, Malaysia
e-mail: amiramatripah@gmail.com

A. Abdul Latif
e-mail: aliza@uniten.edu.my

Z. Che Cob
e-mail: zaihisma@unite.edu.my

S. Mohd Drus
e-mail: sulfeeza@uniten.edu.my

R. Md Anwar
e-mail: mrina@uniten.edu.my

H. Mohd Radzi
e-mail: husni@uniten.edu.my

© The Author(s), under exclusive license to Springer Nature Singapore Pte Ltd. 2023

D.-K. Kang et al. (eds.), *Proceedings of the 9th International Conference on Computational Science and Technology*, Lecture Notes in Electrical Engineering 983,
https://doi.org/10.1007/978-981-19-8406-8_14

and the most used are Neural Network and Support Vector Machine. Similarly, the most common computing tool used is Phyton. The use of social media, as well as computational tools and machine learning algorithm, contributes to current public health efforts to detect any indicators of depression from sources close to patients.

Keywords Depression detection · Machine learning · Social media

1 Introduction

Depression, according to the World Health Organization, is the most prevalent mental illness, affecting more than 250 million people worldwide (WHO). Additionally, depression became a leading cause of suicide in the United States (US), accounting for more than two-thirds of all suicides annually. Contrary to common belief, it is primarily a sickness in affluent countries [1] and Malaysia also is no exception. In fact, depression is the most reported mental illness in Malaysia [2]. It is by far the most severe and treatable disease, with an estimated 2.3 million Malaysians expected to be affected at some point in their life [3]. However, because of their self-denial and lack of understanding of their disease, some patients will continue untreated. Late diagnosis and treatment may aggravate the illness, resulting in a lower quality of life and inability to work, affecting their daily routine or work. Therefore, several studies have found that the data from the social network may be utilized to predict depression. With the rise of social media platforms such as Facebook, Twitter and YouTube, researchers now have access to a large amount of data. These data may be used to gain insight into users' moods and sentiments, including studies that employ semantic analysis on online postings. They believe online sites allow people to express their opinions and sentiments openly and could be valuable in tracking health issues and trends [4]. As a result, using data from these social media to create a prediction model for public health surveillance becomes a viable option. It can help health specialists or other stakeholders involved in public health surveillance and assess communities or individuals at risk by just harnessing online data.

2 Methodology

The Systematic Literature Review began by deciding on a set of research questions that would act as a framework for the research process. The goal of this study is to find an answer to the question, "what might influence the usage of social media as a method for early detection of depression," the following set of research questions are formulated:

- RQ1—Which social media platforms and datasets element is most commonly used in the research of detecting depression signs?

- RQ2—What are the difficulties in using a text-based technique to diagnose depression?
- RQ3—What are the most common machine-learning algorithms for detecting indicators of depression on social media?
- RQ4—What computing technologies are most commonly used to detect depression symptoms on social media?

PRISMA which is a shortform from Preferred Reporting Items for Systematic Reviews and Meta-Analyses will serve as the foundation for better organizing and documenting the complete research process and data. The steps of this inquiry are depicted in Fig. 1.

Many parameters have been established in this paper's technique for collecting reviewed studies, which are explained here. Recent studies published in the last five years, from 2018 to 2022, have been thoroughly evaluated and researched. There have been numerous considerations. One of the considerations is the keyword(s) selection. In this review paper, keywords are classified into four main classes, which are: "depression detection," "machine learning," "mental health," and "social media."

In addition to the first criterion, the published date is based on the most recent ten years. This class's research papers have been thoroughly scrutinized. On the

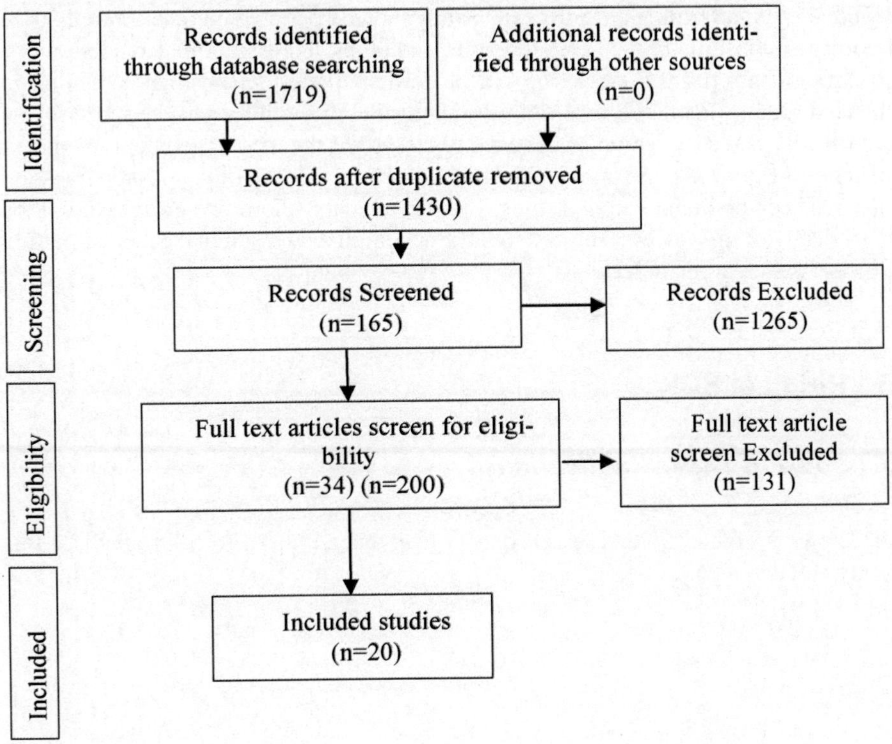

Fig. 1 PRISMA flowchart

Table 1 Selected records

Source	Found	Candidate	Selected
Science Direct	237	88	6
IEEE Explore	402	21	4
PLOS ONE	896	11	0
MDPI	102	22	4
Emerald Insight	24	4	0
SCOPUS	58	19	6
TOTAL	1719	165	20

other hand, the other class has gone so early, and some old study has been examined over the last fifteen years. Still, depression detection utilizing social media has been extensively related to that period. Elsevier, IEEE, PLoS, MDPI, and Emerald are the publishers examined. As a result, ScienceDirect, IEEEXplore, PLOS ONE, MDPI, and Emerald Insight have been the only digital libraries available(s). Another criterion is that the articles must be experimental and published in top-tier journals that Scopus index.

The papers will then be carefully filtered based on a set of parameters, with papers that were not research publications written in English being removed. Other factors that may exclude from inclusion in this study include publication dates before 2018, duplicate publishing of the same research, and other depression-related concerns.

Articles that fulfill the criteria will be further filtered by abstract screening in the next round. First, each record's full-text availability will be checked. If it is, the author will read each record's abstract and classify if the record's content in context matches the review's aim and whether the content can answer the research question that has been predefined in this study. If a record meets those two criteria, it will be considered for this review. The detail of the accomplished result in the record-finding process is shown in Table 1.

3 Result of SLR

3.1 List of Paper Publications

As shown in Table 2, 20 research papers have been chosen to be studied for this study.

Table 2 List of paper publication

Source	Year	Title
Science Direct	2018	"A Large-Scale social media Corpus for the Detection of Youth Depression (Project Note)"
Science Direct	2020	"Predicting Anxiety, Depression and Stress in Modern Life using Machine Learning Algorithms"
Science Direct	2020	"Combining textual features to detect cyberbullying in social media posts"
Science Direct	2021	"An emotion and cognitive based analysis of mental health disorders from social media data"
Science Direct	2021	"An ensemble deep learning technique for detecting suicidal ideation from posts in social media platforms"
Science Direct	2022	"Prediction of generalized anxiety levels during the Covid-19 pandemic: A machine learning-based modeling approach"
IEEE Explore	2019	"Detection of Depression-Related Posts in Reddit social media Forum"
IEEE Explore	2019	"A Novel Co-Training-Based Approach for the Classification of Mental Illnesses Using Social Media Posts"
IEEE Explore	2020	"Predicting Depression Symptoms in an Arabic Psychological Forum"
IEEE Explore	2020	"Utilizing Neural Networks and Linguistic Metadata for Early Detection of Depression Indications in Text Sequences"
MDPI	2022	"Design and Implementation of Attention Depression Detection Model Based on Multimodal Analysis"
MDPI	2022	"Deep Learning for Depression Detection from Textual Data"
MDPI	2022	"Deep Hierarchical Ensemble Model for Suicide Detection on Imbalanced Social Media Data"
MDPI	2022	"Identifying Depression-Related Behaviour on Facebook—An Experimental Study"
SCOPUS	2019	"Detection and Analysis of Stress using Machine Learning Techniques"
SCOPUS	2020	"Deep neural networks detect suicide risk from textual Facebook posts"
SCOPUS	2020	"Predicting depression using deep learning and ensemble algorithms on raw twitter data"
SCOPUS	2022	"A Novel Text Mining Approach for Mental Health Prediction Using Bi-LSTM and BERT Model"
SCOPUS	2022	"Twitter Arabic Sentiment Analysis to Detect Depression Using Machine Learning"
SCOPUS	2022	"Stress detection using natural language processing and machine learning over social interactions"

3.2 Result of Individual Study

In this research step, each record will be reviewed and attempt to collect data from each document listed. The data extraction aims to answer the research questions (RQ) expressed in the first part of this paper.

Table 3 Social media sites and features dataset used

Social media	Features of dataset
Online forum	20,000 posted stories
Reddits	3922 number of posts
	1293 depression-indicative posts and 548 standard posts
	135 depressed users and 752 random users
	95,000 posts
Twitter	3200 tweets
	100,000 tweets
	100,042 tweets
Facebook	144 posts
	403 random entries
	83,292 postings from 1002 users
Sina Weibo	7329 posts

Which platform and datasets are most used in depression sign detection research?

Table 3 shows the most popular social media sites and dataset attributes in depression detection studies from 2018 to early 2022. The social media sites most studied, according to Table 3, are Twitter and Reddit. The work of Tariq et al. [5] was applied to texts in English in the example of Reddit. They use PRAW to get the top thousands posts from each subreddits: r/Depression, r/Anxiety, r/ADHD, and r/Bipolar. In addition, they use Python (API) for Reddit. Users in the subreddit have erased some of the posts, bringing the total number of posts to 3922.

In the example of Twitter, Zaghouani [6] stated that the Twitter API supports the crawling of a user's most recent 3200 tweets and the account's personal information and network information. The above data would be crawled and recorded in a properly structured database. The crawling time for such a collection will likely take 4–6 months.

Furthermore, in a study by Ophir et al. [7], artificial neural network (ANN) models were created to predict suicide risk from social media users' common language. The dataset comprised 83,292 posts written by 1002 verified Facebook users and reliable psychological data about them. We have discovered some social media sites that are exclusive to a specific location. Li et al. [8] investigated Weibo in China, for example. Our findings show that social networking services are becoming more popular worldwide.

What are the difficulties in using a text-based technique to diagnose depression?

The difficulty reported in the selected recordings is shown in Table 4. Though not every publication addresses the difficulty that arises during the research process, it is evident that ethical considerations are the most prevalent. This covers every aspect of data security and privacy protection. Even if most of the data is made public on social networking sites, data about mental health and diseases is considered sensitive. After further investigation, it appears that most of the issues listed in Table 4 are interconnected. According to studies by Uban et al. [9], every technology deployment for mental disease detection should consider potential ethical considerations. If third parties use such technologies, for example, employers that want to screen candidates based on their mental health profile, the subjects' privacy may be jeopardized.

Furthermore, the explicit concern indicated by Tariq et al. [5] is related to results generalization. They presented their findings using a small number of datasets and classifiers. They can incorporate more case studies and classifiers to benchmark the classifier's performance,

Which is the main machine-learning algorithms used for detecting depression signs on social sites?

Table 5 summarises the review of past studies on machine-learning methods utilized in depression sign detection studies in response to this topic.

According to Tariq et al. [5], there has been substantial progress in Machine Learning (ML) to solve real-world problems or introduce automated systems. Machine learning quickly becomes famous for diagnosing mental health disorders because of the large quantity of data available on social media platforms. In addition, he mentioned at several levels, the decision tree technique is excellent for prediction issues since the structure is stable and easy to read. It includes both classification and regression. Furthermore, Renjith et al. [15] say that the Long Short-Term Memory Network (LSTM) can learn long-term dependencies, suggesting that the LSTM can recall information longer. In LSTM, the hidden layers are made up of four gates that determine how much of the past should be remembered and how much of the context should be carried forward. As a result, LSTM becomes a perfect option for recognizing the suicidal context in social media.

Machine-learning techniques improve prediction accuracy in various datasets, which is relevant to our review. Several algorithms are sometimes utilized in a single research project. Support vector machines, Adaboost, SGD, Decision Trees, Random Forest, and k-Nearest Neighbor are all used in a study by Alghamdi et al. [11]. Our review shows that the researchers typically use SVM, Random Forest, or Neural Networks to finish their diagnosis of depression as shown in Table 4. SGD, ADA,

Table 4 Challenges in doing text-based classification

No.	Issue	No. of papers	Study identifiers
1	Limited to one geographical area	3	[5, 10, 11]
2	Limited no of datasets	4	[5, 9, 11, 12]
3	Limited no of classifiers	1	[5]
5	Ethical and privacy concerns	5	[8, 9, 13–15]
6	Severity of errors	1	[14]
7	No means to verify the data validity	2	[9, 16]

Table 5 Machine learning algorithm

No.	Method	Study identifiers
1	Support vector machine (SVM)	[5, 6, 10, 11, 17–20]
2	Multilayer perceptron (MLP)	[17]
3	Random forest	[5, 10, 11, 18, 21, 22]
4	Naïve Bayes	[5, 10, 18, 20, 22]
5	AdaBoost (ADA)	[10, 11]
6	Decision trees (DTs)	[11, 18, 19, 21]
7	k-nearest neighbours (kNN)	[10, 11, 18]
8	Stochastic gradient descent (SGD)	[11]
9	Neural network	[6, 8, 12, 13, 15, 16, 22]
10	Long short-term memory (LSTM)	[15, 22, 23]
11	Logistics regression	[21, 22]
12	Linear SVC	[22]
13	Gradient boosting	[22]

LSTM, linear SVC, k-Nearest Neighbor classifiers, logistics regression, multilayer perceptron, and gradient boosting are some of the less commonly used machine-learning techniques.

What computing technologies are most used to detect depression symptoms on social media?

As shown in Table 6, the authors prioritize Python; for example, Fortunatus et al. [24] create a program in Python 3 that uses Python's NLTK and is run from the command line. A CSV file is expected as input, with the classification result written in an XML file for reading. LIWC (Linguistic Inquiry and Word Count), FastText, and WordNet are second and third positions. FastText is a simple and effective text classification model, as noted by Li et al. [8]. While, in the study by Baheti et al. [20], WordNet was employed for the WSD skip-gram model. It detects whether the blog or text is in Depression, Stress, Normal, Happy, or Relax based on the probabilistic values.

Others include Word2Vec, Gloves, MADAMIRA, CNN, RedditAPI, RStudio, and Weka, all popular but less widely utilized. Musleh et al. [10] used up to three tools to determine classification, including Python, Sklearn, and Grid SearchCV libraries. In contrast, Zaghouani [6] used MADAMIRA, a system for morphological analysis and disambiguation of Arabic and its dialects. Based on the study by Priya et al. [18], machine learning techniques were implemented in the R programming language using Rstudio version 3.5. Based on their symptoms, this will forecast the percentage of persons who suffer from stress, anxiety, or depression.

Table 6 Computing tools

Computing tools	Study
FastText	[8, 13]
Gloves	[11]
Word2Vec	[11]
LIWC	[9, 17]
Python	[5, 10, 23]
MADAMIRA	[6]
TextCNN	[8]
DPCNN	[8]
WordNet	[20, 25, 23]
RedditAPI	[15]
RStudio	[18]
Weka	[19]

4 Discussion

The literature confirms that detecting depression signs using social media data is a developing study area. The data sources employed in different studies may differ. For example, Twitter, Facebook, Reddit, Weibo, and online forums. Users use social media to express themselves based on their desire and the ease with which they can do so. However, we discovered that most of the research is based on the country's most widely used social media. The datasets vary from a few tweets to millions of posts. In our investigation, we discovered that researchers had used a variety of approaches to collect data from tweets or posts posted by people. These approaches can be used in conjunction to achieve better outcomes.

Furthermore, all of the selected studies in the 20 studies examined originated in China, the Arab area, the United States, or other English-speaking countries. Because of cultural and other differences, populations from different nations tend to post at different frequency and with varied content, which might contribute to bias in the diagnosis of depression. The applicability of depression assessment standards across countries and cultures is similarly limited.

In addition, the difficulties while doing the research is addressed in some papers. The most concerning challenges is the ethical and privacy concern. The investigation found that every technology deployment for mental disease detection should consider potential ethical considerations. If third parties use such technologies, for example, employers that want to screen candidates based on their mental health profile, the subjects' privacy may be jeopardized. Further research must be done to address this issue.

Data can be classified and clustered using machine-learning techniques. These tools are beneficial in the process of acquiring accurate findings. Some authors combine these tools to determine which is optimal for the specific topic and data. To process the data, you will need to use computer software. They play an essential role in the classification process by assisting with acquiring natural language information and the translation or processing of the data to be categorized. The computing tools used much vary as there are a lot of tools available.

5 Conclusion

This systematic literature review looks at how a text-based technique for early depression detection is now being used. The study started with 1719 records found, resulting in a total of 165 articles that were further filtered through title and abstract to yield 20 included articles, each of whose content will be analysed to answer the study's research question. According to the findings of this literature review, the three most problematic issues are (1) ethical considerations, (2) a lack of data, and (3) a geographic limitation. This research also provided an overview of the currently used methods for a text-based depression diagnosis. Even though the most popular approaches are (1) Random Forest, (2) Neural Network, and (3) Support Vector Machine, it is intriguing that Naïve Bayes produce the most outstanding results. The current result outperforms any text-based depression detection model. However, more work is needed to increase the model's performance, such as the setup and parameter adjustment.

Acknowledgements We would like to express sincere appreciation to Malaysian Ministry of Higher Education (MOHE) under the Fundamental Research Grant Scheme (FRGS) No. FRGS/1/2020/ICT06/UNITEN/02/2 for providing the fund for this research.

References

1. Latif AA, Cob ZC, Drus SM, Anwar RM, Radzi HM (2021) Understanding depression detection using social media. https://doi.org/10.1109/ICRAIE52900.2021.9703977
2. Aragón ME, Pastor López-Monroy A, González-Gurrola LC, Montes-Y-Gómez M. Detecting depression in social media using fine-grained emotions. Association for Computational Linguistics
3. Leong Bin Abdullah MFI, Ahmad Yusof H, Mohd Shariff N, Hami R, Nisman NF, Law KS (2021) Depression and anxiety in the Malaysian urban population and their association with demographic characteristics, quality of life, and the emergence of the COVID-19 pandemic. Curr Psychol. https://doi.org/10.1007/s12144-021-01492-2
4. Chiong R, Budhi GS, Dhakal S, Chiong F (2021) A textual-based featuring approach for depression detection using machine learning classifiers and social media texts. Comput Biol Med 135. https://doi.org/10.1016/j.compbiomed.2021.104499
5. Tariq S et al (2019) A novel co-training-based approach for the classification of mental illnesses using social media posts. IEEE Access 7:166165–166172. https://doi.org/10.1109/ACCESS.2019.2953087
6. Zaghouani W (2018) A large-scale social media corpus for the detection of youth depression (project note). Procedia Comput Sci 142:347–351. https://doi.org/10.1016/j.procs.2018.10.483

7. Ophir Y, Tikochinski R, Asterhan CSC, Sisso I, Reichart R (2020) Deep neural networks detect suicide risk from textual Facebook posts. Sci Rep 10(1). https://doi.org/10.1038/s41598-020-73917-0
8. Li Z, Zhou J, An Z, Cheng W, Hu B (2022) Deep hierarchical ensemble model for suicide detection on imbalanced social media data. Entropy 24(4):442. https://doi.org/10.3390/e24040442
9. Uban AS, Chulvi B, Rosso P (2021) An emotion and cognitive based analysis of mental health disorders from social media data. Futur Gener Comput Syst 124:480–494. https://doi.org/10.1016/j.future.2021.05.032
10. Musleh DA et al (2022) Twitter Arabic sentiment analysis to detect depression using machine learning. Comput Mater Contin 71(2):3463–3477. https://doi.org/10.32604/cmc.2022.022508
11. Alghamdi NS, Hosni Mahmoud HA, Abraham A, Alanazi SA, García-Hernández L (2020) Predicting depression symptoms in an Arabic psychological forum. IEEE Access 8:57317–57334. https://doi.org/10.1109/ACCESS.2020.2981834
12. Amanat A et al (2022) Deep learning for depression detection from textual data. Electronics (Switzerland) 11(5). https://doi.org/10.3390/electronics11050676
13. Trotzek M, Koitka S, Friedrich CM (2020) Utilizing neural networks and linguistic metadata for early detection of depression indications in text sequences. IEEE Trans Knowl Data Eng 32(3):588–601. https://doi.org/10.1109/TKDE.2018.2885515
14. Kmetty Z, Bozsonyi K (2022) Identifying depression-related behavior on Facebook—an experimental study. Soc Sci 11(3):135. https://doi.org/10.3390/socsci11030135
15. Renjith S, Abraham A, Jyothi SB, Chandran L, Thomson J (2021) An ensemble deep learning technique for detecting suicidal ideation from posts in social media platforms. J King Saud Univ Comput Inf Sci. https://doi.org/10.1016/j.jksuci.2021.11.010
16. Park J, Moon N (2022) Design and implementation of attention depression detection model based on multimodal analysis. Sustainability 14(6):3569. https://doi.org/10.3390/su14063569
17. Tadesse MM, Lin H, Xu B, Yang L (2019) Detection of depression-related posts in Reddit social media forum. IEEE Access 7:44883–44893. https://doi.org/10.1109/ACCESS.2019.2909180
18. Priya A, Garg S, Tigga NP (2020) Predicting anxiety, depression and stress in modern life using machine learning algorithms. Procedia Comput Sci 167:1258–1267. https://doi.org/10.1016/j.procs.2020.03.442
19. Albagmi FM, Alansari A, al Shawan DS, AlNujaidi HY, Olatunji SO (2022) Prediction of generalized anxiety levels during the Covid-19 pandemic: a machine learning-based modeling approach. Inform Med Unlocked 28. https://doi.org/10.1016/j.imu.2022.100854
20. Baheti RR, Kinariwala S (2019) Detection and analysis of stress using machine learning techniques. Int J Eng Adv Technol 9(1):335–342. https://doi.org/10.35940/ijeat.F8573.109119
21. Nijhawan T, Attigeri G, Ananthakrishna T (2022) Stress detection using natural language processing and machine learning over social interactions. J Big Data 9(1). https://doi.org/10.1186/s40537-022-00575-6
22. Shetty NP, Muniyal B, Anand A, Kumar S, Prabhu S (2020) Predicting depression using deep learning and ensemble algorithms on raw twitter data. Int J Electr Comput Eng 10(4):3751–3756. https://doi.org/10.11591/ijece.v10i4.pp3751-3756
23. Zeberga K, Attique M, Shah B, Ali F, Jembre YZ, Chung TS (2022) A novel text mining approach for mental health prediction using Bi-LSTM and BERT model. Comput Intell Neurosci 2022. https://doi.org/10.1155/2022/7893775
24. Fortunatus M, Anthony P, Charters S (2020) Combining textual features to detect cyberbullying in social media posts. Procedia Comput Sci 176:612–621. https://doi.org/10.1016/j.procs.2020.08.063
25. Im TL, San PW, On CK, Alfred R, Anthony P (2013) Analysing market sentiment in financial news using lexical approach. In: 2013 IEEE conference on open systems (ICOS), pp 145–149

Movies and Learning: Does Watching Movies Help Students Improve Learn English?

Aslina Baharum, Nur Faraha Mohd Naim, Rozita Ismail, Farhana Diana Deris, and Noorsidi Aizuddin Mat Noor

Abstract Since English is the most widely used language for communication in the world, mastering English has become essential and important in our lives. Many methods are used to improve the English level, including watching movies. This research aims to study whether watching movies helps students improve their English. This study uses a quantitative method to collect the data through an online questionnaire using Google Forms. The online questionnaire consists of 15 questions with the participation of 250 students from one of the universities in Malaysia, of which each faculty has 25 students for random clustering. The results show that most students usually watch action English movies very often. From the findings, there is no result showing that there is a significant relationship between movie genre preference and English learning. However, it can be concluded that watching movies does help students improve learn English.

Keywords Reading · Writing · Speaking · Vocabulary · Language

---

A. Baharum (✉)
Faculty of Computer and Mathematical Sciences, Universiti Teknologi MARA (UiTM), Jasin Campus, Melaka, Malaysia
e-mail: aslinabaharum@uitm.edu.my

N. F. M. Naim
Faculty of Computing and Informatics, Universiti Malaysia Sabah, Sabah, Malaysia
e-mail: faraha.naim@ums.edu.my

R. Ismail
College of Computing and Informatics, Universiti Tenaga Nasional, Putrajaya Campus, Malaysia
e-mail: irozita@uniten.edu.my

F. D. Deris
Faculty of Social Sciences and Humanities, Universiti Teknologi Malaysia, Johor Bahru, Malaysia
e-mail: diana@utm.my

N. A. M. Noor
UTM CRES, Faculty of Built Environment and Surveying, Universiti Teknologi Malaysia, Johor Bahru, Johor, Malaysia
e-mail: noorsidi@utm.my

© The Author(s), under exclusive license to Springer Nature Singapore Pte Ltd. 2023

D.-K. Kang et al. (eds.), *Proceedings of the 9th International Conference on Computational Science and Technology*, Lecture Notes in Electrical Engineering 983,
https://doi.org/10.1007/978-981-19-8406-8_15

1 Introduction

Communication is always important in daily life because communication is one of the important tools to help us establish connection with people. Based on Liando et al. [1], people need to communicate using language to interact with others and express their feelings or to share ideas. Good communication keeps you in touch with your friends and with those people around you. To communicate, everyone at least learns one language during childhood: the native language. To be able to communicate at an international level, we need to use international languages such as English. English is one of the official international languages because twenty per cent of the world's population speak English as their mother tongue. It is also the most common second language in the world.

Metruk [2] mentioned that it represents principal components of the command of a foreign language to develop language abilities. Listening is the ability to receive language through ears and interpret messages during communication accurately. Listening involves recognizing speech and understanding the speaker's accent or pronunciation, grammar, vocabulary, and meaning. If a person's listening ability is poor, it is easy to misunderstand the message in the communication process. Speaking skills are the skills that express or communicate opinions, feelings, and ideas. Speaking involves the speaker's opinions, feelings, or ideas and explains that speaking is the building process. Speaking is an interactive process in which information is shared, and listeners act when necessary. Therefore, speaking and listening are inseparable between them. Based on Rasyid [3], communication is a two-way process involving the speaker and the listener. Reading skill is the ability to understand meaning from the text, including phonics, word, recognition, vocabulary, decoding, and fluency, and independently read and interact with the message. Reading skills also can be considered as a foundation for writing skills. Reading can enrich our vocabulary and learn new words or terms to improve grammar and spelling when writing. Writing skill is the art of expressing views, ideas, and thoughts. Writing skill also includes all knowledge and abilities related to expressing ideas through words. Good writing skills can acquire knowledge about sentence structure, develop your vocabulary, and build other basic writing skills.

In modern society, the broadness of information obtained through the sources such as the media plays an important role in learning English, so learning foreign languages is not difficult to understand. One of the most effective media to improve English language is by watching movies. According to Aliyev [4], movies are considered authentic materials, and many researchers and language teachers have supported movies in improving the language. Most students think that movies are one of the simplest ways to improve their English. They can attract students' curiosity in learning because the movie can provide audio-visual learning methods. The different types of movies can teach different types of conversation features such as pronunciation and vocabulary. The movie also could help the students to interpret the message by watching the speaker's visual performance and listening to the speaker's voice. The students can improve their English skills while watching movies. Thus, the study presents whether watching movies helps students improve their English and uses sampling techniques to collect data. We can conclude that movies are an integral part of English learning because they can impact the development of English language skills such as reading, listening, writing, and speaking.

2 Related Works

The language learning method may differ from person to person; some prefer to learn a specific language by signing up for classes, and some prefer to learn from books and reading materials. With the progress of modern technology, some can even conduct language learning through all kinds of media. Similar research has been conducted by Tahir [5] to find out the impact of using movies on learning the English language at the University of Halabja. The study has analyzed that movies have an essential role in developing and improving language skills. With the development of technology, the use of movies as a part of technology in teaching has become unavoidable [5]. The data collections of the research are more inclined to personal opinions on whether movies help as a media in language learning. In addition, Al Murshidi [6] showed that movies can be used as an authentic language input to enhance one's English language skills. Through movies, a fun factor provides entertainment for learning. Therefore, students will show interest and listen to the movies. By using conventional teaching methods, this is not achievable [6].

According to Chen [7], while there are many advantages in using movies to teach English, there are also a few disadvantages. First, it is hard to arrange a time to watch a full movie properly. Second, it is also hard to choose a type of movie when students have different preferences. There are several aspects a teacher must pay attention to during a film which are material selections, reasonable arrangement of time and design of good movie teaching steps. The report concluded by agreeing that using movies is an effective teaching method [7]. Other than that, Ebrahimi and Bazaee [8] concluded that movies not only act as a motivation for students, but they could also help students to understand the language in other accents. Subtitles help assist in understanding movies. However, it is assumed that movies with subtitles will only be effective when the students watch the movie more than once. The report ended by mentioning how watching a movie more than once will help students to be able to recognize vocabulary more easily and learn new idioms and expressions [8]. Next, Yaseen and Shakir [9] found that most students prefer to have subtitles in their movies. Foreign subtitles play a huge part in a student's ability to understand the storyboard in a movie. Students can make use of movies as a way of practising their listening, writing and speaking skills [9].

Ebrahimi et al. [10] in their research, watching movies to enhance the students' English language skills are indeed an effective way for students to learn English. Learning processes could not only be done by watching movies entirely, but students could also learn new vocabulary by watching dramas, television shows and programs to improve their English skills. Besides, they also advised that watching movies could benefit students' English writing skills by writing a summary after watching the movies. They concluded that the repetition of watching movies or dramas with a specific theme could also help students learn English more effectively [10]. Furthermore, Iranmanesh and Darani [11] indicated that watching movies could help students improve their speaking and negotiation skills in real life. Students could learn and understand the idioms and expressions from movies and apply them in their daily conversations. Learning English through movies could be more effective than traditional learning methods because students could be motivated and entertained while watching movies and learning English simultaneously. The report ended by recommending that English teachers use movies as teaching material for students [11].

Zhang and Wen [12] in their research, while learning English, students may face some difficulties and pressures during their learning processes, resulting in learning burnout. This phenomenon will reduce the efficiency of the students' English learning progress and affect their academic performances. Thus, the researchers pointed out that using multimedia technology such as movies and documentaries will positively impact students' learning burnout. They also concluded that the use of multimedia technology could help students to understand the English language better [12]. Polat and Eristi [13] also showed that resources such as authentic videos could be a beneficial medium for students to improve their English listening skills. They also discovered that there was a strong relationship between students' listening skills and speaking skills. At the end of the report, the researchers concluded that improving the students' listening skills will be advantageous in developing other language skills such as speaking, writing and reading [13]. In addition, Sadiku [14] found that learning English through movies with subtitles would be an indispensable way for students to strengthen their English language skills. Watching movies with subtitles could help students enrich their English vocabulary and further impact their vocabulary usage during English writing. The researchers also emphasized that using movies as teaching material for students to learn English could help them understand and improve their communication skills in real-life situations [14].

Moreover, Trujillo et al. [15] investigated whether movies with or without subtitles help learners' listening skills and understanding of movies. So, they did a comparative experiment, such as a group of movies with subtitles and a group without subtitles, to test their hearing test results. As a result of their findings, learners watching movies with subtitles can better understand and improve their listening abilities. Watching movies without subtitles, on the other hand, does not improve learners' listening skills. After assessing the statistical findings, experts concluded that watching movies with subtitles can assist learners in better comprehending the film's content and improve their listening abilities [15]. Finally, according to Xing, the researchers believed that most learners from China do not have that good English ability because they lack a real English learning environment. Therefore, they studied whether using movies can help students understand deeper meanings. In addition, Xing [16] stated that the role of teachers is also very important. It is necessary to understand which types of movies students are interested in. Teachers can teach important vocabulary to students through movies. To understand whether the students' English learning ability has been strengthened, they also suggested that the teacher can give the students an exam to test their language ability. That is because teachers can understand whether the effects of teaching through movies can help students learn. From the report, researchers believed that film teaching could help them learn English and enhance their speaking and listening skills [16].

3 Methodology

The participants in the research were 25 students each from 10 faculties in Universiti Malaysia Sabah, with a total of 250 respondents. All the participants are undergraduate students from Universiti Malaysia Sabah. In this research, a descriptive quantitative method was employed. The data collection was done through an online questionnaire which consists of three sections. The questionnaire was adapted from previous studies [1, 4, 17–19]. The first section is the demography of the respondents and Malaysian University English Test (MUET) grade. The second section of the questionnaire is a simple English test from cambridgeenglish.org to examine the English level of the respondents and the English movie's influence on the English test result. Lastly, the third section is the opinion of the respondents on whether watching movies will help improve their English abilities, including vocabulary, listening, pronunciation and writing. The questionnaire is checked for validation and reliability before distribution to the public. The contents of the questionnaire are checked for validation by the expert research and pilot tested by six participants before the real data collection. An online survey was used using the link of the Google Form was shared over to the social media. Data collection was carried out using a random cluster method to avoid any bias during data collection so that students from all years of study at the selected university could complete the questionnaire. The students were required to complete and answer the questionnaire honestly. The duration of the data collection is three weeks, from 11 May 2021 until 07 June 2021. All the responses collected will be kept confidential, remain anonymous and used for academic purposes only. Participation in completing the questionnaire is voluntary. The results from the research were organized and analyzed using Google Forms and Google excel. A demographic of sample data was used to analyze the information provided by the respondents. A Chi-square test, x^2 for independence, was used to evaluate the outcome of this research. This study used a significant value of 5% to indicate statistical significance.

4 Result and Analysis

The summary of the data collected through online questionnaires shows that 55% of the respondents are female and 45% are male. The results also showed that all are undergraduate students. They are from 10 faculties. To ensure the fairness of the results, the number of respondents from each faculty will be the same, which is 25 respondents. Figure 1 indicates the year of study of the respondents. Considering the target respondents being undergraduate students, the range of the year of study is from Year 1 to Year 5. Year 4 students accounted for the least proportion of the respondents, which is 5%. Unsurprisingly, none of the respondents is from Year 5. This could be due to the Year 5 students being only available from FPSK, and the seniors are much occupied with their assignments and internships. Figure 2 shows the data about the MUET grades collected from the 250 respondents. It is clearly shown that more than half of the respondents acquired Band 3 and above in MUET. Only 1% of the respondents get lower than Band 3 in MUET. Since Band 6 is the highest grade and is the grade that is most difficult to achieve in the MUET, the proportion of Band 6 respondents is only 3%. Almost all respondents like to watch English movies, only 1% are not interested in watching English movie. This question is important because this study ought to investigate the relationship between watching movies and the English language level of the respondents. Figure 3 shows the English movie genres that the respondents prefer. Since everyone has different preferences when watching movies, the respondents can choose more than one genre. In total, there were 720 votes from 250 respondents. Out of 9 movie genres, the action movie (Godzilla vs. Kong) is the most popular among the respondents, which was selected by 142 respondents (20%). Figure 4 is the first question from Section 2 in the questionnaire, which asked the respondents to take a simple English test prepared by the University of Cambridge to test their English level before watching the English movie. The results showed that most respondents answered more than half of the questions correctly. This proves that their English proficiency is conformed with their MUET grade. 44% of the respondents answered 16–20 questions correctly out of the total of 25 questions. Of 250 respondents, 30% got 21–25 marks on their English test. Additionally, 23% of the respondents achieve a moderate mark of 11–15 out of 25 questions. However, only 3% answered 6–10 questions correctly, and no respondent obtained less than six marks. Figure 5 shows the number of correct questions increased after watching the short English movie prepared in the questionnaire. To check whether watching English movies can help their English proficiency, the respondents must take the English test once again. Nearly half of the respondents' scores are still the same (45%). At the same time, 30% of the respondents responded that after watching the short English movie, they had some improvement on their test, with the number of correct questions only increasing to less than five questions. Some respondents showed that they had made progress in their test results, with more than five questions answered correctly (19%) and more than ten questions answered correctly (5%). Only 1% of the respondents even got more wrong questions after watching the short English movie.

Table 1 shows the results of the respondent's choices from the online questionnaire. The first question asked about how often the respondents watched English movies. Most respondents said they regularly watched English movies (39%). Question 2 to 5 asked the respondents' opinions on how watching English movies could help them to learn English. From the results, 43% of the respondents agree that they managed to learn English through watching English movies. The results of the

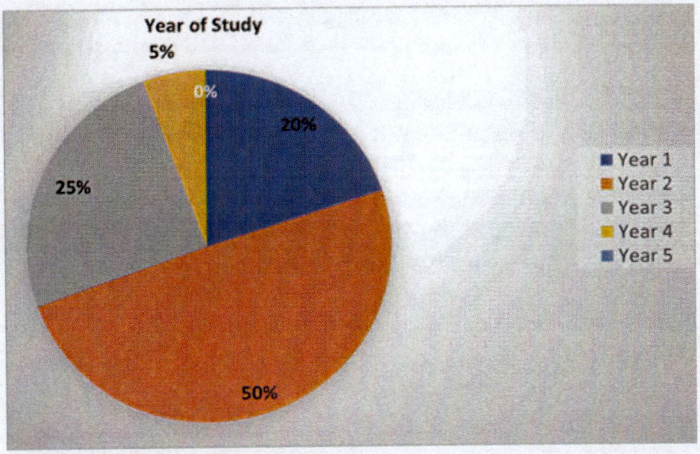

Fig. 1 Year of study of respondent

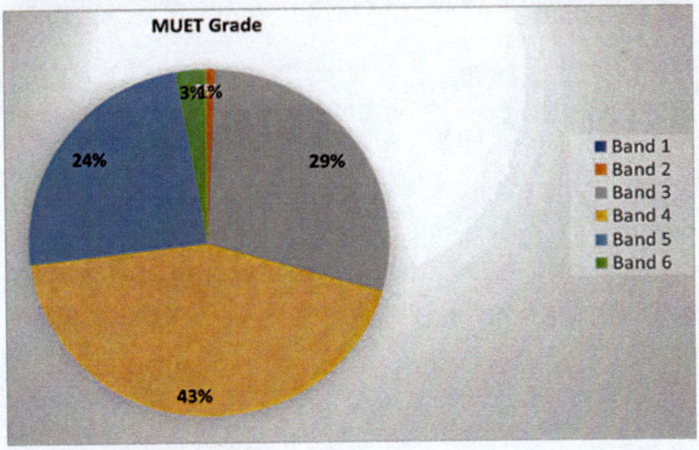

Fig. 2 MUET grade of respondents

respondents' opinion that integrating English movies was useful for developing their writing skills in Question 6. 27% of the respondents strongly agree that integrating English movies was very beneficial for their English writing. Of the 250 respondents, 44% agreed that integrating English movies was useful for developing their writing skills. Next, Tables 2 and 3 are observed and expected results of English test. With a significant value of 5% with a degree of freedom of 4, the critical value is 9.488. The test value falls into the critical region. By using the chi-square test, X^2, it can be concluded that there is a significant relationship between watching movies and English test results.

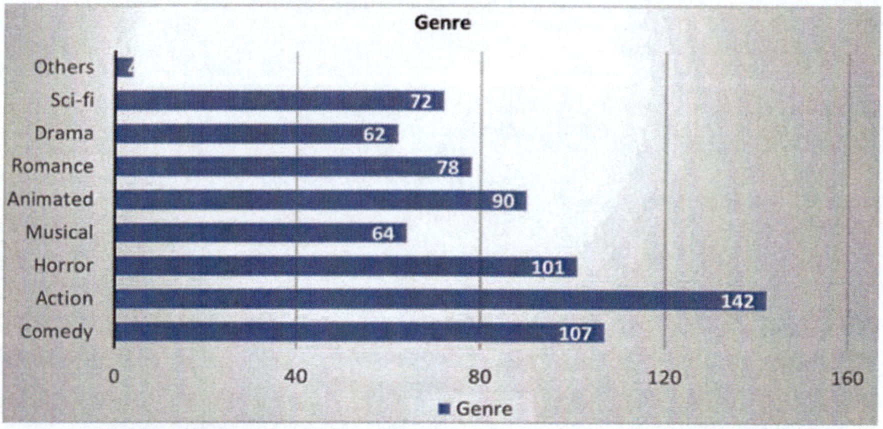

Fig. 3 Movie genres preferred by respondents

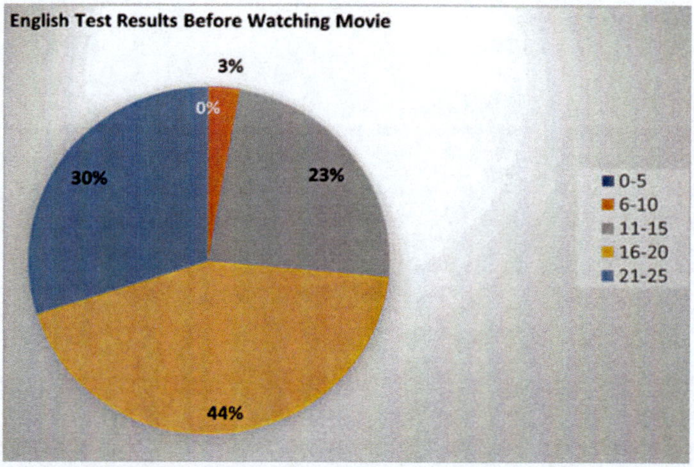

Fig. 4 Before watching movie

5 Findings and Discussion

Results of taking the Simple English Test Before and After watching the Short Video. However, to compare the English level of the respondents before watching the movie, this study provides a Simple English Test prepared by the University of Cambridge. The result of the data collection is that 30% of the respondents got 21 to 25 marks out of 25, whereas more than half of the respondents were in the middle level. In order to compare whether their English level improve after watching the movie, they also need to take the test again after watching the short video provided in the

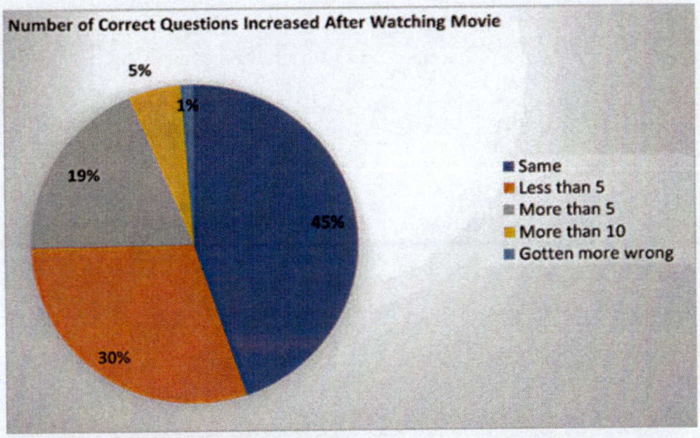

Fig. 5 After watching movie

questionnaire. Based on the results, the data shows that only 30% of the respondents improved their scores for up to 5 questions, while 45% of the respondents did not change their scores, and even 1% made more mistakes. Opinion from watching English Movies. According to data evaluation, 43% of respondents consented that watching movies can enhance their English levels. Many respondents strongly agree that watching English movies can acquire new vocabulary and improve their listening skills. Besides, most respondents also agreed that they could imitate standard pronunciation from movies and improve their speaking skills. From the data collection, a substantial percentage of respondents concurred that watching movies can help them improve their writing skills. On the contrary, a few respondents disagreed that they could find new vocabulary and improve their listening, speaking, and writing skills by watching movies.

The present study investigates whether watching movies helps students to strengthen their English language learning. From the research findings, majority of the respondents think that watching English movies can help them learn English (43% agree + 24% strongly agree). This finding supported the results of a previous study [20] that movies are effective instructional resources that can help students develop their language skills, including reading, listening, speaking, and writing. The majority of the respondents stated that they enjoy watching English movies. This shows that movies can aid in English learning since their favour towards movies will increase their interest in learning English. According to Kabooha [20], incorporating movies into their lectures has significantly enhanced students' motivation to study the English language. From the findings of preference in movie genres, the results show that respondents have different preferences in movie genres. However, no result shows a significant relationship between movie genre preference and English learning in the present study. The findings show that only 30% of the respondents had improved in the test. However, this result does not indicate that watching movies does not help students improve their English. This could happen because 45% of the respondents

Table 1 Number of respondents' choices in percentage

No.	Questions	Strongly disagree/very rare	Disagree/rare	Neutral/sometimes	Agree/often	Strongly agree/very often
1	How often do you watch English movies?	4	9	25	39	24
2	Do you think watching English movies could help you learn English?	0	3	17	43	37
3	Do you think watching English movies could increase your vocabulary?	0.4	2	18	39	41
4	Do you think watching English movies improves your listening ability?	0	0.8	17	38	44
5	Do you agree that watching English movies can help in learning correct pronunciation when speaking English?	0	3	20	39	38
6	Do you agree that integration of the English movie was useful for the development of your writing skills?	0.4	2	25	44	27

Table 2 Observed results of English test results

Movie	Increase in test results after watching movie				
	0–5	6–10	11–15	16–20	21–25
Before	0	7	59	109	75
Then	0	1	9	98	142
		d.f. = 4			

Table 3 Expected results of English test results

Movie	Increase in test results after watching movie				
	0–5	6–10	11–15	16–20	21–25
Before	0	4	34	103	107
After	0	4	34	103	107
$X^2 = 62.873$		d.f. = 4			

obtained the same score on the test before and after watching the short video. There are a few reasons that may cause this result other than not having any improvement; that is, the video is too short, and it is not sufficient to test the effectiveness of watching movies. Aliyev [4] also concluded in her study that the chosen movies should be related to the unit content. Another reason is that the respondents obtained full marks in both tests (before and after watching the short video provided); hence, no difference was observed during the test. From the findings, most of the respondents believe that watching English movies helps develop English language skills. The findings also show that most respondents agree (39% agree + 41% strongly agree) that watching English movies could increase their vocabulary. This could happen because they can find new words from movies and interpret the meaning of vocabulary through the conversation in the movies or from the expression of the actors. This can be supported by a previous study, which claims that English movies could help to improve the English language in developing vocabulary, understanding unfamiliar words, correcting pronunciation, and comprehension skills [21]. The overall respondents also believed (38% agree + 44% strongly agree) that watching movies is highly effective in improving their listening ability. The previous study [18] results also show that watching movies helps to gain more vocabulary and improve other skills like listening. However, movie conversation speed may affect students' listening and comprehension skills. According to Ramadanty [22], students can easily understand the conversation with English subtitles. Most of the respondents agree (39% agree + 38% strongly agree) that watching English movies can help in learning correct pronunciation when speaking English. This finding is supported by the previous study [19] results that students can learn correct pronunciation and better understand unfamiliar accents, dialects, proverbs, and slang from soundtrack movies. Besides, a study also states that movies stimulate students to do better pronunciation in practice and motivate them to apply new vocabulary in daily conversation [1]. Majority of the respondents agree that the integration of English movies was useful for the development of writing skills. This could happen because students learn lots of new vocabulary from movies, which will help in improving writing skills by applying the vocabulary in writing. According to Aliyev [4], watching movies is a useful way to develop ideas for writing essays. In conclusion, the findings of this study show that watching English movies can help students to enhance their English learning. Haq [23] indicates that the advantages of using English movies in English teaching include keeping students interested in learning English and improving student listening and speaking skills, pronunciation, and vocabulary.

6 Conclusion and Recommendations

Based on the findings discussed above, two conclusions are pointed out. The first conclusion is that students can benefit from watching movies to help them learn English. Secondly, students can develop English language skills by watching English movies, which include listening, speaking, reading, and writing. There are several suggestions for further research; first, to obtain more reliable results, future research should be extended to students from different universities or compare public and private institution. Second, since the data collection only used a relatively short time, this study only collected data from the questionnaire for research. Therefore, interviews with respondents can be conducted for future research to obtain more accurate and detailed information. Third, to verify the research results more fairly, different test questions should be arranged. Future research can arrange two different types of tests prepared by the University of Cambridge in the questionnaire to compare whether there is an increase in English proficiency before and after watching the movie. Lastly, future research can also investigate the effectiveness of using subtitles to improve English listening skills and infer if it can improve English proficiency.

Acknowledgements Author would like to thank all students who involve directly in data collection and analysis: Yeo R.Y., Cheng G.X., Leong D., Leong M.Y., Lim Y.X., Loke P.E. This study is currently funded by the SPLB Grant from Universiti Malaysia Sabah (SLB2220).

References

1. Liando NV, Sahetapy RJ, Maru MG (2018) English major students' perceptions towards watching English movies in listening and speaking skills development. Adv Soc Sci Res J 5(6). https://doi.org/10.14738/assrj.56.4627
2. Metruk R (2019) Using English movies and TV programs for developing listening skills of EFL learners. Inf Technol Learn Tools 70(2):227. https://doi.org/10.33407/itlt.v70i2.2488
3. Rasyid S (2016) Using cartoon movie to improve speaking skill. Res Engl Educ (READ) 1(2):161–168

4. Aliyev A (2017) Improving English writing skills of non-native undergraduate learners with the help of movies supported by online technologies. J Educ Black Sea Reg 2(2). https://doi.org/10.31578/jebs.v2i2.38
5. Tahir BJ (2015) The impact of using movies on learning English language at University of Halabja
6. Al Murshidi G (2020) Effectiveness of movies in teaching and learning English as a foreign language at universities in UAE. Psychol Educ 57(6):442–450
7. Chen Y (2015) The application of English movies in English teaching. In: Proceedings of the 2015 international conference on arts, design and contemporary education, pp 764–766. https://doi.org/10.2991/icadce-15.2015.185
8. Ebrahimi Y, Bazaee P (2016) The effect of watching English movies with standard subtitles on EFL learners' content and vocabulary comprehension. J Appl Linguist Lang Res 3(5):284–295
9. Yaseen BH, Shakir H (2015) Movie effects on EFL learners at Iraqi School in Kuala Lumpur. Int J Educ Literacy Stud 3(3):31–36. https://doi.org/10.7575/aiac.ijels.v.3n.3p.31
10. Ebrahimi R, Kargar AA, Zareian A (2018) The effect of input-flood through watching English movies on language productive skills. J Appl Linguist Lang Res 5(2):94–111
11. Iranmanesh A, Darani LH (2018) How can movies be effective in learning English idiomatic and everyday expressions? Int J Learn Teach 10(2):107–115. https://doi.org/10.18844/ijlt.v10i2.3405
12. Zhang M, Wen T (2013) The effect of multimedia technology on undergraduates' learning burnout in college English. In: Proceedings of the 2013 conference on education technology and management science. https://doi.org/10.2991/icetms.2013.66
13. Polat M, Eristi B (2019) The effects of authentic video materials on foreign language listening skill development and listening anxiety at different levels of English proficiency. Int J Contemp Educ Res 6(1). https://doi.org/10.33200/ijcer.567863
14. Sadiku A (2018) The role of subtitled movies on students' vocabulary development. Int J Sci: Basic Appl Res (IJSBAR) 42(1):212–221
15. Trujillo T, Vera-de la Torre A, Cumbe-Coraizaca D (2021) Subtitled and unsubtitled movie and listening comprehension. Educ Eng Future Ind Revolut 1328:241–252. https://doi.org/10.1007/978-3-030-68198-2_22
16. Xing C (2011) How to use English movies to improve a student's listening and speaking ability in Chinese ESL culture learning classrooms. Presented at the Graduate faculty. University of Wisconsin-Plateville
17. Wagner A (2017) Use of movies and television resources in English classes at Asia University. 英語教育研究所紀要 (CELE J) 25:76–98
18. Simamora MWB, Oktaviani L (2020) What is your favorite movie?: A strategy of English education students to improve English vocabulary. J Engl Lang Teach Learn 1(2):44–49
19. Pimsamarn T (2011) A survey of students' opinions on watching English soundtrack movies to enrich listening skill development. Language Institute, Thammasat University
20. Kabooha RH (2016) Using movies in EFL classrooms: a study conducted at the English Language Institute (ELI), King Abdul-Aziz University. Engl Lang Teach 9(3):248. https://doi.org/10.5539/elt.v9n3p248
21. Rojanasuwannachat P, Anekjumnongporn T, Anekjumnongporn P, Chanthaphan S, Paopuka E, James N, Sam-arng C (2021) International program students' attitudes towards using English language movies to develop English communication skills. Asia Soc Issues 14(1)
22. Ramadanty N (2020) The use of movie to learn English in out-of-classroom language learning. Transcript: Rev Engl Teach Learn 2(2)
23. Haq IM (2017) Movie impact in improving English vocabulary

Analyzing the Financial Performance of Transportation Companies with Goal Programming Model

Weng Hoe Lam, Pei Fun Lee, Weng Siew Lam, and Mohd Abidin bin Bakar

Abstract Asset and liability management (ALM) is an important aspect of financial management in all industries. With integrated highways and roadways, land-based transportation has become one of the main modes of moving cargoes due to high accessibility. As a developing country, there are ample rooms of expansion in terms of financial achievement for these companies in Malaysia. A study on ALM of these companies can help to assess the financial stability and highlight possible problems that impede the land-based transportation companies from achieving financial stability. Hence, this paper proposes a mathematical model based on goal programming (GP) to examine the ALM of land-based transportation companies in Malaysia. Key financial indices such as asset, liability, equity, revenue, earnings and optimum management items are also included as the goals in the assessment in this paper. The benchmarks of the goals are compared among the companies in this study. All the companies have successfully obtained excellent results for asset, equity, earnings, and optimum management item goals. With the powerful computational power of GP model, this paper has also been able to point out that the financial stability of land-based transportation companies is largely affected by excess in liabilities, which increased the liquidity and solvency risks of the companies.

Keywords Goal programming · Optimal solution · Transportation · Asset and liability management

W. H. Lam · P. F. Lee · W. S. Lam (✉) · M. A. Bakar
Department of Physical and Mathematical Science, Faculty of Science, Universiti Tunku Abdul Rahman, Kampar Campus, Jalan Universiti, Bandar Barat, 31900 Kampar, Perak, Malaysia
e-mail: lamws@utar.edu.my

W. H. Lam
e-mail: lamws@utar.edu.my

P. F. Lee
e-mail: pflee@utar.edu.my

M. A. Bakar
e-mail: abidin@utar.edu.my

© The Author(s), under exclusive license to Springer Nature Singapore Pte Ltd. 2023

D.-K. Kang et al. (eds.), *Proceedings of the 9th International Conference on Computational Science and Technology*, Lecture Notes in Electrical Engineering 983,
https://doi.org/10.1007/978-981-19-8406-8_16

1 Introduction

The significance of transportation system touches all aspects of life. It is intertwined with the social, political, and economic forces in a society and economy. It has a catalytic effect on a society which affects the movement of goods and people. Transport modes as a whole play an important role in the economic development of a nation. It is critical to ensure that the developments of transport modes are in line with the broad macro-economic objectives of a country [1]. The recent development of technology, societal demand and the growth in e-business and globalization has affected the changes in freight movements. In tandem with the sea-borne transportation, land-based transportation industry also play an important role in the economic development of a nation [2].

Haulage industry becomes the main pivotal for the land-based transportation besides the lorry or trucking operation businesses. Haulage industry is a group of trucking businesses with its core business activities engaging in haulage movements from one place to another. The history of haulage industry in Malaysia began with one prime and three trailers and only one player in the area around Port Klang. Huge market forces especially import and export activities which require containers and other transport unit invited growth in the land-based transportation especially the haulage industry. In the year 2000, the government of Malaysia had proposed an open market concept which ensured that the permits or haulage licenses were available for application to public without much limitation. The total fleets recorded were about 2131 prime movers and 10,701 trailers [3].

Land-based transportation is a massive industry because of the ease of accessibility of roads and highways in Malaysia. However, with Russian invasion since February 2022, the supply of commodities especially oil and gas, has been largely affected which saw a spike in the prices of these commodities [4]. In the meantime, several measures have been in place for sustainable freight transportation to reduce congestions and other environmental impacts. In Malaysia, all commercial vehicles will have to undergo periodic inspections for the safety and sustainability of the vehicles and road users. There is also labor shortage in the road transport sector as companies find it hard to recruit and retain drivers. The shrinking number of drivers will make meeting the delivery deadlines more challenging. Accidents on the road because of driver negligence, vehicle issues, poor weather or geometric roadway design also contribute to the rising logistics costs [5]. All these add to the toll that companies providing road transportation services must bear as part of their operations.

A study by Galkin et al. [6] noted that drivers are more focused when driving vehicles with good ergonomic properties, however, such vehicles tend to cost higher with additional maintenance. Holguín-Veras et al. [7] found that road transportation companies pay exorbitant prices daily which are hard to be passed on to the clients. These impending expenditures include toll fees, parking fees, purchase of environmentally friendly vehicles, off hour deliveries, intelligent transport system, routing and navigation, driver training and changes in delivery schedule or location. Another study by Gonzalez et al. [8] concluded that COVID-19, being an external force, has severely affected the profitability of land transportation companies in Columbia. Izadi et al. [9] studied the possible cost factors in road transportation services. The paper categorized the costs into operational costs such as labor, fuel, maintenance, acquisition, insurance, tax and handling, value of time costs such as travel time and service quality and external costs such as accidental, congestions and environmental costs.

In view of the heightened costs, this paper would like to examine the financial strengths of land-based transportation companies in the logistics industry in Malaysia. Gonzalez-Pascual et al. [10] studied the performance of road freight transportation companies. In this paper, the performances of the companies were evaluated based on the abilities to bring down order management, facility, storage, transportation and logistics management costs. Noting that transportation companies face difficulties to pick up and deliver in town and city areas due to congestions, diverse delivery and limited parking bays, Aljohani and Thompson [11] developed a mathematical cost model to include facility, administrative, personnel and vehicle costs, with the intention of increasing the profitability of these companies. Since existing studies are still limited, the financial strengths of the companies can be assessed using asset and liability management (ALM) in which goal programming (GP) model is a famous, convenient and robust method to assess performances. Financial management is crucial for the companies for sustainability. The past studies on financial management have been done in financial institutions [12–17], construction companies [18, 19], technology companies [20–23] and portfolio investment [24–39]. According to the past studies, there have been no comprehensive studies done on the financial management of transportation companies in Malaysia using GP model.

Goal programming permits researchers to break down multi-criteria decision-making (MCDM) problems with the flexibility to select the objectives and optimization priorities. Constraints will be formulated in the model with the intention of returning minimum deviation from each goal [40]. In MCDM problems, the objective is to determine the best decision alternative by taking multiple decision criteria into consideration such as the evaluation of fast food restaurants [41–43], e-learning methods [44], mobile networks [45–47], jobs [48], machine-tool [49], companies [50–54], course programmes [55], supermarkets [56], library service quality [57] and coffee shops [58]. From this research, the results are expected to contribute to the studied companies by helping them understand the optimum levels of assets, liabilities, equities, revenue, and earnings. This would then balance all the financial indices by identifying the amount needed to reduce excess liabilities or revenue deficiency. This is also in accordance with the aim to maximize shareholders' wealth by creating greater present value of expected future return to a company. With this study, the optimum levels of financial matrices can also enhance the cash flows of the companies for operational, administrative and strategic developmental uses. Moreover, this study will first compare the total values of each goal among the companies and then select the best values for benchmarking for the GP model formulation, which makes this paper distinctive from current literature.

Therefore, this paper carries the aim to optimize the ALM of land-based transportation companies in Malaysia using GP model. The six goals that will be used in this study are asset enhancement, liability reduction, equity accumulation, revenue maximization and earnings amplification. Studies on land-based transportation in Malaysia is very limited, especially in ALM. Zawawi et al. [59] developed a conceptual model to increase road transportation performance in East Malaysia by incorporating factors such as service capability, flexibility, information technology and administration innovation without financial indices. As such, this study hopes to contribute to the literature by offering a robust method to assess the land-based transportation companies financially. With this, the next session will discuss on the methodology used in this paper, followed by results and discussion and conclusion.

2 Methodology

The objective of this research is to optimize the ALM of land-based transportation companies in Malaysia with GP model. The companies are CJCEN, COMPLET and FREIGHT while the assessment period is from 2017 to 2021. Table 1 presents the goals of the ALM of land-based transportation companies in Malaysia.

There are six (6) goals in this research, namely asset enhancement, liability reduction, equity accumulation, revenue maximization, earnings amplification and optimum management item maximization. This implies that the greater the value of each goal (except liability), the company is more optimized in terms of ALM. Companies would prefer lower liabilities to be more stable and have lesser liquidity risk as liabilities could deteriorate the efficient frontiers [60]. A goal is perfectly achieved when there is no existence of positive or negative deviational figures from the benchmark values. In this paper, asset, equity, revenue, earnings, and optimum management items achieve the respective goals when the negative deviational figures are zero because larger values of all these goals are acceptable while values lower than the benchmark values are considered underperformance. Liability goal will be hit only when the positive deviational figures are zero as any excess in liabilities would result in financial stress for the companies [61]. The formulation of GP model is shown below:

Table 1 Goals of the ALM of land-based transportation companies in Malaysia

Goals	Descriptions
1	Asset enhancement
2	Liability reduction
3	Equity accumulation
4	Revenue maximization
5	Earnings amplification
6	Optimum management items maximization

$$\min p = w_1 G_1 + w_2 G_2 + w_3 G_3 + \cdots + w_i G_i \tag{1}$$

subject to:

Constraint-1 (Goal 1):

$$\sum_{v=1}^{u} a_{uv} x_v + d_1^- - d_1^+ = g_1 \tag{2}$$

Constraint-2 (Goal 2):

$$\sum_{v=1}^{u} a_{uv} x_v + d_2^- - d_2^+ = g_2 \tag{3}$$

Constraint-3 (Goal 3):

$$\sum_{u=1}^{v} a_{uv} x_v + d_3^- - d_3^+ = g_3 \tag{4}$$

Constraint-4 (Goal 4):

$$\sum_{u=1}^{v} a_{uv} x_v + d_4^- - d_4^+ = g_4 \tag{5}$$

Constraint-5 (Goal 5):

$$\sum_{u=1}^{v} a_{uv} x_v + d_5^- - d_5^+ = g_5 \tag{6}$$

Constraint-6 (Goal 6):

$$\sum_{u=1}^{v} a_{uv} x_v + d_6^- - d_6^+ = g_6 \tag{7}$$

where

$p =$ objective function,
$w_i =$ weight as $i = 1, 2, 3, 4, \ldots, j$,
$d_i^- =$ negative deviational figure as $i = 1, 2, 3, 4, \ldots, j$,
$d_i^+ =$ positive deviational figure as $i = 1, 2, 3, 4, \ldots, j$,
$a_{uv} =$ weight of goals in year $u = 1, 2, 3, 4, \ldots, v$,
$x_v =$ values of goals in year $u = 1, 2, 3, 4, \ldots, v$,
$g_i =$ benchmark values as $i = 1, 2, 3, 4, \ldots, j$.

After obtaining the values of the financial matrices from 2017 to 2021, the sum of each goal will be taken and compared among the land-based transportation companies. After comparison, the largest value of each goal from the three companies will be used as benchmarks except for the liability goal. All these goals except liability will be attained when there is no value for negative deviational figure. For the benchmark value of the liability goal, the lowest value of all the three companies will be used while this liability goal will be reached only when the positive deviational figure returns zero. The presence of positive deviational figures for maximization goals or negative deviational figures for minimization goals mean that there is potential improvement [62, 63]. The GP model formulation will provide a model value after considering the benchmark value and deviational figure. LINGO is then used for the computation in this paper due to its acceptance and popularity in dealing with optimization issues [64–68].

3 Results and Discussion

Results from the GP model formulation and computation of CJCEN, COMPLET and FREIGHT are being discussed in this section. Table 2 tabulates the ALM of CJCEN, COMPLET and FREIGHT from 2017 to 2021.

Table 2 presents the total amount of asset, liability, equity, revenue, earnings, and optimum management item from 2017 to 2021 for CJCEN, COMPLET and FREIGHT respectively. The largest values are taken as the benchmark values for asset, equity, revenue, earnings and optimum management item. For liability, the minimum value among CJCEN, COMPLET and FREIGHT is used. Using LINGO software, the optimal solutions for CJCEN, COMPLET and FREIGHT are obtained as shown in Tables 3, 4 and 5.

From Table 3, since there is no presence of negative deviational figure, the asset, equity, earnings, and optimum management item goals have been achieved by CJCEN. CJCEN outperformed in equity, earnings, and optimum management item goals as positive deviational figures of 0.5775, 0.0194 and 0.1876 can be found. However, due to the presence of negative deviational figure for revenue, CJCEN failed to achieve the desired revenue. CJCEN has to increase the revenue by 0.2439

Table 2 ALM of CJCEN, COMPLET and FREIGHT (trillion MYR)

Goals	CJCEN	COMPLET	FREIGHT
Asset	3.1906	0.2123	2.3996
Liability	1.3903	0.2115	0.9043
Equity	1.8002	0.7630	1.4953
Revenue	0.4214	0.0879	0.7658
Earnings	0.0251	0.0798	0.0934
Optimum management item	6.8276	2.0582	5.6584

Table 3 Optimal solution of CJCEN (trillion MYR)

Goals	Benchmark	Model value	d_i^-	d_i^+	Achievement
Asset	3.1906	3.1906	0.0000	0.0000	Achieved
Liability	0.2115	0.8129	0.0000	0.6014	Not achieved
Equity	1.8002	2.3777	0.0000	0.5775	Achieved
Revenue	0.7658	0.5219	0.2439	0.0000	Not achieved
Earnings	0.0934	0.1128	0.0000	0.0194	Achieved
Optimum management item	6.8276	7.0152	0.0000	0.1876	Achieved

Table 4 Optimal solution of COMPLET

Goals	Benchmark	Model value	d_i^-	d_i^+	Achievement
Asset	3.1906	3.1906	0.0000	0.0000	Achieved
Liability	0.2115	0.7243	0.0000	0.5128	Not Achieved
Equity	1.8002	2.6818	0.0000	0.8816	Achieved
Revenue	0.7658	0.3993	0.3665	0.0000	Not Achieved
Earnings	0.0934	0.1890	0.0000	0.0956	Achieved
Optimum management item	6.8276	7.1833	0.0000	0.3557	Achieved

Table 5 Optimal solution of FREIGHT

Goals	Benchmark	Model value	d_i^-	d_i^+	Achievement
Asset	3.1906	3.1906	0.0000	0.0000	Achieved
Liability	0.2115	1.1324	0.0000	0.9209	Not Achieved
Equity	1.8002	2.0582	0.0000	0.2580	Achieved
Revenue	0.7658	0.9636	0.0000	0.1978	Achieved
Earnings	0.0934	0.1381	0.0000	0.0447	Achieved
Optimum management item	6.8276	7.4828	0.0000	0.6552	Achieved

trillion MYR to bring the amount up to 1.8002 trillion MYR. CJCEN has also underperformed in liability goal with an excess of 0.6014 trillion MYR. CJCEN has to reduce its liability to 0.2115 trillion MYR to better manage its liabilities and liquidity risks.

Table 4 tabulates the optimal solution of COMPLET whereby COMPLET achieved four out of six goals, namely asset, equity, earnings, and optimum management item. COMPLET has surpluses of 0.8816 trillion MYR in equity, 0.0956 trillion MYR in earnings and 0.3557 trillion MYR in optimum management item. For the non-achievement of revenue goal, COMPLET needs obtain extra 0.3665 trillion MYR in revenue. Liability goal, which is also not achieved, needs to be brought down by 0.5128 trillion MYR to 0.2115 trillion MYR.

As shown in Table 5, FREIGHT has achieved all the goals except the liability goal. Overachievement can be noticed in equity, revenue, earnings and optimum management item goals amounting to 0.2580 trillion MYR, 0.1978 trillion MYR, 0.0447 trillion MYR and 0.6552 trillion MYR respectively. FREIGHT carries a large amount of excess liability (0.9209 trillion MYR) and it is suggested that FREIGHT keeps only 0.2115 trillion MYR in debts.

Table 6 sums up the benchmarks and model values of CJCEN, COMPLET and FREIGHT.

From Table 6, CJCEN, COMPLET and FREIGHT have achieved the asset, equity, earnings and optimum management item goals since all the companies have achieved the benchmarks or reflects surplus amounts. Both CJCEN and COMPLET has not reached the revenue goal since the model values are lesser than the benchmark. With liabilities of greater than 0.2115 trillion MYR, CJCEN, COMPLET and FREIGHT has not been able to attain this goal. Therefore, all three companies should work on minimizing the liabilities to be more financially stable.

Table 7 then provides a comparison of the deviations from benchmark for CJCEN, COMPLET and FREIGHT.

Table 7 shows that excess of 0.6014 trillion MYR, 0.5128 trillion MYR and 0.9209 trillion MYR have been detected in CJCEN, COMPLET and FREIGHT which has caused the non-attainment of this goal in all the companies. Since CJCEN

Table 6 Benchmark and model values of CJCEN, COMPLET and FREIGHT (trillion MYR)

Goals	Benchmark	Model values		
		CJCEN	COMPLET	FREIGHT
Asset	3.1906	3.1906	3.1906	3.1906
Liability	0.2115	0.8129	0.7243	1.1324
Equity	1.8002	2.3777	2.6818	2.0582
Revenue	0.7658	0.5219	0.3993	0.9636
Earnings	0.0934	0.1128	0.1890	0.1381
Optimum management item	6.8276	7.0152	7.1833	7.4828

Table 7 Comparison of deviations from benchmark values (trillion MYR)

Goals	Benchmark	Model values		
		CJCEN	COMPLET	FREIGHT
Asset	3.1906	0.0000	0.0000	0.0000
Liability	0.2115	0.6014	0.5128	0.9209
Equity	1.8002	0.5775	0.8816	0.2580
Revenue	0.7658	-0.2439	-0.3665	0.1978
Earnings	0.0934	0.0194	0.0956	0.0447
Optimum management item	6.8276	0.1876	0.3557	0.6552

and COMPLET must increase their revenues as they are short of 0.2439 trillion MYR and 0.3665 trillion MYR in revenue compared to the benchmark.

4 Conclusion

The objective of this paper is to optimize the ALM of land-based transportation companies in Malaysia from 2017 to 2021 using GP model. The results show that the amount of liabilities that CJCEN, COMPLET and FREIGHT hold are above the optimum value, which will affect their liquidity and solvency risks significantly. These companies could manage their accounts receivables so that they could have more cash flows and show lesser reliance on debts to finance their operations. CJCEN, COMPLET and FREIGHT could also consolidate or restructure their loans to more favorable amounts, tenures, and interest rates. This paper has successfully highlighted the issue that land-based transportation companies may be operating on excess debt in Malaysia due to the rising cost of operation. These companies can then analyze and brainstorm suitable strategies to prevent the spike of liabilities. They could also hedge against possible internal and external events which add pressure to the liquidity and solvency positions of the companies. This paper has also proven that GP model is powerful in handling ALM problems to quantify the financial resilience of these companies amid the COVID-19 pandemic and Ukraine-Russian war. Future studies may then incorporate contemporary indicators such as value-added activities and sustainability indices.

References

1. Zahid MM, Melan M, Sabar R, Prapinit P, Booyarit P (2021) Framework for drop trailer method (DTM) of haulage industry in Malaysia. Turk J Comput Math Educ 12(3):1577–1585
2. Kumar A, Anbanandam R (2019) Development of social sustainability index for freight transportation system. J Clean Prod 210:77–92
3. Melan M, Prapinit P, Boonyarit P, Netsangsee R, Hassan MG (2021) New business segmentation for haulage industry in Malaysia. Turk J Comput Math Educ 12(3):1586–1591
4. Adekoya OB, Oliyide JA, Yaya OS, Al-Faryan MAS (2022) Does oil connect differently with prominent assets during war? Analysis of intra-day data during the Russia-Ukraine saga. Resour Policy 77(102728):1–9
5. Hong J, Park J, Lee G, Park D (2019) Endogenous commercial driver's traffic violations and freight truck-involved crashes on mainlines of expressway. Accid Anal Prev 131:327–335
6. Galkin A, Davidich N, Filina-Dawidowicz L, Davidich Y (2019) Improving the safety of urban freight deliveries by organization of the transportation process considering driver's state. Transp Res Procedia 39:54–63
7. Holguín-Veras J, Leal JA, Sanchez-Diaz I, Browne M, Wojtowicz J (2020) State of the art and practice of urban freight management Part II: financial approaches, logistics, and demand management. Transp Res Part A 137:383–410

8. Gonzalez JN, Camarero-Orive A, González-Cancelas N, Guzman AF (2022) Impact of the COVID-19 pandemic on road freight transportation—a Columbian case study. Res Transp Bus Manag
9. Izadi A, Nabipour M, Titidezh O (2020) Cost models and cost factors of road freight transportation: a literature review and model structure. Fuzzy Inf Eng 1–21
10. Gonzalez-Pascual E, Nosedal-Sanchez J, Garcia-Gutierrez J (2021) Performance evaluation of a road freight transportation company through SCOR metrics. Case Stud Transp Policy 9:1431–1439
11. Aljohani K, Thompson RG (2022) Profitability of freight consolidation facilities: a detailed cost analysis based on theoretical modelling. Res Transp Econ 90(101122):1–18
12. Liew KF, Lam WS, Lam WH (2016) Financial analysis on the company performance in Malaysia with multi-criteria decision making model. Syst Sci Appl Math 1(1):1–7
13. Lam WS, Liew KF, Lam WH (2017) Evaluation on the financial performance of the Malaysian banks with TOPSIS model. Am J Serv Sci Manag 4(2):11–16
14. Naderi S, Minouei M, Gashti H (2013) Asset and liability optimal management mathematical modeling for bank. J Basic Appl Sci Res 3(1):484–493
15. Lam WS, Liew KF, Lam WH (2018) Investigation on the efficiency of financial companies in Malaysia with Data Envelopment Analysis model. J Phys: Conf Ser 995(1):012021
16. Liew KF, Lam WS, Lam WH (2017) An empirical evaluation on the efficiency of the companies in Malaysia with data envelopment analysis model. Adv Sci Lett 23(9):8264–8267
17. Lam WS, Chen JW, Lam WH (2017) Data driven decision analysis in bank financial management with goal programming model. Lect Notes Comput Sci 10645:681–689
18. Lam WS, Lam WH, Jaaman SH, Liew KF (2021) Performance evaluation of construction companies using integrated Entropy-Fuzzy VIKOR model. Entropy 23(3):320, 1–16
19. Lam WS, Liew KF, Lam WH (2019) Investigation on the performance of construction companies in Malaysia with Entropy-TOPSIS model. Earth Environ Sci 385:012006
20. Lam WH, Lam WS, Liew KF (2017) Improvement on the efficiency of technology companies in Malaysia with Data Envelopment Analysis model. Lect Notes Comput Sci 10645:19–30
21. Lam WS, Liew KF, Lam WH (2018) An optimal control on the efficiency of technology companies in Malaysia with Data Envelopment Analysis model. J Telecommun Electron Comput Eng 10(1):107–111
22. Lam WH, Yeoh HB, Lam WS, Chen JW (2018) Analysis on the performance of technology companies with Z score model. Bull Electr Eng Inform 7(4):633–639
23. Lam WH, Yeoh HB, Lam WS, Lai AJX (2020) Evaluation on the financial performance of technology companies in Malaysia with Zmijewski model. Malays J Math Sci 14(S):81–92
24. Lam WS, Jaaman SH, Ismail H (2015) The impact of human behaviour towards portfolio selection in Malaysia. Procedia Soc Behav Sci 172:674–678
25. Lam WS, Jaaman SH, Ismail H (2014) Index tracking modelling in portfolio optimization with mixed integer linear programming. J Appl Sci Agric 9(18):47–50
26. Lam WS, Lam WH (2016) Strategic decision making in portfolio management with goal programming model. Am J Oper Manag Inf Syst 1(1):34–38
27. Lam WS, Lam WH (2016) Mathematical modeling in enhanced index tracking with optimization model. J Numer Anal Appl Math 1(1):1–5
28. Lam WS, Jaaman SH, Ismail H (2014) Comparison between two-stage regression model and variance model in portfolio optimization. J Appl Sci Agric 9(18):36–40
29. Lam WS, Lam WH (2015) Portfolio optimization for index tracking problem with mixed-integer programming model. J Sci Res Dev 2(10):5–8
30. Lam WS, Jaaman SH, Ismail H (2015) An empirical study on the characteristics of high risk aversion behavior in portfolio decision making using regression model. Adv Environ Biol 9(7):17–20
31. Lam WS, Jaaman SH, Ismail H (2015) Investigation on relationship between human behavior and portfolio selection problem in Malaysia using decision making model. Adv Environ Biol 9(7):6–10

32. Lam WS, Jaaman SH, Lam WH (2019) An enhanced Mean-Gini extended model in portfolio optimization with different level of risk aversion. ASM Sci J 12(6):41–46
33. Lam WH, Lam WS (2015) An empirical investigation on portfolio management problem with mean-risk model in Malaysia stock market. Adv Sci Lett 21(5):1293–1294
34. Lam WS, Jaaman SH, Lam WH (2021) Higher moment portfolio optimization with polynomial goal programming approach. J Phys: Conf Ser 1988:012001
35. Lam WS, Jaaman SH, Lam WH (2020) Tracking a benchmark index in portfolio optimization with two-stage mixed integer programming model. J Qual Meas Anal 16(1):53–59
36. Lam WS, Jaaman SH, Lam WH (2019) Mathematical modelling of risk in portfolio optimization with Mean-Gini approach. J Phys: Conf Ser 1212:012031
37. Liew KF, Lam WS, Lam WH (2022) Optimizing the portfolio in Malaysian stock market with mean-variance model. J Posit Sch Psychol 6(3):8382–8388
38. Lam WS, Lam WH, Jaaman SH (2021) Portfolio optimization with a mean-absolute deviation-entropy multi-objective model. Entropy 23(10):1266
39. Lam WH, Jaaman SH, Lam WS (2020) Portfolio optimization of technology companies in Malaysia with Mean-Gini Model. Discov Math 42(1):45–50
40. Defalque CM, da Silva AF, Marins FAS (2021) Goal programming model applied to waste paper logistics processes. Appl Math Model 98:185–206
41. Lam WS, Chen JW, Lam WH (2016) An empirical study on the selection of fast food restaurants among the undergraduates with AHP model. Am J Inf Sci Comput Eng 2(3):15–21
42. Lam WS, Lam WH, Liew KF, Chen JW (2018) An empirical study on the preference of fast food restaurants in Malaysia with AHP-TOPSIS model. J Eng Appl Sci 13:3226–3231
43. Lam WS, Chen JW, Lam WH (2017) Analysis on the preference of fast food restaurants with analytic hierarchy process model. Int J Psychol Cognit Sci 3(6):72–76
44. Lam WS, Lam WH, Liew KF, Bakar MA, Sim JX (2021) Analysis on the e-learning method in Malaysia with AHP-VIKOR model. Int J Inf Educ Technol 11(2):52–58
45. Lam WS, Leong WB, Lam WH (2015) Selection of mobile network operator based on multi-criteria decision making model using Analytic Hierarchy Process. Math Stat J 1(1):12–18
46. Lam WS, Bakar MA, Lam WH, Liew KF (2017) Multi-criteria decision making in the selection of mobile network operators with AHP-TOPSIS model. J Eng Appl Sci 12:6382–6386
47. Lam WS, Lam WH, Bakar M, Sek XY (2018) An empirical evaluation on the preference of mobile broadband in Malaysia with analytic hierarchy process model. Adv Sci Lett 24(11):8573–8577
48. Lam WS, Lee WK, Lam WH (2015) Multi-criteria decision making in job selection problem using analytic hierarchy process model. Math Stat J 1(2):3–7
49. Lam WS, Bishan RS, Lam WH (2017) An empirical study on the mold machine-tool selection in semiconductor industry with Analytic Hierarchy Process model. Adv Sci Lett 23(9):8286–8289
50. Lam WS, Lam WH, Liew KF (2018) Data analysis on the performance of technology sector in Malaysia with Entropy-TOPSIS model. Commun Comput Inf Sci 886:194–203
51. Lam WS, Liew KF, Lam WH (2016) Evaluation on the efficiency of healthcare companies in Malaysia with data envelopment analysis model. SCIREA J Math 1(1):95–106
52. Lam WS, Liew KF, Lam WH (2021) Performance evaluation of construction companies in Malaysia with Entropy-VIKOR model. Eng J 25(1):297–305
53. Lam WS, Lam WH, Bakar MA, Lee PF (2022) Data driven decision analysis on the performance of electronic companies with TOPSIS model. J Adv Inf Technol 13(1):61–66
54. Lam WH, Din MA, Lam WS, Chen JW (2018) Evaluation on the performance of suppliers in Malaysia with TOPSIS model. J Fundam Appl Sci 10(6S):406–415
55. Lam WH, Lam WS, Liew KF, Wong SC (2018) Data driven decision analysis on the selection of course programmes with AHP-TOPSIS model. Int J Supply Chain Manag 7(4):202–208
56. Lam WS, Bishan RS, Lam WH, Liew KF (2018) An empirical evaluation on the preference of supermarkets with analytic hierarchy process model. J Phys: Conf Ser 012037
57. Liew KF, Lam WS, Lam WH (2020) analysis on the library service quality with analytic hierarchy process model. J Phys: Conf Ser 1706:012154

58. Lam WS, Bakar MA, Lam WH, Chen JW, Ma HL (2018) Evaluation on the preference of coffee shops among undergraduate students with analytic hierarchy process model. Int J Supply Chain Manag 7(4):209–215
59. Zawawi NFM, Wahab SA, Yaacob AS, Samy NK, Fazal SA (2016) Measuring the effectiveness of road transportation logistics performance in East Malaysia: a conceptual model. Int J Bus Manag 11(4):110–120
60. Zhang J, Chen P, Jin Z, Li S (2020) Open-loop equilibrium strategy for mean-variance asset-liability management portfolio selection problem with debt ratio. J Comput Appl Math 380(112951):1–17
61. Oliveira WA, Fiorotto DJ, Song X, Jones DF (2021) An extended goal programming model for the multiobjective integrated lot-sizing and cutting stock problem. Eur J Oper Res 295(3):996–1007
62. Cheng J, Feng X, Bai X (2021) Modeling equitable and effective distribution problem in humanitarian relief logistics by robust goal programming. Comput Ind Eng 155(107183):1–13
63. Broz D, Vanzetti N, Corsano G, Montagna JM (2019) Goal programming application for the decision support in the daily production planning of sawmills. Forest Policy Econ 102:29–40
64. Lam WS, Jaaman SH, Ismail H (2015) The impact of different economic scenarios towards portfolio selection in enhanced index tracking problem. Adv Sci Lett 21(5):1285–1288
65. Lam WS, Jaaman SH (2015) Ismail: an empirical comparison of different optimization models in enhanced index tracking problem. Adv Sci Lett 21(5):1278–1281
66. Lee PF, Lam WS, Lam WH (2021) Evaluation and improvement of the efficiency of logistics companies with data envelopment analysis model. Eng J 25(6):45–54
67. Chen TT, Lin C (2022) An FGM decomposition-based fuzzy MCDM method for selecting smart technology applications to support mobile health care during and after the COVID-19 pandemic. Appl Soft Comput 121(108758):1–15
68. Khan MI (2018) Evaluating the strategies of compressed natural gas industry using an integrated SWOT and MCDM approach. J Clean Prod 172:1035–1052

Virtual Reality Vehicle Inspection Routine (VR-VIR): Evaluation of Technology Acceptance

Nor Azura Kamarulzaman, Cheong Soon Nyean, Dendi Permadi, and Mazlan Bin Mahadzir

Abstract This paper proposes a Virtual Reality Vehicle Inspection Routine (VR-VIR) simulation application that teaches students how to perform vehicle inspections, also known as "Rutin Pemeriksaan Kenderaan" in Malaysia, which drivers must learn before obtaining a Malaysian driver's license. The goal of VR-VIR is to provide students with a fresh perspective on learning the vehicle inspection routine module to ensure their vehicle's safety on the road. Users can learn about vehicle inspection routines in a more informative and hands-on manner through the VR-VIR, supplementing existing methods of learning such as books or videos. Users who have never performed vehicle inspections before or are unsure how to do so properly can practice in a safe virtual setting during and post-pandemic. Virtual Reality (VR) is widely used in various fields to aid learning in the virtual world. Many scholars engaged with the Technology Acceptance Model (TAM) to explore user acceptance of new technology. However, there are limited studies on evaluating user acceptance for vehicle inspection routines through VR using TAM. Thus, this research aims to investigate how the VR-VIR affects user acceptance. Davis's TAM final version was used to assess how the VR-VIR was accepted based on elements of perceived ease of use (PEU), perceived usefulness (PU), and behavioral intention (BI). The 28 respondents evaluated the VR-VIR and answered questionnaires through a simple random sampling method. Three experts in IT and VR were involved in VR-VIR construct validation and design evaluation. Data analysis is on descriptive and correlation analysis using SPSS 26.0. The findings showed positive responses from users with medium and high values on PEU (mean = 4.04), PU (mean = 3.72), and BI

N. A. Kamarulzaman (✉) · C. S. Nyean · D. Permadi · M. B. Mahadzir
Multimedia University, 63100 Cyberjaya, Selangor, Malaysia
e-mail: norazurakz@hotmail.com

C. S. Nyean
e-mail: sncheong@mmu.edu.my

D. Permadi
e-mail: dendi.permadi@mmu.edu.my

M. B. Mahadzir
e-mail: mazlan.mahadzir@mmu.edu.my

© The Author(s), under exclusive license to Springer Nature Singapore Pte Ltd. 2023

D.-K. Kang et al. (eds.), *Proceedings of the 9th International Conference on Computational Science and Technology*, Lecture Notes in Electrical Engineering 983,
https://doi.org/10.1007/978-981-19-8406-8_17

(mean = 3.30). Experts' findings also indicated medium and high values in PEU (mean = 3.67), PU (mean = 3.92), and BI (mean = 4.00). The findings on correlation analysis for Hypothesis (H_1, H_2, and H_3) found the correlation coefficient (r) is significant with a value of $r > 0.50$, indicating PEU ($r = 0.63$), PU ($r = 0.59$) and BI ($r = 0.63$) have a strong correlation. Thus, all hypotheses are accepted. The study proved that VR-VIR is accepted by users and can assist them in learning vehicle inspection routines. This research contributes to the vehicle inspection routine in VR transport field, as a virtual learning tool for experiencing the vehicle inspection process. This study can be enhanced by improvements in interaction and application design.

Keywords Virtual reality · Vehicle inspection routine · Technology acceptance model · Vehicles · User acceptance

1 Introduction

Virtual Reality (VR) is a term used to describe technology that replaces sensory ideas with simulated ideas in order to make users believe they are in a simulated world. The definition of VR includes three essential components: an interactive simulation, a presentation method that immerses the user, and a user-centered vision [1, 2]. VR is now widely used in research, learning, invention expansion, training, treatment, entertainment drives, aerospace, automobile, architectural, health, and military applications [3]. In education, VR leads to education that links recent teaching and learning with the future world [4]. In the transport field, VR is applied for design and development, trials, and evaluation with exercise for drivers, and servicing and maintenance [3].

Because of the COVID-19 pandemic, providing virtual service or training for the future unprecedented situation is required. Researchers are investigating ways to provide a better learning experience for students during and after the pandemic using technological advancements such as VR [4]. During the pandemic, driving schools are no longer permitted to operate. Students are unable to enroll in nearby driving schools for driving lessons. This has a significant impact on students attempting to obtain a driver's license. There have been numerous studies on VR in the transportation sector. However, there have been few studies on VR for vehicle inspection routines for driving schools, as well as user acceptance. As a result, the purpose of this paper was to investigate the acceptance of Virtual Reality Vehicle Inspection Routine (VR-VIR) using the Technology Acceptance Model [5]. To fill the disparities, we discovered in the literature on user acceptance of VR in the transportation domain.

2 Literature Review

2.1 Related Work

The growth of virtual reality technology was driven by the need for better computer-user interfaces. VR is broadly used for training in education, architecture, health, and transport [3, 6, 7]. It also provides an efficient and cost-effective way to enhance performance, reduce errors, and learn skills necessary for vehicle inspection [6, 8]. Vehicle inspection is an important procedure by governments globally. The vehicle is checked to confirm the safety rules. Every vehicle driver must thoroughly inspect the vehicle to ensure the overall condition of the vehicle's body is good and should be maintained to avoid on-road car malfunction [8] and endangering public safety [9]. In Malaysia, vehicle inspection routines were implemented by the Road Transport Department (JPJ) for Routine Vehicle Inspection or "Rutin Pemeriksaan Kenderaan (RPK)" to ensure the safety of vehicles on the road. The RPK test is an essential component of the driving license test. The students need to memorize the sequence of activities and make RPK their routine habit before driving. Hence, vehicle inspection has become one of the major aspects of the transport sector.

There are many VR studies found for vehicle inspection and driving, such as surface truck safety training [6], VR experience while driving [10], and virtual vehicle inspection by car rental, maintenance, and insurance companies [11]. However, there is limited VR research found on vehicle inspection routine and evaluation of user acceptance, which becomes the motivation for this study. Hence, the benefit of VR can be extended to vehicle inspection routines in the transport field to enhance student learning.

2.2 Virtual Reality Vehicle Inspection Routine (VR-VIR)

The VR-VIR simulation was created in this study using Unity version 2021.1.17f1 as a VR app that teaches users how to inspect vehicles at home in a fun and educational way, both during and after the COVID-19 pandemic, using head-mounted-devices (HMD) like the Oculus Quest and HTC Vive. The VR-VIR aims to give students a fresh perspective on learning the vehicle inspection routine module, RPK, an official syllabus issued by the Malaysian Road Transport Department, to ensure the safety of their vehicles on the road. VR-VIR enables users to learn about vehicle inspection routines in a more informative and hands-on way, supplementing existing learning methods such as books or videos. When the VI-VIR is launched via HMD, the 3D view of the driving school environment is loaded with the associated digital assets, as shown in Fig. 1. The user is presented with a first-person 3D view of the driving school and is free to navigate it using a pair of controllers. A non-player character representing a driving school officer will greet and introduce users to the learning outcome of the VR-VIR application. Throughout the VR learning

Fig. 1 Driving school environment

Fig. 2 Virtual officer illustrating good and bad conditions of a car part

experience, the virtual officer will walk the user through the vehicle inspection step by step, using photos, videos, audio narrations, and text descriptions via the UI dialogue box. Figure 2 depicts a scenario in which the virtual officer walks the user through the inspection procedures and uses pop-up visuals to show the good and bad conditions of a car part. Users can work their way through the VR-five VIR's sections at their own pace and take a quiz at the end to test their understanding, with their score displayed once they have finished.

2.3 Technology Acceptance Model (TAM)

Technology Acceptance Model (TAM) is broadly applied for determining acceptance [12]. Davis (1989) presented the main TAM, which is reformed from the Theory of Reasoned Action (TRA) established by Fishbein and Ajzen (1975) and the Theory of Planned Behavior (TPB) by Ajzen (1991). These models were effective in foreseeing technology acceptance behavior [13–15]. Much research engaged TAM to explore user acceptance of new technology. The aim of TAM is to describe the acceptance elements to form user behavior through a range of computing and user [15].

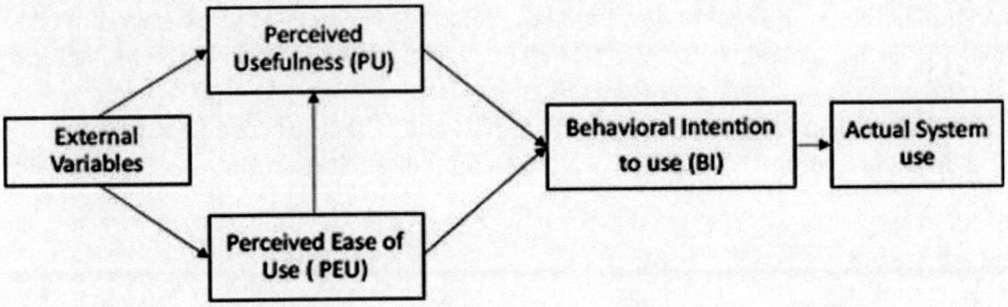

Fig. 3 Technology acceptance model—final version. *Source* Venkatesh and Davis [5]

The revised version of TAM [15] found that PU and PEU had an influence on BI, and removed the attitude element (see Fig. 1). In education, learning ability is vastly influenced by the capability of the application to make the user feel involved [12]. In this research, the final version of the original TAM was adapted for the VR-VIR study as it emphasized the PEU and PU as strong elements when clarifying the BI indirectly to users in using a technology [5] (see Fig. 3). Many studies on user acceptance have used PU and PEU [16]. TAM was chosen because it is consistent and helps explain how people use things from different points of view [17].

Perceived Usefulness (PU). The degree to where people trust using a particular technology will expand their performance. This refers to the definition of usefulness that is proficient to be used. Similar to [18] who viewed the individual when acting a behavior to gain reward. When people recognize technology as useful, they trust it to offer a constructive association [19].

Perceived Ease of Use (PEU). The degree to where people trust using new technology can support to reduce their determinations [20]. It relates to the meaning of ease, free from struggle or great effort. PEU is an element related to the user and the application [20]. It communicates with technology for easily comprehensible, content, ease of knowledge, or easiness of use [19]. Ease of use is interrelated with discovering information, the capability to resolve problems and application functionality [21].

Behavioral Intention (BI) on Actual Usage. The concern of the possibility of a person involved in the behavior of interest [22]. The stronger the intention of behavior, the higher the percentage of the person actually acting out the behavior. Intention to use technology can be described by a user's real use of technology [22]. Various studies investigated the intention to use [16, 23–25]. The actual use will be determined by the users' purpose to use the system [15].

Analysis of Acceptance Elements for Virtual Reality Application. Table 1 summarises the studies on TAM elements for virtual reality. Various studies use two essential principles, PU and PEU, to study user acceptance [16]. The latest version of the original TAM is a common model adopted for virtual reality. It is acknowledged as a strong and dominant model for user acceptance [26–28]. In this study, the final version of Technology TAM developed by Venkatesh and Davis [5] was employed to investigate the acceptance of VR-VIR. This TAM model was developed to achieve an understanding of user acceptance of VR-VIR.

This study applied the elements of PU, PEU and BI as the acceptance of VR-VIR. This literature summarized the acceptance element and description on the variables

Table 1 Summary of studies on TAM for virtual reality application

Studies	Field	Model	Findings	Dimension	References
VR for driver training	Transport	1st modifications Davis, Bgozzi Warshaw (1989)	Attitude as mediator among belief, intention	PU, PEU, attitude, and BI	[28]
VR Driving: TAM	Transport	Final version Venkatesh and Davis (1996)	Excluding Attitude. PU, PEU direct influence BI	PEU, PU, and BI	[10]
VR: Extended TAM	Transport	Final version Venkatesh and Davis (1996)	PU, PEU direct effect on BI	PEU, PU, and BI	[29]
TAM on VR cycling sport	Transport	UTAUT Venkatesh et al. (2003)	6 factors found. 2 used PU, PEU	Realism fidelity visual, audio, interaction	[30]
VR in construction safety training	Construction	Extended TAM, TAM3 Venkatesh and Bala (2008)	PU and PEU are direct predictors of attitude	PEU, PU, BI, attitude, perceived price value, self-efficacy, perceived playfulness	[31]
VR hardware acceptance model	Hardware	VR-HAM Extend and individuating TAM for VR hardware	Parsimonious: acceptance, use, purchase intention. Effect age, curiosity, past use, price	PEU, PU, future use, purchase intention, enjoyment, attitudes, enjoyment	[32]
VR TAM in Classrooms	Classroom	First changes Davis et al. (1989)	Influence on their attitude, intention to use in class	PEU, PU, and BI	[33]
TAM Survey of Metaverse	Metaverse	Extended TAM3 Venkatesh and Bala (2008)	Price, BI, social norm, self-efficiency, pleasure, PU, PEU, attitude	Social norm, self-efficiency, perceived pleasure, curiosity, attitude, PEU, PU, BI	[34]

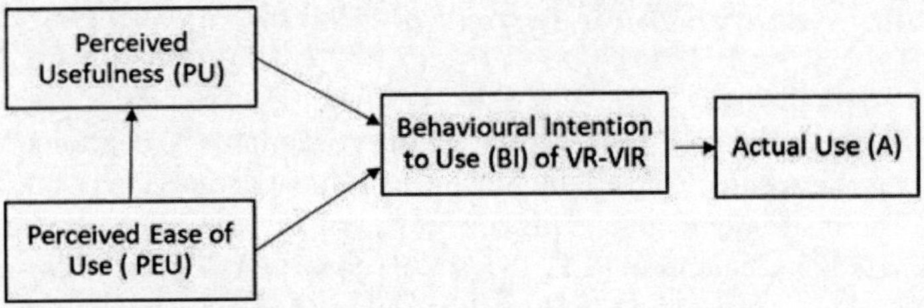

Fig. 4 Technology acceptance model for VR-VIR

and items measurement scale for evaluating the acceptance of VR-VIR from the literature (See Table 1). Figure 4 shows the TAM for VR-VIR.

3 Methodology

This study has reviewed previous studies on user acceptance for virtual reality applications using the technology acceptance model. The articles reviewed between 2018 and 2022 were from online databases, namely Google Scholar, Scopus, and ProQuest, relevant books, and reliable websites. The results summarized the previous studies on VR that employed the TAM. The keyword terms used are "virtual reality", "vehicle inspection routine", "technology acceptance model", "vehicle", and "user acceptance". Then, the acceptance measurement scale for the VR-VIR was proposed. The evaluation of the user acceptance is done through an experimental post method using a close-ended and open-ended questionnaire as the main instrument. The sample of respondents consists of 28 users selected through a simple random sampling method. There were construct validation and evaluation by the experts in VR to validate the instrument and application design. The findings on the acceptance evaluation for VR-VIR were analysed using descriptive and correlation analysis using SPSS 26.0.

4 Result and Discussion

This section discussed the findings of the VR-VIR acceptance evaluation. The quantitative data was tabulated using SPSS version 26.0. To describe demographics and acceptance responses, data were analysed using descriptive statistical analysis. The findings of qualitative data collection and hypothesis testing are also presented.

4.1 Virtual Reality Vehicle Inspection Routine (VR-VIR)

Construct Validation and Reliability on the Instrument for Acceptance VR-VIR. The validity is the accuracy of variables in the instrument to measures what should be measured. The instrument validated by three experts in VR. The questionnaire was developed based on elements of PUE, PU, and BI consisted 12 questions using a five-point Likert scale from 1 to 5 (1 = strongly disagree, 2 = disagree, 3 = moderate, 4 = agree, 5 = strongly agree). Table 3 showed mean, frequency (f), standard deviation (sd), and mean interpretation. Twelve items indicate high mean interpretation level based on mean score interpretation [35]. Experts were agree to strongly agree with PEU (mean = 4.59, sd = 0.58), agree to strongly agree with PU (mean = 4.75. sd = 0.44) and agree to strongly agree with BI (mean = 4.58. sd = 0.58). The overall mean of acceptance was 4.67 which is high. The reliability refers to replication and consistency [36]. The alpha values above 0.70 indicate a high level of reliability and the questionnaire scale are internally consistent [37–39]**.** In this study, **Cronbach's alpha scores show value of 0.93 indicated the questionnaire has high reliability.**

Evaluation on Acceptance of VR-VIR Application (Expert Design). Three experts in IT and VR involved in evaluating the design of VR-IR. There are three components of PEU, PU, and BI. Table 3 shows findings on expert evaluation on acceptance of VR-VIR. Seven items indicate the mean is at a high mean interpretation level (4.01–5.00) and five item is at medium high mean interpretation level (3.01–4.00) based on mean score interpretation by Nunally (1978) [35]. The experts were moderately agree to strongly agree with the PEU (mean = 3.67, sd = 0.73), agree to strongly agree with the PU (mean = 3.92. sd = 0.44 and agree with BI (mean = 4.00. sd = 0.00). The overall mean of acceptance was 3.86 which is medium high mean (see Table 3).

Figures 5 and 6 show the overall view on PEU, PU, BI and recommendation by the expert on VR-VIR. The experts recommended the VR-VIR for learning vehicle inspection routine with some improvement on the design namely, the visibility of the question board, reference image, user interaction, locomotion, and clickable icon in the application.

Evaluation on Acceptance of VR-VIR Application (User). The example data on respondent profile (see Table 2) and user evaluation VR-VIR (see Table 3).

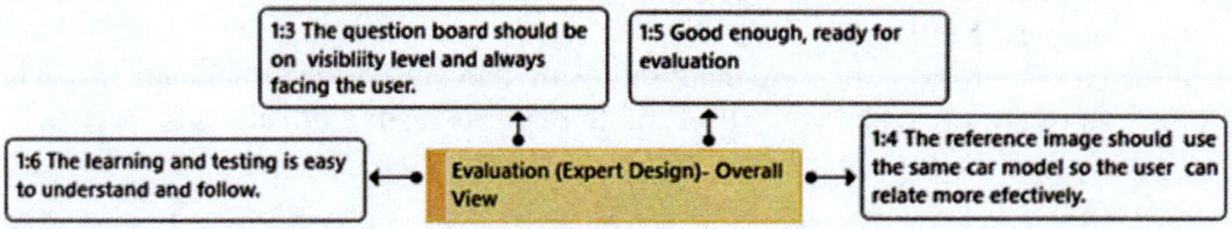

Fig. 5 Evaluation (expert design)—overall view on PEU, PU and BI

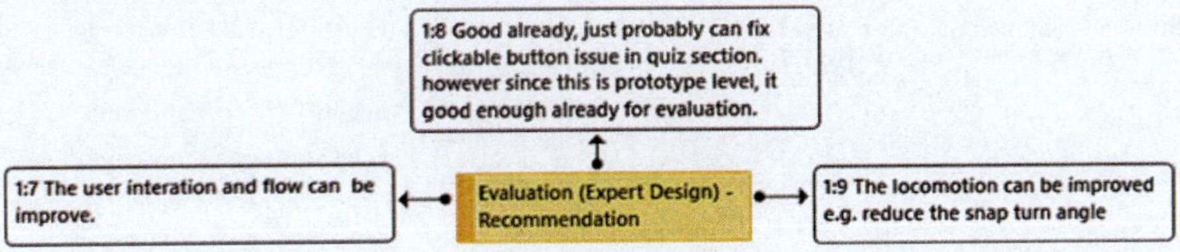

Fig. 6 Evaluation (expert design)—recommendation

Table 2 Respondent profile

Item	Description	Frequency	Percent (%)
Gender	Male	15	53.6
	Female	13	46.4
Age	18–24	27	96.4
	41–64	1	3.6
Highest Education	SPM/SPMV	6	21.4
	STPM/STAM	2	7.1
	Diploma	2	7.1
	Bachelor's degree	18	64
Occupation	Student	25	89.3
	Employed	2	7.1
	Self-employed	1	3.6

Respondent Profile (User). The evaluation was carried out on 28 participants with various background from age 15 to above 65 years. The profile includes gender, age, highest education, and occupation. Data on gender indicated that 53.6% of male respondents (n = 15) outnumbered 46.4% female respondents (n = 13). From 28 respondents, 96.4% (27) were between 18 and 24 years and 3.6% (41) were between 41 and 64. The highest 64.3% of users were studying at bachelor's degree level, followed by 21.4% in SPM/SPMV, 7.1% in STPM/STAM, and 7.1% in Diploma. In occupation, 89.3% of respondents were student, 7.1% were employed, and 3.6% were self-employed.

Table 3 shows mean, frequency, standard deviation, and interpretation of each item. Twelve items were outlined in this study. Five items indicate high mean interpretation level and 7 items are in the medium high mean interpretation level based on mean score interpretation [35]). The findings showed that user was disagreed to agree with PEU (mean = 4.04, sd = 0.82), disagree to strongly agree with the PU (mean = 3.72. sd = 0.93) and disagree to strongly agree with the BI (mean = 3.93. sd = 1.16). The overall mean of the element of acceptance was 3.90 which is medium high mean. Figures 7 and 8 shows overall view on PEU, PU and BI and recommendations for VR-VIR. The users gave the positive feedbacks and suggestions for improvements.

Research Hypothesis. This section presents the result of hypothesis testing on the influence between acceptance of VR-VIR application and behaviour intention to use.

Table 3 Findings on acceptance of VR-VIR (user, expert design, construct validation)

	Item/Scale	User Evaluation		Expert Design		Construct Validation	
		Mean	Sd	Mean	Sd	Mean	Sd
Perceived Ease of Use (PEU) User believe VR-VIR will be able to motivate them in learning vehicle inspection routine [20]	Learning to use VR-VIR would be easy for me	4.04	0.88	3.67	1.16	4.33	0.58
	My interaction with VR-VIR would be clear and understandable	3.93	0.72	3.33	0.58	4.67	0.58
	It would be easy for me to grow into skilful at using VR-VIR	4.04	0.84	4.00	0.00	4.67	0.58
	I find VR-VIR is easy to use	4.14	0.85	3.67	1.16	4.67	0.58
	Perceived Ease of Use	4.04	0.82	3.67	0.73	4.59	0.58
Perceived Usefulness (PU) A level where user believe usage of VR-VIR can decrease a person's work in doing something [20]	Using VR-VIR enhance my learning of vehicle inspection routine	3.75	0.97	4.33	0.58	4.67	0.58
	Using VR-VIR enhances my productivity	3.50	0.96	3.67	0.58	4.67	0.58
	Using VR-VIR enhances my effectiveness	3.75	0.84	4.00	0.00	5.00	0.00
	I believe using VR-VIR is useful in my life	3.89	0.96	3.67	0.58	4.67	0.58
	Perceived Usefulness	3.72	0.93	3.92	0.44	4.75	0.44
Behavioural intention (BI) Thinking of possibility user would be keenly involved in behaviour of concern [39]	There is a high possibility that I will use VR-VIR in future	4.04	1.07	4.00	0.00	4.67	0.58
	I intend to use VR-VIR in future	3.96	1.20	4.00	0.00	4.67	0.58
	I will use VR-VIR in future	4.00	1.12	4.00	0.00	4.67	0.58
	Using VR-VIR in future is important to me	3.71	1.24	4.00	0.00	4.67	0.58
	Behavioural Intention	3.93	1.16	4.00	0.00	4.67	0.58

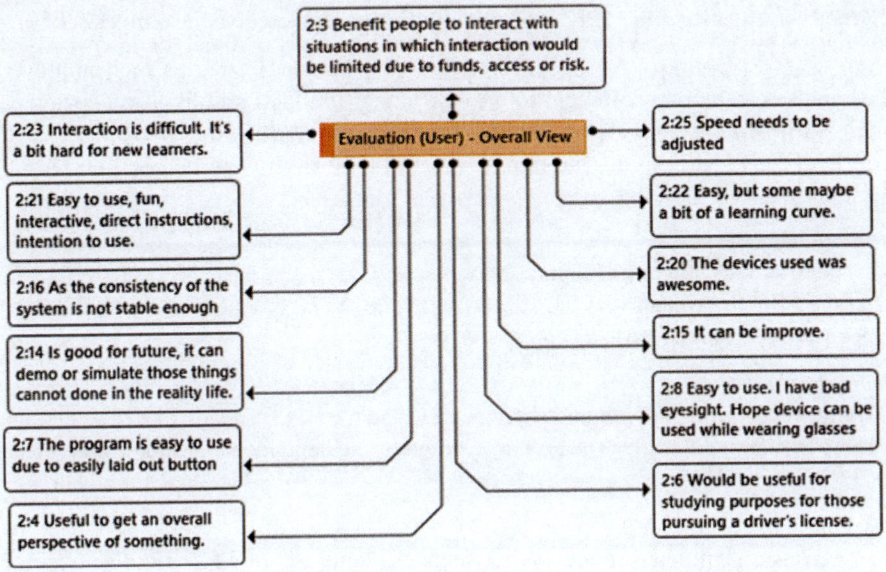

Fig. 7 Evaluation (user)—overall PEU, PU and BI

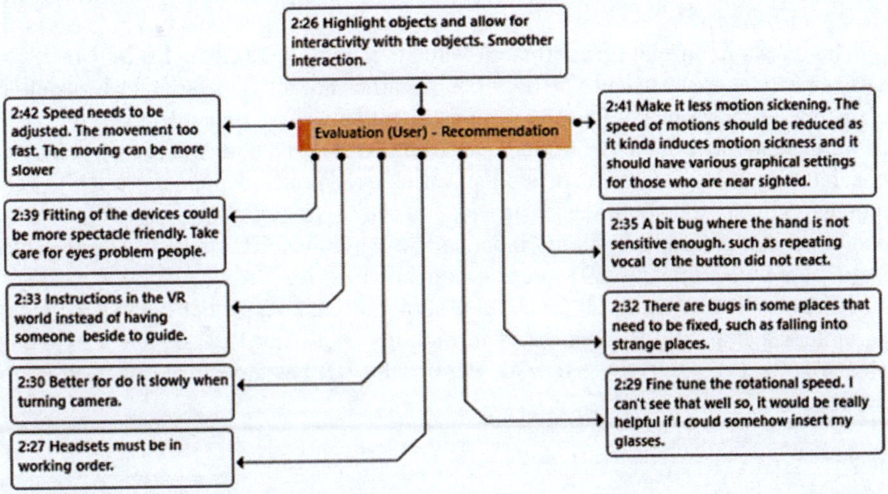

Fig. 8 Evaluation (user)—recommendation

H_1: Perceived ease of use (PEU) positively influences Behavioral Intention (BI) to use Virtual Reality Vehicle Inspection Routine (VR-VIR)

H_2: Perceived usefulness (PU) positively influences Behavioral Intention (BI) to use Virtual Reality Vehicle Inspection Routine (VR-VIR)

H_3: Perceived ease of use (PEU) positively influences Perceived usefulness (PU).

Analysis for Hypothesis (H_1, H_2, H_3). To verify H_1, H_2 and H_3, this research employed Pearson's correlation coefficient to prove whether the PEU and PU is independent, high or low among the independent and dependent variables of the BI to use VR-VIR or PU of VIR-VIR. An interpretation of correlation coefficient is based on Davies (1971) [40].

Hypothesis 1 (H_1).

There is a significant influence between PEU and the BI to use VR-VIR. The average PEU are (mean = 4.04, sd = 0.82). The average dependent variable, BI is (mean = 3.93, sd = 1.16).

Hypothesis 2 (H_2).

There is a significant influence between PU and the BI to use VR-VIR. The average PU are (mean = 3.72, sd = 0.82). The average of the dependent variable, BI is (mean = 3.93, sd = 1.16).

Hypothesis 3 (H_3).

There is a significant influence between PEU and PU of VR-VIR. The average PU is (mean = 3.72, sd = 0.93). The average of the dependent variable, PU is (mean = 3.93, sd = 1.16). Table 4 shows that the Pearson's correlation coefficient between the variables is more than 0.5, showing variables and independent variables have a strong correlation.

The constant correlation coefficient value is among − 1.0 to + 1.0, + 1 ($0 > r > 1$) as a positive connection, − 1 ($0 < r < 1$) signifies among two variables is negative connection. The results of Pearson's correlation analysis for dependent variable (BI) and independent variable (PEU) indicates the correlation coefficient value of 0.634, presents a positive correlation, dependent variable (BI) and independent variable (PU) indicates correlation coefficient with value of 0.673, presents a positive correlation, and dependent variable (PU) and independent variable (PEU) indicates correlation coefficient with value of 0.593, presents a positive correlation. Statistics shows their significance is less than 0.1 ($p \leqq 0.1$), showing the independent and dependent variables have a solid association. The findings show that PEU of VR-VIR has a positive effect on BI to use VR-VIR, PU of VR-VIR has a positive effect on BI to

Table 4 Pearson's correlation analysis (H_1, H_2, H_3)

Variables		PEU	BI	PU
PEU	Pearson Correlation Sig. (2-tailed)	1	0.634** 0.000	
BI	Pearson Correlation Sig. (2-tailed)	0.634** 0.000	1	0.673** 0.000
PU	Pearson Correlation Sig. (2-tailed	0.593** 0.000		1

** Correlation is significant at the 0.01 level (2-tailed). (N = 28)

use VR-VIR, and PEU of VR-VIR has a positive effect on PU of VR VIR. Thus, the hypothesis H_1, H_2, H_3 are accepted.

5 Conclusion

In conclusion, VR-VIR is an innovation for the vehicle inspection routine in a virtual environment. This evaluation was based on findings gathered from literature reviews, user evaluations, and expert feedback. The findings were analyzed through descriptive and correlation analysis. The study received positive feedback from the users, with medium and high values mean for user acceptance. Both PEU and PU have a strong correlation with the BI when using the VR-VIR and have a significant influence on each other. Therefore, all hypotheses are accepted. The results proved that the VR-VIR is accepted by the users. However, some improvement is needed in the user interaction and application design. Practically, this is the new approach in VR related to VIR as a tool to promote VR education in the transport domain. It enriches the VR media for VIR and can assist the JPJ and transport industry with an innovative VR-VIR education. Theoretically, this study enriches the literature and contributes to the body of knowledge in VR for VIR. This study employed the TAM model to evaluate the user acceptance of VR-VIR. It provides a summative and formative evaluation of VR-VIR that can be used as a guideline for researchers and developers. The study contributes to the responses of VR-VIR evaluation based on technology acceptance that includes PEU, PU, and BI. The evaluation and feedback are beneficial for future research and improvement. It is hoped that the proposed VR-VIR can assist and motivate the user to learn a vehicle inspection routine and practice it in the real environment.

Acknowledgements The authors gratefully acknowledge the financial support and VR facilities provided by Faculty of Creative Multimedia, Multimedia University (Cyberjaya campus) to implement the prototype successfully. The authors also extend sincere appreciations to Ms. Nur Afshaa Binti Mohd Ameeruddin, Ms. Raisa Musthofa and Ms. Siti Munirah Binti Razaly in facilitating the data collection.

References

1. Mcneill M, Moshell M, Florida UC, Pausch R, Virginia U, Robinett W, Hill U, Sutherland I, Microsystems S, Urban D, Bishop G, Hill U, Bricken W, Washington U, Brooks F, Hill U, Burbeck C, Hill U, Durlach N, Fuchs H, Hill U, Green M, Alberta U (1992) Research directions in virtual environments. Chapel Hill
2. Muhanna MA (2015) Virtual reality and the CAVE: taxonomy, interaction challenges and research directions. J King Saud Univ Comput Inf Sci. 27(3):344–361
3. Jerald J (2016) The VR book: human-centered design for virtual reality, 1st edn. The VR Book. Morgan & Claypool Publishers, p 223

4. Nalluri SP, Reshma L, Munavalli JR (2021) Evaluation of virtual reality opportunities during pandemic. In: 2021 6th International conference on convergence technology I2CT 2021, pp 1–6
5. Venkatesh V, Davis FD (1996) A model of the antecedents of perceived ease of use: development and test. Decis Sci 27(3):451–481
6. Ennis D, Lucchesu B Oberlander N (2006) Surface mine truck safety training: a VR approach to pre-operational vehicle inspection, vol 1999
7. Jasoren Homepage. https://jasoren.com/4-virtual-reality-applications-in-the-automotive-industry/. Last accessed 11 Aug 2018
8. Driveinmalaysia Homepage. https://driveinmalaysia.com/blog/2021/routine-vehicle-inspection/. Last accessed 11 Aug 2018
9. NST Homepage. https://www.nst.com.my/news/nation/2022/07/815995/review-procedures-vehicle-checks. Last accessed 11 Aug 2018
10. Cosío AA (2019) Virtual reality experiences while driving: a user acceptance model. Technische Universität München
11. Abulkhair M, Sindi H, Barsheed B, Al-Omari M, Al-Shehri R, Al-Basarah R, Al-Harbi W (2015) Car Inspection Syst Procedia Manuf 3:3128–3135
12. Palmas F, Reinelt R, Cichor JE, Plecher DA, Klinker G (2021) Virtual reality public speaking training: experimental evaluation of direct feedback technology acceptance. In: Proceedings—2021 IEEE conference on virtual real 3D user interfaces, VR 2021, pp 463–472
13. Barry M, Jan MT, Islamic I (2018) Factors influencing the use of m-commerce: an extended technology acceptance model perspective. Int J Econ Manag Account. 26(1):157–183
14. Nasri W, Charfeddine L (2012) Factors affecting the adoption of Internet banking in Tunisia: an integration theory of acceptance model and theory of planned behavior. J High Technol Manag Res. 23(1):1–14
15. Davis FD, Bagozzi RP, Warshaw PR (1989) User acceptance of information technology: a comparison of two theoretical models *. Manage Sci 35(8):982–1002
16. Lin C-C (2013) Exploring the relationship between technology acceptance model and usability test. Inf Technol Manag 14:243–255
17. Liu N, Yu R (2017) Identifying design feature factors critical to acceptance and usage behavior of smartphones. Comput Human Behav. 70:131–142
18. Lu HP, Su PYJ (2009) Factors affecting purchase intention on mobile shopping web sites. Internet Res 19(4):442–458
19. Hernandez B, Jimenez J, Jose MM (2009) The impact of self-efficacy, ease of use and usefulness on e-purchasing: an analysis of experienced e-shoppers. Interact Comput 21(1–2):146–156
20. Davis FD (1989) Perceived usefulness, perceived ease of use, and user acceptance of information technology. Manag Inf Syst Res Center Univ Minnesota. 13(3):319–340
21. Pigatto G, Machado JG de CF, Negreti A dos S, Machado LM (2016) Have you chosen your request? Analysis of online food delivery companies in Brazil. Br Food J. 119(3):639–657
22. Ajzen I, Fishbein M (2000) Attitudes and the attitude-behavior relation: reasoned and automatic processes. Eur Rev Soc Psychol 11(1):1–28
23. Shin DH (2009) Towards an understanding of the consumer acceptance of mobile wallet. Comput Human Behav. 25(6):1343–1354
24. Okumus B, Bilgihan A (2014) Proposing a model to test smartphone users' intention to use smart applications when ordering food in restaurants. J Hosp Tour Technol 5(1):31–49
25. Aziz MNA, Harun SN, Baharom MK, Kamaruddin N (2020) The evolution of the technology acceptance model (TAM). Interdiscip Manag Econ Soc Res 2020(August):242–247
26. Aati K, Chang D, Edara P, Sun C (2020) Immersive work zone inspection training using virtual reality. Transp Res Rec 2674(12):224–232
27. Rondan-cataluña FJ, Arenas-gaitán J, Ramírez-correa PE (2015) A comparison of the different versions of popular technology acceptance models a non-linear perspective. Emerald Insight 44(5):788–805
28. Munkvold RI, Robertsen R, Skevik TO (2020) Feasibility study for the use of VR technology for driver training. IRO J Sustain Wirel Syst 2(1)

29. Sagnier C, Loup-Escande E, Lourdeaux D, Thouvenin I, Vallÿry G (2020) User acceptance of virtual reality: an extended technology acceptance model. Int J Hum Comput Interact 36(11):993–1007
30. Mahalil I, Yusof AM, Ibrahim N (2020) A literature review on the usage of technology acceptance model for analysing a virtual reality's cycling sport applications with enhanced realism fidelity. In: 2020 8th International Conference on Information Technology and Multimedia, ICIMU 2020, pp 237–242
31. Zhang M, Shu L, Luo X, Yuan M, Zheng X (2022) Virtual reality technology in construction safety training: extended technology acceptance model. Autom Constr 135(February):104113
32. Manis KT, Choi D (2018) The virtual reality hardware acceptance model (VR-HAM): extending and individuating the technology acceptance model (TAM) for virtual reality hardware. J Bus Res 2019(100):503–513
33. Majid FA, Shamsudin NM (2019) Identifying factors affecting acceptance of virtual reality in classrooms based on technology acceptance model (TAM). Asian J Univ Educ 15(2)
34. Aburbeian AM, Owda AY (2022) A technology acceptance model survey of the metaverse prospects. 285–302
35. Nunnally JC (1978) An overview of psychological measurement. In: Clinical diagnosis of mental disorders. Plenum Press, New York, pp 97–146
36. Mark NKS, Lewis P, Thornhill A (2019) Research methods for business student, 8 edn. Pearson, Harlow, United Kingdom, 833p
37. Al-Maroof RA, Al-Emran M (2021) Recent advances and smart systems in intelligent applications. In: Studies in systems, decision and control, vol. 295. Springer, Al Buraimi, Oman, pp 253–275
38. Taber KS (2018) The use of Cronbach's alpha when developing and reporting research instruments in science education. Res Sci Educ 48(6):1273–1296
39. Venkatesh V, Brown SA, Maruping LM, Bala H (2008) Predicting different conceptualizations of system USE: The competing roles of behavioral intention, facilitating conditions, and behavioral expectation. MIS Q Manag Inf Syst 32(3):483–502
40. Davies JA (1971) Elementary survey analysis. Prentice-Hall, Englewood Cliffs, N.J. 195 pages

Attendance Management System with Half-Covered and Full-Facial Recognition Feature

Kah-Ong Goh, Check-Yee Law, Cu-Kang Tin, Connie Tee, and Yong-Wee Sek

Abstract Manual way of tracking attendance is inefficient and has vulnerability in protecting personal data. Various attendance management systems have been introduced to replace the manual attendance tracking process. This includes the integration of face recognition technique in the attendance management system. However, following the outbreak of pandemic COVID-19, we are strongly advised to always put on a face mask to protect ourselves and others. This practice has caused problems to existing attendance management system with facial recognition. This is because the mask has covered the essential data that can be measured and extracted by the facial recognition algorithms. To overcome this problem, an attendance management system with half-covered and full-facial recognition feature is proposed. Mobile-FaceNet model is used to verify the user identity for a real-time attendance check-in. Users are able to take attendance via the application with or without a face mask.

Keywords Attendance management system · COVID-19 · Face mask · Facial recognition · MobileFaceNets

---

K.-O. Goh · C.-Y. Law (✉) · C.-K. Tin · C. Tee
Multimedia University, Melaka, Malaysia
e-mail: cylaw@mmu.edu.my

K.-O. Goh
e-mail: michael.goh@mmu.edu.my

C.-K. Tin
e-mail: cukang98@gmail.com

C. Tee
e-mail: tee.connie@mmu.edu.my

Y.-W. Sek
Universiti Teknikal Malaysia Melaka, Melaka, Malaysia
e-mail: ywsek@utem.edu.my

© The Author(s), under exclusive license to Springer Nature Singapore Pte Ltd. 2023

D.-K. Kang et al. (eds.), *Proceedings of the 9th International Conference on Computational Science and Technology*, Lecture Notes in Electrical Engineering 983,
https://doi.org/10.1007/978-981-19-8406-8_18

1 Introduction

Attendance management system has widely been used in many organizations. It is used by most companies to keep track of employee working hours and to monitor employee productivity. Many educational institutions also deploy the attendance management system to monitor student attendance to engage students in learning and to minimize the dropout rate. Manual way of tracking attendance is inefficient and has vulnerability in protecting personal data. Therefore, various attendance management systems have been introduced to replace the manual tracking process.

Reference [1] suggested an attendance recording application system based on workflow technology in which the check-in workflow module, the mobile check-in terminal module, the Web check-in service module and the results query module are integrated to track the check-in information. Reference [2] developed student attendance system by using the biometric technology in which attendance is marked when a student place his/her fingerprint on a sensor. Attendance system based on speech biometric had also been proposed [3]. Lukas et al. [4], Salim et al. [5], Mery et al. [6] and Muthunagai et al. [7] proposed the integration of face recognition technique in student attendance management system.

Following the outbreak of pandemic COVID-19, one of the biosafety advice given by the World Health Organization (WHO) to restraint the spread of this infectious virus is to always put on a face mask in public [8]. Such biosafety practice has caused difficulty to some attendance management systems especially those systems that adopt face recognition technique. This is because existing face recognition techniques are designed to recognize an individual via the detection of eyes, nose and mouth. When the face is covered by face mask, some essential data of the nose and mouth are lost. As such, there is a need to develop algorithms for recognizing an individual with a face mask. New strategies must be implemented to increase the robustness of existing systems.

This paper reports the research work of designing and developing a half-covered and full facial recognition attendance management system. Section 2 reviews some related work. Section 3 presents the methodology of the research work in face detection. The prototype design is explained with the context diagram and some user interface screenshots of the mobile application in Sect. 4. Section 5 describes the functional testing results of the proposed system, and the accuracy comparison results for the model chosen in adopting the half-covered and full-facial recognition techniques. The paper is concluded with future work in Sect. 6.

2 Related Work

Various attendance management systems have been developed to track the attendance and performance of employees such as Jibble Tracker, UbiAttendance, and Fareclock. Each of the systems possesses essential features to support different user needs and requirements.

Jibble Tracker that is available in the Google Play is a time clock application for industries to track the attendance of their employees. It is a free-to-use attendance tracking application and is quite popular in the Android application market. A user can check-in and check-out with the application and add a note to it. The employer can then view the attendance log, and the notes or comments made by the employee. Geolocation information is available in this application in which the actual location is shown in Google Map during check-in and check-out. Jibble Tracker allows the employer to turn on "Require Selfie" feature. This requires employees to take a selfie photo whenever they check-in or check-out. However, there is no face recognition feature in this application to verify the users.

UbiAttendance is an attendance tracking system that is available in both Apple Store and Google Play. It comes with Basic and Premium Plan. A user can enjoy a 7-day free trial of the Premium Plan after registered as a user. Some features such as advanced web panel, hourly wages calculation, biometric capture, and geo fencing are not available in the Basic Plan. A user can check their personal attendance logs in a calendar view. Different colours are used to denote different status such as present, absent, off, and missed. This enables a user to easily track their attendance in the calendar view. Various types of attendance records including late comers' log, records of early leavers, and the visitor records are available to view and download as well.

Another popular attendance tracking application is known as Fareclock. Fareclock is available on both the Apple Store and Google Play and can be downloaded and used for free. It is a cloud-based time clock application with facial recognition function that can be used by companies at either a fixed location or any location. All account information is stored by Fareclock in the internet cloud. In the case the device is not connected to the internet, Fareclock can work offline as well. All data will synchronize when an internet connection is established. There are two layers of verification when a user check-in, that is, by using the facial recognition and personal pin. First the application will detect a user's face. Then, the user will be asked to enter the personal pin number to complete the check-in process. This means the person who attempts to check-in must present. However, Fareclock is unable to detect a masked face from the camera input.

Following the outbreak of pandemic COVID-19, we are wearing masks to protect ourselves and others. This practice has caused problems to existing attendance management system with facial recognition. This research work proposes an attendance management system with half-covered and full-facial recognition feature.

3 Methodology

Face detection refers to the technology that can recognize human faces in image or video. Machine learning algorithm is used in face detection to recognize whether there is a face in the image or video. Face detection is a pre-requisite condition to implementing facial recognition technology. As this paper propose the work of half-covered and full facial recognition system, two main pre-requisite work must be accomplished namely, the facial detection and the masked face detection. The former checks if there are any faces in the input images or videos, and the latter aims to recognize the masked face based on the forehead and eyes regions.

A plethora of algorithms have been proposed and used for face detection, including Viola Jones algorithm [9], HOG descriptor algorithm [10], Haar-Like features algorithm [11], VGG-16 face Convolutional Neural Network (CNN) descriptor [12], FaceNet [13], Local Binary Pattern (LBP) [14, 15], MobileFaceNets [16], and many more. Each technique has its own strengths and limitations. Lightweight CNN model that works efficiently on portable devices such as mobile is used in this research work for facial recognition.

MobileFaceNet is proposed by Mohamed et al. [16]. It is a set of Convolutional Neural Network (CNN) models that has lesser parameters (0.99 million) as compared to other pre-trained models. As the MobileFaceNet mobile can be converted to TensorFlowLite file, it requires a very small storage space, that is, around 4.0 Megabyte. So, it is specifically built for real-time facial recognition on embedded or mobile devices with high accuracy. Figure 1 presents the process flow of MobileFaceNet.

MobileFaceNet model takes in an image with 112×112 size as an input and generates a feature map that has 7×7 matrix in size. The Random Forest, RF1 and RF2 correspond to the corner unit and the centre unit of the feature map respectively. The feature map shows that the values of centre units are higher than the corner unit

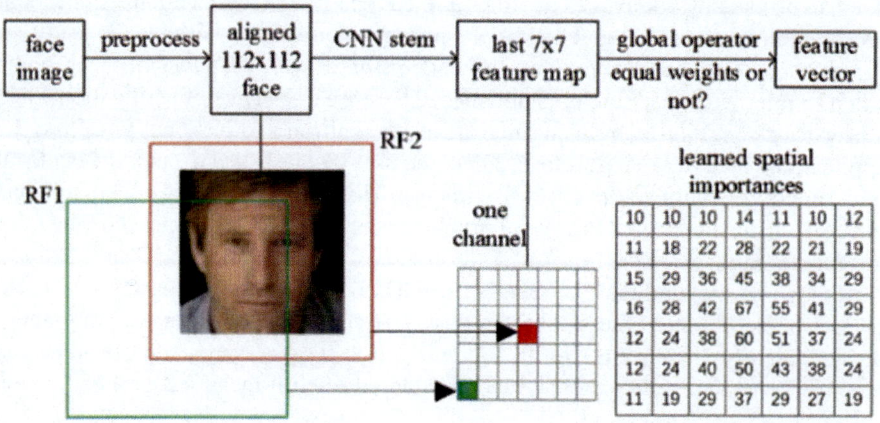

Fig. 1 Process flow of MobileFaceNet

Fig. 2 The architecture of MobileFaceNet

Input	Operator	t	c	n	s
$112^2 \times 3$	conv3x3	-	64	1	2
$56^2 \times 64$	depthwise conv3x3	-	64	1	1
$56^2 \times 64$	bottleneck	2	64	5	2
$28^2 \times 64$	bottleneck	4	128	1	2
$14^2 \times 128$	bottleneck	2	128	6	1
$14^2 \times 128$	bottleneck	4	128	1	2
$7^2 \times 128$	bottleneck	2	128	2	1
$7^2 \times 128$	conv1x1	-	512	1	1
$7^2 \times 512$	linear GDConv7x7	-	512	1	1
$1^2 \times 512$	linear conv1x1	-	128	1	1

since the face is located in the middle of the image. The proposed method uses global depth wise convolutional layers instead of pooling layers so that the model is able to learn from the weights. The Insightface is used as the loss function during the training stage. The architecture of the MobileFaceNet is shown in Fig. 2.

The first convolutional layer takes in an image with the size of 112×112 as an input. The letter t refers to the factor of expansion, c is the output channel that is connected to the next layers, n indicates the number of filters and s represents the stride, that is, the number of pixels that shift over the input matrix. The second last layer, linear GDConv7×7 refers to the global depth wise convolution layer with the kernel size of 7×7.

For the implementation of the system, the features map is transforming into a storable array of numbers and store into the firebase. So, whenever a user signs-in or checks in, the face image will be captured, and the application will search for a similar face in the firebase. This is done by comparing the minimum Euclidean Distances between the face image for each user that has previously been stored in the firebase. Once the user has been determined and the accuracy is above the threshold values, the user data will be returned from the firebase and displayed in the application. The formula for Euclidean Distance is determined as in Eq. 1.

$$d(p,q) = d(q,p) = \sqrt{(q1 - p1)^2 + (q2 - p2)^2 + \cdots + (qn + pn)^2}$$
$$= \sqrt{\sum\nolimits_{i=1}^{n} (qi - pi)^2} \quad (1)$$

When a new user has registered and the face image has been taken, the image that has been processed by the ML Kit face detection API will be inferenced by the tflite model in front-end. Flutter and the returned result, that is the embedded face will be transformed into a storable data structure to be stored in the Firebase. It will then be labelled with the user's name and password so that each user would be identifiable.

4 Prototype Design

Figure 3 shows the context diagram of the proposed half-covered and full facial recognition attendance management mobile application.

First, users would need to register or sign up by entering their personal details including photo shoot of their facial images as in Figs. 4 and 5.

The information will be stored in "user" class in the database. Once done, users would receive a registration confirmation message from the application. Then, they

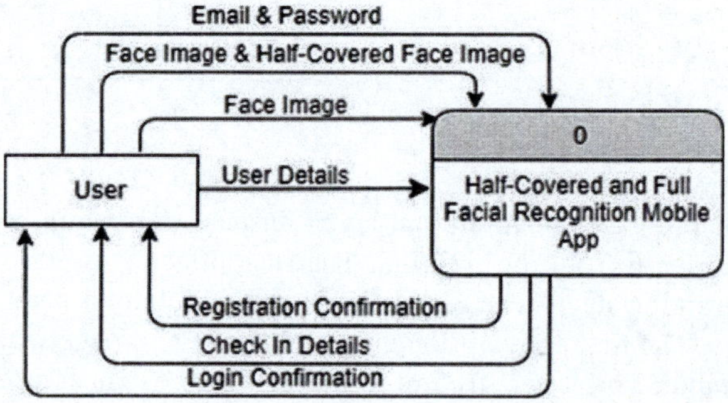

Fig. 3 The context diagram of the proposed system

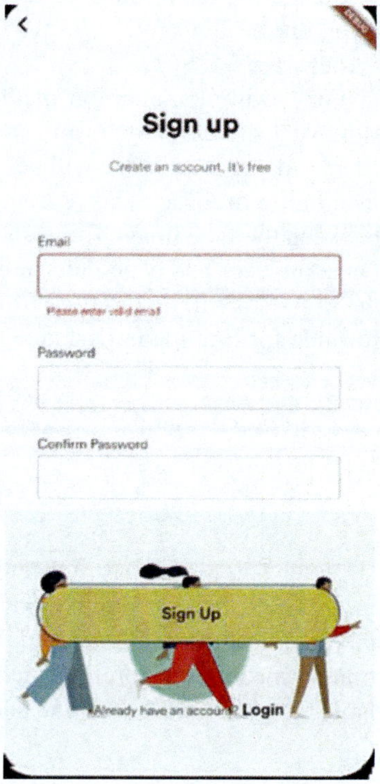

Fig. 4 Sign up page with email and password

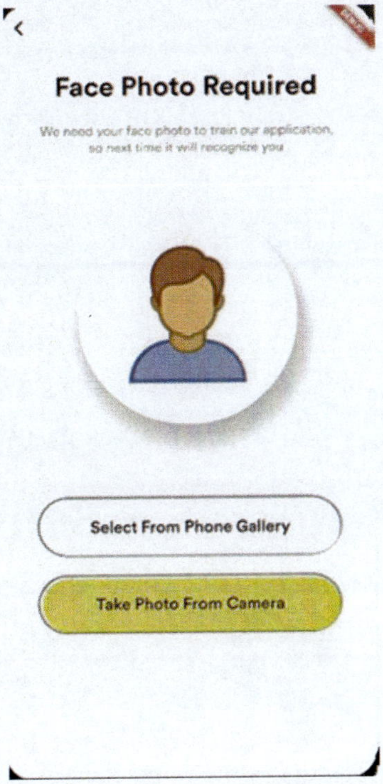

Fig. 5 Sign up page with photo shoot

are able to login by using their email and password (see Fig. 6). Upon login, the application will show location map preview, date and time (see Fig. 7). By confirming the details, users are required to click on "Check In" to proceed to the next page.

The next process is user identity verification via facial recognition. The user's image will be taken as input and the system will search and match the face image with the image data stored in "user" class in the database. The user details will be returned if similar face image is found in the database (see Fig. 8). Otherwise, an error message will be prompted, and the user will be asked to register to use the application. After that, the user is required to enter the password to complete the check-in process. The check-in details will be stored in the database. This enables the users to retrieve the check-in details from the application (see Fig. 9).

5 Prototype Testing

A complete functional testing including testing of the Registration page, Login, Change Password, Check-In and Viewing Attendance Records has been conducted. The application system works well and fulfills the expected results.

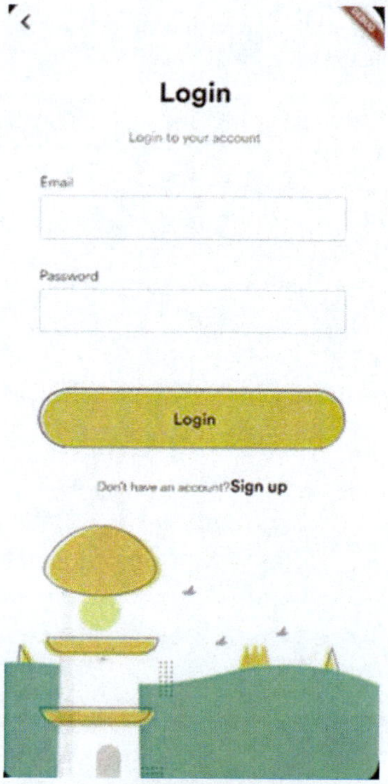

Fig. 6 Login page

Fig. 7 Check In page

Fig. 8 Identity verification via face recognition

Fig. 9 Check-in records

As for the facial recognition model selected to be integrated in the attendance management system, offline model testing has been conducted. Three models are chosen to compare the accuracy. The models are MobileFaceNet, VGG-16, and InceptionResnetV2. The dataset used for the testing were LFW Dataset with mask which is downloaded from Kaggle. The dataset is reduced to 46 classes (person),

Table 1 Accuracy comparison among MobileFaceNet, VGG-16, and InceptionResnetV2

Styles	Train: With masks Test: Without mask	Train: Without masks Test: With mask
Model	Accuracy (%)	
MobileFaceNet	80.9	76.7
VGG-16	56.1	61.5
InceptionResnetV2	52.1	57.6

each class contain 5 images for unmasked face and 5 images for masked face. Table 1 presents the results.

Based on the results, MobileFaceNet has the highest accuracy as compared to VGG-16 and InceptionResnetV2 models with 80.9 and 76.7% for different evaluation criteria. Therefore, MobileFaceNet model is integrated in the proposed attendance management system. For the presentation of this paper, the testing results of two samples with and without face mask are shown in Table 2.

6 Conclusion and Future Work

A half-covered and full-facial recognition attendance management application system has successfully been designed and developed. MobileFaceNet model is used to verify the user identity for attendance check-in in real-time. The application has the capability to recognize an individual with and without face mask. Users can take attendance via the application with or without a face mask. To further improve the model accuracy, future work may include training the model with a larger dataset and testing on different hyperparameter to obtain the best parameter tuning for a better accuracy and system performance.

Table 2 Sample testing results

Test Case	Result (Actual and predicted label)
Without mask 	Actual: Ting Che Siang Predicted: Ting Che Siang
With mask 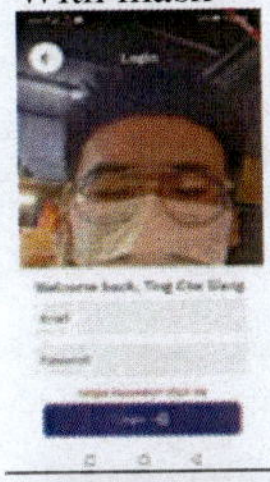	Actual: Ting Che Siang Predicted: Ting Che Siang
Without mask 	Actual: Tin Cu Kang Predicted: Tin Cu Kang
With mask 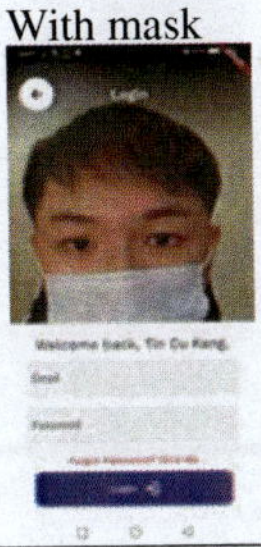	Actual: Tin Cu Kang Predicted: Tin Cu Kang

References

1. Xu K, Yu S, Ding J, Hu R, Zhu Q (2017) Design and implementation of check-in system based on workflow technology. In: 2017 International conference on smart grid and electrical automation (ICSGEA). IEEE, pp 60–63
2. Mohamed BKP, Raghu CV (2012) Fingerprint attendance system for classroom needs. In: 2012 Annual IEEE India conference (INDICON). IEEE, pp 433–438
3. Dey S, Barman S, Bhukya RK, Das RK, Haris BC, Prasanna SRM, Sinha R (2014) Speech

biometric based attendance system. In: Twentieth National conference on communications (NCC). IEEE, pp 1–6
4. Lukas S, Mitra AR, Desanti RI, Krisna D (2016) Student attendance system in classroom using face recognition technique. In: 2016 International conference on information and communication technology convergence (ICTC). IEEE, pp 1032–1035
5. Salim OAR, Olanrewaju RF, Balogun WA (2018) Class attendance management system using face recognition. In: 7th International conference on computer and communication engineering (ICCCE). IEEE, pp 93–98
6. Mery D, Mackenney I, Villalobos E (2019) Student attendance system in crowded classrooms using a smartphone camera. In: IEEE winter conference on applications of computer vision (WACV). IEEE, pp 857–866
7. Muthunagai R, Muruganandhan D, Rajasekaran P (2020) Classroom attendance monitoring using CCTV. In: 2020 International conference on system, computation, automation and networking (ICSCAN). IEEE, pp 1–4
8. World Health Organization Homepage. https://www.who.int/emergencies/diseases/novel-coronavirus-2019/advice-for-public. Last accessed 10 Mar 2022
9. Viola P, Jones M (2001) Rapid object detection using a boosted cascade of simple features. In: 2001 IEEE computer society conference on computer vision and pattern recognition. IEEE, pp I-511–I-518
10. Joshi DK, Rattan K, Rudhra A, Verma N, Pratap M, Vij D (2022) Attendance marking system based on HOG descriptor algorithm. In: 2022 2nd International conference on innovative practices in technology and management (ICIPTM). IEEE, pp 194–199
11. Chaudhari P, Padmane G, Vilhekar R, Dixit S (2020) Attendance management system using face recognition. Int J Adv Res Sci Technol 2(3):15–22
12. Hariri W (2022) Efficient masked face recognition method during the covid-19 pandemic. SIViP 16(3):605–612
13. Anwar A, Raychowdhury A (2020) Masked face recognition for secure authentication. arXiv preprint arXiv:2008.11104
14. Sajida P, Nadeem N, Jherna D (2017) Review on local binary pattern (LBP) texture descriptor and its variants. Int J Adv Res 5(5):708–717
15. Zhang G, Huang X, Li SZ, Wang Y, Wu X (2004) Boosting local binary pattern (LBP)-based face recognition. In: Chinese conference on biometric recognition. Springer, Berlin, Heidelberg, pp 179–186
16. Chen S, Liu Y, Gao X, Han Z (2018) MobileFaceNets: efficient CNNs for accurate real-time face verification on mobile devices. In: Chinese conference on biometric recognition. Springer, Cham, pp 428–438

Employing Smart Educational Aids and Multimedia to Teach Reading Skill to Students with Specific Learning Difficulty: Effectiveness Identification and Task Specificity

Sadeq Al Yaari, Muhammad Alkhunayn, Ayman Al Yaari, Montaha Al Yaari, Aayah Al Yaari, Adham Al Yaari, Sajedah Al Yaari, and Fatehi Eissa

Abstract The study aims at examining the differences in reading comprehension performance that could be seen in students with specific learning difficulty (SpLD) in comparison to typically developing and achieving (TD) children. To that end, different smart educational aids, and multimedia (Computer-assisted language learning/instruction/teaching—CALL/I/T) were used for four months to teach reading skill to a group of fifty-seven preschool children (twenty-seven are SpLD students while thirty are TD controls) of both sexes. Age of all participants ranges between seven and twelve years old. Pre-and-post-tests were administered before and after the course to evaluate participants' reading comprehension. Results of the independent t test between SpLD and control groups and those of Paired t test among SpLD groups indicated that there was a highly statistically significant difference between the groups. Also, SpLD and TD control groups did not differ in gender distribution. Findings reveal that smart educational aids and multimedia play pivotal roles in reading comprehension due to the remarkable effect they leave on student's performance. These outcomes strengthen the evidence that smart educational aids

S. Al Yaari (✉)
Dhamar, Yemen
e-mail: prof.Sadeq@gmail.com

M. Alkhunayn
Department of English, College of Languages and Translation, Al Imam Muhammad Ibn Saud Islamic University, Riyadh, Saudi Arabia

A. Al Yaari · A. Al Yaari
Cairo University, Cairo, Egypt

M. Al Yaari
ACST, Dhamar, Yemen

A. Al Yaari · S. Al Yaari
SOQS, Dhamar, Yemen

F. Eissa
Department of Statistics, Dr. Babasaheb Ambedkar Marathwada University, Aurangabad, India

© The Author(s), under exclusive license to Springer Nature Singapore Pte Ltd. 2023
D.-K. Kang et al. (eds.), *Proceedings of the 9th International Conference on Computational Science and Technology*, Lecture Notes in Electrical Engineering 983,
https://doi.org/10.1007/978-981-19-8406-8_19

and multimedia are effective means for teaching reading skill to preschool students, notably those suffering from SpLD in particular.

Keywords Smart educational aids · Multimedia · Reading · SpLD

1 Introduction

Smart audio-visual aids and multimedia are readily available, mainstream technologies and interactive educative tools that are established within school teaching and learning environments as constructivist learning methods. These knowhow approaches cover all techniques and strategies that are professionally used in the pedagogical system for the purpose of facilitating message encoding and decoding.

Audio-visual as an aid is an indication to any electronic media possessing both sound and visual components, such as slide-tape presentations, films, television programs and other means [1]. Multimedia, on the other side, refers to networked and broadcasting systems and computer-assisted language learning, instruction, or teaching (CALL/I/T). The terms can be broadly extended to include reading, audio and visual means and any other computer-based software programs. Smart audio-visual aids and multimedia in its broader meaning include Professional audio and video technologies, digital devices, electronic multi-use pictures, intelligent games and plays and advanced monitors.

Smart educational aids and multimedia and results of audio-visual educational revolution or multimedia-based education (MBE) new system and the roots of this revolution goes back to Bohemian's ideas introduced in his book Picture of the Sensual World wherein he introduced picture as an example of effective visual means to be used for learning and teaching purposes [2] before other educators extend these ideas to include play materials [3], adopting old philosophy of using pictures as instructional tools [4].

2 Introduction

The technological features of electronic means of reading found to be useful for teaching word recognition [5] but may not be comparable because some techniques do not follow the same distinctive features and other qualities of strategies. For example, the study of [6] on the automatic reading of anti-neutrophil cytoplasmic autoantibody (ANCA-Slides) may be a valuable source of information for refining our understanding of this software that "enables to improve correct pattern recognition" [6]. Blog is another example that found to improve students' reading as reported by Álvarez [7] and this is what was confirmed by Ismail et al. [8] who found them reliable for English as second language (ESL) students. Such outcome has inspired the employment of a support vector machine-based television (T.V) approach to teach English as a foreign language (EFL) student and this supports the literature suggesting a role of visual attention in reading along with that of attention therapy that may be an efficient addition to normal educational practices [9, 10].

For a more precise analysis, the readers' representations of the main protagonist's emotional status in short narratives were investigated in [11] over difficulty levels and the rate of change in performance caused by increasing task difficulty. The significance of such investigation is that it supports the idea that mental models may be of a perceptual nature and may more likely include behavioral elements than emotion labels per se, but how about visual matters? Dexl et al. [12] investigated the safeguards that may help protect children from presbyopia, suggesting the inlay technique in the first year of the child. It is psychoneurolinguistically accepted that augmented benefits and positive feedback, provided by a technical displays and modern educational devices effectively enhance human learning. However, discussion of the reasons for the different impacts of feedback strategies within or between the auditory, visual, and interactive response and the challenges needed to be overcome to provide appropriate feedback in these processes, either in isolation or in combination remains argumentative. Smart audio-visual aids and multimedia found to develop critical thinking and reasoning that are important components of the teaching–learning process as means through which learners are motivated [13]. Recommending this technology for children, the research team has rightly observed: "It has been shown that this technology is a valid and reliable method to evaluate reading speed and reading distance and to calculate reading acuity."

Although adult know how and what topics to select and/or read [14], the specific question of which type of reading (skim/skip/fast/extensive reading vs. scan/slow/intensive reading) one should use remains unclear. Such question motivated [15] to undertake a study on which type of reading one should use. Further results of the types of reading provide positive findings in the integrated use of the best way to read and the right method to use (e.g., reading loudly) [16].

2.1 Related Studies

Smart educational aids and multimedia (Computer-assisted language learning/instruction/teaching—CALL/I/T) are associated with the improvement of language receptive skills and are found to be useful for students with specific learning difficulty (SpLD) [17]. Attentions were first paid to studies investigating SpLD students' deep/surface reading [18] and reading processing of space and shape in blind students' brains [19] before the trend is oriented toward understanding the child's linguistic environment including physicians and nurses and the way they read [20]. Researchers believe that knowing the prevalence of boards, notably for teaching SpLD students, supplements educational interpretation and can overcome the learning difficulties they suffer from [21–23]. Psychoneurolinguistically, smart educational aids and multimedia found to influence SpLD students to understand visual text [24, 25]. In an increasingly smart educational aids and multimedia search, several researchers in the field are concerned with giving crucial answers and gratified justifications to sensitive questions relating to the relationship between human skills and abilities [26] considering means like TV, for example., that found to be well suited to be used in various educational-related measurement settings [27]. Likewise, online computer-based educative programs have been proposed as an accessible means through which reading problems could be solved. Yet, legitimate questions like: What drives people to read? requires factors that, according to [28], not only describe the user motives for consuming the medium, but also predict user's behavior. Also, [29] provided some specific justifications for the major question they posed on why does picture naming take longer than word reading?, stating that "one may speculate post-hoc that the articulation of a smaller chunk (e.g., a word onset in reading) will be swifter and more efficient than the articulation of a complete phonetic program (e.g., a whole word in picture naming), but that subsequent execution will be stretched to compensate for a speedy start."

Similarly, [30] assessed the role of computer laboratory to improve students' ability to overcome reading tasks to emphasize to the importance of this technique in improving students' reading performances like computers, for example, that found to be very useful due to the "the effects of reading from the screen" [31] where "single reading with computer-aided detection (CAD) gave lower sensitivity (87.2% vs 87.7%) and lower specificity (96.9% vs 97.4%)" [32], added. Findings of [29] have opened the door for other researchers to investigate relationships between smart educational aids and multimedia and reading improvement like Christ and his colleagues who, in 2012, suggested a data set that comprises a minimum of 14 curriculum-based measurement of reading [33] is another example of studies that examined the effects of integrating a multicomponent self-monitoring intervention into a targeted reading classroom to determine the presence of a functional relation between a multicomponent self-monitoring intervention and the academic engagement and disruptive behavioral problems. A third example is that of [34] who implemented concept map technique to examine reading proficiency of students before they conclude to recommend it as an influencing strategy.

In fact, the issue of implementing smart educational aids and multimedia for teaching SpLD students reading was explored with the question of whether an absolute or relative educational value [35] can be found to reliably classify the SpLD across different reading rates [36]. Rate changes found to affect the comprehension [37] of children more than adults [38] in such a way that the understanding differences between preschool and secondary school students was greater for slow than normal reading, and for normal than fast reading. Such findings confirmed the belief that smart educational aids and multimedia support learning [39] and are better be used for teaching receptive skills [40] like reading skill [41] and that it should be introduced at early age [42].

Strictly, smart educational aids and multimedia found to improve students' comprehension and attitudes [43] and enhance their understanding of correct terminology and language [44]. In details, the ability to read texts, stories, books, and reports develop over several months (may be years) sense of change [45, 46], future reading plans [47, 48] and integrate into the various environments: Urban and rural [49], educational and scientific channels and networks [50].

Obviously, previous work on the phenomenon under investigated has focused on smart educational aids and multimedia but the target participants were always those belong to one linguistic background. Even under different linguistic environments and/or conditions, the research always lacked the variety in linguistic problems faced by the population sample. Since, however, linguistic difficulties tend to be more open, we have sought to investigate the most common SpLD in preschool students of both sexes to see what effects smart educational aids and multimedia make in their performance in a written skill like reading compared to that of TD controls of the same age. Researchers in the field of psychoneurolinguistics who are unsure of the relationship ought to be drawn between future smart educational aids and multimedia and language productive and receptive skills can expand their thinking about options by reading outcomes of this study.

2.2 *Aims of the Study*

This study is an attempt to understand the roles played by smart educational aids and multimedia (Computer-assisted language learning/instruction/teaching—CALL/I/T) to overcome reading problems in students with specific learning difficulty (SpLD) by investigating the differences in reading comprehension that could be seen in the performance of students with specific learning difficulty (SpLD) compared with typically developing and achieving children (TD).

3 Methodology

Fifty-seven preschool students were recruited in this study from local school of special education and ordinary school. Of them, twenty seven are suffering from specific learning difficulty (SpLD and are distributed as follows: Dyslexia (n = 7 (m = 4; f = 3, f), Dyspraxia (n = 5 (m = 2; f = 3), Attention Deficit Disorder/ Attention Deficit (Hyperactivity) Disorder (n = 4 (m = 2; f = 2), Dyscalculia (n = 5 (m = 3; f = 2) and Dysgraphia (n = 6 (m = 2; f = 4). The rest of the participants are thirty 30 (m = 15; f = 15) typically developing and achieving (TD) students of both sexes and are used here as controls. Both groups (SpLD students and TD controls) were equally matched in IQ ($\geq$75) and age (7–12 years old) and were both involved in a four-month reading course wherein educational aids smart and multimedia (Computer-assisted language learning/instruction/teaching—CALL/I/T) were differently used. A professional speech-language pathologist (SLP) specialist in teaching of reading course has taught the two groups in two separated classes during the time of the course. The course was preceded and followed by pre-and-post-tests that have been administered to evaluate participants' reading comprehension. Results of the two tests were psychoneurolinguistically and statistically assessed. In this regard, a postdoc psychoneurolinguistic researcher has tackled the first part where participants' outcome was investigated considering psychology, cognitive neurosciences and linguistic perspectives, approaches, and theories. The second part has been conducted by a statistician scholar who made use of the best statistical tests like independent sample t test and Paired t test to test the hypothesis of the research and Cohen's d to test the effect size before visualizing the result using the appropriate graphs.

4 Analysis

4.1 Performance of the Subjects in the Pre-and- Post-Test: Comparison Between the Treatment and Control Groups

Different educational aids can be used for teaching reading as a language receptive skill to students with specific learning difficulty (SpLD). Of these aids are smart educational aids and multimedia (Computer-assisted language learning/instruction/teaching—CALL/I/T) that have been employed in this for teaching SpLD students. These include: Three D dimensions models, electronic pictures and drawings, posters, brain maps, smart boards, creative cards, modern graphs and diagrams, electronic forms, menus associated with pictures, mini slides and film strips, educational games, TV educational channels and internet-based interactive programs.

It should be pointed out here that the smart educational aids and multimedia were used for teaching reading comprehension (understanding) to SpLD students through mechanical aspect that involves focusing on training students to learn the roles and functions of discourse markers, expressions, marks (diacritics, punctuation marks and interjections) considering deictic tactics and coherent and cohesive rules. Reading comprehension was given in view of intensive reading (quality, more comprehension, and scanning) and extensive reading (quantity, les comprehension, skimming). Interactive activities include adopting texts and displaying them before practicing them through drills and exercises.

Furthermore, the proportion of comprehension was found to distinguish the SpLD categories across linguistic impairments and disorders where smart educational aids and multimedia proved positive effects on those suffering from dyslexia [51–55], dyscalculia [56], hyperactivity disorder [57–59], dysgraphia [60, 61] and dyspraxia [62–64]. The results support the view that smart educational aids and multimedia remain reliable tools for speech specialists [65] across speech and language pathology.

The effectiveness of the use of smart educational aids and multimedia on the performance of SpLD students was evaluated. Such evaluation included not only the treatment group, but also the control group. As is mentioned in the methodology of this research, pre-and-post-tests were given to the students in questions. The first test was administered at the beginning of the course. Of the fifty-seven participants, eight subjects from the treatment group and four from the control group were diagnosed with lower performance levels (the former scored less than 25 scores; the latter scored less than 50 scores) in the pre-test. The performance of the other forty-five participants from both groups was slightly similar where SpLD students scored less than 50 (below normal average) while TD controls did not reach 80 (upper good). Hence, no one from the SpLD students has passed and no one from the TD controls has reached very well.

In details, the treatment group scored 796 (Mean = 29.4) in the pre-test. Of this number, male students scored 395 (Mean = 30.3) while females scored 401 (Mean = 28.6). Scores of the sub-groups were as follows: Dyslexia = 215 (Mean = 30.7); Dyspraxia = 138 (Mean = 27.6); Hyperactivity Disorder (HD) = 127 (Mean = 31.7); Dyscalculia = 124 (Mean = 24.8); and Dysgraphia = 182 (Mean = 30.3). Note here that no male or female differences were mentioned here because percentages of gender were not equal within sub-groups. For controls, the group scored 1823 (Mean = 60.7). Of this number, male students scored 897 (Mean = 59.8) whereas female students scored 926 (Mean = 61.7).

In comparison to pre-test, the performance of the participants in the post-test was entirely different. Treatment group scored 1232 (Mean = 45.6) wherein male students scored 594 (Mean = 45.6) while female students scored 638 (Mean = 45.5). Since SpLD students were preassigned into five sub-groups, these groups' performances were as follows: Dyslexia = 362 (Mean = 51.7); Dyspraxia = 254 (Mean = 50.8); Hyperactivity Disorder (HD) = 203 (Mean = 50.7); Dyscalculia = 190 (Mean = 38); and Dysgraphia = 286 (Mean = 47.6). Likewise, control group scored 2443 (Mean = 81.4). Of this total number, the male TD students scored 1278 (Mean = 85.2) while female TD students scored 1165 (Mean = 77.6). Table 1 clarifies this picture.

As is seen in Table 1, performances of both groups were remarkably significant, which means that smart educational aids and multimedia were effective. Gender was also investigated to see if it could be used as a key variable for academic performance in both SpLD and TD groups.

The objectives of the study were based initially on the performance differences in understanding written content. Data were obtained from consecutive subjects and healthy volunteers who share similar characteristic features of IQ, educational level, age and manual preference. Differences in performance levels were confirmed by analysis. Gender differences were also considered where participants from both groups (SpLD and TD controls) were matched by gender. Ultimately, both groups did not differ in gender distribution, ($\chi^2 = 0.019$, p-value > 0.05). Also, the Groups did not differ in Age ($t_{55} = -0.796$, p-value $= 0.43$). Table 2 explains the distribution of demographic variables considering gender.

Table 1 Performance of SpLD subjects versus TD controls: pre-and-post-tests outcomes

No.	SpLD Subjects										
	Name	Age	Sex	Diag	Score		Name	Age	Sex	Score	
					Pre-test	Post-test				Pre-test	Post-test
1	M.A	8	F	Dysl	23	41	S.Q.T	7	F	47	71
2	A.S	12	M	Dysl	44	69	M.S.A	7	M	58	88
3	K.N	8	M	Dysl	34	59	A.S.K	9	M	70	92
4	F.A	7	F	Dysl	41	71	S.Y.A	12	F	59	87
5	J.K	9	F	Dysl	18	33	A.M.S	12	F	64	79
6	M.F	11	M	Dysl	22	39	A.Q.N	11	M	71	87
7	W.N	10	M	Dysl	33	50	A.N.A	10	F	41	77
8	W.A	8	F	Dysp	20	36	A.A.S	8	M	55	90
9	H.Z	7	F	Dysp	29	70	M.G.A	9	F	70	89
10	L.F	12	F	Dysp	34	57	T.A.S	12	M	67	88
11	A.A	12	M	Dysp	37	62	A.S.S	7	F	66	78
12	H.A	11	M	Dysp	18	29	T.M.A	8	F	60	80
13	F.A	10	F	HD	19	30	F.A.S	9	F	55	77
14	D.S	9	M	HD	29	41	F.A.A	7	M	53	81
15	M.L	9	F	HD	39	53	S.A.A	11	M	50	69
16	H.A	9	M	HD	40	79	S.A.S	12	M	66	76
17	D.L	9	F	Dysc	17	35	K.A.M	10	M	61	85
18	S.B	11	M	Dysc	33	52	Z.Y.A	10	F	49	75
19	M.R	8	M	Dysc	27	33	M.H.S	10	F	77	94
20	S.A	7	M	Dysc	28	43	H.Y.M	9	M	74	92
21	N.N	8	F	Dysc	19	27	A.B.A	7	F	75	90
22	K.A	7	F	Dysg	25	36	D.A.M	12	F	67	89
23	A.M	7	M	Dysg	31	48	Z.A.M	11	M	76	96
24	N.S	10	F	Dysg	36	55	S.Y.A	8	F	60	88
25	S.N	11	F	Dysg	29	39	M.K.L	9	F	71	91
26	F.Z	8	M	Dysg	25	34	M.L.M	9	M	57	83
27	M.N	8	F	Dysg	36	74	W.N.S	9	M	54	81
28							R.A.N	11	M	73	80
29							T.T.A	10	M	75	90
30							K.W.S	8	F	47	70
Total					796 (A)	1232 (B)				1823 (C)	2443 (D)

A = (M. = 29.4 (Males = 395 (M. = 30.3); Females = 401 (M. = 28.6)
B = (M. = 45.6). Males = 594 (M. = 45.6); Females = 638 (M. = 45.5)
C = (M. = 60.7) (Males = 897(M. = 59.8); Females = 926 (M. = 61.7)
D = (M. = 81.4). Males = 1278 (M. = 85.2); Females = 1165 (M. = 77.6)

Table 2 Demographic variables

Group	N	Gender M/F	Age		
			Mean	SD	Range
SpLD	27	13/14	9.11	1.672	7–12
TD	30	15/15	9.47	1.697	7–12

5 Conclusions and Recommendations

Smart educational aids and multimedia (Computer-assisted language learning/instruction/teaching—CALL/I/T) found to be effective due to that they played pivotal role in overcoming reading difficulties in those who suffer specific learning difficulty (SpLD) of both sexes. They proved to be accountable, reliable, and valid resources for learning how to read and, more importantly, how to understand what you read. The roles of these smart educational aids and multimedia are not restricted to the classical role played by ordinary educational aids as well as audio-visuals aids; rather, their roles extend to go beyond classical means to that of enjoyment by mixing education with entertainment through games, tutorials, simulations, visual realization, and drills.

Again, the smart educational aids and multimedia being investigated in this study found to be useful for reading perception where SpLD preschool students scored remarkable scores in comparison to TD controls. These aids made significant differences, notably in the performance of SpLD preschool students whose dominant language difficulty was reading [55, 66, 67], writing [68, 69], speaking [70, 71] or was related to statistics and/or mathematics [72] or those with attention deficit disorder/attention deficit (Hyperactivity) disorder [73]. They found to be valid, reliable, and accountable for increasing these students' attention and improving their educational performance. These findings provided an educational pattern of performance contrasts among SpLD preschool students.

Roles of these smart educational aids and multimedia found to increase attention of SpLD preschool students who benefitted from them in all their spectrum where score of subgroups in both pre-and-post-tests introduce themselves as strong evidence: Dyslexia (215, (Mean = 30.7) in pre-test versus, (362 (Mean = 51.7) in post-test); dyspraxia (138 (Mean = 27.6) in pre-test versus (254 (Mean = 50.8) in post-test; hyperactivity disorder (127 (Mean = 31.7) in pre-test versus (203 (Mean = 50.7) in post-test; dyscalculia (124 (Mean = 24.8) in pre-test versus (190 (Mean = 38) in post-test; or dysgraphia (182 (Mean = 30.3) in pre-test versus (286 (Mean = 47.6) in post-test. In fact, these aids found to be useful even for typically developing and achieving preschool students whose performance was 1823 (Mean = 60.7) in pre-test versus 1278 (Mean = 85.2) in post-test.

In order to know the distribution of the data, the researchers performed normality tests. Having done so, the mean score data for pre-and-posttests for each group was found to be normally distributed because all *p*-values were > 0.05 in respect to Kolmogorov–Smirnov and Shapiro–Wilk tests. Also, the values of Skewness and

Table 3 Mean score and SD for pre-test and post-test by subject with SpLD and TD

	SpLD		TD	
	Mean	SD	Mean	SD
Pre-test score	29.11	7.910	62.27	10.007
Post-test score	47.96	15.313	83.77	7.403

Kurtosis were in the acceptable intervals. Therefore, to test the hypothesis of the differences between the SpLD and TD groups in the mean score with respect to pre-test and post-test, independent sample t test was performed. Also, Paired t test was conducted to test the hypothesis of the differences between the mean score of pre-test and post-test separately in respect of the SpLD and TD groups. In addition, the effect size was tested by using Cohen's d. Furthermore, the researchers performed the t-test with the bootstrap technique to validate the statistical results. Table 3 shows the mean score and standard deviation of pre-test and post-test regarding to SpLD and TD.

Findings of both SpLD and TD groups in pre-and-post-tests were also examined where, in the pre-test, the results of the independent t test indicated that there was a highly statistically significant difference between the groups ($t_{55} = -13.771$, p-value $< 0.001^{**}$, $d = -3.653$) on the mean score of the pre-test due to the SpLD group's performance at a lower level than of the TD group's (SpLD's mean scores = 29.11, SD = 7.91; TD's mean scores = 62.27, SD = 10.007). Concerning post-test findings, independent t test revealed a highly statistically significant difference between the groups ($t_{55} = -11.417$, p-value $< 0.001^{**}$, $d = -3.029$) on the mean score of the post-test due to the SpLD group's performance at a lower level than of the TD group's (SpLD's mean scores = 47.96, SD = 15.313; TD's mean scores = 83.77, SD = 7.403).

Findings within SpLD groups were also addressed where, in the pre-test, for example, the Paired t test showed a highly statistically significant difference ($t_{26} = -10.457$, p-value $< 0.001^{**}$, $d = -2.013$) between the mean score of pre-test and post-test for the subjects with SpLD. For post-test findings of these groups, results of the Paired t test indicated that there was a highly statistically significant difference ($t_{29} = -17.458$, p-value $< 0.001^{**}$, $d = -3.187$) between the mean score of pre-test and post-test for the TD. Figure 1 below illustrates the mean score and (SEs) recorded by SpLD and TD groups on each test.

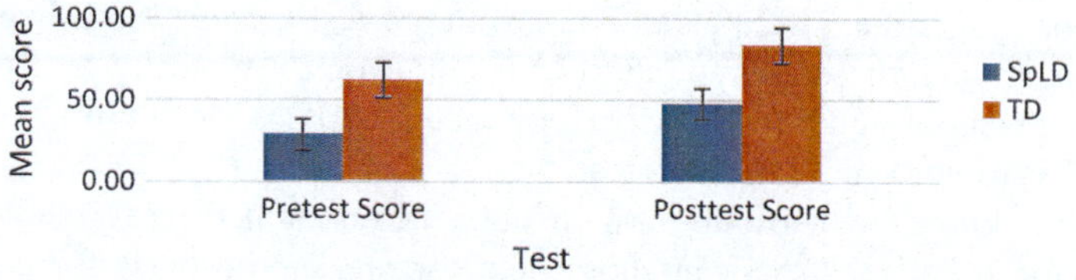

Fig. 1 Mean score and (SEs) recorded by SpLD and TD groups on pre-and-post-tests

As can be clearly seen in Fig. 1, the performance of the participants, notably SpLD subjects is incomparable and the post-test results support that. These outcomes suggested that smart educational aids and multimedia's involvement had a positive effect on the way to overcome difficulties relating to reading comprehension. These findings upkeep those from previous research [6, 7, 21, 26, 68]. The researchers recommend implanting these aids for reading purposes in different psydchoneurolinguistic contexts under different environments. Further research is needed to explore more reading aids for students with various language problems including those suffering from specific language impairments (SLI) and those suffering from autistic spectrum disorders.

References

1. Barman CR (1984) Slide-Tape presentations on a classroom budget. Am Biol Teach 46(6):344–346
2. Aggarwal J (2009) Principles, methods and techniques of teaching. Vikas Publishing House Pvt Ltd., New Delhi
3. Akram S, Sufiana, Malik K (2012) Use of audio-visual aids for effective teaching of biology at secondary schools level. Educ Leadersh 50:10597–10605
4. DeBernardes A, Olsen EG (1948) Audio-Visual and community materials—some recent publications. Educ Leadersh 256–266
5. Ashby J, Yang J, Haver K, Evans C, Rayner K (2012) Eye movements and the perceptual span in silent and oral reading. Atten Percept Psychophys 74(4):634–640
6. Damoiseaux J, Mallet K, Vaessen M, Austen J, Jan-Willem CT (2012) Automatic reading of ANCA-slides: evaluation of the AKLIDES system. Clin Dev Immunol 76(28):74
7. Álvarez G (2012) New technologies in the university context: the use of blogs for developing students' reading and writing skills. RUSC 9(2):185–198
8. Ismail AS, Hamed MA, Abdurrahman GA (2012) Employing reading and writing computer-based instruction in English as a second language in elementary schools. Int J Bus Soc Sci 3(12):3–4
9. Erfani SA (2012) Pictures speak louder than words in ESP, Too! Engl Lang Teach 5(8):164–169
10. Shelley-Tremblay J, Langhinrichsen-Rohling J, John Eyer J (2012) Attention therapy improves reading comprehension in adjudicated teens in a residential facility. J Correct Educ 63(2):49–67
11. Gillioz C, Gygax P, Tapiero I (2012) Individual differences and emotional inferences during reading comprehension. Can J Exp Psychol 66(4):239–250
12. Dexl AK, Seyeddain O, Riha W, Hohensinn M, Rückl T, Hitzl W, Grabner G (2012) Reading performance after implantation of a modified corneal inlay design for the surgical correction of presbyopia: 1-year follow-up. Am J Ophth Almology 153(5):994-1001.e2
13. Mathew N, Alidmat A (2013) A study of the usefulness of audiovisual aids in EFL classroom: implications for effective instruction. Int J High Educ 2:86–92
14. Dalton B, Smith BE (2012) Teachers as designers: multimodal immersion and strategic reading on the internet. Res Sch 19(1):12–25
15. Love J (2012) Reading fast and slow. Am Scholar 81(2):64–72
16. Reynolds M, Mulatti C, Besner D (2012) Reading nonwords aloud: evidence for dynamic control in skilled readers. Psychon Bull Rev 19(6):1135–1141
17. Iram S (2012) Reading needs, facilities and problems of visually impaired people. Pak J Libr Inf Sci 13:1–1
18. Wolf M, Ullman-Shade C, Gottwald S (2012) The emerging, evolving reading brain in a digital culture: implications for new readers, children with reading difficulties, and children without schools. J Cogn Educ Psychol 11(3):230–240

19. Klingenberg OG (2012) Conceptual understanding of shape and space by braille-reading Norwegian students in elementary school. J Vis Impairment Blindness 106(8):453–465
20. Khaliq MF, Noorani MM, Siddiqui UA, Anwar M (2012) Physicians reading and writing practices: a cross-sectional study from civil hospital, Karachi, Pakistan. BMC Med Inform Decis Mak 12(1):76
21. Anson CM, Schwegler RA (2012) Tracking the mind's eye: a new technology for researching twenty-first-century writing and reading processes. Coll Compos Commun 64(1):151–171
22. Henderson JM, Luke SG (2012) Oculomotor inhibition of return in normal and mindless reading. Psychon Bull Rev 19(6):1101–1107
23. Perea M (2012) Revisiting Huey: on the importance of the upper part of words during reading. Psychon Bull Rev 19(6):1148–1153
24. Chen I, Yen J (2013) Hypertext annotation: effects of presentation formats and learner proficiency on reading comprehension and vocabulary learning in foreign languages. Comput Educ 63:416–423
25. Manoli P, Papadopoulou M (2012) Graphic organizers as a reading strategy: research findings and issues. Creat Educ 3(3):348–356
26. Nikolaraizi M, Vekiri I, Easterbrooks SR (2013) Investigating deaf students' use of visual multimedia resources in reading comprehension. Am Ann Deaf 157(5):458–473
27. Notten N, Adam K, Konig RP (2012) Family media matters: unraveling the intergenerational transmission of reading and television tastes. Sociol Perspect 55(4):683–706
28. Gerlich RN, Drumheller K, Sollosy M (2012) The reading motives scale: a uses and gratifications study of what drives people to read. Acad Mark Stud J 16:95
29. Riès S, Legou T, Alario BB, Malfait N (2012) Why does picture naming take longer than word reading? The contribution of articulatory processes. Psychon Bull Rev 19(5):955–961
30. Howe J (2012) Eccentric viewing training and its effect on the reading rates of individuals with absolute central scotomas: a meta-analysis. J Vis Impairment Blindness 106(9):527–542
31. Aydemir Z, Öztűrk E (2012) The effects of reading from the screen on the reading motivation levels of elementary 5th graders. TOJET Turk Online J Educ Technol 11(3):2–5
32. Azavedo E, Zackrisson S, Mejàre I, Heibert-Arnlind M (2016) Is single reading with computer-aided detection (CAD) as good as double reading in mammography screening? A systematic review. BMC Medical Imaging, 12(1):22. Author F (2016) Article title. Journal 2(5):99–110
33. Bruhn A, Watt S (2012) Improving behavior by using multicomponent self-monitoring within a targeted reading intervention. Behav Disord 38(1):3–17
34. Soleimani H, Nabizadeh F (2012) The effect of learner constructed, fill in the map concept map technique, and summarizing strategy on Iranian pre-university students' reading comprehension. Engl Lang Teach 5(9):78–87
35. Linz C (1955) Audio-visual aids and special education. Slow Learn Child 2(2):68–76
36. Magnan A, Ecalle J (2006) Audio-visual training in children with reading disabilities. Comput Educ 46(4):407–425
37. Lundberg I, Leong CK (1986) Compensation in reading disabilities. Adv Psychol 34:171–190
38. Noble L (1990) The use of audio-visual materials in an adult literacy campaign. Concordia University Press, Concordia
39. Sperotto L (2016) The visual support for adults with moderate learning and communication disabilities: how visual aids support learning. Int J Disabil Dev Educ 63(2):260–263
40. Summerfield Q, Foster J (1983) Assessing audiovisual speech-reception disability. In: High technology aids for the disabled. Butterworth-Heinemann, pp 33–41
41. Ortendahl VR (1963) Audio-visual materials and methods for improving the teaching of fourth grade reading. Drake University Press, Drake
42. Hendrickson N, Williams BM (1971) Audio-Visual materials in early childhood education. Young Children, pp 287–289
43. Atmaja D (2018) Using audio–visual aids to improve the students' reading skill. Engl Educ J 6:3
44. Al Aqad M, Al Saqaf M, Muthmainnah M (2021) The impact of audio-visual aids in learning English among MSU third-year students. Engl Franca Acadc J Engl Lang Educ 5(2):201–214

45. Wilson SL, Clements KK (2012) Exploring the professional books: alliances, reading as an expert, and community. N Engl Reading Assoc J 47(2):56–62, 80–81
46. van Steensel R, Oostdam R, van-Gelderen A (2013) Assessing reading comprehension in adolescent low achievers: subskills identification and task specificity. Lang Test 30(1):3-21
47. Breeding M (2013) References and additional reading. Libr Technol Rep 49(1):34–35
48. Ortlieb E (2012) The past, present, and future of reading diagnosis and remediation. J Lang Teach Res 3(3):395–400
49. Kaur P, Jawaid A (2018) A comparative study of urban and rural reading habits. Sarawak Libr J Publ Pustaka Negeri Sarawak 1(1):7–7
50. Loan FA (2012) Reading habits of rural and urban college students in the 21st century. Library Philosophy and Practice, pp 1–10
51. Broadhead M, Daylamani-Zad D, Mackinnon L, Bacon L (2018) A multisensory 3D environment as intervention to aid reading in dyslexia: a proposed framework. In: 2018 10th international conference on virtual worlds and games for serious applications (VS-Games). IEEE, pp 1–4
52. Magnan A, Ecalle J, Veuillet E, Collet L (2004) The effects of an audio-visual training program in dyslexic children. Dyslexia 10(2):131–140
53. Manasa R, Shukla R, Bhattacharyya B, Agarwal R (2020) A comprehensive study of audio-visual aids for dyslexic children. Int J Electr Eng Technol (IJEET) 211–218
54. Pandey S, Srivastava S (2011) Tiblo: a tangible learning aid for children with dyslexia. In: Proceedings of the second conference on creativity and innovation in design, pp 211–220
55. Rüsseler J, Ye Z, Gerth I, Szycik GR, Münte TF (2018) Audio-visual speech perception in adult readers with dyslexia: an fMRI study. Brain Imaging Behav 12(2):357–368
56. Miundy K, Zaman HB, Nordin A (2017) Review on data driven preliminary study pertaining to assistive digital learning technologies to support dyscalculia learners. In: International visual informatics conference. Springer, Cham, pp 233–246
57. Bisch J, Kreifelts B, Bretscher J, Wildgruber D, Fallgatter A, Ethofer T (2016) Emotion perception in adult attention-deficit hyperactivity disorder. J Neural Transm 123(8):961–970
58. Daniel E (2011) Attention deficit hyperactivity disorder–available training module for schoolteachers. J Educ Psychol Couns 2:209–243
59. Heller MD, Roots K, Srivastava S, Schumann J, Srivastava J, Hale TS (2013) A machine learning-based analysis of game data for attention deficit hyperactivity disorder assessment. GAMES FOR HEALTH: Research, Development, and Clinical Applications 2(5):291–298
60. Kamal N, Sharma P, Das R, Goyal V, Gupta R (2022) Virtual technical aids to help people with dysgraphia. In: Approaches and applications of deep learning in virtual medical care, pp. 222–235
61. Tobolcea I, Danubianu M (2010) Computer-based programs in speech therapy of dyslalia and dyslexia-dysgraphia. BRAIN. Broad Res Artif Intell Neurosci 1(2):52–63
62. Al Yaari S (2013) Using audio-visual aids and computer-assisted language instruction (CALI) to overcome learning difficulties of reading in students of special needs. J Study Engl Linguist 1(2). ISSN 2329-7034
63. Bowman SN, Parsons CL, Morris DA (1984) Inconsistency of phonological errors in developmental verbal dyspraxic children as a factor of linguistic task and performance load. Aust J Hum Commun Disord 12(2):109–119
64. Pedro A, Goldschmidt T (2019) Managing dyspraxia: Pre-school teachers' perceptions, experiences, and strategies. J Psychol Afr 29(2):182–186
65. Potter D, Tettemer C (1954) Audio-Visual aids for the speech teacher. Bull Nat Assoc Secondary Sch Principals 38(199):200–204
66. van Laarhoven T, Keetels M, Schakel L, Vroomen J (2018) Audio-visual speech in noise perception in dyslexia. Dev Sci 21(1):12504
67. Yang Y, Yang YH, Li J, Xu M, Bi HY (2020) An audiovisual integration deficit underlies reading failure in nontransparent writing systems: an fMRI study of Chinese children with dyslexia. J Neurolinguistics 54:100884. Ye Z, Rüsseler J, Gerth I, Münte TF (2017) Audiovisual speech integration in the superior temporal region is dysfunctional in dyslexia. Neuroscience 356:1–10

68. Al-Yaari SAS (2013) Using audio-visual aids and computer-assisted language instruction (CALI) to overcome learning difficulties of speaking in students of special needs. J Study Engl Linguist 1(2):231–255
69. Jimale AO, Zainon WMNW, Ahmed YA (2022) Implementation of deaf-blind-dysgraphia communication technique for disabled students in Somalia. In: Proceedings of the 12th National technical seminar on unmanned system technology 2020. Springer, pp 59–68
70. Bodison SC (2015) Developmental dyspraxia and the play skills of children with autism. Am J Occup Ther 69(5):1–6
71. Lundeborg I, McAllister A (2007) Treatment with a combination of intra-oral sensory stimulation and electropalatography in a child with severe developmental dyspraxia. Logoped Phoniatr Vocol 32(2):71–79
72. Yusta N, Karugu G, Muthee J, Tekle T (2016) Impact of instructional resources on mathematics performance of learners with dyscalculia in integrated primary schools, Arusha City Tanzania. J Educ Pract 7(3):1
73. Vahabzadeh A, Keshav NU, Salisbury JP, Sahin NT (2018) Improvement of attention-deficit/hyperactivity disorder symptoms in school-aged children, adolescents, and young adults with autism via a digital smartglasses-based socioemotional coaching aid: short-term, uncontrolled pilot study. JMIR Mental Health 5(2):9631

Educational Video Production: Literature Review on Issues and Challenges of Digital Video Composition

Nur Faraha Mohd. Naim and Ag. Asri Ag. Ibrahim

Abstract This article reviews the issues and challenges of digital video composition to identify the problem in educational video production. The review of related literature discussed four components that lead to the issues and challenges. The discussed components are the insufficient Teaching and Learning (T&L) quality approach in educational videos, recommendation of visual aesthetics application, the video length proxy, and the scant publications on innovative self-directed learning tools, experimental investigation, and engagement measurement which contribute to the issues and challenges in the research study. Therefore, from the review of issues and challenges literature, the paper identifies four problems in educational video production such as the importance of T&L instruments in production value for an educational video, the absence of curiosity on tools and methodological aspects in video production, the pedagogy approach may not be the right concept embedded in an educational video, and the need to provide more control on how learners obtain the information. Hence the paper summarized the application of interactivity as one of the elements in digital video compositing that may create a new perspective on the educational video.

Keywords Educational video production · Digital video compositing · T&L quality approach

N. F. Mohd. Naim (✉)
University Malaysia Sabah Labuan International Campus, Jalan Sungai Pagar, Labuan F. T., 87000 Kota Kinabalu, Malaysia
e-mail: faraha.naim@ums.edu.my

Ag. A. Ag. Ibrahim
Universiti Malaysia Sabah, Jalan UMS, 88400 Labuan, Sabah, Malaysia

© The Author(s), under exclusive license to Springer Nature Singapore Pte Ltd. 2023
D.-K. Kang et al. (eds.), *Proceedings of the 9th International Conference on Computational Science and Technology*, Lecture Notes in Electrical Engineering 983,
https://doi.org/10.1007/978-981-19-8406-8_20

1 Introduction

1.1 Pre-educational Film Development

It has been proven by film and video industry that a video is worth a million words with its improved quality, dramatizes, and provides strong effects on multimedia application. Apart from that, a video has technologically integrated into the business, education, home, and around people for a few decades earlier. For instance, Thomas Edison was one of the enthusiast promoters of an educational film in the 1900s. He formed The Edison Film Library in 1911 that produced a series of educational films on science and history. Though his baseline idea was to promote the company's film products equipment, the school-film market went progressively well between the 1900s and 1920s that caused the formation of other companies in hope to seize the opportunities offered by the school-film market [1]. In 1977, a popular British sitcom Mind Your Language was premiered on TV. It was a show about an English teacher trying his best to help his foreign students learning English in the classroom. The show was mixed with drama, humor, and action but at the same time teaches their audience about how to use an appropriate English language in a conversation. It is a good way to learn English for communication although the show was less approach to reading and writing, yet it still does provide educational value for the audience at that time.

Today, various creative digital videos were seen displayed on TVs, computers, and mobile devices, and some even in a form of edutainment. For instance, the Massive Open Online Course or known as MOOC was first developed in 2012 is a global open course platform that invites educators and academic institutions for community interactions among educators and learners such as discussions, online assignments, educational videos et cetera with the main purpose is for conducting an online session for T&L activities. Therefore, aside from the YouTube channel, another popular educational video accessible on a website for higher education is MOOC. Understandably, educational videos can enhance learners' comprehension of the contents. As stated by Khoo, Khuan, Fatimah Salwah Abd. Hadi and Muhammad Shahbani Abu Bakar about video-based collaborative learning, they noted that the tool helps learners to understand the concepts based on the visualization of the real-world situations that the video provides [2]. The multimedia elements such as audio, graphic, video, text, and animation when effectively combined into one video can contribute to enhancing the way information delivered. Lange and Costley acknowledged the media diversity plays a significant role in increasing the aids for learners to process the information while less diversity leads to limitation in understanding the contents [3].

Beheshti, Taspolat, Kaya & Sapanca also stated that the learning approach via educational video could improve learning results and satisfaction [4]. Adding to the educational video, Sitti Hadijah pointed out that educators need to creatively produce an educational video to cater to their learners' needs, which can contribute to interesting learning experiences among learners and facilitate their learning activities [5].

Hence, the practice of using video as the second medium to deliver education can be widely seen on the internet nowadays via popular online video channels such as YouTube and MOOC. Nonetheless, not all the information shared on the platform produces good quality educational content. Reportedly 58.2% of videos were uploaded to the YouTube platform by dentist professionals, sharing the information on infection control and preventing the COVID-19 from spreading during medical procedures to the dental practitioners. However, the study determined that 43.6% of YouTube videos have poor quality [6].

While educational videos are common nowadays, unfortunately, these creative digital videos are still reserved and not widely used for the T&L classroom by far. Not to mention that interactive video designed for T&L activity is still underdeveloped in the academic area unlike with the industry sector where the feature is utilized in business, marketing, and advertisement strategy. Perhaps in the future, interactive videos might overcome the traditional T&L and slowly modernize the T&L activity inside or outside the classroom. But having said that, what can interactive digital video compositing contribute to educational video? This paper explores the potential of embedding the interactivity element in an educational video and considers the alternative T&L approach (cognitive value) that could blend well with digital video compositing (affective value) to investigate further the possibility of modernizing the traditional T&L approach in educational video.

1.2 Digital Video Compositing

As mentioned by Costello and Youngbloods, compositing is a series of step in mixing two or more videos to form a new motion picture or visual effects. But others describe the compositing in a more specific action [7]. Chen, Zhu, Shamir and Hu describe video composition as extracting a frame sequence from the original video and patch the frame sequence as a foreground onto the other video. They stated that video composition is very useful for motion pictures and entertainment industries thus may also be good for education sector too [8]. Compositing can be described as an evolution of video development with visual effects and thereof emphasizing on the composition layer and technique. For the past recent years, there are studies conducted that focus on the cohesive technical aspect of video compositing can be found.

1. **Illumination**: The blending techniques are applied to the contents between the real captured video camera and the computer-generated imaginary, CGI to reproduce the realism in the virtual set. However, creating the realism of the digital video compositing depends on the consistency of the real-world lighting and the virtual content rendering to reproduce the real-world illumination in the virtual setting [9].

2. **Audio and Video Congruent**: A study was conducted using multiple videos to detect the temporal relationship between audio and video for better perceptual fluency that could increase perceptual pleasure. The experiment focuses on congruent and incongruent audio-visual manipulation. The investigation of subliminal congruency between the two multimedia elements resulted in perceptual pleasure enhancement [10].
3. **Foreground and Background Video Compositing**: Dividing, adjusting the colors, or recoloring then merging the foreground video to the background video. A study was conducted based on this approach to train a classification network by compositing two videos for data augmentation in action recognition which later produce new realistic videos [11].
4. **Image Compositing**: The image compositing method was used in the medical field for static demonstration of surgical operations. The chroma key and tool segmentation were applied in the blending process (mix-blend method) for the fore-ground image and background image, creating a blended image at the end [12].

Other studies also reflect on the mechanical aspect of compositing, and more studies will come. As mentioned by Gyncild, the composition consists of several layers such as image file, video file, an audio file, light layer, camera layer, and other composition to become a new layer. Therefore, creating simple projects may require a single composition but complex projects require several compositions for better organizing on extensive amounts of footage and complicated digital effects [13]. According to Harrington and Jago, working with compositing effects provides new ways of filming and editing arrangement to make it easier in blending visuals [14]. Thus, the key to having impactful results is when compositing is through the combination of filming techniques and dedicated effects.

More so, the video composition is a sequence of elements arranged in a scene that can evoke a specific emotional or intellectual reaction from audience beside from delivering message and purpose. But in what context the video composition could evoke a specific emotional (affective value) and intellectual (cognitive value) reaction from the audience? These questions might be the key to solving the issues and challenges in educational videos.

2 Issues and Challenges of Digital Video Composition in Education

2.1 Review on Related Literature

In 1992, Secules et al. provide results on the effectiveness of videotaped materials (analog video) in a foreign language classroom. They studied on listening comprehension by making a comparison between video-based curriculums using the French in Action video series (native speech conversation) to the traditional curriculum.

They managed to produce improvement in listening comprehension but have no significant difference in reading and writing. They justified the reason is due to the limited capacity of what a video could offer in teaching skills for reading and writing [15]. Thus, the challenges arise from that point during the time. But on closer inspection, the production of this video series was intended for drama concept therefore, it is more into a conversation between actors and actresses thus created a tangible boundary for reading and writing skills. Hence, the video conferencing seems like a good method to switch the conversation between actors and actresses into educators and learners.

In 2003, Fitzgibbon discusses the effectiveness of teaching methods using video conferencing as a delivery mode however she also stated that there are several factors that could render ineffective results. Though the term 'interactive video-conferencing' is usually subjected to technology-based learning discussion, the interactivity will be meaningless if there is no proactive plan from the instructor side. Besides, making students always watching a fixed set camera caused no variety of in-class presentations. In addition to that, eye fatigue toward the visual screen tends to make the audience feeling tired and even irritable. However, the author implied that consideration on colors and fonts selections would help reducing eye strain, foremost, an interactive classroom should be highly maintained to avoid students feeling alienated and neglected. Based on the author's findings, even if video conferencing can offer interactivity but without the variety and visual design consideration in the video production, it will still render the ineffectiveness of the video [16]. Furthermore, video conferencing requires line stability which deters learners with a poor line connection to engage. Though video conferencing can be recorded for them to watch later it deflects the purpose of having interactivity in video conferencing. Even so, using video as a mechanism to convey knowledge and education could still help improve students to understand the message clearer rather than the traditional method of teaching in the classroom. As stated by Rammal, video can be used in a classroom, long-distance learning, and self-study. The video he used as a delivery medium in teaching English as a foreign/second language (EFL/ESL) proven to be effective for both young and adult learners [17]. However, the effectiveness of a video presentation relies on how well instructors manage to utilize the given tools. Thus, the visual aesthetics and interactive learning should be considered in video design and should become significant in educational video production.

In 2006, Zhang, Zhou, Briggs, and Nunamaker conducted an empirical experiment using the LBA System to investigate the effectiveness of e-learning environments on learning outcome and learner satisfaction by examining the influence of interactive video deployment. The LBS system is a prototype that they developed as a sample for the interactive video to see how learners responded. They studied four different settings: Interactive video, non-interactive video, without video, and the traditional classroom. Their research investigation concluded that it may not always be enough by simply incorporating video into an e-learning environment for learning improvement. They suggested that the interactive video can be useful and important to improve learning effectiveness with their ability to give user control to randomly access preferable contents. According to them, this favorable ability in an interactive video may lead to better learning outcomes and higher learner satisfaction [18]. However, their studies do not range to visual effects and the potential bonding between production style, affective value, and cognitive value.

Years later, the educational video concepts are more versatile with the touch of creativity and visual effects, thus, the issues highlighted by researchers, were indirectly solvable through the involvement of creative effects on videos. These videos with visual effects provide more ingenuity in comparison to what can be offered in the past. Furthermore, the MOOC was established in 2012 that serves as a community site for a massive video online learning around the world, provide opportunities for higher-level institutions to utilize and promote educational videos. However, there are still issues, and challenges emerged along with the progressive use of special video effects and video production in MOOC. As mentioned by Kies, Williges, and Rosson that negative attitudes towards educational delivery medium decreased students' enthusiasm, and motivation thus affect their learning. They stated that some researchers claimed that low satisfaction contributes to the lack of motivation, less learning, and high attrition rates [19]. Subtly, this feedback implied to the production value of a video that could be restructured and redesign again using the interactive digital video compositing concept. But what could be achieved by applying interactive digital video compositing in production value? And what is the production value? The production value of a video is a combination of technical skills and qualities of methods and materials to enhance the production of a motion picture. Production value initially is referred to as the cost of overall equipment needed in video production.

The production value was then progressively enhancing where at this stage it includes not only the video equipment but overall aspects for developing a good video such as the idea, storyline, and chunk information and videography skills. Therefore, applying interactive digital compositing in video production might contribute to the enrichment of the production value. The production value enrichment is an expansion of the current production style wherein most educational video seen today are deployed in linear video style. Hansch, Hillers, McConachie, Newman, Schildhauer, and Schmidt provide a typology of video production styles. The typology of production styles is Talking Head, Presentation Slides with Voice-Over, Picture-in-Picture, Text-Overlay, Khan-Style Tablet Capture, Udacity Style Tablet Capture, Actual Paper/Whiteboard, Screencast, Animation, Classroom Lecture, Recorded Seminar, Interview, Conversation, Live Video, Webcam Capture, Demonstration, On Location and Green Screen. They reviewed 20 courses on six major platforms: Coursera, edX, Udacity, iversity, FutureLearn and Khan Academy. These platforms are most widely used in Europe and the United States. Through their participation, they noted a variety of video attributes as they observed how the video was used. The video attributes are video length, video production styles, video quality and audio quality, the level of standardization, production value, and video player interface features [20].

2.2 *Insufficient T&L Quality Approach in Educational Videos*

In general, learning can be defined as a process to obtain knowledge or developing lifelong skills that can be acquired through study, life experience, and being taught by other people. However, as described by Illeris, there are a great number of theories of learning arises, although some still focus on traditional learning concepts yet there are others who try to explore new alternative learning. The author mentioned that these theories are constantly being developed and overlapping but nevertheless worth noting since before, the traditional learning concepts were known as mainly on the acquisition of knowledge and skills, and yet today the concept has evolved and covers a large area that includes emotional, social, and societal dimensional. Thus, the author stated that it is quite difficult to understand a very complex matter and therefore there is no general definition yet that could describe the theory of learning [21].

Illeris statement support Merriam, Caffarella, and Baumgartner explanations in their book regarding the complexity of learning. They mentioned that the theory of learning centralizes human behavior however it is so elusive to comprehend and the fact that there are many people who wrote, thought and investigated the process of learning indicated that this topic indeed complicated to understand [22]. Even so, the learning process has evolved from traditional learning to electronic learning, and thus later, as technology growth is progressively well that it produces a subset of electronic learning, known as mobile learning.

Even so, the traditional T&L quality known in the education sector is a pedagogy approach, however, the evolution of T&L activities and learning environment has constructed the alternative T&L qualities such as andragogy and heutagogy approach. Abraham and Komattil explain the paradigm shift of pedagogy to andragogy to heutagogy. They clarified the pedagogy as teachercentered learning where the teacher has extensive control over the learners, whilst andragogy as expanding the learners' involvement in learning and the teacher's role is to support learners for self-directed learning, and heutagogy as an approach that encourages a self-determined learning and nurturing capability in learners which contribute to flexibility in learning [23].

For video production on MOOC, Breslow, Pritchard, DeBoer, Stump, Ho, and Seaton express their concern on instructional components that they believed is not enough assimilated in MOOC videos. They even stated that the pedagogy goal in the MOOC videos is based on their supposition [24]. Hansch et al. found out most videos in MOOC ignore research on pedagogical strength which contributes weaknesses of lecturing, and the way those video's affordance was underutilized. As a guideline for future reference, they compiled the main production styles that currently used in online learning contexts. However, they encountered that standardizing video production faces challenges due to many limitations, thus, extremely difficult considering that video productions are inclined to personality, abilities, and preferences of the instructor. Hence, Hansch et al. are more interested in pedagogical effectiveness and hope for the future research centralizes on competency measurement that derived from watching a video that specially designed for T&L. They prompted the need to develop new metrics for learning in those contexts [20]. Lange and Costley also supported Hansch et al. claimed, they pointed out the importance of pedagogy implications. Their extant research shows that even if the educational video effectively uses a variety of media but without learning being the main course, it will affect the learning negatively [3].

Kennedy also expresses the need for innovation in MOOC. He hopes for pedagogical research, philosophical and theoretical development, design and facilitation adaptation and implementation in varied context for the future researcher to strive on in MOOC study [25]. The modern educational theory and practices such as principles of cognitive load theory and active teaching practices should become the best approaches for contributors when designing the videos [26]. Hence thereof, video-based learning developed by educators for their learners that embedded with T&L quality is much more supportive and inspiring for T&L 4.0.

2.3 Visual Aesthetics Recommendation

Hansch et al. encouraged course designers to be serious in synchronizing their usage of video with the pedagogical goal. They highlighted that visual aesthetics and production value could influence pedagogical objectives and learning outcomes although the aesthetics recommendation was bound to production styles and affordances [20]. When it comes to video interface, Vieira, Lopes, and Soares identified that the video interfaces on MOOC have significant value to the pedagogical goal [27]. Hansch et al. also provide the same resolution with Vieira findings, that visual aesthetics can affect the pedagogical objectives and learning outcomes. Therefore, the KE methodology can assert the importance of creative visual design aesthetics that could influence the heutagogy goal. Even so, how to apply visual aesthetics to IVML that could trigger the affective and cognitive response? The visual attention concept can be applied in this situation. Referring to Wang, Antonenko, and Dawson, visual attention in instructional videos is the presence of the instructor. Their result study is based on the fixation percentage that shows the instructor attracted the visual attention of the participants and heighten their satisfaction level [28]. According to Kopacz, Banerski, and Biele, a distraction can happen with advanced graphic design such as 3d enhanced video. The viewers are distracted by the complexity of the animation presented during the video narrations. The risk of having a potential communication threat is more inclined to the advanced visualization technologies such as 3D and VR [29]. Another study that focuses on game design interface also stressed out the complexity of visual background may influence the player's performance to complete the tasks [30]. But this depends on the level of difficulty of the message contents. Integrated multimedia can be used for complex information that is difficult to understand, however, if the message is simple information that is easy to understand, avoid using the redundant multimedia materials when presented otherwise the attention might divide [31].

2.4 Video Length Proxy

Besides the pedagogical and visual aspect, other issues are being pointed as well. The disadvantages of video-based learning were often noted as too lengthy [32]. As referred by Buzzetto-More, video length should be a factor in video production. The participants reported that the video length impacted their viewing decision. According to the author findings, the best time for in-person students is between 1.5 and 3 min (48.1%) whilst for online students about 3–7 min (55.6%). Thus, Buzzetto-More findings of the differences between hybrid students and online students in video length preference have brought the urge to know the implication of a video length being segmented by a cue point [33]. By far there are few types of research were conducted focuses on approximate video length for learning engagement and concentration.

Guo, Kim, and Rubin try to ascertain appropriate time duration for an educational video presentation by measuring the best peak time length that holds concentration. Their findings have the same outcome as Buzetto-More. They provide results based on the engagement measurement they conducted which indicated that short video from 0 to 3 min is more engaging and fewer variances in focus, median engagement time is at most 6 min whilst 9 min of video length tend to lose focus halfway through [34].

Vieira et al. discussed the potential benefits and challenges of an educational video in the T&L process at higher education. They claimed that if videos were used creatively, they can be powerful diversity tools that motivate students to embrace knowledge at their own learning pace. They clarified that inside the classroom, the use of educational video can sport on topics debate whilst outside the classroom can be a support for students to self-learning. However, there are challenges ascended according to their research findings. By conducting a test on Massive Open Online Courses (MOOC) video interfaces, Vieira et al. managed to produce results about some design implications with insights that depend "on the visual pedagogical and stylistic properties of the video". They identified issues when designing MOOC interfaces based on feedback such as make short videos, avoid abrupt transitions in videos, offer interactive links and screenshots for highlights, summarize (for selective watchers), and one-click access to crucial parts in video-lectures [27].

2.5 Scant Publications on Innovative Self-directed Learning Tools, Experimental Investigation and Engagement Measurement

Raffaghelli, Cucchiara, and Persico claimed that only a scant experimental investigation concerning the impact of MOOC on pedagogical theory was executed, yet most were inclined to the theoretical and sometimes empirical conclusion that oriented to understanding the concept and provide hypothesis rather than answering the research questions [35]. On the other hand, Kennedy highlighted that not much research was conducted that reflected affective and cognitive values of learners, their behaviors, and the social nature of learning in MOOC [25]. And not to mention, Buzzetto-More clarified that the measurement instruments some researchers used are not concrete enough to be the measurement proxy for engagement [36]. Plus, the engagement could be varied depending on the style of a video presentation. Hansch et al. also concluded that there is scant research going on about the effectiveness of video as a pedagogical tool for online learning yet more toward on educational multimedia from a cognitive science perspective, though little of this research was verified in natural educational surroundings. They doubted the engagement metrics crafted from analysis of clickstream data and viewing statistics implemented by other researchers to be an effective proxy for measuring learning. They stated that the clickstream data analysis and viewing statistics used as engagement metrics for educational videos are not concrete enough to be the proxy [20]. Fournier, Kop, and Durand express the need for powerful tools that could help learners with their self-directed learning. They believed the said tools could contribute to learning quality that is useful for both educators and learners by analyzing the important factors [37]. But what are the important factors that contribute to the quality of learning? According to Chen, Wong, and Chuah the study they conducted on MOOC videos provides insight and evidence of emotions projected in different types of videos. They highlighted that no study related to emotions derived from video style can be found. Hence, they encourage taking learners' learning styles into consideration when examining their emotions toward different types of videos [38]. Moreno-Guerrero, Rodriguez-Jimenez, Gomaz-Garcia, and Navas-Parejo pointed out that the leading role of students should be promoted in T&L in learning theory, and the teachers are required to adopt the procedural standards in T&L activities that included the ICTs to promote positive attitudes in students [39]. Damasceno et al. conducted a study on hyper videos to analyze the Learning Objectives (LOs) by making a comparative analysis of available authoring tools for LOs and the results show the requirements they set are not fulfilled. Nevertheless, they pointed out that recently there are good systematic reviews on interactive videos and hyper videos going on [40].

3 Identifying the Problems in Educational Video Production

3.1 The Importance of T&L Instruments in Production Value for Educational Video

Paper survey was performed to analyze the criteria of an educational video, unfortunately, it is quite difficult to search on interactive digital video compositing designed for T&L. Approximately, many research conducted in relation to the general criteria of educational video that bring about effectiveness and efficiencies to T&L skills, however, their research inclinations are more to a human factor rather than the technical aspect of a video itself and the T&L approach. Only a small number of researches done on the technical instrument of a video that focuses more on the compositing of a good educational video. Nonetheless, a few researchers directed the issues about pedagogy goal deficiency in present video production. Even for online video sessions, researchers recommended using active learning approaches to maintain students' participation [41]. After reviewing the issues and challenges featured in MOOC research articles, most papers had a major mutual significant notion about the importance to implement pedagogical instruments in production value for T&L quality enhancement. Consequently, the notion to investigate the potential of interactive video in line with heutagogy T&L was derived from these issues and challenges in hopes the findings can contribute to enriching the production value in educational video production. And since the time point feature on YouTube has expanded the standard limit of user control, which is the stop, play, and pause buttons offered by online video on a desktop computer, this research delves into the potential of cue point feature in an educational video on mobile devices. Furthermore, the time point feature for mobile devices shows a promising path in interactive video concepts but not much research about the educational video that dives on this area could be found.

3.2 The Absence of Curiosity on Tools and Methodological Aspect

Relatively, the absent of curiosity on tools and methodological aspect become prominence. The same awareness can be applied to the T&L approach behind the MOOC or more precisely, the production value of the current educational videos. The MOOC researchers emphasize that pedagogical influence is important in production value. They are more concern about the absence quality of educational videos as a T&L medium which asynchronous with pedagogy goal. It is a valuable discovery to know that the development of the educational videos inadequate with the T&L instructional design. Their discovery may perhaps be the root of educational video quality deficiency as a T&L medium, thus, conducting an experimental investigation could provide answers that contribute to the new direction of enrichment toward production value for educational video development. However, would it still be applicable to apply pedagogical goal for higher-level education in the twenty-first century? Brame suggested in order creating an effective educational video, the instructor needs to consider three elements for video design and implementation: Cognitive Load, Engagement, and Active Learning [36]. Whereas other authors proposed 4 phases in short instructional video development: Theme and Script, Preparation and Equipment, Filming and Post-Production [42] which are similar to the video production process of pre-production, production, and post-production. It is achievable since there are many tools for creating interactive video and the tutorial in making one can be found on the internet. For instance, one study conducted research on a TikTok media application as a source of knowledge in the medical field, a platform that enables providing short mobile videos, if the video's content were given focused, the vital features were upgraded and the user experience concept was extended in this video platform, the acceptance will be greater in a long term [43]. Additionally, the computer tools can offer the dynamic T&L course construction and manipulation efficiently for module presentation rather than the physical object [44].

3.3 The Pedagogy Approach May Not Be the Right Concept Embedded in Educational Video

Tiernan stated that students preferred short and clear video content that is relevant to the topics. Despite that, there are still some who was unable to concentrate afterward and referring to watching the video as a time to relax. Even though Tiernan suggested that can be addressed by follow-up activities and teaching strategies [45] but at some point, the feedback subsequently has a subtle indication of a T&L method and instructional design of a video. The pedagogy-based instructional design not suitable for self-determined learning and could be fragile to extract affective and cognitive values from learners. Perhaps, the pedagogy approach may not be the right concept embedded in the educational video for higher-level education in the twenty-first century anymore. Researchers also recognize the value of authentic learning experiences for students' engagement in learning activities [46]. Experimental investigation on educational video specifically designed for T&L must be implemented to affirm the validity of pedagogy approach implication toward affective and cognitive values. The reason is to ascertain whether the pedagogy approach applied to the educational video is suitable for self-determined learning and have strong implication to the affective and cognitive thinking. As stated by Raffaghelli et al., many articles and journals offer theoretical and empirical concepts to answer the research questions, yet only a few are based on the actual experimental investigation [35]. Also, a paper survey was conducted on interactive digital video compositing for an educational video which leads to verification that not much articles and journals related to this area could be found.

3.4 Educators Should Give Learners More Control

Regarding the instructional segmentation of an educational video, Lange and Costley stated that allocating the video into segments indeed can reduce the cognitive load. However, they clarify that not every person able to process the information at the same level. They suggested that educators should be giving even more control on how learners obtain the information, for instance by giving them the option of selecting according to their needs and prior knowledge to the content. They concluded that giving the learners option to select, it helps those who needed more time to process the information [3]. Their suggestion can be subtly related to the availability of the cue points feature. An experimental study was conducted on preschoolers (approximate age 3.5 years and above) using interactive apps and noninteractive videos for math learning. The results show that their learning process improved in both interactive apps and the noninteractive video [47] hence subtly insinuating the potential of interactive video mobile learning. Furthermore, according to Mohammed Kamal, interactive digital video (IDV) can provide the cognitive processes to perform faster due to the segmentation of the complex content into several parts while the interactivity allows the active role of the learner and reduces the reading load by engaging visual presentation [48]. For now, the YouTube channel offers the time points feature where instructors can set the time points to cater to learners' need for one-click access to the vital subject. The time points feature embedded in HTML 5 for YouTube platform even enable the instructors to instantly slice down the long video length into a small portion of video segments which help ease the learners to skip other topics and focus onto the intended subject. However, this feature is applicable for e-learning video on electronic devices such as computers and laptops only. The time points are not pertinent to YouTube mobile yet, which deters the usability of m-learning devices such as smartphones and tablets. Thus, remedial action plans to counteract the problem statement need to be constructed. A prototype of interactive video mobile learning should be developed for the sole purpose of research and development as an incentive to the video mobile learning.

4 Conclusions

In summarization, the employment of interactive element in digital video compositing creates a new perspective of educational video and profound insight for other researchers to study current T&L video quality. This review extended the invisible boundary of a traditional educational video where the video is limited only as a linear video timeline, then evolves into something that allows learners to have learning flexibility. Researchers are also invited to conduct a detailed investigation on the potential of interactive video in mobile learning, especially the prospect of it in the education sector and academic research in line with the evolutionary needs of T&L 4.0. The evolutionary needs of T&L 4.0 are referring to the inclusive practice of technology-based learning in education, moving forward from the traditional learning of paper books and pencils. Hence, this paper is only a portion extended from interactive video learning. There are many areas under hyper video where researchers can venture more. The name itself can offer plenty more potential sub-sections for research endeavors such as the incorporation of HCI in video design development, user perspective of UI implanted in video design interface, and the usability of hyper video in accordance with UX. It would be great to expand the standard control available on current educational videos. However, the user perspective may be affected by demographic variables and disparities such as age, gender, education, technological knowledge, geographic, and socioeconomic status [49].

Acknowledgements This paper is of the research grant SDK0058-2018, funded by Universiti Malaysia Sabah. The authors would like to express their gratitude to Universiti Malaysia Sabah.

References

1. Fabos B (2001) Media in the classroom: an alternative history. In: AERA 2001 conference in Seattle, WA, pp 13
2. Khoo YY, Khuan WB, Abd. Hadi FS, Abu Bakar MS (2020) The effect of video-based collaborative learning among economics' Undergraduates in Malaysia. Int J Adv Sci Technol 29(6):279
3. Lange C, Costley J (2020) Improving online video lectures: learning challenges created by media. Int J Educ Technol Higher Educ 17:12
4. Beheshti M, Taspolat A, Kaya SO, Sapanca FH (2018) Characteristics of instructional videos. World J Educ Technol Current Issues 10(1):61
5. Hadijah S (2016) Teaching by using video: ways to make it more meaningful in EFL classrooms. In: International seminar of English language teaching (ISELT 4), pp 307, 308
6. Yuce MO, Adali E, Kanmaz B (2021) An analysis of YouTube videos as educational resources for dental practitioners to prevent the spread of Covid-19. Ir J Med Sci 190:19–26
7. Costello V, Youngblood SA, Youngblood NE (2013) Multimedia foundations: core concepts for digital design. Focal Press, Burlington, p 402
8. Chen T, Zhu JY, Shamir A, Hu SH (2013) Motion-Aware gradient domain video composition. IEEE Trans Image Process 22(7):2532
9. LeGendre C, Ma WC, Pandey R, Fanello S, Rhemann C, Dourgarian J, Busch J, Debevec P (2020) Learning illumination from diverse portraits. In: SIGGRAPH Asia 2020 technical communications (SA'20), vol 7, pp 1–4
10. Lin C, Yeh M, Shams L (2022) Subliminal audio-visual temporal congruency in music videos enhances perceptual pleasure. Neurosci Lett 779:7
11. Gowda SN, Rohrbach M, Keller F, Sevilla-Lara L (2022) Learn2Augment: learning to composite videos for data augmentation in action recognition. https://www.researchgate.net/publication/361253412_Learn2Augment_Learning_to_Composite_Videos_for_Data_Augmentation_in_Action_Recognition. Last accessed 28 June 2022
12. Garcia-Peraza-Herrera LC, Fidon L, D'Ettore C, Stoyanov D, Vercauteren T, Ourselin S (2021) Image compositing for segmentation of surgical tools without manual annotations. IEEE Trans Med Imaging 40(5):1450–1460
13. Gyncild B (2012) Adobe after effects CS6 classroom in a book: the official training workbook from adobe systems. United States of America: Adobe Systems Incorporated, p 15
14. Harrington R, Jago M (2014) Adobe premiere pro CC classroom in a book: the official training workbook from adobe systems. United States of America, Adobe Systems Incorporated, p 366
15. Secules T, Herron C, Tomasello M (1992) The effect of video context on foreign language learning. Mod Lang J 76(4):483
16. Fitzgibbon P (2003) Challenges of video-conferencing teaching and effective teaching methods. Turkish Online J Educ Technol TOJET. 2(3):32
17. Rammal SM (2016) Using video in EFL classroom. http://www.academia.edu/16445198/Using_Video_in_the_EFL_Classroom. Last accessed 13 Nov 2016

18. Zhang D, Zhou L, Briggs RO, Nunamaker JF Jr (2006) Instructional video in e-learning: assessing the impact of interactive video on learning effectiveness. Inf Manage 43:15–27
19. Kies KJ, Williges RC, Rosson MB (1997) Evaluating desktop video conferencing for distance learning. Comput Educ 28(2):85–89
20. Hansch A, Hillers L, McConachie K, Newman C, Schildhauer T, Schmidt P (2015) Video and online learning: critical reflections and findings from the field. HIIG Discussion Paper Series No. 2015-02
21. Illeris K (2009) Contemporary theories of learning: learning theorists in their own words. Routledge Taylor & Francis Group, New York, p 1
22. Merriam BS, Caffarella SR, Baumgartner ML (2007) Learning in adulthood: a comprehensive guide, 3rd ed. United States of America, JosseyBass, p 275
23. Abraham RR, Komattil R (2017) Heutagogic approach to developing capable learners. Med Teach J 39(3):295–298
24. Breslow L, Pritchard DE, DeBoer J, Stump GS, Ho AD, Seaton DT (2017) Studying learning in the worldwide classroom research into edX's First MOOC. Retrieved from http://www.rpajournal.com/dev/wpcontent/uploads/2013/05/SF2.pdf. Last accessed 2 Jan 2017
25. Kennedy J (2014) Characteristics of massive open online courses (MOOCs): a research review, 2009–2012. J Interact Online Learn 13(1):11
26. Velichova D (2022) Visualization in geometry using educational videos https://www.researchgate.net/publication/343419351_VISUALIZATION_IN_GEOMETRY_USING_EDUCATIONAL_VIDEOS. Last accessed 21 June 2022
27. Vieira I, Lopes AP, Soares F (2014) The potential benefits of using videos in higher education. In: Proceedings of EDULEARN14 conference, pp 750–754
28. Wang J, Antonenko P, Dawson K (2020) Does visual attention to the instructor in online video affect learning and learner perceptions? An eye-tracking analysis. Comput Educ 146:1–16
29. Kopacz A, Banerski G, Biele C (2022) Cognitive and visual processing of 3D enhanced disaster risk communication video. Int J Disaster Risk Reduction 75:9
30. Caroux L, Mougine A (2022) Influence of visual background complexity and task difficulty on action video game players' performance. Entertainment Comput 41:1–9
31. Yang J, Li Y, Calic G, Shevchenko A (2020) How multimedia shape crowd funding outcomes: the overshadowing effect of images and videos on text in campaign information. J Bus Res 117:6–18
32. Sablic M, Skugor A (2022) Video-Based learning (VBL)—past, present and future: an overview of the research published from 2008 to 2019. https://www.researchgate.net/publication/342759056. Last accessed 20 June 2022
33. Buzzetto-More NA (2014) An examination of undergraduate student's perceptions and predilections of the use of YouTube in the teaching and learning process. Interdisc J E-Learn Learn Objects 10:28–30
34. Guo PJ, Kim J, Rubin R (2014) How video production affects students engagement: an empirical study of MOOC videos. In: Proceedings of the first ACM conference on learning @ scale conference, p 44
35. Raffaghelli JE, Cucchiara S, Persico D (2015) Methodologies approaches in MOOC research: retracting the myth of protues. British J Educ Technol 46(3):508
36. Brame C (2018) Effective educational videos. https://cft.vanderbilt.edu/guides-sub-pages/effective-educational-videos/. Last accessed 2 Jan 2018
37. Fournier H, Kop R, Durand G (2014) Challenges of research in MOOCs. MERLOT J Online Learn Teach 10(1):14
38. Chen CJ, Wong VS, The CS, Chuah KM (2017) MOOC videos-derived emotions. J Telecommun Electron Comput Eng 9(2–9):139
39. Moreno-Guerrero A, Rodriguez-Jimenez C, Gomez-Garcia G, Navas-Parejo MR (2020) Educational innovation in higher education: use of role playing and educational video in future teachers' training. Sustainability 12(6):2558
40. Damasceno ALB, Busson AJG, Lima TS, Neto CSS (2022) Authoring hypervideos learning objects. https://www.researchgate.net/publication/339621651_Authoring_Hypervideos_Learning_Objects. Last accessed 09 June 2022

41. Castelli FR, Sarvary MA (2021) Why students do not turn on their video cameras during online classes and an equitable and inclusive plan to encourage them To Do So. Ecol Evol 11(8):3565–3576
42. Naik JM, Thieringer FM, Beck-Broichsitter B, Girod S, Naik MN (2022) Shoot and post: the making of educational videos. https://www.researchgate.net/publication/346332034_Shoot_and_Post_The_Making_of_Educational_Videos. Last accessed 10 June 2022
43. Al-Maroof R, Ayoubi K, Alhumaid K, Aburayya A, Alshurideh M, Alfaisal R, Salloum S, Al-Maroof R, Ayoubi K, Alhumaid K, Aburayya A, Alshurideh M, Alfaisal R, Salloum S (2021) The acceptance of social media video for knowledge acquisition, sharing and application: a comparative study among YouTube users and TikTok users' for medical purposes. Int J Data Netw Sci 197–214
44. Thompson CG, Gillern SV (2020) Video-game based instruction for vocabulary acquisition with English language learners: a Bayesian meta-analysis. Educ Res Rev 30:1–23
45. Tiernan P (2015) An inquiry into the current and future uses of digital video in university teaching. Official J IFIP Tech Committee Educ 20(1):87–88
46. Richards JB, Shah NG (2022) What ATS scholar looks for in an educational video. https://www.researchgate.net/publication/347412401_What_ATS_Scholar_Looks_for_in_an_Educational_Video. Last accessed 21 June 2022
47. Griffith SF, Hagan MB, Heymann P, Heflin BH, Bagner DM (2020) Apps as learning tools: a systematic review. Pediatrics 145(1):1–14
48. Kamal M (2020) Effect of interactive video length within e-learning environments on cognitive load, cognitive achievement and retention of learning. Turkish Online J Distance Educ-TOJDE 21(4):68–89
49. Bello-Bravo J, Payumo J, Pittendrigh B (2021) Measuring the impact and reach of informal educational videos on YouTube: the case of scientific animations without borders. Heliyon 7:1–11

Fitweet: Arduino-Based Smartwatch for Early Anticipatory Anxiety Notification System

Muhammad Rahiman Abdulmanab, Nur Faidz Hazirah Nor'Azman, Hasni Meon, Chempaka Seri Abdul Razak, and Siti Hafizah Ab Hamid

Abstract As anxiety becomes increasingly prevalent among youths especially university students, early prevention via anxiety disorder profiling is crucial. Nevertheless, most screening tools to date are not automated, labour-intensive and time-consuming. Mental health detection in online social networks (OSNs) using artificial intelligence (AI) can provide means for capturing overlooked behavioural attributes, but few are centred around anxiety detection. Internet of Things (IoT)'s increasing popularity enables bio-signals monitoring to aid in mental health care. However, minimum research has been devoted to IoT-based assistive technologies for anxiety, which currently are only used by individuals already diagnosed and actively monitored. This study proposes a novel approach to anticipatory anxiety detection for Malaysian university students aged between 18 and 25 inclusive as an early notification system using AI and IoT for bilingual Malay and English tweets and vital signs analysis respectively. Our result shows that Logistic Regression with Bag-of-Words (BoW) is the overall best performing model with 81.06% accuracy. The Arduino smartwatch sensor readings from the MAX30102 pulse oximeter sensor are used to determine different body conditions using a rule-based approach. The comparison test result between the prototype and conventional measuring devices shows the prototype's ability to accurately distinguish vital signs with a relatively low Mean Absolute Percent Error (MAPE) rate of 10.92% and 1.51% for pulse rate and temperature respectively. Together, our proof-of-concept results indicate the potential of AI and IoT in deriving new measures of anxiety, by analysing social data combined with bio-signals monitoring to alert users regarding their anxiety state.

M. R. Abdulmanab · N. F. H. Nor'Azman · S. H. Ab Hamid (✉)
Department of Software Engineering, Faculty of Computer Science and Information Technology, Universiti Malaya, Kuala Lumpur, Malaysia
e-mail: sitihafizah@um.edu.my

H. Meon
Department of Communication, Faculty of Communication, Visual Art and Computing, Universiti Selangor, Bestari Jaya, Malaysia
e-mail: hasni@unisel.edu.my

C. S. A. Razak
School of Pre-University Studies, Taylor's College Lakeside Campus, Subang Jaya, Malaysia
e-mail: chempakaseri.abdulrazak@taylors.edu.my
© The Author(s), under exclusive license to Springer Nature Singapore Pte Ltd. 2023

D.-K. Kang et al. (eds.), *Proceedings of the 9th International Conference on Computational Science and Technology*, Lecture Notes in Electrical Engineering 983,
https://doi.org/10.1007/978-981-19-8406-8_21

Keywords Anxiety · Twitter · Machine learning · Emotion detection · Arduino · Smartwatch · Internet of Things · Wearables

1 Introduction

In Malaysia, the rate of students who suffer from mental health problems doubled in less than a decade (10% in 2011 to 20% in 2016) [1]. Malaysian undergraduates also showed moderate to extremely severe levels of anxiety (51.5–55.0%) [2]. Anxiety is comprised of state anxiety and trait anxiety [3]. State or anticipatory anxiety measures how individuals feel right now and are more likely to feel anxious before certain events while trait anxiety captures the frequency of feeling anxious in general [4].

Mental health detection in Online Social Networks (OSNs) has been widely studied in recent years. Social patterns in OSNs, which make up a critical part of mental health and illness, can be readily observable in the raw Twitter data [5]. As such, social media provides a means for capturing overlooked behavioural attributes [5, 6]. In general, the traditional methods of mental health detection to date consist of face-to-face interviews, self-reporting, or questionnaire distribution which are not always practical as they are not automated, labour-intensive and time-consuming [7]. Thus, previous studies have applied Internet of Things (IoT) technologies, such as wearable sensors and smartphones in healthcare and mental health detection [8, 9]. However, these technologies are typically used by individuals who have already been diagnosed with mental illness and have been monitored over time.

This paper proposes a novel approach to anticipatory anxiety detection for an early notification system using artificial intelligence (AI) and the Internet of Things (IoT) for Malaysian university students aged between 18 and 25 inclusive. To the best of our knowledge, the ability to integrate the emotion detection of a user's tweets with vital signs detection to generate notifications and mobile app recommendations as an early notification system will be a first of its kind. Extensive research has been made on each domain for mental health detection, but none have tried to combine them, unlike our research.

2 Related Works

2.1 *Mental Health Detection in Online Social Networks (OSNs)*

Recently, major advances in machine learning (ML)-centred mental health detection using behavioural and linguistic cues from large-scale social media data have garnered significant attention. Despite its pervasiveness, little research has been made in developing models capable of automatically detecting anxiety specifically compared to other more common mental health disorders such as depression and PTSD [5, 6].

Shen and Rudzicz [10] applied N-gram language modelling, vector embeddings, topic analysis, and emotional norms in detecting anxiety on Reddit posts. For bilingual tweets analysis, Tsugawa et al. [11] estimated signs of depression in individuals through bilingual English and Japanese Twitter posts using Latent Dirichlet Allocation (LDA) or topic modelling inferred from word frequencies, which proved effective in preventing overfitting and improving estimation accuracy. Faryal et al. [12] found that Support Vector Machine (SVM) and Term-Frequency Inverse Document Frequency (TF-IDF) provided the best results on their dataset for Twitter sentiment analysis. Tribedi et al. [13] are among many researchers that investigate ML-based anxiety prediction from public Tweets during COVID-19 using classical ML algorithms with 90% accuracy.

2.2 Mental Health Detection Using the Internet of Things (IoT)

Mental health conditions can be reflected by analysing data linked to behavioural, psychological, and social signals [14]. Several studies [15, 16] have been utilising IoT-based systems in monitoring mental health, assessing, and identifying mental illness based on physiological data especially heart rate variability and pulse rate. Studies from [8, 9] show how IoT has been explored in the healthcare field for its convenience to ensure a seamless process of transferring data and monitoring health conditions. A collective data of pulse rate reading can be observed to monitor anxiety rate and stressed conditions using IoT. In analysing anxiety, increasing heart rate reading during a pressuring situation or anxious condition can be detected using wearable devices [15]. Based on data collected, a study in [16] has come out with a temperature and heartbeat matrix to identify emotions relating to happy (excited), sad, relaxed (normal) and anger. Nevertheless, a standard range of classifying body conditions should be studied more as different people tend to have different resting pulse reading. This can differ according to people's health conditions, age, and lifestyle [17].

3 Methodology

3.1 Data Collection and Labelling

All our tweets are self-collected from more than 20 Malaysian local university students' public Twitter accounts for bilingual Malay and English tweets (excluding retweets and replies) using a customised Python script via Twitter Developers API and Python's *tweepy* library. The raw data set composition comprises a total of 14,417 tweets.

For this study, we labelled the tweets using sentence-level emotion annotation by employing 63 Malaysian local university students who were experts in social science from the Faculty of Communication, Visual Art and Computing, Universiti Selangor (UniSEL) using our custom labelling web application built for this research (http://fitweet.site). We divided the raw data set into 30 subsets of roughly 500 tweets where each will be labelled by 3 annotators based on 3 labels: 'Happy', 'Neither' and 'Worry'. The 'Worry' label represents tweets that exhibit signs of anxiety while the 'Neither' label is intended to filter tweets in which emotion (e.g., Sadness, Anger) does not belong in either 'Happy' or 'Worry'.

3.2 Data Cleaning and Pre-processing

The labelled data set then undergoes our custom data cleaning and pre-processing pipeline. By referring to recommendations on techniques that can yield positive results [18], among the finalised techniques used include conversion of Unicode Emojis as defined by the Unicode Consortium into their text equivalent and contractions expansion since university students tend to use both emojis and contracted words in tweets [19].

Random under-sampling technique is used to balance the data set due to its small size, where the majority class instances ('Happy') are discarded at random until a more balanced distribution is reached. The total amount of tweets in the final data set is 4700 with 2351 and 2349 tweets for the 'Happy' and 'Worry' labels respectively.

3.3 Emotion Detection Model (EDM) Training

An Emotion Detection Model (EDM) is trained with two classes of emotion ('Happy' and 'Worry'). Multiple supervised machine learning classification algorithms such as Multinomial Naïve Bayes (MNB), Logistic Regression (LR), Random Forest (RF) and Support Vector Machine (SVM) are hyperparameter-tuned in Google Collaboratory's *Jupyter* notebook environment using both randomised and grid search cross-validation. They are then trained with different feature extraction techniques such as Bag-of-Words (BoW), Term-Frequency Inverse Document Frequency (TF-IDF) and word embeddings. For word embeddings, we opted to use Google News's *Word2Vec* [20] and Stanford's *GloVe* [21] pre-trained models which leverage large data sets and contain a vast language corpus with billions of words. Stratified K-fold cross-validation with recommended fold value of 10 is used for getting the average performance of the models due to numerous empirical evidence that indicates low variance and bias at said value [22].

Table 1 Pulse rate readings comparison between pulse oximeter, infrared thermometer and Arduino sensor

Pulse oximeter	Arduino pulse sensor (BpM)	Infrared thermometer	Arduino temperature sensor
77	71	36.6	35.351
75	68	36.6	35.347
75	67	36.6	36.026
74	67	36.6	36.1
68	62	36.6	36.453
71	64	36.6	36.523
72	62	36.5	36.437
71	62	36.5	36.312
76	64	36.6	35.687

3.4 Arduino Smartwatch Development

In assembling the smartwatch prototype, ESP8266 NodeMCU v2 is used as a mainboard to connect with the sensors. It combines Wi-Fi and Bluetooth wireless capabilities and dual-core CPU which allows us to send data from the sensors to the web server. The mainboard is connected to a lightweight and low-power OLED screen to display notifications to the users. The vital signs are obtained through MAX30102 Pulse Oximeter by emitting LED as a photon detector and detecting what reflects on the detector as a pulse. An additional component, a rechargeable 3.7 V 350 mAh lithium-polymer battery, is utilised to supply power to the smartwatch and enable it to be used wirelessly for a maximum of 3 h.

To test the prototype's stability and capability, a comparison reading was taken for 9 min with an actual measuring device—an infrared thermometer and pulse oximeter as shown in Table 1. The mean absolute percentage (MAPE) for pulse rate reading is 10.92% while the temperature reading is 1.51%.

3.5 Body Condition's Rule-Based Matrix

An initial test was conducted to determine the range reading and rule-based matrix for body conditions. A controlled test simulating 3 different situations was conducted to study the vital signs pattern. The vital signs were collected for 5 min as a baseline reading while the user was in a resting condition. For the next 7 min, the reading was taken from the user while conducting an interview session to simulate an anxious state. Another 3 min was given to the user to stabilise their body condition before viewing videos that suit the user's liking to simulate an excited state.

From the test, Table 2 shows the pattern of vital signs in a relaxed, anxious and excited state.

Table 2 Vital signs range for each body condition state

Pulse rate (BpM)	Temperature (°C)	Body condition state
58	36.45	Relaxed
63	36.477	
59	34.56	
71	34.751	
67	35.642	
68	35.423	Anxious
65	36.124	
79	36.344	
83	36.1	
112	36.47	
109	36.158	
75	36.5	
74	36.328	Relaxed
67	36.48	
63	36.177	
71	36.82	Excited
68	36.14	
67	36.477	
72	36.71	
71	36.437	

To minimise error and wrong analysis, only normal temperature patterns will be considered to determine body condition. Thus, the temperature reading other than normal temperature, which ranges from 36.3 to 37.2 °C, is labelled as abnormal in the rule-based matrix. From the controlled test conducted, the pulse reading during the anxious state increased and peaked at more than 30 BpM from the average of a relaxed state. The reading of pulse rate during the excited state increased as well but did not exceed 30 BpM from the relaxed state reading. The rule-based matrix is derived by monitoring the range difference from an individual's baseline reading as shown in Table 3.

3.6 Final Rule-Based Matrix

To combine the output of the EDM and the Arduino smartwatch, a custom rule-based matrix is developed to determine the final user emotion and indicate the type of notification released by the system depending on the results of both components as shown in Table 4.

Table 3 Rule-based matrix for body condition

Pulse rate	Increase normal	Increase abnormal	Decrease normal	Decrease abnormal
Below baseline	Invalid	Invalid	Invalid	Invalid
Baseline	Relaxed	Invalid	Relaxed	Invalid
Baseline to (baseline + 30)	Excited	Invalid	Excited	Invalid
(Baseline + 30) to 190	Anxious	Invalid	Anxious	Invalid

There are 4 types of notifications generated by our system: 'Happy', 'Worry', 'No Emotion' and 'System Failure'. The matrix is drafted such that a notification will always be released even if either component fails or is absent. An additional 'Invalid' label is added to both the EDM and Arduino smartwatch output for error handling or missing classifications. For the EDM, a 'No Emotion' label is also included for handling contentious predictions where the final emotion has a narrow probability (between 0.50 inclusive and 0.51 non-inclusive).

Each case in the matrix is represented by assumptions denoted by the letters in the parentheses as explained in Table 5. For example, priorities will be given to the EDM's output over the Arduino smartwatch's if its prediction probability is greater than or equal to 0.51, indicating high emotion intensity. This reflects the rarity and importance of the user's tweets compared to the abundance of their vital signs data. In the scenario of classification conflicts between the EDM and Arduino smartwatch, the notification will default to the 'No Emotion' type on the basis that they will cancel each other out.

4 System Development

4.1 Web-Based System with Emotion Detection Model (EDM)

A web application called *Fitweet* is developed using Python's *Django* back-end framework. Figure 1 shows the interaction overview diagram flow for data retrieval and analysis from both the EDM and Arduino smartwatch for notification generation in the system. The Arduino smartwatch user will first register an account to our web application by signing into their Twitter account. They are required to fill in additional information such as their preferred notification category and basic health information including biometrics and pre-existing health conditions. Users are not required to pair their Arduino smartwatches to use our system for allowing flexibility in exploring other functionalities. Individuals with pre-existing health conditions are also prohibited from registering in our system due to the scope limitation.

Table 4 Final rule-based matrix

		Body condition			
		Excited	Relaxed	Anxious	Invalid
EDM	*Happy*	*Happy* notifications	*Happy* (A) or *No Emotion* (B) notifications	*No Emotion* (C) notifications	*Happy* (D) notifications
	Worry	*No Emotion* (C) notifications	*Worry* (A) or *No Emotion* (B) notifications	*Worry* notifications	*Worry* (D) notifications
	No emotion	*Happy* (B) notifications	*No Emotion* notifications	*Worry* (B) notifications	*No Emotion* (D) notifications
	Invalid	*Happy* (D) notifications	*No Emotion* (D) notifications	*Worry* (D) notifications	*System failure* notifications

Table 5 Final rule-based matrix assumptions

Assumptions	
A	If EDM prediction probabilities ≥ 0.51 (*Happy*/*Worry*), priority will be given to EDM's classified emotion (high emotion intensity)
B	If EDM prediction probabilities < 0.51 (*Happy*/*Worry*), priority will be given to Arduino smartwatch's classified condition (low emotion intensity)
C	If EDM prediction probabilities < 0.51 (*Happy*/*Worry*), priority will be given to Arduino smartwatch's classified condition (low emotion intensity)
D	Missing classification from EDM and/or Arduino smartwatch

When the smartwatch is turned on, it will first connect to the Wi-Fi and check for any paired user from the Firebase Realtime Database. A user can link their account with the smartwatch during registration or when updating account settings using a pre-defined serial number. After first-time pairing, an average resting vital signs reading will be taken for 5 min and saved to the database. If the smartwatch is turned on, the body's vital signs reading will be retrieved and saved for every minute.

By default, all registered users will have their tweets extraction, prediction, and notification release period default to 7 days. The system implements a job scheduler using the *Django APScheduler* library to check each account's next prediction date periodically in a 60-s interval. Users can request for administrators to change the period.

The tweets' emotion analysis result which includes the user's individual and overall emotion label, and intensity will be displayed on their dashboard page in both a chart (word cloud) and tabular manner. For users with their Arduino smartwatch paired to their account, they will also receive a notification message displayed on their dashboard and Arduino smartwatch's OLED screen. The notification message will vary depending on the final emotion predicted and the user's preferred notification category. If their final emotion label is 'Worry', the Arduino smartwatch will display a QR code leading to a page in the web application containing the information of the top 3 most recommended mental health-related mobile applications curated by the system administrators. The *Fitweet* web application can be accessed via the following link (https://fitweet.tech).

5 Results

5.1 *Emotion Detection Model (EDM)*

Based on the prepared data set, Table 6 presents the training results of the EDM algorithms based on their performance in emotion detection of the Twitter users' tweets. The training result showed that Logistic Regression (LR) with Bag-of-Words (BoW) achieved the highest mean score across all performance metrics except for

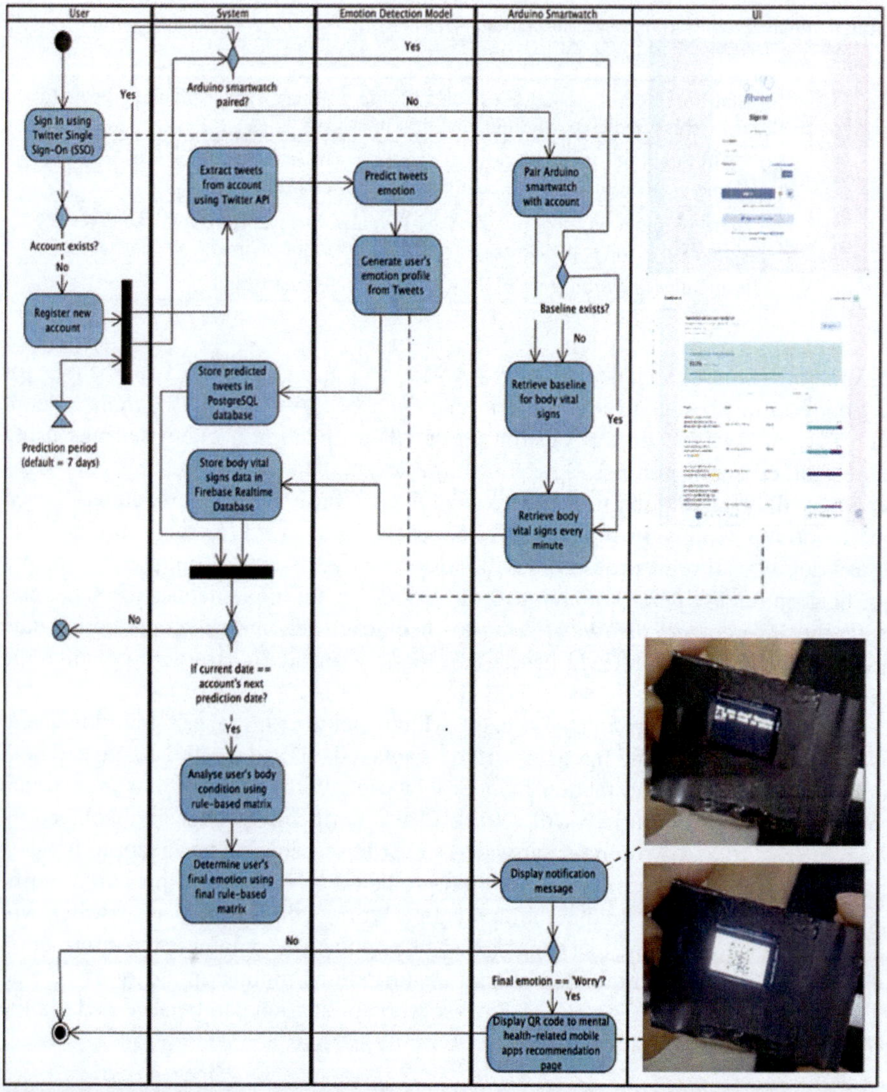

Fig. 1 Interaction overview diagram of *Fitweet* system

Receiver Operating Characteristic (ROC)'s Area Under Curve (AUC) score where Multinomial Naïve Bayes (MNB) with TF-IDF scored the highest. Support Vector Machine (SVM) with Google News' *Word2Vec* [20] scored the lowest in all metrics.

Interestingly, algorithms using pre-trained word embeddings as feature extraction techniques consistently scored poorly compared to BoW and TF-IDF. This could be due to the nature of the tweets in our dataset comprising of mixture between Malay and English languages (mostly informal). Because both *Word2Vec* [20] and *GloVe*

Table 6 Training results for each algorithm

Algorithm	Accuracy	Precision	Recall	F-measure	ROC-AUC
LR_BoW	**0.8106**	**0.8115**	**0.8106**	**0.8105**	0.8786
SVM_TFIDF	0.8078	0.8085	0.8078	0.8077	0.8843
MNB_TFIDF	0.8071	0.8095	0.8071	0.8067	**0.8869**
LR_TFIDF	0.8050	0.8050	0.8050	0.8050	0.8859
MNB_BoW	0.8014	0.8032	0.8014	0.8011	0.8712
LR_GloVe	0.7943	0.7944	0.7943	0.7943	0.8676
RF_BoW	0.7851	0.7970	0.7851	0.7829	0.8669
SVM_BoW	0.7822	0.7826	0.7822	0.7822	0.8469
SVM_GloVe	0.7780	0.7780	0.7780	0.7780	0.8434
RF_TFIDF	0.7780	0.7876	0.7780	0.7761	0.8599
RF_GloVe	0.7766	0.7767	0.7766	0.7766	0.8582
LR_Google	0.7617	0.7617	0.7617	0.7617	0.8298
RF_Google	0.7504	0.7505	0.7504	0.7503	0.8132
SVM_Google	0.7326	0.7329	0.7326	0.7325	0.8114

The bolded values indicate the best performance of the algorithm called LR_BoW. However, MNB_TFIDF algorithm shows the largest ROC-AUC value

[21] are trained with English corpora, they will not include the word representations that did not appear in their trained vocabulary such as Malay words, thus affecting performance.

Word embeddings are also typically used with deep learning algorithms that show significant performance improvements when trained with bigger datasets. In general, small datasets like ours require models that have low complexity (or high bias) to avoid overfitting the model to the data, which may explain why models using word embeddings performed poorly. Finally, although the algorithm using BoW scores the highest overall, models utilising TF-IDF on average scored higher compared to their BoW counterparts across all metrics, proving TF-IDF's effectiveness over BoW.

5.2 *System Testing (UAT)*

System testing is conducted with a 25-year-old female university student from Universiti Kuala Lumpur (UniKL) with no pre-existing health conditions for a duration of 3 days in February 2022. The participant is required to wear the Arduino smartwatch throughout the day with exceptions such as undergoing vigorous activities and sleeping. They are also encouraged to reflect on how they feel at a given time by actively posting tweets on their Twitter account.

Over 60 min of combined vital signs data and 19 total tweets were retrieved during the 3-day monitoring. The average vital signs reading calculated for pulse rate and temperature are 53 BpM and 36.24 °C respectively. Pulse rate readings that exceed 83 BpM for 5 min consecutively will be regarded as an anxious state. Figure 2 highlights the range of anxious and excited states based on the participant's average pulse rate. The reading that falls under the grey area will be considered invalid to indicate device failure.

Because the job scheduler checks each account in the system every 60 s, at precisely midnight on the 3rd day, the system will analyse the emotion of both the participant's tweets and vital signs data and generate the notification based on the final rule-based matrix outlined in Table 4. The emotion analysis result showed that the participant's overall tweets emotion is 'Worry', and their body condition is 'Relaxed'. Therefore, the notification message of the 'Worry' type ("*Are you okay? Remember to go easy on yourself!*") and the QR code for recommended mental health-related mobile applications page is displayed on their Arduino smartwatch as shown in Fig. 3.

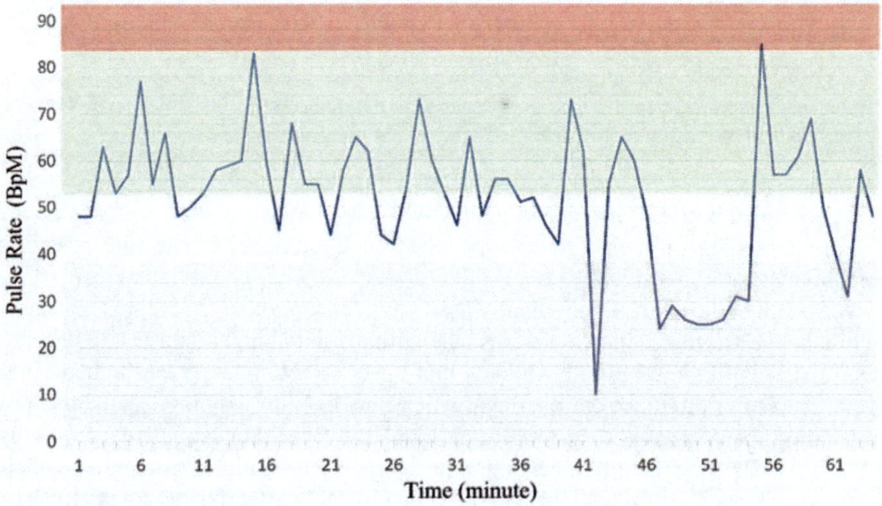

Fig. 2 Range of body condition states is based on the participant's average pulse rate

Fig. 3 Notification message and QR code displayed on Arduino smartwatch's OLED screen

The participant is also required to fill in a Google Form to obtain their feedback and evaluation of our system. A 5-point Likert scale metric is used, where 1 is 'Very Unsatisfied' and 5 is 'Very Satisfied'. Out of 8 total modules in our system, the participant's ratings had a total average score of 4.855. They also remarked the ability to see the analysed emotion of their tweets as the most interesting part of the system. Suggestions for improvements given by them include having internal memory storage in the smartwatch so that it is still accessible in offline mode. The full Google Form survey result can be obtained via the following link (https://bit.ly/39gcLE2).

6 Conclusion

The advent of digital approaches that merges artificial intelligence (AI) and the Internet of Things (IoT) with software engineering provides new opportunities for early detection of one's mental health status. In this paper, we proposed a system that combines the emotion analysis of tweets with vital signs monitoring of temperature and pulse rate using an Arduino smartwatch as an early notification system for anticipatory anxiety.

For improvements, more behavioural data such as sweat, and blood pressure should be detected and a time-series deep learning method such as Long Short-Term Memory (LSTM) should be employed instead of a rule-based approach to increase accuracy. More complex algorithms such as deep learning and transfer learning should be leveraged for EDM training with more emotion labels and a larger dataset. The system testing was only conducted with one participant due to only one Arduino smartwatch prototype being developed for this research. Further testing in longer observation periods with more participants should be conducted in the future.

Together, our proof-of-concept results indicate the potential of AI and IoT in deriving new measures of anxiety, by analysing social data combined with bio-signals monitoring to alert users regarding their emotions. This method is employable by physicians and healthcare agencies to aid in diagnosis by providing nuanced complementary data in transitioning from reactive to preventive care while enabling individuals to be more proactive about their mental health. Nevertheless, we hope that more extensive studies can be conducted to increase the effectiveness of anxiety detection through this method.

References

1. Hezmi MA, Mental health a major concern in coming years among Malaysian students? https://people.utm.my/azril/mental-health-a-major-concern-in-coming-years-among-malaysian-students
2. Gan WY, Taib MNM, Shariff ZM, Saad HA (2011) Disordered eating behaviors, depression, anxiety and stress among Malaysian University students. Coll Student J 45:296
3. Spielberger CD (1983) State-trait anxiety inventory for adults. PsycTESTS Dataset
4. Gruda D, Hasan S (2019) Feeling anxious? Perceiving anxiety in tweets using machine learning. Comput Hum Behav 98:245–255
5. Coppersmith G, Dredze M, Harman C (2014) Quantifying mental health signals in Twitter. Proceedings of the workshop on computational linguistics and clinical psychology: from linguistic signal to clinical reality
6. De Choudhury M, Gamon M, Counts S, Horvitz E (2013) Predicting depression via social media. In: Proceedings of the international AAAI conference on web and social media, vol 7
7. Lin H, Jia J, Qiu J, Zhang Y, Shen G, Xie L, Tang J, Feng L, Chua T-S (2017) Detecting stress based on social interactions in social networks. IEEE Trans Knowl Data Eng 29:1820–1833
8. Plageras AP, Psannis KE (2022) IOT-based health and emotion care system. ICT Express
9. Baig Mohammad G, Shitharth S (2021) Wireless sensor network and IOT based systems for healthcare application. Mater Today Proc
10. Shen JH, Rudzicz F (2017) Detecting anxiety through Reddit. Proceedings of the fourth workshop on computational linguistics and clinical psychology—from linguistic signal to clinical reality, pp 58–65
11. Tsugawa S, Kikuchi Y, Kishino F, Nakajima K, Itoh Y, Ohsaki H (2015) Recognizing depression from Twitter activity. Proceedings of the 33rd annual ACM conference on human factors in computing systems
12. Faryal M, Iqbal MM, Tahreem H (2021) Mental health diseases analysis on Twitter using machine learning. IKSP J Comput Sci Eng 1:16–25
13. Tribedi S, Biswas A, Ghosh SK, Ghosh A (2021) Machine learning based anxiety prediction of general public from tweets during COVID-19. Understanding COVID-19: the role of computational intelligence, pp 291–312
14. Gutierrez LJ, Rabbani K, Ajayi OJ, Gebresilassie SK, Rafferty J, Castro LA, Banos O (2021) Internet of things for mental health: open issues in data acquisition, self-organization, service level agreement, and identity management. Int J Environ Res Public Health 18:1327
15. Roh T, Bong K, Hong S, Cho H, Yoo H-J (2012) Wearable mental-health monitoring platform with independent component analysis and nonlinear chaotic analysis. In: 2012 annual international conference of the IEEE engineering in medicine and biology society
16. Kaur G, Pandey PS (2018) Emotion recognition system using IOT and machine learning—a healthcare application. In: Proceedings of the 23rd conference of open innovations association FRUCT
17. Quer G, Gouda P, Galarnyk M, Topol EJ, Steinhubl SR (2020) Inter- and intraindividual variability in daily resting heart rate and its associations with age, sex, sleep, BMI, and time of year: retrospective, longitudinal cohort study of 92,457 adults. PLOS ONE 15
18. Symeonidis S, Effrosynidis D, Arampatzis A (2018) A comparative evaluation of preprocessing techniques and their interactions for twitter sentiment analysis. Expert Syst Appl 110:298–310
19. Pandarachalil R, Sendhilkumar S, Mahalakshmi GS (2014) Twitter sentiment analysis for large-scale data: an unsupervised approach. Cogn Comput 7:254–262
20. Mikolov T, Sutskever K, Chen K, Corrado G, Dean J (2013) Distributed representations of words and phrases and their compositionality. In: Proceedings of the 26th international conference on neural information processing systems, vol 2, pp 3111–3119

21. Pennington J, Socher R, Manning C (2014) GloVe: global vectors for word representation. In: Proceedings of the 2014 conference on empirical methods in natural language processing (EMNLP)
22. James G, Witten D, Hastie T, Tibshirani R (2017) An introduction to statistical learning: with applications in R. Springer, New York

Student's Mental Health Diagnostic Expert System Using DASS Analysis and Rule-Based Reasoning: A Proposal

Nur Hanani Azami, Faridah Hani Mohamed Salleh, Husni Mohd Radzi, Nor Zakiah Gorment, Ahmad Ikhsan Abdul Aziz, and Yim Jen Nee

Abstract The emergence of coronavirus disease (COVID-19) in December 2019 has led to major economic and social disruption in the new decade of twenty-first century. People from all walks of life are aware of the severity of this COVID-19 pandemic and probably feeling exhausted from the anxiety of it all. It has also brought into focus the mental health of various affected populations including university students. They are known to be the vulnerable group who can be easily exposed to diverse mental health issues such as stress, anxiety, and depression. Hence, this mental health diagnostic expert system called Easy Psychology System is proposed to evaluate the condition of UNITEN student's mental health during COVID-19 pandemic using Depression, Anxiety and Stress Scale (DASS) analysis and Artificial Intelligence (AI) rule-based reasoning method. The goal of this paper is to identify the requirements needed to develop the system in a web-based application platform. A comparative study on existing similar psychology system is also presented. The design and architecture needed to develop the system is also discussed. A review analysis is done to gather the requirements to develop the system. The information gathered will help to develop the system accordingly.

N. H. Azami (✉) · N. Z. Gorment · A. I. A. Aziz
Department of Foundation and Diploma Studies, College of Computing and Informatics, Universiti Tenaga Nasional, Jalan Ikram-Uniten, Kajang, Selangor, Malaysia
e-mail: Nurhanani@uniten.edu.my

N. Z. Gorment
e-mail: Zakiah@uniten.edu.my

A. I. A. Aziz
e-mail: Ikhsan@uniten.edu.my

F. H. M. Salleh · Y. J. Nee
Department of Computing, College of Computing and Informatics, Universiti Tenaga Nasional, Jalan Ikram-Uniten, Kajang, Selangor, Malaysia
e-mail: faridahh@uniten.edu.my

H. M. Radzi
Department of Social Science and Humanities, College of Energy Economics and Social Science, Universiti Tenaga Nasional, Jalan Ikram-Uniten, Kajang, Selangor, Malaysia
e-mail: Husni@uniten.edu.my

© The Author(s), under exclusive license to Springer Nature Singapore Pte Ltd. 2023
D.-K. Kang et al. (eds.), *Proceedings of the 9th International Conference on Computational Science and Technology*, Lecture Notes in Electrical Engineering 983,
https://doi.org/10.1007/978-981-19-8406-8_22

Keywords Mental health · Expert system · DASS analysis · Artificial intelligence · COVID-19 pandemic

1 Introduction

World Health Organization (WHO) announced the coronavirus illness 2019 (COVID-19) epidemic, resulted in severe acute respiratory syndrome (SARS) in 2003, designated an epidemic from March 12, 2020 [1]. This major outbreak brought significant disruptions to education worldwide. The U.N. Educational, Scientific, and Cultural Organization anticipated that 107 states around the world had executed state school and university closures linked to COVID-19 outbreak on March 18, 2020, influencing 862 million students worldwide [2]. According to UNESCO, all the numerous countries have postponed educational activities at educational institutes internationally, as of April 8, 2020, in around 188 countries [3]. To prevent widespread transmission of the COVID-19 virus among students and staff, higher-education institutions across the country have rapidly switched from physical class to online learning [4].

In a short period of time, university students' lives have dramatically changed as they have been asked to leave campus, adjust to new living circumstances, and adapt to online learning platforms. The switch to online learning, particularly in courses that were not originally designed for online delivery likely increases stress among students. Courses designed to include high levels of interaction and hands-on experiences such as practicums, labs, and/or artistic performance have a clear disadvantage regarding the evaluation of students [5]. Some students may struggle to access computers and the internet at home [6]. Additional challenges include concerns about their health, the health of family members, and worry about finances. The pandemic has the potential to negatively impact university student's physically, academically, financially, and psychologically. The mental health of university students has been a rising concern with a significant number of students experiencing psychological distress [7]. This is due to uncertainty regarding academic success, future careers, and social life during university, amongst other concerns [8]. Even before the pandemic, students across the globe experienced increasing levels of anxiety, depressive moods, lack of self-esteem, psychosomatic problems, substance abuse, and suicidality [9]. Therefore, students may need additional resources and services to deal with the physical and mental health repercussions of the disease.

This research will use Depression, Anxiety and Stress Scale (DASS) analysis to gauge UNITEN student's level of stress, anxiety, and depression. It was designed to measure emotional distress in three sub-categories [10] of depression (e.g., loss of self-esteem/incentives and depressed mood), anxiety (e.g., fear and anticipation of negative events) and stress (e.g. persistent state of over arousal and low frustration tolerance) [11]. The Depression scale assesses dysphoria, hopelessness, devaluation of life, self-deprecation, lack of interest/involvement, anhedonia, and inertia. The Anxiety scale assesses autonomic arousal, skeletal muscle effects, situational anxiety, and subjective experience of anxious affect. The Stress scale is sensitive to levels of chronic non-specific arousal. It assesses difficulty relaxing, nervous arousal, and being easily upset/agitated, irritable/over-reactive and impatient [12]. Depending on the score of the DASS analysis, a complete diagnosis of the student's mental health will then be done by using the mental health diagnostic expert system. This will assist them to identify the severity of their mental health as if it was being supervised by a real expert and to suggest a better treatment for them.

2 Related Works

2.1 Review of Existing Applications

Over the years, several similar mental health systems have been developed by various institutions such as JPAN Sabah system's DASS survey form, iCliniq website system, Mental Health America (MHA) website as well as Student Guidance Unit (SGU) Uniten DASS Test's survey form. These systems specialized in mental health with many different types of surveys provided for their users, including the free DASS-21 survey. Among these existing systems, the one related to this research is the survey form belongs to SGU Uniten. All these whiles, they only use Google Form as a platform to conduct DASS analysis. This method has a high risk of security breach as it is difficult for the counsellors to trace a specific respondent's results.

The proposed mental health expert system will have several upgraded features as compared to the existing systems. It will enable the students to answer DASS analysis questions while at a same time assisting the counsellor to keep track of the data and results of their counselee's psychology surveys. It will also keep the record of the student's mental state evaluation. Tables 1 and 2 illustrates the comparison of the existing mental health system.

3 Counsellor's Perception and Experience in Diagnosing Student's Mental Health

3.1 Interview Findings

The counsellors of Student Guidance Unit (SGU) in UNITEN are the expert correspondents for this research. Their perception and opinion are important and very valuable in developing this expert system since they will be the main user and subject matter expert. As such, an interview session was conducted with one of the prominent counsellors of SGU, Dr. Husni Mohd Radzi, to gain her insights and experience in diagnosing student's mental health. The interview technique was chosen for this study because it aims to identify the processes and procedures for accessing and

Table 1 Comparison of existing system based on questionnaire design and weaknesses

System	Questionnaire design	Weakness
JPAN Sabah	• Question is aligned right on PC web browser • Scales and rating of it are placed at every question	• Does not focus on the mental health survey • Difficulty in navigating to the survey page • Button leads to a page that is broken/does not exist • Some questions are aligned left, and some aligned right on mobile screen resolution • No record keeping feature
iClinic depression, anxiety and stress test	• Simple interface • Scales and rating are positioned underneath each question • The questions are split into different parts on the screen • The scale and ratings are placed on top of the table	• Does not record the result of DASS assessment • Might require some programming skills for online communication methods • Website is too simple • There is nothing that attracts users to use it • There is no record keeping
Mental Health America (MHA)	• Answers/scales in the form of button instead	• Search bar does not auto-minimize when user clicks on it and does not use it on mobile screen resolution • Does not save any articles even though there is a function for it • Some button has '+' or '−' after clicking on it which is misleading
SGU UNITEN DASS test	• The scales and ratings are on each question	• Platform is on Google form • There is a delay between user submitting the form and the email sent

diagnosing student's mental health. The findings from this interview session are illustrated in Table 3.

3.2 Summary of Findings from Interview and System Development

This research proposed to have an expert system that tally up the DASS answers and convert all the data into data visualization for easy analysis and viewing. The system

Table 2 Comparison of existing system based on features and special features

System	Features	Special features
JPAN Sabah	• Easy to view the scale rating and questions • Calculates the results of the survey at the end of the survey	
iClinic depression, anxiety and stress test	• Attractive interface • Auto-calculates results • Display the legend of DASS levels for user's references • Has a therapy recommendation section at the end of the survey • Auto calculate results • Has a graph of comparison between the user's scores and the normal scores	
Mental Health America (MHA)	• Split DASS into their own separate survey • Offer different survey other than DASS • Offer support option with affiliated organization • Has a user account • Has informatic articles regarding mental health	• Keep a record of survey done by user • Display a graph based on scores from the survey and test done • Allow user to revisit their answer in the selected test
SGU UNITEN DASS test	• Has a short survey in asking respondent's feelings • Has both English and Malay languages in each question • Platform remember answers if user has not submitted yet • Provide indicator for user's references	• Result and scores are placed in infographic • The results and scores are sent to the respondent's email

would also have a special function where the counselor would be able to record audio or video and saved into the database of the system itself and it will be displayed for the counselee to listen or watch the recording. The use case diagram for the system is presented in Fig. 1.

3.3 Rule-Based Reasoning

The technique of rule-based reasoning is suitable to be chosen in the development of this expert system. This is due to the normal procedures of diagnosing student's mental health which will leads to many possible solutions that can be derived from the diagnostic procedures. It requires specific knowledge to be applied instead of specific technique or algorithms. Relatively, each human experts acquire different level of knowledge base and expertise. The way they handle and conduct the diagnostic procedures will be specifically based on their experience and skills. As such, rule-based reasoning technique is suitable to be used in developing this expert system

Table 3 Interview questions and the responses

Question	Responses
1. How has the SGU been doing its survey especially for DASS? What is the history the unit has when trying to conduct the DASS survey?	We use paper and pencil/pen for manual process. A few years ago, we decided to ask for help from students in UNITEN to convert the form to online format. The students chose Google Form as the platform to do so by using a SGU Gmail account
2. How has the unit been keeping all the data from the survey?	The job is done mostly by the students that handled the Google Form. When we require the overall data from the survey, we will request it from the students who handled the Google Form. The students will then compile everything into an Excel file before sending it
3. Does the unit face any problem when doing the survey in any of the methods?	There are several problems we faced in both methods. With the manual way, it is hard to keep record of all the data as they must manually record it into database and the papers take up spaces. It is also time-consuming as we must wait for the survey respondent to answer and calculate it, which takes up the time of the consultation. For the current method the unit is using, we must wait for the students that handled the Google Form to compile the data into the Excel file
4. What kind of features or improvement would you want to improve on the current system?	There are several things we would like to have on the system. As the Google Form was handled by students, the unit is in a pinch on who to ask for help when the students have already graduated from the university. We would like it if all the features of the system would be in one place. Those features including the survey and the data compilation, so they do not need to do it on another third-party application and can be easily viewed on the system itself. We would also like it if the system had tools that allows them to conduct analysis easily, like the visualization of the data
5. What platform would you want the system to be on?	The platform that would be easily accessible for people to the system is website. Website can be easily adjustable on mobile view and PC view as well

(continued)

Table 3 (continued)

Question	Responses
6. Are there any features do you want the other users (students/staffs/outsiders) to use on the system?	It would be great if the counselee could have their result immediately on the spot, co-relating to the type of system that they want (everything in one system) rather than them having to check the result for their result for their own reference. It would be great if each student/staff has their own profile/account system where we can keep track of their survey result record. It would be good if the color of the system co-relates to psychology colors
7. Regarding the drawing evaluation, how would you evaluate it on the system?	While it would be amazing if it is possible, there are many ways to go through with it. One of them is by letting the counselee to color the shapes provided by the system. With those, the counselor would then evaluate the counselee's mental state through the type of colors they used to color the objects provided. Another is by recording. We hope to have a system where it could record audio or video of them evaluating and summarizing the drawing's significant and meanings and automatically saved it into the system. The counselee can also view the recording on the system itself without having them to download it

since it resembles human reasoning with a blend of heuristic programming. The data can be processed systematically, standardized and the accuracy of the diagnostic treatment can be increased. In rule-based reasoning technique, a set of production rules is made up of many inference rules that are entered as separate rules and are used together to draw the conclusion. For instance, to determine the severity of student's stress level, we can derive several rules such as "*if Y is not adjusting well to his/her study environment AND Y is facing some social problems then the severity of stress is severe*".

4 Identifying the Proposed Mental Health Expert System

4.1 Methodology

A set of interview questions has been set up and the interview session was conducted with one of the prominent SGU counsellor, Dr. Husni Mohd Radzi, to get her insights on the current process and procedures of diagnosing student's mental health. The findings from the interview will be used to develop the system based on Artificial

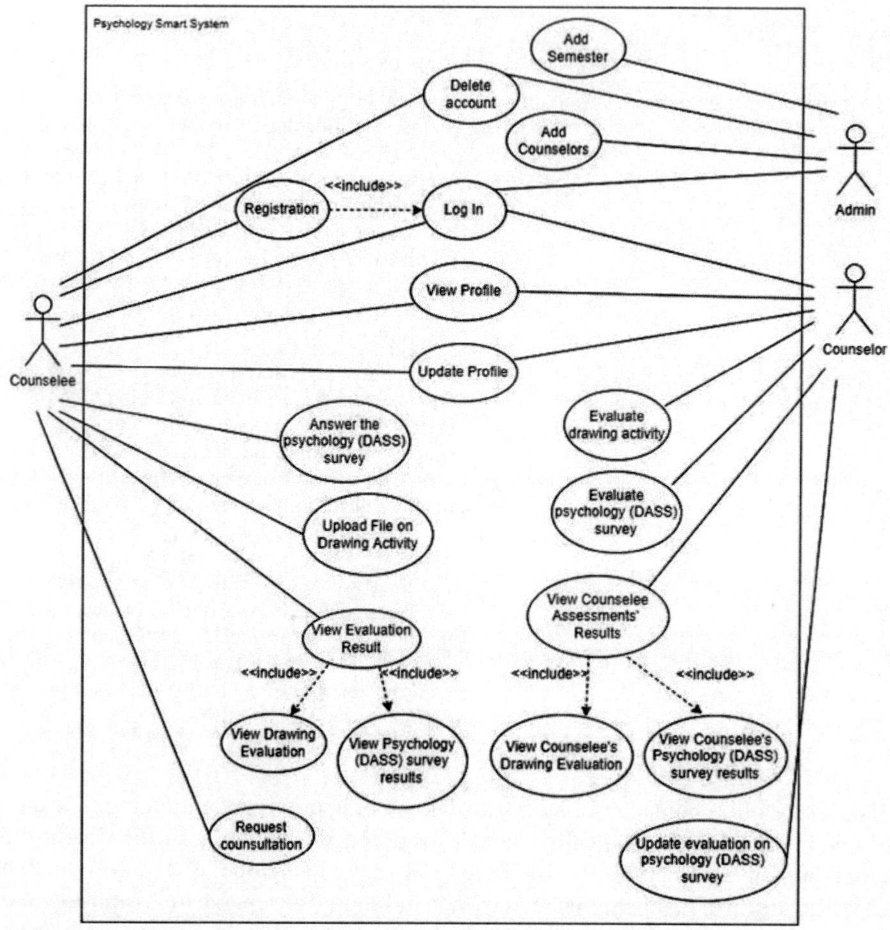

Fig. 1 Use case diagram

Intelligence rule-based reasoning technique and DASS analysis. Then, the link to the website will be dispersed to the students in which they will use the system to answer each DASS analysis questions. The system will then automatically compute the scores gained by the students to determine their mental health state. The counsellor will use the system to check on the student's records and proceed with the expert evaluation. The complete research methodology process can be seen in Fig. 2.

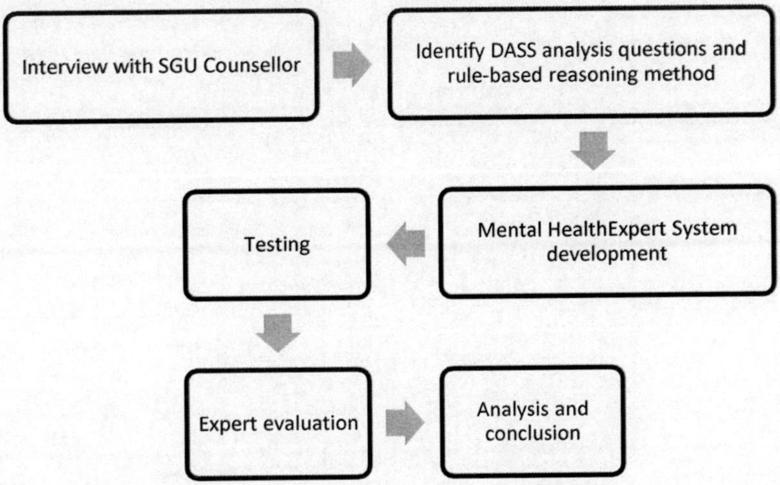

Fig. 2 Research methodology

4.2 *System Architecture*

The proposed mental health expert system will be developed to primarily ease the UNITEN counselor's job of conducting DASS survey and gathering analysis on the mental health of their counselees. The system will also include another aspect of psychological evaluation where the counsellor will be able to view the drawing that the counselee uploaded into the system and able to record their evaluations on the drawing for the counselee. When the counselee answers the DASS survey, the system will automatically calculate the scores and display them in a data visualization format for easy viewing. For the drawing activity, the system will allow counselees to upload their drawing into the system. The system will display the drawing when the counsellors want to view the drawing and record their evaluation of it. The recordings will be saved in database and will always be displayed for viewing on both counselor and counselee's respective page. The system architecture will resemble the standard web application architecture as illustrated by Fig. 3.

5 Conclusion

As a conclusion, students with mental health problems must receive proper treatments and medications so that their conditions get better. If their mental status can be assessed and examined at an early stage, then most probably their mental problems can be detected, controlled, and cured comprehensively. The proposed mental health system will provide the psychology assessment added to the system, a way to keep constant track of the data gathered from the results of assessment as well as

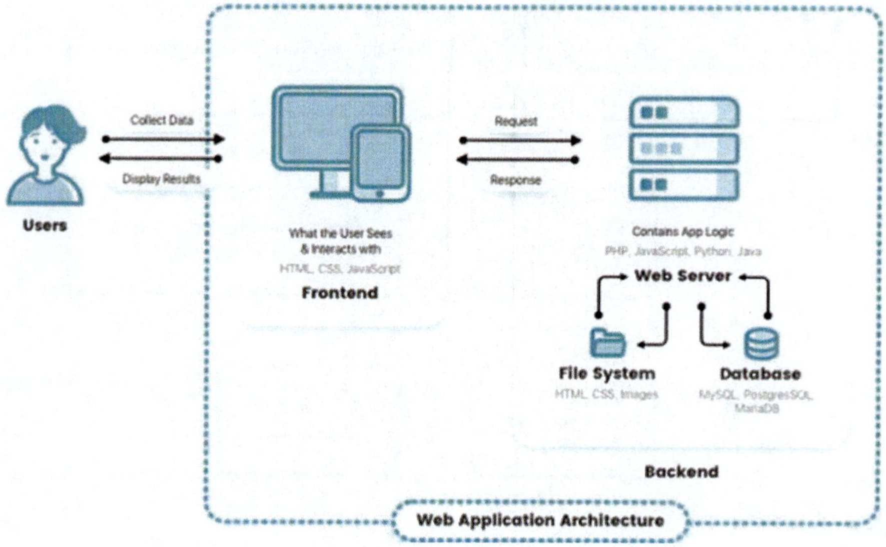

Fig. 3 System architecture

adding a system where the counselor can provide their evaluation on the psychology assessment done online is able to be materialize. This system also provides a good security where users are required to registered through the system before they could access it. It is hoped that this proposed system will be beneficial for SGU UNITEN to help them conduct their standard procedures of accessing and diagnosing student's mental health.

Acknowledgements This work was carried out within the framework of BOLD2021 project and was sponsored by Universiti Tenaga Nasional.

References

1. Tian F (2020) Psychological symptoms of ordinary Chinese citizens based on SCL-90 during the level I emergency response to COVID-19. Psychiatry Res 288:112992
2. Viner RM (2020) School closure and management practices during coronavirus outbreaks including COVID-19: a rapid systematic review. Lancet Child Adolesc Health 4(5):397–404
3. Lee J (2020) Mental health effects of school closures during COVID-19. Lancet Child Adolesc Health
4. Gewin V (2020) Five tips for moving teaching online as COVID-19 takes hold. Nature 580(7802):295–296. pmid:32210377
5. Sahu P (2019) Closure of universities due to coronavirus disease 2019 (COVID-19): impact on Edu
6. Lederman D (2020) How the shift to remote learning might affect students, instructors and colleges

7. American College Health Association (2020) American College Health Association-National College Health Assessment III: Reference Group Executive Summary Fall 2019. Silver Spring, MD
8. Aristovnik A, Kerzic D, Ravselj D, Tomazevic N, Umek L (2019) Impacts of the COVID-19 pandemic on life of higher education students: a global perspective
9. Holm-Hadulla RM, Koutsoukou-Argyraki A (2015) Mental health of students in a globalized world: prevalence of complaints and disorders, methods and effectivity of counseling, structure of mental health services for students. Mental Health Prev 3:1–4
10. Lovibond SH, Lovibond PF (1995) Manual for depression, anxiety and stress scale (DASS), 2nd edn. Sydney Psychology Foundation, Sydney
11. Crawford JR, Henry JD (2003) The depression anxiety stress scales (DASS): normative data and latent structure in a large non-clinical sample. Br J Clin Psychol 42:111–131
12. Jahaz FA (2016) The effectiveness of a counselling programme in relaxing social anxiety related to irrational thinking among Saudi college students. Brunel University, London, Thesis for Doctor of Philosophy

Complementary Preparation for Frame Alignment Algorithms Using PYNQ-Z2 FPGA

Aaron Eldrich Wong, Carlos Louis Ting, Stanley Vincent Lim, Steven Edward Lim, Roger Luis Uy, Rayner Alfred, and Zamhar Iswandono

Abstract FPGAs have been known to run programs at a lower execution time due to its parallelism capabilities. Frame alignment algorithm is a type of sequence alignment that determines the optimal match for a given DNA sequence and reference proteome. However, there are some input data preparations that must be done for the algorithm to work. The PYNQ-Z2 FPGA can speed up conversion processes, which is part of data preparation, to reduce the time it takes to process the data. This research utilizes the computing capabilities of the FPGA as a SIMD paradigm for the conversion process. The tests conducted were successful and favorable as the results showed speedup reaching up to 13,939 times that of non-SIMD implementation.

Keywords Bioinformatics · Data preparation · DNA · RNA · Amino acid · Blosum62 indexing · SIMD · FPGA · PYNQ-Z2

A. E. Wong (✉) · C. L. Ting · S. V. Lim · S. E. Lim · R. L. Uy
Department of Computer Technology, De La Salle University, Manila, Philippines
e-mail: aaron_wong@dlsu.edu.ph

C. L. Ting
e-mail: carlos_ting@dlsu.edu.ph

S. V. Lim
e-mail: stanley_lim@dlsu.edu.ph

S. E. Lim
e-mail: steven_edward_lim@dlsu.edu.ph

R. L. Uy
e-mail: roger.uy@dlsu.edu.ph

R. Alfred
Creative Advanced Machine Intelligence Research Centre, Faculty of Computing and Informatics, Universiti Malaysia Sabah, Kota Kinabalu, Sabah, Malaysia
e-mail: ralfred@ums.edu.my

Z. Iswandono
Higher Colleges of Technology, Abu Dhabi, United Arab Emirates
e-mail: zismail@hct.ac.ae

© The Author(s), under exclusive license to Springer Nature Singapore Pte Ltd. 2023
D.-K. Kang et al. (eds.), *Proceedings of the 9th International Conference on Computational Science and Technology*, Lecture Notes in Electrical Engineering 983,
https://doi.org/10.1007/978-981-19-8406-8_23

1 Introduction

FPGA, short for field programmable gate array, is a device made up of configurable or programmable logics that form matrices [1]. The device offers flexibility in building custom system blocks for specific project usage [2]. In bioinformatics, the primary components of alignment are DNA, RNA, and amino acids [3]. Amino acids formed in a chain are called proteins which are used in gene expressions. Gene expression is an important study as it is used to determine the characteristics of a gene such as function and development. To make the sequence alignment more accurate, possible insertions, deletions, and substitutions to the characters of the sequences must be considered [4]. For this, a frame alignment algorithm can be used to fulfill the sequence aligning process.

The focus of the paper will be the input data preparation aspect of a frame alignment algorithm for it to work. Although the algorithm requires various inputs, only the DNA needs to be processed into something that can be compared to the protein sequence, as well as to give a score to the alignment to get the most optimal match [5, 6]. The first conversion is to make the DNA sequence into an amino acid sequence. After which, the converted amino acid sequence will be converted into an amino acid substitution matrix index which will provide the scores for substituting amino acid x to amino acid y [6]. The substitution matrix that will be used is the Blosum62. A Blosum62 matrix is known for its capability to compare similarities between two amino acids which can be used as a scoring system [7]. To decrease the time for the program, the PYNQ-Z2 was used for the implementation of the two conversion processes, namely DNA sequence to amino acid sequence conversion and amino acid to Blosum62 index conversion. In a paper by El-Wafa et al. [8], it is stated that frame alignment algorithms can be done by either using dynamic programming or heuristics, which only increases accuracy or speed, respectively. By using the Smith-Waterman algorithm which makes use of both dynamic programming and heuristics to speed up the process and provide accurate results. On top of that, El-Wafa also used a modified version of the Smith-Waterman algorithm which allows a parallel computation of the alignment process. The input data preparation was only done sequentially in this paper which means that it was not part of the parallel computation of the alignment. The authors used a ZYNQ-7000 FPGA for the application of the parallel computation of the alignment matrices. The results are then stored in the memory of the hardware. For the rest of the paper, Sect. 2 will be discussing the hardware and how the data flow works. Section 3 will be discussing the two conversion processes and the corresponding conversions of each process. Section 4 for the results and analysis, and the last section will be for conclusion and recommendations.

2 Hardware

2.1 PYNQ-Z2 FPGA

The PYNQ-Z2 FPGA developed by TUL uses an SoC or system on chip called ZYNQ 7000. This unit has two major systems that will be used for the project. First is a dual-core processing system, PS in short, rated at 650 MHz. The second one is the programmable logic or PL. The processing unit reads the PYNQ-Z2 image file which contains the Jupyter Notebook interface for users to access and use. This system can also do basic computations, data processing, and most importantly, serves as a bridge to access the programmable logic. The programmable logic, on the other hand, is the one doing most of the heavy computing and returning the results back to PS for user use or display [9].

2.2 Data Flow

To program the FPGA, a design suite must be used for programming. The design suite that will be used should be best suited for the ZYNQ 7000 series. These third-party softwares are Vivado HLS and Vivado made by Xilinx [10]. The Vivado HLS allows users to program the software that will be integrated into the chosen FPGA. The application has built-in libraries that allow the codes to be automatically run at pipeline when available. Moreover, it uses built-in #PRAGMA directives to tell the device what the parameters for the top function are. After creating the code, users must run the synthesis to make sure that the code is working properly so that it can be exported afterwards. Once synthesis is complete and no errors are found, users can proceed to exporting the RTL to be able to create the custom block diagram. Vivado is used for the creation and customization of the system block diagram. Here, the exported RTL from the Vivado HLS can be used as one of the blocks that will make up the system. The said block will contain the code for the program. The rest of the blocks that will make up the entirety of the system block diagram will be chosen by the user from the list of available parts in the application. Once the required components are added, wiring up the system can be easily done since Vivado has a built-in smart connector that wires the system together correctly. This will allow the data flow from the PS to PL. Once the creation of the custom system block diagram is complete, users can generate bitstream which will create the necessary. The three files that include the bit, .tcl, and .hwh files will be used in order to integrate it to the FPGA. The final step is to place the files generated by Vivado in the Jupyter Notebook interface of the PYNQ-Z2. Here, users will be able to create a notebook that will allow them to initialize the overlay to the system block diagram and pass or receive the data to and from the PL. Note that the Jupyter Notebook is run in the PS of the FPGA. Figure 1 shows a block diagram of the data flow from Vivado HLS to the results.

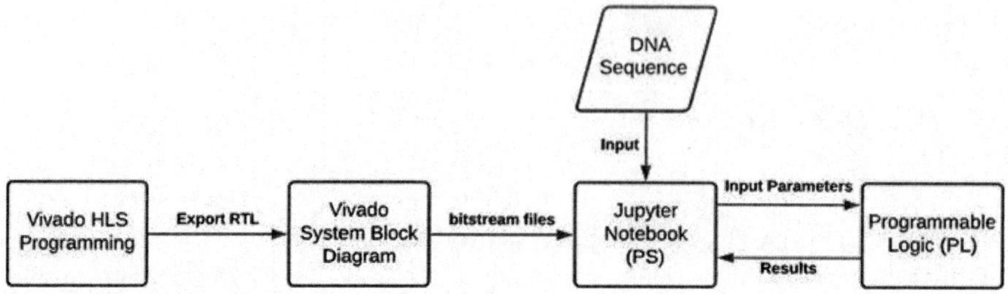

Fig. 1 Data flow block diagram

3 Methodology

The PS of the FPGA will be the one handling the data input of the users, passing the data to the PL, filtering the returned results, displaying the results, and computing the execution time. The only task that PL will be doing is the two conversion processes, which will be done in a SIMD paradigm using the pipeline and unroll functions.

3.1 DNA Sequence to Amino Acid Sequence Conversion

The program for the DNA sequence to amino acid sequence conversion takes in a DNA sequence and converts it into its corresponding amino acid. This is done by taking three characters or codons of the DNA and finding the right conversion in the amino acid hashtable. Ideally, a hashtable or dictionary would be used to speed up the conversion process in C or Python. Unfortunately, Vivado HLS has troubles implementing hash table due to its dynamic nature in memory allocation. Vivado HLS does not allow the use of dynamic memory allocation since it is not sure how the hardware device will react to it. Because of this, an if-else statement will be used instead to make the conversion possible. The execution speed will be compensated with the use of pipelining in Vivado HLS. As an example, Table 1 shows the DNA sequence on top and its corresponding amino acid conversion at the bottom. The conversion process starts at index 1 which will get the codon TGT and give it the amino acid C. This process will iterate until it reaches index N-1 where N is the length of the DNA sequence. This will give the codon AAG the amino acid K as the last codon conversion for this example. Each codon combination will have a corresponding amino acid for this conversion as shown in Table 2.

Table 1 Conversion of codon to amino acid

T	G	T	G	A	T	A	A	G
	C	V	X	D	I	X	K	

Table 2 DNA codons and corresponding amino acid

DNA codon	Amino acid
GTC, GCC, GCA, GCG	A
TGT, TGC	C
GAT, GAC	D
GAA, GAG	E
TTT, TTC	F
GGT, GGC, GGA, GGG	G
CAT, CAC	H
ATA, ATT, ATC	I
AAA, AAG	K
TTA, TTG, CTT, CTC, CTA, CTG	L
ATG	M
AAT, AAC	N
CCT, CCC, CCA, CCG	P
CAA, CAG	Q
CGT, CGC, CGA, CGG, AGA, AGG	R
TCT, TCC, TCA, TCG, AGT, AGC	S
ACT, ACC, ACA, ACG	T
GTT, GTC, GTA, GTG	V
TGG	W
TAT, TAC	Y
TAA, TAG, TGA	X

3.2 Amino Acid Sequence to Blosum62 Index Conversion

As for the amino acid sequence to Blosum62 index conversion, the input would be from the amino acid sequence from the previous conversion. The amino acid will be indexed as a scoring matrix for the sequence alignment algorithm. Again, the scoring matrix that will be used is the Blosum62. With Table 3 as an example, the previously converted amino acid sequence is on top and its corresponding Blosum62 index is at the bottom. Every amino acid in the sequence will be converted into a scoring index and stored in a substitution matrix. Just like the DNA to amino acid conversion, the process is usually to have a hash table for comparison, but again, Vivado HLS will not allow it. The corresponding conversion is shown in Table 4.

Table 3 Conversion of amino acid to Blosum62 index

C	V	X	D	I	X	K
4	19	22	3	9	22	11

Table 4 Amino acid and corresponding Blosum62 index

Amino acid	Blosum62 index
A	0
R	1
N	2
D	3
C	4
Q	5
E	6
G	7
H	8
I	9
L	10
K	11
M	12
F	13
P	14
S	15
T	16
W	17
Y	18
V	19
V	20
B	21
Z	22
X	23

4 Results and Discussion

4.1 Execution Time and Speedup Test

The conversion process was done in both the PS and PL of PYNQ-Z2 to showcase the difference in speed for better understanding of the device. Based on Table 5 which shows the comparison of the runtime of PS and PL, the PS runs take a significantly longer time to execute than the PL runs. Moreover, the execution time of PL is consistently at around 160 μs. A T-test is done to show the significance of the runtime data of the different lengths of input. For this data, a one-tailed T-test is done with a confidence interval of 95%. The null hypothesis is that there is no significant difference between the two data provided; alternative hypothesis is that there is significant difference between the two data provided. The rule is to reject the null hypothesis if p-value is less than or equal to alpha. In this case, the p-value is 0.04

Table 5 Average execution time (in μs) of DNA sequence to amino acid sequence conversion in PS and PL

Length	PS time	PL time
128	1096.75	162.50
256	2233.59	162.20
512	4700.21	160.62
1024	8932.52	159.73
2048	17,513.53	160.85
4096	35,171.03	163.76
8192	70,346.99	159.46
16,384	141,074.16	162.03
32,768	282,012.81	160.89
65,536	572,773.15	158.78
131,072	1159,415.05	160.39
250,000	2233,892.15	160.25

and alpha is 0.05. Therefore, the null hypothesis is rejected and there is significant difference in the data.

Referring to the graph in Fig. 2, the blue dots represent the PS time, and the orange dots represent PL time. This shows that the time it takes to run the program in PS is linearly proportional to the length of the DNA sequence.

The graph in Fig. 3 shows that the speedup in the runtime from PS to PL has an exponential relation to the length of the DNA sequence; meaning that the longer the length is, the higher the speedup will be.

The same can be said for the results of the amino acid to Blosum62 index conversion in a sense that the comparison of the PS and PL runs are also linear with respect to the length of the amino acid sequence. The speedup of the said conversion process is also the same with the DNA sequence to amino acid sequence since the relationship is also exponential with regards to speed up and length. The following table and figures will show the execution time, graph comparison, and speedup test for

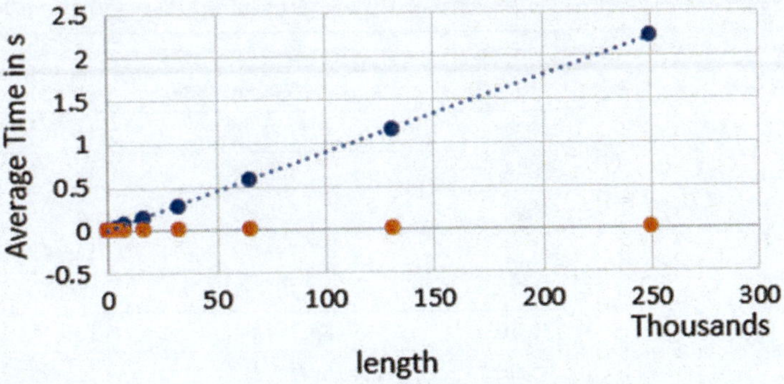

Fig. 2 Graph of DNA sequence to amino acid sequence conversion runtime

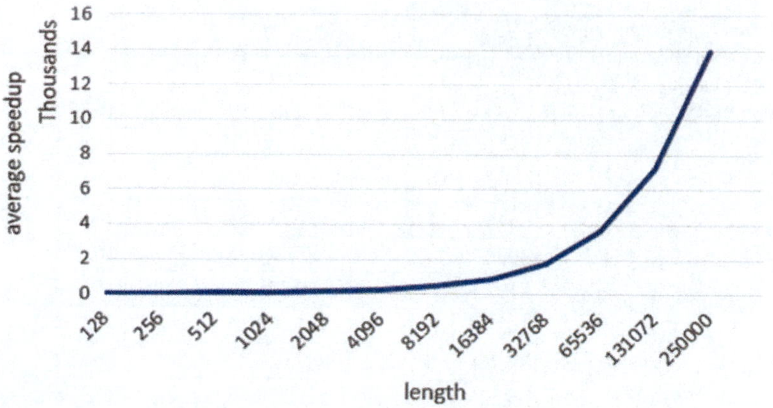

Fig. 3 DNA sequence to amino acid sequence speedup

the amino acid to Blosum62 index conversion. Just like the previous conversion, the same type of T-test was done to see if the data is significant. The same metrics are used, and the p-value calculated is 0.04 which means the null hypothesis is rejected which means there is significant difference in the data.

Table 6 shows that the PS results are much slower than the PL since PL can run more conversions at the same time. Based on observations, the PL runs are, again, consistent at around 158 μs regardless of the length. Figure 4 will be provided for visualization of the speed.

Finally, for the scalability test, the same comparison of PS and PL will be done to see how much of a speedup is present in PL from PS. Like the DNA sequence to amino acid sequence conversion, the speedup graph shows that the relationship of length to speedup is exponential (Fig. 5).

Table 6 Average execution time (in μs) of amino acid sequence to Blosum62 index conversion in PS and PL

Length	PS time	PL time
100	277.93	155.38
128	352.36	155.73
256	699.16	155.37
512	1384.86	160.51
1024	2739.70	156.14
2048	5466.54	155.70
4096	10,826.75	158.05
8192	21,483.72	156.43
16,384	42,902.62	160.49
32,768	85,787.84	156.45
65,000	169,969.97	155.98

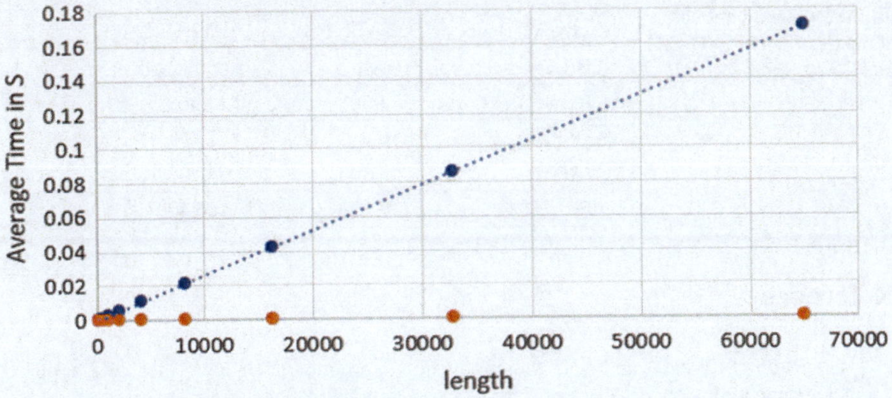

Fig. 4 Graph of amino acid sequence to Blosum62 index conversion runtime

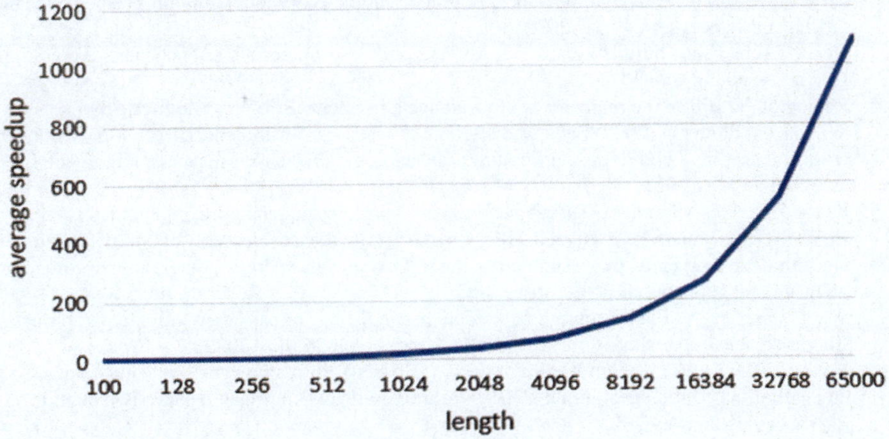

Fig. 5 Amino acid sequence to Blosum62 index speedup

5 Conclusion

In conclusion, the PYNQ-Z2 FPGA is seen to be a device that can process and compute information at a high speed. This is especially seen in the speedup test where it shows that the more input data there is, the better and more efficient the runs will be. Future research can involve using FPGA to speed up the frame alignment algorithm [8]. But, the PYNQ-Z2 has limited resources which is not suitable for the alignment algorithm due to its resource demand. Alternate FPGAs such as the GENESYS2 will more likely have the power and resources to run the alignment algorithm. Table 7 shows the comparison of the PYNQ-Z2 and the GENESYS2 FPGA [11, 12].

Table 7 Specifications comparison for PYNQ-Z2 and GENESYS2

FPGA	GENESYS2	PYNQ-Z2
BRAM	16 Mb	630 KB or ~ 5 Mb
Logic slices	50,950	13,300
DSP slices	840	220
LUT	203,800	53,200
FF	407,600	106,400

References

1. Keim R (2018) What is an Fpga? An introduction to programmable logic—technical articles. All about circuits
2. What is FPGA? FPGA Basics, Applications and Uses: Arrow.com. https://www.arrow.com/en/research-and-events/articles/fpga-basics-architecture-applications-and-uses. Last accessed 26 May 2022
3. Brown T (2022) Transcriptomes and proteomes. https://www.ncbi.nlm.nih.gov/books/NBK21121/. Last accessed 24 May 2022
4. Gene Expression. https://www.sciencedirect.com/topics/materials-science/gene-expression. Last accessed 24 May 2022
5. Batzoglou S (2005) The many faces of sequence alignment. Briefings Bioinform 1(6)
6. Sequence Alignment (2022) Sequence alignment—an overview|ScienceDirect Topics. https://www.sciencedirect.com/topics/agricultural-and-biological-sciences/sequence-alignment. Last accessed 25 May 2022
7. Zhang Z (1997) Aligning a DNA sequence with a protein sequence. J Computat Biol 4(3)
8. El-Wafa WA, Seliem AG, Hamed HFA (2016) Hardware acceleration of Smith-Waterman algorithm for short read DNA alignment using FPGA. In: Proceedings—international computer software and applications conference, vol 2
9. What is an FPGA? Field Programmable gate array Xilinx. https://www.xilinx.com/products/silicon-devices/fpga/what-is-an-fpga.html. Last accessed 26 May 2022
10. Vivado 2019.1 (2022) High-level synthesis (C based). https://www.xilinx.com/support/documentation-navigation/design-hubs/2019-1/dh0012-vivado-high-level-synthesis-hub.html. Last accessed 26 May 2022
11. Genesys 2 kintex-7 FPGA Development Board, Diligent. https://digilent.com/shop/genesys-2-kintex-7-fpga-development-board/. Last accessed 26 May 2022
12. TUL (2022) https://www.tulembedded.com/FPGA/ProductsPYNQ-Z2.html. Last accessed 26 May 2022

Identification and Analysis of Maize Leaf Diseases and Insect Pests Based on Machine Learning

Xu Ming, Siti Arpah Ahmad, and Shafaf Ibrahim

Abstract Maize is the most productive food crop in the world, and its leaves are extremely susceptible to disease. However, the yearly crop yields are reduced by the disease and its interference to corn leaf diseases. Therefore, finding a way to identify corn leaf diseases more quickly and accurately is a current concerning matter in agriculture. Traditional machine learning extracts the crop disease features such as texture, color and visual as the basis for machine learning model classification. Yet, many disease images are needed to train the machine learning algorithms. It is difficult to collect disease samples evenly and meticulously during production as well. With the progress of smart agriculture, in-depth learning, and the introduction of agriculture, especially in the field of crop disease diagnosis, CNN in-depth learning technology has made varying degrees of progress in identifying crop diseases and pests. The experimental results of many scholars have verified that various CNN models are very effective in identifying leaf diseases. These improved models can better extract the characteristics of the target and improve the accuracy of identifying corn leaf diseases.

Keywords Machine learning · Maize leaf · Pest identification

X. Ming (✉) · S. A. Ahmad
Faculty of Computer and Mathematical Sciences, Universiti Teknologi MARA, 40450 Shah, Malaysia
e-mail: okthere@126.com

S. A. Ahmad
e-mail: arpah@tmsk.uitm.edu.my

S. Ibrahim
Faculty of Computer and Mathematical Sciences, Universiti Teknologi MARA Cawangan Melaka Kampus Jasin, 77400 Merlimaun, Malaysia
e-mail: shafaf2429@uitm.edu.my

© The Author(s), under exclusive license to Springer Nature Singapore Pte Ltd. 2023
D.-K. Kang et al. (eds.), *Proceedings of the 9th International Conference on Computational Science and Technology*, Lecture Notes in Electrical Engineering 983,
https://doi.org/10.1007/978-981-19-8406-8_24

1 Introduction

Corn is the most productive food crop in the world, and the crop with the greatest yield increasing potential among cereal crops. It has many uses, such as food, feed, industrial raw materials, crops, etc. Improving the yield and quality of corn is of great significance to solving the world food problem. Malaysia imports a wide range of corn from around the world every year for breeding.

Corn is not only an important raw material for food processing and feed processing, but also the highest yielding crop among world-class cash crops and cereal crops. The yield of some experimental quality corn in China has reached 1500 kg per mu. Corn is used in various fields, such as the feed industry, poultry industry, and the beverage industry. It has become a major food source in many parts of the world, and its total output has exceeded that of wheat. However, due to the influence of microorganisms, maize planting methods and different regions, the types of maize leaf diseases are also increasing accordingly. The diseases in maize leaves are one of the important reasons that limit the development of maize yield [1].

In the growth cycle of maize, various diseases may be encountered every year in the process of planting it. At the same time, the problem is that the leaves of maize are extremely susceptible to diseases. Due to the interference of corn leaf diseases, the yield and quality of crops in the current year will be reduced to varying degrees, which will cause huge economic losses to the farmers planting corn. In recent years, the occurrence of corn leaf diseases such as corn rust, corn leaf spot, corn leaf blight, and corn borer disease has become more and more widespread. The annual yield of corn will continue to decline due to the occurrence of these diseases. When the disease is serious, it will affect the quality of corn, and the degree of damage is also increasing.

Corn diseases tend to increase year by year. The impact of diseases has seriously restricted the yield and quality of corn. Generally, the yield of corn can be reduced by about 20%, and in serious cases, it can lead to a 50% reduction. Identifying different types of corn leaf diseases is the basis for crop disease prevention and control. Only when these diseases are found in time can farmers who grow crops prevent and control them in time. Corn leaf diseases mainly include the disease images of corn mosaic, gray leaf spot, rust, and leaf spot. A problem to be solved in agriculture is how to identify maize leaf diseases more quickly and accurately. Currently, most of the work is focusing on certain type of corn disease such as rust and leaf spot only. Maize leaf disease are many, thus many future investigations can be done.

2 Application of Traditional Machine Learning

Machine learning has been widely used in agriculture in recent years. Because the symptoms of infected crops mainly appear in the visual characteristics of crop leaves, computer vision is an effective and fast method used to identify different crop leaf diseases. Establishing a mathematical model, obtaining different levels of information related to different image features, and usage of different classifiers are needed to identify different types of diseases and pests [2]. Experts are required to segment the disease spots and extract some features of the disease spots. The extracted disease features are used to train the machine learning model. These crop disease features extracted manually are used as the basis for machine learning model classification. The crop disease features often used for training mainly include texture features [3, 4], color features [5], visual features [6]. Feature construction methods [7, 8] can also be used further to construct more relevant features to improve the data modelling in identifying some different types of crop diseases.

Researchers studying machine learning technology have also tried various methods to identify some different types of crop diseases. The classical machine learning methods for crop disease identification mainly include support vector machine [3, 4], K-means clustering [9], simple linear iterative clustering [10], distance measurement [11]. Syarief et al. [12] used fuzzy least square vector machine (FLSVM) to classify maize leaf disease. The grid method is used to locate feature points, the speed up robot feature (surf) is used for feature extraction, the K-means is used to cluster the visual vocabulary of words, and the support vector machine (SVM) is used for classification. Classification data is the corn leaf image from plantvillage dataset.

Chodey et al. [13] introduced practical digital image processing into neural network technology. The moving frames of video are preprocessed, and the object motion is tracked by K-means clustering. From the segmented image, a new feature evaluation called distributed intensity-based LBP feature (DI-LBP) is extracted, as well as edge and color. In addition, the features are classified using an optimized neural network (NN).

These research methods require manual extraction and selection of crop disease features in the identification of maize leaf diseases. However, the difference between the color and texture of different maize leaf diseases is small. The manual extraction of features is subjective and exploratory, and the preprocessing of maize leaf images is complex. Thus, the result of feature extraction has a serious impact on the recognition accuracy of traditional schemes for corn leaf diseases to a certain extent. The low-level features extracted from the diseased corn leaf images have limited ability to describe the low-level features. The performance of these methods cannot meet the needs of large-scale identification of corn field diseases. Therefore, in the subsequent research on the algorithm for corn leaf disease, we should study how to automatically obtain the image features of different types of corn leaf diseases to reduce the subjectivity of manual feature extraction and make the algorithm adapt to the recognition of corn leaf disease in different scenarios. The relationship between maize leaf disease and pets is lacking. Identifying both factors in early stage could contribute to better corn production.

3 Application of Deep Learning in Crop Disease Identification

With the development of GPU integrated intelligent hardware, agricultural intelligent applications related to automatic learning are growing day by day, leading to the progress of new methods and models [14, 15]. When using conventional machine learning approaches for agricultural disease picture recognition, the feature parameters needed to build the model must be manually designed, which is heavily dependent on the prior knowledge of specialists. This prior function design is unnecessary when utilizing a deep learning framework based on convolutional neural networks (CNN) [16]. The larger the training sample data is, the more accurate the recognition result is.

With the progress of smart agriculture, deep learning has become a method to realize task transfer. The introduction of in-depth learning into agriculture, especially in the field of crop disease diagnosis, has only begun in recent years, but CNN in-depth learning technology has made varying degrees of progress in identifying crop diseases and pests [17, 18]. Hu et al. [19] An optimization model obtained by fine tuning the GoogLeNet pre-training network, adjusting the optimizer, learning rate, and other parameters is proposed. The average recognition accuracy of corn diseases of corn leaf rust, corn rust, corn North leaf blight, and healthy leaves reaches 97.6%, and the highest accuracy is improved by 5.9% compared with the original GoogLeNet model. Lv et al. [20] designed the convolution neural network of DMS robust to identify the disease of corn leaves. Firstly, a maize leaf feature enhancement framework with the ability to enhance maize features in complex environments was designed. On this basis, extended convolution and multi-scale convolution layers are combined to improve the feature extraction ability of the network. Through batch normalization to avoid over adaptation of the network, the robustness of the model is enhanced, and the expected results are shown in the classification task of crop diseases.

Xu et al. [21] improved the accuracy of maize disease identification by adding a multi-scale feature extraction module to a network and combining global pooling. Firstly, based on the Alexnet model, multi-layer convolution operation is added, and then the migration learning method is used to solve the over-fitting problem caused by insufficient sample data. This improved model can better extract the characteristics of the target and improve the accuracy of identifying corn leaf diseases. Bansal et al. [22] used Bayesian optimization technology to fine tune the super parameters of the DCNN model. The model is trained and tested on the open data set of the maize disease sub data set. The accuracy of the DCNN model trained by optimized parameters is 96.05%. The experimental results verify that the proposed neural network based on the recognition model is very effective in identifying leaf diseases.

Sm et al. [23] used intelligent equipment to provide automatic identification and diagnosis of corn to effectively improve the yield of corn diseases and reduce the impact of diseases on corn output. The performance of the neural network is improved by changing the parameters and pooling on the GPU system.

Mazzia et al. [24] and other scholars put forward a lc & cc deep learning model, which combines the cyclic neural network (RNN) and convolutional neural network (CNN) to reduce the manual feature stage modeling. The author also tested and compared other widely used traditional machine learning algorithms. The overall accuracy of the proposed R-CNN method is 96.5%.

Sun et al. [25] applied the transfer learning technology to the problem of corn leaf disease identification and used the CNN model with an open database for disease identification and diagnosis. A new full connection layer is designed based on the existing network. Then, the network sends the weights trained on the Imagenet image data set to the new model, and the final recognition accuracy of the network is 95.33%.

4 Comparison and Analysis

Machine learning can learn useful features from many plants disease images to improve recognition accuracy [26]. However, most deep learning network models have very deep network structures [27]. These models cannot accurately identify crop diseases with limited data because they require many parameters and data to calculate. However, field experts focusing on corn leaf disease are still in the minority. Many aspects, such as extraction of disease characteristics in changing environments, uneven illumination, reflection of incident light sources, etc., will affect the diagnosis of maize leaf diseases. It is worthwhile to take on more problems in the identification of maize leaf diseases. The research of maize leaf disease identification will be worthy of further challenges. Because the actual shooting environment of maize leaf disease is often very complex and the characteristics of maize leaf disease are diverse, the structure of different networks may affect the recognition rate of the algorithm.

5 Conclusion

Numerous disease image samples are needed to train machine learning algorithms. However, in practice, it is extremely difficult to collect disease samples in a precise and uniform manner. Traditional machine learning methods was found to be difficult in studying the identification and analysis of corn disease. The segmentation is mainly based on the threshold of corn disease spots, and the recognition is carried out with a convolution neural network model. Most scholars use GPUs to find the best parameters and speed up the development of deep learning models. The colour and texture of the sample image are also critical for training and recognition purposes. In the complex field environment, illumination affects the colour of a sample image, which in turn affects the extraction of features. Thus, the current machine learning algorithm should be improved in future research to study image data collected, particularly in the field environment and the uneven number of samples.

References

1. Priyadharshini RA, Arivazhagan S, Arun M, Mirnalini A (2019) Maize leaf disease classification using deep convolutional neural networks. Neural Comput Appl 31(12)
2. Wu Y, Wu J, Hu G (2020) MMFS: a grape disease recognition method based on multi-feature fusion and SVM. In: ICCBDC'20: 2020 4th international conference on cloud and big data computing
3. Gd A, Vk B, Hdj B (2019) A novel computer vision based neutrosophic approach for leaf disease identification and classification. Measurement 135:782–794
4. Pantazi XE, Moshou D, Tamouridou AA (2019) Automated leaf disease detection in different crop species through image features analysis and one class classifiers. Comput Electron Agric 156:96–104
5. Wu Y (2021) Identification of maize leaf diseases based on convolutional neural network. J Phys Conf Ser 1748(3):032004
6. Yigit E, Sabanci K, Toktas A et al (2019) A study on the visual features of leaves in plant identification using artificial intelligence techniques. Comput Electron Agric 156(1):369–377
7. Alfred R (2008) DARA: Data summarisation with feature construction. In: Proceedings—2nd Asia international conference on modelling and simulation, AMS 2008, art. no. 4530583, pp 830–835. https://doi.org/10.1109/AMS.2008.131
8. Alfred R (2010) Feature transformation: a genetic-based feature construction method for data summarization. Comput Intell 26(3):337–357. https://doi.org/10.1111/j.1467-8640.2010.00362.x
9. Hameed JS, BB Stünda (2020) Analysis of using k-means clustering with convolutional neural network architectures for automatic plant leaf disease recognition. Talent Develop Excellence 12(2s):2250–2264
10. Sadeghi-Te Hran P, Virlet N, Ampe EM, Reyns P, Hawkesford MJ (2019) Deepcount: in-field automatic quantification of wheat spikes using simple linear iterative clustering and deep convolutional neural networks. Front Plant Sci 10:1176–1179
11. Dhingra G, Kumar V, Joshi HD (2019) Basil leaves disease classification and identification by incorporating survival of fittest approach. Chemom Intell Lab Syst
12. Syarief M, Prastiti N, Setiawan W (2019) Maize leaf disease image classification using bag of features. J Infotel 11(2):48
13. Chodey MD, Shariff CN (2021) Neural network-based pest detection with k-means segmentation: impact of improved dragonfly algorithm. J Inform Knowl Manag 20(3):2150040
14. Li Y, Wang H, Dang LM, Sadeghi-Niaraki A, Moon H (2020) Crop pest recognition in natural scenes using convolutional neural networks. Comput Electron Agric 169:105174
15. Wang J, Li Y, Feng H, Ren L, Wu J (2020) Common pests image recognition based on deep convolutional neural network. Comput Electron Agric 179(1):105834
16. Sz A, Sz B, Cz C, Xw A, Yun SA (2019) Cucumber leaf disease identification with a global pooling dilated convolutional neural network. Comput Electron Agric 162:422–430
17. Geetharamani G, Pandian A (2019) Identification of plant leaf diseases using a nine-layer deep convolutional neural network. Comput Electr Eng 76:323–338
18. Zeng W, Li M (2020) Crop leaf disease recognition based on self-attention convolutional neural network. Comput Electron Agric 172:105341
19. Hu R, Zhang S, Wang P, Xu G, Qian Y (2020) The identification of corn leaf diseases is based on transfer learning and data augmentation. In: CSSE 2020: 2020 3rd international conference on computer science and software engineering
20. Lv M, Zhou G, He M, Chen A, Hu Y (2020) Maize leaf disease identification based on feature enhancement and dms-robust alexnet. IEEE Access 99:1–1
21. Xu Y, Zhao B, Zhai Y, Chen Q, Zhou Y (2021) The maize disease identification method is based on a multi-scale convolutional global pooling neural network. IEEE Access 99:1–1
22. Bansal S, Kumar A (2021) Automatic deep neural network hyper-parameter optimization for maize disease detection. IOP Conf Series Mater Sci Eng 1022(1):012089 (9pp)

23. Sm A, Rs A, Dr A (2020) Deep convolutional neural network-based detection system for real-time corn plant disease recognition. Proc Comput Sci 167:2003–2010
24. Mazzia V, Khaliq A, Chiaberge M (2020) Improvement in land cover and crop classification based on temporal features learning from sentinel-2 data using recurrent-convolutional neural network (r-cnn)
25. Sun X, Wei J (2020) Identification of maize diseases based on transfer learning. J Phys Conf Series 1437:012080
26. Quan L, Feng H, Lv Y, Wang Q, Yuan Z (2019) Maize seedling detection under different growth stages and complex field environments based on an improved faster r–cnn. Biosys Eng 184(1):1–23
27. Mj A, Lei ZB, Qw C (2020) Automatic grape leaf diseases identification via unitedmodel based on multiple convolutional neural networks. Inf Process Agric 7(3):418–426

Information Technology Model for Health Sector in Oman

Louay Al Nuaimy and Younis AL-Mawali

Abstract So far, information technology development takes the boom part which has not been widely absorbed in developing countries. Health care systems may be transformed by information systems. The main drivers of globalization are this type of system. Information technology development is the degree to which it is easier to bring information into organizational exercises. This reflection relates to knowing the level and sophistication of information within the health care sector in the Sultanate of Oman and deducing the emergence of information technology for the prevailing IT brands used in Oman (computers, flexible phones, internet technology) and consequently their effects on health care in the system of transfer of care in educational clinics in the Sultanate of Oman. Oman This reflection relates to knowing the level and development of information in the health care sector in the Sultanate of Oman and deducing the emergence of information technology for the dominant IT brands used in Oman (computers, flexible phones, internet technology) and thus their correctness. Effects of care in the care transfer system in teaching clinics in the Sultanate of Oman. In addition, data were collected on the basis of studies and oral interviews. The results of the investigation were like that revealed of among similar studies in health care sectors of developing countries.

Keywords Information technology · Health sector · Organizational exercises

1 Introduction

The last decade of the twentieth century saw the rapid and dynamic implementation of information technology, transforming the world into a global village. As a result of this evolving trend, countries have experienced unprecedented productivity gains and

L. Al Nuaimy · Y. AL-Mawali (✉)
Oman College of Management and Technology, Halban, Sultanate of Oman
e-mail: 201912037@omancollege.edu.om

L. Al Nuaimy
e-mail: loay.alneimy@omancollege.edu.om

© The Author(s), under exclusive license to Springer Nature Singapore Pte Ltd. 2023
D.-K. Kang et al. (eds.), *Proceedings of the 9th International Conference on Computational Science and Technology*, Lecture Notes in Electrical Engineering 983,
https://doi.org/10.1007/978-981-19-8406-8_25

continued economic growth and development. But this commendable development was not easy. In fact, it has focused only on the technologically superior countries, and this has held back the creation of IT in growing countries like Nigeria. This is particularly evident in health in Oman, where the use of computers, cell phones and the Internet to force character and collective efforts to improve the productivity of the efforts of health professionals is not always recognized. As a result, many vital processes that need to be carried out slow down to a very rapid rate, sometimes ending in death [1]. Due to the lack of net supplies, many health care experts are unable to improve their skills and keep pace with new features in global health care, and it is well known that the economic power of a country now does not depend on natural resources. but not enough staff. Productivity can be hampered when people don't have access to world-class healthcare. Statistical technologies were identified as a real device to improve the excellence of health structures and the competence of health specialists in each developed and growing international location. The above problem requires finding ways to improve the performance and high quality of fitness structures in developing countries, as this could also allow the fitness neighborhood to evolve with population growth. Information technologies have changed the face of the scenario in which we find ourselves. Data generation enables direct communication between people around the world, quick access to international libraries, record assets, and many different opportunities [2]. Information technologies can also improve health systems. So, it can be described as one of the main drivers of globalization. Fact technologies can also enhance physical structures. Therefore, it can be described as one of the essential engines of globalization. The age of statistics includes the manufacture of communication technologies ranging from radio, film, television, printing, and the cell phone to offerings throughout information technology includes the manufacture of communication technologies ranging from radio, film, television, printing, and the telephone to services such as the Internet, cell phones, and digital video, and this research is limited to determining the extent to which technology can or can reach the statistics. suitable grade. It was integrated into the fitness zone of Oman, the use of teaching hospitals in the universities of Oman as a case study [3].

2 Modern Technology

Time period recording includes any tool, networked machine, or device subsystem that is used to obtain, store, system, control, transmit, manipulate, display, alter, broadcast, or transmit statistics or facts to obtain of. This includes computers, add-on systems, software programs, firmware, procedures, and similar offerings (along with support services), and related resources [4]. There may not be a universally general definition of ICT, as ICT methods, standards and packages evolve almost daily. however, ICT can be considered a document processing technology. Furthermore, ICT includes all products that store, retrieve, process, transmit or receive data electronically in virtual form. The rise of the age of statistics is gaining momentum across all lifestyles and more and more international places like Oman are taking advantage of this opportunity in various sectors including health device improvement [5]. The introduction of registration and communication technologies in Oman's training hospitals is the result of applied health work through the Sultan Qaboos University Computing Center, Sohar University and Nizwa University Teaching Medical Institution within the discipline of modern technology [6].

IT leakage is defined as the volume at which various IT teams are brought into organizational sports. More primarily, IT leakage refers to the frequency of use of generation, the full use of utility capabilities, the level of integrated and complementary use of various technologies, and the use of generation for organizational functions compared to the standard techniques of facts and verbal exchange. (Digital equipment offers many new opportunities [7]. Spurred on by many authors, Huber notes that the conversational capabilities of advanced record technology allow people to: (a) communicate easily and cheaply across time and location geographic image, (b) speak faster and more appropriately to audiences, (c) reliably and cost-effectively collect and index the content and nature of release activities, and (d) gain admission and engagement selectively in a network or communications management event similarly, the author states that data generation can aid decision making because it enables people to: maintain and retrieve large amounts of data more quickly, access it selectively, and acquire data and reconfigurations faster and better as it should be to generate new statistics and also be more reliable and less expensive to register and r retrieve information [8]. The nature and nature of organizational transactions, as well as trends in mobile technology in recent years, have enabled executives to use and access those electronic media almost anytime, anywhere, and to transmit statistics in almost any form (Such as text, audio, and photography). Before long, data generation will increase the richness, complexity, and mobility of experience and statistics due to increased speed of verbal exchange, data availability, bandwidth, connectivity, remote access, and portable storage. Facts are a major asset in the ultra-modern marketplace as they enable better decision making and problem solving. Similarly, theories and empirical studies suggest that the era of records will increase according to personal effectiveness, improve center control in decision making, improve verbal exchange techniques and, consequently, the work done. Infusion has been defined within the literature as the technique of deep and extensive incorporation of IT packages into the work system of a man or a woman or a company [9]. The amount to which general innovativeness is integrated into the work or control structures of the company and the intensity of this integration is very great. Cooper and zoomed suggest that infusion is the final level of a six-grade sequential version of an IT application. The first stage, initiation, is the system of exploring organizational problems and possibilities and implementing IT solutions and is related to the solution level. The next two stages, adoption, and adaptation, represent Lewin's section of exchange. Accreditation method that receives organizational support for IT implementation, while customization consists of its development, installation, and protection [10]. The remaining three levels, Fixation, Routine, and Infusion can be set to the Leen Refreeze stage. Cooper and Zoom describe the adoption section as an attempt to get a company's contributors to commit to using the technology, and the routine represents the changes that occur within the business machine of a laptop account so that it doesn't looks like new. Subsequently, leakage occurs as statistical technology becomes generalized within the work structures of a commercial firm, groups, or individuals. There is some empirical evidence to suggest the possibility of leakage, that is essential. Cooper and zoomed studied the integration of MRP structures in a control assignment in 62 manufacturing companies. Low disapproval rates (for example, the use of only a few preliminary MRP functions) have been observed to depend on the health (or coincidence) between interest and era. However, the best levels of IT (i.e., the use of many MRP capabilities) may also depend more on personal interest and political behavior than on the interaction of functions and the right technology. Its effects imply that regulatory modifications are required even for low stages of MRP infusion. However, those changes are especially positioned within the manufacturing and warehouse departments. Alternatively, larger infusion volumes require modifications involving the entire age [11].

3 Injecting Information Technology into the Health Sector

Adoption of data technology in the updated volume of the health area until the date on which the era of statistics within the health sector is incorporated or universalized. The mix of fact generation in health care plays the following roles: Provision of consultation, analysis, and telemedicine through telemedicine. They are increasingly used in growing countries, providing up-to-date patient records and experts, who then offer recommendations on appropriate remedies [12]. Internet communications and satellite TV for PCs are also used for collaboration and collaboration among physicians. It allows you to exchange knowledge on new treatment strategies, update each specific case, obtain up-to-date expert advice, and ensure the most effective remedy. It is not more effective for treatment, but also for health research and training. Aptitude studies are greatly facilitated through the established order of library networks and different information corporations, thus increasing the access of university students and up-to-date researchers and their transmission of information from the field up-to-date evaluation in remote research institutes and the exchange of ideas on similar up-to-date research areas among researchers from extraordinary institutes. Up-to-date ICT schooling mechanisms are increasingly being used to teach health workers. They have been proven to facilitate the transmission of facts could be very beneficial in spreading epidemic surveillance activities. The Internet can be used for epidemiological surveillance and typical planning of vaccination campaigns. This is actively achieved in many sub-Saharan African countries. The benefits of ICT in the field of health are numbers ooze many of the many blessings of statistical technology in the health field is the effective and efficient exchange of information on fitness and skills among health experts. Spread without difficulty the treatment of the disorder and manipulate the facts. Reduce the costs and inconvenience of transporting patients. Data on disease surveillance and intervention updated by hand all. HMIS database to make effective plans and modify policies. While IT brings many advantages, it

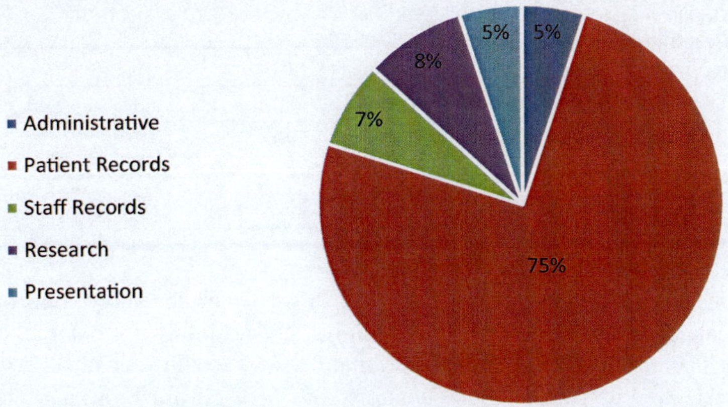

Fig. 1 Pie chart showing PC utilization in percentage

also brings challenges, and successfully recognize the upgraded benefits, you want to upgrade them. The main up-to-date healthcare IT challenges in Oman consist of poor IT skills among healthcare professionals and occasional literacy in Oman. Attitudes towards the use of information technologies (the fear factor) [13]. Lack of information technology density in health facilities. Low-level technicians and computer technicians. There is no standard format for record keeping. There are no standard labels.

4 Methodology

This study was conducted to arrive at a pump model for the fitness records era area within the Sultanate of Oman. As a team, questionnaires and private interviews have been used. This trial uses 4 selected training hospitals in Oman; University hospitals have been visited to decide the extent of reputation and the effect of fact generation indicators on healthcare structures sometime in the last 5 years. Oman University Hospitals is the product of a health project by the Computing Center of Sultan Qaboos University, Sohar University, Nizwa University Hospital for Modern Technology and Oman Health Intermediate School (Fig. 1; Table 1).

5 Results

The results showed that among all the university hospitals visited; The nearest university hospital in Sohar is connected to the Internet. Different university hospitals use the connection to the school network. Internet use the test showed that most health

Table 1 Mobile application in university hospitals

Year	Percentage of users (%)
2017	10
2018	80
2019	90
2020	95
2021	97

professionals use external Internet facilities. In particular, the study confirmed that approximately 33% of Internet users in university hospitals use external Internet facilities. The accumulated data showed that 0.9% of health specialists do not use the Internet, 71% use it for email functions, 38.7% use it for browsing, and 0.4% use it for clinical studies and electronic medicine. Parent 1 shows the net use among forensic doctors in Omani university hospitals. Mobile phones most health workers in Oman's university hospitals use mobile phones. The survey found that around 89.32% of forensic doctors use mobile phones and the best 0.6% do not. The survey also found that the use of mobile phones in Amman teaching hospitals has increased in recent years. Desk 1 suggests the trend of mobile phone use in educational hospitals. Respondents said that the use of cell phones significantly improved employee productivity through the following methods that helped communicate and disseminate vital statistics on occasion. He helped me with phone calls and consultations with my old and young colleagues, as well as with the sick. Help convey accurate information for rapid analysis and treatment of patients. More importantly, health workers outside the clinic can be reached on their cell phones in an emergency [14].

6 Discussions

The result of this study showed that statistical technology is not always fully adopted in Oman compared to other developed countries. While cell phones are owned by most healthcare workers, having access to the Internet and a computer is horrible. This may be related to the lack of funding within the clinic and insufficient facilities along with private workplaces for forensic doctors. In this study, we not only analyzed the acceptability or degree of combination of the three major (or unusual) indicators of information technology in the healthcare district in Oman, but we were also able to extend mathematical models of IT transformation in the Oman health district. However, these fashions lose their validity as saturation tactics. Large deviations from the above derived mathematical gradients are expected to appear once more than 65% of the highest leakage stage is reached.

7 Conclusion

After a detailed examination of the implications of this, it can be concluded that data and conversation technology contribute to the productivity efforts of medical experts, in addition to the prosperity of fitness equipment in training clinics within the Sultanate of Oman. Colleague. In the end, the contribution of the authorities and control of the clinic was requested in the purchase of more ICT centers for hospitals. The hospital administration also periodically organizes training courses for its employees. This will be an important contribution to worker productivity and the growth of the health system.

References

1. Echendu AJ, Peter Claver CO (2021) Smart city technology: a potential solution to Africa's growing population and rapid urbanization? Develop Stud Res 8(1):82–93. https://doi.org/10.1080/21665095.2021.1894963
2. Al-Shorbaji N (2000) Health Informations and telematics with reference to the work of WHO/EMRO. World Health Organization, Alexandria, Egypt. www.who.sci.eq
3. Srivastava A (2005) Accelerating delivery of health services. UN World Food Programme, New Delhi
4. Wager KA et al (2017) Health care information systems: a practical approach for health care management. John Wiley & Sons, Incorporated. ProQuest Ebook Central. https://www.proquest.com/legacydocview/EBC/4815068?accountid=195692
5. Tran DM, Thwaites CL, Jennifer Ilo VN, McKnight J, An PL, Paton C, Phil Vietnam ICU Translational Applications Laboratory (VITAL) 5 (2022) Digital health policy and programs for hospital care in Vietnam: scoping review. J Med Internet Res.https://doi.org/10.2196/32392
6. Request for information (RFI) (2006) Improving health and accelerating personalized health care through health information technology and genomic information in population- and community-based health care delivery systems. Washington: Federal Information & News Dispatch, LLC. Retrieved from Research Library https://www.proquest.com/reports/request-information-rfi-improving-health/docview/190177570/se-2?accountid=195692
7. Heponiemi T, Gluschkoff K, Vehko T, Kaihlanen A-M, Saranto K, Nissinen S et al (2021) Electronic health record implementations and insufficient training endanger nurses' well-being: cross-sectional survey study. J Med Internet Res. https://doi.org/10.2196/27096
8. Fine SL, Kane JC, Spiegel PB, Tol WA, Ventevogel P (2022) Ten years of tracking mental health in refugee primary health care settings: an updated analysis of data from UNHCR's health information system (2009–2018). BMC Med 20:1–16. https://doi.org/10.1186/s12916-022-02371-8
9. Ward BW, Sengupta M, DeFrances CJ, Lau DT (2021) COVID-19 pandemic impact on the national health care surveys. Am J Public Health 111(12):2141–2148.https://doi.org/10.2105/AJPH.2021.306514
10. Gingold DB, Liang Y, Stryckman B, Marcozzi D (2021) The effect of a mobile integrated health program on health care cost and utilization. Health Serv Res 56(6):1146–1155. https://doi.org/10.1111/1475-6773.13773
11. Martin CW, Savai S, Mokaya B, Mbugua S, Ribeka N, Cholli P, Yeung A (2021) mUzima mobile electronic health record (EHR) system: development and implementation at scale. J Med Internet Res. https://doi.org/10.2196/26381

12. Yeng PK, Nweke LO, Bian Y, Muhammad AF, Snekkenes EA (2021) Artificial intelligence-based framework for analyzing health care staff security practice: mapping review and simulation study. JMIR Med Inform 9(12). https://doi.org/10.2196/19250
13. Concetta Tania DI, Carinci F, Oderkirk J, Smith D, Siano M, de Marco DA et al (2021) Assessing data protection and governance in health information systems: a novel methodology of privacy and ethics impact and performance assessment (PEIPA). J Med Ethics 47(12):e23. https://doi.org/10.1136/medethics-2019-105948
14. Leijdesdorff S, Klaassen R, Wairata D-J, Rosema S, Therese vA, Popma A (2021) Barriers and facilitators on the pathway to mental health care among 12–25 year olds. Int J Qual Stud Health Well-Being 16(1). https://doi.org/10.1080/17482631.2021.1963110

A Simulation Approaches in Modelling the Assessment for Evacuation Model

Amir Haikal Abdul Halim, Khyrina Airin Fariza Abu Samah, Lala Septem Riza, and Raihah Aminuddin

Abstract In the event of an emergency evacuation, the total time taken to reach the main exit happens critical and important. The time taken reflected the barriers such as blockages and queuing process. Although the Uniform Building By-Law has mandatory declared that each building must follow the standard evacuation preparedness procedure, the building owner is unaware of the best evacuation model to be implemented. Besides, the owner has a problem with the suitability and high cost incurred if they directly implement the model. Therefore, this study proposes to model the simulation approaches in assessing evacuation models to identify the best model based on the building layout. Pathfinder, AnyLogic, and PedGo are the three-simulation software used in this study. The comparison has been made by conducting six simulation studies using 50, 100 and 150 occupants as the design parameter, with six attributes implemented in each simulation software. A high-rise 14th-floor plan was used as the case study to evaluate the actual scenarios of an evacuation. As a result, the Pathfinder evacuation model shows the lowest average time taken percentage, with the lowest mean value of 300.9983, ANOVA test with a significant value of 0.024, which is less than 0.05.

Keywords Assessment algorithm · Evacuation model · Simulation

A. H. A. Halim · K. A. F. A. Samah (✉) · R. Aminuddin
Faculty of Computer and Mathematical Sciences, Universiti Teknologi MARA Cawangan Melaka Kampus Jasin Melaka, Merlimau, Malaysia
e-mail: khyrina783@uitm.edu.my

A. H. A. Halim
e-mail: ameerhaical@gmail.com

R. Aminuddin
e-mail: raihah1@uitm.edu.my

L. S. Riza
Department of Computer Science Education, Universitas Pendidikan Indonesia, Bandung, Indonesia
e-mail: lala.s.riza@upi.edu

© The Author(s), under exclusive license to Springer Nature Singapore Pte Ltd. 2023

D.-K. Kang et al. (eds.), *Proceedings of the 9th International Conference on Computational Science and Technology*, Lecture Notes in Electrical Engineering 983,
https://doi.org/10.1007/978-981-19-8406-8_26

1 Introduction

The population of big and medium-sized cities worldwide is growing despite the constrained space in urban areas. High-rise structures have sprung up due to the need to handle this increase [1]. As a result, number of fire cases in the high-rise buildings have increased frequently during the past several decades and impacted major cities' skyline [2]. Rapid evacuation of large groups of people in high-rise buildings has become a worldwide issue and is being researched. Rapid evacuation has long been a critical yet challenging challenge because of people's various floors, heights, and concentration in high-rise buildings [3]. The Fire and Rescue Service Department together with Occupational, Health and Safety Environment advise people to leave the premises once the emergency alarm sounds, as soon as possible within three minutes. There is a need to evaluate time optimization in building evacuation to reduce the risk of human fatalities [4].

Several studies on crowd evacuation have been performed. Designing suitable building amenities, grasping the shortest time to evacuate, and implementing proper guiding methods might help reduce casualties and losses during crowd evacuations [5]. Due to the requirement to design buildings for rapid evacuation, evacuation model usage has increased over time. Similarly, evacuation models may effectively improve emergency planning and incident management in major outdoor fire situations [6]. Researchers have developed many indoor evacuations models. Most evacuation models are divided into microscopic and macroscopic models. The microscopic model focuses on individuals' interactions, whereas the macroscopic model defines crowds as a whole [7]. Some researchers have also suggested mesoscopic models, which explore how varied individual behaviours might affect collective dynamics [8]. Nevertheless, the macroscopic and microscopic models have frequently simulated pedestrian traffic in evacuation studies [9]. Thus, this study does not consider the mesoscopic model as it does not consider how people connect with others [10].

Comparison between macroscopic and microscopic models, microscopic models have become increasingly popular in recent years [11]. It is due to macroscopic models being created to mimic pedestrians. Although the microscopic model does not fully understand individual interactions and human behaviours [12], it covers other factors of individuals' qualities. It includes age, fatness, stoutness, and velocity, which are considered in the microscopic evacuation models. The in-depth or speculative information can be converse and shared by the evacuees [12]. Thus, this research will focus mainly on the microscopic model.

Uniform Building By-Law (UBBL) requires each building to prepare evacuation preparedness in choosing the best evacuation model. The owner must apply an evacuation model to the building. They must choose a suitable evacuation model to be implemented. Once they have decided, they need to apply the evacuation model. However, a missing process is found during the decision-making process: assessing the suitable evacuation model [13, 14]. Thus, we focus on assessing the suitable evacuation model because not every building can use the same evacuation plan.

Therefore, since the actual process involved is time-consuming and has a high-cost incurred, we propose the simulation approach to overcome the problem. In addition, this study focuses on high-rise buildings' emergency evacuation using three microscopic evacuation models. We also set the three design parameters as the standard procedure to ensure the acceleration of the evacuation.

2 Evacuation Model

The aims of evacuation models have two basic objectives. First, to determine the appropriateness of the architectural plan and exit capabilities for the building. Second, in the event of a fire or other calamity, it is important to make sure that evacuees spend as little time as possible in the safety zone, providing for the best escape route. A model of evacuation was developed to comprehend how quick flows and human experiences affect the evacuation duration [15].

Furthermore, there is three main task focuses in the evacuation model: (1) go towards the closest exit, (2) go into the zone with few people, and (3) try to prevent any obstacles during evacuation. The three tasks can become complicated with the bigger size, capability, and architecture of a high-rise building floor plan. Three primary models that are typically used in the microscopic model are the agent-based model (ABM), cellular automata (CA), and social force model (SFM) [16]. This study used three simulation software that represents the three microscopic models. These include Pathfinder, represented by ABM; PedGo represented by CA; and AnyLogic, represented by SFM. The simulation software is used to compare the simulation results.

ABM is more relied on because it can handle a higher degree of system complexity and simulate agent's interactivity on one or another level. The derivation from experience can be used to model social institutions, behaviour, properties and social characteristics [17]. SFM was innovated by [18]. Generally, its main contribution was to human behaviour, which associated motion with the laws of pedestrian behaviour. This model has been generally adopted for pedestrian dynamics and primarily explains pedestrian activity and simulations [19]. Furthermore, CA has become an area of study at the national and international frontiers as it can involve the characteristics of the human, the psychological reaction to disasters, relationships between people, and the effect on the evacuation behavior of the structure and climate [20].

Table 1 Software requirements

Software	Description
Microsoft word document	Project documentation
AutoCAD	To create a 2D line to be used in evacuation simulation
Pathfinder	An ABM evacuation simulation used to get data
AnyLogic	An SFM evacuation simulation used to get data
PedGo	A CA evacuation simulation used to get data

Table 2 Hardware requirements

Hardware	Description
Laptop model	MSI GF63 9SCSR-888 Gaming Laptop
Processor	Intel(R) Core (TM) i5-9300H CPU @ 2.40 GHz
Operating system	64-bit operating system
Windows version	Windows 10

3 Modelling Simulation Methodology

3.1 Software and Hardware Requirements

Both software and hardware are essential to run the simulation and fulfill the research requirements. Table 1 shows software requirements, and Table 2 shows the hardware requirements.

3.2 Level 14 Floor Plan Layout

The floor chosen for this research building is a high-rise building on level 14 of Yayasan Melaka's Building. It is an office layout with numerous rooms and dividing partitions that could make evacuation difficult. Each storey of the building has two entrances, and stairs are considered an escaping route. Because the use of elevators is restricted and windows cannot be utilised to escape due to the height of the structure, neither option is permitted as a means of escape. Figure 1 shows the level 14 floor plan of the Yayasan Melaka building.

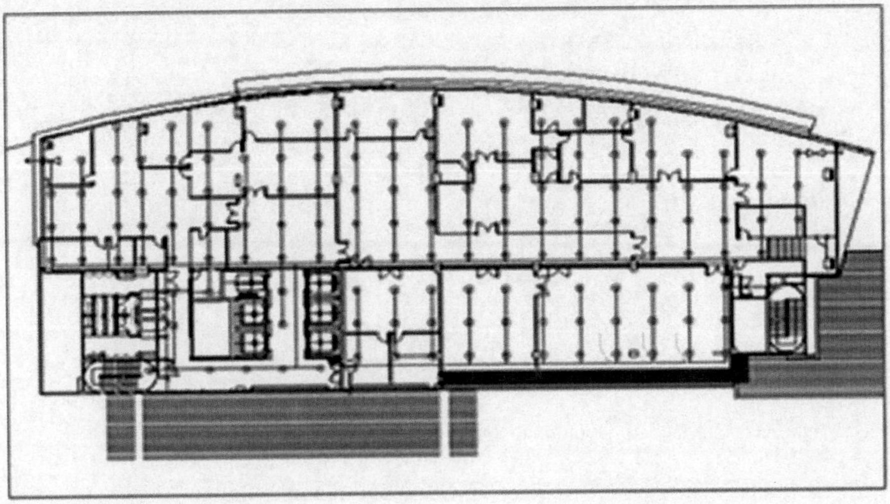

Fig. 1 Level 14 floor plan of Yayasan Melaka

3.3 Mapping Floor Plan

Once the layout has been drawn in AutoCAD and saved into .dwg format, the layout will be implemented in each simulation software, Pathfinder, PedGo, and AnyLogic, to obtain simulation results. The layouts will be mapped and drawn according to the simulation studies. The mapping procedure will adhere to a few set rules. These are the guidelines:

1. For each of the simulations, the pathways are fixed.
2. For each simulation software, the agents are placed in the same room.

Such requirements must be monitored because each simulation software's functional capabilities differ. Due to that, certain simulations require manual mapping for the agent's path from the starting position to the last position so they can move throughout the simulation. For each simulation software, fair simulations are made possible by agents scattered over the same rooms. Similar procedures were used for each simulation software in layout mapping but not identical. Figure 2 shows the Pathfinder interface to map the layout, Fig. 3 shows the PedGo interface to map the layout, and Fig. 4 shows the AnyLogic interface to map the layout.

3.4 Level 14 Simulation Studies

Once settled in the mapping process, we proceed with the simulation design parameter and attributes setting that aims to assess and identify the most effective evacuation

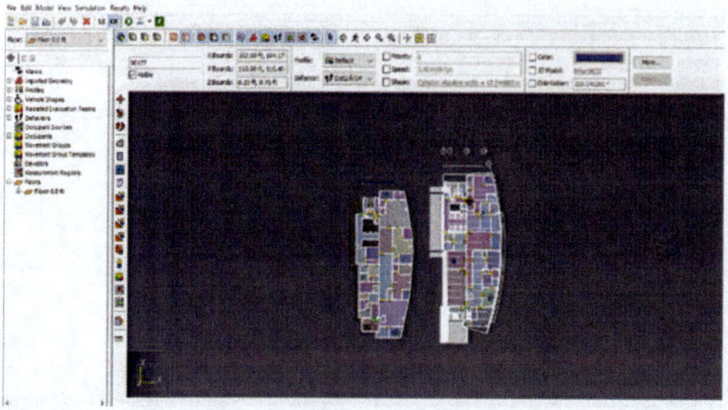

Fig. 2 Pathfinder interface for layout mapping

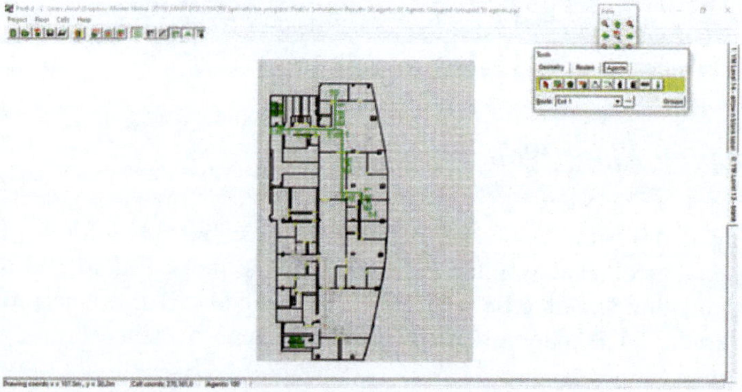

Fig. 3 PedGo interface for layout mapping

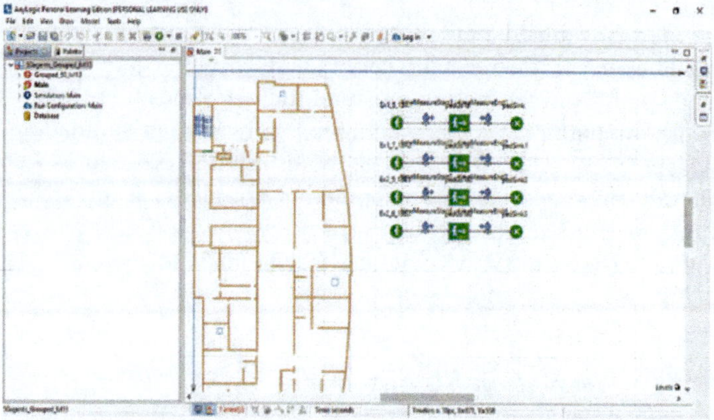

Fig. 4 AnyLogic interface for layout mapping

Table 3 Level 14 simulation studies

Simulation study	Number of agents	Agents' behaviour	Room size, ft^2	No. of doors	No. of staircases	Blockage condition	No. of exits
SS14-1	50	Group	1.5*E*	13	14	Yes	2
SS14-2	50	Scattered	1.5*E*	13	14	Yes	2
SS14-3	100	Group	1.5*E*	14	14	Yes	1
SS14-4	100	Scattered	1.5*E*	14	14	Yes	1
SS14-5	150	Group	1.5*E*	20	14	No	2
SS14-6	150	Scattered	1.5*E*	20	14	No	2

plan for the specific structure, as in Table 3. We consider the building's appropriateness when choosing the value. We set six simulation studies consisting of three design parameters. The number of agents' design parameters initially starts at 50. It increases by 50 every time until there are 150 agents altogether.

At the same time, six attributes have been used for the simulation: agents' behaviour, size of the room, doors, staircases, condition of blockage, and exit number. Each simulation study involves different behaviour (group or scattered) that distinguishes the agent. The original layout for room size is set at $1.5Eft^2$. The total number of doors used by the agents determines the number of doors, and the total number of staircases depends on the structure (two or four). This study utilises a mean overall time taken of 0.44 m/s using stairs [21] with a length of 7.384 m. It calculates how long it takes agents to escape via stairs [22]. Thus, it shows that an agent will take about 16.78 s to go down one staircase. Since there are 14 staircases from level 14 to ground level, it takes about 234.92 s for an agent to go down. Thus, the overall calculation for the time taken for one agent to escape requires that the time taken produced from the simulation will be added to the time taken during going down the stairs. It will ensure the accuracy of the results produced. One or two exits are chosen depending on whether a blockage is necessary.

4 Result and Discussion

4.1 Level 14 Simulation Studies Results

Based on the simulation studies, we record the execution time for each simulation software and plot the line graph to illustrate the difference between the simulation. Figure 5 shows the simulation result for SS14-1 for 50 agents with the time taken range distribution within 240–345 s. Figure 6 shows similar trends for SS14-2, but the distribution time taken is within 240–365 s. Followed the second design parameter of 100 agents, Fig. 7 shows the SS14-3 time taken distribution is between 240 to 440 s. The same pattern of line graph distributed for SS14-4 with a range between

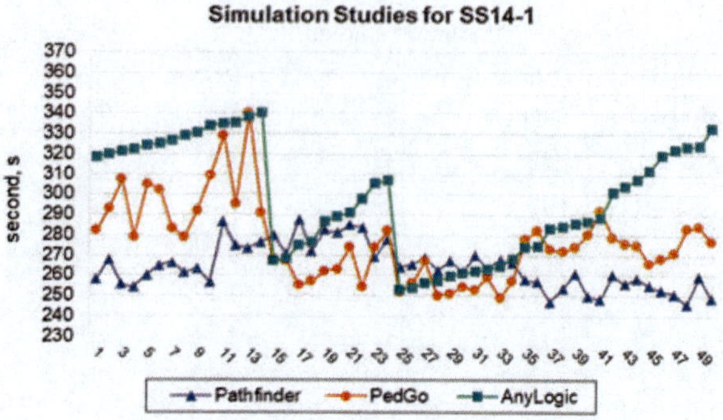

Fig. 5 Simulation results for SS14-1

240–460 s as in Fig. 8. For the last design parameter, 150 agents, the simulation result for SS14-5, as in Fig. 9, depicted the time taken between 240–560 s. Finally, Fig. 10 shows the simulation results for SS14-6, which have a range time taken from 240 to 570 s.

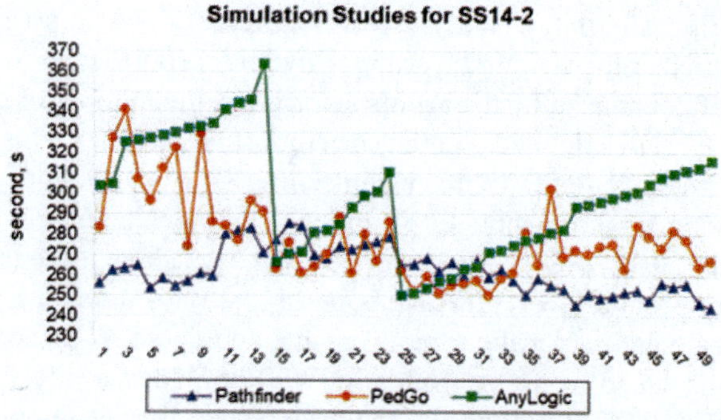

Fig. 6 Simulation results for SS14-2

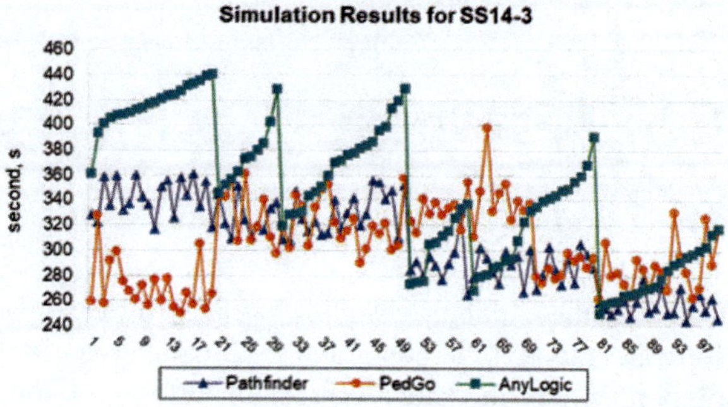

Fig. 7 Simulation results for SS14-3

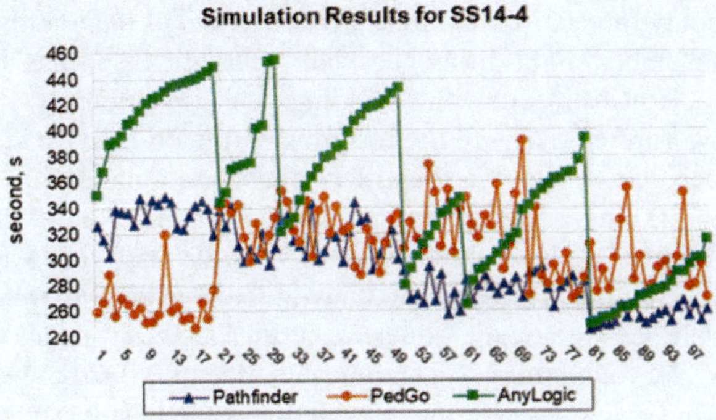

Fig. 8 Simulation results for SS14-4

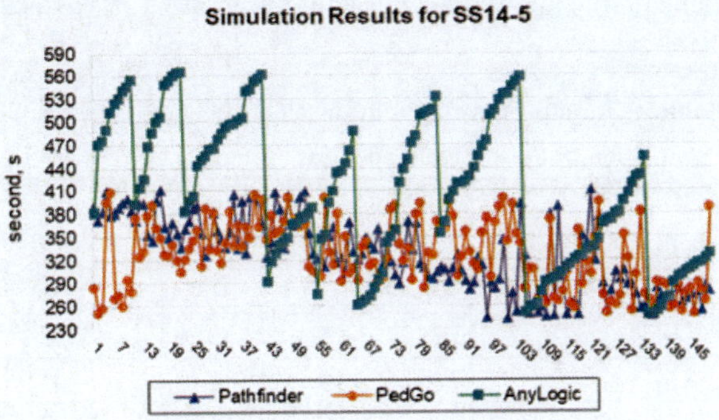

Fig. 9 Simulation results for SS14-5

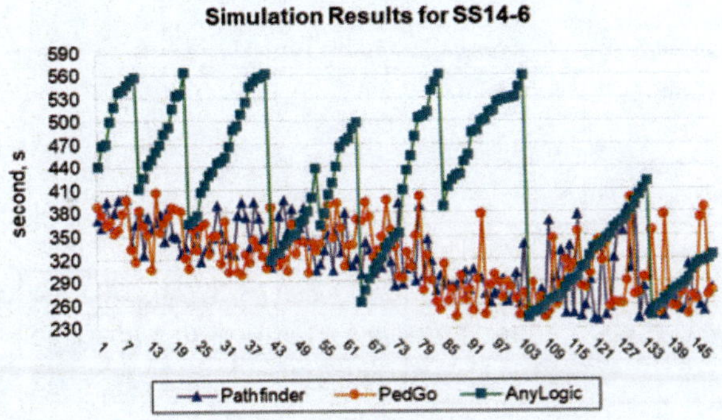

Fig. 10 Simulation results for SS14-6

4.2 Analysis of the Lowest Average Time Taken Results

The comparison among the three-simulation software continued by calculating the average time taken for each simulation study on the average lowest time taken. Various attributes can influence the result of the minimum time taken to evacuate from level 14. The time-taken results produced from the simulations are then examined and contrasted. Table 4 provides the result of the average time taken for the simulation study for each simulation software. It shows that Pathfinder and PedGo average time taken has slight differences, but AnyLogic result has significant differences.

However, since one building can implement only one evacuation model, thus we further analysed the result using the descriptive statistics test in Table 5. In the descriptive test, we found Pathfinder has a mean of 295.4400, PedGo with a mean of 300.9983 and AnyLogic 351.3717. Next, we did the ANOVA test as in Table 6 because we have three simulations to be analysed. We found a significant value of 0.024 which is less than 0.05 with the F value of 4.858. Based on the ANOVA result that has significant value, we further analysed using the Post-Hoc test as in Table 7. We found there is a significant value of 0.032 between Pathfinder and AnyLogic. Thus, we conclude that Pathfinder is the best evacuation model due to the lowest mean value. Therefore, it may be concluded that Pathfinder represents the ABM as the suitable and the best evacuation model for Yayasan Melaka.

Table 4 Results of average time taken of simulation studies for level 14

Simulation study	Pathfinder (s)	PedGo (s)	AnyLogic (s)	Chosen simulation
SS14-1	264.52	276.32	297.72	Pathfinder
SS14-2	262.48	277.90	297.57	Pathfinder
SS14-3	304.84	304.43	346.55	PedGo
SS14-4	295.72	303.96	352.96	Pathfinder
SS14-5	326.36	324.49	405.96	PedGo
SS14-6	318.72	318.89	407.47	Pathfinder

Table 5 Descriptive statistics table for level 14

Time	N	Mean	Std. deviation	Std. error	95% Confidence interval for mean		Minimum	Maximum
					Lower bound	Upper bound		
Pathfinder	6	295.4400	26.94012	10.99826	267.1681	323.7119	262.48	326.36
PedGo	6	300.9983	20.17459	8.23624	279.8264	322.1703	276.32	324.49
AnyLogic	6	351.3717	48.83810	19.93807	300.1192	402.6241	297.57	407.47
Total	18	315.9367	41.29034	9.73223	295.4035	336.4699	262.48	407.47

Table 6 ANOVA test for level 14

Time	Sum of squares	df	Mean square	F	Sig
Between groups	11,393.438	2	5696.719	4.858	0.024
Within groups	17,589.723	15	1172.648		
Total	28,983.162	17			

Table 7 Post-hoc test

Dependent variable: time

Tukey HSD

(I) ChosenSimulation	(J) ChosenSimulation	Mean difference (I-J)	Std. error	Sig	95% confidence interval	
					Lower bound	Upper bound
Pathfinder	PedGo	– 5.55833	19.77075	0.957	– 56.9123	45.7956
	AnyLogic	– 55.93167*	19.77075	0.032	– 107.2856	– 4.5777
PedGo	Pathfinder	5.55833	19.77075	0.957	– 45.7956	56.9123
	AnyLogic	– 50.37333	19.77075	0.055	– 101.7273	0.9806
AnyLogic	Pathfinder	55.93167*	19.77075	0.032	4.5777	107.2856
	PedGo	50.37333	19.77075	0.055	– 0.9806	101.7273

*The mean difference is significant at the 0.05 level

5 Conclusion

The aim of this research is to model the indoor evacuation assessment using a simulation approach since actual implementation requires a high cost. Simulation helps identify the best evacuation followed the specific design parameter set for the simulation. We found that modelling the assessment evacuation model can be implemented using the simulation approach by properly setting the methodology with the design parameters and attributes. Using the case study, we conclude that the ABM model is suitable for level 14 in high-rise buildings such as Yayasan Melaka. The suitability of the current evacuation models depends on the building's architecture. Different high-rise structures call for different kinds of evacuation models. For future study, we plan to assess other high-rise buildings using the other simulation with other design parameters.

Acknowledgements The research was funded by Universiti Teknologi MARA Cawangan Melaka under the TEJA Grant 2022 (GDT 2022/1-19).

References

1. Ding N, Chen T, Zhu Y, Lu Y (2021) State-of-the-art high-rise building emergency evacuation behavior. Phys A Stat Mech Appl. https://doi.org/10.1016/j.physa.2020.125168
2. Ding N, Chen T, Zhang H (2017) Simulation of high-rise building evacuation considering fatigue factor based on cellular automata: a case study in China. Build Simul 10:407–418

3. Zhang X (2017) Study on rapid evacuation in high-rise buildings. Eng Sci Technol an Int J 20:1203–1210
4. Sharbini H, Sallehuddin R, Haron H (2021) Crowd evacuation simulation model with soft computing optimization techniques: a systematic literature review. J Manag Anal 8:443–485
5. Li Y, Chen M, Dou Z, Zheng X, Cheng Y, Mebarki A (2019) A review of cellular automata models for crowd evacuation. Phys A Stat Mech its Appl 526:120752
6. Ronchi E (2021) Developing and validating evacuation models for fire safety engineering. Fire Saf J. https://doi.org/10.1016/j.firesaf.2020.103020
7. Li X, Zhou J, Chen F, Zhang Z (2018) Cluster risk of walking scenarios based on macroscopic flow model and crowding force analysis. Sustain. https://doi.org/10.3390/su10020385
8. Zhu Z, Zhou L, Zhang C, Lin B, Cui Y, Che M (2018) Modeling of macroscopic building evacuation using IFC data. ISPRS Int J Geo-Inform. https://doi.org/10.3390/ijgi7080302
9. Shi M, Lee EWM, Ma Y (2018) A novel grid-based mesoscopic model for evacuation dynamics. Phys A Stat Mech its Appl 497:198–210
10. Ming R, Peng X (2017) Study on the social force model of personnel evacuation in large stadiums. In: 14th international conference on services systems and services management ICSSSM 2017—proceedings, pp 1–5
11. Li Y, Chen M, Zheng X, Dou Z, Cheng Y (2020) Relationship between behavior aggressiveness and pedestrian dynamics using behavior-based cellular automata model. Appl Math Comput. https://doi.org/10.1016/j.amc.2019.124941
12. Ndiaye IA, Neron E, Jouglet A (2017) Macroscopic evacuation plans for natural disasters: a lexicographical approach for duration and safety criteria: Lex((Q|S) Flow). OR Spectr. https://doi.org/10.1007/s00291-016-0451-1
13. Samah KAFA, Halim AHA, Ibrahim Z (2022) An assessment algorithm for indoor evacuation model. Int J Inform Vis 6:168–176. https://doi.org/10.30630/joiv.6.1-2.933
14. Halim AHA, Samah KAFA (2022) Validation of evacuation assessment algorithm in finding the best indoor evacuation model. Int J Adv Comput Sci Appl 13:340–347. https://doi.org/10.14569/ijacsa.2022.0130540
15. Gaire N (2017) A study on human evacuation behavior involving individuals with disabilities in a building. Utah State University
16. Fayez L (2017) Modeling family behaviours in crowd simulation. Qatar University
17. Jumadi CS, Quincey D (2017) A conceptual design of spatio-emporal agent-based model for volcanic evacuation. Systems 5:53
18. Helbing D (1998) A fluid dynamic model for the movement of pedestrians, pp 1–23
19. Zhou J, Guo Y, Dong S, Zhang M, Mao T (2019) Simulation of pedestrian evacuation route choice using social force model in large-scale public space: comparison of five evacuation strategies. PLoS ONE 14:1–25
20. Yang Y, Deng J, Xie CC, Jiang YT (2014) Design and implementation of fire safety evacuation simulation software based on cellular automata model. Proc Eng 71:364–371
21. Kuligowski ED, Peacock RD, Reneke PA et al (2015) Movement on stairs during building evacuations. Natl Inst Stand Technol Tech Note 1–213
22. Hyun-seung H, Jun-ho C, Won-hwa H (2011) Calculating and verifying the staircase-length for evacuation analysis. Pedestr Evacuation Dyn. https://doi.org/10.1007/978-1-4419-9725-8

Gamified Augmented Reality Mobile Application for Tourism in Kuching

Chung Ang Tay, Bryan Benjamin Dominic, Hai Inn Ho, Norahima Annuar, and Nur Elana Mohd Saferinor

Abstract Augmented Reality (AR) for tourism utilizing mobile application mostly features scanning prefixed image marker by tourist in order to retrieve additional multimedia content for learning. This paper proposes the use of similar AR application for tourism that extends beyond learning to include gamification for the enhanced tourist's experience. Majority of tourism AR uses marker-based method like scanning QR code to load virtual objects on the screen. The current study will explore markerless method using spatial anchor. Furthermore, functionality for site operator, an aspect often neglected in literature, like the ability to dynamically add content without relying on developer, will be categorically examined. Platforms, frameworks and services used include Unity, AR Core or AR Foundation and Azure cloud services. This would culminate in implementation and evaluation of an AR mobile application prototype that benefits both tourists and site operators towards proliferation of gamified AR tourism in Kuching city of Sarawak state in Malaysia.

Keywords Augmented reality · Mobile application · Tourism · Gamification · Spatial anchor

C. A. Tay (✉) · B. B. Dominic · H. I. Ho · N. Annuar · N. E. M. Saferinor
i-CATS University College, Kuching, Sarawak, Malaysia
e-mail: chungang.tay@icats.edu.my

B. B. Dominic
e-mail: 2018311527@icats.edu.my

H. I. Ho
e-mail: hai.inn@icats.edu.my

N. Annuar
e-mail: norahima@icats.edu.my

N. E. M. Saferinor
e-mail: elana@icats.edu.my

© The Author(s), under exclusive license to Springer Nature Singapore Pte Ltd. 2023

D.-K. Kang et al. (eds.), *Proceedings of the 9th International Conference on Computational Science and Technology*, Lecture Notes in Electrical Engineering 983,
https://doi.org/10.1007/978-981-19-8406-8_27

1 Introduction

A typical mobile augmented reality (AR) would involve the use of a mobile device scanning the tourism site for markers like a QR code (marker-based AR) or a natural trait of the physical environment (marker-less based AR) resulting in loading of virtual objects such as a video clip, a text display or similar multimedia object being overlaid over the site's environment as perceived through the mobile's device screen [1–4]. As pointed out in [1] and confirmed by an actual study closer to home in Sarawak [5], marker-based AR is widespread, but the less intrusive marker-less based AR entices further interest and study.

The studies of tourism AR in Malaysia like [5] and [6], seem to suggest that the application is mainly to educate the tourist by providing more information on the sites, the former for Sarawak and the latter for UNESCO sites in Malaysia. A survey by [7] found a scarcity of AR tourism research in Malaysia and Asia continent (but primarily South Korea) at 15% compared to Europe at 73%. In fact [7] only attributed [6] as the only research from Malaysia. It further highlighted that AR research at its inception in 2005 delve on design and development and eventually developed into newer themes of user experience and gamification in 2015 onwards. It is observed that [5] and [6] seem to concur with this evolving trend with exception of gamification in AR. In fact, gamification was adopted in tourism AR in Asia [8] and Europe [9, 10]. Thus, it can be surmised more studies are required for gamified AR application in Malaysia and particularly Sarawak.

Hence this study calls for development of a gamified AR mobile application for tourism in Kuching, Sarawak with aims of exploring marker-less AR technologies, the potential of gamification and a better understanding of AR application in this region.

2 Literature Review

Using a similar approach by [11], related works on AR on tourism are surveyed with attention on few added dimensions: gamification, the AR type (is it marker-based, marker-less, or location-based [1]), contribution to tourism and the study's location:

It can be gleaned from Table 1 that majority of the educational AR applications favor marker-based AR due to its accuracy and reliability [1]. Thus, the current study attempts to fill this gap by creating mobile tourism AR using marker-less technique and with a different technology from that used in [12]. The current study also uses the marker-based AR. Next AR works involving gamification are considered.

All the games mentioned in Table 2 are educational and so the non-game context qualifies their association with the term gamification [16]. Such games instill fun and positive experience for tourists. Thus, the current study would incorporate gaming elements.

Table 1 List of educational tourism AR without gamification

Reference	AR type/contribution to tourism	Technology	Location
Saragih [11]	Marker-based method. Loads model and information upon marker detection presumably found on site Helps tourism through active participation of tourists	Kudan AR SDK	Batam, Indonesia
Demir and Karaarslan [12]	Location-based (GPS) and marker-less (relying on special algorithms to detect unique features of an image) to detect scenery and loads information accordingly. Also links to and extracts information from social media like displaying reviews	OpenCV and Phyton	Turkey
Safitri et al. [13]	Marker-based to scan map and loading model and additional multimedia objects	Unity 3D Vuforia	Indonesia
Obeidy et al. [14]	Location-based. The smart glass uses context aware AR engine to create interesting experience for tourists on sites	BeyondAR	Malaysia
Sidyawati et al. [15]	Marker-less as the application can scan the original masks (the museum's artefacts on exhibit) and load virtual objects on the screen's gadgets. Promotes visits to museums	Vuforia	Malaysia
Lee et al. [5]	Marker-based. It is more concerned with ease of integrating multimedia objects on mobile	ARToolKitPlus OpenGLES	Sarawak, Malaysia

Table 2 List of gamified tourism AR

Reference	AR type/contribution to tourism	Technology	Location
Paliokas et al. [9]	Marker-based using QR code is mentioned to trigger AR objects (like 3D content overlay on artefacts exhibited in museums). The game is a quiz that involves answering by scanning actual artefacts on display in the museum. It seems, in this case, a marker-less technique is used to detect the correct artefact. Fun and informal learning	Unity and ARCore	Greece
Koo et al. [8]	An impressive marker-less method that involves scan mode: taking a picture using mobile device which is then binarized and compared with matching stored binarized images. When a match is found, AR objects would be loaded. Examples: (1) scanning an informative panel in the scene will load AR object that overlays the same panel. (2) scanning along a timeline on a wall panel will result in automatic loading of different AR objects corresponding to the timeline resulting in a smoother experience compared to abrupt marker-based scanning The game: user can choose to construct a structure (like roof, floor, wall) using given materials via AR and in the process learning something about the fortress construction: the tourism site	Unity and Unity ARkit plugin	Republic of Korea
Nobrega et al. [10]	Location-based AR that load virtual objects for storytelling based on GPS tracking (of where the tourist is) The AR game involves the tourist as a sailor of a traditional boat in the area and the mission being to transport as many wine barrels as possible along the river. Fun and informal learning	Unity 3D	Portugal

In gamified AR, it seems that marker-less method and the choice of Unity may not be coincidental. The marker-less method is possibly less obstructive [8] compared to having to scan assigned markers like QR code at intervals. The same literature also seems to suggest possible problem in getting tourists to scan images accurately for comparison. And such challenge may explain the choice for location-based AR (using GPS sensor) as in [10], instead of relying on computer vision systems alone. The current study allays such concern by using spatial anchor [17] to be used together with ARKit/ARCore-enabled Android devices. Much guidance is found in [2] (a mixed reality application using head mount display and HoloLens to train surgeons) that uses Azure Spatial Anchors that relies on ability to deduce precise location through perception of relationships between people, spaces, and devices within a physical space. No other sensors are required. Azure's cloud database ensures persistency of such anchors making them discoverable in future and by others (https://azure.microsoft.com/en-us/services/spatial-anchors/). Unity and Microsoft HOLOLENS are used in [2], but the current study will use ARCore as in [17] as it would be a mobile AR application.

It seems that literature does not discuss AR tourism beyond its role as mere publication platform. Using the web technology as a benchmark, the web literally proliferates only when it becomes web 2.0 and beyond: enabling active participation by other players [18]. As such, the current study would consider building a mobile application catering for both tourists and the site operators. In addition to the above review, the present study may address in a limited way the following gaps identified by [7] namely: domination of study in Europe, domination of PLS-SEM methods and limited study on gamification.

3 Research Method

Though stages of data collection and analysis may not be apparent, Hasan [19] argues that information systems development ought to be recognized as a valid research method so long as contribution to knowledge is present. It suggests knowledge creation to be inherent in the system development itself. Thus, the knowledge contribution and system development method must be scrutinized.

The contributions in this study, considering previous literature review are: (i) software artefact: gamified tourism AR mobile application for Kuching, (ii) implementation of marker-less using Microsoft Azure's spatial anchor, (iii) gamification and (iv) an app that is more participatory involving both tourists and related actors than merely a publication platform.

The system development method is based on Game Development Life Cycle (GDLC) proposed by [20] as follows (Table 3).

Table 3 Game development life cycle

GDLC step	Description	Application/adaptation
Initiation	Conceptual form of the game	Develop use case: in consultation with Dean of the Faculty of Tourism and Hospitality Conception of Mata Kuching AR Tour Guide mobile app
Pre-production	Game Design Document (GDD) and prototype	Second author's final year project thesis as GDD and Mata Kuching App
Production	Development of assets and coding	Development of Mata Kuching App
Testing	Internal testing for playability and usability	One to one testing with Dean of the Faculty of Tourism and Hospitality (FTH) playing the role as tourist. Final Year Project Presentation by second author to panel assessors from Faculty of Science and Technology. Due to COVID-19 related precautions, the lobby of the university would be the tour site
Beta	Third party or external testers	The Mata Kuching app made available for download for 1 week period by any walk-in staff at the university's lobby
Release	Ready to be released to the public	Not done. The GDLC is iterative in nature and the findings from both testing and beta may suggest a return to pre-production phase

4 Result and Discussion

The use case is as follows (Fig. 1).

The Mata Kuching app is developed using Unity. AR Foundation/ARCore is required to support AR for mobile devices currently limited to Android devices. Figure 2 shows screenshots of the Mata Kuching AR Tour app video demo (https://youtube.com/playlist?list=PLcxi50r736YcjURGaabIQMO-HHXahgJWk): Fig. 2 illustrates the screen of managing users and Fig. 3 illustrates the screen of Creating exhibit and uploading upload audio and 3D model (Figs. 4, 5, 6, 7 and 8).

The Headhunter event is a game that the tourist can play on the tour site. The manager can place an anchor for the hidden skull. The tourists when choosing to play the game will be given hints where to look for the skull. This game is based on the past practice of certain Dayak warriors in Borneo engaging in beheading their enemies and collecting heads as trophies (Fig. 9).

The study successfully produced a gamified tourism AR app prototype capable of supporting marker-less and marker-based methods. The Dean was impressed though much guidance was required initially and finding a compatible mobile phone was an

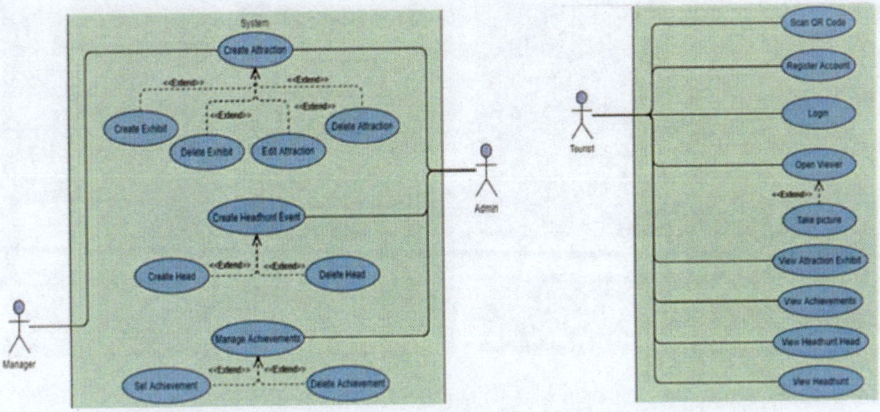

Fig. 1 The Manager refers to the tour site operator or the tour guide. The other actors being admin referring to the developer or future IT support and of course the tourist

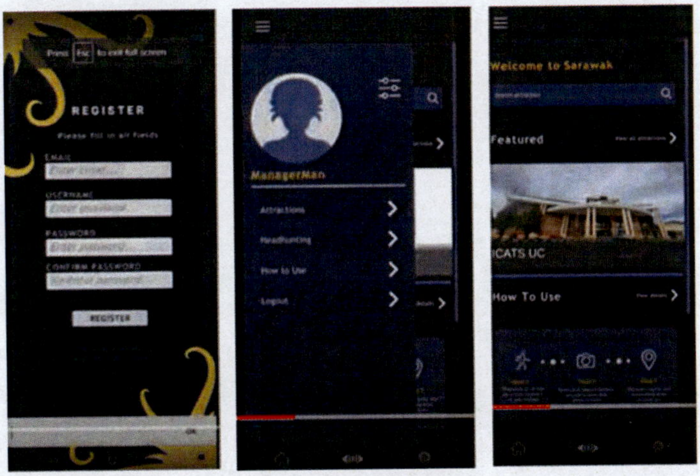

Fig. 2 Managing users

issue. During the Final Year Presentation, intermittent interruptions to Azure services were observed. The Beta results are as follows.

Despite limited number of responses, due to COVID-19 restrictions, the following can be observed from Table 4: (1) A very favorable rating for UI design suggests AR can easily impress. This agrees with first-hand impression by the Dean (FTH)as well as during demonstration before final year project panel (despite reported technical challenges during the demonstration), and thus validating this observation. (2) The mixed review for 'app difficulty' suggests technical problem, as noted previously, problems with initial installation of app (which requires AR Foundation/AR Core

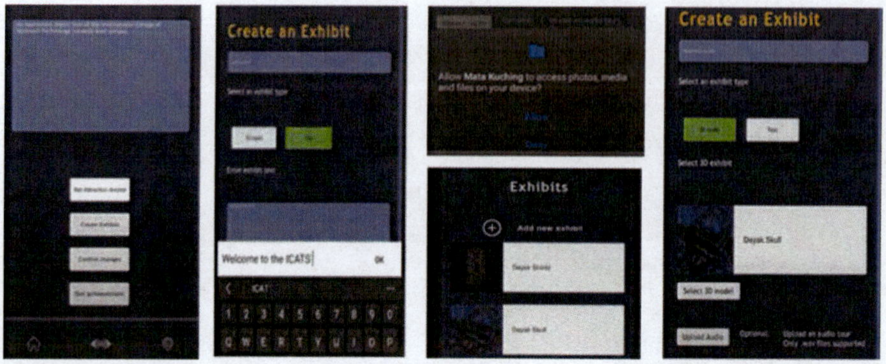

Fig. 3 Create exhibit. Upload audio and 3D model

Fig. 4 Placing anchor in the lobby and upload to Azure Cloud database. Note that the text/audio clip and the skull 3D model use two different anchors at different locations in the lobby floor

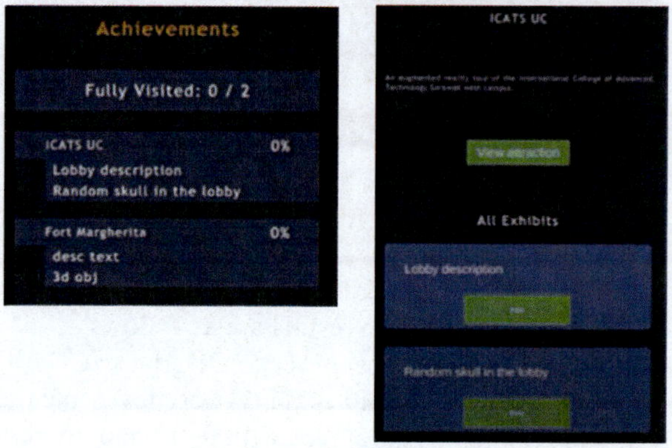

Fig. 5 Achievements to encourage visits and showing available exhibits

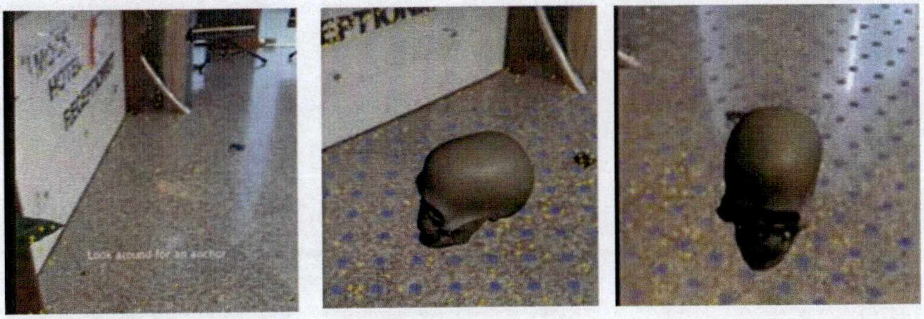

Fig. 6 The maker-less technique (not using QR code). Once the invisible anchor is detected, the 3D model is loaded virtually on the lobby floor

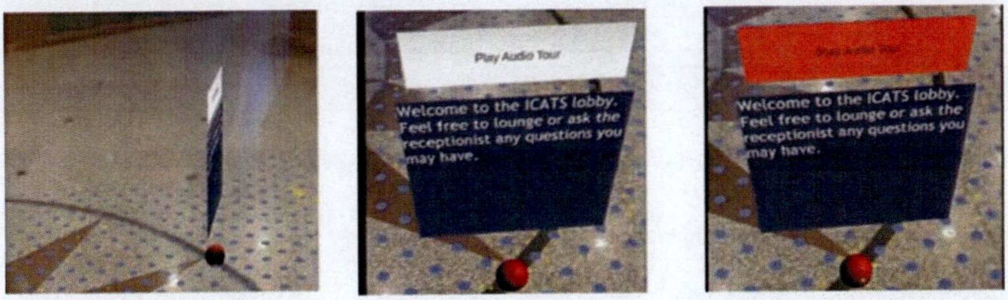

Fig. 7 Once the appropriate anchor is detected, a virtual text panel and virtual button to play sound clip are loaded. The button flashes red when pressed

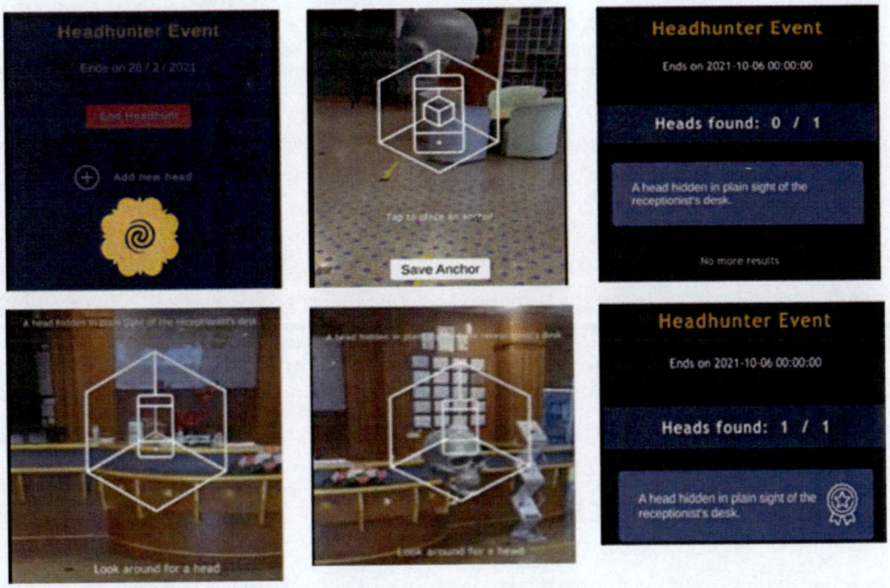

Fig. 8 Screenshots of Mata Kuching app

Fig. 9 The Dean (FTH) with the second author during testing

Table 4 Beta results based on walk-ins in the lobby for 1 week trial period (1 Oct–8 Oct 2021)

Area	Very poor	Poor	Average	Good	Very good
Overall speed	–	–	1	4	1
Overall performance	–	–	1	4	1
Overall UI design	–	–	–	1	5
App difficulty	–	2 (hard)	2 (normal)	2 (easy)	
Likeliness of using app	–	–	1	1	4
Likeliness of app replacing tour guide	–	–	2 average	2 quite likely	2 very likely
Total	–	2 (5.6%)	7 (19.4%)	14 (38.9%)	13 (36.1%)

installation) and initial instruction on how to use the app. (3) A high score for 'likeliness of using app' suggests the potential for such app is readily perceived and thus this project is justified. (4) A mixed result for its potential to replace a tour guide may suggest that the respondents may not fully understand all the functions of the app especially those use cases pertaining to manager (i.e., the tour site operator). Besides the choice of lobby as the trial site and lack of sample content may make it difficult for respondents to envisage the app's full potential.

A combined score of 75% (sum of Good and Very Good in Table 4) in the independent beta results validate the study in the sense that it is perceived as achieving its aims and in agreement with approval from both Dean (FTH) and final year project presentation panel. However, all three would suggest areas of improvements as expected in GDLC and discussed above.

5 Comparison with Other Works

Peculiar in this study is the use of spatial anchor [2, 17]. While [17] relied on custom built LAN anchor server (custom Python app on CentOS7), this study relied on Azure cloud services. This poses connectivity issue (i.e., the intermittent interruptions mentioned earlier) and is to be rectified by ensuring the AR session was tracking before starting the Azure session. The app is now improved with a loading screen before the start of Azure session with an additional prompt to remind users to be in a well-lit environment. Azure spatial anchor is also used in [2] but also includes Azure Cosmo DB and Azure Digital Twins, both of which are part of mixed reality tool kit. A mixed reality is not content with just loading virtual objects in real world but using them to navigate it. But the current study is primarily concerned with AR using Android device. Both [17] and [2] uses HoloLens.

The choice of using spatial anchor, makes this project stand apart from similar AR tourism applications that aim to implement a marker-less method using either location-based approach [10] and image processing [8]. This study demonstrates a better approach to AR tourism in Sarawak compared to current marker-based adopted in [5].

Furthermore, there is no gamification in [5, 14] and [15], all of which are AR tourism in Malaysia. The app produced in this study has a simple culturally based headhunting game, though lacking in finesse to those found in Table 2.

Finally, none of the AR tourism apps gleaned from literature seem to address the co-creation mentality as discussed in [18] using the evolvement of web 2.0. This mean presumably the content creators would have to rely on developers whenever a new content is required and the AR apps function mainly as viewing platform by the tourists. Thus, the dual function of Mata Kuching AR Tour app to enable managers to upload content on their own including determining where such content will be loaded, as well as a viewing platform by tourists, is a significant trailblaze.

6 Conclusion and Future Works

As indicated by the beta testing, the Mata Kuching app can still be further enhanced, and this is expected in GDLC. The prototype fulfills the expected contributions: a localized app, use of spatial anchor, gamification and is more than a publishing platform. Future works include: (1) implementing the Mata Kuching app in an actual tourist site in Kuching and with extensive beta testing. This cannot be done in the current study due to COVID-19 restrictions. (2) to conduct empirical quantitative findings or qualitative findings based on the app.

References

1. Tian F, Xu F, Fu J (2013) Augmented reality technology overview for tourism app development. In: 2013 IEEE international conference on machine learning and cybernetics, vol 4, pp 1483–1489
2. Mamdouh R, El-Bakry HM, Riad A, El-Khamisy N (2020) Using Azure to construct recent architecture for visualize training in real-time. Ind J Sci Technol 13(23):2355–2370

3. Ming C, On CK, Rayner A, Guan TT, Patricia A (2018) The determinant factors affecting cloud computing adoption by small and medium enterprises (SMEs) in Sabah, Malaysia. J Telecommun Electron Comput Eng 10:83–88
4. Kim JJ, Wang Y, Wang H, Lee S, Yokota T, Someya T (2021) Skin electronics: next-generation device platform for virtual and augmented reality. Adv Func Mater 31(39):2009602
5. Lee LS, Ng GW, Tan KY, Shaharuddin SS, Wan-Busrah SF (2018) Integrating interactive multimedia objects in mobile augmented reality for Sarawak tourism. Adv Sci Lett 24(2):1017–1021
6. Obeidy WK, Arshad H, Huang JY (2017) An acceptance model for smart glasses based tourism augmented reality. In: 2017 AIP conference proceedings, vol 1891, no 1, p 020080
7. Jingen Liang L, Elliot S (2021) A systematic review of augmented reality tourism research: what is now and what is next? Tourism Hospitality Res 21(1):15–30
8. Koo S, Kim J, Kim C, Kim J, Cha HS (2019) Development of an augmented reality tour guide for a cultural heritage site. J Comput Cult Heritage 12(4):1–24
9. Paliokas I, Patenidis AT, Mitsopoulou EE, Tsita C, Pehlivanides G, Karyati E, Tsafaras S, Stathopoulos EA, Kokkalas A, Diplaris S, Meditskos G (2020) A gamified augmented reality application for digital heritage and tourism. Appl Sci 10(21):7868
10. Nóbrega R, Jacob J, Coelho A, Weber J, Ribeiro J, Ferreira S (2017) Mobile location-based augmented reality applications for urban tourism storytelling. In: 2017 IEEE 24° Encontro Português de Computação Gráfica e Interação (EPCGI), pp 1–8
11. Saragih RE (2020) Development of interactive mobile application with augmented reality for tourism sites in Batam. In: 2020 IEEE fourth world conference on smart trends in systems, security and sustainability, pp 512–517
12. Demir ÖF, Karaarslan E (2018) Augmented reality application for smart tourism: GökovAR. In: 2018 IEEE 6th international istanbul smart grids and cities congress and fair (ICSG), pp 164–167
13. Safitri R, Yusra DS, Hermawan D, Ripmiatin E, Pradani W (2017) Mobile tourism application using augmented reality. In: IEEE 2017 5th international conference on cyber and IT service management (CITSM), pp 1–6
14. Obeidy WK, Arshad H, Huang JY (2018) TouristicAR: a smart glass augmented reality application for UNESCO world heritage sites in Malaysia. J Telecommun Electron Comput Eng (JTEC) 10(3–2):101–108
15. Sidyawati L, Sudarmanto JA, Prasetyo AR (2019) Berahim EMHB: Nusantara mask heritage Malaysia: infographic application development of masks of Malaysian indigenous tribes at the museum of Asian art Malaysia based on augmented reality as media of tourism education. Jurnal IPTA (Industri Perjalanan Wisata) 7(2):163–173
16. Yong E, Baharum A, Tanalol SH, Deris FD, Noor NAM (2021) Gamification for tourist guide application. Trans Sci Technol 8(3–2):335–342
17. He W, Xi M, Gardner H, Swift B, Adcock M (2021) Spatial anchor based indoor asset tracking. In: 2021 IEEE virtual reality and 3D user interfaces (VR) pp 255–259
18. Zhang X, Liu H, Yao P (2021) Research jungle on online consumer behaviour in the context of web 2.0: traceability, frontiers and perspectives in the post-pandemic era. J Theor Appl Electron Commer Res 16(5):1740–1767
19. Hasan H (2003) Information systems development as a research method. Australas J Inf Syst 11(1):4–13
20. Ramadan R, Widyani Y (2013) Game development life cycle guidelines. In: 2013 IEEE international conference on advanced computer science and information systems (ICACSIS) pp 95–100

Prediction of CO_2 Breakthrough Time in Packed Bed Adsorption Column Using Artificial Neural Network

Mohd Hardyianto Vai Bahrun, Norasikin Othman, Awang Bono, and Muhammad Abbas Ahmad Zaini

Abstract Carbon dioxide is a component that constitutes a significant fraction of natural and biogas after methane and needs to be removed to increase its energy content for efficient combustion. A packed bed adsorption column is often utilized for separating gas constituents, and its dynamic performances are determined for successful cyclic adsorption design. Experimental procedure and phenomenological modeling are often used to determine or predict non-linear packed bed dynamic behavior. A simplistic surrogate model is a powerful tool for generalizing non-linear problems. In this work, a simple artificial neural network (ANN) was applied to predict the breakthrough time of CO_2 in a packed bed adsorption column. The phenomenological packed bed column mathematical model was validated prior to dataset generation for ANN model training. The ANN architecture with a 4-10-1 structure is used. Several data pre-processing techniques and activation functions were analyzed to improve the learning ability of the ANN. The results show that the data pre-processing transformation techniques could increase the learning ability of the ANN model. The log transformation normalization (LTN) with *elu* activation function in the hidden layer is the best choice and is used to train the ANN. The trained ANN model can predict the breakthrough time with 99.99% accuracy

M. H. V. Bahrun · N. Othman · M. A. A. Zaini (✉)
Department of Chemical Engineering, Faculty of Chemical and Energy Engineering, Universiti Teknologi Malaysia, 81310 Johor Bahru, Johor, Malaysia
e-mail: abbas@cheme.utm.my

M. H. V. Bahrun
e-mail: mohdhardyiantovai@graduate.utm.my

N. Othman
e-mail: norasikin@cheme.utm.my

N. Othman · M. A. A. Zaini
Centre of Lipids Engineering and Applied Research (CLEAR), Ibnu Sina Institute for Scientific and Industrial Research (ISI-SIR), Universiti Teknologi Malaysia, 81310 Johor Bahru, Johor, Malaysia

A. Bono
GRISM Innovative Solutions, Kota Kinabalu, Sabah, Malaysia

© The Author(s), under exclusive license to Springer Nature Singapore Pte Ltd. 2023

D.-K. Kang et al. (eds.), *Proceedings of the 9th International Conference on Computational Science and Technology*, Lecture Notes in Electrical Engineering 983,
https://doi.org/10.1007/978-981-19-8406-8_28

between the actual and ANN model values. This demonstrates that the trained ANN model can be used to predict the breakthrough time of CO_2 with excellent accuracy.

Keywords CO_2 removal · Natural gas · Biogas · Data pre-processing · Adsorption modeling · Adsorption simulation

1 Introduction

The demand for the quest for renewable alternative fuels has increased significantly. It is expected to keep growing due to the non-renewable fossil fuels that we used today being utilized at a tremendous rate, as well as to mitigate environmental issues and minimize the dependency on non-renewable reserves [1]. Natural gas, an example of fossil fuels other than oil and coal, is the fastest-growing fossil fuel, accounting for about a quarter of global electricity generation [2]. However, due to natural gas being a non-renewable resource, an alternative to natural gas, which is renewable, is necessary. The natural gas substitute, known as biogas, is one of the alternative fuels that are very similar to natural gas and has a high content of methane and carbon dioxide upon formation. The future of biogas as a sustainable fuel is bright, with significant market potential since there is a vast abundance of biogas sources [3]. Since the natural or biogas contain CO_2, it must be removed up to a satisfactory methane level, so the gas is suitable for fuel application. And one of the promising gas separation approaches is the adsorption process due to its minimal energy penalty, low installation and operating cost, and simple yet safe operation [4].

A practical adsorption mode used in the industry for gas separation is usually in a continuous operation mode by utilizing a packed bed column. The successful design of a column adsorption process requires a description of the dynamic behavior of the adsorbate in a packed bed column, referred to as a breakthrough curve (BTC). The BTC can provide information on the adsorbate concentration of a system's dynamic response with a function of elapsed time [5]. From the BTC, information on breakthrough and saturation times can be obtained, which is one of the key elements in evaluating the feasibility of the column adsorption system and is essential for practical scaling up [6]. Also, the BTC investigation is the premise and study basis of cyclic adsorption cycle design and the foundation of the theoretical verification method through the dynamic mathematical model [7, 8].

An artificial neural network (ANN) is a computational model that is widely used for dealing with highly non-linear problems. It is a powerful modeling technique with high generalization ability without understanding the underlying phenomenological mechanisms behind a process [9]. Its ability to model complex functional relations between various parameters makes the ANN one of the most important modeling tools in the field of adsorption for predicting its dynamic performances [9, 10]. Developing the ANN model is bound to many parameters' adjustment so that the learning process can be optimized. Improving the learning capability of ANN has been a widely researched topic, and many works have been reported [11–14]. However, the effect of different data pre-processing techniques and activation functions in the hidden layer on the learning capability of the ANN is less studied.

Therefore, this work aims to apply neural network prediction to the breakthrough time of CO_2 adsorption in a packed bed column. The packed bed column mathematical model validation is performed to ensure the reliability of the dataset generated by the model. Then, from the generated dataset, an ANN model is built to predict CO_2 breakthrough time by taking four input parameters (system pressure, feed volume flowrate, bed column height, CO_2 feed mole fraction). Several dataset pre-processing transformation techniques and activation functions are evaluated to explore the effect of different techniques and activation functions on the learning capability of the ANN model.

2 Packed Bed Column Mathematical Model

A mathematical model describing the breakthrough curve is required based on mass, momentum, and energy balances, including the appropriate initial and boundary conditions. In this work, an isothermal packed bed column adsorption (no energy balance equation required) is solved using Aspen Adsorption using Euler numerical integration method. The governing mathematical model equations are listed in Table 1, and the underlying assumptions of the model can be found in Ref. [1].

The model parameters used in this work are taken from ref. [7] in order to validate the simulated breakthrough curves presented in this work with the experimental data from the previously cited reference. The adsorbent's physical properties and the specification of the packed bed adsorption column are shown in Table 2.

The Langmuir–Freundlich isotherm model describes the pure component adsorption relationship of CO_2 and CH_4. Their respective isotherm parameters are given in Table 3.

3 ANN Architecture for CO_2 Breakthrough Time Prediction

The ANN model is developed to predict the CO_2 breakthrough time in a packed bed adsorption column. The ANN is written in Python using Spyder IDE, employing several open-source libraries, among others are Tensorflow/Keras, NumPy, Pandas, Matplotlib, Scipy, and Sklearn. The maximum epoch of training cycles is set as 5000. The default settings of ANN model training are used unless specified.

Table 1 Mathematical model equations for conservation of mass and momentum, and correlations used to estimate unknown transport properties [1, 15, 16]

Items	Equations
Conservation of mass	*Component i mass balance* $-\varepsilon_b D_{ax}\frac{\partial^2 c_i}{\partial z^2} + \frac{\partial (v_g c_i)}{\partial z} + \left[\varepsilon_b + (1-\varepsilon_b)\varepsilon_p\right]\frac{\partial c_i}{\partial t} + (1-\varepsilon_b)\rho_s\frac{\partial q_i}{\partial t} = 0$
	Overall mass balance $\frac{\partial (v_g c)}{\partial z} + \left[\varepsilon_b + (1-\varepsilon_b)\varepsilon_p\right]\frac{\partial c}{\partial t} + (1-\varepsilon_b)\rho_p\sum\frac{\partial q_i}{\partial t} = 0$
Conservation of momentum	*Ergun's equation* $\frac{\partial P}{\partial z} = -\left(\frac{1.50\times 10^{-3}\mu(1-\varepsilon_b)^2}{\varepsilon_b^3(2r_p\psi)^2}v_g + 1.75\times 10^{-5}\frac{(1-\varepsilon_b)M\rho_g}{2r_p\psi\varepsilon_b^3}v_g^2\right)$
	Correlations
Binary molecular diffusivity, D_m	$D_m = \frac{1.00\times 10^{-3}T^{1.75}\left(\frac{1}{M_j}+\frac{1}{M_i}\right)^{0.5}}{P\left[\left(\sum_A V_i\right)^{\frac{1}{3}} + \left(\sum_B V_i\right)^{\frac{1}{3}}\right]^2}$
Axial dispersion coefficient, D_{ax}	$D_{ax} = \frac{D_m}{\varepsilon_b}\left(0.3\varepsilon + \frac{1}{2}\frac{ScRe}{1+\frac{3.8}{ScRe}}\right)$
External film mass transfer resistance, $k_{f,i}$	$Sh = 2 + 1.1Sc^{1/3}Re^{0.6}$ for $3 < Re < 10^4$
Knudsen diffusivity, $D_{K,i}$	$D_{K,i} = 97r_{mac}\sqrt{\frac{T}{M_i}}$
Macropore diffusivity, $D_{p,i}$	$\frac{1}{D_{p,i}} = \tau\left(\frac{1}{D_{K,i}} + \frac{1}{D_{m,i}}\right)$
LDF mass transfer coefficient, $k_{LDF,i}$	$\frac{1}{k_{LDF,i}} = \frac{r_p K_i}{3k_{f,i}} + \frac{r_p^2 K_i}{15\varepsilon_p D_{p,i}}$, where $K = \frac{dq_0}{dC_0}\cdot\frac{\rho_s}{\varepsilon_b}$

Table 2 Parameters of packed bed adsorption column and adsorbent [7]

Parameter	Value	Parameter	Value
Bed column characteristics		*Adsorbent characteristics*	
Bed height, H_b (m)	0.1185	Adsorbent	Pine sawdust AC
Inner bed diameter, d_b (m)	0.013	Solid density, ρ_s (kg/m^3)	255.7
Bed porosity, ε_b (–)	0.581	Particle diameter, d_p (mm)	2.96
Process conditions		Particle porosity, ε_p (–)	0.681
Feed flowrate, F (mL/min)	30	Mean pore radius, r_{pore} (nm)	5.0*
Feed mole fraction, y_F (–)	Equimolar	Shape factor, Ψ (–)	0.91
Pressure, P (bar)	1.2, 3, 5, 10	Particle tortuosity, τ (–)	3*
Temperature, T (K)	303		

Values with * are assumed

Table 3 The Langmuir–Freundlich isotherm parameters for CO_2 and CH_4 onto pine sawdust AC [7]

Adsorbent	Pine sawdust AC	
Adsorbate	CO_2	CH_4
$IP_1 \times 10^3$ (kmol/kg)	6.107	5.864
$IP_2 \times 10^4$ (1/bar)	6.312	6.030
$IP_3 \times 10^1$ (–)	7.315	7.831
$IP_4 \times 10^{-3}$ (K)	2.108	1.864
$IP_5 \times 10^4$ (1/bar)	8.507	7.688
$IP_6 \times 10^{-3}$ (K)	1.982	1.847

3.1 CO_2 Breakthrough Time Dataset Generation and Preprocessing for ANN

The artificial neural network (ANN) is used to predict the CO_2 breakthrough time of the dynamic packed bed adsorption column. Four inputs have been chosen for the ANN model, including the system's pressure, feed volume flowrate, bed column height, and CO_2 feed mole fraction. The datasets are generated using the validated packed bed column mathematical model described in Sect. 2, by running the model at pre-defined experimental conditions, and the result in terms of CO_2 breakthrough time is recorded (time at which the outlet CO_2 component mole fraction reaches 0.05).

To initialize the training of the ANN, the dataset is pre-processed to normalize or standardize the dataset. The effect of different data pre-processing transformation techniques is investigated to explore different techniques on the learning performance of ANN. The data pre-processing transformation techniques (Table 4) studied are min–max scaler normalization (MMN), z-score normalization (ZSN), decimal scaling normalization (DSN), Box-Cox transformation normalization (BCN), and log transformation normalization (LTN). Then, the datasets are divided into training (80%) and testing/validation (20%) performance sets.

3.2 Topology and Activation Functions for ANN Model

The structure of the artificial neural network (ANN) is a significant key that determines the accuracy of the outputs. In this work, an ANN of feed-forward neural network is applied to model the CO_2 breakthrough time with four inputs. Figure 1 shows the ANN architecture used in this study. The ANN comprises three layers, including an input, an output, and one hidden layer with ten neurons.

Different activation functions give a different performance of ANN model training. Several activation functions (Table 5) in the hidden layer are investigated by

Table 4 Different data pre-processing transformation techniques equation

Data pre-processing transformation	Expression
MMN	$X_{MMN\_ scaled} = \frac{X - X_{min}}{X_{max} - X_{min}}$ where X_{min}, X_{max} are data's value minimum and maximum
ZSN	$X_{ZSN\_ scaled} = \frac{(X - \mu)}{SD}$ where μ = mean, SD = standard deviation
DSN	$X_{DSN\_ scaled} = \frac{X}{10^{i}}$ where i is the smallest integer such that max $\lvert X_{DSN\_scaled} \rvert < 1$
BCN	$X_{BCN\_ scaled} = \frac{X^{\lambda} - 1}{\lambda}$ where λ = constant
LTN	$X_{LTN\_ scaled} = \log_{10} X$

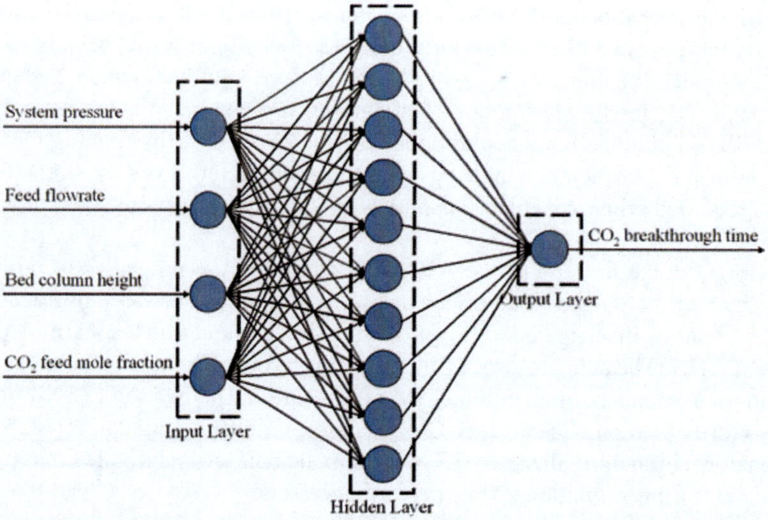

Fig. 1 Artificial neural network architecture applies for the CO_2 packed bed adsorption column modelling

testing different functions, including *sigmoid*, *tanh*, *relu*, *elu*, and *selu*, at a different data pre-processing transformation technique. The output layer activation function is fixed using a *linear* transfer function. An extended version of the stochastic gradient descent method, the Adam optimization algorithm, is used to train the neural network, with a learning rate of 0.1.

Table 5 Respective equation of different activation functions

Activation functions	Equation
Sigmoid	$f(x)=\frac{1}{1+e^{-x}}$
Tanh	$f(x) = \frac{e^x-e^{-x}}{e^x+e^{-x}}$
ReLU	$f(x) = \begin{cases} 0, & x \le 0 \\ x, & x > 0 \end{cases}$
ELU	$f(x) = \begin{cases} \alpha(e^x - 1), & x \le 0 \\ x, & x > 0 \end{cases}$ where $\alpha = 1.0$
SeLU	$f(x) = \begin{cases} \beta x, & x \le 0 \\ \alpha\beta(e^x - 1), & x > 0 \end{cases}$ where $\alpha = 1.67326324$, $\beta = 1.05070098$
Linear (output function)	$f(x) = x$

4 Results and Discussion

4.1 *Packed Bed Column Mathematical Model Validation*

Figure 2 compares the experiment (points) and the model (solid lines) results at different pressures, ranging from 1.2 to 10 bar. Overall, the simulation outputs are the same as the experimental results reported by [7]. Thus, the packed bed mathematical model is said to be well-correlated with the experimental results. Though, they appear to show a slight deviation, most probably due to the difference in the overall mass transfer coefficient and/or the axial dispersion coefficients that were predicted by the correlations.

The validated packed bed column mathematical model is used to generate a different combination of experimental conditions. A total of 30 simulated datasets are used to train and test the ANN model. Table 6 shows the descriptive statistics of the input and output model parameters used for the ANN model. The X_{min}, X_{max}, μ, and SD represent the data's minimum, maximum, mean, and standard deviation.

4.2 *Effect of Data Transformation Techniques and Activation Functions on the Learning Ability of ANN*

The aim of data pre-processing is to facilitate the learning process of the ANN by transforming and scaling the dataset in a proper way. The effect of different data pre-processing transformation techniques and activation functions in the hidden layer

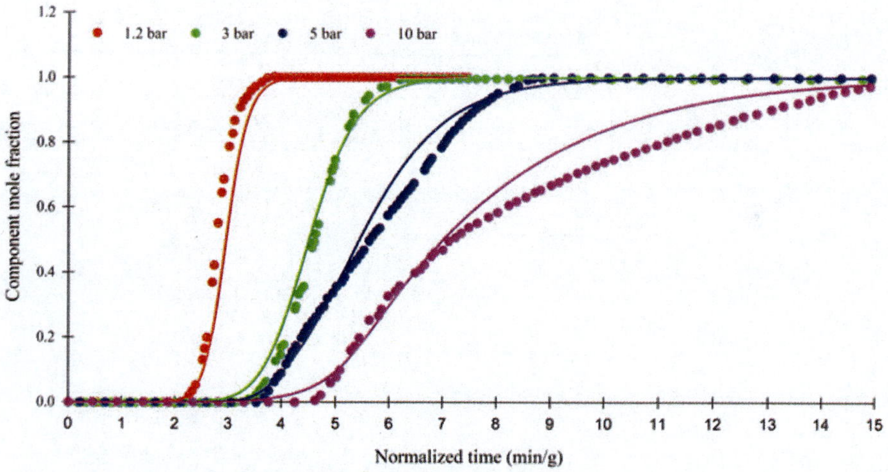

Fig. 2 Comparison between simulation results (solid lines) with experimental data (points) of breakthrough curves for CO_2/CH_4 adsorption. The CH_4 breakthrough curve are not shown

Table 6 Statistical indices of the input and output ANN model parameters of the dataset

Parameters	Unit	X_{min}	X_{max}	μ	SD
Input model parameters					
System's pressure	bar	1	10	5.5	3.5453
Feed volume flowrate	mL/min	20	100	60	31.5135
Bed column height	m	0.1	1	0.55	0.3545
CO_2 feed mole fraction	-	0.2	0.8	0.5	0.1182
Output model parameter					
CO_2 breakthrough time	min	2.0526	358.7903	67.6231	85.7527

on the ANN model prediction performance were established to select the optimal option for the ANN model training and validation. Figure 3 displays the kernel density function of raw and pre-processed datasets (output values only) using different transformation techniques. The raw dataset is positively skewed, as shown by the extended right-hand tail of the plot in Fig. 3a.

Figure 3 indicates that the BCN and LTN techniques able to normalize the dataset and follow nearly the bell-shaped normal distribution (skewness close to 0), as compared to the raw dataset before pre-processing. The normality is not achieved by MMN, ZSN, and DSN, as can be seen from the skewness, however the data values were narrowed to a certain range. The details of each data pre-processing transformation technique are presented in Table 7, with their respective root mean square error (RMSE) and R^2 at different activation functions used in the hidden layer.

Table 7 shows the RMSE and R^2 obtained with several data pre-processing techniques (MMN, ZSN, DSN, BCN, LTN) using different activation functions in the

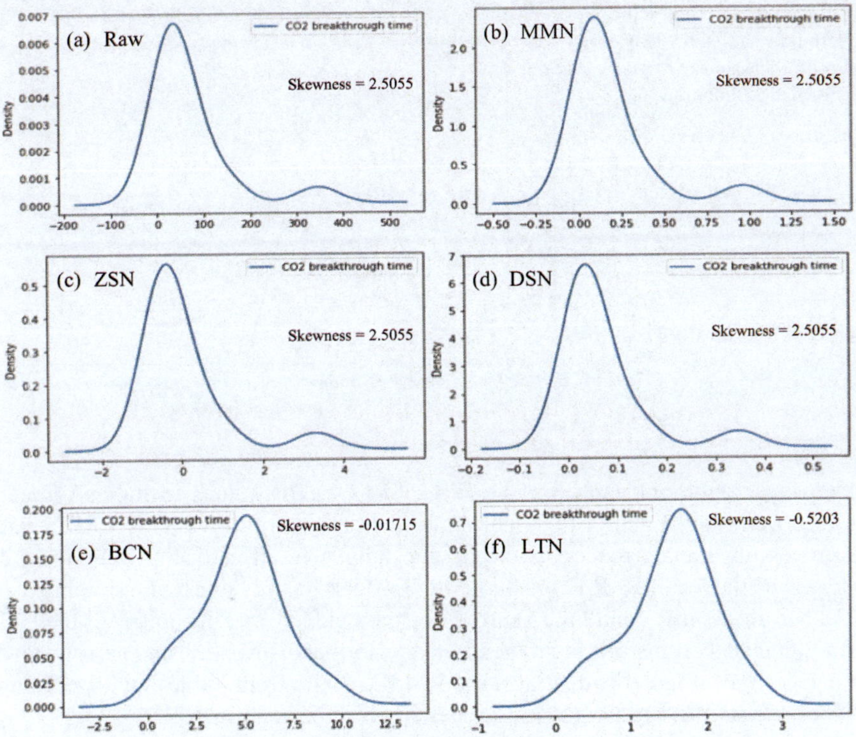

Fig. 3 Kernel density function of the output raw and pre-processed datasets

Table 7 RMSE and R^2 on different data transformation techniques and activation functions in the hidden layer. The RMSE and R^2 were calculated after the data de-normalization

Data transform	Root mean square error (R^2)				
	Sigmoid	Tanh	ReLU	ELU	SELU
Raw (control)	84.8157 (− 83.3368)	84.8157 (− 83.3368)	6.1667 (0.9946)	*3.3941 (0.9984)*	5.9636 (0.9950)
MMN	3.8534 (0.9978)	4.4584 (0.9973)	3.6890 (0.9981)	*3.4228 (0.9984)*	5.3281 (0.9960)
ZSN	8.5035 (0.9896)	6.7623 (0.9937)	4.9806 (0.9965)	*3.1780 (0.9986)*	8.0155 (0.9913)
DSN	*3.3691 (0.9984)*	8.0959 (0.9910)	21.4899 (0.9311)	13.4822 (0.9754)	11.9340 (0.9790)
BCN	4.6083 (0.9968)	28.1184 (0.8358)	*0.8804 (0.9999)*	1.7928 (0.9996)	4.1695 (0.9976)
LTN	1.1864 (0.9998)	0.9503 (0.9999)	8.6558 (0.9907)	*0.8523 (0.9999)*	1.7685 (0.9996)

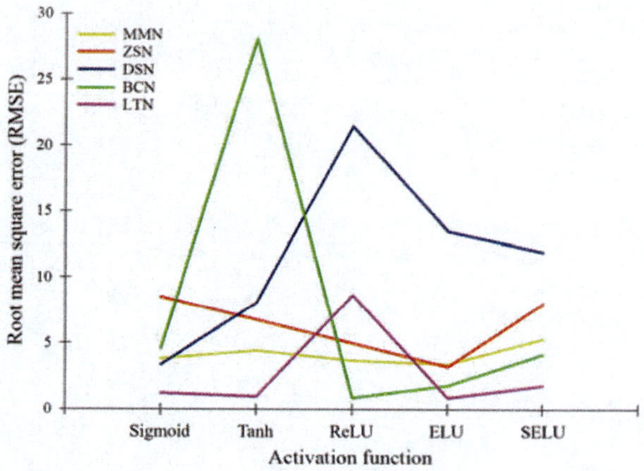

Fig. 4 RMSE variation with activation functions for five data pre-processing transformation techniques

hidden layer (sigmoid, tanh, ReLU, ELU, SELU). The values in italics indicate the best activation function for a particular data transform. Overall, the LTN data pre-processing transformation is the best technique in all activation functions, as indicated by the lower RMSE and higher R^2 (> 0.99), mainly due to the suitability of the LTN for a positive- and high-skewed raw dataset [17], and the ability of the LTN to simultaneously transforming to/nearly to the normal distribution and narrow down the dataset [12]. Figure 4 illustrates the RMSE results from Table 7 with different activation functions for all data pre-processing techniques tested.

4.3 CO_2 Breakthrough Time Prediction by Artificial Neural Network

The LTN data pre-processing transformation technique with *elu* activation function was selected due to its excellent performance in predicting the CO_2 breakthrough time. Figure 5 compares the data sequence (30 data points) of CO_2 breakthrough time between the actual and ANN model predicted. The figure observed nearly perfect overlapping between the actual and ANN model predicted data, suggesting a strong agreement between the trained ANN model and the actual data. Also, according to Table 8, it was proven that the mean difference of CO_2 breakthrough time between the actual and ANN model values is 0.08751 (95% CL [− 0.23446, 0.40948]), and there is no significant average difference between the actual and ANN model values ($t_{29} = 0.556$, $p > 0.05$). The *p*-value = 0.583 (> 0.05) indicates that there is no difference in the data values between the actual and ANN model. Therefore, these two means, the actual and ANN models are not significantly different. In other words, the ANN model prediction of CO_2 breakthrough time is similar to the actual values obtained through the phenomenological modeling through Aspen Adsorption, which suggests the strong predicted values of the ANN model.

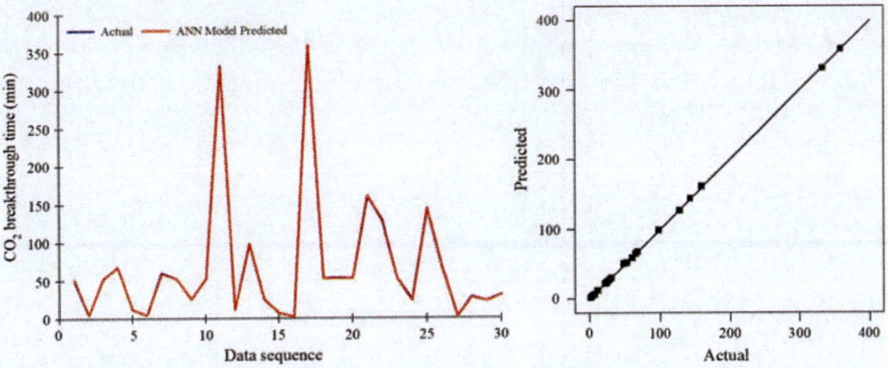

Fig. 5 Agreement between the ANN model predicted and actual values for the chosen ANN model (data de-normalized)

Table 8 Paired samples test between the actual data versus ANN model (LTN-*elu*)

Pair	Paired differences					t	df	Sig. (two-tailed)
	μ	SD	Std. error mean	95% confidence level				
				Lower	Upper			
Actual and ANN	0.08751	0.86226	0.15743	– 0.23446	0.40948	0.556	29	0.583

To predict the CO_2 breakthrough time using the trained ANN model, the values in Table 9 are used. The table reports the weights and biases of the trained ANN model, whereby the model's accuracy is over 99.99%. However, the weights and biases in the table are values for the ANN model trained using LTN normalized datasets. To use them, the ANN's raw input model parameters (system pressure, feed volume flowrate, bed column height, CO_2 feed mole fraction) need to be pre-processed using the LTN technique. Then, the obtained output (CO_2 breakthrough time) needs to be de-normalized back by inversing the LTN, so that the output value is at the appropriate unit.

5 Conclusion

ANN is a reliable model for predicting the CO_2 breakthrough time of packed bed column adsorption. The ANN model with a 4-10-1 structure provides excellent model prediction with an accuracy of 99.99% in excess. The LTN is the best technique for data pre-processing, which showed an R^2 of over 0.99 for all the tested activation functions. When combined with *elu* activation function, the LTN-*elu* ANN model achieved over 99.99% accuracy. It is important to pre-process the dataset to narrow

Table 9 Weights and biases for the ANN model (LTN—*elu*)—$R^2 = 99.99\%$

Input layer weight matrix				Hidden layer bias vector	Hidden layer weight vector	Output layer bias vector
W_{ij}				b_1	W_2	b_2
j = 1	j = 2	j = 3	j = 4			
– 0.228156939	0.395489514	0.055413138	0.948005319	– 0.028703336	– 0.002705482	0.777417898
0.198885605	0.086720623	1.107950091	0.321181089	0.314799041	0.004857629	
0.461167037	– 0.347541213	0.326638877	– 0.348954499	0.298380971	– 0.006402044	
0.627989709	– 0.292449713	0.677941561	0.072305761	0.395670116	0.012166242	
– 0.026199397	1.302408457	– 1.72740829	– 0.154439509	– 1.367269874	– 0.753220499	
– 0.604172587	0.568156004	2.117257357	0.39304021	0.151778728	– 0.083618067	
– 0.277662665	– 0.491306037	0.961406112	1.752529383	– 0.682884336	– 1.075071096	
– 3.537484169	1.08609724	0.013839924	0.267211616	– 0.123217568	– 0.093533047	
– 1.879289985	– 1.353732109	1.346389294	– 0.302497894	– 0.766480446	– 0.967149615	
0.677423835	– 0.691293776	0.553026795	– 0.320476055	0.725408792	0.042563438	

the value ranges and reduce the skewness. The results indicate that the data pre-processing transformation techniques increase the learning capability of the ANN model. This work demonstrates that the ANN model can provide an excellent prediction of chemical processes, in this case, the dynamic adsorption performance prediction in terms of breakthrough time. Furthermore, the ANN model paves the way for rapid design and optimization of dynamic packed bed adsorption column, particularly on the more practical gas separation associated with it, which is the cyclic adsorption process i.e. the pressure swing adsorption (PSA), unlike the laborious and/or costly experimental procedure or numerical simulation.

Acknowledgements The authors are grateful for the scholarship awarded by Universiti Teknologi Malaysia under UTM PhD Zamalah Scholarship and UTM Zamalah Research Fund No. 00N08. Extended thanks also to GRISM Innovative Solutions for technical support.

References

1. Bahrun MHV, Bono A, Othman N, Ahmad Zaini MA (2022) Carbon dioxide removal from biogas through pressure swing adsorption—a review. Chem Eng Res Des 183:285–306
2. IEA: Gas. https://www.iea.org/fuels-and-technologies/gas. Last accessed 02 July 2022
3. Mhadmhan S, Ngamcharussrivichai C, Hinchiranan N, Kuchonthara P, Li Y, Wang S, Reubroycharoen P (2022) Direct biogas upgrading via CO_2 methanation to high-quality biomethane over NiMg/CNT-SiO_2 fiber catalysts. Fuel 310:122289
4. Grande CA (2012) Advances in pressure swing adsorption for gas separation. ISRN Chem Eng 2012:1–13
5. Hodaifa G, Alami SBD, Ochando-Pulido JM, Víctor-Ortega MD (2014) Iron removal from liquid effluents by olive stones on adsorption column: breakthrough curves. Ecol Eng 73:270–275
6. Bahrun MHV, Kamin Z, Anisuzzaman SM, Bono A (2021) Assessment of adsorbent for removing lead (Pb) ion in an industrial-scaled packed bed column. J Eng Sci Technol 16:1213–1231
7. Durán I, Rubiera F, Pevida C (2022) Modeling a biogas upgrading PSA unit with a sustainable activated carbon derived from pine sawdust. Sensitivity analysis on the adsorption of CO_2 and CH_4 mixtures. Chem Eng J 428
8. Ma S, Tong L, Ye F, Xiao J, Pierre B, Chahine R (2019) Hydrogen purification layered bed optimization based on artificial neural network prediction of breakthrough curves. Int J Hydrogen Energy 44:5324–5333
9. Chittoo BS, Sutherland C (2020) Column breakthrough studies for the removal and recovery of phosphate by lime-iron sludge: Modeling and optimization using artificial neural network and adaptive neuro-fuzzy inference system. Chin J Chem Eng 28:1847–1859
10. Bai S, Li J, Ding W, Chen S, Ya R (2022) Removal of boron by a modified resin in fixed bed column: breakthrough curve analysis using dynamic adsorption models and artificial neural network model. Chemosphere 296
11. Nawi NM, Atomi WH, Rehman MZ (2013) The effect of data pre-processing on optimized training of artificial neural networks. Procedia Technol 11:32–39
12. Bowden GJ, Dandy GC, Maier HR (2003) Data transformation for neural network models in water resources applications. J Hydroinform 5:245–258
13. Plumb AP, Rowe RC, York P, Brown M (2005) Optimisation of the predictive ability of artificial neural network (ANN) models: a comparison of three ANN programs and four classes of training algorithm. Eur J Pharm Sci 25:395–405

14. Aggarwal R, Kumar R (2015) Effect of training functions of artificial neural networks (ANN) on time series forecasting. Int J Comput Appl 109:14–17
15. Fuller EN, Schettler PD, Giddings JC (1966) A new method for prediction of binary gas-phase diffusion coefficients. Ind Eng Chem 58:18–27
16. Gutiérrez Ortiz FJ, Barragán Rodríguez M, Yang RT (2019) Modeling of fixed-bed columns for gas physical adsorption. Chem Eng J 378:121985
17. Feng C, Wang H, Lu N, Chen T, He H, Lu Y, Tu XM (2014) Log-transformation and its implications for data analysis. Shanghai Arch Psychiatry 26:105–109

A Deep Learning Model with Name Attention to Predict the Stock Trend from News Headline

Tien-Ping Tan, Phaik Ching Soon, Marwan Aleasa, Huah Yong Chan, and Keng Hoon Gan

Abstract Market sentiment is the attitude or feeling that traders or investors have about a company, industry, or financial market. Day traders and technical analysts rely on market sentiment, as it influences the price movement of a stock. Sentiment analysis is a natural language processing approach to analyze market sentiment. In this work, we analyze the effect of the news on stock price trends on the Kuala Lumpur Stock Exchange or Bursa Malaysia. Specifically, we analyze the financial news headlines involving one or more listed companies and determine whether it will move the stock price up, down, or flat. We propose a deep neural network architecture with name attention to model the relationship between a news headline and the stock price trend. The main idea of the approach is to focus on the target company in the headline when modeling the relationship between a headline and the stock trend using attention. The experiment result shows that the proposed approach performs better than other baseline approaches in terms of accuracy, precision, recall, and f1.

Keywords Name attention · Deep neural networks · Stock price trend

1 Introduction

Market sentiment is the attitude or feeling that traders or investors have about a company, industry, or financial market, due to an exterior event such as the announcement of the quarterly report, the changes in the management of a company, the price movement of a stock, the release of the economic data, the changes in the government policy, pandemics, etc. Market sentiment is also called "investor sentiment".

T.-P. Tan (✉) · P. C. Soon · M. Aleasa · H. Y. Chan · K. H. Gan
School of Computer Sciences, Universiti Sains Malaysia, 11800 USM, Penang, Malaysia
e-mail: tienping@usm.my

H. Y. Chan
e-mail: hychan@usm.my

K. H. Gan
e-mail: khgan@usm.my

© The Author(s), under exclusive license to Springer Nature Singapore Pte Ltd. 2023
D.-K. Kang et al. (eds.), *Proceedings of the 9th International Conference on Computational Science and Technology*, Lecture Notes in Electrical Engineering 983,
https://doi.org/10.1007/978-981-19-8406-8_29

Day traders and technical analysts rely on market sentiment, as it influences the price movement of stocks. Sentiment analysis is a natural language processing approach to analyze market sentiment. Sentiment analysis in the financial market simplifies the analysis of complex events such as the analysis of the quarterly report, macroeconomic data, and news events that often required professional expertise. Often, the output of the sentiment analysis is in the form of categories (e.g., positive, negative, or neutral), or the sentiment analysis can be a part of the stock price prediction module.

In this work, we analyze the effect of the news on stock price trends on the Kuala Lumpur Stock Exchange or Bursa Malaysia. Specifically, we analyze the financial news headlines involving one or more listed companies and determine whether it will move the stock price up, down, or flat. Instead of evaluating the sentiment and then predicting the effect of the sentiment on a stock price, we model the relationship between a piece of news and the stock price directly. In general, financial news involving a company can be divided into the following topics: quarterly results, financial analyst report, and news events. Some examples of financial news headlines from The Edge Markets are shown below.

Quarterly Result

- `2019-10-30, Caring Pharmacy 1Q net profit rises 2.27% on higher sales from 16 new stores`
- `2019-11-18, Cash balance grows but liquidity still an issue at AirAsia X.`

Financial analyst Report

- `2019-08-30, Carimin Petroleum may rise higher, says RHB Retail Research`
- `2019-10-15, CGS-CIMB Research raises target price for MISC to RM8.61.`

Events

- `2017-10-02, CPO prices soften outlook for Kim Loong's 2H`
- `2019-12-05, Gamuda expects to sign Penang Transport Master Plan PDP agreement in 'next few weeks', says MD`
- `2020-01-15, Celcom eyes higher non-consumer mobile business revenue contribution.`

2 Related Works

News events can affect the stock market. For example, when a company announces an unexpectedly good result, the stock price may increase. Lim et al. [1] investigate different machine learning techniques such as decision trees, naive Bayes, SVM, neural networks, etc., in predicting the news sentiment in news headlines. Song et al. [2] used news sentiment to predict stock price movement. Their study found that different news may affect an investor's reaction toward financial markets differently, and the extreme changes in the news sentiment may result in better signals when it comes to predicting the cumulative market return and implied volatility of the S&P 500 index. Sentiment shocks are abnormal changes in the prices that may be caused by the release of unexpected macroeconomic data, financial report results, corporate actions, etc. They showed that extreme positive sentiment is a reliable clue to predicting the potential increase in stock prices. The finding contradicts the study from [3] that found that investors often underreact to the official news (e.g., quarterly report results) and overreact to external good or bad news. The approach [2] gave a higher return than the simple buy-and-hold S&P500 strategy. Their approach proposed buying stocks when the sentiment is extremely positive and selling stocks when the sentiment is extremely negative. However, the approach using sentiment trend tracing has shown to provide a better cumulative return compared to using sentiment shock. Riordan et al. [4] studied the impacts of positive, negative, and neutral messages on price discovery and trading activity. They have shown that negative news produces stronger reactions compared to positive news.

Li et al. [5] developed an approach that combines technical indicators from stock prices and news sentiments to predict stock movement. For sentiment analysis, dictionaries were used to generate sentiment score. The sequential representation constructed from technical indicators and sentiment vectors were fed into a two-layer LSTM neural network followed by a dense layer. Jing et al. [6] extracted textual data from a major stock forum to analyze the sentiment. They proposed a sentiment analysis model based on the Convolutional Neural Network (CNN). It was integrated to an LSTM neural network for stock price prediction.

3 Methodology

We propose a deep neural network architecture to model the relationship between headlines and stock price trends. The main idea of the approach is to focus on the target company in the headline when modeling the relationship between a headline and the stock trend. The architecture consists of two sentence encoders. A sentence encoder performs sentence embedding on the headline, and another sentence encoder embeds the target company name to the sentence vector. An attention distribution is calculated from the output of both encoders. The purpose of the attention layer is to focus on the name of the company in question in the headline as a headline can have more than one company. A convolutional layer and max-pooling layer are then used to extract high-level features. The features are then classified into one of three possible trends: up, down, or flat.

Figure 1 shows our proposed architecture. Given a headline, x_h, and the name of a company, x_c, the texts are tokenized and normalized. The tokens in the sentences are then converted to corresponding embedding vectors, v_h, and v_c. The word embedding vectors can be pretrained using word embedding algorithms such as word2vec [7], GloVe [8], etc. from a text corpus. The word vector v_h is input to a long-short term memory, $LSTM_h$, while the vector v_c is input to an $LSTM_c$. At each timestep, an LSTM outputs a state. The $LSTM_h$ outputs hidden states h, while the $LSTM_c$ outputs hidden states s. The attention scores, e, are calculated using Eq. (1). The scores are then normalized using softmax function with Eq. (2) and produce attention distribution, α [9]. The attention distribution is then concatenated to h. If the length of the headline and the company name is N and M respectively, then an attention matrix of size NxM will be produced.

$$e(t,i) = \left[s(t)^T h(i=1), s(t)^T h(2) \ldots, s(t)^T h(N)\right] \quad (1)$$

$$\alpha(t,i) = softmax(e(t,i)) \quad (2)$$

$$u(t) = [h(t); \alpha(t)] \quad (3)$$

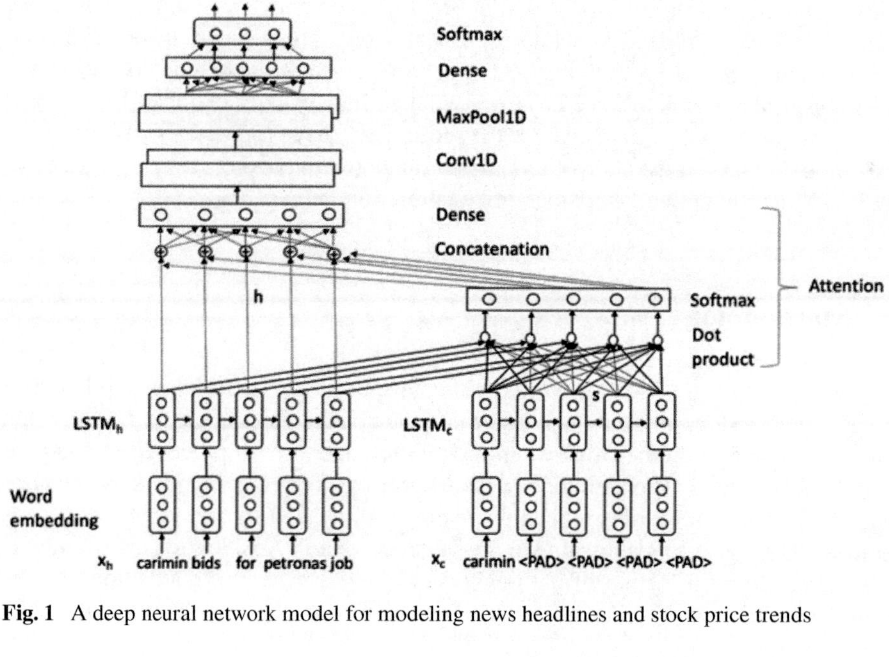

Fig. 1 A deep neural network model for modeling news headlines and stock price trends

The u matrix is then passed to a dense layer. The output of the dense layer then goes into the convolutional layer. The convolutional layer extracts high-level features from the input matrix. The max-pooling layer that comes after the convolutional layer then selects the prominent features from the matrix. The matrix is flattened to become a single dimension vector. The vector is input to a dense layer of neurons. The output layer consists of a dense layer with three neurons followed by a softmax layer that classifies the headline as up, down, or flat.

4 Experiments

Bursa Malaysia stock data were extracted from Yahoo Finance. News headlines from 2017 to 2021 were extracted from The Edge Market news websites. We used the news articles from The Edge Market because the news is current and up to date. Thus, it is one of the most trusted financial news publishers in Malaysia [10]. About 58 thousand news headlines and their publication date and time were extracted. Regular expressions were used to match the names of the listed company in the headlines. A news headline may have zero or more matching companies. News headlines without any matches will be discarded. For each matching company, the stock prices for the company will be retrieved based on the publication date using the rules in Table 1.

For instance, consider the following news headline: "`Dialog` awarded `Petronas` master service agreement", with the publication date and time of `2019-07-01T18:41:56 + 08:00`. The matching companies for the headlines are Dialog and Petronas. Thus, the close prices, Price_1 for Dialog and Petronas on 1 July 2019 and Close prices, Price_2 for Dialog and Petronas on 2 July 2019 were retrieved. The data was annotated automatically. There are three categories: uptrend, downtrend, and flat. If the difference between Price_2 and Price_1 (Price_2-Price_1/Price_1) is more than $+$ 1%, then it is "uptrend", and annotated as 2. But if the difference is lower than $-$ 1%, then it is annotated as 0, or "downtrend", else it is annotated as 1, "flat". A total of 24,388 samples were generated. Table 2 presents the trend distribution of the dataset.

The dataset was split according to the ratios 0.7, 0.1, and 0.2 to construct the train set, test set, and validation set respectively. The texts were tokenized using NLTK English word tokenizer [11]. We evaluated the conventional machine learning

Table 1 Rules for retrieving stock prices based on the published time

Published time	Price_1	Price_2
Day i, Weekday, 9 am–5 pm	Day i Open price	Day i Close price
Day i, Weekday (except Friday and holiday), 5 pm–Day i + 1, 9 am	Day i Close price	Day i + 1 Close price
Day i, holiday–Day j, weekday, 9 am	Day i-n Close price	Day j Close price
Day i, Friday, 5 pm–Day j, Monday, 9 am	Day i Close price	Day j Close price

Table 2 Distribution of the dataset

Trend	Count	Distribution
Down	6802	0.2789
Flat	9268	0.38
Up	8318	0.3411

algorithms such as decision tree, Naïve Bayes, and SVM in predicting the trend. Table 3 shows the accuracy, $\text{Precision}_{\text{macro}}$, $\text{Recall}_{\text{macro}}$, and F1_{macro} of the algorithms when evaluated on the test set. SVM produces the best result in terms of accuracy, $\text{Precision}_{\text{macro}}$, $\text{Recall}_{\text{macro}}$, and F1_{macro} compared to decision tree and Naïve Bayes.

For the proposed approach, word embedding vectors were pretrained using the word2vec algorithm. The news headlines and news articles were used as input text for learning the word embedding vectors. The minimum word count was set at 2, and the size of the word vector was set at 64. The setting for the proposed model was as follows: the timestep for LSTM_h was set to 100, while the timestep for LSTM_c was set to 5. The number of hidden states for both LSTMs was set to 128. The number of hidden states for the first dense layer was set to 32. The size of the filter and the size of the kernel for the convolutional layer were set to 32 and 3 respectively. The max-pooling size was set at 2, and the number of hidden states for the second dense layer before the output layer was set at 16.

For training the model, NLTK English word tokenizer was first used to tokenize the texts. The words were then mapped to vectors using the word embedding vectors pretrained earlier using word2vec [7], and the vectors were then input to the LSTM layers. Adam was used as the optimizer for training the model. The batch size was set at 16, and the maximum number of epochs was 8. At every 100 steps, the current model will be evaluated using the validation set. If the model scores a higher accuracy than the previous model, the model will be saved. Figure 2 shows the changes in the accuracy of the proposed model for the training set and validation set. The training stops after it completed the 4th epoch. From the figure, we can see that the accuracy of the model improves over time. We evaluated the test set using the best model stored at the validation step.

Besides the conventional machine learning algorithms, we also evaluated some baseline deep learning classifiers, such as the LSTM model and the bidirectional LSTM model. For both the LSTM layer, the number of hidden states was set at

Table 3 Baseline classification results

Classification approach	Accuracy	$\text{Precision}_{\text{macro}}$	$\text{Recall}_{\text{macro}}$	F1_{macro}
Decision tree	0.4190	0.4008	0.3999	0.3992
Naïve Bayes	0.2956	0.3716	0.3461	0.2699
SVM	**0.4510**	**0.4324**	**0.4136**	**0.4084**

The bold text represents the result of using SVM to classify the sentiment of the headlines. The accuracy, precision, recall, and F1 are 0.4510, 0.4324, 0.4136, and 0.4084 respectively

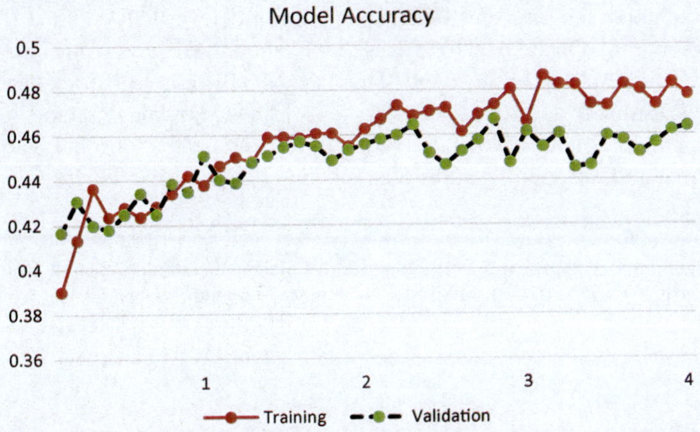

Fig. 2 Changes in the accuracy of the training set and validation set during model training

Table 4 Comparing the proposed model result with baseline models

Classification approach	Accuracy	$\text{Precision}_{\text{macro}}$	$\text{Recall}_{\text{macro}}$	F1_{macro}
SVM	0.4510	0.4324	0.4136	0.4084
LSTM	0.4474	0.4492	0.4183	0.3898
Bidirectional LSTM	0.4559	0.4528	0.4360	0.4292
LSTM + CNN + Name attention (proposed)	**0.4758**	**0.4762**	**0.4573**	**0.4467**

128. After the LSTM layer, it is followed by two dense layers with the number of hidden states set at 32 and 8 respectively, before connecting to the output layer. The same tokenizer, and word embedding vectors used the proposed model were applied when training the baseline models. Table 4 shows the results of our proposed model compared to other baseline models. The result shows that the result of the proposed approach is better than the baseline models in term of accuracy, precision, recall and f1. Bidirectional LSTM has a slightly better results compared to SVM. Our proposed model achieved about 0.2 absolute improvement in term of accuracy, precision, recall, and f1, compared to the bidirectional LSTM.

5 Conclusions

This paper proposes a deep neural network model with attention mechanism to focus on the company in the headline during the modeling of the relationship between headlines and stock price trends. The proposed approach performs better than other baseline approaches in term of the accuracy, precision, recall, and f1. By comparing the results of the proposed approach and the distribution of the dataset, it shows that the proposed approach performs better than random. Nevertheless, the difference is not by huge margin. One reason for this is that by the time a news come out, the stock price has already reacted to the event. This is especially true for stock news related to quarterly results. In addition, a stock price is influence by many factors and thus it is a challenging problem. The future work will look at using additional features such as stock price and introducing additional attributes to improve the prediction of the stock trend.

Acknowledgements This research was supported by the Fundamental Research Grant Scheme FRGS/1/2021/ICT02/USM/02/4 from Ministry of Higher Education Malaysia.

References

1. Lim SLO, Lim HM, Tan EK, Tan T-P (2020) Examining machine learning techniques in business news headline sentiment analysis. In: Alfred R, Lim Y, Haviluddin H, On C (eds) Computational science and technology. Lecture notes in electrical engineering, vol 603. Springer, Singapore
2. Song Q, Liu A, Yang SY, Deane A, Datta K (2015) An extreme firm-specific news sentiment asymmetry-based trading strategy. In: 2015 IEEE symposium series on computational intelligence, pp 898–904. Cape Town
3. Barberis N, Shleifer A, Vishny R (1998) A model of investor sentiment. J Finance Econ 49(3):307–343
4. Riordan R, Storkenmaier A, Wagener M, Zhang SS (2013) Public information arrival: price discovery and liquidity in electronic limit order markets. J Bank Finance 37(4):1148–1159
5. Li X, Wu P, Wang W (2020) Incorporating stock prices and news sentiments for stock market prediction: a case of Hong Kong. Inf Process Manag 57(5):102212
6. Jing N, Wu Z, Wang H (2021) A hybrid model integrating deep learning with investor sentiment analysis for stock price prediction. Expert Syst Appl 178
7. Mikolov T, Sutskever I, Chen K, Corrado GS, Dean J (2013) Distributed representations of words and phrases and their compositionality. In: Advances in neural processing information systems. Nevada, pp 3111–3119
8. Pennington J, Socher R, Manning CD (2014) GloVe: global vectors for word representation. In: Proceedings of empirical methods in natural language processing (EMNLP). Doha, pp 1532–1543
9. Luong M-T, Pham H, Manning CD (2015) Effective approaches to attention-based neural machine translation. In: Proceedings of conference on empirical methods in natural language processing. Lisbon, Portugal, pp 1412–1421
10. Im TL, San PW, On CK, Alfred R, Anthony P (2013) Analysing market sentiment in financial news using lexical approach. IEEE Conf Open Syst (ICOS) 2013:145–149
11. Bird S, Klein E, Loper E (2009) Natural language processing with Python: analyzing text with the natural language toolkit. O'Reilly Media, Inc.

Hand Gesture Signature Recognition with Machine Learning Algorithms

Azra Izni Anis binti Azlin, Pang Ying Han, Khoh Wee How, and Ooi Shih Yin

Abstract Hand gesture recognition is regarded as an imperative Human–Computer Interaction technology for non-verbal communication and interaction between human and computers. Inspired by hand gesture recognition systems, hand gesture signature recognition is proposed to replace the existing online handwritten signature recognition which requires a stylus pen and tablet as middleware between users and the systems. The possibility of germs contamination on these shared devices alerts public concerns, especially during this COVID-19 pandemic. Hence, an in-air signature recognition, termed as iSign, that utilizes hand gesture signing movement is proposed in this paper. In iSign, a Kinect Sensor is adopted to detect and capture the hand movement of the users while signing. This mode of data acquisition does not require users for any direct touch on any devices. Next, the captured video data is transformed into feature template. Various machine learning algorithms, i.e. Bayesian network, Naïve Bayes, Decision Tree, Random Forest etc., are examined to classify the hand gesture signatures. Empirical results demonstrate that Random Forest performs the best among the machine learning algorithms with accuracy of 93%.

Keywords Hand gesture · Signature recognition · Machine learning · Random forest

A. I. A. binti Azlin · P. Y. Han (✉) · K. W. How · O. S. Yin
Faculty of Information Science and Technology, Multimedia University, 75450 Ayer Keroh, Melacca, Malaysia
e-mail: yhpang@mmu.edu.my

A. I. A. binti Azlin
e-mail: 1171201765@student.mmu.edu.my

K. W. How
e-mail: whkhoh@mmu.edu.my

O. S. Yin
e-mail: syooi@mmu.edu.my

© The Author(s), under exclusive license to Springer Nature Singapore Pte Ltd. 2023

D.-K. Kang et al. (eds.), *Proceedings of the 9th International Conference on Computational Science and Technology*, Lecture Notes in Electrical Engineering 983,
https://doi.org/10.1007/978-981-19-8406-8_30

1 Introduction

Signatures have been counted as an identity authentication form in the society for hundreds of years. Signature recognition is the most used approach in commercial and legal industries for personal authentication. There are two mode of signature analysis methods: offline and online [1]. In offline signature mode, the signature patterns are signed on papers and scanned. The digitized signature data is further processed and analysed for verification. Online signature system requires data acquisition devices, such as stylus pen and tablet to capture the static and dynamic signing data such as geometrical shapes, pressure sensitivity, velocity, acceleration, etc.

In these recent years, the world has been acutely affected by Coronavirus disease (COVID-19). This global virus is an infectious disease and has a catastrophic effect worldwide causing nearly 5.4 million deaths [2]. There are almost 280 million confirmed cases globally. This COVID-19 pandemic has accelerated the use of touchless technology due to hygiene desires. Hence, the classical offline and online hand signature systems which involve shared or touch-based properties (i.e., paper, pen, stylus pen, and tablet) should be replaced to minimize the spread of the virus through the contaminated shared/common device. One of the alternatives is to use touchless handwritten signature recognition technology.

Dynamic hand gesture is a pattern of a temporal sequence of points concerning the hand centroid of the individual while performing a hand motion [3]. Hand gesture recognition is a popular Human–Computer Interaction technology for non-verbal touchless communication and interaction between humans and computers. Inspired by the hand gesture system, hand gesture signature recognition is proposed to replace the existing handwritten signature recognition as a means of touchless personal identity verification technology. In this paper, an in-air handwritten signature recognition, termed as iSign, that utilizes hand gesture signing movement is proposed. Signing using hand gestures offers several discriminative characteristics for classification such as signature shape, hand motion orientation, position or trajectory, and motion speed or direction. Orientation is how the hand is being rotated whether in left or right directions; the signature shape is regarding how the signature is formed due to the unique signing behaviour of individuals; position or trajectory is about how users move their hands while signing; motion speed or direction is how fast the hands' movements are when the users are signing.

The iSign system is non-intrusive and reliable since the hand gesture signature is utterly untraceable. The touchless characteristic of iSign resolves the hygiene issue and diminishes the likelihood of the outspread of contagious viruses. In iSign, there are four main phases: signing data acquisition, data pre-processing, feature extraction/projection and data classification. The overview of iSign is illustrated in Fig. 1. During the data acquisition, a Microsoft Kinect Sensor is adopted to detect and capture the hand signing movement into a video format. This mode of data acquisition does not require users for any direct touch on any devices. Next, the captured video data is pre-processed for enhanced quality before data analysis. In this phase, pre-processing processes such as region of interest localization, background subtraction,

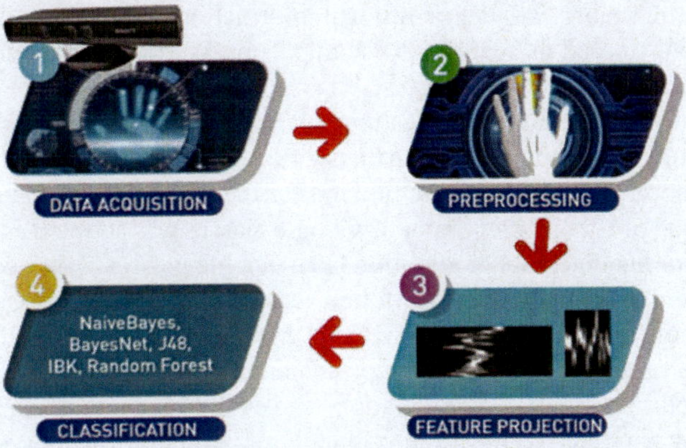

Fig. 1 Overview of the proposed system

denoising, etc. are performed. Then, this pre-processed data is transformed into feature template. Lastly, various machine learning algorithms, i.e. Bayesian network, Naïve Bayes, Decision Tree, Random Forest etc., are examined to classify the hand gesture signatures.

2 Related Works

Since Microsoft Kinect sensor is the main acquisition device in iSign, a brief description of this sensor is discussed here. The affordable price and ample availability of Kinect sensor propagate the application of Kinect sensor in the gesture recognition [4]. In recent years, Kinect camera is gradually replacing time-of-flight camera and stereo camera in Human–Computer Interaction community for visual computation. This is because Kinect camera is not only able to capture colour information, but it also affords depth information as an additional knowledge for computer vision processing [5]. Hand gesture signature recognition can be grouped as an online signature recognition. The core difference between both is that the latter is using a stylus pen and digital tablet for signature acquisition; whereas, the former is using camera, i.e., Kinect sensor, to capture the signing movement of the hand. In literature, Dynamic Time Warping is one of the popular techniques used in online signature recognition [6, 7]. Tian et al. [8] proposed in-air password recognition using a Kinect camera. Password signing trajectory was computed by associating the user's fingertip in each video frame. Then, Dynamic Time Warping was employed to analyse and classify the signing trajectory. There is another similar work that utilizes the depth information captured by the Kinect sensor for hand gesture signature recognition [9]. In this work, the positional trajectory was determined by connecting the fingertip and palm mass centre points. Besides, additional information, such as velocity and acceleration, was considered as features. Dynamic Time Warping was used as distance classifier on the extracted features.

Furthermore, a personal identification for home entertainment based on hand gesture signature was proposed [10]. In this work, point distribution model and neighbourhood components analysis were utilized for measuring the signature trajectory distance. Instead of using Kinect camera, Google glasses was adopted as a data acquisition device for hand gesture recognition [11]. Google glasses was used to track the finger movement of the hand gesture. Motion vectors were computed by using Optical Flow calculation. Besides, Gaussian Mixture Modelling and Principal Component Analysis were used for binary mark representative feature. Lastly, Dynamic Time Warping is used for signature classification. Sun et al. [12] presented a hand gesture recognition for personal verification to unlock mobile phone. This system utilized three-dimensional (3D) acceleration information. A template updating strategy was proposed for persistent template update to tackle the physiological variations of the hand gestures.

Besides, an imagery-based hand gesture signature recognition system was proposed [13]. This system manipulated both temporal and spatial information of the signing hand gesture. To be specific, hand movement orientation and hand shape were examined to learn the hand gesture signature. This rich feature analysis resulted an encouraging performance in personal identification and verification. Additionally, the authors further improved their hand gesture signature recognition with multi-view spatiotemporal analysis based on spectral histogramming [14]. In this system, data feature learning is performed via filter convolution and nonlinear feature analysis through a series of feature binarization, block-wise histogramming and histogram concatenation for a global descriptor. In recent years, deep learning methods have been explored in hand gesture recognition. For instances, a Long Short-Term Memory (LSTM) had been applied in a hand gesture recognition based on trajectories features [15]. This proposed deep learning model was constructed for self-learning the spatial, temporal, and Doppler features of the hand motions. Yu et al. [16] proposed a dynamic hand gesture recognition based on 2D convolutional neural network and feature fusion. The authors adopted a fractional-order model to extract optical flow frames of the data video and then incorporated the fractional-order into the neural network. Empirical results showed a promising performance of this deep learning approach.

3 The Proposed System

As aforementioned, iSign system comprises of four phases: signing data acquisition, data pre-processing, feature extraction/projection and data classification. Microsoft Kinect camera is employed to capture the hand gesture signature data. The captured data is in a video format, a sequence of depth images which describes the hand motion of a subject while signing. Next, this raw data is pre-processed to segment the region of interest (ROI). The nearest point from the camera (i.e., the nearest fingertip from

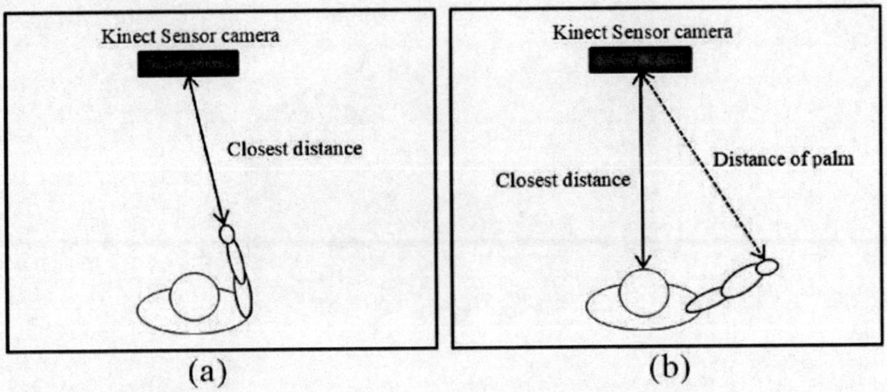

Fig. 2 (Left) the detection of the nearest point, i.e., signing fingertip. (Right) the loss of signing motion tracing, i.e., the detection of another body part as the nearest point (the picture is adopted from [13])

the camera) is utilized to trace the signing motion. There is always a challenge that the signing hand may be moving backward during the signing. Hence, an abrupt change of the nearest point from the fingertip to other parts such as nose tip or chest could be possible. This results the loss of signing motion tracing, as illustrated in Fig. 2. A predictive palm segmentation method is proposed to assure the hand allocation precision. More explanation of the method can be found in this paper [17].

After the pre-processing phase, the resultant data is a 3D binary spatio-temporal imagery sequence, as illustrated in Fig. 3. From the figure, we can observe that the resultant data embraces the detected hand location, hand shape, signing motion and speed. Next, feature projection is performed to produce a compact but informative feature template. In this work, *x*-profile projection is computed by projecting the 3D binary imagery sequence along the *y*-axis. This projection unveils the pattern of the time and signing motion horizontally in a compact projected template. Besides, *y*-profile projection is calculated by projecting the 3D sequence along the *x*-axis, exhibiting the pattern of the time, and signing motion vertically. The *x*-profile and *y*-profile projections are illustrated in Fig. 4.

After the feature projection phase, the projected templates, i.e., *x*-profile and *y*-profile, are further analysed using multiple machine learning algorithms for signature classification. In this work, the performances of the classification algorithms, such as Bayesian network, Naïve Bayes, Decision Tree, Random Forest and nearest-neighbour classifier, in recognizing hand gesture signatures are analysed using WEKA software, a java-based machine learning tool.

Bayesian Network is a probabilistic graphical model, composing modes and directed edges, captures conditionally dependent and independent relationships between random variables. It adopts Bayesian inferences for probability computations. Naïve Bayes is an effective classifier algorithm based on the class conditional independence assumption. This classifier estimates the likelihood of a feature vector being connected with a label. J48 decision tree is a statistical classifier that could

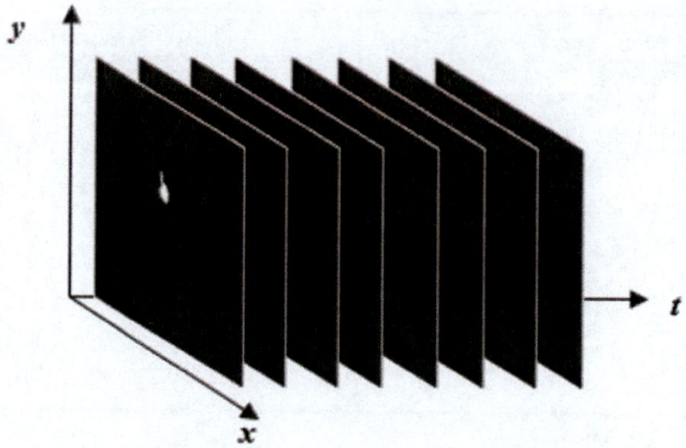

Fig. 3 3D binary spatio-temporal imagery sequence of a signing process

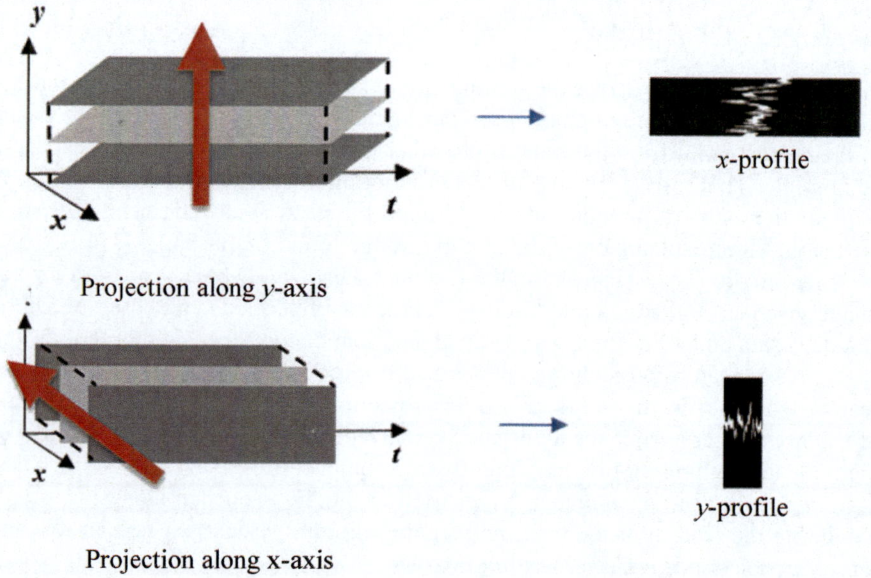

Fig. 4 x-profile (upper) and y-profile (lower) projections

categorize new unlabeled data based on its similarity to those previously classified data. Random Forest is a collection of decision trees in such a way that the tree results are aggregated into a unit result. In other words, this classifier creates a "forest" out of a cluster of decision trees. Nearest-neighbour classifier uses a distance metric to find the nearest occurrences in the training data for each test case.

4 Database and Experimental Protocol

In this work, the proposed system is assessed by using a self-collected Hand Gesture Signature (HGS) database [13]. By using a Microsoft Kinect camera, image sequences of hand signing movement are recorded in a video format. There were 100 volunteers participated in this data collection with age ranging from 18 to 40 years. Each participant contributed 20 HGS samples and these HGS samples would be separated into training and testing sets depending on the implemented testing protocol. Training set is used to train the system model; whereas testing samples are new and unseen to the system model, i.e., there is no overlap in samples between the training and testing sets and used for system performance evaluation.

Two testing protocols were carried out in this work: cross-validation protocol and testing split protocol. In the first protocol, tenfold cross-validation was performed. The HGS database was partitioned into ten equal-sized subsets. A single subset was held for testing the model, and the remaining nine subsets were integrated and used for model learning. The cross-validation process was repeated for 10 folds with each of the 10 subsets was used once as the testing sample set. Then, the obtained ten results from the folds were averaged for performance indicator. In the testing split protocol, the HGS database was randomly split into two subsets: testing and training sets. In this work, different ratio of splitting was performed with 60:40, 70:30 and 80:20 (ratio of training set: testing set).

5 Performance Results and Discussion

In this work, several performance measure metrics were used to assess the system performance, such as True Positive (TP) rate, False Positive (FP) rate, Mean Absolute Error (MAE) and Root Mean Squared Error (RMSE). In cross-validation testing, x-profile, y-profile, and concatenation of x-profile with y-profile were assessed. Tables 1 and 2 record the performance results of *x*-profile and *y*-profile feature sets, respectively. Table 3 records the performance result of the concatenation of *x*-profile and *y*-profile feature sets, known as *xy*-profile.

Table 1 Performance results of iSign on *x*-profile feature set

Classifier	Accuracy (%)	TP rate	FP rate	MAE	RMSE
Bayesian network	54	0.540	0.005	0.0092	0.0948
Naïve Bayes	42.05	0.421	0.006	0.0116	0.1073
J48 decision tree	50	0.500	0.005	0.0104	0.0954
Random forest	76.9	0.769	0.002	0.0117	0.0675
Nearest-neighbor classifier	72	0.720	0.003	0.0063	0.073

Table 2 Performance results of iSign on *y*-profile feature set

Classifier	Accuracy (%)	TP Rate	FP Rate	MAE	RMSE
Bayesian network	62.9	0.629	0.004	0.0073	0.0839
Naïve Bayes	55.05	0.551	0.005	0.0088	0.0934
J48 decision tree	57	0.570	0.004	0.0088	0.0872
Random forest	78.1	0.781	0.002	0.0104	0.0634
Nearest-neighbor classifier	70.95	0.710	0.003	0.0064	0.0736

Table 3 Performance results of iSign on *xy*-profile feature set

Classifier	Accuracy (%)	TP rate	FP rate	MAE	RMSE
Bayesian network	84.75	0.848	0.002	0.0029	0.0538
Naïve Bayes	72.8	0.728	0.003	0.0052	0.0723
J48 decision tree	69.9	0.699	0.003	0.0062	0.0733
Random forest	93	0.930	0.001	0.0083	0.0507
Nearest-neighbor classifier	90.1	0.901	0.001	0.0028	0.0428

From the tables, we can notice that *xy*-profile showing superior performance. It can be deduced that the integrated features (i.e., *xy*-profile) is able to provide richer information for signature hand gesture recognition. To be specific, the *xy*-profile feature delivers the time and signing motion pattern characteristics in both horizontal and vertical directional analysis. Among the classifiers, Random Forest shows the best performance in this application. Random Forest improves on bagging since it could decorrelate the individual tree with the splitting on a random subset of features. In other words, at each tree split, Random Forest only considers for a small subset of features, instead of the entire features of the model.

Next, the system performance was evaluated using testing split protocol with a ratio of M:N training and test samples by using xy-profile. In this work, different ratio of splitting was conducted with 60:40, 70:30 and 80:20 (ratio of training set: testing set). Tables 4, 5 and 6 summarize the performance of iSign with ratios of 60:40, 70:30 and 80:20, respectively. The purpose of this experiment is mainly to study the influence of training samples towards the feature learning efficiency. From the experimental results, it is observed that the ratio of 80:20 training-test samples shows higher accuracy rates. More training samples always supply extra information which helps machines to learn the underlying data structure and make predictions. In other words, the utilization of more training samples can produce better recognition. It is because different kinds of data characteristics are available and extracted from multiple data samples to provide richer information for feature analysis. More training samples, more opportunities for the learning algorithms to learn and understand the underlying mapping of the input data to the output.

Table 4 Performance with 60:40 training-test samples in testing split protocol

Classifier	Accuracy (%)	TP rate	FP rate	MAE	RMSE
Bayesian network	78.63	0.786	0.002	0.0042	0.0639
Naïve Bayes	67.75	0.678	0.003	0.0062	0.0787
J48 decision tree	61.13	0.611	0.004	0.0079	0.0836
Random forest	89.13	0.891	0.001	0.0098	0.0577
Nearest-neighbor classifier	87.13	0.871	0.001	0.0038	0.0484

Table 5 Performance with 70:30 training-test samples in testing split protocol

Classifier	Accuracy (%)	TP rate	FP rate	MAE	RMSE
Bayesian network	80.5	0.805	0.002	0.0037	0.0607
Naïve Bayes	68.83	0.688	0.003	0.006	0.0774
J48 decision tree	59.5	0.595	0.004	0.0081	0.0848
Random forest	89.17	0.892	0.001	0.0094	0.0558
Nearest-neighbor classifier	87.33	0.873	0.001	0.0036	0.0481

Table 6 Performance with 80:20 training-test samples in testing split protocol

Classifier	Accuracy (%)	TP rate	FP rate	MAE	RMSE
Bayesian network	83.25	0.833	0.002	0.0032	0.0565
Naïve Bayes	71.25	0.713	0.003	0.0055	0.0743
J48 decision tree	64.25	0.643	0.004	0.0072	0.0802
Random forest	92.25	0.923	0.001	0.0088	0.0529
Nearest-neighbor classifier	88.75	0.888	0.001	0.0032	0.0455

6 Conclusion

This paper presents a contactless hand gesture signature recognition by using machine learning technology. In this system, there are four stages to recognize the detected in-air hand gesture handwritten signatures: data acquisition, data pre-processing, feature transformation and data classification. Data acquisition is to detect and capture the signing hand movements using Microsoft Kinect. In data pre-processing, the quality of the captured data is enhanced. Next, the pre-processed data is transformed to the desired form. In this stage, directional projections are performed vertically and horizontally to generate x-profile and y-profile. Lastly, classification is performed. In this work, multiple machine learning algorithms are explored, such as Random Forest, Nearest-neighbour classifier, Bayes Net, Naïve Bayes, and J48 Decision Tree. Empirical results show that Random Forest classifier is able to achieve the best hand gesture signature recognition performance among the other machine learning classifiers.

Acknowledgements This research is supported by Fundamental Research Grant Scheme (FRGS), FRGS/1/2021/ICT02/MMU/03/3. Authors acknowledged the Ministry of Higher Education Malaysia (MOHE).

References

1. Karouni A, Daya B, Bahlak S (2011) Offline signature recognition using neural networks approach. Procedia CS 3:155–161. https://doi.org/10.1016/j.procs.2010.12.027
2. Covid flight cancellations delay holiday travel—BBC News. https://www.bbc.com/news/world-59798682. Last accessed 28 Feb 2022
3. Giang D, Vu D, Hai V, Hai T (2016) Dynamic hand gesture recognition using spatial-temporal features. https://doi.org/10.15625/vap.2015.00158
4. Khoh WH, Pang YH, Teoh ABJ, Ooi SY (2021) In-air hand gesture signature using transfer learning and its forgery attack. Appl Soft Comput 113. [108033]. https://doi.org/10.1016/j.asoc.2021.108033
5. Han J, Shao L, Xu D, Shotton J (2013) Enhanced computer vision with Microsoft Kinect sensor: a review. IEEE Trans Cybern 43(5):1318–1334. https://doi.org/10.1109/TCYB.2013.2265378
6. Sharma A, Sundaram S (2016) An enhanced contextual DTW based system for online signature verification using vector quantization. Pattern Recogn Lett 84. https://doi.org/10.1016/j.patrec.2016.07.015
7. Durrani M, Khan S, Khalid S (2019) VerSig: a new approach for online signature verification. Cluster Comput 22. https://doi.org/10.1007/s10586-017-1129-4
8. Tian J, Qu C, Xu W, Wang S (2013) KinWrite: handwriting-based authentication using Kinect. NDSS
9. Jeon JH, Oh BS, Toh KA (2012) A system for hand gesture-based signature recognition. In: 2012 12th international conference on control, automation, robotics and vision, ICARCV 2012 [6485153], pp 171–175. https://doi.org/10.1109/ICARCV.2012.6485153
10. Mendels O, Stern H, Berman S (2014) User identification for home entertainment based on free-air hand motion signatures. IEEE Trans Syst Man Cybern Syst 44(11):1461–1473 [6842691]. https://doi.org/10.1109/TSMC.2014.2329652
11. Sajid H, Cheung SS (2015) VSig: hand-gestured signature recognition and authentication with wearable camera. In: 2015 IEEE international workshop on information forensics and security (WIFS), pp 1–6
12. Sun Z, Wang Y, Qu G, Zhou Z (2016) A 3-D hand gesture signature based biometric authentication system for smartphones. Sec Commun Netw 9, 11:1359–1373. https://doi.org/10.1002/sec.1422
13. Khoh WH, Pang Y, Teoh AB (2018) In-air hand gesture signature recognition system based on 3-dimensional imagery. Multimedia Tools Appl 78:6913–6937
14. Khoh WH, Pang YH, Ooi SY, Yap HY (2021) Spatiotemporal spectral histogramming analysis in hand gesture signature recognition. J Intell Fuzzy Syst 40(3):4275–7286
15. Yang Z, Zheng X (2021) Hand gesture recognition based on trajectories features and computation-efficient reused LSTM network. IEEE Sens J 21(15):16945–16960
16. Yu J, Qin M, Zhou S (2022) Dynamic gesture recognition based on 2D convolutional neural network and feature fusion. Sci Rep 12(1):1–15
17. Khoh WH, Pang YH, Teoh AB, Ooi SY (2021) In-air hand gesture signature using transfer learning and its forgery attack. Appl Soft Comput 113, Part A, 108033. ISSN 1568-4946. https://doi.org/10.1016/j.asoc.2021.108033

Optimization of Convolutional Neural Network in Paddy Disease Detection

Thyfanney David, Rayner Alfred, Joe Henry Obit, Florence Sia Fui, Januarius Gobilik, Zamhar Iswandono, and Haviluddin Haviluddin

Abstract In Sabah, agriculture is an important economic sector. The situation has recently worsened due to paddy cultivation and rice production management issues. A traditional form, such as detecting disease with the naked eye, is susceptible to high error rates and incorrect classification. To improve the long-term sustainability of the paddy industry, disease detection is critical. The collaboration between Universiti Malaysia Sabah and other government agency has opened new research opportunities in agricultural programs in Sabah and this has sparked the initiative to use advanced technology in Smart Farming. It allowed researchers to utilize digital technologies to jumpstart sustainable and competitive agriculture development. This paper proposes a Paddy Leaf disease detection and classification algorithm that applies deep learning approach. The proposed solution can later assist farmers to diagnose the frequently occurring disease automatically. CNN architectures have successfully been developed to solve various prediction and classification tasks. Due

T. David · R. Alfred (✉) · J. H. Obit · F. S. Fui
Creative Advanced Machine Intelligence Research Centre, Faculty of Computing and Informatics, Universiti Malaysia Sabah, Kota Kinabalu, Sabah, Malaysia
e-mail: ralfred@ums.edu.my

T. David
e-mail: bi18110123@student.ums.edu.my

J. H. Obit
e-mail: joehenry@ums.edu.my

F. S. Fui
e-mail: florence.sfs@ums.edu.my

J. Gobilik
Faculty of Sustainable Agriculture, Universiti Malaysia Sabah, Sabah, Malaysia
e-mail: jgobilik@ums.edu.my

Z. Iswandono
Higher Colleges of Technology, Abu Dhabi, United Arab Emirates
e-mail: zismail@hct.ac.ae

H. Haviluddin
Universitas Mulawarman, Samarinda, Indonesia
e-mail: haviluddin@unmul.ac.id

© The Author(s), under exclusive license to Springer Nature Singapore Pte Ltd. 2023

D.-K. Kang et al. (eds.), *Proceedings of the 9th International Conference on Computational Science and Technology*, Lecture Notes in Electrical Engineering 983,
https://doi.org/10.1007/978-981-19-8406-8_31

to its excellent performance, the aim of this paper is to formulate a deep learning approach to recognize and detect paddy disease based on the paddy leaf images. In this work, an optimized deep convolutional neural network model will be utilized and assessed to diagnose the health of the paddy based on its leave condition. Based on the results obtained, the number of epochs and the dropout rate have a great influence on the performance of the CNN models. For instance, having a high number of epoch's value and smaller percentage of dropout rate, the proposed CNN model is able to classify the type of paddy disease with performance accuracy of 64.80%.

Keywords Paddy leaf disease · Classification · Convolutional neural network · Deep learning approach

1 Introduction

Paddy disease has a significant economic impact on the paddy industry. The most damaging diseases that can cause economic damage to paddy are blast and rust, bacterium, and stem rot. The decrease in return would eventually have an impact on profit, lowering productivity. Agriculture is an important economic sector in Sabah, but the quantity and quality of paddy production have recently declined due to management issues associated with paddy cultivation and rice production. A traditional methodology, such as naked eye disease detection, is also susceptible to high error rates and incorrect classification. The time period complexity is substantial and laborious due to the presence of paddy disease on the plant, reflected on leaves by displaying certain unique disease signs. It is difficult to correctly classify the disease and predict its infection on a large agricultural scale. Manual inspection prevents the disease from being diagnosed too early. The disease would then spread to other crops. Convolutional Neural Network (CNN) is one of the deep learning approaches that has been widely used in machine learning and pattern recognition because of its ability to extract useful features. Another important aspect to consider is that deep neural networks are flexible and adaptive and is able to perform well when dealing with fine details in images. It is also possible that the model will be able to learn useful functionality from the dataset. Using convolution images and filters, CNN is able to generate invariant features passed on to the next layer of the neural network. To produce more consistent and abstract qualities, the following layer features are convoluted with different filters, and the process is repeated until the final part is obtained. Recent researches have suggested that CNN architectures have satisfactorily evolved to solve a wide variety of image classification tasks. For instance, Krizhevsky used a deep CNN to conduct object classification and eventually won first place in the ImageNet Large Scale Visual Recognition Challenge 2012, sponsored by the National Science Foundation [9]. When it comes to computer vision tasks such as object or image classification, CNNs are extremely effective and accurate. According to Chouiekha's research findings, deep convolutional neural networks significantly outperform other traditional machine learning algorithms such as random forest, support vector machines, and gradient boosting classifiers, especially in terms of accuracy [9].

Disease identification is critical in agriculture to improve the sustainability of the paddy industry. With the outstanding performance of the CNN algorithm, farmers will be able to diagnose the frequently occurring disease [5]. It is an end-to-end learning solution that eliminates the need for preprocessing images and directly extracts relevant high-level features from raw images. This method then allows farmers to minimize their efforts while still taking the necessary action based on the outcome. Additionally, the new farmer can benefit from this method by learning about the various causes of paddy disease.

Identifying a plant disease based on its appearance and characteristics can be a tedious and time-consuming task. The conventional methods of determining the plant disease and information through book and catalog searches required considerable time and even confusion. The primary issue is that farmers require a much more convenient and rapid identification of a plant disease, particularly a paddy. This is because paddy cannot be planted anywhere. It is hard to find someone who is an expert on identifying its disease without any intervention of technologies.

Hence, the aim of this paper is to propose and assess the proposed deep learning approach to classification and detection of paddy diseases based on the paddy leaf images. In this work, the performance of the proposed CNN model will also be investigated by varying the deep learning parameters, that include the number of epoch and the dropout rate, to get the optimized parameter settings of the CNN model that can determine the health of the paddy with higher accuracy based on its leaf condition. Section 2 provides the detail of the literature review or related works. Section 3 presents the proposed approach, Sect. 4 presents the results and discussion and finally Sect. 5 concludes this paper.

2 Related Works

In the field of machine vision, plant disease detection is a crucial research topic. It is a technology that employs machine vision equipment to acquire photos to determine whether the gathered plant photos contain illness [15]. Plant disease detection play a significant role in determining the quality or the plants or crops. On the other hand, plant disease is difficult to discern due to the vast range of variances in structure, size, texture, color, environment, arrangement, and imaging light. Deep learning approach has achieved significant advances in digital image processing in recent years, significantly outperforming other machine learning methods. As a result, deploying a deep learning technology to detect plant disease has becoming a hot topic among researchers. As a result, the deep learning method can significantly contribute to plant disease detection with a well-known deep convolutional neural network (CNN).

Deep neural network models are being used in image recognition to automate feature extraction from high-dimensional feature space, which offers significant advantages over traditional manual design feature extraction methods [1–4, 23]. CNN's robust feature extraction capabilities have become a classification network most widely used in pattern classification for plant disease. However, few past projects have been executed involving plant disease detection using a Convolutional Neural Network [6, 17, 21, 24].

As a developing country with the second-largest population, India has a high demand for food as it increases daily as its population grows. On the other hand, the increasing population leads to a lousy climate that contributes to global warming and creates a bad environment for plants to grow and survive as they are exposed to new diseases and infections. Traditional methods such as naked-eye detection are time-consuming and necessitate a large workforce to monitor plants in farms continuously. As a result, alternative approaches to detect a plant disease are developed to implement a deep learning classification model to help Indian farmers detect disease early. Image preprocessing techniques have been combined with machine learning algorithms to produce a better result. The CNN model is used to detect plant disease cultivated in Punjab, Maharashtra, Karnataka, Gujarat, West Bengal, and Andhra Pradesh [20].

A study on the detection of plant leaf disease using CNN has been conducted and the aim was to design a module that can predict plant disease for people who had no prior programming experience to learn and obtain information about the plant disease [8]. The database was created to keep track of all different plant leaf diseases. The module was also trained repeatedly to achieve the highest level of accuracy. If an input image is provided to the module, the feature of the image will be compared to the features that had been trained in the database. Thirty-eight images have been collected as samples made up of different plants with various diseases such as tomatoes, grapes, healthy leaves, etc. The CNN model was trained and tested using this leaf disease dataset. The dataset went through preprocessing that includes image reshaping, resizing, and conversion to an array format. A dataset that contains 38 different plant leaf diseases is compiled, and any image from this dataset can be used to test the input image. The module was able to extract the leaf features from a new input image provided after being created and then proceeds to go through the CNN model. The feature set is then compared to a dataset that has already been trained in the database. The module will then determine whether the input image has been affected by a disease or a healthy input image.

A machine learning based technology was developed to detect paddy crop disease that employed two principal phases: training the model and detecting the disease image [22]. The dataset includes healthy and diseased leaf image datasets. Two thousand images of each rice blast, bacterial blight, and healthy paddy leaves were gathered and downloaded from `Kaggle.com`. The Convolutional Neural Network was then used to train these images.

A disease symptom identification in paddy leaf using image processing has also been conducted [19]. In this work, a software prototype approach was proposed for paddy disease identification based on infected images of diverse rice plants. Images of infected leaves were captured using a digital camera, which were processed utilizing preprocessing, segmentation, feature extraction, and picture classification. In this study, the researchers run the input image through techniques like acquisition,

Fig. 1 Common image processing system configuration

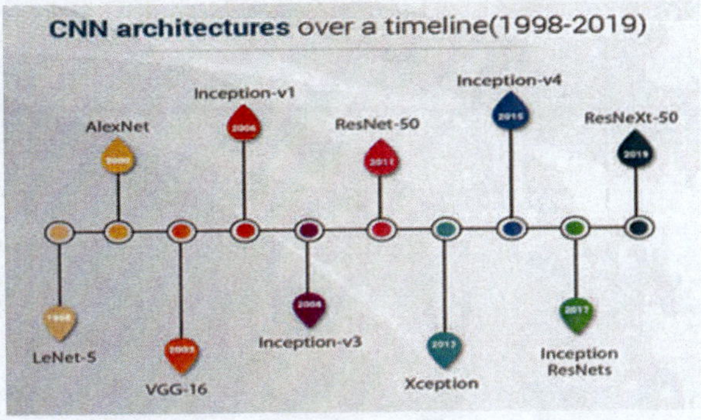

Fig. 2 CNN architectures over a timeline (1998–2019) [10]

segmentation, preprocessing, and then extract the feature from the image. Then, it is classified into two stages using SVM to decide whether it is defective or not, based on lesion kind, border color, spot on the leaf, and paddy leaf condition.

A study on finding the best automated convolutional neural network-based approach for paddy leaf disease detection [13]. This study aimed to deliver the best disease detection results possible using deep learning to get the highest accuracy. The suggested detection model was combined from the benchmarked methodology with customized deep learning. The four models used were VGG-19, ResNet-101, Inception-Res-NetV2 and Xception deep CNN models. In another research, he conducted a study on developing a model for detecting leaf disease using images of corn, peach, grape, potato, and strawberry developed using computer science and engineering and machine learning, specifically deep learning [14]. Several popular food items, such as mangoes and strawberries, can be found in abundance in Bangladesh. Corn, peach, grape, potato, and strawberry farmers face many disorder and insect attractions, and thus a significant loss. This study aimed to prevent these losses and provide an immediate cure by utilizing image processing and CNN model.

A study on rice blast, which is the most common paddy disease in the world was conducted by creating a dataset for rice blast disease and using that dataset to train and test a CNN model for disease classification [16] (Figs. 1 and 2).

Fig. 3 Brown spot rice disease

3 Methods

3.1 Datasets

In this work, the dataset was collected from the kaggle owned by Do [12]. The dataset is a collection of paddy leaves containing 400 images of four classes of rice leaf conditions. The classes include three classes of rice diseases which are Brown Spot (see Fig. 3), Hispa (see Fig. 4) and Leaf Blast (see Fig. 5) rice diseases and the fourth class includes all images of healthy rice leaf (see Fig. 6).

The dataset is split into two, one for training and one for testing. Each class consists of 100 images in which 60 images will be used for training the CNN model, while 40 images of each class are used for testing the performance of the CNN model.

3.2 Methods

The CNN model consists of three essential layers, namely the Convolutional, Pooling and Fully Connected layers. Each layer of the CNN model will be built by using the Keras libraries.

Convolutional Layer There are two ways of initializing a CNN, and this work, the sequence layers technique is used to initialize the CNN model. The sequential

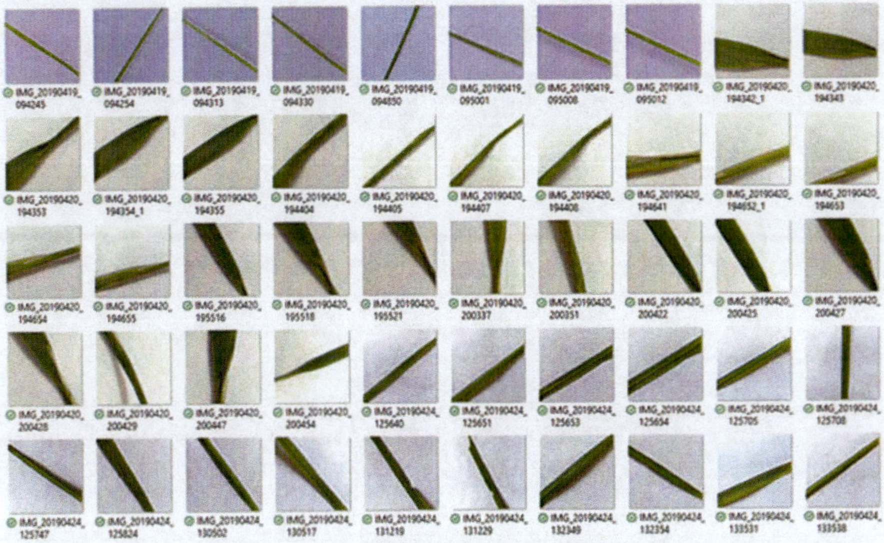

Fig. 4 Hispa rice disease

Fig. 5 Leaf blast rice disease

Fig. 6 Healthy rice

from the `Keras` model is imported to initialize the CNN model. Then, the `Conv2D` function will be used to add the convolution layer. The `Con2D` will be taking four arguments which are listed below.

- The number of filters.
- The shape of each filter.
- Input shape and the type of images.
- Activation function.

In this work, the values of the parameters of the CNN model are set as follows; Steps = 100, Learning Rate = 1e−3, Batch Size = 32, Width = 256, Height = 256 and Depth = 3.

Pooling Layer The pooling operation must be performed on the feature maps. The `MaxPooling2D` packages will be implemented to assist this process. This pooling function is applied to decrease the image size and the total number of nodes. Besides that, the `Flatten` packages will also be implemented in this process. Flattening is used to transform pool images into continuous vectors, which is necessary because it alters the image pixels into a single-dimensional vector.

Fully Connected Layer After the pooling layer process is done, the fully connected layers are created from the connecting set of nodes. The Dense package will also be applied to these layers to complete the process.

Figure 7 shows the codes used to create the CNN model for classifying the type of rice diseases based on the condition of the leaf. The first Conv2D layer is added with a total of 32 filters in which the size of each filter is 3×3. A ReLU activation function will be used in this layer. The usage of ReLU helps to prevent the exponential growth in the computation required to operate the neural network. If

```
model = Sequential()
inputShape = (WIDTH, HEIGHT, DEPTH)
chanDim = -1
if K.image_data_format() == "channels_first":
    inputShape = (DEPTH, HEIGHT, WIDTH)
    chanDim = 1
model.add(Conv2D(32, (3, 3), padding="same",input_shape=inputShape))
model.add(Activation("relu"))
model.add(BatchNormalization(axis=chanDim))
model.add(MaxPooling2D(pool_size=(3, 3)))
model.add(Dropout(0.25))
model.add(Conv2D(64, (3, 3), padding="same"))
model.add(Activation("relu"))
model.add(BatchNormalization(axis=chanDim))
model.add(Conv2D(64, (3, 3), padding="same"))
model.add(Activation("relu"))
model.add(BatchNormalization(axis=chanDim))
model.add(MaxPooling2D(pool_size=(2, 2)))
model.add(Dropout(0.25))
model.add(Conv2D(128, (3, 3), padding="same"))
model.add(Activation("relu"))
model.add(BatchNormalization(axis=chanDim))
model.add(Conv2D(128, (3, 3), padding="same"))
model.add(Activation("relu"))
model.add(BatchNormalization(axis=chanDim))
model.add(MaxPooling2D(pool_size=(2, 2)))
model.add(Dropout(0.25))
model.add(Flatten())
model.add(Dense(256))
model.add(Activation("relu"))
model.add(BatchNormalization())
model.add(Dropout(0.3))
model.add(Dense(n_classes))
model.add(Activation("softmax"))
```

Fig. 7 Initializing the proposed CNN model using sequential method

the CNN scales in size, the computational cost of adding extra ReLUs increases linearly. Empirically, early papers observed that training a deep network with ReLU tended to converge much more quickly and reliably than training a deep network with sigmoid activation. After that, a batch normalization is added to make computation efficient by reducing values between 0 and 1. Next, a max pooling with the size of 3 × 3 is added to reduce the spatial dimensions and the percentage of dropout is set to 0.25. The second and third Conv2D layers are added with a total of 64 filters in each layer with the size of each filter is 3 × 3. Next, a max pooling with the size of 2 × 2 is added with the percentage of dropout is set to 0.25. The fourth and fifth Conv2D layers are added with a total of 128 filters in each layer with the size of each filter is 3 × 3. Next, a max pooling with the size of 2 × 2 is added with the percentage of dropout is set to 0.25. Next, a `Flatten()` function is used to return a copy of an array collapsed into one dimension. Then, a 256-neuron output layer is added by writing Dense `(256)` and finally a 4-neurons output layer is added that represents four types of classes of data that will be modelled in this work. Softmax activation function is used in this layer, as the activation for the last layer of a classification network because the result is interpreted as a probability distribution.

After building the CNN model is completed, then the Adam Optimizer is initialized with learning rate and decay parameters. The type of loss and metrics for the model are chosen and compiled for the model training.

3.3 Optimizing CNN Parameters: Epochs and Dropout Rate

The experiment design in this work was outlined to investigate the effects of varying the values of the number of epochs [7] and the percentage of dropout [11]. The parameters that are varied in this experiment includes the number of epochs and percentage of dropout. The number of epochs will be varied with the following values, epochs = 100, 300, 600 and 900 and the percentage dropouts will be varied with the following values, dropout = 0.25, 0.5, and 0.75. The experiments that are conducted in this work is outlined to investigate the effects of varying the number of epochs and percentage of dropout of the CNN model.

The Number of Epochs Epochs is a cycle of data trained in the neural network. The number of Epochs indicates how many cycles the dataset were trained in the model. The number of Epochs must be set in some value to define the hyperparameters of the Convolutional Neural Network model architecture before the training starts. There is no fixed value for setting up the number epochs to define a good model. This is because the compatibilities for number of epochs on the Convolutional Neural Network depend on the dataset that will be trained [7].

The Percentage of Dropout The function of dropout in the CNN model is to prevent the model from overfitting during the training progress. By setting a percentage of dropout in the model layer, it will deactivate some of the neuron layers to prevent the model from overfitting thus nullifying the output [18].

3.4 Efficiency Evaluation of the CNN Model Based on Accuracy

The test dataset will be used to evaluate the performance of the trained model. The model must be able to detect the input image whether it is infected or not. If the input image is infected, it must be able to show what types of disease it belongs to, and if it is healthy, it will tell the user that the input image is healthy. The performance of the proposed CNN model will be made based on the accuracy of the model to classify rice leaf diseases [25].

Table 1 CNN accuracy of classification during the training phase

Epochs	Dropout %	Accuracy %
100	0.25	41.50
	0.50	42.25
	0.75	40.80
300	0.25	62.50
	0.50	64.25
	0.75	63.65
600	0.25	63.50
	0.50	64.00
	0.75	62.75
900	0.25	65.00
	0.50	64.50
	0.75	62.50

4 Results and Discussion

Based on the results shown in Table 1, it shows that by using large number of epochs (e.g., 900) and 0.25% or 0.50% of dropout rate, the constructed CNN model can provide the highest result which is 77.0% for Training Accuracy. This combination of values for the number of epochs and percentage of dropout will be used as a default setting during the testing phase for the paddy disease classification model.

Figure 8 depicts the training and validation accuracy and loss for the CNN model for the highest accuracy of 42.25% when the number of epochs is set to 100 and the percentage of dropout to 0.5. In contrast, Fig. 9 depicts the training and validation accuracy and loss for the CNN model for the highest accuracy of 64.25% when the number of epochs is set to 300 and the percentage of dropout to 0.5. Next, Fig. 10 depicts the training and validation accuracy and loss for the CNN model for the highest accuracy of 64.00% when the number of epochs is set to 600 and the percentage of dropout to 0.5. Finally, Fig. 11 depicts the training and validation accuracy and loss for the CNN model for the highest accuracy of 65.00% when the number of epochs is set to 900 and the percentage of dropout to 0.25.

Finally, the CNN model is trained again using the same parameter settings that was concluded to have the best accuracy performance (e.g., number of epochs is 900 and the 0.25% of dropout rate) and also the training dataset that was defined earlier. Then, the trained CNN model is then tested using the test dataset which is predefined earlier. This is done for 10 times and all the 10 accuracy results are averaged to get the average accuracy. The proposed CNN model is able to classify the paddy disease based on its leaf conditions with an average of 64.80%.

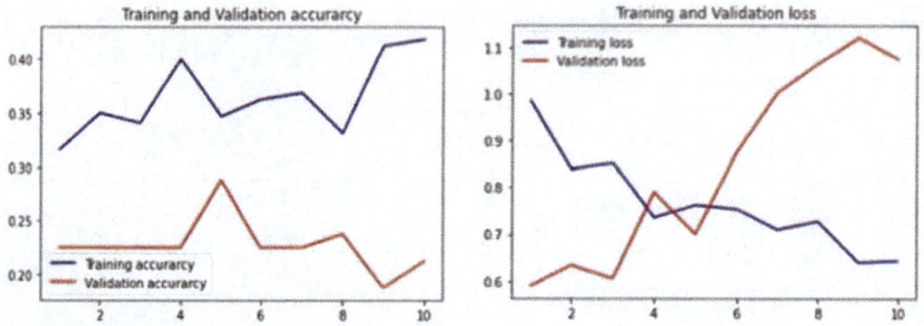

Fig. 8 Training and validation graph for accuracy and loss for 100 epochs and 0.5% dropout

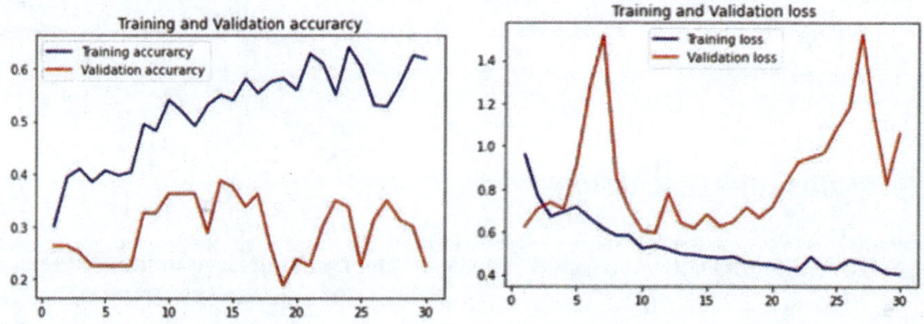

Fig. 9 Training and validation graph for accuracy and loss for 300 epochs and 0.5% dropout

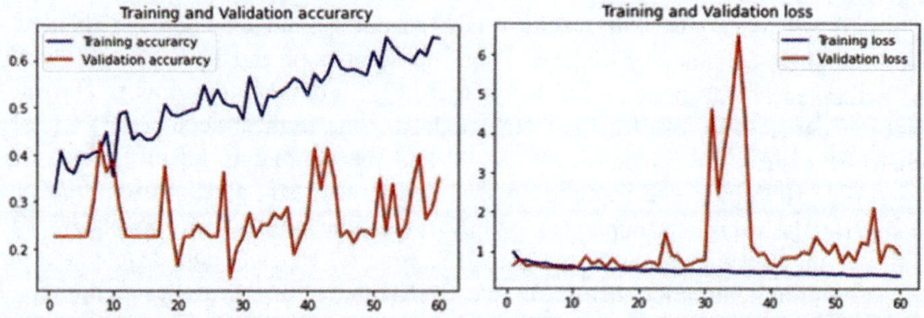

Fig. 10 Training and validation graph for accuracy and loss for 600 epochs and 0.5% dropout

5 Conclusion

This work investigated the effectiveness of applying a deep learning approach to perform a multi-class classification task to classify paddy rice disease based on its leaf conditions. The number of epochs and the percentage of dropout rate of the CNN model can have a great influence on the accuracy performance of the CNN

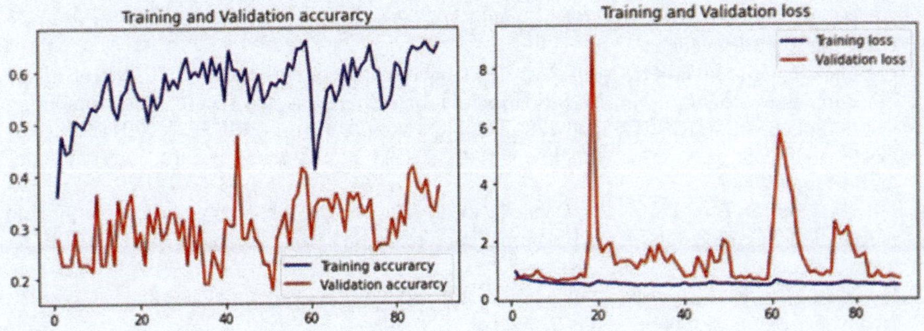

Fig. 11 Training and validation graph for accuracy and loss for 900 epochs and 0.25% dropout

models. For instance, having a high number of epoch's value of 900 and a smaller percentage of dropout rate value (e.g., 0.25), the CNN model can produce accuracy performance of 64.8%. For future work, the model can be improved by adding a masking process for the dataset. The leaf images in the dataset could be separated from its background so it can be classified with more effectively by reducing noise and data leakage. These image dataset could then be used to train the CNN model more effectively thus increasing the testing accuracy.

Acknowledgements This work was supported in part by research grants from Universiti Malaysia Sabah (DFK2004).

References

1. Alfred R (2007) The study of dynamic aggregation of relational attributes on relational data mining. In: Lecture notes in computer science (including subseries lecture notes in artificial intelligence and lecture notes in bioinformatics), vol 4632, pp 214–226. https://doi.org/10.1007/978-3-540-73871-8_21
2. Alfred R (2008) DARA: data summarisation with feature construction, pp 830–835. https://doi.org/10.1109/AMS.2008.131
3. Alfred R (2009) Optimizing feature construction process for dynamic aggregation of relational attributes. J Comput Sci 5(11):864–877. https://doi.org/10.3844/jcssp.2009.864.877
4. Alfred R (2010) Feature transformation: a genetic-based feature construction method for data summarization. Comput Intell 26(3):337–357. https://doi.org/10.1111/j.1467-8640.2010.00362.x
5. Alfred R, Obit JH, Chin CPY, Haviluddin H, Lim Y (2021) Towards paddy rice smart farming: a review on big data, machine learning, and rice production tasks. IEEE Access 9:50358–50380. https://doi.org/10.1109/ACCESS.2021.3069449
6. Appalanaidu MV, Kumaravelan G (2021) Plant leaf disease detection and classification using machine learning approaches: a review. In: Saini HS, Sayal R, Govardhan A, Buyya R (eds) Innovations in computer science and engineering. Springer Singapore, Singapore, pp 515–525
7. Aradhya AM, Ashfahani A, Angelina F, Pratama M, de Mello RF, Sundaram S (2022) Autonomous CNN (AutoCNN): a data-driven approach to network architecture determination. Inf Sci

8. Bharath S, Kumar KV, Pavithran R, Malathi T (2020) Detection of plant leaf diseases using CNN. Int Res J Eng Technol (IRJET)
9. Chauhan R, Ghanshala KK, Joshi R (2018) Convolutional neural network (CNN) for image detection and recognition. In: 2018 first international conference on secure cyber computing and communication (ICSCCC), pp 278–282. https://doi.org/10.1109/ICSCCC.2018.8703316
10. CNN architectures over a timeline (1998–2019). https://www.aismartz.com/blog/cnn-architectures/
11. Dileep P, Das D, Bora PK (2020) Dense layer dropout based CNN architecture for automatic modulation classification. In: 2020 national conference on communications (NCC). IEEE, pp 1–5
12. Do HM (2019) Rice diseases image dataset. https://www.kaggle.com/datasets/minhhuy2810/rice-diseases-image-dataset
13. Islam MA, Shuvo MNR, Shamsojjaman M, Hasan S, Hossain MS, Khatun T (2021) An automated convolutional neural network based approach for paddy leaf disease detection. Int J Adv Comput Sci Appl 12(1)
14. Islam MT (2020) Plant disease detection using CNN model and image processing. Int J Eng Res Technol (IJERT) 9
15. Lee SH, Chan CS, Mayo SJ, Remagnino P (2017) How deep learning extracts and learns leaf features for plant classification. Pattern Recognit 71:1–13. https://doi.org/10.1016/j.patcog.2017.05.015
16. Liang WJ, Zhang H, Zhang GF, Cao HX (2019) Rice blast disease recognition using a deep convolutional neural network. Sci Rep 9(1):1–10
17. Mohanty SP, Hughes DP, Salathé M (2016) Using deep learning for image-based plant disease detection. Front Plant Sci 7. https://doi.org/10.3389/fpls.2016.01419. https://www.frontiersin.org/articles/10.3389/fpls.2016.01419
18. Nandini GS, Kumar AS, Chidananda K (2021) Dropout technique for image classification based on extreme learning machine. Glob Transit Proc 2(1):111–116. https://doi.org/10.1016/j.gltp.2021.01.015. https://www.sciencedirect.com/science/article/pii/S2666285X21000157. In: 1st international conference on advances in information, computing and trends in data engineering (AICDE-2020)
19. Narmadha R (2017) Disease symptoms identification in paddy leaf using image processing
20. Nidhis AD, Pardhu CNV, Reddy KC, Deepa K (2019) Cluster based paddy leaf disease detection, classification and diagnosis in crop health monitoring unit. In: Peter JD, Fernandes SL, Eduardo Thomaz C, Viriri S (eds) Computer aided intervention and diagnostics in clinical and medical images. Springer International Publishing, Cham, pp 281–291
21. Pothen ME, Pai ML (2020) Detection of rice leaf diseases using image processing. In: 2020 fourth international conference on computing methodologies and communication (ICCMC), pp 424–430. https://doi.org/10.1109/ICCMC48092.2020.ICCMC-00080
22. PrajwalGowda B, Nisarga M, Rachana M, Shashank S, Raj BS (2020) Paddy crop disease detection using machine learning. Int J Eng Res Technol 8(13)
23. Sainin MS, Alfred R (2011) A genetic based wrapper feature selection approach using nearest neighbour distance matrix, pp 237–242. https://doi.org/10.1109/DMO.2011.5976534
24. Singh JP, Pradhan C, Das SC (2020) Image processing and machine learning techniques to detect and classify paddy leaf diseases: a review. In: Swain D, Pattnaik PK, Gupta PK (eds) Machine learning and information processing. Springer Singapore, Singapore, pp 161–172
25. Trivedi J, Shamnani Y, Gajjar R (2020) Plant leaf disease detection using machine learning. In: International conference on emerging technology trends in electronics communication and networking. Springer, pp 267–276

Method Fragments for Service Oriented Architecture Realization: Review and Future Research Directions

Supriya Pulparambil and Basel Bani-Ismail

Abstract As an architectural style, service-oriented architecture (SOA) improves the efficiency of developing information systems and enables a rapid response to ever-changing business needs. SOA adoption is a complex process, with different levels of maturity that require know-how in terms of practices. Organizations require methods to engineer services and redesign the existing business processes. The business processes have to be managed in real-time to provide better services. A business process is composed of one or more individual services. Besides the business process composition steps, various supporting activities are required to realize business process services. This research explores methods or a part of the method which supports the organization to achieve the promised benefits of SOA. This paper also discusses the future research directions in the context of adopting SOA to provide heterogeneous IoT services and automated business processes using blockchain technologies.

Keywords SOA · Method · Composition · Method fragment · Service discovery · Service selection · Service negotiation · Change management · Service monitoring · Service measuring

1 Introduction

SOA adoption is an incremental process, with different levels of abstraction; in fact, it requires many methods adopted at each level to realize SOA. Advanced organizations implement SOA to achieve interoperability and business flexibility [1]. With the increasing popularity of SOA and cloud paradigm, methods for constructing high-quality service-based software are demanded [2, 3]. Adequate methodical guidance

S. Pulparambil (✉) · B. Bani-Ismail
Department of Computer Science and MIS, Oman College of Management and Technology, Muscat, Oman
e-mail: supriya.pulparambil@omancollege.edu.om

B. Bani-Ismail
e-mail: basel.baniismail@omancollege.edu.om

© The Author(s), under exclusive license to Springer Nature Singapore Pte Ltd. 2023

D.-K. Kang et al. (eds.), *Proceedings of the 9th International Conference on Computational Science and Technology*, Lecture Notes in Electrical Engineering 983,
https://doi.org/10.1007/978-981-19-8406-8_32

that is obtained from successful experiences of service-oriented system development is essential for an enterprise to move to a higher level of SOA adoption maturity.

The SOA adoption methods have already been explored, specifically in the service engineering domain [4]. A new discipline for designing and developing service-based systems is emerged to support enterprises in SOA adoption and is widely called service-oriented software engineering (SOSE) methods [3, 5]. SOSE methods are mainly concerned with service engineering and compositions [3, 6] to develop candidate services and composite services with a combination of multiple services. These methods focus on developing different types of services such as technical services, business services, or business process services [7]. The realization of technical and business services is mainly through a legacy to SOA migration or implementing SOSE methods, whereas the business processes are realized through composition. Although a systematic SOSE method cannot always guarantee high-quality service-based systems; it can guide various service development activities. By considering the different stages of SOA adoption, the studies claimed that the existing SOSE methods don't reflect the features and processes that would progressively achieve the SOA benefits [8]. These methods contain a set of activities and guidelines; the activities generate some work products as output and take a few work products as inputs based on some guidelines. The activities of a method can be grouped and referred to as method fragment which represents a part of the method. The composition of business process services requires method fragments for service discovery, selection, negotiation, and composition [5]. Besides these method fragments, enterprises require supporting method fragments for successful SOA adoption [6]. Therefore, this research reviews the existing composition methods to identify activities for SOA realization. Though SOA offers several promises of service integration and interoperability, the configuration and manageability of IoT-based services face several challenges in their integration [1] and composition [2]. The literature review helped to identify additional task patterns for service composition and these task patterns are used to extract method fragments. The remainder of this paper is structured as follows: Sect. 2 presents the literature review. Section 3 discusses the results of the review. Section 4 presents the future research directions and Sect. 5 concludes the paper by discussing the impact and limitations of the study.

2 Literature Review

SOA is an approach for software system development based on services and service-orientation design principles [3]. Service engineering methods or SOSE methods support the systematic approach to the design, development, and maintenance of service-oriented systems. In literature, there are several SOSE methods available; some of the well-known methods are SOAF [4], SODD methodology [9], SOMA [5], and SOAD [3]. The service engineering lifecycle begins with service candidate identification, followed by service modelling for the definition and design of services. The service development phase takes the service model as input and develops new services. The deployed services are continuously monitored. A combination of simple candidate services makes a complex composite service using orchestration or choreography styles [6]. The results from the monitoring phase determine the changes required, and in a few cases, the services might be retired, and the next iteration begins.

The realization of business process services is achieved through composition methods. Considering this, a literature review has been conducted to identify the SOA realization methods that focus on composition. 19 proven methods are examined in detail to extract the method fragments based on task patterns. The studied methods are given in Table 1. The task patterns are used to extract reusable method fragments from selected methods. In [7], four task patterns are already defined as a legacy to SOA migration, service discovery, selection, negotiation, and composition. Besides these patterns, additional task patterns are identified in the literature such as change management, service monitoring, service measurement, and error handling (resilience). On preliminary analysis, two method fragments are extracted based on these patterns to support SOA realization. In addition, this study has proposed two new method fragments for candidate service measuring and monitoring.

Table 1 Selected methods

Method	Focus
[8]	Specification-based service discovery, selection, and composition
[10]	Input/output-based semantic service discovery and composition
[11]	Heterogeneous composition
[12]	Change management in service composition
[13]	Tuple-based service description modeling for composition
[14]	QoS-aware service selection and composition
[15]	QoS-aware service selection and composition
[16]	Resilience measurement in collaborative business process
[17]	Resilient service composition
[18]	QoS-aware dynamic composition
[19]	Service recommendation for composition
[20]	Measuring service adaptability for composition
[21]	Collaborative service monitoring
[22]	QoS-aware service discovery and selection
[23]	Pre-contractual service negotiation
[24]	Fuzzy truth proposition-based service negotiation
[25]	Reinforcement learning for QoS negotiation
[26]	Ontology-based service discovery
[27]	Probabilistic topic modelling for service selection

3 Results and Analysis

This section presents the method fragments selected from the literature review. The SOA realization approaches can be static or dynamic. In static approaches, the interactions between services are predefined whereas dynamic approaches adopt automated composition techniques. The different methods or method fragments that come under static approaches are presented in Fig. 1.

The business process services are realized through composition. Most of the selected studies focus on how to discover, select, and recommend services; however, the activities differ in each study. In addition to the method fragments for service discovery, selection, and negotiation the review also identified four supporting method fragments. The supporting fragments are (i) change management in service composition, (ii) resilience measuring, (iii) candidate service monitoring, and (iv) service measuring. The method fragments can be formally presented using software process engineering metamodel [28] with a set of activities, roles involved, and work products produced. In this paper, the key activities are presented for each method fragment.

3.1 Service Discovery

The heterogeneous nature of services and the demands from users have made service discovery an open issue [29], hence the discovery and selection of existing services is a concern for enterprises. Service discovery is generally based on keyword-based matching and can be classified as grammar-based discovery (input/output matching), semantic-based discovery, and QoS-aware discovery [22]. The service descriptions play a vital role in classifying services, service discovery, service selection, matchmaking, and dynamic service composition [13]. There are different methods for service discovery such as specification-based service discovery [8], input–output based service discovery [10], ontology-based service discovery [26, 29], input/output and context similarity–based service discovery [30], QoS aware service discovery [22]. The key activities for specification-based, input–output-based, ontology-based, and QoS-aware service discovery are presented as method fragments in Table 2.

Fig. 1 Service realization methods

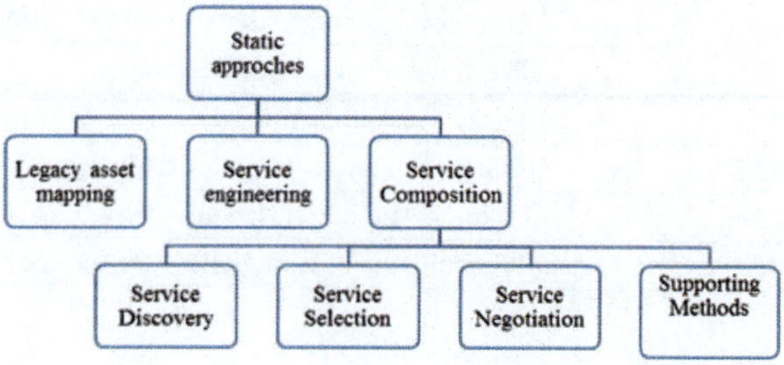

Table 2 Service discovery method fragments

Method fragment	Activities
Specification-based service discovery [8]	1. Requirement acquisition to get the clues for service discovery 2. Organize the requirements to a structured format including the non-functional requirements as constraints 3. A textual matching through keyword-based search recursively
Input/output based service discovery [10]	1. Matchmaking according to inputs and outputs through semantic similarity 2. Define functions for compatible matchmaking with different degrees 3. Find all services with a partial match to use available information 4. Find the matched services for both inputs and outputs
Ontology-based service discovery [26]	1. Advertise the services with semantic annotations 2. Context-aware ontology definition to use during service matching process 3. Define an ontology base to store application and domain ontologies 4. Define a context-aware service discovery algorithm 5. Service matching and ranking using semantic matching and checking the consistency of user requests based on defined ontologies defined
QoS-aware service discovery [22]	1. Service discovery is based on a support vector machine according to the quality requirements of the user 2. Dynamic selection of services based on service quality perception using the QoS model

3.2 *Service Selection*

After service discovery, methods are needed to select the best services. The service descriptions represent the functionality of services, and the service search engine retrieves the services efficiently on how well the descriptions are modeled [27]. The service selection methods available in the existing literature are mostly based on service quality [14, 31, 32]. The service selection can be done using different techniques such as specification-based [8], QoS [33], and functional clustering [15]. The emergence of the internet of things (IoT) and cloud services have redefined the QoS parameters for service selection by considering network and device capabilities [32]. This has resulted in new methods for service selection using deep and reinforcement learning [33]. A few method fragments for service selection are presented in Table 3.

Table 3 Service selection method fragments

Method fragment	Activities
Specification-based service selection [8]	1. Filter the candidate services and obtain the most appropriate service for each function defined in the informal specification 2. Rank the services associated with a given function 3. Decompose the given function into sub functions and proceed with the service ranking for subfunctions 4. Formalize the association between informal specification function and the services using formal languages such as BPMN 5. Dynamic testing based on specification 6. Develop semi-formal specifications as process flow diagrams 7. Transform the process flow diagram to executable language
QoS correlation aware service selection [14]	1. Pre-process the discovered services for the service tasks in the service plan 2. Find the service with an optimal value for the given QoS attribute and remove services with no correlations 3. Select the services with adjacent task correlation and non-adjacent task correlation as per the service plan 4. Select the optimal QoS composing service plan
Functional clustering based service selection [15]	1. Define the matchmaking methods for the services in the service plan using semantic and functional compliance 2. Group services according to locations, data centre, city, and host 3. Merge the services according to the functionality graph 4. Unify the cluster 5. Resolve service dependencies based on locations 6. Determine backup for service clusters 7. Determine service backup for clusters and sub-clusters
Probabilistic topic modelling based service selection [27]	1. Generate the topic vector by extracting the topics 2. Group services based on service topics and form service clusters 3. Apply clustering techniques like K-means 4. Select the right cluster and select the service from that cluster iteratively

Table 4 Service selection negotiation method fragments

Method fragment	Activities
Pre-contractual service negotiation [23]	1. Define negotiation objectives 2. Define rules for communication and message formats 3. Propose a decision-making model by considering the strategies of provider and consumer 4. Define functions to accept or reject negotiation proposals and prepare counterproposals 5. Define strategies for negotiation and coordination
Fuzzy truth proposition based service negotiation [24]	1. Define negotiator profile with a set of primitive fuzzy propositions 2. Generate the strategies for the negotiation proposal 3. Specify negotiation process including negotiation proposal, counter proposal, acceptance/rejection degree 4. Define strategies for negotiation proposal including time and resource variables 5. Create the negotiation proposal
Reinforcement learning based QoS negotiation [25]	1. Identify the QoS parameters for negotiation 2. Define the negotiation protocol and strategies for negotiation 3. Model the QoS negotiation as a set of n Markov decision processes

3.3 Service Negotiation

After service selection, the service provider and consumer negotiate and reach an agreement on certain values for service parameters as they have different preferences. The service level agreement between provider and consumer is just like a contract. The negotiation methods are presented in Table 4.

3.4 Supporting Method Fragments

There are different supporting method fragments required for successful service composition. A few method fragments are presented in Table 5. This includes two new method fragments proposed in this study for service measuring and candidate service monitoring.

The first aspect is change management, which is a major concern in service composition. The workflow-based processes have a long life and have a business committed to its customers for a long time [12]. The change management processes in such services are critical. Either the owner of the business process or the changes initiated by the providers of the individual services can begin a change management process. The top-down changes are normally initiated by the process owner. The changes include new business requirements, government policy changes, and regulations. This requires a central mechanism to manage the change requests. The activities are presented in Table 5.

The second aspect is resilience measurement. The business processes should be resilient to withstand unexpected failures [16]. Usually, the agreement-based collaborative approach is followed in enterprises. Resilience can be looked at from different

Table 5 Supporting method fragments

Method fragment	Activities
Change management [12]	1. Provide a mechanism to specify change requests 2. Define a layered architecture for change management with a schema layer and an instance layer 3. Define a change reaction process to change on the schema level and the instance level 4. Define the data flow and control flow of the schema graph 5. Categorize the change operators such as functional or non-functional changes 6. Define a process for change enactment at the schema level 7. Generate composition instance
Resilience measuring [16]	1. Handle all the error conditions for the individual process model and ensure the flexibility of the process model design to be more resilient 2. Identify and analyze the unspecified factors of the process model such as events and conditions which will impact the process instance 3. Identify the conditions for the services which are commonly used and critical. This is to analyze the impact of service replacement in case of any failures to keep the process instance active 4. Implement automatic service replacement mechanisms based on a service backup plan 5. Provide resource elasticity by accessing a pool of resources
Candidate service monitoring	1. Define intervals to collect statistics, this is to collect runtime information required for system monitoring 2. Define interfaces to access statistical information of services from the enterprise service bus middleware 3. Set the SLA alerts for the individual service level and the process level 4. Set alerts for errors in the message flow 5. Implement mechanisms to send a status message at critical steps of the business process 6. Implement a real-time viewer to correlate the messages based on message semantics with services in the business process to get a complete view of SLA management and problem determination 7. Enable business activity monitoring

(continued)

Table 5 (continued)

Method fragment	Activities
Service measuring	1. Define a plan for measuring SOA adoption to define a plan and scoring framework for SOA adoption to measure success 2. Define metrics for design principles to measure SOA adoption success (a) Define metrics for service reuse (b) Define metrics for business process variance (c) Define metrics for change management 3. Track measuring mechanisms 4. Develop a scoring framework to evaluate SOA investments against the SOA adoption roadmap defined in an enterprise 5. Create guidelines for contribution to SOA 6. Score the projects based on guidelines and the scoring framework

perspectives such as service level, process model level, instance level, organizational level, and resource level. In the context of collaborative business process services, resilience focuses on minimizing the interruption and reducing the impact of interruption (if any). Table 5 presents the activities. The resilience metrics and composed business process model are the main work products involved.

In addition to the above method fragments, two additional method fragments are proposed in this study for service monitoring and measuring. Candidate service level monitoring is required to get a complete view of service-level agreement (SLA) management and problem determination. Traditional tracking is not enough for SOA and monitoring is needed at both the service layer and business events. The runtime data collection and log data of services identify abnormal events [21]. The activities are presented in Table 5. The service measuring is done with different types of metrics. The metrics and policies are generally covered under the governance dimension of SOA adoption. The metrics are required to measure the effectiveness and standard compliance of services. A scoring framework has to be developed. Different attributes of services need to be measured to ensure the quality of services [20]. The activities are listed in Table 5. The quality team leader and the solution architect are involved in defining service level metrics. The metrics, scoring framework, and guidelines are the main work products of the service measuring methods.

4 Research Directions and Future Aspects

The findings of this review are more focused on adopting SOA to provide traditional enterprise services. However, these method fragments can be customized to compose IoT-based services. The additional method fragments discussed in this work support static business process composition. Based on these findings, there is an opportunity to investigate further on method fragments for dynamic and secure composition especially when multiple devices are involved in service composition.

Similarly, the recent advances in 4th industrial revolution technologies such as cloud computing [34], IoT, and blockchain technology solutions and their integration with SOA are studied limited. The IoT platform generally contains heterogeneous service providers and the service composition and provisioning differ from traditional service-oriented systems. Hence, integration of the traditional process model with SOA for IoT service development or adopting existing SOSE methods must be further investigated in terms of methodical requirements. The selection of services and composition in the IoT environment is a major concern due to resource constraints. More specifically, QoS parameters and energy consumption needs to be balanced in service selection for IoT systems.

Recently, enterprises started integrating blockchain technologies to ensure trusted relations. The SOA-based automated business process management solutions must ensure their trustworthiness; rather than a mere technology change, the enterprises must adopt the required methods for accelerating the business process composition with verification of the quality of service. Such integration requires specific rules and directions described through methods.

A detailed analysis of existing SOA methods helps to extract the task patterns that need to be customized for 4th industrial revolution technologies. These technologies such as IoT changes the way of service interaction and the types of service providers. To summarize, the composition methods need to be further analyzed in the context of a more dynamic environment.

5 Conclusion

Considering the growing complexity of the IT infrastructure and the increased maintenance cost of legacy systems, SOA is most relevant to enterprises. Service engineering or SOSE methods support the systematic approach to the design, development, and maintenance of service-oriented systems. The candidate services are composed to develop complex services. This research studies some of the existing composition methods to identify the supporting method fragments for SOA realization. The study has identified method fragments for service discovery, selection, negotiation, and supporting activities. The new developments in service discovery and selection are primarily based on QoS attributes. On preliminary analysis, 15 method fragments are discussed in this study. Among the identified method fragments, 13 are extracted from the existing literature and two are newly proposed in this study. The four method fragments identified to support SOA realization are (i) change management in service composition, (ii) resilience measuring, (iii) candidate service monitoring, and (iv) service measuring. This study has analyzed only 19 methods for static composition to identify method fragments which is a limitation of this research. This review can be further extended as a systematic literature analysis in the context of SOA adoption methods for automated business process management and heterogeneous IoT services.

References

1. Mishra SK, Sarkar A (2021) Service-oriented architecture for Internet of Things: a semantic approach. J King Saud Univ Comput Inf Sci no. xxxx
2. Adewuyi AA, Cheng H, Shi Q, Cao J, Wang X, Zhou B (2022) SC-TRUST: a dynamic model for trustworthy service composition in the Internet of Things. IEEE Internet Things J 9(5):3298–3312
3. Erl T (2008) Service-oriented architecture: concepts, technology & design. Prentice Hall, Pearson
4. Erradi A, Anand S, Kulkarni N (2006) SOAF: An architectural framework for service definition and realization. In: SCC'06. IEEE international conference on services computing, pp 151–158
5. Arsanjani A, Ghosh S, Allam A, Abdollah T, Ganapathy S, Holley K (2008) SOMA: a method for developing service-oriented solutions. IBM Syst J 47(3):377–396
6. The Open Group (2009) Soa Source Book (TOGAF series). Van Haren Publishing
7. Gholami MF, Sharifi M, Jamshidi P (2014) Enhancing the OPEN process framework with service-oriented method fragments. Softw Syst Model 13(1):361–390
8. Miao W, Liu S (2013) A formal engineering framework for service-based software modeling. IEEE Trans Serv Comput 6(4):536–550
9. Papazoglou MP, Van Den Heuvel W-J (2006) Service-oriented design and development methodology. Int J Web Eng Technol 2(4):412–442
10. Rodriguez-Mier P, Pedrinaci C, Lama M, Mucientes M (2016) An integrated semantic web service discovery and composition framework. IEEE Trans Serv Comput 9(4):537–550
11. Lee J, Lee SJ, Wang PF (2015) A framework for composing SOAP, Non-SOAP and non-web services. IEEE Trans Serv Comput 8(2):240–250
12. Liu X, Bouguettaya A, Wu J, Zhou L (2013) Ev-LCS: a system for the evolution of long-term composed services. IEEE Trans Serv Comput 6(1):102–115
13. Yong-Yi Fanjiang J-YK, Syu Y, Ma S-P (2017) An overview and classification of services description approaches in automated services composition. IEEE Trans Serv Comput 10(2):176–189
14. Deng S, Wu H, Hu D, Zhao JL (2016) Service selection for composition with QoS correlations. IEEE Trans Serv Comput 9(2):291–303
15. Wagner F, Ishikawa F, Honiden S (2016) Robust service compositions with functional and location diversity. IEEE Trans Serv Comput 9(2):277–290
16. de Vrieze P, Xu L (2016) Resilience analysis of collaborative process management systems. IFIP Adv Inf Commun Technol 480(1):124–133
17. Palade A, Sci M (2019) Stigmergic QoS optimisation for flexible service composition in mobile environments by University of Dublin, Trinity College
18. Jiang W, Zhang C, Huang Z, Chen M, Hu S, Liu Z (2010) QSynth: a tool for QoS-aware automatic service composition. In: ICWS 2010—2010 IEEE 8th international conference on web services, pp 42–49
19. Liu J, Tang M, Zheng Z, Liu XF, Lyu S (2016) Location-aware and personalized collaborative filtering for web service recommendation. IEEE Trans Serv Comput 9(5):686–699
20. Botangen KA, Yu J, Han Y, Sheng QZ, Han J (2020) Quantifying the adaptability of workflow-based service compositions. Futur Gener Comput Syst 102:95–111
21. El Ioini N, Garibbo A, Sillitti A, Succi G (2013) An open source monitoring framework for enterprise SOA. In: IFIP advances in information and communication technology, vol 404

22. Zhang N (2021) Service discovery and selection based on dynamic QoS in the Internet of Things. Complexity 2021
23. Hashmi K, Malik Z, Najmi E, Alhosban A, Medjahed B (2016) A web service negotiation management and QoS dependency modeling framework. ACM Trans Manag Inf Syst 7(2):1–33
24. Yao Y, Ma LML (2008) Automated negotiation for web services. In: 2008 11th IEEE Singapore international conference on communication systems, pp 1436–1440
25. Udoh IS, Kotonya G (2020) A reinforcement learning QoS negotiation model for IoT middleware. In: IoTBDS 2020—proceedings of the 5th international conference on Internet of Things, big data and security, no. IoTBDS, pp 205–212
26. Zeshan F et al (2017) Ontology-based service discovery framework for dynamic environments. IET Softw 11(2):64–74
27. Bukhari A, Liu X (2018) A web service search engine for large-scale web service discovery based on the probabilistic topic modeling and clustering. Serv Oriented Comput Appl 1–14
28. Omg (2008) Software & systems process engineering meta-model specification V2.0, p 236
29. Al-Sayed MM, Hassan HA, Omara FA (2020) An intelligent cloud service discovery framework. Futur Gener Comput Syst 106:438–466
30. Lu C (2021) A novel web service discovery method combining semantic interface similarity and context similarity. IOP Conf Ser Earth Environ Sci 693(1):0–12
31. Ranjan A, Sahoo B (2020) Web service selection mechanism in service-oriented architecture based on publish-subscribe pattern in fog environment. Adv Intell Syst Comput 1056(January):269–281
32. Tong E et al (2021) A hierarchical energy-efficient service selection approach with QoS constraints for Internet of Things. IEEE Trans Green Commun Netw 5(2):645–657
33. Kondo Y, Moustafa A (2022) Service selection for service-oriented architecture using off-line reinforcement learning in dynamic environments 1(Icaart):64–70
34. Ming C, On CK, Rayner A, Guan TT, Patricia A (2018) The determinant factors affecting cloud computing adoption by small and medium enterprises (SMEs) in Sabah, Malaysia. J Telecommun Electron Comput Eng 10:83–88

Identification of Banana Leaf Diseases and Detection

Wassem I. A. E. Altabaji, Wooi-Haw Tan, Chee-Pun Ooi, and Yi-Fei Tan

Abstract In modern agriculture, there is a growing interest and attention concentrated on the diagnosis of plant diseases. Recognizing early signs and symptoms of a disease can help to keep outbreaks of the disease on crops to a minimum. Aside from that, most currently available solutions rely on manual identification by a small number of experts, which waste both time and money. Certain artificial intelligence technologies are now being used to provide faster detection and identification of diseases while requiring less work and expenses. This research presents a proposed methodology for the analysis and identification of banana leaf diseases that makes use of Convolutional Neural Networks (CNN). The experimental results demonstrated that the suggested method is capable of providing satisfactory detection and classification of three major banana leaf diseases, namely Cordana, Pestalotiopsis, and Sigatoka, with the validation accuracy of up to 96.24%.

Keywords CNN · Feature extraction · MobileNet · Xception · InceptionV3 · Banana leaf

1 Introduction

In the evolution of moderately active human civilization, agriculture was a key achievement because it gives humans the ability to live in the cities by obtaining surplus food from large organisms. Diseases have consistently ravaged crops around the world, resulting in huge productivity and economic losses for the agricultural industry. Crop production for human use often necessitates large-scale agricultural

W. I. A. E. Altabaji · W.-H. Tan (✉) · C.-P. Ooi · Y.-F. Tan
Faculty of Engineering, Multimedia University, Cyberjaya, Malaysia
e-mail: twhaw@mmu.edu.my

C.-P. Ooi
e-mail: cpooi@mmu.edu.my

Y.-F. Tan
e-mail: yftan@mmu.edu.my

© The Author(s), under exclusive license to Springer Nature Singapore Pte Ltd. 2023
D.-K. Kang et al. (eds.), *Proceedings of the 9th International Conference on Computational Science and Technology*, Lecture Notes in Electrical Engineering 983,
https://doi.org/10.1007/978-981-19-8406-8_33

activities. Tropical and temperate locations around the world have been severely affected by a variety of environmental elements and unpredictably shifting climates and atmospheres. Consequently, crop yields could be significantly impacted by these factors [1].

Food insecurity can occur if plant diseases are not detected in time and can have a negative impact on agricultural production. For food security and agricultural output, rice, maize, and other staple crops are essential. Plant disease prevention and control can only be successful if early detection and forecasting are in place. In third world countries, plant disease detection has traditionally relied on visual observations by experienced farmers, but this requires constant monitoring by experts, which can be prohibitively expensive on large farms [2].

Recent years have seen a dramatic rise using deep learning in multiple fields such as the field of digital image. Researchers have been looking into how to use deep learning technology to analyze and identify plant diseases. Using deep learning models such as CNN in a wide range of Computer Vision (CV) applications, including traffic detection, medical picture recognition, scenario text detection, facial expression recognition, and other types of facial recognition. Several regional and international businesses have developed several deep learning-based photo identification software systems for detecting plant diseases in real-world agricultural operations. An intelligent system that learns from experience to detect plant diseases would be a valuable tool not only for academics, but for industry as well [3].

CNNs are neural networks that have a complex system architecture and can conduct convolutional operations on input data. In addition to being composed of several layers, the model also features powerful autonomous learning and feature expression capabilities. It is capable of autonomously extracting image features for the purposes of picture classification and identification. The CNN model is consisting of the following layers: the input, the convolutional, the pooling, and the full connection layers then the output layer. The convolution and pooling layers alternate inside the model, and when their neurons are linked, feature extraction occurs, sending the output to the fully connected layer for classification. The CNN model is one of the most extensively used in deep learning the core structural features, which provide large model capacity and intricate information [4].

Transfer learning is a sort of machine learning in which CNN that have been learned for one task are used as the basis for a model that is being developed for another task. Instead of starting from scratch and randomly initializing the weights, the weights can be initialized using a pre-trained network on large, labelled datasets, such as public picture datasets, which can save time and effort. Over this, it is recommended that models that have already been trained on the large typical dataset ImageNet and then moved to the task specified on the target dataset to be used. Networks for classifying plant diseases that have been established, such as the VGGNet, ResNet, Inception V4, DenseNets, and SqueezeNet, are significantly reliant on neural network architectures [5].

2 Related Work

Banana leaves were classified using machine learning as following: healthy leaves and diseased leaves. Based on the seven models that have been constructed, it has been determined that color and shape are the most useful attributes for classifying banana leaves. The AUC (Area Under Curve) metric was used to evaluate the performance of the model. The obtained result of the classification shows that 0.96 for the classification of black Sigatoka disease, 0.91 for Bacterial Wilt disease, and 0.99 for healthy leaves. Moreover, because the classification is only performed on healthy leaves or infected leaves, the proposed method classification shows disease with an accuracy of 96.2%. It should be noted that the images utilized in this research were derived from this work of literature [6].

In order to detect diseases of banana leaves such as Black Sigatoka and Fusarium Wilt, deep learning models are being created. The Inception V3 model, which achieved 95 percent accuracy in a real-world context, was used to construct the best model that was discovered. Furthermore, VGG-16 design was shown to be one of the most effective of the five architectures tested; as a result, this model was selected for comparison with the suggested model and transfer learning technique [7].

With the Support Vector Machine (SVM) technique, the author concentrates on the classification of four different types of diseases. SVM is an algorithm that analyzes data for the purposes of classification and regression. An image acquisition, image pre-processing, segmentation, feature extraction, and classification system are the proposed five steps. Furthermore, a collection of contrast enhancement approaches, including histogram equalization and K-means segmentation, has been compiled together [8].

The researchers utilized a VGG16 model with a transfer learning strategy to identify the disease that was present in the millet crop. This study gathered 124 different leaf images and categorized them as either healthy or affected by mildew. The VGG16 model was found to be accurate 95% of the overall [9].

A proposed technique for detecting diseases in coffee leaf utilizing Fuzzy Logic Based Expert System, CNN with Data Augmentation, RBF Neural Network, and CNN with Data Augmentation and Transfer Learning using Inception V3 model. This research makes use of two different kinds of datasets, the first of which consists of actual leaf images, while the second consists of selected symptom parts taken from leaf pictures. Every one of the datasets includes five distinct classifications of leaf images. A satisfactory result of 97.61% was produced by the IncepionV3 model [10].

The author presented a technique that combines image segmentation and soft computing to detect and categorize plant leaf diseases automatically. A genetic algorithm was used to accomplish this strategy. For classification purposes, the K-mean clustering method with the lowest distance criteria was first used, however this was later abandoned. The system's accuracy has been assessed to be 86.54%. By using the SVM algorithm, the accuracy of the recommended methodology was boosted to 93.63%, while the accuracy of the SVM technique was improved to 95.71% [11].

Fig. 1 Banana leaf sample

Table 1 Train-validation-split of the dataset

Dataset	Classes	Train	Validation
Banana leaf	Cordana	122	40
	Pestalotiopsis	133	40
	Sigatoka	433	40
	Healthy	89	40

A ResNet50 CNN model was utilized in order to identify the diseases that were present in the tomato leaf images. The dataset contains 3,000 images, which may be broken down into three categories. This model has a precision that is equal to 98.0% [12].

3 Dataset Description

Plant Village Dataset is available for research purposes on Kaggle that is an open-source data repository. The datasets contain 55,000 well-labeled photos of healthy and diseased leaves from apples, blueberries, cherries, grapes, peaches, peppers, oranges, tomatoes, and potatoes. The dataset comprises healthy and sick apple, blueberry, cherry, grape, peach, pepper, orange, tomato, and potato leaves. Each type of leaf disease in a crop is classified separately. The banana Leaf dataset from the Kaggle repository was chosen and divided into 4 classes, healthy and another 3 common diseases which are Cordana, Pestalotiopsis, and Sigatoka. Figure 1 show samples of the banana leaf.

The number of images in each class varies, ranging from 89 to 433. For the classification, only banana leaf images were considered. Table 1 shows the train-validation split of the dataset.

4 Proposed Approach

Deep learning model will be implemented in this research to categorize and detect banana leaf diseases for early-stage diagnosis (Fig. 2).

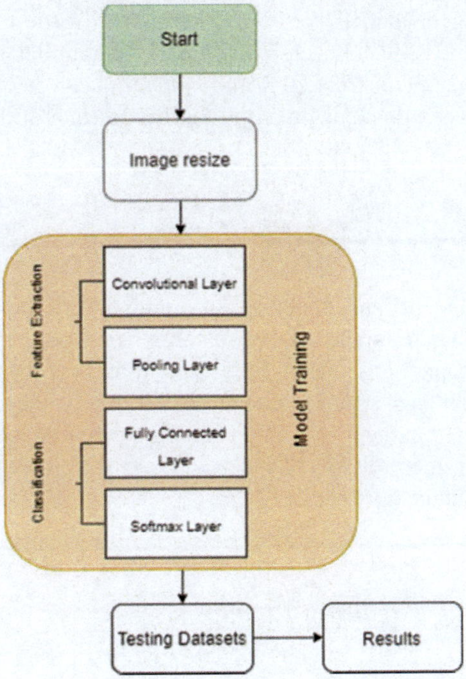

Fig. 2 Flow chart of the proposed system

4.1 Image Resize

Since CNN architectures may only accept images with a fixed dimension, all images were reduced to a fixed resolution of 224 X 224 pixels.

4.2 Convolutional Neural Network

CNN are multilayer networks where the architecture is indicative of their performance. It is made up of four types of layers: convolution, pooling, fully connected, and Softmax layer. The first two types of layers merge to produce a feature extractor, while the third layer performs the function of a classification algorithm. The pooling layer reduces the dimensionality of the features returned by the convolutional layer, which is beneficial. The fully connected layer is followed by the Softmax layer, which categorizes the images based on the feature that was taken from the images. A sequence of learnable filters is used by the convolution layer to extract information from the input image throughout the processing stage. When using a sliding window, the 2-D feature map is constructed by taking the dot product of each filter and the raw picture pixel and putting them together. Rectified linear units (ReLUs) are a sort of activation function that is quite popular. The max pooling layer is a subsampling layer that reduces the overall size of the feature map. Once this is accomplished, the fully connected layer makes a complete connection to each feature map that has been constructed. Softmax layer categorizes the images in a multiclass problem by assigning a decimal probability to each of the categories.

4.3 Convolution Layer

The primary objective of a convolution layer is to locate and map the presence of local conjunctions of features from the preceding layer to a feature map. Multiple convolution layers with many filters will be utilized to build the output feature maps for each specified model. Additionally, these maps record the location of the feature within the image and its compliance with the filter. As a result, each filter is spatially trained relative to the volume to which it is applied, and each filter detects certain image attributes.

4.4 Pooling Layer

The design of convolutional neural networks includes convolutional layers followed by subsampling layers. Each sub-sampling layer reduces the size of the convolution maps and introduces invariance for any (small) rotations or transformations in the input. The maximum pooling layer's output is defined by the maximum activation value for each feature map's sub-sampling in the input layer. Max-pooling minimizes the feature map's size.

4.5 Fully Connected Layer

The fully connected layers are the final layers of the network architecture. A flattened version of the output from the final Pooling or Convolutional Layer is fed into the fully connected layer. The classification operation is carried out in the fully connected layers. Within this fully connected layer, the features are integrated to form a model.

4.6 Softmax Layer

The output layer makes use of the Softmax activation function, which exponentially normalizes the output of the fully connected layer(s) and provides a probability distribution over the different classes of objects.

Table 2 Model sizes

Models	Size (MB)	Number of layers
Xception	88	71
InceptionV3	92	48
MobileNetV2	14	53

4.7 Model Training

Several deep neural network models will be investigated in this research utilizing the dataset that has been chosen. The deep neural network models that will be used in this research are MobileNetV2, InceptionV3, and Xception. One of these models is chosen based on the model size, which is advantageous because it reduces the amount of memory required during the training process. The transfer learning method will be utilized to increase the generalization of these models. Transfer learning is the process by which whatever is learned in one setting is used to improve generalization in another setting. Various data augmentation techniques, such as rotation, mirroring, and rescaling, will be used during the training phase. The sizes and number of layers for each model are listed in Table 2.

4.8 Training Settings

Some settings such as the optimizer, learning rate, metrics, batch size, epochs and callbacks have an impact on how well the proposed network will perform during training. The Adam optimizer is used to minimize the loss function for the models that have been proposed. It started out with a learning rate of 0.01. The batch size is defined as the number of pictures that are fed into the model at a given time. All the models training parameters are depicted in Table 3.

Table 3 Model settings

Parameters	Value
Optimizer	Adam
Initial learning rate	0.01
Batch size	10
Loss	Categorical crossentropy
Metrics	Accuracy
Epochs	50

4.9 Evaluation Tool

Google Colab is used as an evaluation tool for the selected models to monitor the performance of each model.

5 Results and Discussion

The dataset was divided into two parts, 80% for training and 20% for validation. After multiple training attempts for each model, the results are different based on the model architecture. Each model was trained without any data augmentation first to observe the performance of the model. After adding few data augmentations and finalized the parameters, the results were recoded to observe the performance of each model. Table 4 shows the results of each model.

The results in Table 4 show that each model perform differently while the parameters are the same. This is due to the model layers architecture where the feature extraction behaves differently. All the selected models were pre-trained using transfer learning technique. The parameters in Table 3 were used in the implemented CNN architectures mentioned in the preceding section. In comparison to InceptionV3, Xception and MobileNetV2, Xception achieved the highest accuracy. Many factors were utilized to evaluate performance, such as performance accuracy, training loss, and time per epoch. The obtained results show that Xception model provide the optimal performance with accuracy of 96.24%. The better accuracy with less loss of the model after the training will be considered a significant difference which will provide a better comparison among the models. The obtained results are validated by training the model using the banana leaf dataset with the validation dataset. Further wok will be implemented in the future to compare between the proposed method and the remaining methods.

Figure 3 illustrates the loss of the Xception model during the training process which has a very low loss compared to the validation loss at the beginning of the process.

Figure 4 depicts the accuracy of the Xception model were the training accuracy and validation accuracy are almost the same. The obtained results show the performance of the Xception model and how close the train accuracy and validation accuracy which lead to better detection in future test. The deep-learning architecture based on convolutional neural networks detected banana leaf diseases with high success

Table 4 Models performance

Model	Training accuracy (%)	Training loss (%)	Val accuracy (%)	Val loss (%)
Inception V3	77.48	0.7	69	1.07
MobileNetV2	98.32	0.068	83.74	0.66
Xception	99.87	0.019	96.24	0.20

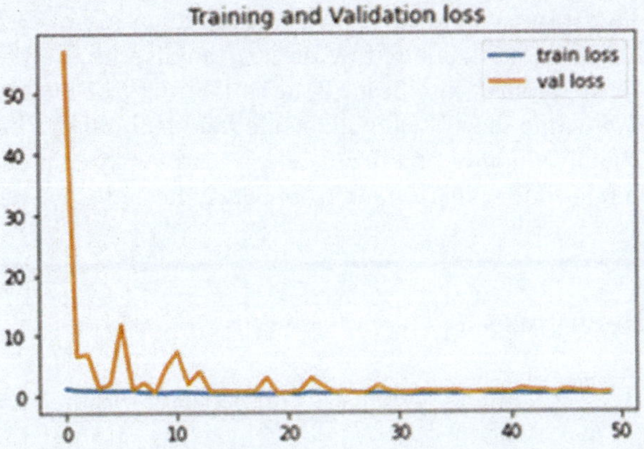

Fig. 3 Training and validation loss for Xception model

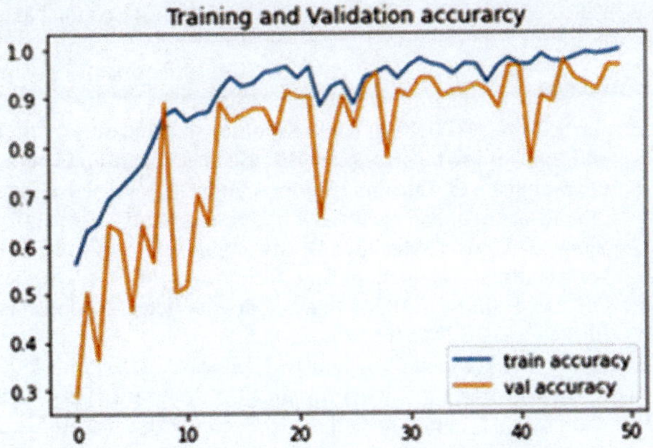

Fig. 4 Training and validation accuracy for Xception model

rates; nonetheless, it has significant limits and there is potential for future research. The deep-learning algorithm misclassified the sample photos due to some noise in the images.

6 Conclusion

A developed automated system to classify diseases in banana leaves such as Cordana, Pestalotiopsis, and Sigatoka, was used in this study. The classification accuracy of the system over the validation dataset was 96.24%, demonstrating that transfer learning can be used to improve classification accuracy. The developed system is still in its early stages, but it has the potential to be utilized to identify and detect diseases. Furthermore, the aim of the research to have a fast, easy, and high-performance system for plant diseases detection and identification has been achieved. Future studies will entail assessing and improving performance on noisy images. However, the dataset should be expanded to include more terrain areas and illness images. Aerial images acquired by drones can also be used to increase image quality and view. Another significant issue is that the test images are from the same dataset. Evaluating the system with actual field pictures is a critical and difficult challenge. The invention of an effective deep learning system capable of identifying diseases in real-time scenarios and from data collected from various datasets is required.

References

1. Chakraborty S, Newton AC (2011) Climate change, plant diseases and food security: an overview. Plant Pathol 60(1):2–14
2. Chen J, Chen J, Zhang D, Sun Y, Nanehkaran YA (2020) Using deep transfer learning for image-based plant disease identification. Comput Electron Agric 173
3. Liu J, Wang X (2021) Plant diseases and pests detection based on deep learning: a review. Plant Methods 17(1)
4. Jadhav SB, Udupi VR, Patil SB (2021) Identification of plant diseases using convolutional neural networks. Int J Inf Technol 13(6):2461–2470
5. Barbedo JGA (2018) Impact of dataset size and variety on the effectiveness of deep learning and transfer learning for plant disease classification. Comput Electron Agric 153:46–53
6. Owomugisha G, Owomugisha G, Quinn JA, Mwebaze E, Lwasa J, Automated vision-based diagnosis of banana bacterial wilt disease and black sigatoka disease
7. Sanga S, Mero V, Machuve D, Mwanganda D (2020) Mobile-based deep learning models for banana diseases detection. Eng Technol Appl Sci Res 10(3):5674–5677
8. Oo YM, Htun NC (2018) Plant leaf disease detection and classification using image processing. Int J Res Eng 5(9):516–523
9. Coulibaly S, Kamsu-Foguem B, Kamissoko D, Traore D (2019) Deep neural networks with transfer learning in millet crop images. Comput Ind 108:115–120
10. Kumar M, Gupta P, Madhav P, Sachin (2020) Disease detection in coffee plants using convolutional neural network, pp 755–760
11. Singh V, Misra AK (2017) Detection of plant leaf diseases using image segmentation and soft computing techniques. Inf Process Agric 4(1):41–49
12. Di J, Li F, Yang Y, Yu S (2020) A tomato leaf diseases classification method based on deep learning. In: Proceedings of 32nd Chinese control decision conference CCDC 2020, pp 1446–1450

Development of a Detection System for Endangered Mammals in Negros Island, Philippines Using YOLOv5n

John Alfred J. Castañeda, Angelo L. De Castro, Michael Aaron G. Sy, Nouar AlDahoul, Myles Joshua T. Tan, and Hezerul Abdul Karim

Abstract Many species have gone extinct as a result of human neglect and various environmental influences. Monitoring these species has proven to be a challenge due to their small population, remote habitats, and evasiveness, among other reasons. Nonetheless, they can be routinely tracked by using CCTV cameras. This project made use of a transfer learning approach to detect the presence of Malayan civets (*Viverra tangalunga*), Visayan leopard cats (*Prionailurus javanensis sumatranus*), Visayan spotted deer (*Rusa alfredi*) and Visayan warty pigs (*Sus cebifrons*) in the forests of Negros Island. We also developed a web application to record camera data and assign timestamps. The setup consisted of a Raspberry Pi 3B+ and a Raspberry Pi camera module powered by a solar power bank. The method used YOLOv5n, a lightweight object detection algorithm, to detect the four species. The trained model yielded 91% mAP, 64% mAP@0.5:0.95, and the following average precisions: 94% (Visayan warty pig), 91% (Malayan civet), 88% (Visayan leopard cat), and 91% (Visayan spotted deer) with 4.5 GFLOPs.

Keywords Animal detection · Transfer learning · Endangered species · Negros Island · Object detection · Philippines · Raspberry Pi · YOLOv5n

Supported by Multimedia University, Malaysia. J. Castañeda, A. De Castro, and M. Sy contributed equally and share first authorship.

J. A. J. Castañeda · A. L. De Castro · M. A. G. Sy · M. J. T. Tan (✉)
Electronics Engineering, University of St. La Salle, Bacolod, Philippines
e-mail: mj.tan@usls.edu.ph

N. AlDahoul (✉) · H. A. Karim (✉)
Faculty of Engineering, Multimedia University, Cyberjaya, Malaysia
e-mail: nouar.aldahoul@live.iium.edu.my

H. A. Karim
e-mail: hezerul@mmu.edu.my

M. J. T. Tan
Natural Sciences, University of St. La Salle, Bacolod, Philippines

Chemical Engineering, University of St. La Salle, Bacolod, Philippines

© The Author(s), under exclusive license to Springer Nature Singapore Pte Ltd. 2023

D.-K. Kang et al. (eds.), *Proceedings of the 9th International Conference on Computational Science and Technology*, Lecture Notes in Electrical Engineering 983,
https://doi.org/10.1007/978-981-19-8406-8_35

1 Introduction

Animal tracking is a tool that helps in understanding how species move within a local area and migrate across land masses and bodies of water [1]. The information that are gathered from this tool help in identifying movement patterns, which are key factors in the study of environmental issues, such as climate and land change, and biodiversity loss; and help combat the destructive impacts of deforestation and illegal animal poaching on wildlife. Currently, although there are many animal tracking systems available in the market, most systems cost hefty amounts. Thus, we endeavored to find a way around these costs by developing a more budget-friendly system that would cost less to maintain and to develop. The high costs of setting up animal tracking and observation systems have made it very difficult for small animal conservation institutions, especially in the Philippines, to acquire these systems and employ them in their various efforts [2].

Object detection was the main idea of this project. We utilized the nano version of the You Only Look Once (YOLOv5n) [3] object detection model that we trained to detect the four endangered species endemic to Negros Island in the Philippines. We gathered an initial set of 300 images for each of the species, all of which were labeled using Tzutalin's labelImg [4] and were later split into 70% for training and 30% for testing. This is the first part of a larger project whose objective is to develop a budget-friendly and performance-efficient animal tracking system with superior detection performance. The four endangered species the were of interest to this project were the Malayan civet (*Viverra tangalunga*), the Visayan leopard cat (*Prionailurus javanensis sumatranus*), the Visayan spotted deer (*Rusa alfredi*) and the Visayan warty pig (*Sus cebifrons*).

2 Related Literature

2.1 Object Detection Algorithms

Deep learning models for object recognition and extraction are generally referred to as compositional models, where the object is depicted as a multilayered structure of image primitives [5]. Conventional object detection involves three steps. It begins by utilizing movable window frames of varied sizes and dimensions as viable regions on the input image. Then, it extracts features from each potential region's local data using classical elements, such as color, structure, size; and middle-level or high-level semantic characteristics. Finally, it can use classifiers like support vector machine models to yield the output [6]. Object detection frameworks based on convolution neural networks typically belong to one of two major groups. The first group is the Region-based convolutional neural network (R-CNN) [7], which is a two-stage approach (region proposal network [RPN] and region of interest [RoI] pooling), and the other group is the You Only Look Once (YOLO) model [8], which is a regression-based target detection system.

The R-CNN is a deep learning architecture that employs regions to recognize an image [9]. It involves a selective search, which combines an exhaustive search, followed by segmentation. Faster R-CNN detection, on the other hand, uses a region proposal network (RPN) to develop region suggestions instantaneously in the system. The RPN uses anchor boxes for object detection, which creates faster and more accurate region recommendations [10].

2.2 *Computer Vision Systems in Animal Detection*

The development of software that involve deep learning (DL) models are computationally expensive and the necessary development and deployment hardware are likewise, costly. However, they are necessary in wildlife monitoring. A paper described a study that used YOLOv5 to categorize images of animals in Poland [11]. They used the open-source TRAPPER software that organized and harmonized camera traps and large-scale data with an 85% F1-score in detecting 12 common species.

Global food security is an important societal concern and in 2050, the population will reach 9.2 billion people [12]. Daily food protein consumption usually comes from beef and chicken, so precision livestock farming employs this technology to rear these animals, as well. It analyzes animal behavior objectively, minimizing human biases. Non-invasive computer vision technologies could be made affordable and less intrusive. Cameras and other recording devices deployed on fixed (ceilings, passageways) or movable (rail systems, ground robots, drones) equipment can be used to monitor these animals (cattle, sheep, pigs, poultry, etc.) [13].

3 Methodology

3.1 *Proposed System*

The project made use of the nano version (YOLOv5n) of the pre-built YOLOv5 model. Its design is based on the Faster R-CNN model, which uses a bounding box regression and an anchor box network structure. YOLO carries out bounding box regression and classification at the same time, while the Faster R-CNN does these two steps separately, giving YOLO an advantage over the Faster R-CNN model in real-time object detection [15].

The model can label the categories and surrounding boxes on an image. It then uses a CNN to calculate each region's attributes. It would then predict its categories and bounding boxes by using the characteristics of each defined region. Prediction determines accuracy, speed, and cost of the static camera. We created a flowchart (Fig. 1) to show how the system works based on our objectives.

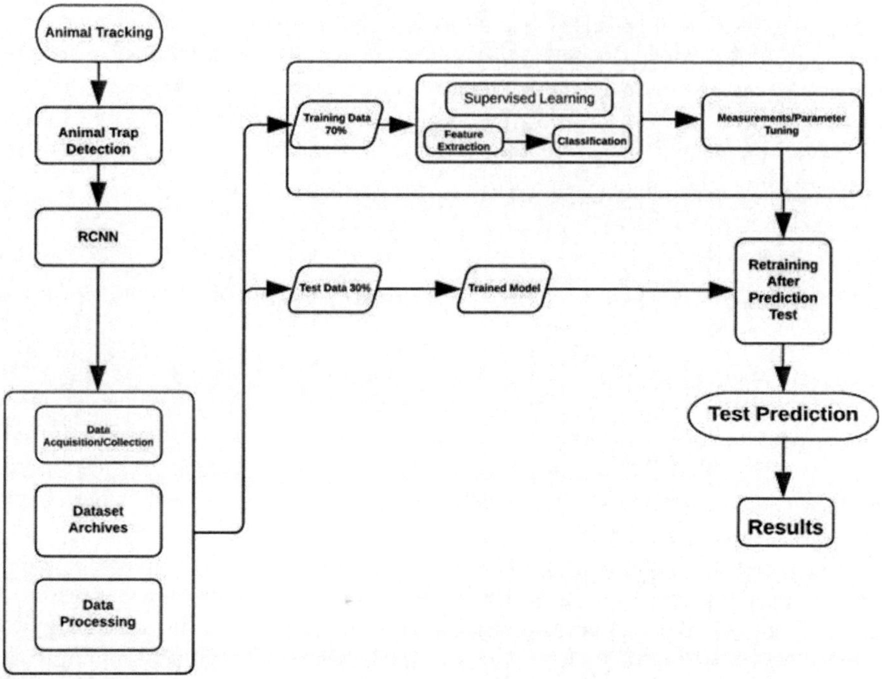

Fig. 1 System flowchart for the endangered mammal detection system

3.2 Proposed Dataset

Initially, we aimed to acquire data from Kaggle and Data World. However, because of the lack of images for each of the four mammal species, we downloaded more images from Google Images and manually labeled each image using a tool called labelImg [4]. The proposed dataset included 548 Visayan warty pigs, 300 Visayan spotted deer, 298 Visayan leopard cats, and 300 Malayan civets. 70% of the data were utilized for training and 30% for testing. The model was trained and validated to enhance accuracy and reduce losses.

3.3 Augmentation

The dataset was rather small so image augmentation was used to increase the amount of data available for training. The Albumentations tool was used to perform blur, randombrightnesscontrast, randomgamma, and CLAHE functions. This was followed by copying and rotating the photos.

3.4 Data Preparation

Because the model used was pre-built, no actual data preparation, such as changing the image size and image type, was necessary. The YOLO model handled all the fitting of data and we only had to split the dataset into training and test sets. However, the only data fitting that YOLO carried out automatically was resizing images into 640×640. Splitting is an important part of training because if the split is not good enough or not thoroughly randomized, model accuracy might be affected. The dataset was split using the random.randint function from Python 3 [16].

3.5 YOLOv5 Architecture

YOLOv5 is known as a single-stage object detector, so it does one pass through the neural network and predicts all the bounding boxes at the same time. This is the reason for its detection speed. This is also one of its downfalls as it could become less accurate than some two-stage object detectors. Similar to other single-stage object detectors, YOLOv5 also has three important parts, which are the model backbone, model neck and the model head [3].

The YOLOv5 employs Cross Stage Partial Networks (CSP) as its model backbone, which gives rich gradient combinations and relevant features from an input image while reducing computation time. The backbone of YOLOv5 is a Cross Stage Partial DenseNet (CSPDenseNet), which combines DenseNet and CSPNet. It first uses the CSPNet procedure, where the dense block is split into two pieces by the channel, with one going to the stage's end and the other going through the dense block [17].

Path Aggregation Network (PA-Net) employs bottom-up path augmentation to enhance and minimize the feature pyramid and to rectify low-level signal localization in the YOLOv5 model neck. It uses FPN to provide feature map layers and ResNet as its basic structure. Adaptive feature pooling is necessary because it employs low feature levels, resulting in smaller proposals. ROIAlign collects tiny feature maps from each area of interest and combines the feature grids from each level. On each proposal's pooled feature grid, a tiny fully convolutional network (FCN) predicts the mask. With a pixel mask, it predicts each class [18].

The final part of the YOLOv5 architecture is the model head which is responsible for the final detection of the model. It applies anchor boxes on features and generates final output features using the training done by the backbone and neck of the model [3]. Figure 2 summarizes the YOLOv5 architecture.

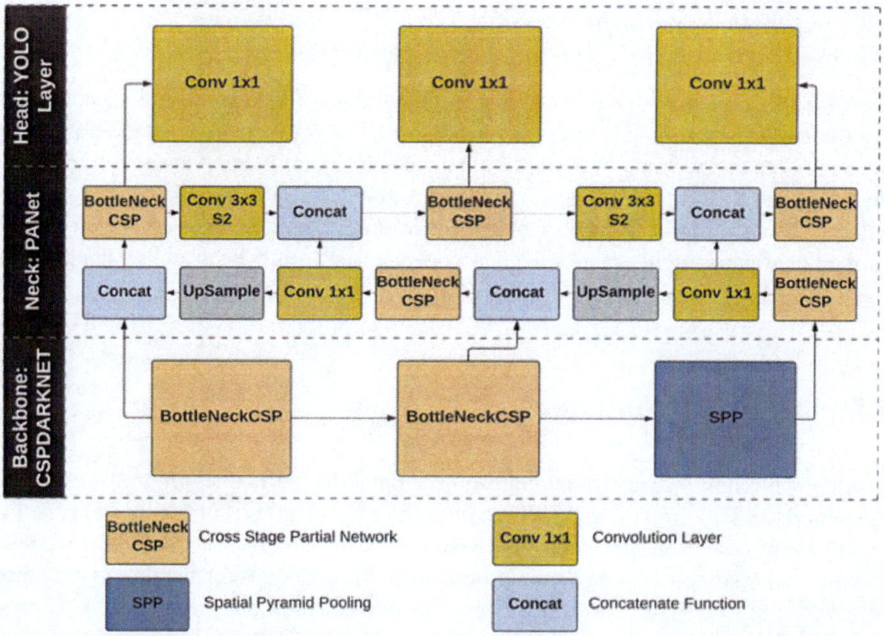

Fig. 2 Full architecture of YOLOv5

3.6 *Activation Function*

In every deep neural network, the choices of activation functions are the most critical factor. In YOLO v5's hidden layers, the Leaky ReLU activation function was applied, whereas the sigmoid activation function was utilized in the last detection layer [19].

3.7 *Optimization Function*

Stochastic gradient descent (SGD) is the standard optimization function for training in YOLOv5. Thus, we used this optimization technique. It is utilized in neural network models to determine the model metrics that provide the optimum match between projected and actual outputs [20].

3.8 *Loss Function*

A compound loss is produced for the YOLO family predicated on objectness score, class probability score, and bounding box regression score. We utilized the Binary

Cross-Entropy with the Logits Loss function from PyTorch for class probability and object score loss computation [21].

4 Metrics

In order to evaluate the success of the machine learning model, a variety of performance indicators were employed. Precision, Recall, F1-score, mAP, and losses are all key criteria to examine when validating a detection model, and these were all taken into account in this project. The following is a succinct list of performance outcomes (TP = true positive, TN = true negative, FP = false positive, FN = false negative).

4.1 Precision

It is a metric that calculates the number of samples that were correctly detected as a particular species out of all samples that were detected as that species, both correctly and incorrectly.

$$Precision = \frac{TP}{TP + FP}$$

4.2 Recall

It is the percentage of all correctly detected animals of a particular species out of all animals in the dataset that belong to that species.

$$Recall = \frac{TP}{TP + FN}$$

4.3 F1 Score

It is calculated as the harmonic mean of precision and recall.

$$F1\,Score = \frac{2 \times precision \times recall}{precision + recall} = \frac{2TP}{2TP + FP + FN}$$

4.4 Mean Average Precision (mAP)

Mean Average Precision (mAP) is the recent standard metric for evaluating object detection algorithms. mAP improves both precision and recall.

$$mAP = \frac{1}{n}\sum_{i=1}^{i=n} AP_i$$

mAP@0.5:0.95 is a variation of the mAP, which averages it into various IoU thresholds from 0.5 to 0.95.

4.5 Losses

Box loss (box_loss) indicates how effectively the system can detect an object's center and how accurately the projected bounding box encompasses it. Objectness loss (obj_loss) measures the probability of the occurrence of an object in a particular region of interest. Classification loss (cls_loss) measures how effectively the system can predict the correct object class.

5 Experiment

In this section, we discuss the details of the experiment, how YOLOv5n is more reliable than other object detection algorithms, and we show the results of training.

5.1 Experimental Parameters

Our model dynamically resizes all training set images to 640 × 640. We employed augmentation to improve our dataset. Our images were also resized to 640 × 640 for testing purposes. Table 1 shows the experimental hyperparameters of our model.

5.2 PR Curve

When plotting a trained model's precision (y-axis) against its recall (x-axis), we can extract the mean average precision (mAP) from the area under this curve. As seen in Fig. 3, the trained model yielded a high mAP with all classes having an average precision of 91% at an IOU threshold of 0.5, indicating the model should be accurate for detecting animals.

Table 1 Experimental model hyperparameters

Parameters	Value
Learning rate	0.01
Learning rate decay	0.999
Learning rate decay step	1
Weight rate decay	5E−4
Momentum	0.937
Batch size	16
Number of iterations	100

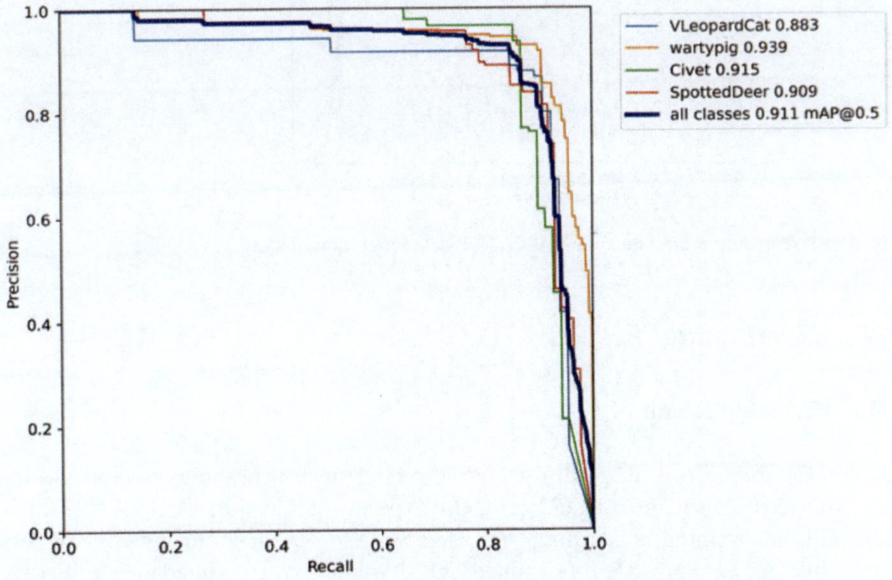

Fig. 3 Precision versus recall curve of the trained model

5.3 *F1 Score*

Precision and recall can reflect a model's performance at several confidence levels. Consequently, the F1-score is particularly valuable for determining the ideal degree of confidence for a model's accuracy and recall. There is a confidence interval of 0–1. From a model's F1-score sequence, a single value evaluation parameter may be derived to help determine its optimal quality.

In Fig. 4, the F1-score confidence rating that balances accuracy and recall is 0.278. A higher confidence level is beneficial in some situations. Because our model's F1-score is approximately 0.89, a 0.45 confidence level may be optimal. Following precision and recall measures at 0.6 confidence level may be a suitable starting point. When recall exceeds 0.45, it declines while accuracy remains high.

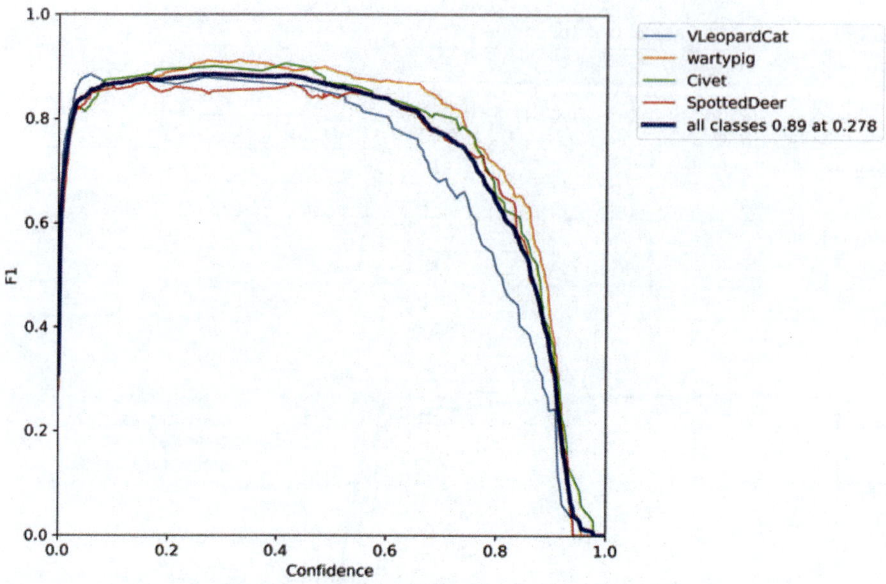

Fig. 4 F1-score curve for a custom YOLOv5 object recognition model

5.4 *Experimental Results*

5.4.1 Training Results

Google Colaboratory was used to train the model since it provides free access to sophisticated GPUs without installation. Our experiment ran on an NVIDIA® Tesla® K80 GPU for training and testing. We used SGD to optimize our network. Web-accessible YOLOv5 provides pre-trained COCO weights. We modified the number of training epochs for the model's upper layers to detect the four species in question. 19 epochs took 65 min. In Fig. 5, it was shown how our model's performance improved with the number of epochs.

As seen in Fig. 5, accuracy, recall, and mAP peaked after around 100 training epochs. After epoch 50, classification losses rise to 0.01 while validation data objectness falls until epoch 20. To get the top 100 epoch results, we utilized YOLOv5's function to save the last and best results.

In Table 2, the three architectures evaluated to train the custom dataset are shown. YOLOv5's mean average precision (mAP), mAP@0.5:0.95 intersection over union (IoU), and average precision (AP) for each species were significantly higher than those of R-CNN and NanoDet. The species' AP also differed. YOLOv5n uses a small number of GFLOPs. This means that it is a lighter version that can operate on Raspberry Pi 3B+, which runs on 5.1 GFLOPs. YOLOv5n performed significantly better than Faster R-CNN despite using significantly less GFLOPs.

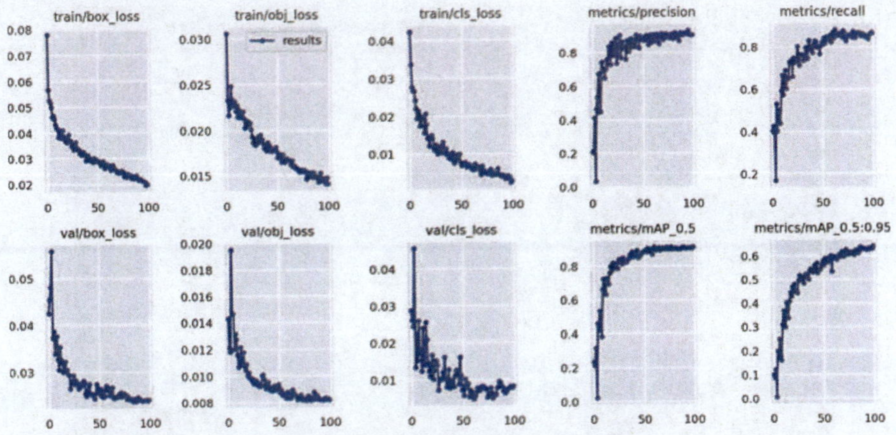

Fig. 5 Various evaluation results of our YOLOv5n model

Table 2 Comparison of performance results between YOLOv5n, faster R-CNN, and NanoDet

Model	mAP		Average precision (AP)				GFLOPs
	mAP	mAP @0.5:0.95	V. warty pig	M. civet	V. leopard cat	V. spotted deer	
YOLOv5n	0.91	0.64	0.94	0.91	0.88	0.91	4.5
Faster R-CNN [14]	0.67	0.44	0	0.64	0.63	0.50	850
NanoDet [22]	0.58	0.51	0.73	0.67	0.52	0.54	1.20

Figure 6 shows examples of correctly detected mammals. Some images show instances where multiple animals of the same species are detected within the same frame.

6 Conclusions

It was our hope to raise awareness of the tragic state of biodiversity in Negros Island in the Philippines, and to contribute, in some way, to the solution. Having shown that we were able to come up with a model that could accurately detect the four endangered mammal species in the island and that could be run on a Raspberry Pi 3B+ board, the next step for us would be to develop the deployment hardware.

Fig. 6 Examples of correctly detected animals: **a** Malayan civet, **b** Visayan leopard cat, **c** Visayan spotted dear, **d** Visayan warty pig

References

1. Clark PE et al (2006) An advanced, low-cost, GPS-based animal tracking system. Rangel Ecol Manag 59(3):334–340. https://doi.org/10.2111/05-162R.1
2. Ani PAB, Castillo MB (2020) Revisiting the state of Philippine biodiversity and the legislation on access and benefit sharing. Food and Fertilizer Technology Center for the Asian and Pacific Region (FFTC-AP)
3. Jocher G et al (2020) yolov5. Code repository. https://github.com/ultralytics/yolov5
4. Tzutalin D (2015) LabelImg. GitHub Repository 6
5. Szegedy C et al (2013) Deep neural networks for object detection. Advances in neural information processing systems, vol 26
6. Malisiewicz T et al (2011) Ensemble of exemplar-SVMs for object detection and beyond. In: 2011 international conference on computer vision. IEEE, pp 89–96
7. Girshick R et al (2015) Region-based convolutional networks for accurate object detection and segmentation. IEEE Trans Pattern Anal Mach Intell 38(1):142–158. https://doi.org/10.1109/TPAMI.2015.2437384

8. Redmon J et al (2016) You only look once: unified, real-time object detection. In: Proceedings of the IEEE conference on computer vision and pattern recognition, pp 779–788. https://doi.org/10.1109/CVPR.2016.91
9. Girshick R (2015) Fast R-CNN. In: Proceedings of the IEEE international conference on computer vision, pp 1440–1448. https://doi.org/10.1109/ICCV.2015.169
10. Ren S et al Faster R-CNN: Towards real-time object detection with region proposal networks. IEEE Trans Pattern Anal Mach Intell (2017). https://doi.org/10.1109/tpami.2016.2577031
11. Choiński M et al (2021) A first step towards automated species recognition from camera trap images of mammals using AI in a European temperate forest. In: International conference on computer information systems and industrial management. Springer, pp 299–310. https://doi.org/10.1007/978-3-030-84340-3_24
12. Yitbarek MB (2019) Livestock and livestock product trends by 2050. IJAR 4:30
13. Li G et al (2021) Practices and applications of convolutional neural network-based computer vision systems in animal farming: a review. Sensors 21(4):1492. https://doi.org/10.3390/s21041492
14. Wu Y et al (2019) Detectron2
15. Nguyen N-D et al (2020) An evaluation of deep learning methods for small object detection. J Electr Comput Eng 2020. https://doi.org/10.1155/2020/3189691
16. Random—generate pseudo-random numbers—Python 3.10.4 documentation
17. Wang C-Y et al (2020) CSPNet: a new backbone that can enhance learning capability of CNN. In: Proceedings of the IEEE/CVF conference on computer vision and pattern recognition workshops, pp 390–391. https://doi.org/10.1109/CVPRW50498.2020.00203
18. Liu S et al (2018) Path aggregation network for instance segmentation. In: Proceedings of the IEEE conference on computer vision and pattern recognition, pp 8759–8768. https://doi.org/10.1109/CVPR.2018.00913
19. Nwankpa C et al (2018) Activation functions: Comparison of trends in practice and research for deep learning. arXiv preprint arXiv:1811.03378
20. Ruder S (2016) An overview of gradient descent optimization algorithms. arXiv preprint arXiv:1609.04747
21. Zhang Z, Sabuncu M (2018) Generalized cross entropy loss for training deep neural networks with noisy labels. Advances in neural information processing systems, vol 31
22. NanoDet Authors (2021) NanoDet

A Study on the Performances of the SPRT Control Chart When Estimating Process Parameters

Jing Wei Teoh, Wei Lin Teoh, Laila El-Ghandour, Zhi Lin Chong, and Sin Yin Teh

Abstract The sequential-probability-ratio-test (SPRT) control chart is well-known for its outstanding detection ability towards a wide range of process mean shifts. In the existing literature, the SPRT chart has been regularly designed from the standpoint that the process parameters are known. Nonetheless, industrial practitioners reveal that these process parameters are hardly known. They often estimate these process parameters using a modest number of Phase-I samples. This in turn leads to different outcomes obtained across practitioners. In this paper, we present detailed insights about the distribution of the conditional ATS when the process parameters are estimated. Also, some statistical properties of the ATS through Monte Carlo simulation are presented in this paper. Particularly, the standard deviation of the ATS is examined in this paper to evaluate the practitioner-to-practitioner variability. Results show that a large Phase-I sample, and a large desired maximum mean shift in optimization design, are required to reduce volatilities in the average and standard deviation of the ATS values. Finally, we provide recommendations for the Phase-I sample sizes tailored to specific objectives of the SPRT chart when the process parameters are estimated.

J. W. Teoh (✉) · W. L. Teoh
Heriot-Watt University Malaysia, Putrajaya, Malaysia
e-mail: t.jing_wei@hw.ac.uk

W. L. Teoh
e-mail: wei_lin.teoh@hw.ac.uk

L. El-Ghandour
Heriot-Watt University, Edinburgh, UK
e-mail: l.el_ghandour@hw.ac.uk

Z. L. Chong
Universiti Tunku Abdul Rahman, Kajang, Selangor, Malaysia
e-mail: chongzl@utar.edu.my

S. Y. Teh
Universiti Sains Malaysia, Penang, Malaysia
e-mail: tehsyin@usm.my

© The Author(s), under exclusive license to Springer Nature Singapore Pte Ltd. 2023
D.-K. Kang et al. (eds.), *Proceedings of the 9th International Conference on Computational Science and Technology*, Lecture Notes in Electrical Engineering 983,
https://doi.org/10.1007/978-981-19-8406-8_36

Keywords Average time to signal · Estimated parameters · Quality control · SPRT control chart · Standard deviation of the average time to signal

1 Introduction

Statistical Quality Control (SQC) is an essential procedure to ensure that the quality of products and services conforms to a specific standard. The implementation of SQC can be found in a myriad of settings, such as those in manufacturing industries, banking sectors, clinical laboratory testing, etc. Contemporarily, rapid advancement in technology has led to the acceleration of revolutionary changes within the quality control regimen. For instance, manual inspection has been gradually replaced by automated quality management systems in factories. Cloud computing infrastructures have made cost-effective storage possible without compromising data security, while supporting a broad range of analytical tools in generating business insights. Among the tools in SQC, the control chart is unequivocally the most popular statistical method. It can alert the personnel whenever a process drifts off in a relatively short time. To date, numerous research articles on quality control have investigated the sequential-probability-ratio-test (SPRT) control chart, which has proven advantages in applications involving expensive or destructive testing [1]. The SPRT chart applies a hypothesis test at each sampling interval by taking a variable number of observations. Each observation is used to compute the control statistic of the SPRT chart with two control limits.

The SPRT chart was first developed based on the idea of sequential analysis [1–3]. Unlike control charts with fixed subgroups, the SPRT chart samples a random number of observations in each sequential test until a conclusion is attained [3]. Existing work has shown that the SPRT chart presents superior performance over the variable-sample-size-variable-sampling-interval (VSS-VSI) Shewhart $\overline{X}$ and cumulative sum (CUSUM) charts in detecting both small and large process mean shifts [3, 4]. In addition, [5, 6] investigated the detection ability of the SPRT chart towards shifts in process variability. Apart from variable-type data, the SPRT chart has also been studied for its usefulness in monitoring attribute-type data, such as proportions and count data [7, 8]. To further boost the SPRT chart's performance, Ou et al. [9] proposed an optimization design of the SPRT chart by minimizing the average extra quadratic loss (AEQL). It is shown that the optimal SPRT chart outperforms the basic SPRT chart by a substantial degree in terms of the overall detection effectiveness and out-of-control performances towards moderate and large mean shifts [9].

Most of the control charts have been designed under the setting that the process parameters (i.e., the mean and standard deviation) are known. In real situations, not only are these parameters unknown, but they are often estimated from in-control samples that are set aside for Phase-I chart's calibration. These Phase-I parameter estimates, by nature, are subject to a certain degree of variability due to differences in sampling procedures and environment. As a result, practitioners may obtain outcomes that elicit different conclusions about an underlying process. This instability in charts' performances were first recognized by [10, 11]. References [12–14] showed that the commonly used performance metrics, i.e., the average run length (ARL) and the average time to signal (ATS), are extremely inconsistent under parameter estimation with a small Phase-I data set. It is also revealed that the conditional in-control performances of the control charts with estimated process parameters are largely undesirable [12]. Specifically, traditional control charts with estimated process parameters tend to produce a huge number of false alarms even when the process is in-control. This is deemed endangering to businesses in the long run. To that end, several researchers have established research work encompassing control charts with modified control limits [15–21]. A review of several methods for the Shewhart, exponentially weighted moving average (EWMA) and CUSUM control charts with estimated process parameters can be found in [22].

Since none of the existing work considers the impact of parameter estimation on the SPRT chart, this paper conducts a preliminary study on the performances of the SPRT chart with estimated process parameters. In this paper, we investigate the conditional ATS (CATS) distribution, the average of the ATS (AATS), and the standard deviation of the ATS (SDATS) of the SPRT chart with estimated process parameters. In Sect. 2, we describe the architecture of the SPRT chart with known process parameters. We also provide formulae for the average sample number (ASN) and the ATS of the SPRT chart with known process parameters. In Sect. 3, we study the distribution of CATS of the SPRT chart when the process parameters are estimated. Also, we provide approximate results for the AATS and SDATS through Monte Carlo simulation. Last but not the least, we present a summary and some suggestions in Sect. 4.

2 The Working Mechanism of the SPRT Chart

Let $(Y_1, Y_2, \ldots, Y_n)$ be a sequence of Phase-II samples of an SPRT, where n represents the random sample number of the SPRT. Assume that each observation Y_i $(i = 1, 2, \ldots, n)$ is sampled from a normal distribution with mean μ_0 and standard deviation σ_0. . Suppose further that all the observations are sampled independently. To compute the control statistic of the SPRT chart, each measurement must be standardized against the process parameters (i.e., μ_0 and σ_0), i.e.,

$$Z_i = \frac{Y_i - \mu_0}{\sigma_0}. \tag{1}$$

The standardized observation is then inserted in the control statistic T_i of the upper one-sided SPRT chart, which is calculated from the following recursive relation, i.e.,

$$T_i = T_{i-1} + Z_i - k, \tag{2}$$

for $i = 1, 2, \ldots, n$, where Z_i is obtained from (1), and $k > 0$ is a reference parameter. Note that $T_0 = 0$ is set in this paper as the fast-initial-response feature is not considered. Each control statistic leads to one of the following three conclusions, i.e., to accept, to reject, or to be indecisive of the hypothesis that the process is in-control. Suppose that the SPRT chart has a lower control limit g and an upper control limit h ($g < h$). When the control statistic T_i falls below g, the process is accepted as in-control. When T_i rises beyond h, the process is indicated as out-of-control, and investigations for the root causes are conducted promptly. When T_i lies between g and h, no conclusion can be made and sampling resumes. Each SPRT is conducted at a fixed time interval, $d > 0$, until the operation of the SPRT chart is suspended due to an out-of-control event.

Stoumbos and Reynolds [3] introduced the use of the ASN and ATS to evaluate the expected sample size and the time-sensitive ARL, respectively, of the SPRT chart. Both formulae can be derived by means of the Markov chain approach and are consistent with the integral equations presented in [3]. We present both the ASN and ATS formulae using the Markov chain approach in this paper. Suppose that the region between g and h is split into l subintervals, where each subinterval has width $\Delta = (h - g)/l$. To apply the concept of absorbing Markov chain, one needs to first construct the transition probability matrix $\mathbf{R}$ with specified mean shift $\delta = (\mu_{\text{new}} - \mu_0)/\sigma_0$, where μ_{new} is the shifted mean in a process. The transition probability matrix $\mathbf{R}$ is given as

$$\mathbf{R} = \begin{pmatrix} r_{1,1} & r_{1,2} & \cdots & r_{1,l} \\ r_{2,1} & r_{2,2} & \cdots & r_{2,l} \\ \vdots & \vdots & \ddots & \vdots \\ r_{l,1} & r_{l,2} & \cdots & r_{l,l} \end{pmatrix}, \tag{3}$$

where each entry $r_{a,b}$ is evaluated as

$$r_{a,b} = \Phi((b - a + 0.5)\Delta + k - \delta) - \Phi((b - a - 0.5)\Delta + k - \delta), \tag{4}$$

for $a, b = 1, 2, \ldots, l$. Note that $r_{a,b}$ is the transition probability moving from state a to state b, while $\Phi(\cdot)$ in (4) is the cumulative distribution function (cdf) of the standard normal distribution. The formula for the ASN of the SPRT chart is provided by [3] as

$$\text{ASN} = 1 + \mathbf{C}^{\text{T}}(\mathbf{I} - \mathbf{R})^{-1}\mathbf{1}, \tag{5}$$

where $\mathbf{R}$ is quoted directly from (3), $\mathbf{I}$ is the $l \times l$ identity matrix, and $\mathbf{1}$ is a vector with entries equal to one. Also, $\mathbf{C}$ in (5) is a vector of initial transition probabilities, with the ath entry c_a, which is given by $\Phi(g + a\Delta + k - \delta) - \Phi(g + (a - 1)\Delta + k - \delta)$.

The ATS of the SPRT chart can be obtained using a similar argument. To evaluate both the in-control and out-of-control steady-state ATS, one needs to first recognize that the time to signal is a geometric random variable with probability equal to $1 - P(\delta)$. Here, $P(\delta)$ is the probability of accepting the process, which is a function of the mean shift δ. By means of the Markov chain approach, $P(\delta)$ is evaluated as

$$P(\delta) = P_0 + \mathbf{C}^{\mathrm{T}}(\mathbf{I} - \mathbf{R})^{-1}\mathbf{Q}, \tag{6}$$

where P_0 is computed as $\Phi(g + k - \delta)$ and $\mathbf{Q}$ is a vector with the ath entry q_a, which is given by $\Phi((0.5 - a)\Delta + k - \delta)$.

Using the properties of a geometric distribution, [3] derived the in-control ATS (ATS_0) and out-of-control steady-state ATS (ATS_1) of the SPRT chart with known process parameters as

$$\text{ATS}_0 = \frac{d}{1 - P(0)} \tag{7}$$

and

$$\text{ATS}_1 = \left(\frac{1}{1 - P(\delta)} - 0.5\right) \times d, \tag{8}$$

respectively, where $P(0)$ (i.e., $\delta = 0$) and $P(\delta)$ (i.e., $\delta \neq 0$) in (7) and (8), respectively, can be computed directly from (6).

When the process parameters μ_0 and σ_0 are known, the ATS provides a sound measure of the performance of the SPRT chart at any specified mean shift δ. Since the process parameters are barely known in practical situations, we need to estimate them from a set of validated in-control Phase-I samples. Denote $(X_1, X_2, \ldots, X_m)$ as the Phase-I data set of size m. The common estimators adopted to estimate the mean μ_0 and the standard deviation σ_0 are

$$\hat{\mu}_0 = \overline{X} = \frac{1}{m}\sum_{s=1}^{m} X_s \tag{9}$$

and

$$\hat{\sigma}_0 = \sqrt{\frac{1}{m-1}\sum_{s=1}^{m}\left(X_s - \overline{X}\right)^2}, \tag{10}$$

respectively. By substituting $\hat{\mu}_0$ and $\hat{\sigma}_0$ in (1), the performance metric (i.e., ATS) may be subject to variation, depending on the number of Phase-I samples acquired. Recognizing that the ATS has become a conditional random variable of $\hat{\mu}_0$ and $\hat{\sigma}_0$., we investigate the CATS distribution, AATS and SDATS values in the following section to gauge the impact of practitioner-to-practitioner variability upon the performances of the SPRT chart with estimated process parameters. These CATS distributions, AATS and SDATS values are computed through Monte Carlo simulation.

3 Results and Discussions

In this section, we explore the distribution of CATS for the SPRT chart with estimated mean and standard deviation given in (9) and (10), respectively. Saleh et al. [12] showed that the run-length properties of a control chart with estimated process parameters could be extremely unstable when the Phase-I sample size is small. Hence, we demonstrate the effect of parameter estimation on the AATS and SDATS over a range of values of $m \in \{100, 200, 400, 600, 1000, 1500, 2000, +\infty\}$. Note that $m = +\infty$ represents the case in which the process parameters are known and require no estimation.

The SPRT chart has five optimal charting parameters, i.e., the in-control ASN (ASN_0), k, d, g, and h. In this paper, we adopt four different combinations of optimal charting parameters $(ASN_0, k, d, g, h) \in \{(2.122, 0.360, 0.707, 0.147, 6.194), (1.500, 0.410, 0.500, 0.561, 5.481), (2.260, 0.250, 0.753, 0.438, 11.439), (1.500, 0.320, 0.500, 0.804, 9.170)\}$ obtained from Table 3 of [9]. These optimal parameters are computed by minimizing the AEQL metric when the process parameters are known. See Sect. 2 and [9] for the detailed formulae and optimization algorithm, respectively, of the SPRT chart with known process parameters. It is important to note that the first two combinations of charting parameters yield $ATS_0 = 250$, whereas the last two combinations yield $ATS_0 = 1500$. All the SPRT charts are designed with an inspection rate equal to three. The four combinations of optimal charting parameters will be used to compute the distribution of CATS, AATS and SDATS values presented in Sects. 3.1, 3.2 and 3.3.

In Sect. 3.1, we provide boxplots for the distribution of conditional ATS_0 ($CATS_0$) values for all the four charting combinations stated in the preceding paragraph, specifically when $m = 100$. We first formulate the problems encountered in the analyses of the distribution of $CATS_0$. Then, we reveal some of the treatments implemented in the remainder of Sect. 3. In Sect. 3.2, we tabulate the values of AATS and SDATS for the four charting combinations in Tables 1, 2, 3 and 4, when $m \in \{100, 200, 400, 600, 1000, 1500, 2000, +\infty\}$ and $\delta \in \{0.00, 0.25, 0.50, 0.75, 1.00, 1.25, 1.50, 1.75, 2.00\}$. Note that all the results are computed using Monte Carlo simulation with 100,000 simulation runs. The two values in each cell of Tables 1, 2, 3 and 4 show the AATS (the first row of each cell) and SDATS (the second row of each cell) values. In Sect. 3.3, we provide the values of the in-control AATS ($AATS_0$) and SDATS ($SDATS_0$) of the SPRT chart with estimated process parameters over a range of values of m in Tables 5 and 6, respectively. We highlight the minimum m required in each SPRT chart to ensure that the specified criteria are met.

Table 1 AATS and SDATS values for (ASN_0, k, d, g, h) = (2.122, 0.360, 0.707, 0.147, 6.194), $ATS_0 = 250$, $m \in \{100, 200, 400, 600, 1000, 1500, 2000, +\infty\}$, and $\delta \in \{0.00, 0.25, 0.50, 0.75, 1.00, 1.25, 1.50, 1.75, 2.00\}$

m	100	200	400	600	1000	1500	2000	$+\infty$
δ	AATS	AATS	AATS	AATS	AATS	AATS	AATS	ATS
	SDATS	SDATS	SDATS	SDATS	SDATS	SDATS	SDATS	SDATS
0.00	672.07	395.65	311.84	289.07	272.39	264.51	260.69	249.64
	2397.42	532.16	238.70	169.76	118.24	91.67	77.39	0
0.25	36.31	25.83	22.24	21.22	20.46	20.09	19.91	19.39
	76.06	23.70	12.15	9.00	6.47	5.09	4.33	0
0.50	4.50	3.93	3.70	3.63	3.58	3.56	3.54	3.51
	3.97	1.78	1.06	0.82	0.61	0.49	0.42	0
0.75	1.54	1.48	1.46	1.45	1.44	1.44	1.44	1.43
	0.52	0.31	0.20	0.16	0.12	0.10	0.09	0
1.00	0.90	0.88	0.88	0.88	0.87	0.87	0.87	0.87
	0.16	0.10	0.07	0.06	0.04	0.04	0.03	0
1.25	0.65	0.64	0.64	0.64	0.64	0.64	0.64	0.64
	0.07	0.05	0.03	0.03	0.02	0.02	0.02	0
1.50	0.52	0.52	0.52	0.52	0.52	0.52	0.52	0.52
	0.04	0.03	0.02	0.02	0.01	0.01	0.01	0
1.75	0.45	0.45	0.45	0.45	0.45	0.45	0.45	0.45
	0.02	0.02	0.01	0.01	0.01	0.01	0.00	0
2.00	0.41	0.41	0.41	0.41	0.41	0.41	0.41	0.41
	0.01	0.01	0.01	0.01	0.00	0.00	0.00	0

3.1 *Distribution of the CATS for the SPRT Chart with Estimated Process Parameters*

Figure 1 shows four boxplots for the distribution of $CATS_0$ obtained from the SPRT chart with estimated process parameters when $m = 100$. Note that 10,000 $CATS_0$ values are generated for each boxplot. The two vertical dotted lines in Fig. 1 represent pre-specified in-control ATS, i.e., $ATS_0 = 250$ and $ATS_0 = 1500$, respectively. We denote all the four combinations of charting parameters (ASN_0, k, d, g, h) $\in$ {(2.122, 0.360, 0.707, 0.147, 6.194), (1.500, 0.410, 0.500, 0.561, 5.481), (2.260, 0.250, 0.753, 0.438, 11.439), (1.500, 0.320, 0.500, 0.804, 9.170)} as Comb 1, Comb 2, Comb 3, and Comb 4, respectively.

From Fig. 1, it is observed that the medians for the distribution of $CATS_0$ are very close to their corresponding nominal ATS_0 values (i.e., 250 or 1500). Clearly, the $CATS_0$ distributions under Comb 3 and Comb 4 are much more variable compared to those under Comb 1 and Comb 2. This may be explained by the fact that the control limits are wider under Comb 3 and Comb 4. Hence, more variability is

Table 2 AATS and SDATS values for $(\mathrm{ASN}_0, k, d, g, h) = (1.500, 0.410, 0.500, 0.561, 5.481)$, $\mathrm{ATS}_0 = 250$, $m \in \{100, 200, 400, 600, 1000, 1500, 2000, +\infty\}$, and $\delta \in \{0.00, 0.25, 0.50, 0.75, 1.00, 1.25, 1.50, 1.75, 2.00\}$

m	100	200	400	600	1000	1500	2000	$+\infty$
δ	AATS	AATS	AATS	AATS	AATS	AATS	AATS	ATS
	SDATS	SDATS	SDATS	SDATS	SDATS	SDATS	SDATS	SDATS
0.00	569.72	367.35	301.20	282.70	268.97	262.43	259.25	249.99
	1557.45	427.08	206.46	150.05	106.24	83.02	70.36	0
0.25	44.58	33.26	29.19	28.00	27.11	26.68	26.47	25.85
	78.03	27.92	15.00	11.26	8.17	6.46	5.51	0
0.50	6.38	5.53	5.20	5.10	5.02	4.98	4.96	4.91
	5.77	2.71	1.62	1.26	0.94	0.75	0.65	0
0.75	1.99	1.89	1.85	1.83	1.82	1.82	1.82	1.81
	0.83	0.49	0.32	0.26	0.19	0.16	0.14	0
1.00	1.04	1.02	1.01	1.00	1.00	1.00	1.00	1.00
	0.25	0.16	0.11	0.09	0.07	0.06	0.05	0
1.25	0.68	0.68	0.67	0.67	0.67	0.67	0.67	0.67
	0.11	0.08	0.05	0.04	0.03	0.03	0.02	0
1.50	0.51	0.50	0.50	0.50	0.50	0.50	0.50	0.50
	0.06	0.04	0.03	0.02	0.02	0.02	0.01	0
1.75	0.41	0.41	0.41	0.41	0.41	0.40	0.40	0.40
	0.04	0.03	0.02	0.01	0.01	0.01	0.01	0
2.00	0.35	0.35	0.35	0.35	0.35	0.35	0.35	0.35
	0.02	0.02	0.01	0.01	0.01	0.01	0.00	0

present in the distribution of CATS_0, leading to greater uncertainty. Besides, it is noticed that in all the four boxplots, approximately 50% of the CATS_0 values are less than the recommended in-control ATS. For instance, the first two boxplots clearly show that 50% of the CATS_0 values fall below 250, whereas the last two boxplots show that more than 50% of the CATS_0 are lower than 1500. Since almost half of the practitioners are expected to suffer more frequent false alarms than the recommended rate of 1/250 or 1/1500, this marks a red flag for practitioners.

Based on the interpretations, one quickly recognizes two key issues concerning the SPRT chart with estimated process parameters. First, the variability in the distribution of CATS_0 is huge, as the CATS_0 values range from zero to at least four times the specified $\mathrm{ATS}_0 \in \{250, 1500\}$. Second, the number of false alarms is large. This is deemed highly intolerable to businesses aiming at achieving cost and production efficiency. It is possible to increase the number of Phase-I samples used in parameter estimation to reduce the stated variability [12]. This will be discussed in the next two subsections.

Table 3 AATS and SDATS values for $(ASN_0, k, d, g, h) = (2.260, 0.250, 0.753, 0.438, 11.439)$, $ATS_0 = 1500$, $m \in \{100, 200, 400, 600, 1000, 1500, 2000, +\infty\}$, and $\delta \in \{0.00, 0.25, 0.50, 0.75, 1.00, 1.25, 1.50, 1.75, 2.00\}$

m	100	200	400	600	1000	1500	2000	$+\infty$
δ	AATS	AATS	AATS	AATS	AATS	AATS	AATS	ATS
	SDATS	SDATS	SDATS	SDATS	SDATS	SDATS	SDATS	SDATS
0.00	33,007.52	5899.17	2862.75	2289.12	1925.40	1769.18	1696.67	1499.27
	2,339,817.69	30,983.01	5072.00	2748.83	1571.47	1110.43	896.16	0
0.25	114.29	41.12	28.44	25.59	23.66	22.78	22.37	21.20
	2383.09	95.85	26.36	16.75	10.82	8.12	6.75	0
0.50	4.28	3.67	3.47	3.42	3.38	3.36	3.35	3.32
	6.31	1.52	0.85	0.65	0.48	0.39	0.33	0
0.75	1.67	1.62	1.60	1.59	1.59	1.59	1.59	1.58
	0.45	0.28	0.19	0.15	0.12	0.09	0.08	0
1.00	1.04	1.03	1.02	1.02	1.02	1.02	1.02	1.02
	0.17	0.12	0.08	0.06	0.05	0.04	0.04	0
1.25	0.76	0.76	0.75	0.75	0.75	0.75	0.75	0.75
	0.09	0.06	0.04	0.03	0.03	0.02	0.02	0
1.50	0.61	0.61	0.60	0.60	0.60	0.60	0.60	0.60
	0.05	0.04	0.02	0.02	0.02	0.01	0.01	0
1.75	0.52	0.52	0.52	0.51	0.51	0.51	0.51	0.51
	0.03	0.02	0.02	0.01	0.01	0.01	0.01	0
2.00	0.46	0.46	0.46	0.46	0.46	0.46	0.46	0.46
	0.02	0.01	0.01	0.01	0.01	0.00	0.00	0

3.2 *AATS and SDATS of the SPRT Chart with Estimated Process Parameters*

Tables 1, 2, 3 and 4 display the values of AATS and SDATS of the SPRT chart with estimated process parameters, for four combinations of charting parameters. Note that AATS is used as the performance measure for the SPRT charts with estimated process parameters, whereas ATS is used in the case of known process parameters. Since the ATS of the SPRT chart with known process parameters is unaffected by parameter estimation, the $SDATS_0$ and out-of-control SDATS ($SDATS_1$) are equal to zero.

In all the four tables, one observes that both the AATS and SDATS values decrease as the Phase-I sample size m increases. It is noticed that the AATS values in the cases of estimated process parameters approach the ATS value in the case of known process parameters ($m = +\infty$) for large values of m. When δ is zero or small ($\delta < 0.50$), one perceives considerable differences between the AATS and ATS values. On the contrary, when δ is large, there is hardly any visible distinction between chart's

Table 4 AATS and SDATS values for $(ASN_0, k, d, g, h) = (1.500, 0.320, 0.500, 0.804, 9.170)$, $ATS_0 = 1500$, $m \in \{100, 200, 400, 600, 1000, 1500, 2000, +\infty\}$, and $\delta \in \{0.00, 0.25, 0.50, 0.75, 1.00, 1.25, 1.50, 1.75, 2.00\}$

m	100	200	400	600	1000	1500	2000	$+\infty$
δ	AATS	AATS	AATS	AATS	AATS	AATS	AATS	ATS
	SDATS	SDATS	SDATS	SDATS	SDATS	SDATS	SDATS	SDATS
0.00	12,595.87	3908.91	2366.60	2023.19	1791.00	1687.01	1637.76	1500.11
	268,963.50	11,710.21	3066.60	1883.79	1178.34	868.25	715.02	0
0.25	147.66	70.01	51.86	47.35	44.17	42.70	41.99	39.98
	1245.62	121.07	43.60	29.42	19.76	15.07	12.63	0
0.50	7.06	5.69	5.25	5.12	5.03	4.99	4.96	4.90
	11.77	3.23	1.73	1.31	0.96	0.76	0.65	0
0.75	2.08	1.99	1.95	1.94	1.93	1.92	1.92	1.91
	0.80	0.47	0.31	0.25	0.19	0.15	0.13	0
1.00	1.15	1.13	1.12	1.11	1.11	1.11	1.11	1.11
	0.26	0.17	0.12	0.10	0.07	0.06	0.05	0
1.25	0.77	0.76	0.75	0.75	0.75	0.75	0.75	0.75
	0.13	0.09	0.06	0.05	0.04	0.03	0.03	0
1.50	0.57	0.56	0.56	0.56	0.56	0.56	0.56	0.56
	0.07	0.05	0.04	0.03	0.02	0.02	0.02	0
1.75	0.45	0.45	0.45	0.45	0.45	0.45	0.45	0.45
	0.05	0.03	0.02	0.02	0.01	0.01	0.01	0
2.00	0.38	0.37	0.37	0.37	0.37	0.37	0.37	0.37
	0.03	0.02	0.01	0.01	0.01	0.01	0.01	0

performances at any level of m. For instance, in Table 1, the AATS values evaluated at $\delta = 2.00$ are all equal to 0.41. When the process is in-control ($\delta = 0$), the $SDATS_0$ values in all the four tables are extremely large, especially when m is small (i.e., $m = 100$). One can see that the $SDATS_0$ values in Tables 1 and 2 are both greater than 1000 (i.e., four times of $ATS_0 = 250$), whereas those in Tables 3 and 4 exceed 200,000 (i.e., about 133 times of $ATS_0 = 1500$). In the out-of-control cases ($\delta > 0$), the $SDATS_1$ values decrease rapidly to zero for moderate and large δ. This indicates that the out-of-control AATS ($AATS_1$) values are rather concentrated around their mean when the process shift is large. As m increases, the SDATS values seem to decay in an "exponential" trend towards zero.

Another interesting feature worth highlighting is the difference in magnitudes of $AATS_0$ and $SDATS_0$ between different combinations of charting parameters, when the desired ATS_0 is the same. For instance, it is observed that, for all values of m, the $AATS_0$ and $SDATS_0$ values in Table 1 are higher than those in Table 2, even though both charts are deliberately designed to achieve $ATS_0 = 250$ and an inspection rate of three. Here, the hidden feature is the maximum mean shift, δ_{max}, considered in the

Table 5 $AATS_0$ values for four combinations of charting parameters (ASN_0, k, d, g, h) for different values of m

ATS_0	250	250	1500	1500
(ASN_0, k, d)	(2.122, 0.360, 0.707)	(1.500, 0.410, 0.500)	(2.260, 0.250, 0.753)	(1.500, 0.320, 0.500)
(g, h)	(0.147, 6.194)	(0.561, 5.481)	(0.438, 11.439)	(0.804, 9.170)
m	$AATS_0$	$AATS_0$	$AATS_0$	$AATS_0$
1000	272.39	268.97	1925.40	1791.00
1300	266.91	264.42	1815.60	1718.21
1500	264.51	**262.43**	1769.18	1687.01
1800	**261.96**	260.30	1720.45	1653.98
2100	260.15	258.80	1686.61	1630.86
2400	258.80	257.67	1661.74	1613.78
2700	257.76	256.80	1642.69	1600.65
3000	256.94	256.11	1627.64	1590.24
3300	256.26	255.55	1615.45	1581.78
3600	255.70	255.08	1605.37	**1574.77**
3900	255.23	254.68	1596.90	1568.87
4200	254.82	254.34	1589.69	1563.83
4500	254.47	254.05	1583.46	1559.48
4800	254.17	253.79	1578.04	1555.69
5000	253.98	253.64	**1574.80**	1553.42
$+\infty$	250.00	250.00	1500.00	1500.00

AEQL optimization design of the SPRT chart with known process parameters. The AEQL is calculated by taking the weighted average of ATS across a range of mean shifts $\delta \in (0, \delta_{max})$. By employing different values of δ_{max}, practitioners may obtain different optimal SPRT charts with slightly different performances. In this paper, both optimal charting parameters used in Tables 1 and 3 are designed with $\delta_{max} = 2$, whereas those used in Tables 2 and 4 are designed with $\delta_{max} = 4$. Based on the results, we deduce that the SPRT charts designed using $\delta_{max} = 2$ are more susceptible to large variation in performances compared to those designed using $\delta_{max} = 4$.

3.3 *Sample Size Recommendations for the SPRT Chart with Estimated Process Parameters*

Tables 5 and 6 show the values of $AATS_0$ and $SDATS_0$, respectively, for the SPRT chart under different combinations of charting parameters and values of m. The

Table 6 $SDATS_0$ values for four combinations of charting parameters (ASN_0, k, d, g, h) for different values of m

ATS_0	250	250	1500	1500
(ASN_0, k, d)	(2.122, 0.360, 0.707)	(1.500, 0.410, 0.500)	(2.260, 0.250, 0.753)	(1.500, 0.320, 0.500)
(g, h)	(0.147, 6.194)	(0.561, 5.481)	(0.438, 11.439)	(0.804, 9.170)
m	$SDATS_0$	$SDATS_0$	$SDATS_0$	$SDATS_0$
10,000	32.58	29.89	339.87	284.12
13,000	28.47	26.14	295.32	247.52
15,000	26.47	**24.30**	273.79	229.75
17,000	**24.83**	22.80	256.37	215.32
20,000	22.86	21.00	235.52	198.01
23,000	21.30	19.57	219.05	184.29
26,000	20.02	18.39	205.61	173.08
29,000	18.94	17.41	194.38	163.70
32,000	18.02	16.56	184.80	155.69
35,000	17.23	15.83	176.51	**148.75**
38,000	16.53	15.19	169.25	142.67
41,000	15.91	14.62	162.81	137.27
43,000	15.53	14.27	158.91	134.00
46,000	15.01	13.80	153.55	129.50
49,000	14.54	13.37	**148.69**	125.42
$+\infty$	0	0	0	0

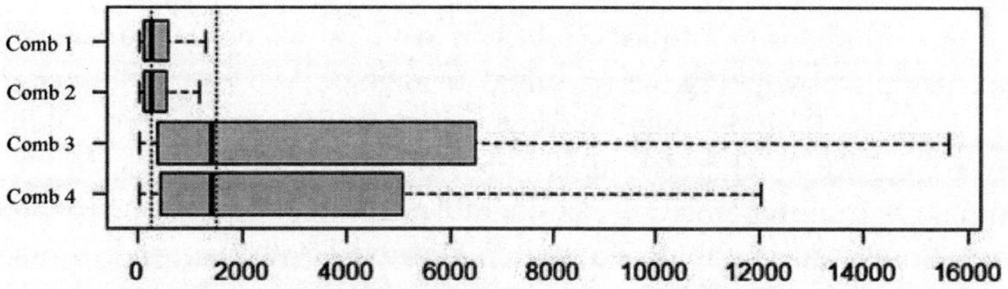

Fig. 1 Boxplots for the distribution of $CATS_0$ for the SPRT chart when $m = 100$ for four combinations of charting parameters

bolded values represent the minimum m required to achieve the desired levels of in-control performances. Particularly, to reduce practitioner-to-practitioner variability in the chart's performances, we propose the minimum m based on two criteria [12]:

(i) the $AATS_0$ value lies within 5% errors of the corresponding ATS_0 value, or
(ii) the $SDATS_0$ value falls below 10% of the corresponding ATS_0 value.

Note that criterion (i) aims to enhance the accuracy of the $CATS_0$ values, whereas criterion (ii) improves the precision of the $CATS_0$ values.

In Table 5, it is noticed that the minimum m required to achieve $AATS_0$ values within 5% of the ATS_0 value are 1800, 1500, 5000, and 3600 for the SPRT chart designed under Comb 1, Comb 2, Comb 3, and Comb 4, respectively. It is observed that large values of m are required to guarantee consistent performances across practitioners when a large ATS_0 is used in control chart's design. Referring to the optimization design of Ou et al. [9], results show that large Phase-I sample sizes are needed to control the $AATS_0$ values when a small δ_{max} is used. For example, the SPRT charts designed under Comb 1 and Comb 3 (with $\delta_{max} = 2$) require larger Phase-I sample sizes compared to those designed under Comb 2 and Comb 4 (with $\delta_{max} = 4$), respectively, to achieve 5% errors from their corresponding ATS_0 values.

On the other hand, in Table 6, it is observed that the minimum m required to achieve $SDATS_0$ values within 10% of the desired ATS_0 value are 17,000, 15,000, 49,000, and 35,000 for the SPRT chart designed under Comb 1, Comb 2, Comb 3, and Comb 4, respectively. As an example, to determine the minimum m for the SPRT chart designed under Comb 1, we increase m gradually until the $SDATS_0$ value falls just below 25 (i.e., 10% of the $ATS_0 = 250$). Note that when m is increased indefinitely, the $SDATS_0$ should decrease to zero. However, it is usually sufficient to control the $SDATS_0$ value within a small percentage of the desired ATS_0 to guarantee stable performances across practitioners [12]. One interesting pattern to note is that the minimum values of m required in Table 6 are almost ten times greater than those required in Table 5 for all combinations of charting parameters. In fact, such a large m is almost always deemed impractical. Therefore, it seems that sample size recommendations based on the $SDATS_0$ may not be completely feasible.

4 Conclusion

In this paper, we first present the distribution of $CATS_0$ for a modest number of Phase-I samples, followed by an in-depth analysis of the statistical properties of the ATS, i.e., AATS and SDATS. Results show that, with a large value of the desired ATS_0, the distribution of $CATS_0$ is likely to be extremely unstable for small values of m. Besides, it is revealed that almost 50% of the $CATS_0$ values fall below the nominal ATS_0, signifying undesirably large proportions of false alarms. Furthermore, we show that the AATS values converge to the ATS values as m increases. We also demonstrate that the SDATS values decrease to zero with a sufficiently large m. It is observed that different values of δ_{max} in the optimization design may have different effects on the $AATS_0$ and $SDATS_0$ of the SPRT chart with estimated process parameters. Specifically, we notice that setting a large δ_{max} helps to reduce volatilities in the $AATS_0$ and $SDATS_0$ values.

In the final section, we provide sample size recommendations based on the $AATS_0$ and $SDATS_0$ values of the SPRT chart with estimated process parameters. We recommend $m > 1500$ for the SPRT chart designed with $ATS_0 = 250$, and $m > 3600$ for the SPRT chart designed with $ATS_0 = 1500$, to guarantee that $AATS_0$ is at most 5% away from the nominal ATS_0 value. On the other hand, recommendations based on the $SDATS_0$ values may be infeasible to be implemented, since they require m to be at least 15,000, even for the best performing chart, to attain an $SDATS_0$ value that is lower than 10% of the nominal ATS_0 value. In view of this issue, future research may explore the concept of modified control limits in the design of the SPRT chart with estimated process parameters, while retaining suitably large numbers of Phase-I samples.

Acknowledgements This work was supported by the Ministry of Higher Education (MOHE) Malaysia and Heriot-Watt University Malaysia under Fundamental Research Grant Scheme (FRGS), no. FRGS/1/2021/STG06/HWUM/02/1.

References

1. Wald A (1947) Sequential analysis. Dover Publications, New York
2. Siegmund D (1985) Sequential analysis: tests and confidence intervals. Springer-Verlag, New York
3. Stoumbos ZG, Reynolds MR (1997) Control charts applying a sequential test at fixed sampling intervals. J Qual Technol 29(1):21–40
4. Ou Y, Wu Z, Tsung F (2012) A comparison study of effectiveness and robustness of control charts for monitoring process mean. Int J Prod Econ 135(1):479–490
5. Godase DG, Mahadik SB (2019) The SPRT control chart for process dispersion. Qual Reliab Eng Int 35(6):1878–1889
6. Mahadik SB, Godase DG, Teoh WL (2021) A two-sided SPRT control chart for process dispersion. J Stat Comput Sim 91(17):3603–3614
7. Reynolds MR, Stoumbos ZG (2001) Monitoring a proportion using CUSUM and SPRT control charts. Frontiers in statistical quality control 6. Springer-Verlag, Berlin, pp 155–175
8. Yu FJ, Low CY, Cheng SS (2003) Design for an SPRT control scheme based on linguistic data. Int J Prod Res 41(6):1299–1309
9. Ou Y, Wu Z, Lee KM, Chen S (2011) A highly effective SPRT control chart for monitoring process mean. Technical report, School of Mechanical and Aerospace Engineering, Nan-yang Technological University
10. Chakraborti S (2006) Parameter estimation and design considerations in prospective applications of the chart. J Appl Stat 33(4):439–459
11. Chen G (1997) The mean and standard deviation of the run length distribution of charts when control limits are estimated. Stat Sin 7(3):789–798
12. Saleh NA, Mahmoud MA, Keefe MJ, Woodall WH (2015) The difficulty in designing Shewhart and X control charts with estimated parameters. J Qual Technol 47(2):127–138
13. Mim FN, Khoo MBC, Saha S, Castagliola P (2021) Revised triple sampling $\overline{X}$ control charts for the mean with known and estimated process parameters. Int J Prod Res 60(16):4911–4935
14. Johannssen A, Chukhrova N, Castagliola P (2022) The performance of the hypergeometric np chart with estimated parameter. Eur J Oper Res 296(3):873–899
15. Sarmiento MGC, Jardim FS, Chakraborti S, Epprecht EK (2020) Design of variance control charts with estimated parameters: A head to head comparison between two perspectives. J Qual Technol 54(3):249–268
16. Khusna H, Mashuri M, Ahsan M, Suhartono S, Prastyo DD (2020) Bootstrap-based maximum multivariate CUSUM control chart. Qual Technol Quant M 17(1):52–74

17. Jaime M, Francisco A, Eugenio EK (2021) Guaranteeing acceptable in-control and out-of-control performance of joint-S control charts with estimated parameters. Qual Technol Quant M 18(6):701–717
18. Hu XL, Castagliola P, Tang A, Zhong J (2021) Guaranteed conditional performance of the median run length based EWMA chart with unknown process parameters. Commun Stat-Simul C 50(12):4280–4299
19. Lee MH, Khoo MBC (2021) Double sampling np chart with estimated process parameter. Commun Stat-Simul C 50(8):2232–2250
20. Ming C, On CK, Rayner A, Guan TT, Patricia A (2018) The determinant factors affecting cloud computing adoption by small and medium enterprises (SMEs) in Sabah, Malaysia. J Telecommun Electron Comput Eng 10(3-2):83–88
21. Li J (2022) Adaptive CUSUM chart with cautious parameter learning. Qual Reliab Eng Int 38(6):3135−3156
22. Does RJMM, Goedhart R, Woodall WH (2020) On the design of control charts with guaranteed conditional performance underestimated parameters. Qual Reliab Eng Int 36(8):2610–2620

Exploration and Showcase of FSMs and HFSMs Traffic in Low-Cost Educational Driving Simulator

Yun-Quan Cheng, Sarina Mansor, Ji-Jian Chin, Hezerul Abdul Karim, and Ban Kar-Weng

Abstract Finite state machine (FSM) is a model of computation that executes an exact finite number of states at any given time where Hierarchical Finite State Machines (HFSM) can group multiple FSMs and execute as one state. Given these techniques can be used to exclusively execute certain states, it is widely used in games. In this paper we will explore and showcase the implementation of FSMs and HFSMs to create traffic non-playing characters (NPC) in our low-cost educational driving simulators. We will also discuss possible techniques that can be use. It is also in the interest to push for more understanding in any gap for traffic NPCs in similar low-cost educational driving sims in the future.

Keywords Driving education · Driving simulator · Finite state machines · Hierarchical finite state machines · Low-cost

1 Introduction

Finite state machines have been one of the oldest techniques used in games such as old as classic PACMAN [14] and as new as the controversial hit Cyberpunk 2077. Throughout the years in games, these techniques share a same goal, which is to make non player characters (NPCs) more intelligent [14]. By reacting to certain conditions

Y.-Q. Cheng (✉) · S. Mansor · H. A. Karim
Faculty of Engineering, Multimedia University, Cyberjaya, Selangor, Malaysia
e-mail: tylerchengyq@gmail.com

S. Mansor
e-mail: sarina.mansor@mmu.edu.my

H. A. Karim
e-mail: hezerul@mmu.edu.my

J.-J. Chin · B. Kar-Weng
Faculty of Computing and Informatics, Multimedia University, Cyberjaya, Selangor, Malaysia
e-mail: jjchin@mmu.edu.my

© The Author(s), under exclusive license to Springer Nature Singapore Pte Ltd. 2023

D.-K. Kang et al. (eds.), *Proceedings of the 9th International Conference on Computational Science and Technology*, Lecture Notes in Electrical Engineering 983,
https://doi.org/10.1007/978-981-19-8406-8_37

with certain states that is pre-programmed, these NPCs could seem like they are reacting to the player's actions.

The gaming industry has come a long way in terms of NPC intelligence. With the rise in popularity for educational/serious games, these technologies could benefit educational usage in a way never explored before.

As driving education starts to turn to simulators, the benefits were visible, in a paper by Chan et al. [3], the authors argued driving simulators can be effective tools for evaluating novice drivers [3, 4]. Kandhai et al. presented their analysis on the use of immersive driving simulation for driving education and reported positive results on favourable learning experiences [4, 15]. With these papers stating its viable, a push for a study here is more than welcomed. With a study done by Ismail et al. [4, 13] that surveyed the readiness of Malaysian drivers' for using a driving simulator as a tool to improve driving skills. Results from the survey were positive and the majority would utilise such a tool if provided. In more recent studies that were done by Pierro et al., the authors ran a Pilot Project to Validate the Transfer of Training of Driving Skills Learned on a High Fidelity Driving Simulator to On-Road Driving and concluded that there was positive transfer of driving skills to road exam performance [11] and encouraging evidence that driving simulator-based training (DBST) had positive impact on driving infractions on the first two years of unsupervised driving [11]. Another study on Transfer of Skills Learned on a Driving Simulator to On-Road Driving Behaviour done by Pierro et al. as well also show results that few hours of DBST for on-road training is associated with reduced infractions in the first 2 years of unsupervised driving after licensing [12]. In addition to all above, driving simulators are easy to control and repeat [18], and widespread entry level racing sims are seen throughout the globe [4, 18] driving costs down. The stage has been set for such technologies to be exploited for mass-educational purposes.

Source codes and techniques that are used in games are usually secretive, for example, Warner Bros Interactive Entertainment recently secured patent for its Nemesis System in the game Shadow of Mordor [2]. Even the techniques are widely used and common place in many games, due to profitability and trade secrets, usually implementation of the NPCs not released to the public. Even with the book Game AI Pro, are mostly conceptual.

In this article, we set our major objective as to explore and showcase the implementation of FSMs and HFSMs for NPCs in our educational driving simulator project and its future direction. As our low-cost educational driving simulator is built based on Malaysian Standardised License Examination and Malaysian traffic. The purpose for these NPCs is to mimic certain traffic conditions found in Malaysia, where we reside.

The major contribution is summarised as below.

1. We explore and showcase the use of FSM and HFSM in our own project and display our implementations.
2. We discuss other possible techniques that can be used in the future to help create realistic the NPCs.

The organisation of this article is as follows. Section 2 explains the basis of implementation of NPCs. Section 3 explains our methodology between FSMs and HFSMs. Section 4 would be our discussion on the technologies and potential effects of the implementation. Section 5 conclude the article.

2 Basis of Implementation

As we are developing an educational driving simulator, one of our objectives are to help learner drivers to gain more hands-on experience and engrain certain muscle memory and reflexes for real world driving. Hence the format of standardised driving tests will not suffice and clicking to identify hazards with photos [9] is not engaging. With this reason, we implemented some NPCs into our driving simulator to mimic Malaysian traffic conditions to challenge our learner drivers. As this challenges can help to create a great learning experience with interactive instructions and stimulation [4, 7, 10]. Since there is a link with positive results on specific skill-oriented aspects of driving and simulation [1, 17], there is no reason to shrug off an implementation of such NPCs. The "World's Fastest Gamer", James Baldwin, has also proven that racing and driving skills are transferable from virtual to reality by winning his British GT debut after winning the E-sport racing championship "World's Fastest Gamer" 2019 [8]. With multiple studies and a real life example, it is hard not to implement some NPCs to challenge learner drivers to develop necessary driving skills.

3 Methodology

3.1 Educational Driving Simulator

Our educational driving simulator is built on Unity Engine, the engine came with some basic programming that requires setup and some further codes to create a functioning NPC. The purpose of building these NPCs in the driving sim is to act as traffic, these NPCs will be taking the form of cars, roaming around the streets or creating situations based on certain conditions. Such NPCs is able to mimic Malaysian traffic and provide our users with challenges to react to, we then can implement tests based on these NPCs (Fig. 1).

3.2 Finite State Machine

We used a simple finite state machine in the NPC cars to allow them to follow a set route. The NPCs were programmed to go to a set checkpoint, after it reaches a

Fig. 1 A few NPC cars as traffic

checkpoint, it will look for the next checkpoint and starts its journey to it, this is done repeatedly without halt. In this sense, whenever the conditions of reaching its latest check point is fulfilled, it will check and reset its look ahead to the next latest check point, then repeats this simple cycle (Figs. 2 and 3).

With this simple state FSM, we were able to create a realistic static roaming traffic.

3.3 Hierarchical Finite State Machine

To create a continuous flow of traffic without halt, we designed a simple HFSM using our simple FSM as base. The HFSM simply checks if the NPC has reached its final point and resets its position back to an initial point. This simple HFSM creates a sense of uninterrupted continuous flow of traffic for the user (Figs. 4, 5, 6 and 7).

We also designed a simple NPC car which would cut into the users' lane and would cut back to its lane after the user fulfilled a certain condition. The condition is still in progress as we are still deciding what would be the best condition to trigger the action. The purpose of having this NPC is to create challenges for the user (Figs. 8, 9 and 10).

With the use of such HFSMs, we were able to create and mimic real traffic to a certain extent based on users' input, position and condition.

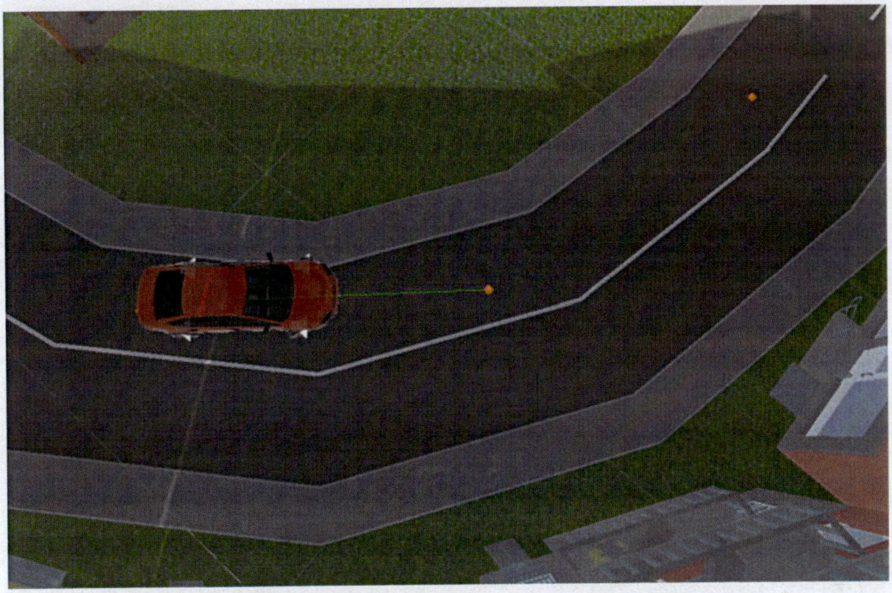

Fig. 2 An NPC car following the next checkpoint

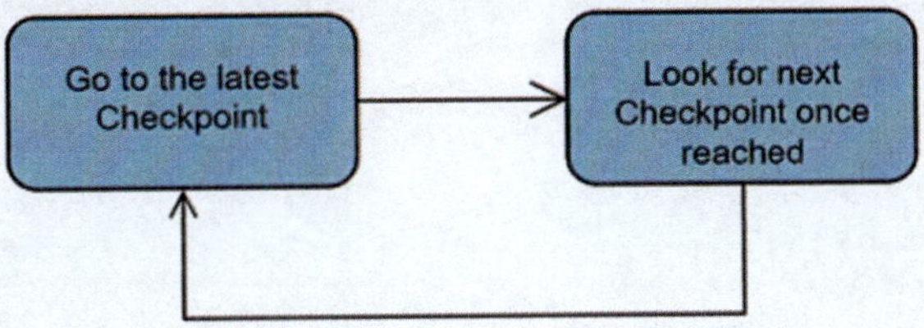

Fig. 3 State diagram of the basic NPC's

4 Discussion

4.1 Technology

With our project using FSMs and HFSMs in the traffic NPCs, we did imagine our project to use more advanced techniques like Artificial Intelligence and Machine Learning. There is a desire to create such NPCs that function like actual cars in the real world based on autonomous driving (AD) technology. This development path, however, is extremely complex. Although, this is not impossible as shown by BMW. BMW Autonomous Driving department has built and tested their AD technology in unity [5, 6], they used such technology to run more than 240 million virtual kilometers. Their virtual AD cars can drive themselves in any various traffic situations with a variety of weather conditions such as night driving.

Fig. 4 Traffic NPCs from initial point

Fig. 5 The initial point of the NPCs

The prospect of bringing such technology into low-cost educational driving sims is appealing. With independent NPC cars that roam around and will react to the users' action without pre-programmed conditions or reactions will provide a sense of real-world randomness to the user as we have difficulty predicting what these AD NPC's actions.

The advantages of AD driving tech in driving sims are noticeable compared to FSMs and HFSMs but it has its own disadvantages that make FSMs and HFSMs more approachable. We tabulated a simple table below to discuss their comparison (Fig. 11).

Fig. 6 The end point for NPCs

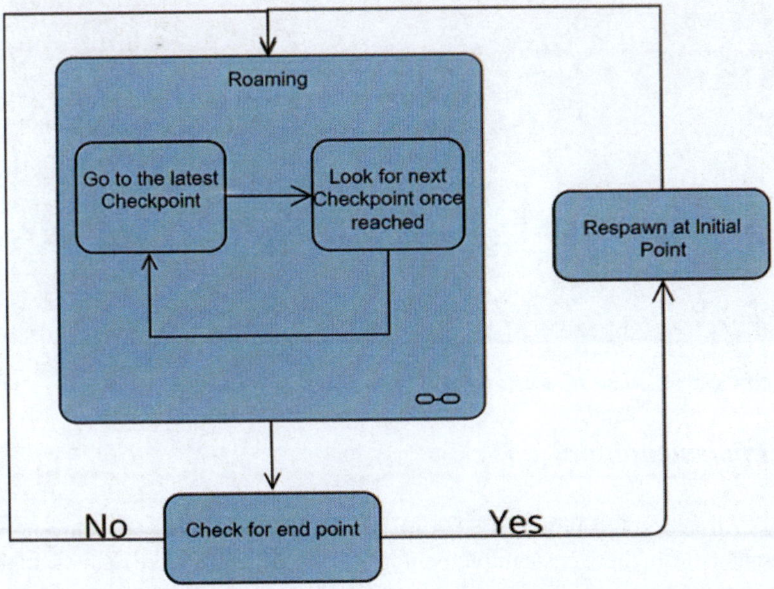

Fig. 7 State diagram of traffic NPCs

Shown in the comparison chart, it is realistic that we chose FSMs and HFSMs for our project as we can design and tailor these systems for our NPC to suit every situation that we design. Although AD can react and adapt on the fly, their behaviour is unpredictable. Developing HFSMs are also more approachable compared to designing AD as they are immensely complex and require a lot of manpower.

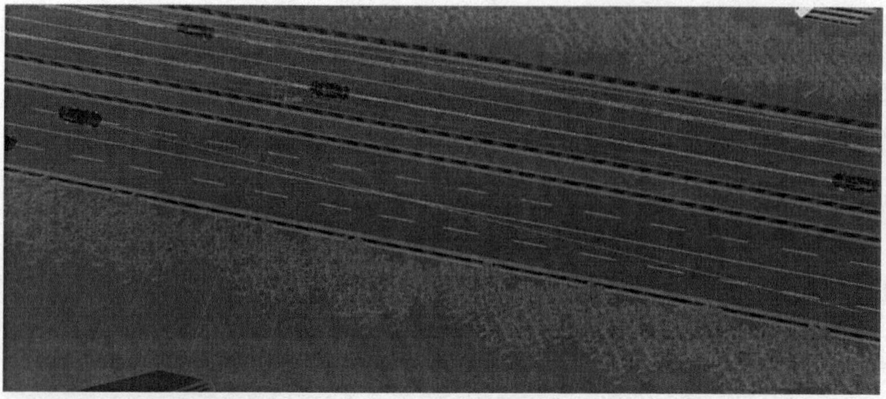

Fig. 8 The path of NPC

Fig. 9 NPC cutting in lane from user point of view (POV)

4.2 Implementation

The basic structural FSMs and HFSMs are already implemented in the current prototype version. From these basic implementations, we noticed a more realistic highway in one of our scenarios.

We plan to add more features such as built-in traffic light adhering systems, lane follow, and adaptive car follow systems into our final version. These systems would potentially provide a more realistic feel to the traffic around the user as they can adhere to traffic lights, follow and adapt to the user on highway situations.

For AD, we did not plan to implement this into our final version as it is too complex for our scale. Although, having AD cars as independent agents in an educational driving sim such as ours would bring great value to low-cost driving sims and the learner drivers.

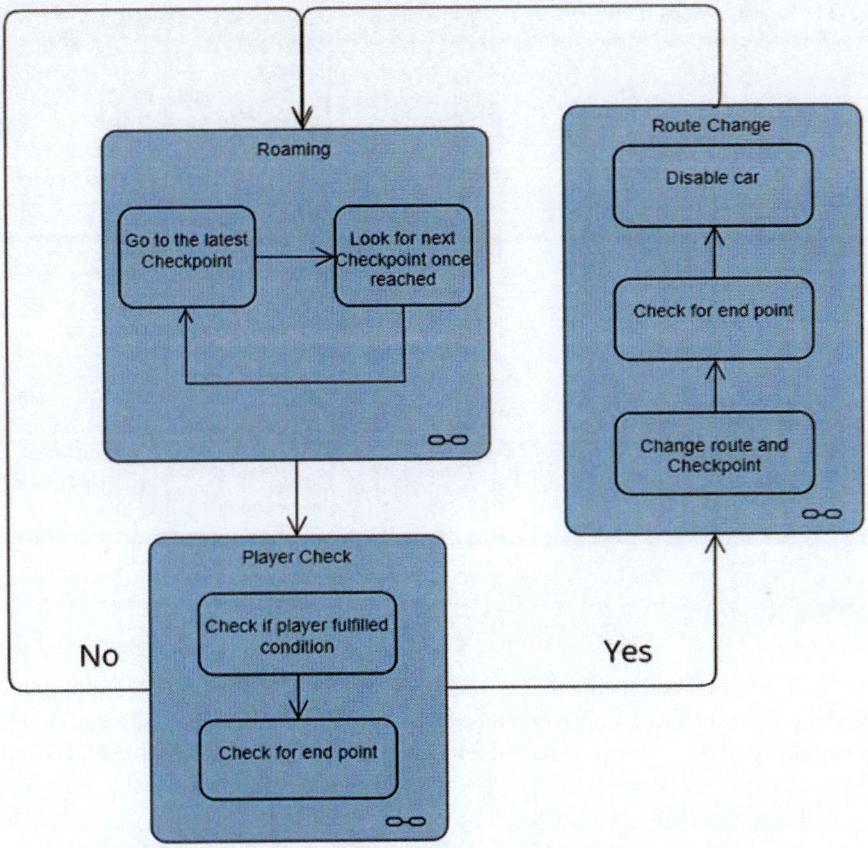

Fig. 10 State diagram of lane cutting NPC

4.3 Potential Effects

As these implementation have yet to be tested. We predict the potential effects of the implementation would be able to provide challenges and really put the users' abilities to the test. The majority of users found adaptive AI in racing games to have a positive effect on their experiences [16]. Implementing HFSMs that can mimic adaptiveness that can provide challenges can possibly have a positive effect on the user's experience. It also can create a sense of realism as the streets and highways will be filled with traffic that roam around independently that we believe will positively impact the user's experience.

Fig. 11 Comparison chart of FSM, HFSM, and AD

Technology	Advantages	Disadvantages
FSMs	Simple to design and tailored to our use case	Too many FSMs for a certain condition will add complexity
HFSMs	Can contain multiple states of FSM in one state	More complex than FSMs, require fine tuning of sub-states
AD	Independent Car agents, able to react and adapt to their surroundings	Complex to design and develop. Unpredictable behaviour of agents

5 Conclusion

We conclude with showcasing that using FSMs and HFSMs in our current project is sufficient to create and mimic a certain extent of Malaysian traffic as shown. Future works related to this project is the refinement of the use of HFSMs in the NPCs and may include the exploration of use of AD technology or any other techniques that can create independent agents that can be unpredictable in behaviour. We would also like to see the gap here in which use of advanced techniques and technologies is used for low-cost educational driving sims and serious games can be explored.

Acknowledgements The authors would like to thank the Ministry of Education of Malaysia in providing financial support for this work through the Fundamental Research Grant Scheme (FRGS/1/2020/SS0/MMU/02/1).

References

1. Backlund P, Engström H, Johannesson M, Lebram M (2010) Games for traffic education: an experimental study of a game-based driving simulator. Simul Gaming 41:145–169. https://doi.org/10.1177/1046878107311455
2. Batchelor J (2021) Warner bros finally secures patent for shadow of Mordor's nemesis system. https://www.gamesindustry.biz/articles/2021-02-08-warner-bros-finally-secures-patent-for-shadow-of-mordors-nemesis-system
3. Chan E, Pradhan A, Pollatsek A, Knodler M, Fisher D (2010) Are driving simulators effective tools for evaluating novice drivers' hazard anticipation, speed management, and attention maintenance skills. Transp Res Part F Traff Psychol Behav 13:343–353. https://doi.org/10.1016/j.trf.2010.04.001

4. Cheng YQ, Mansor S, Chin JJ, Karim HA (2022) Driving simulator for drivers education with artificial intelligence traffic and virtual reality: a review. In: Alfred R, Lim Y (eds) Proceedings of the 8th international conference on computational science and technology. Springer, Singapore, pp 483–494
5. Davis N (2020) BMW employs unity across its automotive lifecycle. https://blog.unity.com/manufacturing/visualizing-bmws-self-driving-future
6. Davis N (2020) BMW's autonomous driving journey with unity. https://blog.unity.com/manufacturing/the-road-to-240-million-virtual-kilometers-bmws-autonomous-driving-journey-with-unity
7. Erhel S, Jamet E (2013) Digital game-based learning: impact of instructions and feedback on motivation and learning effectiveness. Comput Educ 67:156–167. https://doi.org/10.1016/j.compedu.2013.02.019
8. Evans A (2020) "World's fastest gamer" wins on British GT debut
9. Fisher D, Pollatsek A, Pradhan A (2006) Can novice drivers be trained to scan for information that will reduce their likelihood of a crash? Inj Prev: J Int Soc Child Adolesc Inj Prev 12(Suppl 1):i25–i29. https://doi.org/10.1136/ip.2006.012021
10. Guillén Nieto V, Alesón-Carbonell M (2011) Serious games and learning effectiveness: the case of it's a deal! Comput Educ 58:435–448. https://doi.org/10.1016/j.compedu.2011.07.015
11. Hirsch P, Bellavance F (2016) Pilot project to validate the transfer of training of driving skills learned on a high fidelity driving simulator to on-road driving. https://doi.org/10.13140/RG.2.1.5092.6483
12. Hirsch P, Bellavance F (2017) Transfer of skills learned on a driving simulator to on-road driving behavior. Transp Res Rec 2660(1):1–6
13. Ismail R, Md Saad MH, Mohd Nor J, Rosli R. Driver's readiness to use car driving simulator as a tool to improve driving skill: a preliminary study. J Teknol 78:137–142. https://doi.org/10.11113/jt.v78.9161
14. Jagdale D (2021) Finite state machine in game development, pp 384–390. https://doi.org/10.48175/IJARSCT-2062
15. Kandhai K, Smith M, Kanneh A (2011) Immersive driving simulation for driver education and analysis. In: 2011 16th international conference on computer games (CGAMES), pp 288–292
16. MacKinnon B (2012) Adaptive AI in a racing simulator: a challenging and fun environment for any user. Comput Games J 1(1):71–89. https://doi.org/10.1007/BF03392329
17. Uhr M, Felix D, Williams B, Krueger H (2020) Transfer of training in an advanced driving simulator–comparison between real world environment and simulation in a manoeuvring driving task
18. Wang Y, Zhang W, Wu S, Guo Y (2007) Simulators for driving safety study—a literature review. In: Shumaker R (ed) Virtual reality. Springer, Berlin, Heidelberg, pp 584–593

An Investigation of the Relationship Between Agricultural Price and Its Determinants in Malaysia

Wong Weng San, Chen ZhiYuan, and Teo Wing Leong

Abstract Prices for major food commodities such as grains and vegetable oils have risen sharply in recent years which have triggered the interest of researchers to investigate the factors of price movement. Understanding agricultural commodity price relationships can help both government and farmers to raise the awareness regarding the price volatility in agricultural market, production costs and other potential risks in the future market. This paper examines the relationships between the agricultural commodities price and the potential price determinants such as exchange rate, temperature, rainfall and covid 19 cases in Malaysia. The investigation is examined by using Granger Causality test and Johansen co-integration test. The estimated results provide evidence a unidirectional causal relationship running from the covid-19 cases to chicken price. Furthermore, there is also empirical evidence of cointegrating vectors exists between temperature and chicken. Overall, the findings have significant implications for predicting the future price of agricultural commodities in Malaysia.

Keywords Agricultural commodities · Price determinants · Granger causality test · Johansen cointegration test

1 Introduction

Even though the price of agricultural products have risen recently, its contribution to Malaysia's GDP declined from 10% in 21st century to 7.54% in 2018. However, agriculture sector is still an important sector in Malaysia's economy. Many

W. W. San · C. ZhiYuan (✉)
School of Computer Science, University of Nottingham Malaysia, Semenyih, Malaysia
e-mail: zhiyuan.chen@nottingham.edu.my

W. W. San
e-mail: khcy5wws@exmail.nottingham.edu.my

T. W. Leong
School of Economics, University of Nottingham Malaysia, Semenyih, Malaysia
e-mail: wingleong.teo@nottingham.edu.my

© The Author(s), under exclusive license to Springer Nature Singapore Pte Ltd. 2023
D.-K. Kang et al. (eds.), *Proceedings of the 9th International Conference on Computational Science and Technology*, Lecture Notes in Electrical Engineering 983,
https://doi.org/10.1007/978-981-19-8406-8_38

exogenous factors and variables could be the main causes of such a trend. Hence, risks management has been important in agriculture business, owing to the fact that the uncertainties of agriculture inherent in natural phenomena, weather, government policy, the handling of living organism and other aspect that can impact the price of agricultural products [7]. For decades, price volatility in agricultural market becomes an ongoing issue. Price volatility occurs when the price fluctuations of a commodity become excessive and unpredictable over a short period of time. The instability of agricultural commodities price due to disease outbreaks and adverse weather events, such as flood and droughts will cause supply volatility, impacting producer incomes, endanger the accessibility to food by consumer and market trades [22]. All these impacts may lead to food insecurity, malnutrition, food wastage and starvation [9]. As a consequence, farmers, policy-makers as well as other participants have the urgent needs to have more professional skills and advice in both production and business management to anticipate and mitigate the potential agricultural risks [15]. Studies such as Chen et al. [5] and Chen et al. [26] perform time series analysis and develop forecasting model to predict future events of agricultural markets in Malaysia.

In this paper, Granger Causality Test and Johansen Cointegration Test have been conducted to examine the relationship between rainfall, temperature, exchange rates and Covid-19 positive cases and the selected agricultural prices namely chicken, chili and tomato.

2 Literature Review

There are many variants that can lead to market volatility. Understanding the factors that contribute to the commodity price is significant for the government to evaluate new policies for better plan for plantation and breeding. Determinants of agricultural commodity prices can be categorised into demand-side factors and supply-side factors [4].

From the supply side, the factors are supply shocks, adverse change in weather conditions, climate changes, rising production cost, and high price of energy and energy resources. Besides, demand-side factors include population growth, economic development, income growth, exchange rate changes, speculation and increased production of liquid biofuel [4].

Due to the high dependence of agricultural production on natural phenomena, especially crop production, adverse changes in climate and weather condition can have significant implication to the agricultural market. Climate and temperature changes can affects many aspects in agriculture, such as crops, livestock and fishery. There is optimal temperature for any particular crop for growth and reproduction. If the surrounding temperature exceeds the optimal temperature, crop production will decline [11].

In literature, there are studies investigate the relationship between climate or temperature change and agricultural commodity price. Using VAR methods, Marini

[16] find that increase in temperature is likely to affect the commodities price to increase. Bhanumurthy et al. [3] performed cointegration and causality analysis and showed that the expected climate condition, rainfall has significant negative impact on both commodity spot price and future price. A global dynamic computable general equilibrium model proposed by Bandara and Cai [2] to investigate the relationship between changes in crop productivity on food prices as well as food security in five selected countries. The agriculture sectors in these countries were predicted to be unfavourably influenced by the productivity changes due to climate changed, leading to a food shortage by 2030. Mirzabaev and Tsegai [17] used an innovative estimation method with Feasible Generalized Least Squares (FGLS) panel regression to investigate the impact of weather shocks on the prices of agricultural commodity in Central Asia for the period of 2000–2010. Analysis performed suggested that the impact of weather shocks on the prices of wheat and potato in Central Asia is crucial. Tran et al. [23], extend a storage models and find decreasing yield and increasing yield variability caused by climate change results in the increase of commodity price level and volatility.

Exchange rate is one of the demand-side factors for agricultural product prices. By definition, exchange rate is the flow of the currency between two countries. Due to the fact that US dollar is the world dominant currency, the foreign currency that most of the countries used for trading in market is USD. The movement in exchange rate can influence agricultural import and exports, agricultural production and investment. There are many studies in the literature that determine the linkage between exchange rate and the price of agricultural commodities.

Frank and Garcia [6] applied VAR and VECM procedures on weekly cash data between 1998 and 2009 to investigate the dependency between agricultural commodity price, crude oil prices and exchange rates. The results indicate that shocks in exchange rate have a small impact on the price of agricultural commodities, the corn and livestock. In recent times, agricultural markets seems to be more dependent on exchange rates and less dependent on oil prices.

Rezitis [21] uses panel cointegration and error correction models and the results show that crude oil price has a positive impact on commodity food prices while exchange rate impact negatively on food prices. Causal analysis done by Nazlioglu and Soytas [19] implies that spot oil prices, and exchange rate can help to improve the forecasting agricultural price. Using long run level model and Short-Run Error Correction Mechanism (ECM) Models, Hatzenbuehler et al. [12] find evidence that agricultural prices are responsive to exchange rate under the conditions when the stocks is low and there is a policy shifts.

The year of 2020 has ushered in a global pandemic caused by Coronavirus Disease (COVID-19). It has spread to many countries in the world and has influenced to many sectors, especially agricultural sector because of the implementation of government policy to handle the pandemic situation caused by the disease such as the movement's control order (MCO) in Malaysia. Travel restrictions during this period has disconnected farm production to the consumer. The emergence of covid-19 seriously disrupt market demand and supply of agriculture industries [24]. Negative impacts of the pandemic on supply and demand for food might lead to price volatility and food insecurity. Therefore, many recent works have investigated the impacts on agriculture and solutions to deal with the potential situations.

Wang et al. [25] finds the impact of COVID-19 on the cross-correlation of multi-fractality between selected agricultural future markets and Brent Crude Oil. Nugroho et al. [20] find that the Covid-19 pandemic did not have significant impact on the Robusta coffee spot price movement. Granger causality test done by Gök et al. [8] shows that covid-19 mortality rates granger-cause the prices of corn, rapeseed, and oats. Meanwhile, the movements in the mortality rate could not help to predict soy-beans prices in neither the short-run nor long-run.

In the literature, there are studies on determinants of agricultural commodity prices in Malaysia. Ibrahim [14] uses a nonlinear autoregressive distributed lags (NARDL) model and suggests that there exists a cointegrating relationship between food price, oil price and GDP. Another work by Hasan and Masih [10] also employs Nonlinear ARDL (NARDL) technique to identify the relationship between food price, crude oil price, GDP and exchange rate. From the experimental result, ARDL and NARDL finds that there is cointegrating relationship between food price and the factor variables, while Vector Correction Model (VECM) supports that exchange rate have the most significant impact on food price. The authors suggest that the linkage between food price and interest rate, trade balance as well as government policy can be considered for future research.

Factors determining the agricultural commodity prices deserve great attention from farmers, investors and policy makers in Malaysia. Early preparation such as propose new control policies and make adjustment to the plantation plan to cope with all the potential situations that will impact the agricultural market. As a result, information about the determinants of commodities price is beneficial for all participants in the market to create a better and more effective risk management plan in agriculture sector.

3 Data Collections

The data collection is targeting the time series data based on monthly basis that could start from the year of 2015 which should includes agricultural commodity price and other variables data reflecting exogenous factors affecting price.

The dataset was collected from FAMA official portal. One challenge here is to extract all the data from the weekly report downloaded from the website and combine them into one single csv file. The whole process was done by using excel tools and processed with python code.

In addition, price determinants such as temperature, rainfall, and crude oil price were collected to study potential factors that could affect the agricultural commodity prices in Malaysia. These data set will then combine with the agricultural price for convenience purpose.

4 Methods

4.1 Granger Causality Tests

Theoretically, the aim of Granger Causality Test is to find out the causality relationship between two time series data and improve the forecasting. It is a statistical test for identifying the usefulness of one variable to forecast another. If one series x is helpful for forecasting another time series y, it is said to be granger-cause the series y. In other words, x contains significant information that can help to predict the future values of y. If a time series is found to be useful for forecasting, then it can be considered adding to the learning model to help reduce the prediction errors.

The steps for the test are simple, which are as follows:

1. Identify null hypothesis and alternate hypothesis. Example for null hypothesis, y(t) does not Granger-cause x(t).
2. Determine the lags. The lag value can be determined after several experiments or the criterion approach.
3. Calculate the f-statistic.
4. If the F statistic is greater than the f-value, reject the null hypothesis.

4.2 Co-integration Tests

Cointegration test is used to examine the existence of the correlation between two or more non-stationary time series in long term. In the literature, both [13, 18] implement cointegration test to find the relationship between the exchange rates and commodities prices. In this research, cointegration test are used to find evidence to support the relationship present between the commodity price and the influential factors in Malaysia. Johansen's test is an improved method to address the limitations seen in the Engle-Granger test. It allows multiple cointegrating relationship. There are two forms for this tests:

1. Trace Tests
   The trace test aims to determine the number of cointegration vectors in a time series (i.e. K). The null hypothesis are described as follows:
   Null Hypothesis,H_0: K is equal to K_0
   Alternate Hypothesis, H_1: K greater than to K_0
   Initially, K_0 is set to 0, if null hypothesis is rejected, then at least one cointegrating relationship present in the time series.
2. Maximum Eigenvalue test
   This test is similar to trace test but the hypothesis is different:
   Null Hypothesis, H_0: K is equal to K_0
   Alternate Hypothesis, H_1: K is equal to $K_0 + 1$

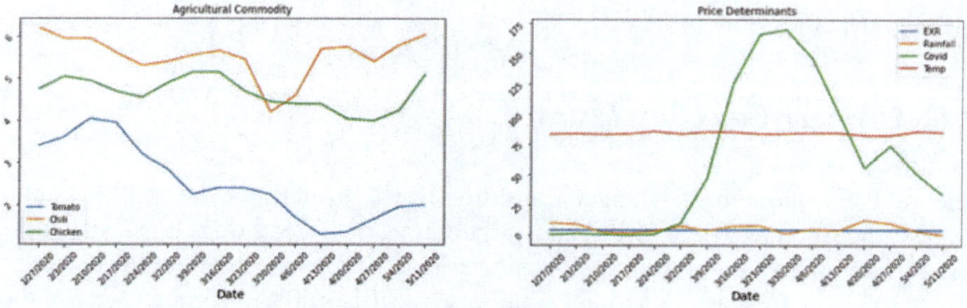

Fig. 1 Time series plot for both agricultural commodities price and price determinants

Initially, K_0 is set to 0, if null hypothesis is rejected, then there exists only one possible combination of the variables to produce a stationary process.

5 Results

5.1 Data Preparation

This section aims to preprocess the data before the Granger Causality test. This is because the prerequisite for running the test is that time series data is required to be stationary which means it should contain constant mean and variance. Transformation is necessary when a series is confirmed non-stationary. The most common method that helps to solve the non stationarity problem in time series is difference transform, either first-order or second-order differencing.

Most popular statistical test to check the stationarity is Augmented Dicker-Fuller Test (ADF). In this result, the price of agricultural products were obtained from the official website of Malaysia's Federal Agricultural Marketing Authority (FAMA) [1] The data obtained is on a weekly basis, starts from the year of 2009 until August 2020 and the price value is expressed in Ringgit Malaysia per kilogram. Figure 1 shows the time series plot for FAMA dataset and the potential agricultural price determinants. By visual inspection, tomato and Covid series are not stationary. However, results from the ADF test indicate that, *p*-values of both chili and rainfall variables are greater than significant value (0.05), which means both of the series are not stationary. In other words, chili and rainfall data are required to perform a difference transform, while the rest of the variables data can proceed to Granger Causality test as they are stationary series. Figure 2 shows the result of 1st difference transform for both chili and rainfall and they seems to become stationary. This is then confirmed by a second ADF test, and now all the series are ready for Granger Causality Test.

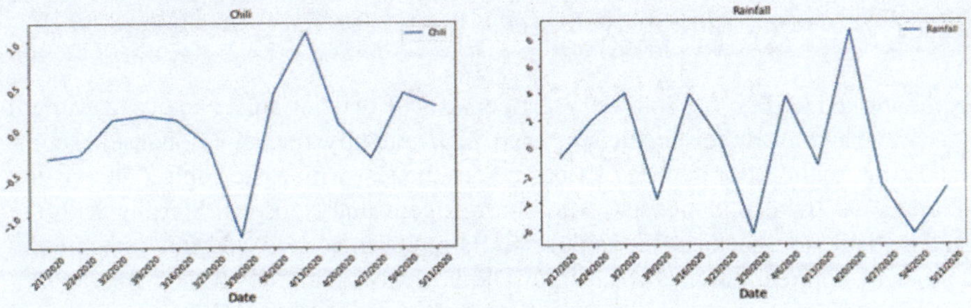

Fig. 2 Plot for chili and rainfall after 1st differencing

Table 1 Summary of Granger causality test

Variables	Commodity	F test	p-value
Exchange rate	Tomato	11.3522	0.0114
	Red chili	2.5986	0.1648
	Chicken	0.3265	0.8071
Rainfall	Tomato	0.5249	0.6951
	Red chili	0.3823	0.9555
	Chicken	0.8481	0.5523
Covid 19 cases	Tomato	4.6895	0.0646
	Red chili	4.1684	0.0792
	Chicken	8.3166	0.0218
Temperature	Tomato	0.5545	0.6672
	Red chili	0.3654	0.7816
	Chicken	0.6243	0.6295

5.2 *Granger Causality Test*

This apply the grangercausalitytests() function to perform a Granger-Causality test to determine if the price determinants are the factors that affect agricultural commodity prices. In the experiment, test are conducted using three lags. Table 1 summarise F Test and p-value between price determinants and commodity prices at lag = 3.

In general, p-values for all the combination of price determinant series and agricultural price are considerably high thus the price determinants do not Granger cause the price of agricultural commodities. There are only 2 exceptions which are Exchange Rate (EXR) and tomato as well as Covid and Chicken. The null hypothesis can be rejected since the p-value is less than 0.05. From the analysis, we conclude that the information of EXR is useful in the prediction of tomato future price. Similarly, we can also conclude that the number of covid-19 new cases Granger causes the chicken price, so knowing the number of covid cases can help to predict future price of chicken.

5.3 *Johansen Cointegration Test*

As mentioned earlier, the Johansen test consists of two variants, namely maximum eigenvalue and the trace statistic approach. Conveniently, the coint_johansen() command from python library allows to carry out the test with ease. Table 2 depicts the summary of Trace statistics and Maximum Eigenvalue statistics. Starting with the trace statistic approach, in the table, rank is represented by r and this indicate the number of cointegrating relationships. The critical value at level 0.05 for r = 0 is 15.4943, while r = 1, the critical value is 3.8415. When r = 0, EXR-Tomato and Temperature-Chicken have trace values greater than critical value, which are 16.099 and 17.079 respectively (>15.4943). This means that the null hypothesis which suggests that for $r > 0$ can be rejected, there exists some cointegration. When r = 1, the trace result for Temperature-Chicken is 7.0414 > 15.4943. This again reject the null hypothesis. Therefore, it can be concluded that there are at most 2 cointegrating relationships present between Temperature and Chicken price.

Table 2 Summary of Johansen cointegration test

Variables	Commodity	Null hypothesis	Trace	Max eigenvalue
Exchange rate	Tomato	r = 0	16.098740	18.912242
		r = 1	2.813502	2.813502
	Red chili	r = 0	7.536802	9.928124
		r = 1	2.391322	2.391322
	Chicken	r = 0	7.406226	8.074343
		r = 1	0.668117	0.668117
Rainfall	Tomato	r = 0	10.416764	14.213786
		r = 1	3.797022	3.797022
	Red chili	r = 0	8.940592	13.939558
		r = 1	4.998967	4.998967
	Chicken	r = 0	11.987107	16.728982
		r = 1	4.741874	4.741874
Covid 19 cases	Tomato	r = 0	14.057658	21.507241
		r = 1	7.449583	7.449583
	Red chili	r = 0	13.263342	18.556147
		r = 1	5.292805	5.292805
	Chicken	r = 0	8.966830	13.866019
		r = 1	4.899189	4.899189
Temperature	Tomato	r = 0	11.457112	18.943942
		r = 1	7.486830	7.486830
	Red chili	r = 0	12.895255	22.528762
		r = 1	9.633507	9.633507
	Chicken	r = 0	17.078686	24.120091
		r = 1	7.041405	7.041405

Next is the test using maximum eigenvalue variant. The critical value at level 0.05 for r = 0 is 14.2639, while r = 1 is 3.8415. From Table 2, There are total 7 combinations have eigenvalue greater that the critical value, and suggests that there is some cointegration present for $r > 0$. When r = 1, 6 out of 7 combinations have eigenvalue greater than critical value (3.8415). This again reject the null hypothesis. Overall, it can be concluded that there are at least two cointegrating relationships present for both Covid-19 and Temperature with agricultural prices. By now, it is very clear to see the relationships between all the factors and commodity prices. It can indeed deepen the understanding in how the agricultural market are affected and use this knowledge to improve the predictions.

6 Conclusion

In summary, the investigation by using Granger Causality test and Johansen cointegration test provide evidence that there exists a unidirectional causal relationship running from the covid-19 cases to chicken price. Furthermore, there is also empirical evidence of cointegrating vectors exists between temperature and chicken. In general, the obtained results provide some important information which is useful for relevant parties such as policymaker and farmers to better anticipate the future events in agricultural market. So, they can decide and construct more suitable strategies and policies to avoid potential risk. Since the production of agricultural commodities is highly dependent on temperature conditions in surrounding environment, the disease like Covid-19 will severely impact the economy of Malaysia which heavily depend on agricultural market. As found causality and cointegrating relationship between the variables, early plans or actions could be taken accordingly to avoid unexpected risks in the future. Thus, Malaysia's agricultural development can be improved and more effective in the industry such as cropping, transport, production and trade of agricultural commodities.

Acknowledgements The authors would like to acknowledge the Ministry of Education Malaysia FRGS Grant with reference code FRGS/1/2018/ICT02/UNIM/02/1.

References

1. https://sdvi.fama.gov.my/index.html
2. Bandara J, Cai Y (2014) The impact of climate change on food crop productivity, food prices and food security in South Asia. Econ Anal Policy 44. https://doi.org/10.1016/j.eap.2014.09.005
3. Bhanumurthy N, Dua P, Kumawat L (2013) Weather shocks and agricultural commodity prices in India. Clim Change Econ 4(03):1350011
4. Borychowski M, Czyżewski A (2015) Determinants of prices increase of agricultural commodities in a global context. Management 19(2):152–167

5. Chen YC, Rogoff K, Rossi B (2010) Predicting agri-commodity prices: an asset pricing approach. Available at SSRN 1616853
6. Frank J, Garcia P (2010) How strong are the linkages among agricultural, oil, and exchange rate markets?
7. Girdžiūtė L (2012) Risks in agriculture and opportunities of their integrated evaluation. Procedia Soc Behav Sci 62:783–790. https://doi.org/10.1016/j.sbspro.2012.09.132
8. Gök R, Kara E (2020) Impacts of the covid-19 pandemic on the agricultural prices: new insights from CWT granger causality test. Ekonomi Politika & Finans Araştırmaları Dergisi 5:76–96. https://doi.org/10.30784/epfad.810558
9. Harvey CA, Rakotobe ZL, Rao NS, Dave RB, Razafimahatratra H, Rabarijohn RH, Rajaofara H, MacKinnon JL (2014) Extreme vulnerability of smallholder farmers to agricultural risks and climate change in Madagascar. Philos Trans Roy Soc B: Biol Sci 369
10. Hasan AN, Masih M (2018) Determinants of food price inflation: evidence from Malaysia based on linear and nonlinear ARDL. MPRA Paper 91517, University Library of Munich, Germany. https://ideas.repec.org/p/pra/mprapa/91517.html
11. Hatfield J (2015) Temperature extremes: effect on plant growth and development. Weather Clim Extremes 10:WACED1400046. https://doi.org/10.1016/j.wace.2015.08.001
12. Hatzenbuehler P, Abbott P, Foster K (2016) Agricultural commodity prices and exchange rates under structural change. J Agric Resour Econ 41:204–224. https://doi.org/10.22004/ag.econ.235153
13. Hudson D, Harri A, Nalley L (2009) The relationship between oil, exchange rates, and commodity prices. J Agric Appl Econ 41. https://doi.org/10.1017/S1074070800002959
14. Ibrahim MH (2015) Oil and food prices in Malaysia: a nonlinear ARDL analysis. Agric Food Econ 3:1–14
15. Kahan D et al (2008) Managing risk in farming. Food and Agriculture Organization of the United Nations
16. Marini A (2019) The impact of weather on commodity prices: a warning for the future. Discussion Papers 1902, University of Exeter, Department of Economics. https://EconPapers.repec.org/RePEc:exe:wpaper:1902
17. Mirzabaev A, Tsegai D (2012) Effects of weather shocks on agricultural commodity prices in central Asia
18. Musunuru NM (2017) Causal relationships between grain, meat prices and exchange rates. Int J Food Agric Econ (IJFAEC) (1128-2018-044). https://doi.org/10.22004/ag.econ.266439
19. Nazlioglu S, Soytas U (2012) Oil price, agricultural commodity prices, and the dollar: a panel cointegration and causality analysis. Energy Econ—ENERG ECON 34. https://doi.org/10.1016/j.eneco.2011.09.008
20. Nugroho W, Astuti A, Astutik S (2021) Vector error correction model to forecasting spot prices for coffee commodities during covid-19 pandemic. J Phys: Conf Ser 1811:012076. https://doi.org/10.1088/1742-6596/1811/1/012076
21. Rezitis AN (2015) Empirical analysis of agricultural commodity prices, crude oil prices and US dollar exchange rates using panel data econometric methods. Int J Energy Econ Policy 5(3):851–868
22. Tedesco I (2018) A holistic approach to agricultural risk management for improving resilience. Platform for Agricultural Risk Management, Working Paper 5
23. Tran A, Welch J, Lobell D, Roberts M, Schlenker W (2012) Commodity prices and volatility in response to anticipated climate change
24. Waiho K, Fazhan H, Ishak SD, Kasan NA, Liew HJ, Norainy MH, Ikhwanuddin M (2020) Potential impacts of covid-19 on the aquaculture sector of Malaysia and its coping strategies. Aquac Rep 18:100450

25. Wang J, Shao W, Kim J (2020) Analysis of the impact of covid-19 on the correlations between crude oil and agricultural futures. Chaos Solitons Fractals 136:109896. https://doi.org/10.1016/j.chaos.2020.109896
26. Yuan CZ, San WW, Leong TW (2020) Determining optimal lag time selection function with novel machine learning strategies for better agricultural commodity prices forecasting in Malaysia. In: Proceedings of the 2020 2nd international conference on information technology and computer communications, pp 37–42

Hybrid Wireless Home Security System Using Bluetooth and IoT Technologies

Gwo Chin Chung, Jun Jiat Tiang, Soo Fun Tan, and Khan Vun Teong

Abstract Home security systems are a must-have for any home in any part of the world to help protect our homes and other places alike. The main threats people face include intruders, break-ins, robberies, and fires, to mention but a few. While some of these threats, such as fires, can be contained by automatic actuators, others, such as break-ins and robberies, can't be prevented by automatic systems. The implementation of technologies such as surveillance cameras and the Internet of Things (IoT) is able to prevent such criminals, but they all require internet connection, high power consumption, and a high cost of setup and maintenance, which cannot be afforded by low-income earners. Hence, this paper intends to address some of the issues as stated above by proposing a hybrid wireless home security system that is able to provide low-end functions without the internet ability as well as high-end functions with the internet connection. For the low-end setup, Bluetooth is chosen. Meanwhile, IoT is implemented for high-end purposes. The proposed system includes the functions of breaching and fire alerts, as well as the control of light for energy saving. A new method has also been proposed in order to reduce the breach detection errors from the infrared (IR) sensors. Finally, a final prototype for the performance demonstration has been successfully built to showcase the functions of Bluetooth and IoT technologies.

Keywords Home security · Bluetooth · Internet of Things · Infrared sensor

---

G. C. Chung (✉) · J. J. Tiang
Faculty of Engineering, Multimedia University, 63100 Cyberjaya, Malaysia
e-mail: gcchung@mmu.edu.my

J. J. Tiang
e-mail: jjtiang@mmu.edu.my

S. F. Tan · K. V. Teong
University Malaysia Sabah, 88400 Kota Kinabalu, Malaysia
e-mail: soofun@ums.edu.my

K. V. Teong
e-mail: nichtkv@ums.edu.my

© The Author(s), under exclusive license to Springer Nature Singapore Pte Ltd. 2023

D.-K. Kang et al. (eds.), *Proceedings of the 9th International Conference on Computational Science and Technology*, Lecture Notes in Electrical Engineering 983,
https://doi.org/10.1007/978-981-19-8406-8_39

1 Introduction

Today, many people will have their homes broken into without their knowledge. According to the latest statistics [1], even a well-developed country such as the USA has a burglary happening on an average every 30 s. This situation is even worse in most non-developed countries [2]. In many of these break-ins, the intruders are overly cautious and go unnoticed, causing various damages to the homeowner and, in some cases, the loss of valuable property and/or life. In other cases, the intruder will be noticed by the homeowner but will have very little time to react in self-defense if he isn't alerted by the system at the moment of intrusion. Therefore, the few seconds lost at the time of intrusion could mean life or death for the homeowner if he didn't expect it at all. As such, it is very crucial to have systems that provide real-time and current information to improve the situational and positional awareness of homeowners as regards to the latest status of their home or the position of any intruders so as to take a better course of action.

This can be accomplished through a live internet feed with cameras strategically placed as needed, or through the use of Global System for Mobile (GSM) capable systems that feed the user with data as needed, as reported in [3, 4], where the technologies were used to track the intrusion of both animals and humans into the farms [3] and to detect robberies that occurred in the house [4]. The camera was also used to perform face recognition for a door lock system as suggested in [5]. There are also projects being developed that do not use cameras, such as [6], which uses PIR sensors for home surveillance. More advanced technologies, such as GSM and GPS, were combined in [7, 8] to improve the security of women. Internet of Things (IoT) was also employed in [8] and in [9] for women's security and babies' monitoring system. While all of these mentioned systems will do the job well, they all require internet, or Wi-Fi, which requires high power consumption, which can also be disadvantageous due to certain factors such as, poor connections, undedicated data lines that lose valuable time, and above all, many of these systems seem to be expensive and can't be afforded by low income earners in many nations, such as countries in Africa, South East Asia, etc. According to the 2022 statistics, some countries have less than 50% of internet users per population [10] and less than 50% of electricity coverage per population [11]. This forms one of the core values of this research, which is to provide an alternative way for home security systems without relying on the internet, reduce power usage, cut down the cost, but still maintain good awareness.

Thus, this paper proposes a hybrid wireless home security system that implements low-end functions without internet connectivity as well as high-end functions with internet connectivity. For low-end systems, Bluetooth, which is always dedicated to serving intended master and slave devices only [12], is low cost, very low power usage, and wireless control of the security system within any part of the home is used to benefit the homeowner in the event of a break-in. Using available technologies, the homeowner will be able to request and receive data from the security system, such as the detection of intruders and fire incidents, about the current situation of each room and perimeter fence side. All these can be achieved using a sensor-based microcontroller with sensors such as infrared (IR), light, and fire sensors. Even better, the system can be hidden out of view or from easily accessible areas so that it is not physically destroyed by the intruder to disable it, with only the display and keypad being in direct view as their destruction does not degrade the overall performance of the system. Finally, the IoT is also implemented in this paper as an alternative mode for offering high-end functions such as mobile applications.

This paper has the following organization. In Sect. 2, the architecture of the proposed system is described. Then, the result of the demonstration of the hardware prototype and software implementation is presented in Sect. 3. Finally, the contributions of this paper as well as future recommendations are summarized.

2 System Architecture

2.1 Targeted Home Concept

Figure 1 is the description of the home targeted by this project, and it has the following characteristics:

- the home has a fence (perimeter wall) around it (marked red).
- the fence is clear of any obstacles for at least 0.5 m inwards to have a clear infrared transmission (marked yellow).
- the shape of the house is not a matter, and it can have any number of fence sides. Only four sides are showcased in this project.
- the fence cannot be a curve, or else infrared technology cannot be applied.
- the home can be with or without internet connectivity, but it must have a minimum electricity supply, such as a battery.

Fig. 1 Description of the targeted home

2.2 Hardware Description

Table 1 summarizes the functions of each hardware component used in this paper. In general, a microcontroller named as, Arduino Mega serves as the brain of the system. All the sensors and the other components used in this project will be connected to the microcontroller for activity controlling and coordinating purposes. The Arduino Mega 2560 is chosen for this research since it has the capability of storing data in its memory and consists of several interrupt pins, which makes it viable for such a project that requires multitudes of sensors.

There are two operation modes in this proposed system: Bluetooth and Wi-Fi connection. For a low-end setup, a Bluetooth transmitter is used to communicate between the Android devices and the Arduino. The data arrives on the serial line of the Arduino processor, from which it is picked and processed. The maximum range of transmission is 10 m, which is reasonable for an average home. On the other hand, for the high-end functions with internet connectivity, a Wi-Fi module is used for communication purposes as well as to push specific data to the cloud server using the IoT platform. Using this wireless capability, home users will be able to control their devices from almost any part of the home that falls within this range.

The main objective of this paper is achieved by installing multiple infrared line sensors on the fence around the house. Due to budget constraints, these sensors are made up of an IR transmitter and an IR receiver, which have only a 5 cm range of detection with low power usage. Whenever there is an intruder breaching the fence and the IR sensors are turned on, the alarm system will be triggered. However, the drawbacks of using IR line detection are the false alarms due to various non-human objects such as birds, dogs, and randomly flying objects such as papers, polythene, etc., which could trigger false and unwanted alarms. Hence, a solution is proposed in

Table 1 Functions of the components

Component		Function
Arduino Mega 2560	Low-end/high-end	To centralise data collection and control activities
Bluetooth transmitter (Bluebee/SKXBEE)	Low-end	To transmit data and commands between the Android devices and the microcontroller
Wireless module (ESP-01)	High-end	To transfer data to the IoT server via the internet
IR line sensor	Low-end/high-end	To detect movement of an object bypassing the fencing of an area
Light sensor module (SN-LIGHT)	Low-end/high-end	To detect light density of the environment
Fire/flame sensor	Low-end/high-end	To sense fire in the house
Keypad and liquid crystal display (LCD)	Low-end/high-end	To input simple commands and display basic messages

Fig. 2 Elimination of undesired IR detection

this project, whereby on each fence side, three sensors will be placed along the height as shown in Fig. 2. A minimum of two consecutive sensors along the height must be activated to trigger the alert system since an adult human will have an average height higher than the non-human objects.

To detect the breaching values of the IR sensors, the sensor analogue value must be determined. Figure 3 shows the sensitivity of the IR sensor along the length of the fence. Therefore, the equation of the analogue value, Y is formulated as:

$$Y = 106.6X + 315 \tag{1}$$

where X is the actual fence distance entered by the users during the initial setup of the system. When Y is obtained, the system will trigger the breach alert when the analogue sensor value exceeds the Y value.

As an add-on, the light sensor module is also offered to keep the home's exterior always lit automatically while conserving power whenever the lights aren't required. This light sensor module will be used to switch on a series of lights in this project along the fence of the house model. In a real-world implementation, the output of this sensor would be used to drive a higher voltage switching circuit, controlling all the exterior lighting of homes. This module serves two main purposes in the home. One is to reduce the energy consumption by automatically switching off the many fence

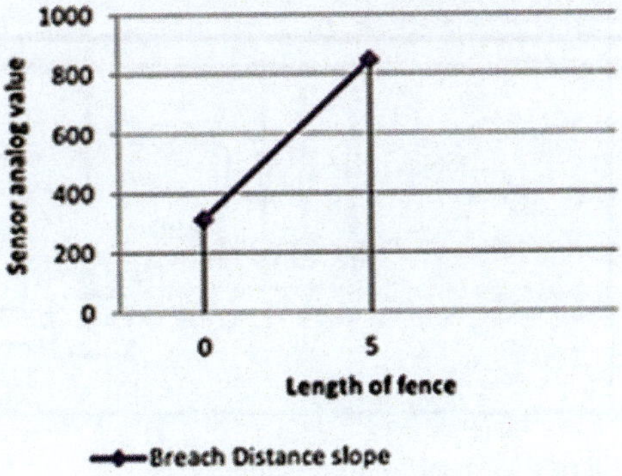

Fig. 3 Breach slope of IR sensor

lights when they're not needed, especially during the daytime. The second purpose is that it improves the general security of the home when the owner is not around.

Besides that, a fire alert sensor is added to further enhance the security of the home. It can comprise various flame sensors placed in the desired or most vulnerable rooms, such as kitchens, car garages, or as desired. The sensor used in this project has the capability to detect fire or other light in the wavelength range of 760–1100 nm. The sensor has both analogue and digital outputs, and the choice of output depends on the application. In this project, the digital output is preferred to indicate the presence of light in the above-stated range in which the fire flame is most common as light.

Finally, other hardware such as the LCD, keypad, and buzzer are also required in the system. The LCD is very important in communicating with the processor via the keypad, especially during the first run of the software when setting the password and home parameters, as it provides feedback to the user who may not have any engineering knowledge or background. The full schematic diagram of the proposed prototype is shown in Fig. 4. Only one of the IR line sensors is shown in the diagram since it consists of multiple sensors that are unable to be fully displayed in the above schematic. A multiplexer (MUX) is used since the goal of this project is to provide a home security system to all areas, with or without internet access. The function of MUX is to act as a switch between Bluetooth and Wi-Fi modules, which share the same connection to the serial port of the Arduino. This switching mode is set during the setup of the system and the MUX operations are illustrated in Table 2.

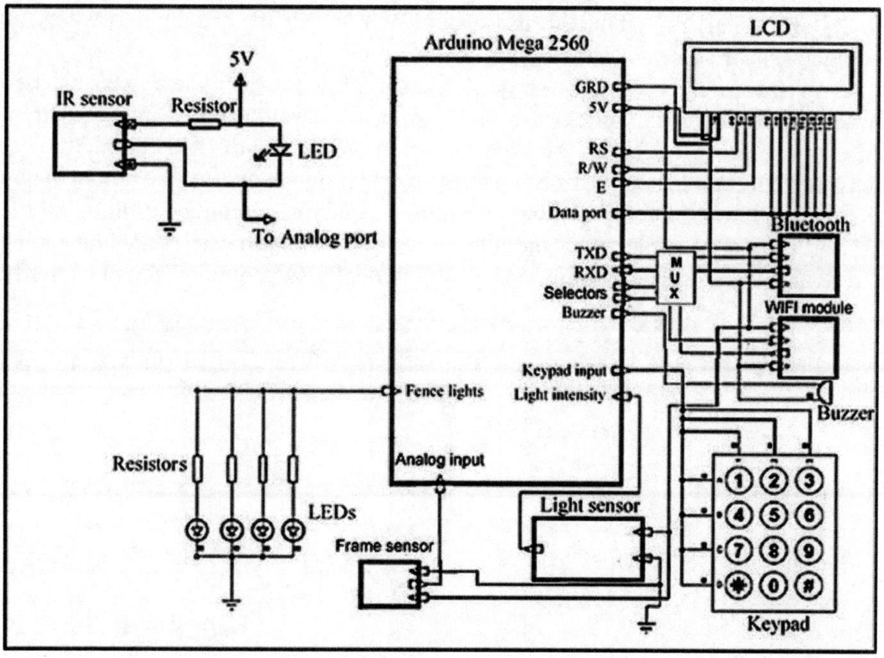

Fig. 4 Schematic of the circuit

Table 2 Multiplexer operations

Selectors	Mode
00	Bluetooth is selected to transmit signal to or receive signal from the Arduino
01	No operation
10	No operation
11	Wi-Fi is selected to transmit signal to or receive signal from the Arduino

2.3 *Software Description*

Figure 5 shows the general flow of the execution of the security software, which is summarised as below:

- On initial boot, the system will check if the functional parameters, i.e., password and home fence parameter, as well as the operation mode of the hardware (Bluetooth or Wi-Fi) have been set.

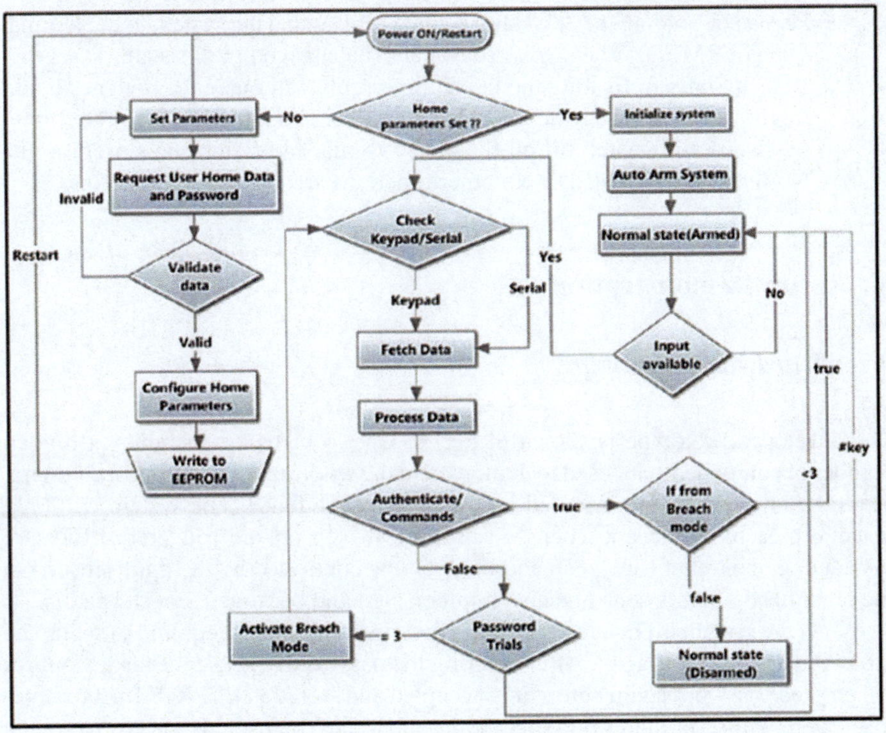

Fig. 5 Flow chart of the system

- If the functional parameters have been set, then the system will initialize, set up security conditions, and then wait to enter disarmed mode, or else it will request the user to insert the data as requested.
- Once the user enters the data, it will be saved in the EEPROM, the internal memory of the processor, and referenced for the subsequent boots. This data can't be changed by the user afterwards, unless by authorised personnel.
- Entering the password 3 times wrongly will trigger the breach condition, which will thus require a correct password to go back to the armed Normal state. Notice that there are three states, i.e., Normal and Breach states. The Normal state, however, has two sub-states, i.e., Armed and Disarmed sub-states, which are explained as below.

  (a) Normal State (ARMED): In this state, the system will enable all breach interrupts, which will trigger the breach condition when the appropriate conditions are satisfied. Entering the correct password will switch the system to the Normal State (DISARMED).
  (b) Normal State (DISARMED): In this state, all the breach interrupts (only fire and light interrupts run in both normal states) are disabled, and therefore the home is not being monitored for any intrusions. Pressing the "#" key on the keypad or sending the "#" via Bluetooth will switch the system to the Normal State (ARMED). The password can also be changed in this state.
  (c) Breach State: Activating the breach interrupts will make the system go into this breach state where the alert systems will be activated. Entering the correct password will switch off all the alert systems and return the system to the Normal State (ARMED). No other functions are available in this state.

3 Result Demonstration

3.1 Hardware Prototype

The developed prototype is shown in Fig. 6. Due to budget constraints, a limited number of components are used to demonstrate the workings of the proposed system. Three IR line sensors are installed on the posts of the fence. The system will only detect a breaching incident when two or more IR sensors are triggered. LEDs are installed on the top of each post, and they will be controlled by the light sensor. On the other hand, the fire sensor is also implemented and packaged in a casing box.

When the system is powered on for the first time, the user is required to set up the password as shown in Fig. 7. After entering the password, the user needs to confirm the entered password again before it is accepted and stored in EEPROM. At the end, the system will restart after pressing the reset button. The user also needs to set the home parameters, such as fence perimeter based on the actual perimeter of the house, and also the operation modes of the system, through the keypad input. If not, the system will sense the incorrect breaching message and notify the user.

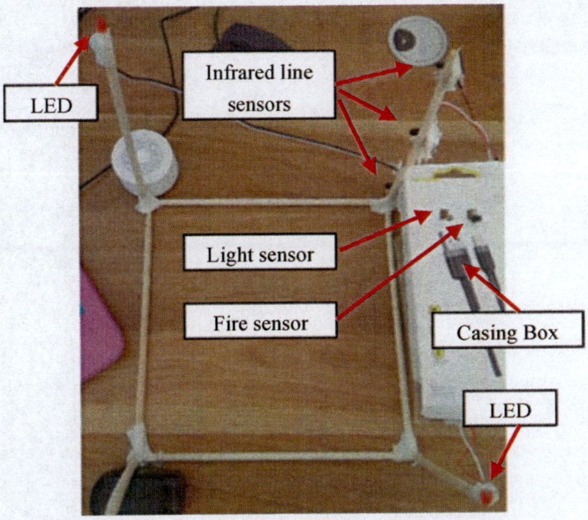

Fig. 6 Hardware prototype

Fig. 7 Password setup

3.2 Bluetooth Application (Low-End Function)

In Bluetooth mode, if a mobile device such as a tablet or phone is to be used, the device must be first paired with the developed prototype via Bluetooth. To receive messages from the system, a Bluetooth serial port profile (SPP) [13] application is used in this project. Figure 8 displays all possible messages that can be viewed on the application, and some of the common messages have been summarised in Table 3.

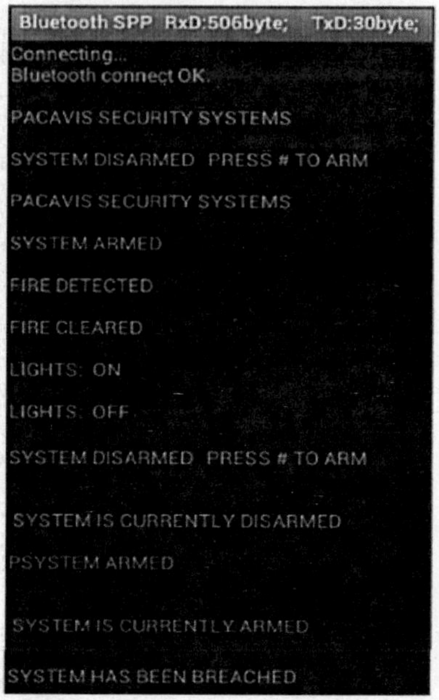

Fig. 8 Messages received on the Bluetooth application

Table 3 Definition of the received messages

Message	Definition
Fire detected/cleared	A fire incident has been detected and it will only be cleared after putting the fire out
Lights on/off	The fence light will turn on or off
System is currently armed	The system is armed, and it will monitor all security parameters
System is currently disarmed	The system is currently disarmed and idle. You are not protected from intruders in this state. However, the fire and light systems still work as usual
System has been breached	An intruder has been detected past the fence of the house property

3.3 IoT Platform (High-End Function)

In Wi-Fi mode, the alert messages are sent to the smart devices through the cloud server using the IoT platform if the internet service is available to the users, as shown in Figs. 9 and 10. The platform used in this project is named Blynk [14], which is a free open-source platform. A simple mobile application is developed using Blynk in this project for the purpose of receiving notifications from the Arduino and displaying them on a smart phone to alert the users of possible fire detection and breaching activities that may happen whenever the owners are not at home. Users can also view the surroundings of the house via a surveillance camera using the developed mobile

Fig. 9 Notification on fire detection

Fig. 10 Notification on breach detection

application if one has been installed at the house. These are the advanced functions offered by the IoT technology whenever the internet is available.

4 Conclusion

A limited budget for prototyping means only a partially scaled model showing all intended functions will be built. Therefore, this research is intended to showcase the wireless capability and advantages of a remotely controlled and secured system within a home using already available technologies such as micro-controllers, Android, Bluetooth, and IoT. With such motivation, a final prototype has been successfully showcased and all the proposed system functions, such as breach and fire detection, control of light intensity, Bluetooth, and IoT, have been working properly. The hardware and software can be extended to other platforms such as Apple, Windows, and any other capable platform, with other microcontrollers and even better graphical displays as needed. The wireless functionality in this project also allows additional functions to be added, such as remotely locking doors; cutting power to specific rooms; locking gates; activating fire extinguishers, etc.

References

1. Rebecca E (2022) 8 surprising home burglary facts and stats. https://www.safewise.com/blog/8-surprising-home-burglary-statistics/. Last accessed 1 Mar 2022
2. Burglary rate in selected countries worldwide in 2018. https://www.statista.com/statistics/1238258/burglary-rate-country/. Last accessed 1 May 2021
3. Muhammad Y, Kolo JG, Ohize OH (2018) Development of GSM based farm security system using motion sensors and IP camera. In: Proceedings of national engineering conference 2018 (ABU NEC 2018), pp 227–232
4. Shankar K, Neelima N, Naveen S, Sai Vardham B (2021) Design security system based on Arduino using motion tracking camera. J Eng Sci (JES) 12(6):515–521
5. Sandar S, Oo SAN (2019) Development of a secured door lock system based on face recognition using Raspberry Pi and GSM module. Int J Trend Sci Res Develop (IJTSRD) 3(5):357–361. https://doi.org/10.31142/ijtsrd25280
6. Praveen Reddy P, Prudhvi Pavan P, Devika M (2019) Home surveillance without using CCTV camera. Int Res J Eng Technol (IRJET) 6(3):4536–4541
7. Sunehra D, Sreshta VS, Shashank V, Kumar Goud BU (2022) Raspberry Pi based smart wearable device for women safety using GPS and GSM technology. In: IEEE international conference for innovation in technology (INOCON), pp 1–5. https://doi.org/10.1109/INOCON50539.2020.9298449
8. Prakash N, Udayakumar E, Kumareshan N, Gowrishankar R (2021) GSM-based design and implementation of women safety device using Internet of Things. In: Peter J, Fernandes S, Alavi A (eds) Intelligence in big data technologies—beyond the hype. Advances in intelligent systems and computing, vol 1167. Springer, Singapore, pp 978–981. https://doi.org/10.1007/978-981-15-5285-4_16
9. Chung GC, Teong KV, Tan SF, Sundarajoo RA (2022) IoT-based autonomous baby monitoring system using mobile app control. In: Alfred R, Lim Y (eds) Proceedings of the 8th international conference on computational science and technology (ICCST). Lecture notes in electrical engineering, vol 835. Springer, Singapore, pp 183–193. https://doi.org/10.1007/978-981-16-8515-6_15
10. World Internet Users and 2022 Population Stats. https://www.internetworldstats.com/stats.htm#google_vignette. Last accessed 1 Mar 2022
11. Access to electricity (% of population). https://data.worldbank.org/indicator/EG.ELC.ACCS.ZS?most_recent_value_desc=false. Last accessed 1 Mar 2022
12. Zeadally S, Siddiqui F, Baig Z (2019) 25 years of Bluetooth technology. Future Internet 11(9):1–24. https://doi.org/10.3390/fi11090194
13. Rachael L (2021) Bluetooth in Android. https://scholarworks.iupui.edu/bitstream/handle/1805/12496/LIBBY_ECET491_POSTER.pdf;jsessionid=07F2B302E0F42D3264DD3A9B2DFC5437?sequence=3. Last accessed 1 Jan 2021
14. Blynk IoT platform: for businesses and developers. https://blynk.io/. Last accessed 1 Mar 2021

The Perspectives on Developing a Conceptual Model for Exploring Emergent Behaviors in Complex Systems

Sa'adah Hassan, Novia Indriaty Admodisastro, Hazura Zulzalil, and Mohd Hafeez Osman

Abstract In systems development, it is necessary to think thoroughly about any emergent behaviors that might occur and create strategies to prevent the undesirable emergent behavior from occurring or at least mitigate them. Often dealing with that at later development stages is costly. Emergent behaviors are the main concern, particularly in the development of complex systems. However, identifying and predicting them is challenging in complex systems as it involves lots of components and interactions. Prior to making wise design decisions, developers need to elucidate and understand the emergent behaviors. For this reason, a conceptual model that highlights the presence of emergent behavior at an early stage is essential in the systems development. The conceptual model is useful in achieving higher coverage as it is correlated with the probability of identifying more emergent behaviors. Therefore, this paper goes through the existing work focusing on effort in modeling emergent behavior and presents perspectives on developing a conceptual model that is able to inform the presence of emergent behavior.

Keywords Complex systems · Conceptual model · Emergent behavior · Software development

S. Hassan (✉) · N. I. Admodisastro · H. Zulzalil · M. H. Osman
Faculty of Computer Science and Information Technology, Universiti Putra Malaysia, 43400 Serdang, Selangor, Malaysia
e-mail: saadah@upm.edu.my

N. I. Admodisastro
e-mail: novia@upm.edu.my

H. Zulzalil
e-mail: hazura@upm.edu.my

M. H. Osman
e-mail: hafeez@upm.edu.my

© The Author(s), under exclusive license to Springer Nature Singapore Pte Ltd. 2023

D.-K. Kang et al. (eds.), *Proceedings of the 9th International Conference on Computational Science and Technology*, Lecture Notes in Electrical Engineering 983,
https://doi.org/10.1007/978-981-19-8406-8_40

1 Introduction

Emergent behaviors or emergent properties are the main concern in the development of complex systems. Complex systems have more elements and hierarchy levels compared to simpler systems which can cause greater interaction between components leading to a large potential for emergent behaviors [1]. As an example, work by Nguyen et al. [2] highlights the challenge in social networks and the impact of unwanted emergent behaviors. They mentioned that this issue is extremely important for many complex systems in practice.

Commonly, the characteristics and desirable behaviors of the proposed system are identified at the early stage of the development process. Besides, any undesirable behaviors should also need to be considered by the developers. They need to create strategies to prevent them from occurring or mitigate them. Though, the expected or unexpected emergent behaviors that are beneficial will give positive surprise [3]. In case of the expected undesirable emergent behavior, it can be avoided or mitigated with appropriate system design. Still, the problematic case is the unexpected undesirable emergent behavior that can be harmful to the system. Grieves [4] categorizes the system behaviors into four categories: the Predicted Desirable (PD), the Predictable Undesirable (PU), the Unpredicted Undesirable (UU), and the Unpredicted Desirable (UD). The PU and UU are the behaviors that we want to eliminate or at least decrease. Yet, the traditional approach to system development is still unable to address the unpredicted undesirable behaviors (UU) [4].

Most researchers agree that the occurrence of emergent behavior is from the interactions (e.g., [5–11]). Vinerbi et al. [12] identify components, interactions, and context as sources of emergence. The interactions between sub-systems, components, and environment, can be the key contributors to emergence. Rouff et al. [13] also mentioned that the combination of various behaviors can become more complex than each individual network in isolation and can result in unexpected emergent behaviors. It also can occur from the interactions with external agents including other systems, humans, and the environment [7]. Besides, Grieves [4] also highlights that many of issues leading to undesirable and unpredicted emergent behavior are due to human interaction. Szabo et al. [9] mentioned that the approaches using interaction as a measure of emergence have the potential to alleviate in identifying the existence of emergent behavior, in which they analyse absolute and relative methods that use interaction as a measure of emergence.

According to Szabo et al. [9], understanding the behavior of complex systems is becoming a crucial issue as systems grow, and the interconnection and geographical distribution of their components diversifies. Yet very few practical approaches for the identification of emergent behavior exist, and many are unfeasible to implement [9]. Pomorova and Hovorushchenko [10] mentioned that the available approaches are theoretical and only explain the nature of emergent properties and those approaches does not provide the prediction of the emergence and evaluation of the influence of emergent properties in the software system. Besides, in System-of-Systems (SoS), Thiago et al. [14] conducted a study to elicit and analyze the state-of-the-art on emergent behavior in SoS and obtained results show that the studies do not deeply investigate emergent behavior. While Raz et al. [11] also mentioned about the need to properly understood the interactions and their effects on design dealt with. However, they added that how to understand and control emergent behavior are still poorly understood.

For this reason, a model that highlights the presence of emergent behavior is essential in system development. A recent study [15] highlights issues with using multiple conceptual models. They mentioned that models have been used to represent the state and process behaviors of a system and the more complex the system becomes, the more it demands multiple models to be simultaneously utilized and integrated among each other. However, the existing methods are not able to accurately model all the interactions between subsystems and components [11]. Hence, a conceptual model of emergent behavior seems useful to address the issues. The conceptual model will act as a tool used at the early stage of development to investigate the potential occurrence of emergent behavior. However, there are a few perspectives that need to be considered in developing the conceptual model; hence it will be presented in this paper.

The structure of this paper is organized as follows: investigates the related work for modeling emergent behavior in Sect. 2. Section 3 provides an overview of research on modeling emergent behavior. Section 4 presents the perspectives on the development of a conceptual model for exploring emergent behavior. Section 5 concludes the paper.

2 Related Work on Modelling Emergent Behavior

This section presents researchers effort towards modeling emergent behaviors. We gain insight into the current research by reviewing the existing work that address topic on modeling emergent behavior/property. However, there are very limited numbers of work that cover the topic, thus we search in emergent behavior/property modeling and emergent behavior/property identification/detection/prediction related work.

McDonald and Weir [6] was introduced a conceptual model of emergence based on meta classes of emergence. Figure 1 shows the proposed conceptual model that consists of meta classes derived from the previous literature. The proposed work offers a useful framework for investigation of emergence and its causation in real complex systems.

Work by Rouff et al. [13, 16] discusses a mathematical modelling technique for prediction and detection of emergent behavior, e.g., in social networks using Semi-Boolean Algebra [16]. In 2011, Ferreira and Tejeda [8] proposed an ontology model for characterizing emergent properties from the system engineering perspective. Other work by Ferreira and Tejeda [7] discussed on the evolving ability to predict and detect emergent properties that is associated to stages and processes in the development lifecycle. For example, using use case model or scenario to predict emergent properties at requirements analysis stage. In addition, a use case to predict

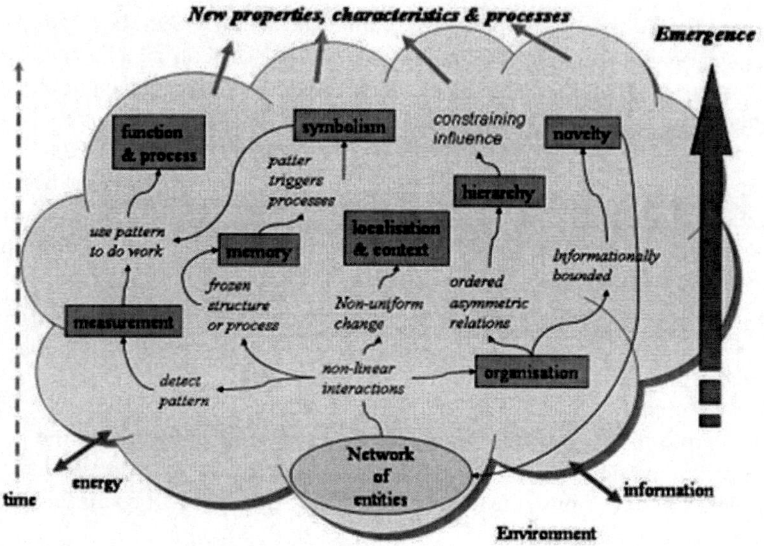

Fig. 1 A conceptual model/framework of emergent [6]

undesirable emergent properties is also proposed, this use case represent functionality of an expert system or a decision support system that allows a user to predict emergent properties and the use case can be realized by using available methods and techniques. However, to predict the emergent properties, some information is needed and should be available as preconditions, including physical architecture and design models. Thus, it's likely appropriate at later stages in software development lifecycle. They further discuss the ideas for decision support system to support system architects in discovering undesirable emergent properties and defining candidates for root causes of the properties in a system's physical architecture [17]. In addition, in 2013 they proposed an approach to predict undesirable emergent properties by means of risk management [1]. In their work, they discussed emergent property taxonomy and how an organization might consider addressing the risk of undesirable emergent behaviors throughout the system lifecycle. They mentioned that detecting and predicting emergent behaviors is highly based on the knowledge about the systems, its requirement and traceability model.

Multi-agent systems (MAS) and Complex Adaptive Systems (CAS) are examples of complex systems, in which, issues of emergent properties are also get a lot of attention from researchers. This including research for detecting and predicting emergent properties in CAS to avoid possible risks to the systems. Existing methods to handle emergence have focused on attempting to predict it through modeling and simulation of large MAS (e.g., [18–20]). However, it is mentioned that such approaches are limited in their effectiveness as the CAS requires detection at runtime. Whereas Birdsey and Szabo [21] proposed software architecture for identifying emergent behaviour in MAS as it happens, using interval-based snapshots and emergent behaviour metrics. Szabo et al. [22] introduce a grammar-based approach to formalize and identify the existence and extent of emergence without the need for prior knowledge of emergent properties. They employed formal grammars to capture agent interactions in the forms of words written on a common tape.

In software product line (SPL), Mauricio et al. [23] focuses on the manifestation of the interaction and is applicable where unexpected and undesired behaviour occurs. Where, the feature interactions could be analysed in Domain Engineering to anticipate unwanted interactions between SPL features and during the selection of features for specific products in Application Engineering. Mauricio et al. [23] came out with a catalogue (Fig. 2) to help in detecting potential feature interactions in SPL. While, Siegmund et al. [24] propose an approach to predict a product's non-functional properties based on the product's feature selection. However, the approach is only applicable for all quantifiable non-functional properties.

In Lucas et al. [25] work, they proposed Process Risk Indicator (PRI) methodology for analyzing and evaluating emergent properties early in the development cycle. They focus on process risk management using the artifacts from the development process to capture information about emergent properties. They analyze the artifacts to quantify process risks that may lead to higher system risk with the assumption that risk mitigation processes reduce system risks.

SoS is an example of a complex system that is formed by the information interactions and physical interactions among constituent systems (CSs) [3]. The unexpected behaviors might occur after the SoS has been deployed. Work by Chiprianov et al. [26] focusing on modeling and analyzing security and performance of SoS. While Mori et al. [27] present a framework to support an evolutionary threat analysis by considering the emergent phenomena originating from an evolving SoS. They argue that by analyzing emerging properties of a SoS, it is possible to naturally capture threats possibly leading to the violation of security requirements, thus, supporting the prevention of malicious attacks and the recovery from SoS damages. Axelsson and Kobetski [28] presents a risk analysis method that is based on systems thinking for deriving requirements on the CSs to reduce the emergent risks on the SoS as a whole. The modeling activity in their proposed work aims at deriving a system-theoretic model of the SoS for risk analysis and adopting control diagram (Fig. 3) as the basis. The modeling activity is an analysis of the SoS description document, as defined in STECA [29] where the analyst scans the text in the document and identifies concepts that should appear as elements in the model. And, by using the document and domain knowledge to identify information flows between the elements. A SoS control model consists of functional elements, such as controllers, controlled processes, sensors, and actuators. The arrows represent control flows, i.e., information transfers, between the functional elements. A recent article by El Hachem, et al. [30] proposed a method called Systems-of-Systems Security (SoSSec) to address the challenges of security modeling and analysis during the architecture phase of the SoS development life cycle.

Having ability to handle the emergent behaviors is one of the good characteristics of a resilient system. Resilience is about the delivery of desired capabilities by the

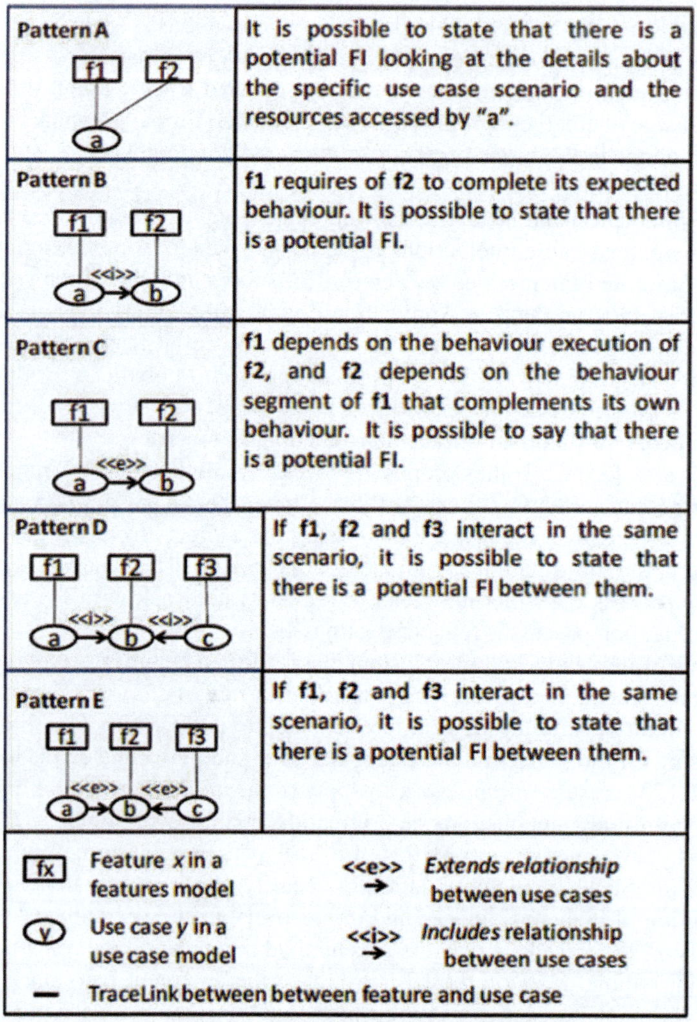

Fig. 2 Patterns catalog to detect potential feature interactions [23]

system even if in the face of difficulty. Brtis and McEvilley [31] shared the modeling information that needs to be captured during the various lifecycle stages to support the effective development and documentation of resilience scenarios and resilience requirements. They mentioned that to establish the resilience needs, properties, and constraints, it should start early in the development lifecycle. Thus, they discuss activities that need to be added in software engineering methodology to address system's resiliency. However, the adversities need to be identified prior to start their proposed activities.

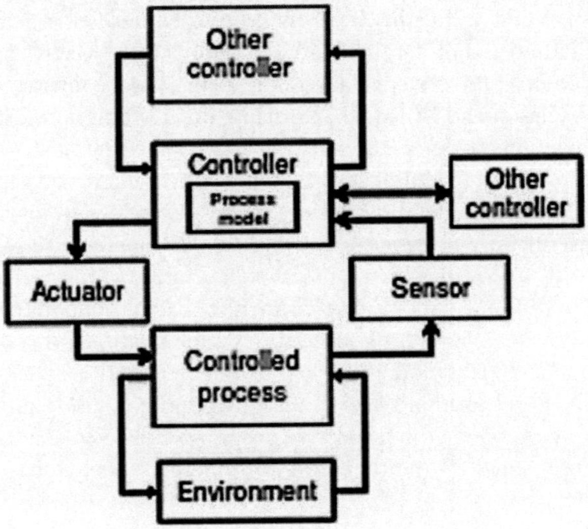

Fig. 3 Control diagram concepts for modelling a system [28]

We also bring to light efforts on modeling the complex computing environment for different purposes but highlight the need of model to facilitate further investigations. For example, work by Bychkov et al. [32] focuses on modeling heterogeneous distributed computing environment. In which, they mentioned that representation the heterogeneous knowledge of subject domains in a MAS is still a challenging problem. Therefore, they proposed an approach to the design of the conceptual model of a heterogeneous distributed computing environment based on multi-agent management. An important component of the proposed model is algorithmic knowledge, which provides the support of planning, resources allocation and dynamic decomposition of a problem to sub-problems. A recent study by Lana et al. [33] report on the state-of-the-art requirements modeling to support system that are formed of heterogeneous and independent constituent systems. Yet, based on their study, the current modeling approaches, particularly concerning on emergent properties, are insufficient to support the development of such systems.

3 Discussion

In this section, we provide an overview from the analysis of the related work on modelling emergent behavior.

If we look in term of the coverage, most of the proposed work are focusing on certain emergent behavior, for example security (e.g., [27, 30]) and quantifiable emergent properties (e.g., performance, footprint) and memory consumption (e.g., [24]). Basically, those are the assessable emergent behaviors.

While as for the artefacts used to explore emergent behavior, we can conclude that mostly, researchers are using model like UML diagrams, e.g., use case model, activity, and sequence diagrams, to analyse the interactions (e.g., [23]). Additionally, Axelsson and Kobetski [28] using control diagram. Those are useful to analyze the functionality of a complex system by looking at it from an abstract level view to foreseeing potential interactions. While class and state chart diagrams are more dedicated to detailed design phases as they are closer to an implementation, thus useful for detection. Yet, using formal language (e.g., [13, 16]) provides accurate results. However, formal method has rarely been applied in software development as it is difficult and only those who have background in formal language can understand it. Besides, it is hardly used for communications; stakeholders must be familiar with mathematical modelling. While there are work that use concrete artifact, e.g., source code. For example, in SPL, it may require not only domain knowledge, but also the source code of the respective program. However, source code not available yet at the early stage, so the prediction of emergent behavior is likely conducted at the later stages.

As an overview of the existing work, there is lack of effort to probe for plausible emergent behavior at the early stage. Even though they acknowledge that it is important to recognize it at early stages. Hence, a conceptual model would be useful for exploring emergent behavior that is hard to predict in the ill-defined computing situation. Study of related work leads to the understanding that there are few perspectives need to be considered to build the conceptual model for exploring emergent behavior. The perspectives are discussed in detail in the following section.

4 Perspectives for Developing Conceptual Model

A conceptual modeling approach allows a useful representation of salient features pertaining to the issues under investigation [34]. In which, such a model "provides a working strategy, a scheme containing general, major concepts and their interrelations. It orients research towards specific sets of research questions." [34]. Valuable information for building the conceptual model was derived by examining the previous related work to see what should be usefully included in the model. Consequently, we present perspectives for consideration to build a conceptual model for exploring the emergent behavior of a system.

4.1 Types of Emergent Behavior

We can conclude that there are four kinds of emergent behavior (i.e., predicted, and unpredicted desirable emergent, predicted and unpredicted undesirable emergent). For both the predicted desirable and undesirable emergent, since they are predictable, there are many techniques and strategies for the prediction that have been proposed (e.g., [35, 36]). While it is not an issue for the unpredicted desirable emergent as it is the behavior that gives benefits to the user. Yet the unpredicted undesirable emergent behavior is the behavior that needs to be addressed. Knowing the different kinds of emergent it will help to expand the coverage as it is correlated with the chances of predicting further emergent behaviors.

4.2 Potential Source of the Existences

Based on the literature, most researchers agree that the existence of emergent behavior is from interactions. Thus, interactions need to be properly understood so that the designer will investigate their effects and deal with it. To facilitate that, we have to know the source of the interaction that we categorize into two; internal and external. Internal is where the components within the system interact, while external is where the system interacts with the computing environment (i.e., human, nature and other systems). Figure 4 illustrates the two categories of the interaction. The boxes represent elements or components in the system as well as in the environment.

Internal Interaction. It would be easier to predict the interaction within the system (i.e., internal) and its potential consequences. Here, we suggest a *Component Interaction Matrix* that would be useful to identify possible behavior that might occur. Table 1 is an example of the Component Interaction Matrix.

This matrix allows stakeholders to predict potential emergent behavior of a system that can occur from the interaction of the components. Each intersection in the matrix

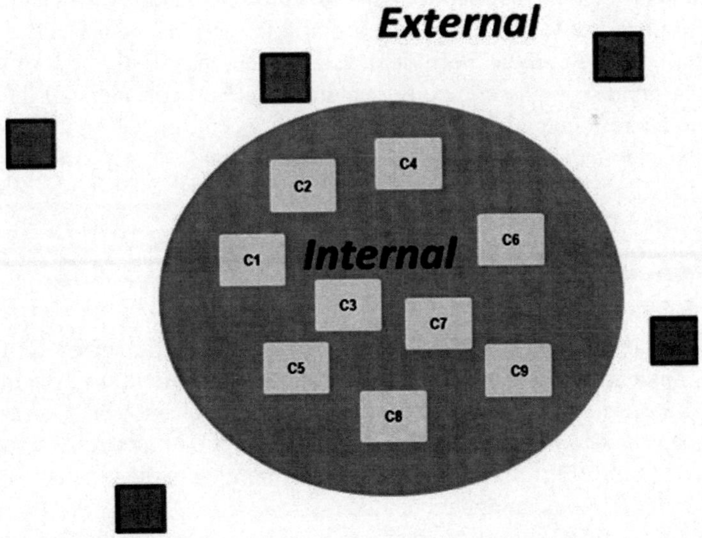

Fig. 4 Categories of interaction

Table 1 An example of component interaction matrix

Component	C1	C2	...	Cn
C1	nil	*C1;C2*	*C1;C...*	*C1;Cn*
C2	*C2;C1*	nil	*C2;C...*	*C2;Cn*
...	*C...;C1*	*C...;C2*	nil	*C...;Cn*
Cn	*Cn;C1*	*Cn;C2*	*Cn;C...*	nil

represents an interaction of the components, here, both planned and unplanned interaction will be measured. Each intersection can be comprehended to include information such as:

- Information that involved (i.e., interaction messages, data)
- The risks of unavailability
- The consequences
- The hazards (uncertainty about probabilities).

External Interaction. External interactions show an interaction of the system with many kinds of external elements, including humans (i.e., intended users, unintended users), nature (e.g., disaster, weather), and other systems. Information about the system environment should be available. For the intended or planned interaction to achieve certain functionality, the emergent behavior whether desirable or undesirable is likely can be predicted. For example, an interaction with an intended user or other systems that involve information flows to achieve a desirable system's behavior. While the undesirable emergent behavior or misbehavior can be predicted by exploring potential stimulus that can outgrowth from the interaction. The proposed interaction matrix can be applied to assist this process. Though, we cannot outline unpredicted undesirable emergent behavior at this early stage. Typically, it can be detected once the system has been deployed. Hence, that would be useful to have a model that represents the computing environment and considers all the external elements to address potential behavior that might occur, including the undesirable emergent behavior.

4.3 The Related Artifacts

It would be easier to predict and detect emergent behavior towards the end of the software development lifecycle as more detailed information about the system is available [7]. As the system development evolves, more concrete models, such as detailed class diagrams and component diagrams, are available to observe or discover emergent behavior (refer Fig. 5). Particularly for predicted emergent behavior whether desirable or undesirable emergent behavior.

Nevertheless, if the emergent behaviors have been addressed at the early stage, developers carried out the remainder of the development activities will handle the

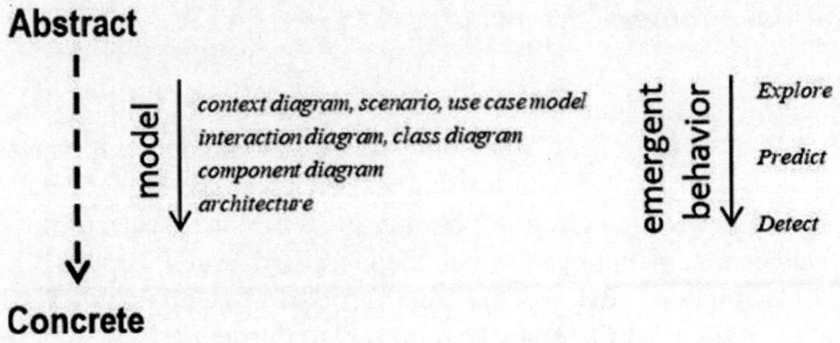

Fig. 5 Model evolvement and discoverability of emergent behavior

emergent behaviors considerations with minimal changes. Thus, our focus is on developing a conceptual model at the early stage.

Requirements engineering is the first stage in software development lifecycle where the requirements are identified. Use case, scenario and context diagram are usually developed to model the requirements. System context, boundary and scope were also clearly demonstrated at this stage. In addition to that, we suggest that a conceptual model that able to address emergent behavior should also be useful at this early stage. Much of the artifacts from requirements model can be used for this purpose. Context diagram and use case diagram are useful to picturize the interaction of system with the environment. Even though, there is an emergent behavior that cannot be predicted, but at least we able to address the potential occurrences of the behaviors. For example, additions of the use case, the misuse case, to explore potential security threat and its mitigation.

Emergent behaviors are behaviors that arise when the system is integrated and does not exist in the individual system's components. It's in line with systems thinking, an approach that offers whole-part hierarchy and can deal with varieties of elements, including humans and software. The fundamental idea in systems thinking is that the system cannot be analyzed by looking just at the individual elements but must be seen as a whole [37], thus it is useful to capture these emergent behaviors. A focus of systems thinking is on understanding the links and interactions among the elements for the entire system's composition. System thinking is often used in problems in which the relationships among components are non-linear [38]. An example of system thinking is Soft System Methodology (SSM). Many researchers have applied SSM in systems development to explore ill-defined problems (e.g., [39, 40]).

Once additional detail that shows the architecture and structure of the system becomes available, we can adopt other proposed approaches such as the meta classes of emergence proposed by McDonald and Weir [6] to detailing out the model. Besides, a catalogue pattern proposed by Mauricio et al. [23] would also be useful to detect further potential interactions and predict emergent behaviour at this stage.

4.4 The Stakeholders' Knowledge

In developing the conceptual model, knowledge from the stakeholders is required to explore further the emergent behavior. Various stakeholders will involve in the development and their involvement is usually at different stages in the development lifecycle. Their knowledge is useful, possibly in addressing issues of the proposed system as well as treatment or solution to the identified issues.

To know what they can offer, we can refer to article by Mylopoulos [41] where they briefly examine the kinds of knowledge represented during the system development process. The knowledge is classified into four *worlds*: subject, system, usage, and development. The *subject world* comprises of the subject matter for the system. The *system world* describes the system being develop at several layers may range from requirements, to design and an implementation. The *usage world* defines the environment where the system functions and consists of agents, activities, tasks, projects, users, user interfaces (with the system) and so forth. The *development world* describes the process and people involved in the system development including, the adopted methodology, schedule, and the design decisions along with the justifications behind those. All this knowledge is relevant during the initial development of the system but also later on after deployment.

Stakeholders with different background and knowledge will be able to offer various viewpoints, then if possible, to utilize and integrate the viewpoints and represented in the conceptual model. Besides, the model should be easily interpreted by stakeholders as it is intended for understanding the situation and communication.

5 Conclusion

Identifying and predicting emergent behavior is challenging particularly in complex systems as it involves greater interactions. It would be useful to have a conceptual model that is able to address emergent behavior at an early stage. Essentially, addressing the emergent behavior at the early stage can leverage a successful design in which the negative effects or risks due to undesirable emergent behavior will be reduced. In consequence, it will lead to better decision-making in developing a resilient system. However, there has been limited investigation of how the emergent behaviors of computing systems can be explicitly modeled and lack of effort to probe for plausible emergent behavior at the early stage. Thus, this paper presents the perspectives to be considered for building a conceptual model for exploring emergent behavior. In the future, we intend to develop a conceptual model by considering those perspectives. Further work would be an evaluation of the proposed model by applying it in a suitable case study.

Acknowledgements This work was supported by the Ministry of Higher Education under Fundamental Research Grant Scheme (FRGS/1/2020/ICT01/UPM/02/1).

References

1. Ferreira S, Faezipour M, Corley HW (2013) Defining and addressing the risk of undesirable emergent properties. In: 2013 IEEE international systems conference (SysCon), Orlando, FL, pp 836–840
2. Nguyen HT, Nguyen NP, Vu T, Hoang HX, Dinh TN (2017) Transitivity demolition and the fall of social networks. IEEE Access 5:15913–15926
3. Kopetz H, Höftberger O, Frömel B, Brancati F, Bondavalli A (2015) Towards an understanding of emergence in systems-of-systems. In: 2015 10th system of systems engineering conference (SoSE), pp 214–219
4. Grieves M, Vickers J (2017) Digital twin: mitigating unpredictable, undesirable emergent behavior in complex systems. In: Kahlen J, Flumerfelt S, Alves A (eds) Transdisciplinary perspectives on complex systems. Springer, Cham
5. Johnson CW (2006) What are emergent properties and how do they affect the engineering of complex systems. Reliab Eng Syst Saf 91(12):1475–1481
6. McDonald DM, Weir GRS (2006) Developing a conceptual model for exploring emergence. In: International conference on complex systems (ICCS2006), 28–31 May 2006, Reading, UK
7. Ferreira S, Tejeda J (2011) An evolving understanding for predicting emergent properties. In: Proceedings of the IEEE international systems conference, April 4–7, Montreal, Quebec, Canada
8. Ferreira S, Tejeda J (2011) An ontology model for characterizing emergent properties from a system engineering perspective. In: Proceedings of the 9th conference on systems engineering research (CSER), 15–16 April, Los Angeles, CA
9. Szabo C, Teo YM, Chengleput GK (2014) Understanding complex systems: using interaction as a measure of emergence. In: Proceedings of the 2014 winter simulation conference (WSC '14). IEEE Press, pp 207–218
10. Pomorova O, Hovorushchenko T (2015) The way to detection of software emergent properties. In: The 8th IEEE international conference on intelligent data acquisition and advanced computing systems: technology and applications, 24–26 Sep 2015, Warsaw, Poland
11. Raz AK, Llinas J, Mittu R, Lawless W (2019) Engineering for emergence in information fusion systems: a review of some challenges. In: 2019 22nd international conference on information fusion (FUSION), pp 1–8
12. Vinerbi L, Bondavalli A, Lollini P (2010) Emergence: a new source of failures in complex systems. In: 2010 third international conference on dependability
13. Rouff C, Haglich P, Pullum L (2010) Detecting emergence in social networks. In: 2010 IEEE second international conference on social computing (SocialCom 2010). The second IEEE international conference on privacy, security, risk and trust (PASSAT 2010) (SOCIALCOM), Minneapolis, MN, pp 693–696
14. Inocêncio TJ, Gonzales GR, Cavalcante E, Horita FEA (2019) Emergent behavior in system-of-systems: a systematic mapping study. In: Proceedings of the XXXIII Brazilian symposium on software engineering (SBES 2019). Association for computing machinery, New York, NY, USA, pp 140–149
15. Ong D, Jabbari M (2019) A review of problems and challenges of using multiple conceptual models. In: Proceedings of the 27th European conference on information systems (ECIS), Stockholm & Uppsala, Sweden, 8–14 June 2019
16. Rouff C, Hinchey M, Truszkowski W, Vanderbilt A, Rash J (2004) Properties of a formal method for prediction of emergent behaviours in swarm-based systems. NASA technical documents
17. Ferreira S, Tejeda J (2011) Ideas for a decision support system for predicting undesirable emergent properties. In: Proceedings of the 9th conference on systems engineering research (CSER), 15–16 April, Los Angeles, CA
18. Kub´ık A (2003) Toward a formalization of emergence. Artif Life 9(1):41–65
19. Moshirpour M, Mousavi A, Far BH (2012) Detecting emergent behavior in distributed systems using scenario-based specifications. Int J Softw Eng Knowl Eng 22(06):729–746

20. Teo YM, Luong BL, Szabo C (2013) Formalization of emergence in multi-agent systems. In: Proceedings of the 2013 ACM SIGSIM conference on principles of advanced discrete simulation—SIGSIM-PADS '13, p 231
21. Birdsey L, Szabo C (2014) An architecture for identifying emergent behavior in multi-agent systems. In: Proceedings of the 2014 international conference on autonomous agents and multi-agent systems (AAMAS '14). International foundation for autonomous agents and multiagent systems, Richland, SC, pp 1455–1456
22. Szabo C, Teo YM (2015) Formalization of weak emergence in multiagent systems. ACM Trans Model Comput Simul 26, 1, Article 6, 25 p
23. Alférez M, Moreira A, Kulesza U, Araújo J, Mateus R, Amaral V (2009) Detecting feature interactions in SPL requirements analysis models. In: Proceedings of the first international workshop on feature-oriented software development (FOSD '09). Association for Computing Machinery, New York, NY, USA, pp 117–123
24. Siegmund N, Rosenmüller M, Kästner C, Giarrusso PG, Apel S, Kolesnikov SS (2013) Scalable prediction of non-functional properties in software product lines: Footprint and memory consumption. Inf Softw Technol 55(3):491–507
25. Layman L, Basili VR, Zelkowitz MV (2014) A methodology for exposing risk in achieving emergent system properties. ACM Trans Softw Eng Methodol 23, 3, Article 22, 28 p
26. Chiprianov V, Falkner K, Gallon L, Munier M (2014) Towards modelling and analysing non-functional properties of systems of systems. In: 2014 9th international conference on system of systems engineering (SOSE), pp 289–294
27. Mori M, Ceccarelli A, Zoppi T, Bondavalli A (2016) On the impact of emergent properties on SoS security. In: 2016 11th system of systems engineering conference (SoSE), Kongsberg, pp 1–6
28. Axelsson J, Kobetski A (2018) Towards a risk analysis method for systems-of-systems based on systems thinking. In: 12th annual IEEE international systems conference, SysCon 2018, pp 1–8. Institute of Electrical and Electronics Engineers Inc.
29. Fleming CH, Leveson N (2015) Integrating systems safety into systems engineering during concept development. In: 25th annual INCOSE international symposium
30. El Hachem J, Chiprianov V, Ali Babar M, Al Khalil T, Aniorte P (2020) Modeling, analyzing and predicting security cascading attacks in smart buildings systems-of-systems. J Syst Softw 162:110484
31. Brtis JS, McEvilley MA (2019) Systems engineering for resilience, MITRE technical report, The MITRE Corporation
32. Bychkov I, Oparin G, Tchernykh A, Feoktistov A, Bogdanova V, Gorsky S (2017) Conceptual model of problem-oriented heterogeneous distributed computing environment with multi-agent management. Proc Comput Sci 103:162–167
33. Lana CA, Guessi M, Antonino PO, Rombach D, Nakagawa EY (2019) A systematic identification of formal and semi-formal languages and techniques for software-intensive systems-of-systems requirements modeling. IEEE Syst J 13(3):2201–2212
34. Järvelin K, Wilson TD (2003) On conceptual models for information seeking and retrieval research. Inf Res 9(1)
35. Beurier G, Simonin O, Ferber J (2002) Model and simulation of multilevel emergence. In: Proceedings 2nd IEEE international symposium on signal processing and information technology, pp 231–236
36. Gore R, Reynolds PF, Tang L, Brogan DC (2007) Explanation exploration: exploring emergent behaviour. In: 21st international workshop on principles of advanced and distributed simulation (PADS'07), pp 113–122
37. Axelsson J (2015) Systems-of-systems for border-crossing innovation in the digitized society: a strategic research and innovation agenda for Sweden. SICS Technical Report T2015:07
38. Petkov D, Edgar-Nevill D, Madachy R, O'Connor R (2008) Information systems, software engineering, and systems thinking: challenges and opportunities. Int J Inf Technol Syst Approach (IJITSA) 1(1):62–78

39. Hassan S, Bustard DW, McSherry DMG (2006) Soft systems methodology in autonomic computing analysis. In: Proceedings of UKSS international conference, Oxford, pp 106–115
40. Too CW, Hassan S (2015) The adoptions of USEPs in identifying usability requirements in SSM-based framework: a case study. In: Fujita H, Selamat A (eds) Intelligent software methodologies, tools and techniques communications in computer and information science, vol 513. Springer International Publishing, Switzerland, pp 71–82
41. Mylopoulos J (1992) Conceptual modelling and telos. In: Loucopoulos P, Zicari R (eds) Conceptual modelling, databases and CASE: an integrated view of information systems development. McGraw Hill

Performance Analysis of Transformerless Photovoltaic Three-Phase Inverter Using Sinusoidal Pulse Width Modulation (SPWM) Method

M. Irwanto, N. A. E. M. Foze, and Y. T. Nugraha

Abstract This paper focuses on the simulation and analysis of transformerless photovoltaic three-phase inverter using the sinusoidal pulse width modulation (SPWM) method. The photovoltaic (PV) inverter typically applies a transformerless circuit configuration due to its high performance. The low output of inverter may cause the distortion in the output waveform and inverter will shut down. To prevent this problem from occurring, several proposals were use that are the three-phase power supply of photovoltaics in series connection to ensure the high output voltage for inverter. Next, the MOSFET operate at high frequencies, it may also be used for speedy switching applications with low turn-off losses. Lastly, using the SPWM method to reproduce a sine wave by producing one or more square pulses of voltage per half cycle and keep high performance in an inverter to avoid the distortion on output waveform. As a result, the inverter system is high efficiency, long life and fast response to load change. While for the output waveform is in the sine wave so the total harmonic distortion for current and voltage are low compared to square wave.

Keywords Pulse width modulation · Three-phase inverter · Total harmonic distortion

M. Irwanto (✉) · Y. T. Nugraha
Department of Electrical Engineering, Universitas Prima Indonesia (UNPRI), Medan, Indonesia
e-mail: muhammadirwanto@unprimdn.ac.id

Y. T. Nugraha
e-mail: yogatrinugraha@unprimdn.ac.id

M. Irwanto · N. A. E. M. Foze
Fellow of Center of Excellence for Renewable Energy (CERE), Universiti Malaysia Perlis (UniMAP), Perlis, Malaysia

© The Author(s), under exclusive license to Springer Nature Singapore Pte Ltd. 2023

D.-K. Kang et al. (eds.), *Proceedings of the 9th International Conference on Computational Science and Technology*, Lecture Notes in Electrical Engineering 983,
https://doi.org/10.1007/978-981-19-8406-8_41

1 Introduction

Compared to traditional transformer-based inverters, transformerless photovoltaic (PV) inverters are becoming more popular owing to their lower cost, smaller size, and higher efficiency. Increasing the quantity of grid-connected transformerless PV systems might pose a variety of technical issues. An inverter is a power electronic device that transforms energy from one form to another at the needed frequency and voltage, such as direct current (DC) to alternating current (AC). This may be classed based on the source of power as well as the design of the power circuit.

The inverters are classified as either voltage source inverter (VSI) or current source inverter (CSI). At the inverter's input terminals, the VSI type inverter has a reduced impedance DC voltage source. The CSI type inverter uses a DC current source with a high impedance. An inverter without a transformer is called a transformerless inverter. A computerized multi-step procedure and electronic components that mimic a transformer are used by transformerless inverters to provide the voltage stepping function [1]. The voltage regulator, MOSFET switch, and a few more minor electronics components make up the solar inverter.

The inverter output waveform as shown in Fig. 1a is square waveform that is an inexpensive and easy inverter option. Due to its poor quality of output voltage and excessive harmonics, it is often not utilized commercially. As a prominent kind of inverter on the consumer market today, modified sine wave inverters as shown in Fig. 1b may be used for a wide range of electrical loads, and they can be utilized in solar systems where waveform shape isn't critical. It is provider in terms of AC voltage waveform quality and total harmonic distortion (THD) [2, 3]. Thyristor output stages chop and invert the DC input positive power to form a square wave alternating positive to negative AC output signal that is filtered to resemble a sine wave as shown in Fig. 1c and remove undesirable harmonics from the output signal, hence the term Inverter [2, 4].

In the circuit design of transformerless PV inverter, to provide the sine wave for output inverter is by using the sinusoidal pulse width modulation (SPWM) method.

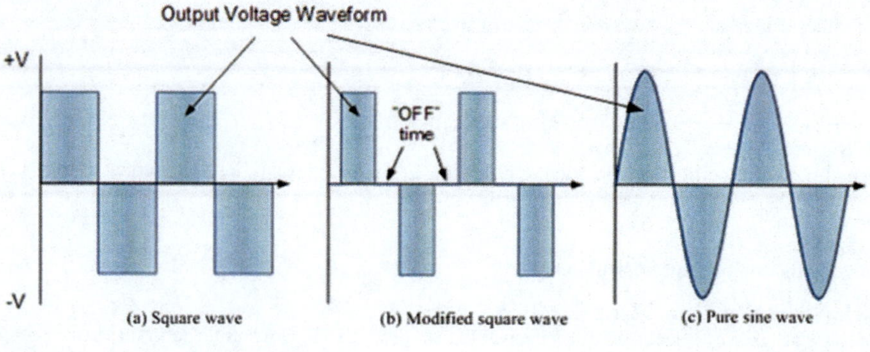

Fig. 1 The output waveform of inverter

Table 1 The comparison of the existing method and proposed method

Parameters	Existing methods	Proposed method
Power supply	Most researcher use the DC voltage as the power supply	The photovoltaic as the power supply that in series connection
Switching	Insulated-gate bipolar transistor (IGBT)	Metal-oxide semiconductor field-effect transistor (MOSFET)
Load	Resistance load	Resistance Inductance load

Next, the problem of inverter's performance is a low output voltage because the output inverter will determine the phase of inverter. Then, to get the high output inverter, the circuit must insert the MOSFET switch and photovoltaics in series connection because this connection can provide the high output voltage while the MOSFET can ability to operate at high frequencies [5–7].

From the research of the previous works, the existing method still lack in the system. Hence, there are several parameters will be proposed to improve the system. Table 1 shows the comparison of the method used by previous work and the proposed method that will be used as an improvement to the system.

The reasons for opting the parameters for the proposed method, first is the photovoltaics in series connection because when many panels are connected in series, the output voltages add up and the output current stays the same compare connection in parallel, their output currents add up, but their output voltages remain constant. Second, the switching MOSFET because the switching frequency of a MOSFET is greater than IGBT. While, MOSFET has a positive temperature coefficient, which means it doesn't become too hot. It can have substantially lower on-state losses than a bipolar component due to its on-state resistance having no theoretical limit. The MOSFET also has a body-drain diode, which comes in helpful when dealing with currents that are restricted in free-wheeling. Lastly, the implement of the resistance inductance at the load represents the operation of the motor in the actual system because the output voltage and current are smooth than resistance load.

2 Methodology

The research procedure had been discussed; this project will achieve the voltage input as a 345 V as a three-phase input for inverter. This will be done by implementing a suitable methodology to develop a better system with the best technique, which is the most important for the project to be done successfully. The wrong methodology chosen will cause failure to the whole future developed system. That why the working flow planning is important to make sure that have taken the correct method, to get right result and then get a good grade for this project. It will be discussing details in this chapter. All the methods are used to meet this study's objective to determine the simulation analysis of transformerless photovoltaic three-phase inverter using sinusoidal pulse width modulation (SPWM) method. This section also will discuss the materials are used in more detail. The flowchart in Fig. 2 is the steps that will used to complete this project.

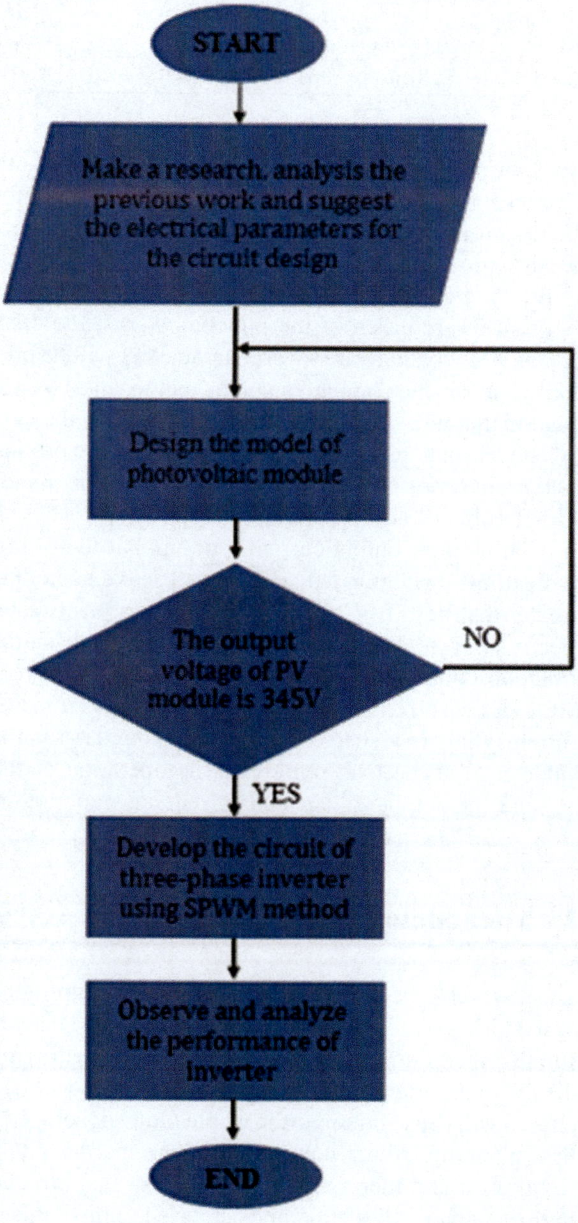

Fig. 2 The flow chart of the project

Table 2 Electrical parameters of Yingli Solar brand

Parameters	Symbol	Value
Power output (W)	Pmax	530
Voltage at Pmax (V)	VPmax	41.30
Open-circuit voltage (V)	Voc	49.15
Short-circuit current (A)	Isc	13.76
Power output tolerance (W)	ΔPmax	0/+ 5
Module efficiency (%)	NPmax	20.48
Current at Pmax (A)	IPmax	12.84

By following the flowchart, there are many solar cells are linked in series and parallel to make a PV module, as is widely known. The nonlinear and multivariable properties of a solar cell's output curves are dictated by the solar cell parameters that characterize its model under various operating situations. Several models, however, have been proposed and shown to be successful in modelling the behavior of solar cell systems by considering a variety of physical characteristics. The photogenerated current (Iph), saturation current (ID), series resistance (Rs), shunt resistance (Rsh), and diode ideality factor are the most important photovoltaic metrics (n). For this project, the Yingli Solar brand with a data sheet in Table 2 is used because of their advantages are doubled strength for multiplied reliability and high product quality.

The PV module's datasheet includes information such as open-circuit voltage (Voc), short circuit current (Isc), peak power (Pmax), voltage and current at maximum peak point (Vmpp, Impp) under various indoor settings such as the standard test condition (STC) and nominal operating cell temperature (NOCT). The key factors that influence PV module characteristics are weather conditions, kinds of materials utilized, and production technology. The environment in which PV systems are installed differs from that of STCs. Using analytical methods, extraction of PV module parameters can be achieved using data in the manufacturer's datasheet. To know the number of photovoltaics that will used in the simulation, can use the circuit of photovoltaic performance shown in Fig. 3. Also, the transformerless series connection photovoltaic three-phase inverter is shown in Fig. 4.

3 Results and Discussion

This section describes the functional test results for every development process involved to complete the application. For this project must provide the circuit of transformerless photovoltaic three-phase inverter using SPWM and use eight photovoltaics as a voltage source to generate the three-phase of the inverter. This is because each photovoltaic will supply 43.98 V, so the target for the three-phase inverter is using the eight photovoltaics that can supply 351.83 V. The methods and design

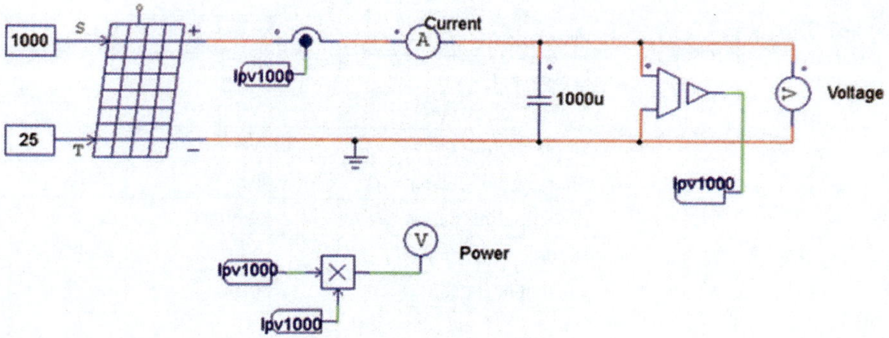

Fig. 3 The circuit of photovoltaic performance

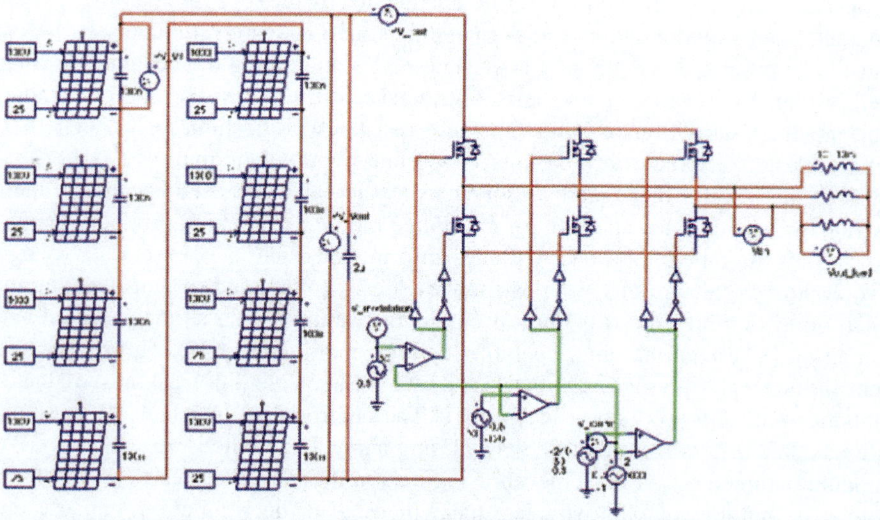

Fig. 4 The transformerless series connection photovoltaic three-phase inverter

will be explained. The result of the photovoltaic and inverter performances will be presented based on the result of simulation by using the PSIM software.

3.1 Output Performance of Photovoltaic

By refer the Fig. 3 and photovoltaic datasheet, after running the simulation the output voltage of photovoltaic is 49.109 V, the output current is 13.999 A as shown in Fig. 5, and the output power is 504.502 W as shown in Fig. 6.

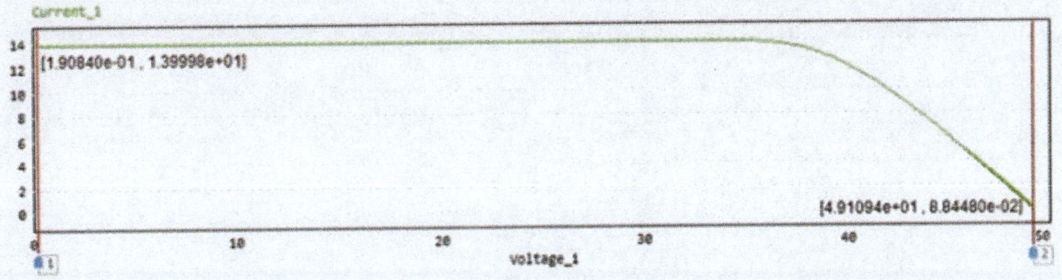

Fig. 5 The current versus voltage of photovoltaic performance

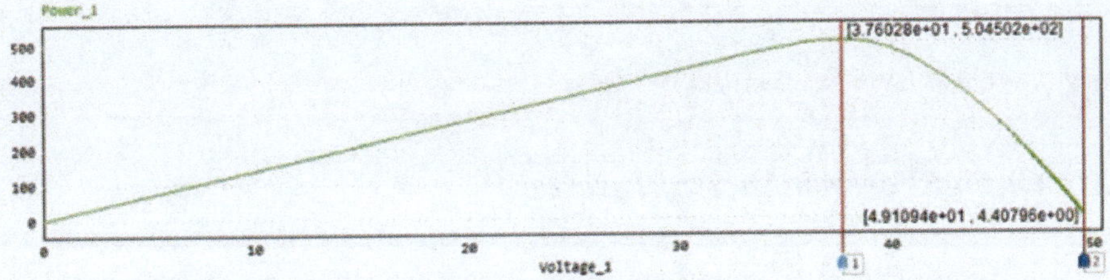

Fig. 6 The power versus voltage of photovoltaic performance

Figure 7 shows the output current of PV module when connected to the inverter. The output current of photovoltaic is 13.0665 A that means the current is suitable for the electronics equipment that is because if the current flow is high, it will make the equipment not durable and easily damaged.

Then, the output voltage of photovoltaic is 392.457 V as shown in Fig. 8 and meet the required criteria expected result for three-phase as the input of inverter. The reason why output voltage is higher than 345 V because it should install the photovoltaic panels more than seven panels otherwise the output photovoltaic will less than 345 V. Based on the results of output power photovoltaic as shown in Fig. 9, their value is high because the output voltage is high. This is called output power is directly proportional to the output voltage.

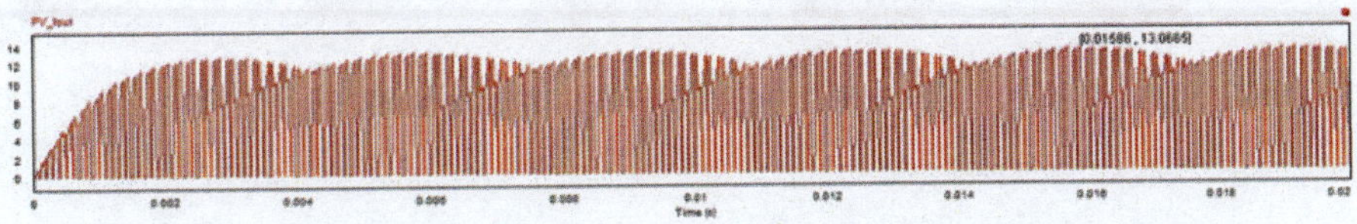

Fig. 7 The output current waveform of PV

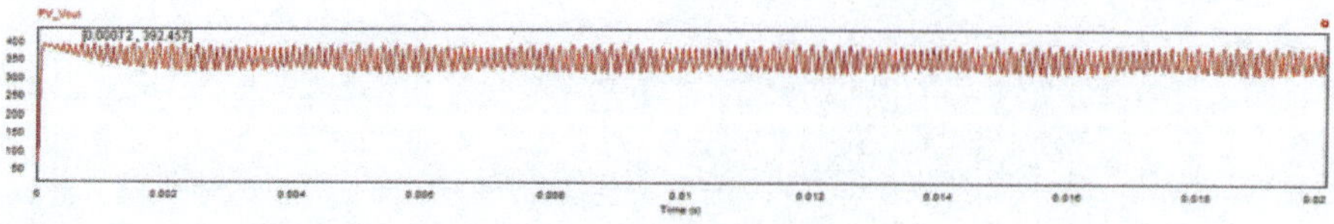

Fig. 8 The output voltage waveform of PV

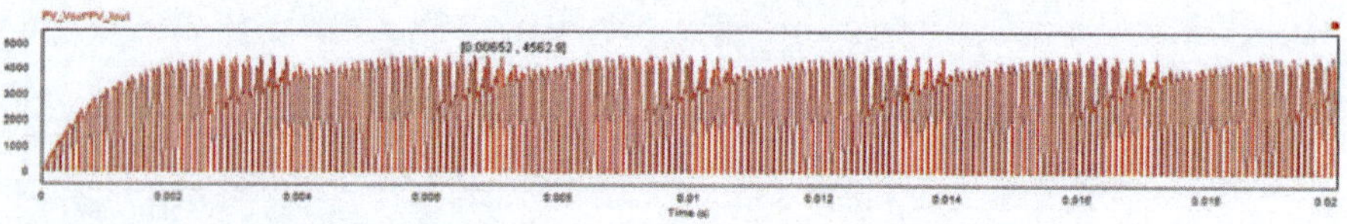

Fig. 9 The output power waveform of PV

3.2 Output Performance of Inverter

The waveform of inverter output voltage is square wave that has fade to become sine because of PWM technique with the distortion with the peak is 391.337 V as shown in Fig. 10. The output current as shown in Fig. 11 is in sine wave (sinusoidal) because of the PWM technique with sinusoidal reference signals have 120° phase. The output power of the inverter as shown in Fig. 12 is taken by the value of the power factor because the concept of alternating current (AC) to get the output power is Power = V × I × pf. For the simulation, to get the waveform of output power by combining the output voltage with output current. Then, select analysis and choose power factor (pf) at the above on menu.

The SPWM refers to the generation of PWM output with a sine wave as the modulating signal. The sinusoidal reference signals have a 120° phase for each load. But their starting is different, for the RL1a starts with the negative peak. While RL1b

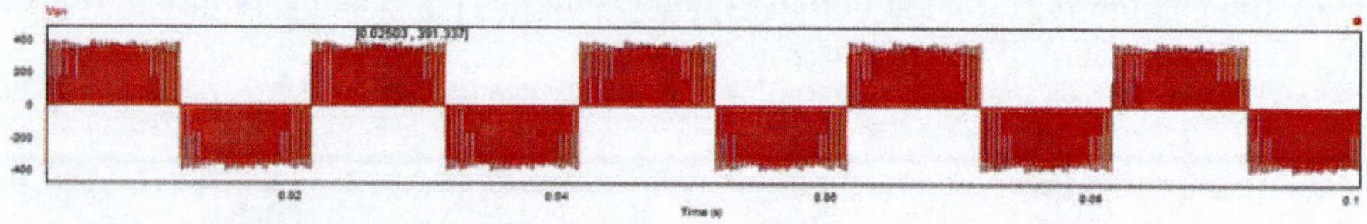

Fig. 10 The output voltage waveform of inverter

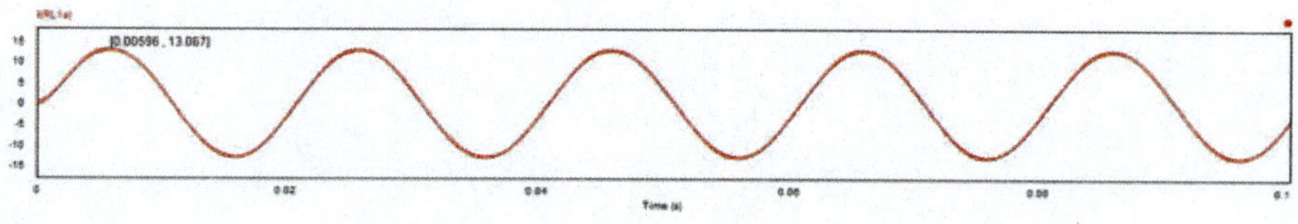

Fig. 11 The output current waveform of inverter

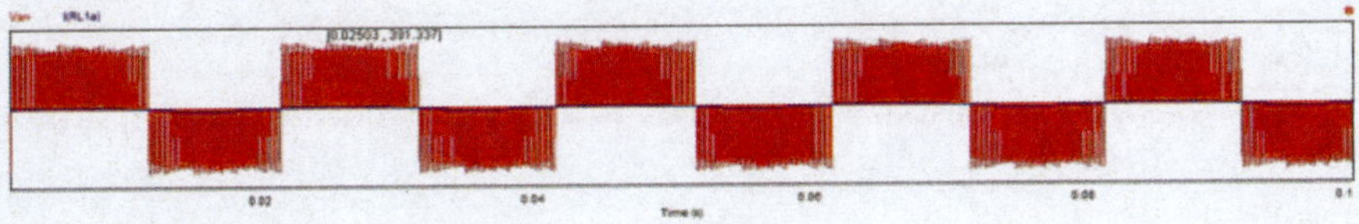

Fig. 12 The output power waveform of inverter

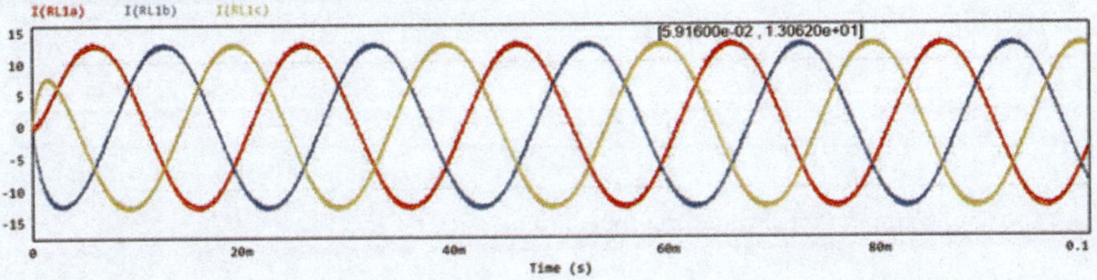

Fig. 13 The output current of three phase inverter

and RL1c start with a positive peak as shown by the output current of three phase inverter in Fig. 13.

3.3 *Output Performance of Inverter*

Based on the current load waveform of the inverter in Fig. 14, the total harmonic distortion value is 0.0178 at a fundamental frequency 50 Hz, and this value of THD is the best value for current THD that is less than 20%.

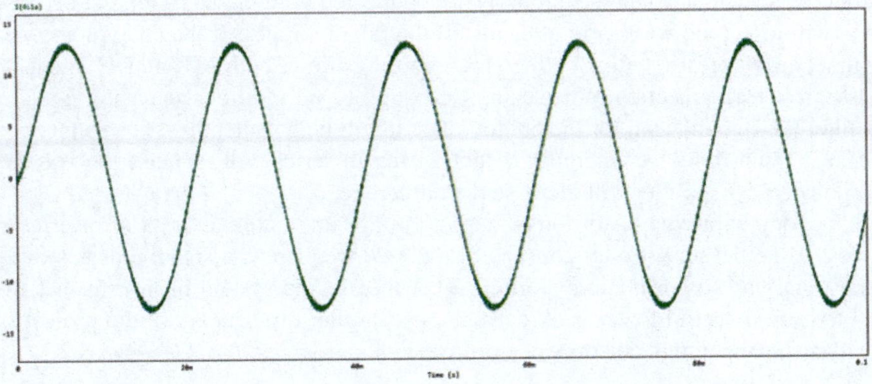

Fig. 14 The waveform of the current load

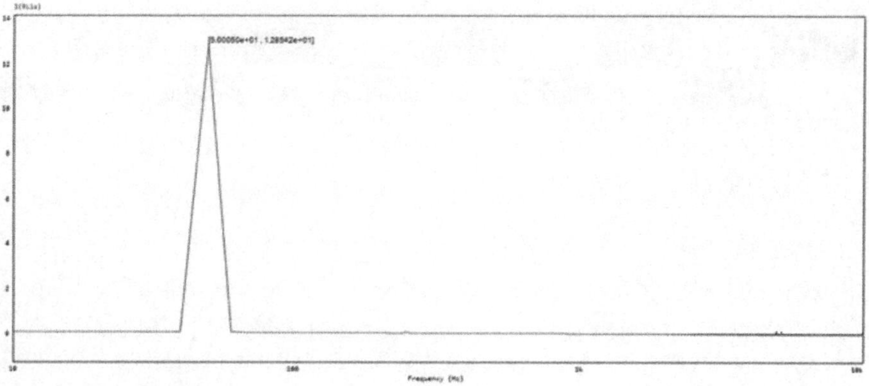

Fig. 15 The spectrum of the inverter's output current

Because the current load is a purely sinusoidal waveform so has no harmonic distortion; by referring the Fig. 15, the spectrum waveform, there is no peak after the fundamental frequency 50 Hz peak. Also, it cannot prove the value of THD by using THD calculation to compare with the measured value from the simulation. Leakage is the technical term for this kind of FFT mistake. When the discrete data capture does not end at the same point in the sine wave as it began, leakage occurs. In theory, leaking would not be an issue if an unlimited number of discrete data points were collected.

4 Conclusion

In conclusion, based on the theoretical analysis from the previous research, several problems will arise when the circuit is not connected with the DC coupling because as it will affect the waveform after install the photovoltaics. If the DC voltage as a source, the waveform will not affect. DC coupling allows both AC and DC signals to pass through a connection. When using DC coupling, no additional capacitor is added to filter the signal. Next, for the photovoltaic it will be installed in series connection because main thing to remember is that wiring in series will increase the voltage, while wiring in parallel will increase the amperage.

The total harmonic distortion of output voltage and output current of inverter in a best value because it only consists of a single frequency, a signal that is strictly sinusoidal has no harmonic distortion. That means, the system of inverter has the higher power factor, lower peak currents, and higher efficiency. So, the project of analysis having a high performance of inverter.

Acknowledgements Thank you very much to Indonesian Ministry of Education, Culture, Research and Technology for a grant of University Superior Basic Research with a grant number of 63/LL1/LT/K/2022.

References

1. Three Phase Inverter: Circuit, Working and Its Applications. https://www.elprocus.com/what-is-a-three-phase-inverter-working-itsapplications/. Accessed 04 Dec 2021
2. (PDF) A study of harmonics in pulse width modulation power inverter as a motor drive. https://www.researchgate.net/publication/280024746_A_Study_of_Harmonics_in_Pulse_Width_Modulation_Power_Inverter_as_a_Motor_Drive. Accessed 04 Dec 2021
3. Development and modelling of three phase inverter for harmonic improvement using sinusoidal pulse width modulation (SPWM) control technique. Int J Recent Technol Eng 8(4):1897–1902, 2019. https://doi.org/10.35940/IJRTE.C4624.118419
4. Connecting solar panels together for increased power. https://www.alternative-energy-tutorials.com/solar-power/connecting-solarpanels-together.html. Accessed 12 Jun 2022
5. Zahira R, Fathima AP, Muthu R (2014) SPWM technique for reducing harmonics in three-phase non-linear load. In: Proceedings of the 2014 IEEE 2nd international conference on electrical energy systems, ICEES 2014, pp 56–60.https://doi.org/10.1109/ICEES.2014.6924141
6. Sine pulse width modulation (SPWM) and its working. https://www.watelectronics.com/sine-pulse-width-modulation-spwm-working/. Accessed 12 Jun 2022
7. Xie H, Li H, Liu H, Chen Q (2013) Harmonic wave analysis and suppression research on three-phase SPWM inverter. Inf Technol J 12(12):2366–2373. https://doi.org/10.3923/ITJ.2013.2366.2373

Long COVID Emotion Analyzer: Using Machine Learning Approach

Kalaimagal Ramakrishnan, Vimala Balakrishnan, Gan Joo Han, and Ng Kee Seong

Abstract COVID-19 has scattered around the world since end of year 2019. In the past two years, countless people who have recovered from COVID-19 will still have sequelae. The sequelae are called as long-COVID. Although many people recover without treatment, unfortunately they come across long-COVID effects for weeks or even months. This paper aims to find out the emotions of people towards long-COVID using a machine learning approach based on the data obtained from one of the most popular social media—Twitter. Furthermore, the study deployed the proposed model by developing a prototype named Long COVID Emotion Analyzer. Twitter data were collected using the hash tag Long COVID for a total of five months (i.e., May–September 2021), resulting in 97,098 clean tweets. IBM Watson Tone Analyzer was used to label the emotion of the tweets. (i.e., sadness, joy, fear, anger, analytical, tentative, confident, neutral). Several machine learning algorithms were used to train and test the dataset. Results indicate Logistic Regression with Unigram of Bag of Words to be the best predicting model, with accuracy = 88%, F1-score = 88%, Area Under Curve = 97%.

Keywords Emotion · Machine learning · Twitter · Long COVID

1 Introduction

Most infected persons with COVID-19 recover within weeks after being unwell. However, several people experiencing post-COVID symptoms. Post-COVID syndromes are a slew of new, recurring, or chronic health issues that can strike for weeks or more after a person is initially infected with the COVID-19 virus. Even patients who showed no signs of COVID-19 at the beginning of the infection may

K. Ramakrishnan (✉)
University of Nottingham Malaysia, Semenyih, Malaysia
e-mail: Kalaimagal.ramakrishnan@nottingham.edu.my

V. Balakrishnan · G. J. Han · N. K. Seong
University of Malaya, Kuala Lumpur, Malaysia
e-mail: vimala.balakrishnan@um.edu.my

© The Author(s), under exclusive license to Springer Nature Singapore Pte Ltd. 2023

D.-K. Kang et al. (eds.), *Proceedings of the 9th International Conference on Computational Science and Technology*, Lecture Notes in Electrical Engineering 983,
https://doi.org/10.1007/978-981-19-8406-8_42

experience post COVID complications. These diseases might exhibit themselves as a range of health issues throughout time [1].

Post-COVID-19 syndrome, post-acute COVID-19, or chronic COVID are some of the terms used to describe long COVID syndrome [1]. It was first identified in social support groups, then in scientific and medical circles. COVID-19 survivors of all ages are afflicted by this long-term COVID illness, which is unclear. The most common symptoms are weariness and trouble breathing, which can remain for months after exposure to COVID-19. Long COVID can cause cognitive and mental deficits, joint pain, chest pain, palpitations, smell and taste dysfunctions, cough, headache, cardiac and gastrointestinal problems (Yong, 2021). On the Internet, a growing number of public emotions are freely ex- pressed. Through social media posts such as Tweets and Facebook, people expressed their concerns about the spread of COVID and the long COVID symptoms. Even though individuals may convey their thoughts and feelings through audio and video files, majority of people still communicate and engage in everyday life through text on social media platforms. When it comes to communicating their feelings about others, people still prefer words [2].

2 Literature Review

During recent years, the usage of social media during times of COVID-19 crisis has increased dramatically. Using Twitter APIs, several academics from all around the world are building COVID-19 datasets [3 4]. Mostly of the tweets are related to keywords such as COVID, corona, pandemic, and quarantine. There is still has limited study which related to our research about long COVID. Few would have predicted that the COVID-19 pandemic, which was declared by the World Health Organization (WHO) in March 2020, would be chronic [5]. The new acute respiratory syndrome coronavirus 2 is the cause of COVID-19 (SARS-CoV-2) [6]. This longer-lasting COVID-19 situations started gaining attention and the term long COVID or post COVID syndrome or long-haul COVID-19 started to gain acknowledgment in the communities of medical and scientific [7]. The potential persistent symptoms of long COVID include myalgia, chest and joint aches, palpitations, cognitive and mental impairments, cardiac difficulties etc. [8]. The symptoms are still not discovered fully, and further research is expected to prove the aspects cause long COVID.

To perform emotion analysis, the study has investigated a few emotion models to select the suitable model. Emotion is a concept that describes personal, or group feelings [9]. Scherer's circumplex continuum model assumes that all emotions exist in a two-dimensional space continuum with four dimensions [10]. This includes valence which study the positive or negative of an emotion, the excitement to study the degree of excitement connected with the emotion, the coping potential that controls the feeling to control someone in certain situations, and goal attainment which study the capability to achieve a goal in the circumstances. It is difficult to detect emotion from the language and semantics of these texts. Although there are many studies accessible on tweets classification, there are only few studies focused on the emotion using COVID-19 related tweets. There is no related work on emotion detection related to long COVID tweets. Some related to emotion analysis studies are reviewed. Kabir and Madria [11] created a neural network model and trained it with manually labelled data to automatically recognize numerous emotions in COVID-19 tweets at fine-grained labelling. They also built a customized Q&A RoBERTa model to obtain terms from tweets that are most accountable for the associated emotions. Their classification model outperformed the other systems, with a Jaccard score of 64.75% and an accuracy of 89.51%. The custom RoBERTa Q&A model surpasses other models with a Jaccard score of 78.65%. Kulai et al. [12] fetched tweets using Twitter APIs for peak COVID months and perform emotion analysis by using Facebook's FastText. Facebook's FastText is an NLP library normally used for text representation and classification.

Hassan et al. [13] applied social media data to research and detect emotions from text using emotion theories, NLP, and machine learning approaches to determine a person's depression degree. They concluded by using a voting model and feature selection approach to compare ME, SVM and NBclassifiers. Their findings revealed that SVM performed better than NB and ME classifiers. SVM has a 91% accuracy, Naive Bayes has an 83% accuracy, and ME has an 80% accuracy. Sentiment analysis systems have been developed that uses an ensemble of classifiers to recognize emotions in text automatically [2, 14]. They used ensemble of Naive Bayes, maximum entropy learner, and knowledge-based tools and the best performance of a sole classifier is achieved by the Naïve Bayes classifier. Da Silva et al. [15] developed a method that uses classifier ensembles and lexicons to automatically classify the sentiment of tweets. They used Multinomial Naive Bayes, Logistic Regression, Random Forest, Support Vector Machines (LibSVM), and ensembles and they found that ensembles revealed better accuracy rates than single base classifiers for all the datasets. Hassan et al. [13] proposed Emotex that inevitably classify the text messages of individuals to imply their emotional states. They utilized the well-established Russell's Circumplex model by considering four major classes of emotions which are Happy-Active, Happy-Inactive, Unhappy-Active, and Unhappy-Inactive. For classifying different emotions categories of Twitter tweets, they examined the accuracy of numerous Machine Learning (ML) methods, including SVM, Decision Tree, KNN and Naive Bayes. Their method achieved a high accuracy of over 90% and is robust across different learning algorithms. Desmet and Hoste [16] used natural languages processing techniques and machine learning methodology such as SVM for fine-grained emotion detection in suicide notes.

However, there is no available studies on tweets emotion extraction related to long COVID. There is also lack of publicly available datasets. This study scraped the raw tweets related to long COVID from Twitter and provide emotion label to the data. Supervised machine learning approach was used to train an emotion predictor model and deployed it as a data product which is a web application hosting on Heroku platform.

3 Issue to Be Addressed

Every country has instituted lockdown measures since the emergence of COVID-19, and people seldom venture out to mingle with others. This incident shows that some people, including long COVID patients, are unable to express their emotions face-to-face with their friends or family. This is because everyone is locked up at home. So, it is difficult to know what the emotions of patients with long COVID experience is when they suffer from long COVID symptoms. It is also important to determine the kind of feelings people have about long COVID. Everyone is more inclined to vent their emotions on social media, including one of the biggest social networking platforms Twitter has 353 million active users [17].

The goal of this study is to create a long COVID emotion analyzer for persons who are concerned about long COVID issues or who are long COVID patients. The web application could analyze how people react to the long COVID and what emotions they own. The application can find out the trends of keywords used by people regarding the long COVID issues. The application can also analyze the emotion from the text. The scope of the study includes searching for the dataset, training the model, and deploying a model on a web application for the users to view the tweets analysis results about long COVID and predict the emotion of text entered by users. At the end of the study, a web application is deployed and hosted online, and it can be accessible anywhere from a computer. It also can support other web browsing enabled devices which having smaller screens such as smartphones and tablets, but it is not recommended for a better visual experience. The objectives of this study are to explore the trend of keywords across the different emotions, to determine the emotions using supervised algorithms, to assess the effectiveness of the emotion model by using evaluation metrics and to develop a product for the emotion analyzer.

4 Methodology

Tweets were crawled and collected using Twitter streaming API service between May 1, 2021, and September 30, 2021, based on one hashtag, namely, #LongCOVID. The initial number of tweets gathered were 130,005. The dataset is stored in a CSV file. Thus, a series of preprocessing tasks were performed to improve the dataset's quality. Several common criteria were used, namely Tweets in any other language other than English were removed, all URLS, emails, emoticons and emojis were removed, special characters such as @, #, & etc., multiple spaces, punctuations and digits were removed, Tweets containing fewer than three words (e.g., good day, sleep) were removed, user handles (e.g., @Maguire) and hashtags were removed, all re-tweets and duplicate entries were removed, convert upper case to lower case, and finally stop words were identified and removed [18]. By implementing these steps, a total of 32,908 tweets were reduced, leaving 97,098 tweets for further analysis.

Then the lemmatization is applied to the clean tweets instead of stemming because stemming operates on a single word without knowledge of the context (e.g., change, changing, changes, changed, changer → chang) while lemmatization determines the lemma of a word based on its intended meaning. The process converts words into their basic forms. Unlike stemming, lemmatization depends on correctly identifying the intended part of speech and meaning of a word in a sentence (e.g., change, changing, changes, changed, changer → change; drive, drives, drove, driven → drive) [19]. It is necessary to choose the emotion model before exploring the tweets emotion. This study utilizes the Scherer's circumplex model by considering 7 emotions and classify tweets as 'joy', 'sadness', 'anger', 'fear', 'analytical', 'tentative', 'confident' or no emotion which is 'neutral'. The circumplex model of this study is shown in Fig. 1.

The emotions were labelling using the IBM Watson Tone Analyzer which calculates the emotion score between a range of 0–1. The highest score in "Document-level" is picked. The example is shown in Table 1.

The way of text is evaluated and how it is characterised is crucial for the execution of a machine learning approach. In this study, bag-of-words (BOW) and term frequency inverse document frequency (TF-IDF) representation techniques are used for the representation of natural language text. BOW is the simplest form of text representation in numbers and is often used in text mining applications. TF-IDF refers to the numerical statistic that indicate how crucial a word is to a document in a compilation. Besides that, N-grams feature extraction method also were experimented in the vectorization process. N-grams basically is a collection of word tokens from a text document such that these tokens are contiguous and occur in a sequence.

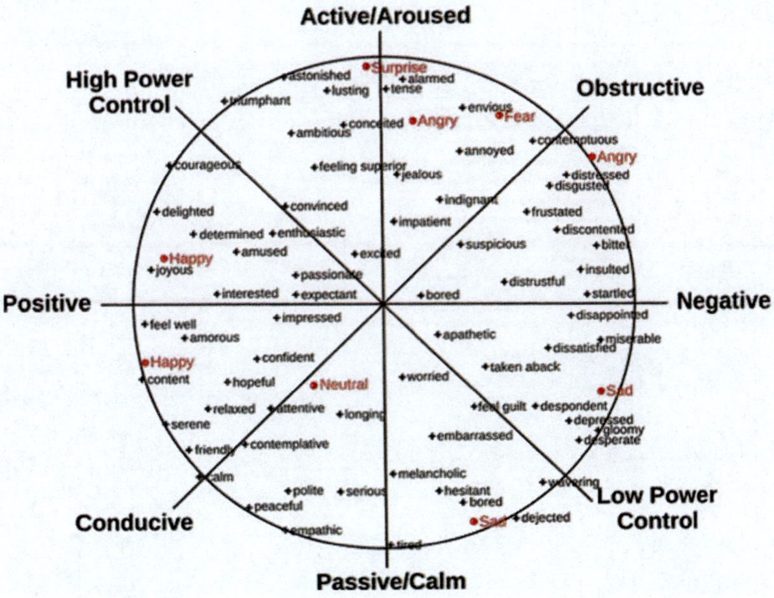

Fig. 1 Circumplex model for this study

Table 1 Labelling the example Tweets

Example Tweets	Confidence score
"Migraine like headaches weeks week stopped amp haven't come wonder we've covidish thing long covid knowing we've covid point pos antibody test remember ill"	Sadness: 0.65 Analytical: 0.62 Tentative: 0.61 ∴Label: sadness
"Favorite aunt covid bad thing believed amp spread conspiracy theories vaccine pacemaker long term effect vaccine wouldn't"	Sadness: 0.54 Analytical: 0.88 Tentative: 0.81 ∴ Label: analytical
"Stiking thinking antiopioid tardsdoctors prescribing opioids covid longhaulers raising addiction fears"	Fear: 0.92 Analytical: 0.65 ∴ Label: fear
"Rare bullshit friends w long covid hospitalized dont know cases cfs infections friends' cfslike cases covid year"	Anger: 0.64 ∴ Label: anger
"Number young healthy people got case covid tiny people experiencing longterm symptoms tiny fraction laughable be- lieves longer bc look obviously false"	No result ∴ Label: neutral

In this study, BOW with unigram and bigram and TF-IDF with unigram and bigram are used. All the tasks were accomplished using Python (Fig. 2).

Exploratory data analysis is carried out to find some initial insights and more understand about the data. Result of the IBM Watson Tone Analyzer labelling revealed 36,530 (37.62%) analytical tweets, 19,822 (20.41%) sadness tweets, 18,031 (18.57%) neutral tweets, 7959 (8.20%) tentative tweets, 5957 (6.14%) joy tweets,

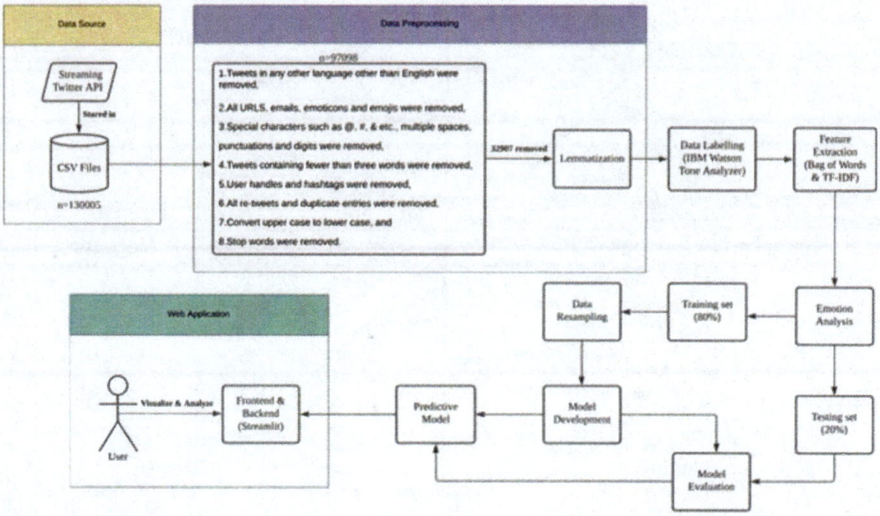

Fig. 2 Overall pipeline

4791 (4.93%) confident tweets, 2610 (2.69%) fear tweets, and 1398 (1.44%) anger tweets. The analytical category is the highest followed by sadness category while the anger category is the lowest. Thus, people have an analytical and sadness emotion towards the topic of long COVID. The emotions classes are imbalanced and needed to be balanced before passing into machine learning algorithms. This is because imbalanced dataset will cause the model biased to the major category class. Besides that, we also find out some of the most common words in the dataset which can be considered as stop words and need to be removed before proceeding to model building part. This is because these common stop words exist in all categories of emotions and will not help in the model building, even will worsen the performance of model. Round 2 of data cleaning step is taken to remove these common stop words such as 'covid', 'long', 'vaccine', 'know', 'people', 'amp', 'time', 'need', 'like', 'year', 'term', 'risk', 'vaccinate', 'symptom', 'work', etc. The others visualization results will be shown in Sect. 5. The dataset is divided into 80% of training set and 20% of testing set. The training set is applied for model training while the testing set is used for model evaluation. From the previous emotion analysis, the problem of imbalanced data is found. The training set is resampled to become balance before passing into machine learning algorithms for training. Majority classes such as analytical, sadness, and neutral are down-sampled while minority classes such as tentative, joy, confident, fear, and anger are up-sampled. The task is to build a model that could predict the emotion from the user's input text which is a multiclassification problem. There are a variety of algorithms to choose from, each with its own set of benefits and drawbacks. Exploration of the various algorithms is essential to determine the best algorithms to be used to build the model for web application. Thus, six machine learning algorithms were used in this study, i.e., Multinomial Naïve Bayes, Logistic Regression, Random Forest Classifier, K-Nearest Neighbors, Support Vector Machine, and Decision Tree. The parameters for each algorithm used in this study are Multinomial Naïve Bayes: alpha = 0.315789, fit_prior = True; Logistic Regression: C = 10,000, solver = 'saga', n_jobs = −1; Random Forest Classifier: n_estimators = 200, n_jobs = −1; K-Nearest Neighbors: Default; Support Vector Machine: loss = 'log', penalty = 'l2', alpha = 1e−3, random_state = 42, max_iter = 5, tol = None and Decision Tree: criterion = 'gini', max_depth = None. Performance evaluation is a crucial step in analyzing a model's performance by comparing the model's predictions to the actual labels. The evaluation metrics which are used for the model evaluation are accuracy, F1-score (macro), Precision (macro), Recall (macro) and Area Under Curve (AUC). The best performing model which is logistic regression with unigram is used as the predictive model of the web application. The architecture of the web application being used is Streamlit framework. Streamlit is an open-source app framework which all the code are in Python. No front-end and back-end experience required to build a powerful web application using Streamlit. Then, the web application is hosted by using Heroku, a website that provided free web hosting services. Heroku is chosen because it provided free web hosting services to host the web application. Heroku is also user friendly for developer to deploy their application.

5 Results

Based on the results, the highest emotion is analytical, followed by sadness and neutral, while anger is the lowest. This shows that people are having an analytical emotion to- wards the long COVID topic as majority of the people are not knowledgeable about long COVID. So, they maintain an analytical mood to look at and study about long COVID. People do not have enough information about the long COVID situation. However, the results shows that the sadness category is the second highest emotion in this dataset. This shows a group of people feel sad because of long COVID. The reasons could be they are the victims who suffering from long COVID, or their family, parents, partner, siblings, or friends are the victims of long COVID.

Word cloud is used to visualize the most popular words used in each category of emotion. Words such as "effect", "think", "longterm", etc. are most frequently used in analytical category. Words like "fact", "think", "study", "understand", "cause", "impact" can indicate people's reasoning and analytical attitude about long COVID topic. Words such as "die", "death", "suffer", "bad", "lose", etc. are most frequently used in sadness category. These words can indicate people's feeling of loss and disadvantage. Words such as "new", "case", "test", "health", "day", etc. are most frequently used in neutral category. These words can indicate people feel indifferent and nothing in particular. Words such as "hope", "likely", "chance", "believe", "maybe", "wonder", "probably", "hopefully", "possible", "unknown", "guess", etc. are most frequently used in tentative category. These words can indicate a person feels hesitant or unsure about something, might be questionable, doubtful, or debatable. Words such as "good", "great", "thank", "love", "happy", etc. are most frequently used in joy category. These words can indicate a person has shades of satisfaction, enjoyment, and pleasure. Joy brings a sense of well-being, inner peace, love, safety, and contentment. Words such as "sure", "actually", "absolutely", "exactly", "completely", etc. are most frequently used in confidence category. A person who feels confidence might be perceived as assured, collected, hopeful, or egotistical. Words such as "fear", "worry", "scary", "afraid", "scar", etc. are most used in fear category. These words can indicate a person has an unpleasant emotion caused by the threat such as danger, pain, or harm. It may be a moderate caution or an extreme phobia. words such as "idiot", "shit", "angry", "hate", "bullshit", "stupid", etc. are most frequently used in anger category. These words can indicate a person has a strong feeling of annoyance, displeasure, or hostility. This emotion may evoke due to conflict, injustice, humiliation, betrayal or negligence. If anger is active, the person attacks the target, verbally or physically. If anger is passive, the person silently sulks and feels tension and hostility (Table 2).

N-grams are widely used in text mining and natural language processing. It is a continuous sequence of N elements from a given sample of text. The N value must be an integer and is set to be a unigram n = 1 (one or characters), a bigram n = 2 (two words or characters), or a trigram n = 3 (three words or characters) [20]. In this study, we used N-grams to explore the top words in the dataset. The top 10

Table 2 Abbreviations used in Tables 3, 4, 5 and 6

MNB	Multinomial Naïve Bayes
LR	Logistic regression
RF	Random forest
KNN	K-nearest neighbors
SVM	Support vector machine
DT	Decision tree

most popular one word in the dataset. Words like "effect", "die", "death", "think", and "case" are the most often used single word. Words like "longterm effect", "wear mask", "natural immunity", "delta variant", and "mild case" are the most often used two words. Words like "longterm care facility", "post viral syndrome", "post viral fatigue", "chronic fatigue syndrome", and "longterm care home" are the most often used three words. Tables 3, 4, 5 and 6 shows the model performance assessment of all classifiers based on the results of the accuracy, precision, F1-score, recall and Area Under Curve (AUC) model evaluation results. These results were achieved using the grid's best parameter. The dataset was split once it was collected and pre-processed: 80% of the data was allocated for training, while 20% was allocated for testing.

Only the confusion metric report and ROC curves of the best performing classifier which is Logistic Regression with BOW and unigram will be displayed (Figs. 3 and 4).

By assessing the performance of the six classifiers using N-grams, BOW and TF-IDF, Logistic regression with BOW and unigram gained the highest accuracy rate at 87.79%, and scored 87.55%, 87.52%, 87.72% and 97.69% for F1-score, precision, recall, and AUC, respectively. The classifier that showed the least accuracy was KNN with TF-IDF and unigram, with a rate of 43.43%. Thus, the best classifier which is Logistic regression with BOW and unigram is deployed in the web application.

Table 3 Experiment result of bag of words (unigram)

Bag of words (unigram)					
Model	Accuracy	F1-score	Precision	Recall	AUC
MNB	72.93	72.07	72.18	72.87	94.59
LR	**87.79**	**87.55**	**87.52**	**87.72**	97.69
RF	84.95	84.41	84.51	84.82	**98.04**
KNN	57.40	57.01	62.69	57.11	87.22
SVM	72.11	71.90	72.24	72.02	93.90
DT	81.40	80.72	80.58	81.27	89.35

The bold signifies the best value for the Accuracy, F1-score, Precision, Recall and AUC for all the models listed

Table 4 Experiment result of bag of words (bigram)

Bag of words (bigram)					
Model	Accuracy	F1-score	Precision	Recall	AUC
MNB	75.38	74.49	74.58	75.30	95.68
LR	**77.89**	**78.19**	79.98	**77.73**	**96.16**
RF	76.37	76.78	**80.55**	76.15	94.88
KNN	52.07	51.78	63.18	51.81	82.61
SVM	70.06	69.79	70.79	69.93	91.20
DT	75.71	75.92	78.66	75.51	89.74

The bold signifies the best value for the Accuracy, F1-score, Precision, Recall and AUC for all the models listed

Table 5 Experiment result of TF-IDF (unigram)

TF-IDF (unigram)					
Model	Accuracy	F1-score	Precision	Recall	AUC
MNB	72.57	71.58	72.01	72.52	94.49
LR	**86.39**	**86.10**	**86.01**	**86.32**	97.43
RF	85.48	85.01	85.04	85.37	**98.20**
KNN	43.43	42.48	78.72	43.00	84.37
SVM	64.33	63.86	64.77	64.30	91.25
DT	81.63	80.96	80.81	81.49	89.45

The bold signifies the best value for the Accuracy, F1-score, Precision, Recall and AUC for all the models listed

Table 6 Experiment result of TF-IDF (bigram)

TF-IDF (bigram)					
Model	Accuracy	F1-score	Precision	Recall	AUC
MNB	75.39	74.34	74.60	75.31	95.61
LR	**77.76**	**77.57**	77.63	**77.65**	**96.13**
RF	76.27	76.42	**79.10**	76.07	94.64
KNN	52.09	51.38	61.95	51.80	82.26
SVM	68.11	67.91	74.46	68.21	91.68
DT	75.04	74.95	77.23	74.84	89.01

The bold signifies the best value for the Accuracy, F1-score, Precision, Recall and AUC for all the models listed

Fig. 3 Confusion metric of LR with BOW and unigram

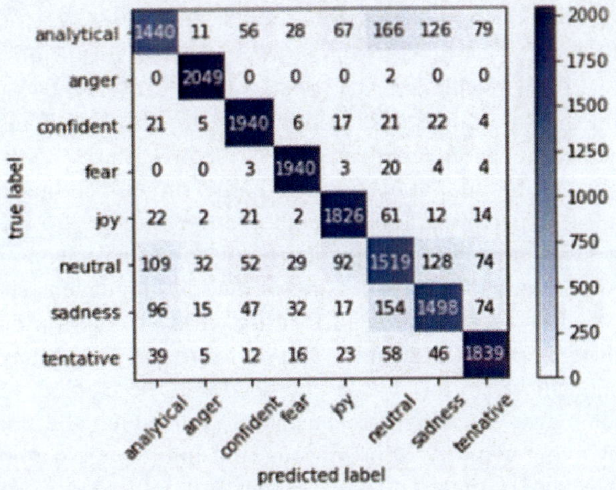

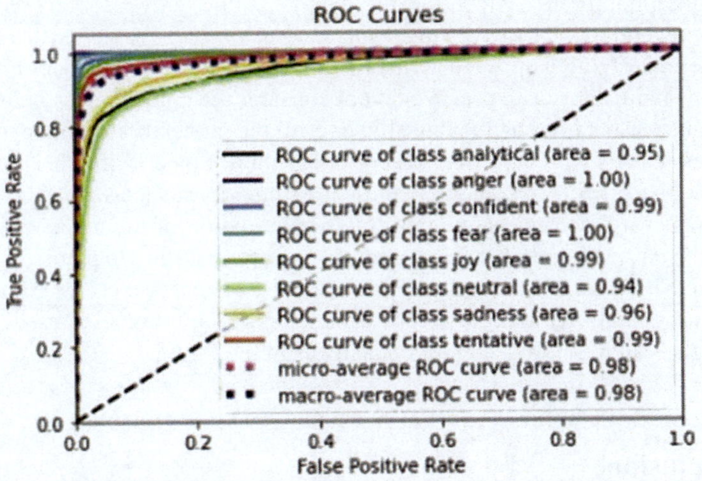

Fig. 4 ROC curves of LR with BOW and unigram

6 Discussion

The findings of emotional analysis and all supervised machine learning models can be seen in previous mentioned figures and tables. The machine learning classifiers perform greatly in terms of accuracy, precision, F1-score, AUC and recall except k-Nearest Neighbors which always perform the worst in all result. Logistic Regression outperformed the other classifiers on all methods. Overall, Logistic Regression with Bag of Words and Unigram is the best performing models with accuracy, F1-score, precision, recall, and AUC of 87.79, 87.55, 87.52, 87.72 and 97.69%. The second-best model is Logistic Regression with TF-IDF and unigram with accuracy, F1-score, precision, recall, and AUC of 86.39%, 86.10%, 86.01%, 86.32%, and 97.43%. While the third-best model is Random Forest with TF-IDF and unigram with accuracy, F1-score, precision, recall, and AUC of 85.84, 85.01, 85.04, 85.37 and 98.20%. Besides that, we found that bi-grams of both word embedding methods which are BOW and TF-IDF performed poorly compared to unigram for this dataset.

The most common challenge of emotion analysis is that we cannot ignore the importance of the processing of natural languages. The granularity of the data set, which is created after NLP in the case of emotion analysis, is directly related to the experiment's accuracy and performance [21]. Many irregularities, diversity, and subjectivity in the dataset are needed to be tackled while dealing with the natural language. There is also has limitation which the data created on Twitter keeps evolving with new abbreviations and emojis. For example, "tbh" which is frequently used lately means "To Be Honest", "fyi" means "For your information", "wfh" means "work from home". There is also some noisy words which is hard to clean such as "haaapppppyyyy" which means "happy". Although we already reclean the dataset for second round by removing some new customize stop words, there are still have some underlying stop words which can be removed before model training to improve the model performance. We do not consider the emoticons and hashtags of the tweets because it can obstruct the efficiency of the model. For example, users can post sadness tweets with happy emoticons or hashtags. Since Twitter become one of the biggest platforms for people to express their emotions, sarcasm has become the most popular way of expression now. Another limitation of the model is it unable to detect the emotion underlying the sarcastic tweets, resulting in predict the wrong emotion. Positive sentences with terrible feelings, or vice versa, are examples of sarcastic viewpoints. Training a model with some of the most often used sarcastic words might lead to improved emotion analysis in the future.

7 Conclusion

In conclusion, based on the findings discussed, people are having an analytical emotion towards the long COVID topic. After pre-processing the dataset, six different supervised machine learning text classifiers were used. These algorithms were used to analyze and predict the emotion from the tweets. Furthermore, the models have been evaluated in the aspects of accuracy, F1-score, precision, recall, and AUC. The greatest performance of this study was obtained by LR with BOW and unigram with 87.79% accuracy.

References

1. Centers for Disease Control and Prevention (2021, September 16) Post-COVID Conditions. Retrieved from https://www.cdc.gov/coronavirus/2019-ncov/long-termeffects/index.html
2. Im TL, San PW, On CK, Alfred R, Anthony P (2013) Analysing market sentiment in financial news using lexical approach. IEEE Conference on Open Systems (ICOS) 2013:145–149
3. Banda JM, Tekumalla R, Wang G, Yu J, Liu T, Ding Y, Chowell G (2021) A large-scale COVID-19 Twitter chatter dataset for open scientific research—an international collaboration. Epidemiologia 2(3):315–324
4. Chen E, Lerman K, Ferrara E (2020) COVID-19: the first public coronavirus Twitter dataset. arXiv e-prints. arXiv preprint arXiv:2003.07372
5. World Health Organization (2021) Coronavirus disease (COVID19) pandemic. Retrieved from https://www.who.int/emergencies/diseases/novel-coronavirus2019
6. Hu B, Guo H, Zhou P, Shi Z-L (2021) Characteristics of SARS-CoV-2 and COVID-19. Nat Rev Microbiol 19(3):141–154
7. Callard F, Perego E (2021) How and why patients made long COVID. Soc Sci Med 268:113426
8. Yong SJ (2021) Long COVID or post-COVID-19 syndrome: putative pathophysiology, risk factors, and treatments. Infect Dis 1–18(2021):1924397
9. Ortony A, Clore GL, Collins A (1990) The cognitive structure of emotions: Cambridge university press
10. Scherer KR (2005) What are emotions? and how can they be measured? Soc Sci Inf 44(4):695–729
11. Kabir MY, Madria S (2021) EMOCOV: machine learning for emotion detection, analysis and visualization using COVID19 tweets. Online Soc Netw Media 23:100135
12. Kulai A, Sankhe M, Anglekar S, Halbe A (2021) Emotion analysis of COVID tweets using fasttext supervised classifier model. In: Paper presented at the 2021 international conference on communication information and computing technology (ICCICT)
13. Hassan AU, Hussain J, Hussain M, Sadiq M, Lee S (2017) Sentiment analysis of social networking sites (SNS) data using machine learning approach for the measurement of depression. In: Paper presented at the 2017 international conference on information and communication technology convergence (ICTC)
14. Perikos I, Hatzilygeroudis I (2016) Recognizing emotions in text using ensemble of classifiers. Eng Appl Artif Intell 51:191–201
15. Da Silva NF, Hruschka ER, Hruschka ER Jr (2014) Tweet sentiment analysis with classifier ensembles. Decis Support Syst 66:170–179
16. Desmet B, Hoste V (2013) Emotion detection in suicide notes. Expert Syst Appl 40(16):6351–6358
17. Sailunaz K, Dhaliwal M, Rokne J, Alhajj R (2018) Emotion detection from text and speech: a survey. Soc Netw Anal Min 8(1):1–26. https://doi.org/10.1007/s13278-018
18. Chekima K, Alfred R (2016) An automatic construction of Malay stop words based on aggregation method. In: International conference on soft computing in data science, Springer, Singapore, pp 180–189
19. Müller T, Cotterell R, Fraser A, Schütze H (2015) Joint lemmatization and morphological tagging with lemming. In: Paper presented at the proceedings of the 2015 conference on empirical methods in natural language processing
20. Majumder P, Mitra M, Chaudhuri B (2002) N-gram: a language independent approach to IR and NLP. In: Paper presented at the international conference on universal knowledge and language
21. Gupta P, Kumar S, Suman R, Kumar V (2020) Sentiment analysis of lockdown in India during COVID-19: a case study on Twitter. IEEE Trans Comput Soc Syst

A Study on Strain-Based Elements for Solid Mechanics: Initial Formulation

Logah Perumal, Wei Hao Koh, and Palanisamy Chockalingam

Abstract Formulation of a strain-based triangular element with three nodes and nine degrees of freedom by modification of existing functions is presented. The strain-based element is then tested by examining two benchmark problems. Although the current formulation yields better results as compared to the other elements considered in this study, the current element holds drawbacks due to equilibrium conditions that are not met.

Keywords Strain-based element · Compatibility · Continuity · Triangular element

1 Introduction

Strain-based elements were first introduced in the early 1970s [1–3]. These elements have certain advantages, such as flexibility in the element shape, interpolants that consist of simple polynomials, high convergence, and accuracy with coarse mesh. Formulation of strain-based elements starts with the description of strain variables (ε_x, ε_y and γ_{xy}) within the problem domain using suitable functions. Usually, polynomial functions are used to describe the strain variables. Over the years, the method has been studied and improved further. Some researchers have presented new techniques to formulate the strain-based elements [4, 5], while others have proposed the functions for the strain variables through trial and error by modifying the initial functions that were proposed in [6] or by other means. A list of strain-based finite elements developed through the latter ways can be seen in [7].

L. Perumal (✉) · W. H. Koh · P. Chockalingam
Faculty of Engineering and Technology, Multimedia University, Jalan Ayer Keroh Lama, 75450 Melaka, Malaysia
e-mail: logah.perumal@mmu.edu.my

P. Chockalingam
e-mail: palanisamy.chockalingam@mmu.edu.my

© The Author(s), under exclusive license to Springer Nature Singapore Pte Ltd. 2023
D.-K. Kang et al. (eds.), *Proceedings of the 9th International Conference on Computational Science and Technology*, Lecture Notes in Electrical Engineering 983,
https://doi.org/10.1007/978-981-19-8406-8_43

In order to achieve convergence, functions for the strain variables should satisfy the compatibility relation below.

$$\frac{\partial^2 \varepsilon_x}{\partial y^2} + \frac{\partial^2 \varepsilon_y}{\partial x^2} - \frac{\partial^2 \gamma_{xy}}{\partial x \partial y} = 0 \tag{1}$$

Apart from the compatibility relation, the equilibrium conditions should also be satisfied in order to achieve optimal performance [4]. Therefore, both compatibility and equilibrium conditions are being considered during the formulation of new strain-based elements [8, 9].

In this work, a set of functions are obtained for the strain variables by modifying the existing functions. The compatibility relation is satisfied in the current formulation. However, the equilibrium conditions are not satisfied. This paper is arranged as follows. Related works are described in Sect. 2. Formulation of the strain-based element through modification of existing formulations is explained in Sect. 3. The simulation setup is provided in Sect. 4. Section 5 is used for the presentation of simulation results and discussion. Finally, the paper is concluded in Sect. 6.

2 Related Works

The principal formulation for strain-based elements starts with expressing the strain variables using suitable functions. One of the early formulations for strain-based elements (to be used for plane elasticity) can be seen in [6], for a triangular element with nine degrees of freedom. Since then, new formulations have been proposed by others by using the methods mentioned in Sect. 1. Current work focuses on modification of the existing formulations for triangular elements with a total of nine degrees of freedom, and related works are briefly stated in this section. The early formulation in [6] is given as:

$$\varepsilon_x = a_4 + a_5 y + a_7 x \tag{2a}$$

$$\varepsilon_y = a_6 + a_7 x + a_5 y \tag{2b}$$

$$\gamma_{xy} = a_8 + a_9 (x + y) \tag{2c}$$

It is seen that the formulation is based on first-order polynomial functions from the Pascal triangle. Later, a new formulation is obtained through trial and error in [10], which is known as SBTREIR:

$$\varepsilon_x = a_4 + a_5 y + a_9 \left(\frac{y^2}{4} \right) \tag{3a}$$

$$\varepsilon_y = a_6 + a_7 x - a_9\left(\frac{x^2}{4}\right) \tag{3b}$$

$$\gamma_{xy} = a_8 + a_9(x + y) - a_5\left(\frac{x^2}{4}\right) + a_7\left(\frac{y^2}{4}\right) \tag{3c}$$

Higher order terms have been introduced into the formulation, and some of the original terms from Eqs. (2a)–(2c) are still present and can be identified easily. Another formulation with higher order terms is seen in [11], which is given as:

$$\varepsilon_x = a_4 + a_5 y \tag{4a}$$

$$\varepsilon_y = a_6 + a_7 x \tag{4b}$$

$$\gamma_{xy} = a_8 + a_9\left(x^2 + y^2\right) \tag{4c}$$

It can be noticed that the last terms in Eqs. (2a) and (2b) have been omitted, and the order of the polynomial terms in Eq. (2c) has been increased by one. Subsequently, another formulation known as SBTE was proposed in [12]:

$$\varepsilon_x = a_4 + a_5 y + a_9 y^2 \tag{5a}$$

$$\varepsilon_y = a_6 + a_7 x + a_9 x^2 \tag{5b}$$

$$\gamma_{xy} = a_8 + 2a_9(x + y)^2 \tag{5c}$$

The degrees of freedom for the normal strains are assigned in a similar way as in Eqs. (3a) and (3b), whereas differences are seen in the equations for the shear strain among Eqs. (3c) and (5c). The first two terms for the normal strains and the first term for the shear strain remain the same for all the formulations shown above. These formulations were developed by making sure that the compatibility equation was satisfied.

3 Methodology

The modified functions (for current element) for the strain variables (straining mode) are given as follows:

$$\varepsilon_x = a_4 + a_5 y + a_9 y^2 \tag{6a}$$

$$\varepsilon_y = a_6 + a_7 x + a_9 x^2 \quad (6b)$$

$$\gamma_{xy} = a_8 + 2a_9(x+y)^2 - a_5\left(\frac{x^2}{4}\right) + a_7\left(\frac{y^2}{4}\right) \quad (6c)$$

Functions in Eq. (6) are obtained by a slight modification of strain functions of the SBTE element. The two additional terms $\left(-a_5\left(\frac{x^2}{4}\right) + a_7\left(\frac{y^2}{4}\right)\right)$ that are introduced for the shear component γ_{xy} in Eq. (6c) are taken from the formulation of the SBTREIR element. The strain displacement relationships for in-plane elasticity are given by:

$$\varepsilon_x = \frac{\partial U}{\partial x} \quad (7a)$$

$$\varepsilon_y = \frac{\partial V}{\partial y} \quad (7b)$$

$$\gamma_{xy} = \frac{\partial U}{\partial y} + \frac{\partial V}{\partial x} \quad (7c)$$

The displacement functions U and V for the straining mode are then obtained by substituting the strain functions (Eqs. (6a) and (6b)) into strain displacement relationship formulas (Eqs. (7a) and (7b)) and performing the integration. Functions that arise from the integration of Eqs. (7a) and (7b) are substituted into Eq. (7c), and an expression for γ_{xy} is obtained, which will be in terms of unknown functions (resulted from the integrations). The other expression for γ_{xy} is provided by Eq. (6c). These two different expressions for the same shear strain γ are then compared and solved for the unknown functions that resulted from the earlier integrations.

The displacement fields for the rigid body mode are given by [10, 13]:

$$U = a_1 - a_3 y \quad (8a)$$

$$V = a_2 + a_3 x \quad (8b)$$

The rotational degree of freedom, θ, is obtained by using the following equation:

$$\theta = \frac{1}{2}\left(\frac{\partial V}{\partial x} - \frac{\partial U}{\partial y}\right) \quad (9)$$

The complete expressions (containing both rigid body and straining modes) for the displacement fields U and V, as well as the rotational degree of freedom, θ are given as:

$$u = a_1 - a_3 y + a_4 x + a_5 xy - a_7\left(\frac{y^2}{2} - \frac{y^3}{12}\right) + a_8\left(\frac{y}{2}\right) + a_9\left(xy^2 + \frac{2y^3}{3}\right) \quad (10a)$$

$$v = a_2 + a_3 x - a_5\left(\frac{x^2}{2} + \frac{x^3}{12}\right) + a_6 y + a_7 xy + a_8\left(\frac{x}{2}\right) + a_9\left(x^2 y + \frac{2x^3}{3}\right) \quad (10b)$$

$$\theta = a_3 - a_5\left(x + \frac{x^2}{8}\right) + a_7\left(y - \frac{y^2}{8}\right) + a_9(x^2 - y^2) \quad (10c)$$

The stiffness matrix is obtained by using the formula:

$$K_e = (Q^{-1})^T\left[\iint_{\Omega}(B^T DB)d\Omega\right]Q^{-1} \quad (11)$$

where Q is the total nodal displacement interpolation matrix (geometric matrix), B is the three-by-nine strain interpolation matrix, D is the constitutive matrix, and Ω is the element domain.

4 Simulation Setup

The strain-based triangular elements are tested by performing computer simulations. Two benchmark problems are considered. The first benchmark problem is a deep cantilever beam with a shear load at the free end (the other end is fully constrained). Properties of the beam are length = 10 m, height = 4 m, thickness = 0.0625 m, modulus of elasticity = 100 GPa and Poisson's ratio = 0.2. Magnitude of the uniformly distributed shear load, F, is 100 kN. The problem geometry and boundary conditions are shown in Fig. 1a. A mesh of 2 by 5 (20 elements) is shown in Fig. 1a. Convergence and accuracy of the elements in calculating the deflection of point A (midpoint of the free end) are obtained. Analytical solution for the deflection at point A is 1.105 mm [10].

The second benchmark problem is the thick curved beam with a shear load at the free end (the other end is fully constrained). Properties of the beam are: $R_1 = 10$ m, $R_2 = 15$ m, thickness = 1 m, modulus of elasticity = 1000 Pa, and Poisson's ratio = 0. Magnitude of the uniformly distributed shear load, F, is 600 N. The problem geometry and boundary conditions are shown in Fig. 1b. Convergence and accuracy of the elements in calculating the deflection of point A (bottom corner of the free end) are obtained. The analytical solution for the deflection at point A is 90.1 m [4].

Performance of the modified strain functions given in Eq. (6) is compared with the conventional constant strain triangular elements (CST), SBTE, and SBTREIR elements.

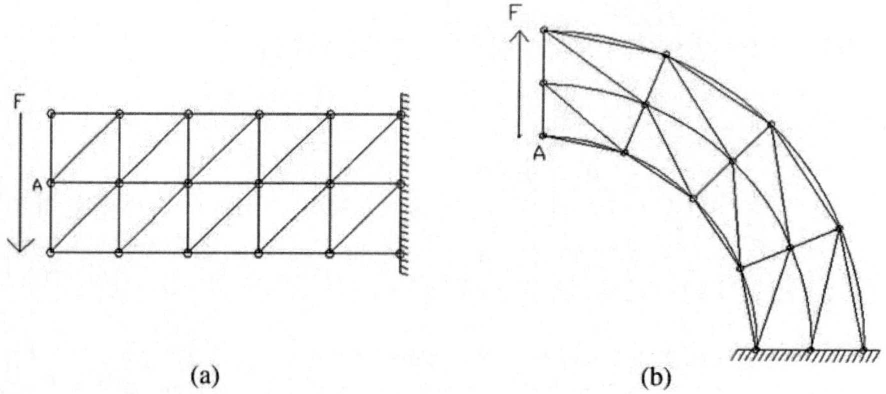

Fig. 1 **a** A deep cantilever beam with a shear load **b** a thick curved cantilever beam with a shear load

5 Results and Discussion

The modified functions for the current element are shown in Eqs. (6) and (10). The functions are obtained by distributing complete and balanced polynomial terms from the Pascal triangle among the strain variables. Modifications to the functions caused cubic terms to be present (in the expressions for displacements U and V) among more degrees of freedom (a_5, a_7, and a_9), as shown in Eq. (10). Simulation results for the benchmark problems are shown in Tables 1 and 2, as well as in Figs. 2 and 3. As shown in the figures, all strain-based elements converge towards a steady-state solution when the number of elements in the mesh is increased. Convergence of the solutions is possible due to the fulfilment of the compatibility relation. However, the accuracy of these elements varies. Current element shows the most rapid convergence rate and the most accurate results among all the elements considered in this work. Furthermore, for both benchmark problems, the deflection predicted by the current element has the slightest deviation from the exact solution (-0.1% and 2.1%, respectively) compared to the other elements, as shown in Tables 1 and 2.

Table 1 Simulation results for the first benchmark problem (deep cantilever beam)

Number of elements	Deflection of point A in mm (relative error in %)			
	SBTREIR	Current element	CST	SBTE
20	0.786 (−28.9)	0.913 (−17.3)	0.632 (−42.8)	1.173 (6.1)
80	0.988 (−10.6)	1.028 (−7.0)	0.922 (−16.6)	1.290 (16.8)
120	1.024 (−7.3)	1.052 (−4.8)	0.978 (−11.5)	1.302 (17.9)
180	1.056 (−4.4)	1.076 (−2.7)	1.015 (−8.1)	1.326 (20.0)
320	1.086 (−1.8)	1.104 (−0.1)	1.053 (−4.7)	1.341 (21.3)
Exact solution	1.105			

Table 2 Simulation results for the second benchmark problem (thick curved beam)

Number of elements	Deflection of point A in m (relative error in %)			
	SBTREIR	Current element	CST	SBTE
8	65.7 (−27.1)	44.5 (−50.7)	22.0 (−75.6)	65.1 (−27.7)
16	80.3 (−10.9)	54.3 (−39.7)	31.1 (−65.5)	70.3 (−21.9)
32	90.1 (0.0)	63.4 (−29.7)	48.8 (−45.8)	81.8 (−9.2)
128	104.9 (16.4)	80.2 (−10.9)	74.1 (−17.8)	91.4 (1.4)
320	114.7 (27.3)	91.5 (1.5)	81.7 (−9.4)	96.6 (7.2)
512	111.6 (23.9)	92.0 (2.1)	85.6 (−5.0)	96.0 (6.5)
Exact solution	90.1			

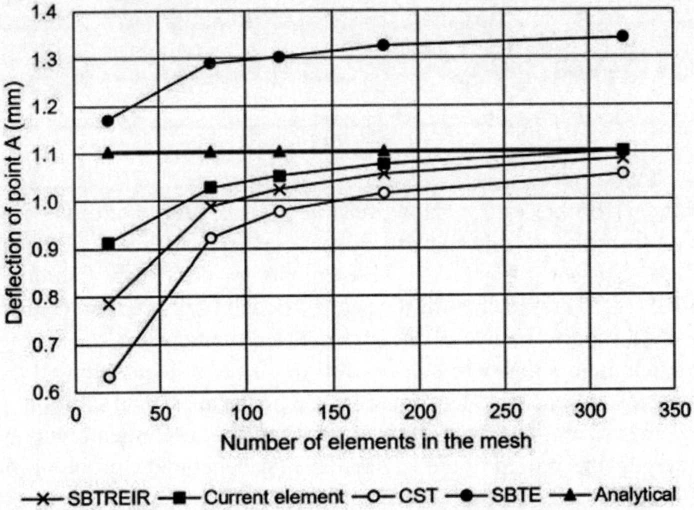

Fig. 2 Simulation results for the first benchmark problem (deep cantilever beam)

Also, it should be noted that the assignment of the degrees of freedom for the current formulation is exactly like the SBTREIR element. However, the polynomial terms and the constants accompanying these degrees of freedom have been changed. For instance, the constants and polynomial terms accompanying the degree of freedom a_9 are different in these two formulations. These modifications turn out to have a prominent effect on the convergence rate and accuracy, as shown in both figures; it has been found that modification of existing formulations can improve convergence and accuracy. This improvement may be due to an increase in the order of the polynomial term accompanying a_9 in the shear component of the strain functions (Eq. (6c)), and higher-order terms assigned to more degrees of freedom could

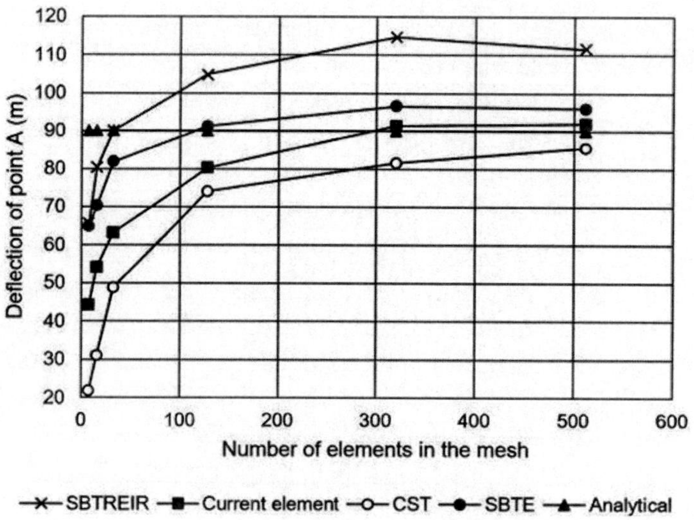

Fig. 3 Simulation results for the second benchmark problem (thick curved beam)

lead to better performance [13]. However, the equilibrium conditions are not met in the current formulation, which can lead to poor performance for other problems and applications, as shown by the SBTE elements for the first benchmark problem (Fig. 2) and SBTREIR elements for the second benchmark problem (Fig. 3).

Judging by the performance of the strain-based elements shown in Figs. 2 and 3, the current element appears to behave well in different beam geometries, and the computed result shows an excellent agreement with the analytical solution. However, while most other strain-based elements, except the SBTE element, display comparable accuracy to the present element for the first benchmark problem, this is not the case for the second problem. For example, although the error in the cantilever beam's deflection computed with the SBTREIR element is within an acceptable range (−1.8% for 320 elements), the predicted behavior of the thick curved beam is much more exaggerated than the theory. Moreover, as shown in Table 2, the relative error for the SBTREIR is maximum (27.3%) when the mesh has 320 elements, and a finer mesh results in a lower deviation from the analytical solution, suggesting that a high number of SBTREIR elements is needed for achieving convergence and accurate results. Besides, compared to the first problem, the SBTE element produces much more accurate results for the second problem. Therefore, the SBTREIR and SBTE elements may not yield consistent results when applied to different geometries.

6 Conclusion

A strain-based element has been formulated by modification of the existing element. The compatibility relation is satisfied in the current formulation. However, the equilibrium conditions are not satisfied. The work presented here is a work in progress. The ongoing works include incorporating compatibility and equilibrium conditions into the element formulation and extending the method to general polygons.

Acknowledgements The authors would like to thank the Research Management Centre (RMC) of Multimedia University, Malaysia, for the support provided through the grant: MMUI/220016.

References

1. Ashwell DG, Sabir AB (1971) Limitations of certain curved finite elements when applied to arches. Int J Mech Sci 13(2):133–139
2. Ashwell DG, Sabir AB, Roberts TM (1971) Further studies in the application of curve finite elements to circular arches. Int J Mech Sci 13(6):507–517
3. Ashwell DG, Sabir AB (1972) A new cylindrical shell finite element based on simple independent strain functions. Int J Mech Sci 14(3):171–183
4. Pajand MR, Moghaddam NG, Ramezani M (2021) Review of the strain-based formulation for analysis of plane structures part II: evaluation of the numerical performance. Iran J Numer Anal Optim 11(2):485–511
5. Himeur M, Zergua A, Guenfoud M (2015) A finite element based on the strain approach using Airy's function. Arab J Sci Eng 40(3):719–733
6. Sabir AB (1983) A new class of finite elements for plane elasticity problems. In: CAFEM7 7th international conference on structural mechanics in reactor technology, Chicago
7. Pajand MR, Moghaddam NG, Ramezani M (2021) Review of the strain-based formulation for analysis of plane structures part I: formulation of basics and the existing elements. Iran J Numer Anal Optim 11(2):437–483
8. Fortas L, Belounar L, Merzouki T (2019) Formulation of a new finite element based on assumed strains for membrane structures. Int J Adv Struct Eng 11(Suppl 1):9–18
9. Belarbi MT, Bourezane M (2005) On improved Sabir triangular element with drilling rotation. Eur J Environ Civ Eng 9(9–10):1151–1175
10. Tayeh SM (2003) New Strain-based triangular and rectangular finite elements for plane elasticity problems. Master Thesis. The Islamic University
11. Belarbi MT, Bourezane M (2014) An assumed strain based on triangular element with drilling rotation. Mail Sci Tech Knowl 6(6):117–123
12. Rebiai C (2018) Finite element analysis of 2-D structures by new strain based triangular element. J Mech 35(3):305–313
13. Perumal L, Chockalingam P (2022) A study on formulation of strain-based triangular elements. J Eng Technol Appl Phys 4(1):25–29

A Blockchain-Based E-Voting System

Mohammad Hadiff bin Hamidey and Swee-Huay Heng

Abstract E-voting methods have become increasingly popular in recent years due to the rapid growth of internet and computer networks. Many e-voting schemes have been proposed by researchers that use cryptographic tool such as blind signature technique and homomorphic encryption technique. However, most of these proposed schemes use a centralised database as their data storage medium and it causes various security risks, such as data alteration by a third-party in the network due to the use of central database system. Therefore, it is a need to have a trusted entity to manage the voting process. This brings the involvement of blockchain technology in this proposed e-voting scheme which can provide a decentralised model that makes the network flexible, reliable and able to support real-time services. This study focuses on the development of a secure blockchain-based e-voting system. Finding shows that the proposed scheme is secure and efficient as it outperforms some of the existing schemes with respect to transaction cost.

Keywords Blockchain · Ethereum · Smart contract · E-voting

1 Introduction

Around the world, e-voting is becoming more common as the time passed. The technology of e-voting has evolved from the use of punched card, optical scan, cryptographic tools, and the latest being the blockchain technology that provides better security. The evolution of e-voting system is majorly affected by the attempts to make the system more secure, transparent, and verifiable. Due to its importance, various works have been proposed to achieve the objectives of making a particular e-voting system more efficient and durable. For e-voting systems to be widely used, they must comply to specific benchmark requirements.

M. H. Hamidey · S.-H. Heng (✉)
Faculty of Information Science and Technology, Multimedia University, Melaka, Malaysia
e-mail: shheng@mmu.edu.my

© The Author(s), under exclusive license to Springer Nature Singapore Pte Ltd. 2023
D.-K. Kang et al. (eds.), *Proceedings of the 9th International Conference on Computational Science and Technology*, Lecture Notes in Electrical Engineering 983,
https://doi.org/10.1007/978-981-19-8406-8_44

E-voting system is designed to give an edge to voting process. However, even with the e-voting system itself has its own drawbacks. One of the main concerns regarding e-voting system is the confidentiality of data. An e-voting system that does not keep voter's ballot as confidential will increase the opportunity of vote buying and extortion [1]. Other than that, there is no physical ballot used in e-voting system. Instead, digital ballot is applied, and it may easily be tampered with, given the unsecured nature of computer networks. Next, e-voting system that does not properly check the authenticity of voters may cause problems such as fake votes by ineligible voter and double voting done by a single voter [1]. Another concern about nowadays voting process is verifiability. The voters cannot check and verify whether their votes were properly received and have been included in final tally. Lastly, most of the voting systems are using centralised database to store the data and votes of users. The use of a central database system may cause data modification by the third-party in the network [1].

The main objective of this paper focuses on the aspects in countering the existence of problems or security issues in e-voting system.

1.1 Security Prerequisites

In most e-voting systems, the process of election usually includes four main stages, namely, preparation stage, registration stage, voting stage, and tallying stage. The following are some critical security requirements for e-voting schemes [2]:

- Authenticity: Only users who have fulfilled the requirements should be able to vote.
- Integrity: Integrity of data should be preserved, and all the votes must be properly counted during tallying process.
- Verifiability: During the final counting stage, a voter must be able to verify that his vote is counted and tallied properly.
- Anonymity: The link between the voter and their vote should always be meaningless.
- Confidentiality: Votes data should remain confidential and secret.
- Non-repudiation: A voter cannot deny the casting of vote.
- Fairness: No party will be able to have any information regarding the election result until the process ends to ensure a fair election.
- Long-term security: E-voting system should be able to withstand any kind of attack or risk throughout years to preserve voter's privacy.

Table 1 Drawbacks of cryptographic based E-voting system

Cryptographic scheme	Disadvantage	Description
Mix-net	Does not provide vote validation	Since votes are decrypted individually, it is possible to examine whether each encrypted vote follows the election rules, and if it does not, it will be considered as an invalid vote [4]. Therefore, modifying the system to cast invalid vote might be seen as an attack against the system
Homomorphic	Does not provide total vote verifiability	The voters can confirm that their vote was cast. However, they cannot confirm that their vote was cast for their choice of candidate [5]
Blind signature	Must have trusted third-party included in the voting process	The inclusion of third-party in voting process such as to count total number of votes will give threats to the status of validity and authenticity of election result [2]

2 Literature Review

2.1 Cryptographic Based Approach

Most of the existing schemes of e-voting system is based on the underlying concept of cryptographic techniques. The e-voting schemes can be mainly categorised into three cryptographic approaches, which are mix-net, homomorphic encryption, and blind signature. These approaches use mathematical algorithm to fulfil the security requirements needed in e-voting system. This section will discuss the security issues found in these three cryptographic schemes used in e-voting system. Please refer to Table 1 for a brief comparison on the existing e-voting approaches. Interested readers may refer to [3] for a more comprehensive review.

2.2 Blockchain Technology

Nakamoto first released his bitcoin white paper in 2008 which transformed the realm of cybersecurity. With the well-known technology known as "Blockchain", he integrated certain hashing techniques to acquire the immutability of data as well as giving a secure protection in transferring and storing data in a distributed ledger. It also helps in securely exchanging transactions of assets such as funds, votes, and data access rights. Currently, it is impossible to conduct a fair election without risking rigging and manipulation of the results [6]. However, the advent of the blockchain, which opened a plethora of possibilities, prompt researchers to investigate and conclude that based on their findings, blockchain is suitable foundation for e-voting. Some of the most significant benefits of implementing e-voting system based on blockchain are anonymity and accuracy as well as able to defend against DoS and distributed DoS attacks at the highest level of performance [6].

Related Works on Blockchain E-Voting System. The scheme proposed by Cruz and Kaji [7] used Bitcoin protocol along with blind signature techniques to develop an e-voting platform. This scheme was able to provide anonymity and individual verifiability through the use blind signature technique. It is also able to provide universal verifiability using bitcoin blockchain. However, the proposed scheme still requires external entity which is counter that acts as the transaction verifier of voting process. This may cause problem such as malicious attempt by a corrupted counter to fake the result of election. Besides, the use of bitcoin blockchain also has its own drawbacks. According to [7], a block containing transactions is validated and confirmed every 10 min on average in bitcoin. To achieve finality, bitcoin recommends that a block needs to be confirmed for at least 60 min (six blocks been mined). Thus, this may cause delay in each transaction of voting process and may cause problem especially in large-scale election.

The scheme proposed by Bistarelli et al. [8] was based on bitcoin protocol in which votes calculation is done by adding each candidate's tokens on the bitcoin blockchain. In terms of security requirements, this scheme can address eligibility and authentication issue in the pre-voting phase. They also provided individual and universal verifiability using bitcoin blockchain. Bitcoin blockchain also helps in maintaining the integrity of data as well as preserving the anonymity of voters' vote during election process. However, according to [8], the bitcoin consensus mechanism is time-consuming and computationally costly causing the scalability performance of the system to be limited. This scheme also requires the involvement of third-party to help in conducting the election which might increases the chance of data tampering and data leakage.

The scheme proposed by Shukla et al. [9] used Ethereum technology to implement a secure e-voting system. Given the security nature of Ethereum, it is difficult to change the source code defined in the smart contract. Thus, it will be able to maintain the secrecy as well as the integrity of data. This proposed scheme was able to address some important security features in e-voting system such as votes transparency, integrity of data, and non-repudiation of votes [9]. However, the proposed scheme does not properly provide authenticity measure for eligible voter. Other than that, the system also does not provide verifiability feature for voters to ensure that they can verify their votes upon completion of election process. Thus, the voters cannot verify whether their choice of vote has been included in the final tally.

The scheme proposed by Lai et al. [10] requires a low level of confidence between the participants. They used Ethereum protocol as their main architecture to build the voting system which can eliminate trust on third-party as well as maintaining the integrity of data. They also provided anonymity to user's information by using ring signature technique. However, the use of ring signature makes it difficult to organise multiple signer entities. They also utilised proof of work consensus model which consumes a lot of energy. For example, miners' supercomputers from all around the world will need to execute and monitor millions of computations every second. As a result, it requires a huge number of cost and energy because of the necessity of having a lot of computing power [10].

The scheme proposed by Shahzad and Crowcroft [11] ensures reliability in their system as they used proof of completeness as well as process model to describe the system's architecture in detail. This scheme was able to provide privacy and anonymity as well as addressing security issues in a small-scale election. However, a few issues have been raised. As an example, this scheme requires the use of third-party to help in conducting the election. This increases the chance of data tampering, data leakage, and unfair tabulation of result in which could affect end-to-end verification [11]. Besides, the scheme also does not provide the voters with verifiability feature to ensure that their votes have been submitted and counted. Lastly, in terms of a larger scale election, polling procedure may experience some delay because of the block generation and sealing process.

The scheme proposed by Yi [12] was based on blockchain which also included approaches for enhancing security requirements of an e-voting system in a peer-to-peer network. This scheme employs distributed ledger that helps to prevent falsification and forgery of votes. They also used Elliptic Curve Cryptography (ECC) as their user credential model to provide non-repudiation and authentication features. However, this scheme found difficulty to implement a counter-measure assaults. This system requires the engagement of third-party to help in conducting the election. Thus, it does not properly suit a system with centralised usage involving multiple agents [12].

3 Proposed Solution

3.1 Ethereum

In this project, Ethereum technology will be used because it is one of the widely used and open-source platforms for developing decentralised-based application. Ethereum is one of the blockchain technology that utilises the peer-to-peer network without the need for central authority or any intermediary parties. Blockchain technology could be defined as cryptographically secure because of the nature of generated digital currency which is protected by complicated and complex mathematical algorithms. The algorithms used are incredibly difficult to crack and thus could be considered as a type of firewall. The act of cheating the system such as creating fake transactions and erasing transactions is almost impossible to be done. As a result, this degree of authority provided by blockchain is ideal for a variety of high-security projects such as e-voting.

Ethereum is a distributed network of computers running software or also could be known as nodes that validate blocks and transaction data. Node can be referred as a running piece of client software. To ensure Ethereum network's security and accuracy, client will act as the validator of all transactions in each block. In any related Ethereum project, client is needed in the computer to run node. In this proposed scheme, the voter will act as the client who verifies and validates the transaction occurs in the network.

There are two types of Ethereum network. The first type of network is public network. On a public blockchain, anybody may view or make transactions, and every transaction could be validated by everyone on the network. The second type of network is private network which works similarly to a public network except that its nodes are not connected to the main network. One type of private network is development network. In development network, developers test and run their Ethereum application to see how it works before deploying it. In this proposed scheme, private development network will be used to ensure smooth development of blockchain-based application.

The specification of the voting system is defined in smart contracts written in Solidity programming language. After the generation of smart contract, the compilation process will take place with the use of Solidity compiler tool. Then, migration tool will be used to deploy smart contract into the Ethereum network. Each voter in this system must have their own machine to use the voting system such as smartphone or tablet. The machine used must have the capability to access the internet to use the voting system that utilises Web3 and Truffle for interacting with the deployed voting contract. After that, voter will also need to have an Ethereum wallet or Metamask account to store and spend some amount of ETH to create a new vote block. This vote block will be connected to the previous existing vote block using Merkle Hash root function, which is one of the core concepts of Ethereum. Any update to the vote block will be updated to all the Ethereum nodes. The amount of ETH spent will be depending on the size of data within each created block. Every voter will be given different Ethereum address by the administrator. To use the system, they will be required to login to the system and Metamask wallet using the given address. After the election ends, the tallying process will take place and the result will be shown to all participants. The architecture of the proposed Ethereum based e-voting system is as depicted in Fig. 1.

3.2 *Tools and Languages*

Metamask. Metamask is a cryptocurrency wallet that allows a user to store ETH and use it to do any transaction in Ethereum network. When users first created their Metamask account, they will be provided with a pair of public and private key. The

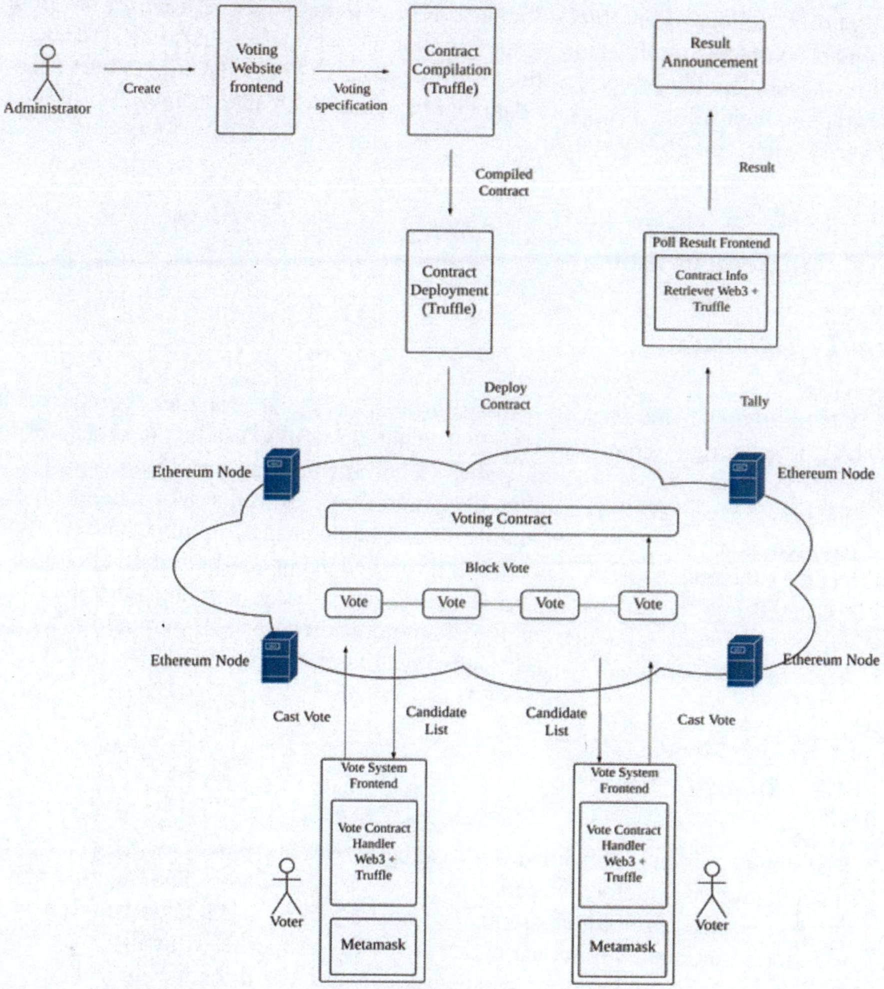

Fig. 1 System architecture of the proposed scheme

public key represents the voter's identity in the network while the private key acts as the access controller to stored ETH and any authentication process when using smart contracts.

Ganache. Ganache is part of the truffle suite ecosystem which acts as a high-end development tool used to run a local blockchain for Ethereum decentralised application development. In this project, Ganache user interface will be used to serve as a local blockchain node. Instead of downloading the whole blockchain node which require a high cost and time, development process will be much easier with the use of Ganache.

Solidity. Solidity is an object-oriented programming language used for developing smart contract. Simply said, a smart contract is a program that runs on Ethereum blockchain. In this project, Solidity will be used to develop voting smart contract which include all the necessary data and functions such as to vote as a voter and to verify voter's eligibility status.

4 Implementation

4.1 Database

The first part of this project is to ensure that only eligible voters can access and use the e-voting system. Thus, MySQL database will be used to store data of user. Two functions for authentication process are written in JavaScript language. The first function is to register user which will be done by the administrator. Three values are required which are user's Ethereum address, name, and IC number. The IC number and the Ethereum address of voter will be hashed using bcrypt function before being registered into the database. The second function is the authentication part. Voter will be required to input their name, Ethereum address, and IC number. If any of the required input is not valid, user's authentication process will fail.

4.2 Smart Contract

The smart contract is divided into several parts. The first part is related to the authentication process which required the use of MySQL database. The function will take three input from user which are name, phone number, and IC number. The address of voter is automatically set using one of solidity properties which is "msg.sender". It is the address on the user's Metamask account, and it cannot be changed since every transaction will need to be signed using the said address. If any of the input values does not match with the registered details on the database, the authentication process will fail. If the authentication succeeds, all the details will be stored in the struct of voter. The verified status is set to true indicating that the verification process is successful. For each new registered voter, the voter's count will be incremented by 1.

The second function is to set election details which could only be executed by administrator. The function will take five input values which are election title, election location, organisation title, start time, and end time. All these information will be updated to election details struct and election time struct. The start flag will be set to true indicating that the election has started, and the end flag will be set to false as its initial value as in Fig. 2.

```
function setElectionDetails(
    string memory _electionLocation,
    string memory _electionTitle,
    string memory _organizationTitle,
    string memory _timeStart,
    string memory _timeEnd
)
    public
    // Only admin can add
    onlyAdmin
{
    electionDetails = ElectionDetails(_electionLocation, _electionTitle, _organizationTitle);
    electionTime = ElectionTime (_timeStart, _timeEnd);
    start = true;
    end = false;
}
```

Fig. 2 Function to set election details

```
// Vote
function vote(uint256 candidateId, string memory voteTime, string memory voteReceipt) public {
    require(voterDetails[msg.sender].hasVoted == false);
    require(voterDetails[msg.sender].isVerified == true);
    require(start == true);
    require(end == false);
    candidateDetails[candidateId].voteCount += 1;
    voterDetails[msg.sender].hasVoted = true;
    voterDetails[msg.sender].vote_Time = voteTime;
    voterDetails[msg.sender].vote_Receipt = voteReceipt;
}
```

Fig. 3 Function to vote for candidate

The last part of this smart contract is the voting function. There are some requirements which needs to be fulfilled for the function to be executed successfully. The first requirement is voter's vote status should only be false which means that only voters who have not voted yet can vote for their choice of candidate. The second requirement is voter's verified status should only be true meaning that only voters who have already verified can proceed to do the voting process. The next requirement is the start flag should be true and the end flag should be false indicating that the election process is still on going. If all the requirements are fulfilled, the vote count for voter's chosen candidate will be incremented by 1. The vote status of the voter will be changed to true indicating that they have already voted and cannot vote again. The vote time and vote ID will also be updated to the struct of voter. Please refer Fig. 3.

4.3 Ganache and Metamask Setting

Ganache will provide 10 accounts with each account is prefilled with 100 amounts of Ether. The first address listed in Ganache will automatically be assigned to administrator and used to deploy the contract to blockchain. An instance of local blockchain will run on the computer with the use of Ganache. The private development network is configured on Metamask wallet and will be connected to Ganache. The configuration is as follows. The name of the network will be set as Election. The RPC URL server is set to https://localhost:7545. Lastly, chain ID is set to 1337. After configuring the network, accounts from Ganache can be imported into Metamask wallet by copying and pasting the private key. The account can be used throughout election process.

5 Result and Findings

Voters can do several functions such as authentication, voting, and view result. When users first enter the system, they need to go through authentication process to verify their eligibility status. Then, voters can proceed to the voting section and vote for their choice of candidate. List of candidates will appear along with their information such as name, organisation, and slogan. Every transaction requires the user to sign the message using their private key before it can be submitted. Figure 4 shows an example where voters need to sign the message transaction before submitting their vote. After the election has ended, voter can navigate to the result section to view the election result. The winner is shown in box along with their details on top of the page.

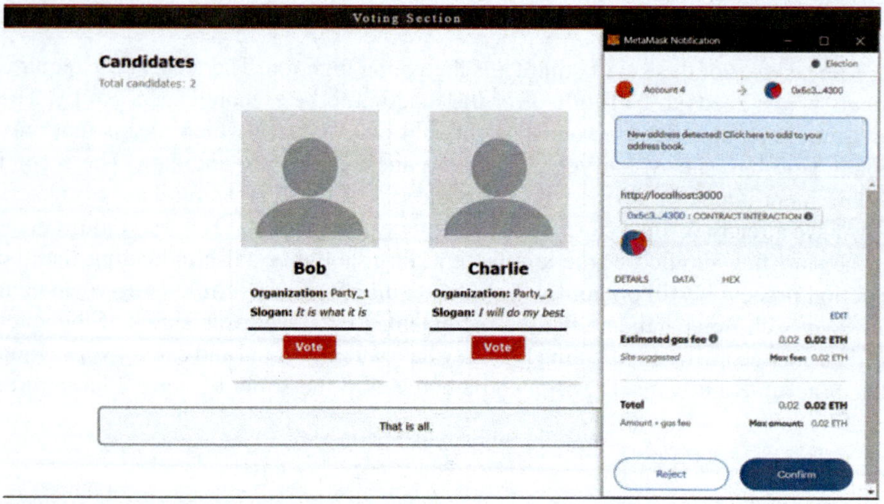

Fig. 4 Signing message before submitting vote

5.1 Security Fulfillment

This part will discuss the security fulfilment by the e-voting system using six properties which are authenticity, integrity, verifiability, anonymity, confidentiality, and non-repudiation. The security fulfilment of this proposed scheme will also be compared with the existing schemes reviewed in the Sect. 2.2.1 earlier. Please refer Table 2 for the summary of comparison.

The authentication is achieved through the use of MySQL database and unique Ethereum address. Every user in the system will be given unique Ethereum address before the start of election process. Before using the system, voter needs to sign in to the Metamask account with the address given by the administrator. During registration or authentication process, voter needs to input three values which are name, IC number, and Ethereum address. The address is automatically set by the system following the address on user's Metamask account. If any value of name, IC number or Ethereum address does not match with the registered details in MySQL database, authentication process will fail. Thus, authentication is achieved in this project.

Integrity of data can be maintained with the use of blockchain technology. It is known that blockchain is an immutable storage ledger which records and stores every transaction in a decentralised way. The technology able to maintain the integrity of data because of the way it chains each node hash value to each other in sequential order using Merkle root hash technique. Any change in the bottom node will also cause the change of hash in the node above it. Eventually, the root hash of the tree will also change. Thus, integrity of data is maintained in this Ethereum e-voting system.

The implementation of Ethereum blockchain in this project also helps to ensure universal verifiability and individual verifiability. The state of transparency in blockchain network will allows any transaction to be visible to all user on the network. Therefore, validation of transactions could be done by anyone presents on the Ethereum network. Each user could also check their transaction history to verify that the transaction is successful and recorded in the blockchain. Thus, verifiability feature is provided.

Voter anonymity is preserved using unique Ethereum address. Every user on the system is represented by different unique public key cryptography identifier. Thus, voter's information such as name and IC number is kept secret.

Blockchain technology has helped to ensure integrity and availability in the CIA triad. However, blockchain technology does not implement confidentiality as strictly as compared to the integrity and availability of the data stored. Due to the nature of transparency and decentralised of blockchain, the elements of data are viewable to all users on the blockchain network. Therefore, the secrecy component of blockchain technology is hard to be maintained. However, only user with valid access can view the data element thus maintaining the confidentiality of data.

Non-repudiation is achieved with the availability of digital signature in Ethereum transaction. Whenever transaction is sent by an account to Ethereum network, it needs to be sent along with a digital signature produced by the private key corresponds to

Table 2 Comparison of existing schemes and proposed scheme

Scheme	Protocol	Security requirements						
		Authenticity	Integrity	Verifiability	Anonymity	Confidentiality	Non-repudiation	No 3rd party involvement
[7]	Bitcoin	✓	✓	✓	✓	✓	✓	✗
[8]	Bitcoin	✓	✓	✓	✓	✓	✓	✗
[9]	Ethereum	✗	✓	✗	✓	✓	✓	✓
[10]	Ethereum	✓	✗	✓	✓	✓	✓	✓
[11]	Bitcoin	✓	✓	✗	✓	✓	✓	✗
[12]	Bitcoin	✓	✗	✓	✓	✓	✓	✗
Proposed scheme	Ethereum	✓	✓	✓	✓	✓	✓	✓

that Ethereum address. Since elliptic curve cryptography allows the verification of valid transaction by checking if the digital signature matches the transaction details and the Ethereum address to which access is being requested, users cannot deny the transaction comes from them. Therefore, non-repudiation is achieved.

5.2 *Smart Contract Evaluation*

In this part, the evaluation of the smart contract is carried out. Unlike previous practice, the evaluation process will use real public test network which is Rinkeby. This is to give a more accurate evaluation result in real-life environment than just evaluating it in development network. Remix platform will be used for the evaluation process. The cost required for voting transaction will be evaluated and compared with existing schemes.

The fee of the transaction of the proposed scheme for voting process is 0.000,117 ETH. The detail of transaction from Etherscan is depicted in Fig. 5. In comparison with other existing schemes proposed in [13] and [14], they require 0.000652 ETH and 0.000288 ETH, respectively to submit vote. This shows that the proposed scheme has the better efficiency as it requires the least ETH amount of computational cost to submit the vote transaction, as depicted in Table 3.

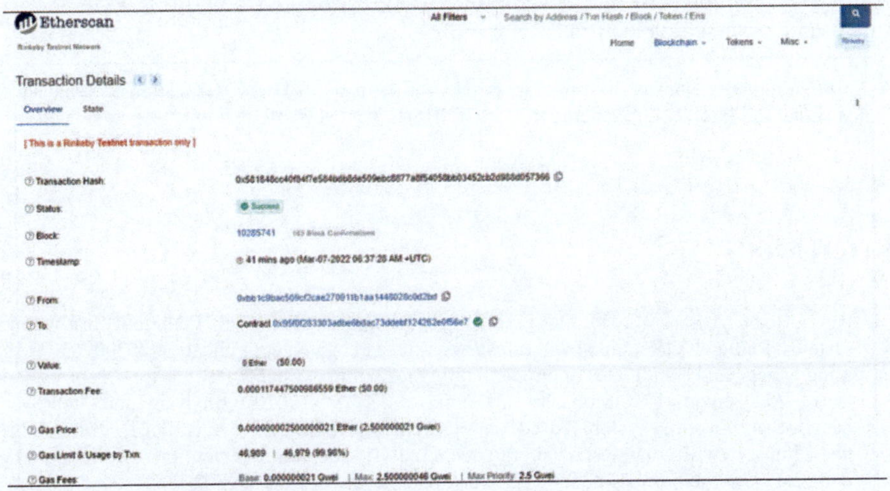

Fig. 5 Transaction detail for voting process

Table 3 Comparison of cost to submit vote between proposed scheme and existing schemes

Scheme	Cost to submit vote (ETH)
[13]	0.000652
[14]	0.000288
Proposed scheme	0.000117

6 Conclusion and Future Works

We proposed an e-voting scheme that utilises blockchain technology specifically the power of smart contract in Ethereum technology. This study argues that blockchain technology might have the potential to bring voting process to another level of secureness, thus fulfilling all the security requirements needed in e-voting system. In conclusion, this study can theoretically demonstrate and address the security requirements in an e-voting system such as authenticity, integrity of data, verifiability, voter's anonymity, confidentiality of data, and non-repudiation.

In continuation of this work, biometric mechanism will be used to provide a more accurate and stronger authentication method. As this study only focuses on the security requirements of e-voting system, it does not take scalability and reliability issue into consideration. Thus, future work will include more factors such as scalability, processing overhead, and energy efficiency to design a more secure and reliable e-voting system. Other blockchain-based application frameworks such as Hyperledger will also be explored to increase the possibility of adopting e-voting system in a small- and large-scale election.

Acknowledgements This work was supported by the Ministry of Higher Education of Malaysia's Fundamental Research Grant Scheme (FRGS/1/2018/ICT04/MMU/01/01).

References

1. Pan H, Hou E, Ansari N (2011, May) Ensuring voters and candidates' confidentiality in e-voting systems. Research Gate; unknown. https://www.researchgate.net/publication/252009636_Ensuring_voters_and_candidates'_confidentiality_in_Evoting_systems
2. Doost M, Kavousi A, Mohajeri J, Salmasizadeh M (2020) Analysis and Improvement of an e-voting system based on blockchain. Retrieved 16 Oct 2021, from undefined. https://www.semanticscholar.org/paper/Analysis-and-Improvement-of-an-E-voting-System-on-Doost-Kavousi/d11b3666841b59aa6b2f53ecc0a40bd159099302
3. Kho Y-X, Heng S-H, Chin J-J (2022) A review of cryptographic electronic voting. Symmetry 14(5):858. https://doi.org/10.3390/sym14050858
4. Bibiloni P, Escala A, Morillo P (2015) Vote validatability in mix-net-based e voting. Lecture notes in computer science, pp 92–109. https://doi.org/10.1007/978-3-319-22270-7_6
5. Aziz A (2019) Coercion-resistant e-voting scheme with blind signatures. In: 2019 Cybersecurity and cyberforensics conference (CCC). https://doi.org/10.1109/ccc.2019.00009
6. Al-Rawy M, Elci A (2019). Secure i-voting scheme with blockchain technology and blind signature. J Digit Sci (1):3–14. https://doi.org/10.33847/2686-8296.1.1_1

7. Cruz JP, Kaji Y (2017, March 21) E-voting system based on the bitcoin protocol and blind signatures. Retrieved 16 Oct 2021, from Research Gate website: https://www.researchgate.net/publication/317100187_E-voting_System_Based_on_the_Bitcoin_Protocol_and_Blind_Signatures
8. Bistarelli S, Mantilacci M, Santancini P, Santini F (2017, April 3). An end-to-end voting-system based on bitcoin. Retrieved 21 March 2022, from Research Gate website: https://www.researchgate.net/publication/317161233_An_end-to-end_voting-system_based_on_bitcoin
9. Shukla S, Thasmiya AN, Shashank DO, Mamatha HR (2018) Online voting application using ethereum blockchain. In: 2018 International conference on advances in computing, communications, and informatics (ICACCI). https://doi.org/10.1109/icacci.2018.8554652
10. Lai W.-J, Hsieh Y, Hsueh C.-W, Wu J.-L (2018) DATE: a decentralized, anonymous, and transparent e-voting system. In: 2018 1st IEEE International conference on hot information-centric networking (HotICN). https://doi.org/10.1109/hoticn.2018.8605994
11. Shahzad B, Crowcroft J (2019) Trustworthy electronic voting using adjusted blockchain technology. IEEE Access 7:24477–24488. https://doi.org/10.1109/access.2019.2895670
12. Yi H (2019, May 28). Securing e-voting based on blockchain in P2P network. Retrieved 21 March 2022, from ResearchGate website: https://www.researchgate.net/publication/333439952_Securing_e-voting_based_on_blockchain_in_P2P_network
13. Angsuchotmetee C, Setthawong P, Udomviriyalanon S (2019) BlockVOTE: an architecture of a blockchain-based electronic voting system. In: 2019 23rd International computer science and engineering conference (ICSEC). https://doi.org/10.1109/icsec47112.2019.8974826
14. Canessane RA, Srinivasan N, Beuria A, Singh A, Kumar BM (2019) Decentralised applications using ethereum blockchain. In: 2019 fifth international conference on science technology engineering and mathematics (ICONSTEM). https://doi.org/10.1109/iconstem.2019.8918887

Equipment Failure Analysis for Oil and Gas Industry with an Ensemble Predictive Model

ZhiYuan Chen, Olugbenro. O. Selere, and Nicholas Lu Chee Seng

Abstract In the oil and gas industry it has become of primary importance that assets are properly used and maintained as the cost of failure for these assets is extremely expensive. An ensemble predictive model has been proposed in this paper, which aims at improving the classification accuracy of the Sequential Minimal Optimization (SMO) Algorithm for equipment failure analysis in the oil and gas industry. SMO is used to train a Support Vector Machine (SVM) in properly classifying failure and normal instances from oil and gas industry equipment data. Recent applications of failure analysis have made use of the LibSVM technique for SVM implementation and in this paper we show that the proposed solution can perform much better when using the SMO. The proposed solution implements the ensemble approach to combine different classifiers so that the performance of the SVM training classifier is improved. The improved SMO approach is also a solution for the underperformance issue when dealing with imbalanced dataset. From the experiments, the best Model III: Stacked SMO using PART, Multilayer perceptron and Naïve Bayes correctly classified instances at 98.3018%. The classification performance of this predictive model is good, performing considerably better than the LibSVM and the other classifiers when applied to the pre-processed dataset with the unsupervised learning approach.

Keywords Artificial Intelligence · Sequential minimal optimization (SMO) · Support Vector Machine (SVM) · Ensemble learning · Class imbalance

Z. Chen (✉) · Olugbenro. O. Selere
School of Computer Science, University of Nottingham Malaysia, 43500 Semenyih, Malaysia
e-mail: Zhiyuan.chen@nottingham.edu.my

Olugbenro. O. Selere
e-mail: Khcx4oor@nottingham.edu.my

N. L. C. Seng
Tentacle Technologies, 50470 Kuala Lumpur, Malaysia
e-mail: nicklu@tentacletech.com

© The Author(s), under exclusive license to Springer Nature Singapore Pte Ltd. 2023
D.-K. Kang et al. (eds.), *Proceedings of the 9th International Conference on Computational Science and Technology*, Lecture Notes in Electrical Engineering 983,
https://doi.org/10.1007/978-981-19-8406-8_45

1 Introduction

In the oil and gas industry it has become of primary importance that assets are properly used and maintained as the cost of failure for these assets is extremely expensive. Failure or incidents are events that occur in rarity but when it happens, the consequence can be very harmful. Proper understanding of which way of using an asset can lead to failure is vital, as Oil and Gas industries need to address Health, Safety and Environment issues. Equipment monitoring at oil and gas plants can be a very difficult task, because of the varying conditions plant operators must face as regards to the specific equipment.

The purpose of this work is to develop a solution using machine learning techniques for data-driven failure analysis. The solution is to properly classify failure and normal events from oil and gas equipment data. The main classifier used towards implementing this solution is the Sequential Minimal Optimizer (SMO) which is used to train a support vector machine. This classifier was chosen because of the faster approach taken towards the quadratic programming used by the original SVM created by Vapnik (1992), and because the dataset used presented a class imbalance problem.

This class imbalance was a result of having more normal class labels in the data than the failure instances, and because of this, the SVM training classifier (SMO) needed to be optimized to handle such data. The optimization approach led to the development of the ensemble model using the ensemble learning approach called stacking. The ensemble approach was taken because of the need to ensure that the classification algorithm selected did not attempt to over-fit the data during training. Through the stacking approach, the SVM training classifier (SMO) would learn from the prediction approach used by the base classifiers to better predict the normal and failure events in the resulting dataset.

Three classifiers were selected for the stacking approach with the SMO: the random forest decision tree algorithm, the partial-decision trees (PART) rule-based algorithm, and the multilayer perceptron algorithm. These algorithms were selected based on their predictive performance with the pre-processed imbalanced dataset. The pre-processing was done with either oversampling using the synthetic minority oversampling technique (SMOTE) or under-sampling technique.

Through the stacking approach, the SVM training classifier (SMO) would learn from the prediction approach used by the base classifiers to better predict the normal and failure events in the resulting dataset. The resulting model was a better performer than the SMO on its own after conducting tests on the dataset without any sampling or cost sensitive learning technique applied to the dataset.

2 Literature Review

2.1 Predictive Analytics

Predictive analytics is a branch of machine learning. The notion behind predictive analytics involves making future predictions about events after conducting techniques encompassing machine learning and data mining on current and historical data [1]. The core of predictive analytics relies on capturing the possible relationships existing between variables and the predictive variables from past occurrences, to predict future outcomes [2]. Predictive modelling draws from statistics and optimization techniques to extract accurate information from large volumes of data. From such modelling can oil and gas personnel produce interpretable information that can be used to understand the implications of events, enabling them to act based on these implications [3].

In the oil and gas industry the system in place for conducting asset maintenance was the enterprise asset management system (EAM). Predictive analytics builds on this system by combining real-time data from sensor placed in equipments with historical data to predict potential asset failures. This then enables the move from a reactive mode of maintenance whereby equipment must fail before fixed, to proactive maintenance whereby equipment can be maintained before failure [3].

2.2 Applications of Failure Analysis

A review of literature revealed some recent application of failure analysis in the oil and gas industry. The work proposed by Saybani et al., shows how a machine learning technique called anomaly detection can be used in the detecting failure in sensor data and to predict the next occurrence of a sensor failure [4]. They highlight the fact that currently, human experts try to detect sensor failures or anomalies; they do this by watching and studying the data during production process, but they have limits to how much they can observe properly. Reducing equipment downtime, increasing reliability and availability of the equipments are considered as the most important strategical objectives, which can optimize the life cycle of the equipment [5]. It was considered that the costs associated with manufacturing design as fixed and predetermined, and therefore they suggested, to be competitive in the open market, the users have no other choice than optimizing life cycle of engines during their operation and maintenance. In the context of this paper, "engines" here corresponds to the equipment used in a refinery [5].

The focus of the work proposed by Patri et al., is to demonstrate how a time series analysis approach can be applied to failure detection and failure prediction from the streams of sensor data [6]. This method involves identifying shapelets which are short instances that are particularly distinct in the streams of sensor data. The streams of data obtained are from electrical submersible pumps, each instrumented with sensors that continually measure electrical properties of the pump. The shapelets approach is particularly appropriate for large oil and gas enterprise datasets because the algorithm does not access of the historical data. This greatly reduces the amount of data that needs to be stored for analysis. Moreover, unlike model-based approaches, shapelet-based analysis does not make any assumptions about the underlying nature of the data, making it practical for applications where a detailed physical model of the pump is not available. Examples of such are conducting experiments in a simulated laboratory [7]. The results from failure prediction and detection were then compared with other classifiers, and that showed that the time-series shapelets approach was a better performer in failure detection than the other classifiers but in failure prediction, outperformed by the SVM (rbf), AdaBoost, J48 and RF (random forest) classifiers. The method can detect failures only from intake pressure measurements with a precision of 89 and 100% recall. The accuracy of predicting failures is lower with a precision of 78% at 78% recall. Towards better prediction and detection, multiples measurement attributes can be considered.

Also, in the work proposed by Orrù et al., they proposed a machine learning-supervised algorithm for the fault diagnosis of rotating machinery in the oil and gas industry. The proposed system works with real-life historical data from process and equipment sensors mounted on the selected machinery. To validate the ML model, two different algorithms—the Support Vector Machine (SVM) and the Multilayer Perceptron (MLP) have been compared. This model performs rather well that an overall accuracy of 98.2% has been achieved [8]. Tewari et al. presented ensemble methods to combine Adaboost and Random Forest with the data re-sampling techniques to handle the complex drill bit selection process. K-nearest neighbors, naïve Bayes, multilayer perceptron, and support vector machine have been evaluated and the results show the random forest with bootstrap has given the most impressive performance for bit type selection with accuracy ranges from 92 to 99% [9].

3 Methodology

3.1 Unsupervised Learning

Supervised learning entails using a function to learn a mapping between a set of input variables and an output variable and applying this mapping function to predict the outputs for unseen data [10]. Unlike supervised learning, unsupervised learning involves drawing inference from unlabeled data. The method of unsupervised learning under focus is clustering analysis. Cluster analysis involves the creation of different categories containing objects of similar characteristics. In other words, cluster analysis aims at placing data objects into different groups depending on the degree of association between the data objects to determine if they belong to that group or not.

Various algorithms exist for conducting clustering analysis such as the popular k-means algorithm, but the clustering algorithm considered is the expectation maximization (EM). The EM algorithm approaches the clustering technique in a different way from the k-means algorithm by estimating the cluster to which an observation belongs by computing probabilities based on one or more probability distributions. The overall aim of this algorithm is to maximize the cluster an observation would belong to the given number of clusters to assign observations to. Because of this uniqueness from the k-means, the EM algorithm can also be applied to both continuous and categorical variables.

The EM algorithm requires us to iterate through the following two steps:

- The Expectation Step: Using the current best guess for the parameters of the data model, we construct an expression for the log-likelihood for all data, observed and unobserved, and then, marginalize the expression with respect to the unobserved data. This expression will be shown to depend on both the current best guess for the model parameters and the model parameters treated as variables.
- The Maximization Step: Given the expression resulting from the previous step, for the next guess we choose those values for the model parameters that maximize the expectation expression. These constitute our best new guess for the model parameters.

The output of the Expectation Step codifies our expectation with regard to what model parameters are most consistent with the data actually observed and with the current guess for the parameters provided we maximize the expression yielded by this step. We stop iterating through the two steps when any further change in the log-likelihood of the observed data falls below some small threshold [11].

3.2 Sequential Minimal Optimization

The sequential minimal optimization algorithm created by Platt [12] is a simple algorithm that can quickly solve the SVM Quadratic Programming (QP) problem without any extra matrix storage and without using numerical QP optimization steps at all. SMO decomposes the overall QP problem into QP sub-problems, using Osuna's theorem to ensure convergence.

To solve the QP problem below

$$\max_{\alpha} W(\alpha) = \sum_{i=1}^{l} \alpha_i - \frac{1}{2}\sum_{i=1}^{l}\sum_{j=1}^{l} y_i y_j k(\vec{x_i}, \vec{x_j})\alpha_i \alpha_j$$

$$0 \le \alpha_i \le C, \quad \forall i,$$

$$\sum_{i=1}^{l} y_i \alpha_i = 0.$$

The SMO aims to satisfy the Karush-Kuhn-Tucker (KKT) conditions [13]. Platt when describing KKT says "a point is an optimal point if and only if the Karush-Kuhn-Tucsker (KKT) conditions are fulfilled and $Q_{ij} = y_i y_j k(\overrightarrow{x_i}; \overrightarrow{x_j})$ is positive semi-definite. Such a point may be a non-unique and non-isolated optimum" [14]. These KKT conditions which are particularly simple, allow for the QP to be solved when for all i:

$$\begin{aligned} \alpha_i = 0 \quad &\Rightarrow y_i f(\overrightarrow{x_i}) \geq 1, \\ 0 < \alpha_i < C &\Rightarrow y_i f(\overrightarrow{x_i}) = 1, \\ \alpha_i = C \quad &\Rightarrow y_i f(\overrightarrow{x_i}) \leq 1. \end{aligned}$$

Unlike the methods for solving the SVM QP problem, SMO looks to solve the smallest possible optimization problem at every step. To do this the SMO chooses two Lagrange multiplier which it jointly optimizes, after which it finds the optimal values for these multipliers and updates the SVM with the new values obtained from the optimization step. The reason for selecting two Lagrange multipliers at every step is because in the standard SVM QP problem, the smallest possible optimization problem involves two Lagrange multipliers, and these multipliers must obey a linear equality constraint [14] (see Fig. 1).

For each method, three steps are illustrated. The horizontal thin line at every step represents the training set, while the thick boxes represent the Lagrange multipliers being optimized at that step. A given group of three lines corresponds to three training iterations, with the first iteration at the top.

What gives the SMO the advantage over other SVM training methods is the analytical way in which it solves the tow Lagrange multipliers. This analytic method makes the SMO avoid the numerical QP optimization step. This is because the SMO solves as many optimizations as possible sub-problems which makes it solve the overall QP problem very quickly [14]. Also, because the SMO doesn't use matrix algorithms, it is less likely to be susceptible to numerical precision problems [14].

The two Lagrange multipliers as shown in Fig. 1 [14] must fulfil all the constraints of the full problem. The inequality constraints cause the Lagrange multipliers to lie in the box. The linear equality constraint causes them to lie on a diagonal line.

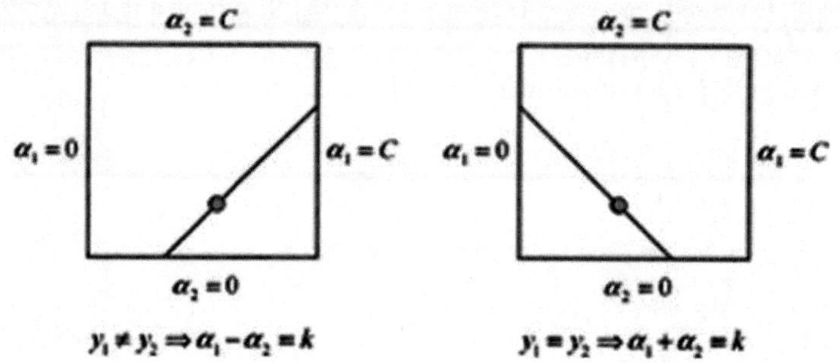

Fig. 1 Two lagrange multipliers

Therefore, one step of SMO must find an optimum of the objective function on a diagonal line segment [14].

3.3 Ensemble Learning

Ensemble learning is the process that combining classifiers to solve a computational intelligence problem. It is only used to improve the predictive performance of a model, or in other instances to reduce the likelihood of having selected a wrong model [15].

There are three common methods of ensemble learning which are boosting, bagging, and stacking. Bagging is one of the most intuitive and simplest to implement and tends to perform considerably well [16]. In boosting, resampling is strategically geared to provide the most informative training data for each consecutive classifier.

Stacking is the method of involving an ensemble of classifiers and its outputs are used as inputs to a second-level meta-classifier to learn the mapping between the ensemble outputs and the actual correct classes. In essence, when a particular classifier incorrectly misclassifies instances from a particular region because of incorrectly learning the region of the feature space, a second-level meta-classifiers can then learn from this behavior and that of other classifiers to correct such improper training [15]. Figure 2 illustrates the stacked generalization approach, where classifiers are trained using training parameters θ1 through θT to output hypotheses h1 through hT. The outputs of these classifiers and the corresponding true classes are then used as input/output training pairs for the second level classifier, CT + 1, as shown in Fig. 2.

4 Experimental Setup

4.1 Dataset Description

The dataset used contained 9 attributes and 875,000 instances. The data was collected from an online source for a study conducted on oil and gas pipes to determine the hazard levels. The dataset was unlabeled prompting the need for a machine learning approach to give class labels to the data before conducting predictions. Also, the dataset comprised of numeric attributes and the class label as the nominal attribute. There was no case of missing data in the dataset and the dataset contained continuous variables. The data recordings were taken for 15 days starting from 16/12/2014 to 31/12/2014, with features such as operating temperature ((°C), working pressure (psi), flow rate (cc/min) etc.

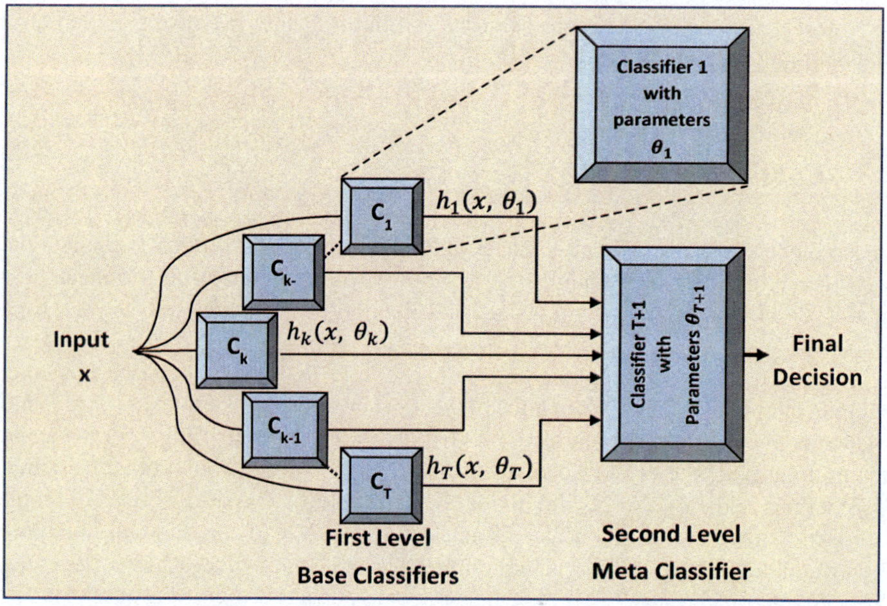

Fig. 2 Stacking technique

4.2 Classifiers Selected

Classifiers were selected based on the literature review and the reduction of variance and bias when learning from the training set. Therefore, Multilayer perceptron (MLP), Naïve Bayes (NB), PART, LibSVM, J48 and the Random Forest (RF) were selected. These classifiers were then used for comparison with the SVM training classifier (SMO) to measure the performance.

4.3 Data Preprocessing, Training and Evaluation

As the dataset was unlabeled, an approach was used to assign class labels to the data. This approach was an unsupervised clustering approach discussed in Sect. 3.1 known as the expectation maximization algorithm. The number of clusters was set to two for the failure and normal events and the resulting class distribution displayed a class imbalance problem with the normal class labels taking 87% of the data, and the failure class labels taking 13% of the data.

Considering the class imbalance problem, sampling techniques such as oversampling using SMOTE technique and under-sampling using the random sub-sample technique, were used to deal with the imbalance presented in the dataset. Another imbalance technique considered was the cost-sensitive learning technique.

Since the dataset containing over 870,000 instances is quite large, the dataset is split into training and test set using a 66% split. This means that the classifiers would be learning on about 574,200 instances and conducting classification on the remaining 295,800 instances. The split was selected to allow for the classifiers to learn more from the training set and have enough instances to conduct proper classification on. Cross Validation was considered for the evaluation of the dataset but because of hardware constraints which would lead to high computational complexity, the training and test set split seemed a much better option.

5 Results

The following Tables 1, 2, 3 and 4 show the results of the performance measure of the classifiers when oversampling and under sampling were applied to the dataset. Also, the tables show the performance measures after cost-sensitive learning using a cost matrix was applied to the dataset, and when no sampling or cost-sensitive learning was applied on the dataset.

The second set of experiments conducted to compare the performance of the ensemble models created when trained on the dataset without sampling and cost-sensitive learning technique applied to the dataset (Table 5). The models created are:

Table 1 Performance comparison of ML techniques with no sampling

Measure	J48	PART	MLP	NB	RF	LibSVM	SMO
TP rate	0.983	0.983	0.981	0.979	0.982	0.971	0.980
FP rate	0.090	0.086	0.057	0.032	0.081	0.028	0.059
Precision	0.983	0.983	0.981	0.981	0.982	0.975	0.981
Recall	0.983	0.983	0.981	0.979	0.982	0.971	0.981
F-measure	0.983	0.983	0.981	0.980	0.982	0.972	0.981
ROC	0.990	0.998	0.991	0.998	0.998	0.972	0.961

Table 2 Performance comparison of ML techniques with oversampling

Measure	J48	PART	MLP	NB	RF	LibSVM	SMO
TP rate	0.982	0.982	0.981	0.978	0.983	0.969	0.980
FP rate	0.039	0.031	0.022	0.012	0.034	0.011	0.014
Precision	0.982	0.982	0.982	0.980	0.983	0.973	0.981
Recall	0.982	0.982	0.981	0.978	0.983	0.969	0.980
F-measure	0.982	0.982	0.981	0.979	0.983	0.970	0.981
ROC	0.994	0.995	0.991	0.999	0.999	0.979	0.961

Table 3 Performance comparison of ML techniques with under sampling

Measure	J48	PART	MLP	NB	RF	LibSVM	SMO
TP rate	0.981	0.981	0.980	0.978	0.981	0.971	0.978
FP rate	0.019	0.019	0.020	0.021	0.019	0.029	0.021
Precision	0.981	0.981	0.980	0.979	0.981	0.972	0.979
Recall	0.981	0.981	0.980	0.978	0.981	0.971	0.978
F-measure	0.981	0.981	0.980	0.978	0.981	0.971	0.978
ROC	0.989	0.988	0.993	0.998	0.998	0.971	0.990

Table 4 Performance comparison of ML techniques using cost-sensitive learning with a cost matrix

Measure	J48	PART	MLP	NB	RF	LibSVM	SMO
TP rate	0.983	0.983	0.981	0.981	0.983	0.958	0.975
FP rate	0.111	0.113	0.071	0.072	0.102	0.180	0.115
Precision	0.983	0.983	0.981	0.981	0.983	0.957	0.975
Recall	0.983	0.983	0.981	0.981	0.983	0.958	0.975
F-measure	0.982	0.982	0.981	0.981	0.982	0.957	0.975
ROC	0.991	0.982	0.991	0.998	0.998	0.889	0.991

Table 5 Performance of ensemble models

Measure	Model I	Model II	Model III	Model IV	Model V
TP rate	0.983	0.982	0.983	0.983	0.982
FP rate	0.112	0.081	0.086	0.114	0.085
Precision	0.983	0.982	0.983	0.983	0.982
Recall	0.983	0.982	0.983	0.983	0.982
F-measure	0.982	0.982	0.983	0.982	0.982
ROC	0.936	0.998	0.998	0.934	0.993

- Model I: Stacked SMO using j48 and Multilayer perceptron
- Model II: Stacked SMO using Random Forest and Naïve Bayes
- Model III: Stacked SMO using PART, Multilayer perceptron, and Naïve Bayes
- Model IV: Stacked SMO using Random Forest and PART
- Model V: Stacked SMO using Random Forest, Naïve Bayes and Multilayer perceptron.

Upon comparison of the best performing ensemble model, the model is then compared with the other classifiers on the dataset when no sampling technique or cost-sensitive learning has been applied (Table 6).

Table 6 Performance of ensemble model compared with other machine learning techniques

Measure	J48	PART	MLP	NB	RF	LibSVM	SMO	Model III
TP rate	0.983	0.983	0.981	0.979	0.982	0.971	0.980	0.983
FP rate	0.090	0.086	0.057	0.032	0.081	0.028	0.059	0.086
Precision	0.983	0.983	0.981	0.981	0.982	0.975	0.981	0.983
Recall	0.983	0.983	0.981	0.979	0.982	0.971	0.981	0.983
F-measure	0.983	0.983	0.981	0.980	0.982	0.972	0.981	0.983
ROC	0.990	0.998	0.991	0.998	0.998	0.972	0.961	0.998

6 Discussion

The results from the experiments conducted show the performance of some classifiers to be above the rest when sampling and cost sensitive learning techniques have been applied to the dataset. The PART and Random Forest classifier perform the best when initially trained on the raw dataset. With oversampling and under-sampling applied the Naïve Bayes and Random Forest technique are best performing with regards to the ROC measure, although the Naïve Bayes has a lower true positive rate (tpr) on both occasions. When applying cost-sensitive learning to the classifiers the true positive rate of the Naïve Bayes increases even as the ROC measure remains the same indicating that this classifier is a strong predictor. The same can also be said of the Random Forest classifier although this classifier has a tendency for overfitting the positive (normal) instances.

The prompt to compare different stacking models was brought about by the lack of any significant distance between the performance measurements for the classifiers, apart from the LibSVM model which was the underperformer in all experiments conducted. From the Table 5 the two best performing models were model II and model III.

The better model was the model III as it correctly classified instances at 98.3018% compared to the 98.2312% from model II. The difference is not statistically significant because the model II has a lower false negative rate compared to model III. The reasons for model III having more correctly classified instances are because of the extra base classifier, because the more prediction outputs for the meta-classifier to learn from, the better the decision making. This is made evident when the ensemble model is compared against the other classifiers as shown in Table 6 when no sampling or cost-sensitive learning technique has been applied on the dataset. The model performs much better to the original SMO with a particular increase in the ROC value from 0.961 to 0.998. This increase can be attributed to the use of the stacking approach to improve the prediction capability of the SMO when learning from other classifiers.

7 Conclusion

In conclusion the review of literature has given insight into the notion of failure analysis and how it is being conducted by oil and gas personnel. Also, the recent applications of failure analysis in various industries have been discussed. The methodology has reviewed the various approaches considered towards the creation of the ensemble model such as the expectation maximization clustering technique for assigning class labels, the sampling and cost-sensitive learning techniques for addressing the class imbalance, and the ensemble learning approach used for the creating the ensemble model.

The objectives of this research have been achieved where it has been proved that the SMO technique can be used for analysis of oil and gas data than the libSVM technique. The created ensemble learning model using the stacking approach has also shown to be a better improved variant of the SMO technique. This clarifies the notion that ensemble models are better performers than original models on their own. The SMO training algorithm also shows it capability for use in failure analysis in the oil and gas industry.

Regarding any future implementation, although the classification has been addressed using the sample dataset obtained, the focus would be on applying this model to large oil and gas data, to determine if this model would be best used for analysis. This is because oil and gas data can be large, so a future approach would be to use a window-based approach to take segments of the oil and gas dataset and then conducting prediction on it.

Another future implementation would be the use of the SMO for the creation of a pre-emptive incident engine, which can then be used to build various predictive models to make predictions on real-time oil and gas dataset. Also, inference can also be drawn from the confusion matrix built from the predictive model as for how oil and gas experts can consider normal, failure events, and events which can lead to failure.

Acknowledgements We would like to thank the Tentacle Technologies Sdn Bhd Big Data Analytics Lab for the assistance of conducting experiments.

The preprint version of this paper has been published in [17] https://arxiv.org/abs/2012.15030. Doi = https://doi.org/10.48550/arXiv.2012.15030. Licensing terms: CC BY 4.0.

References

1. Strickland J (2015) Predictive analytics using R. Lulu.com
2. Eric S (2013) Predictive analytics: the power to predict who will click, Buy, Lie, or Die. Wiley
3. Nicholson R, Madden C, Feblowitz J, Bigliani R (2010) The role of predictive analytics in asset optimization for the oil and gas industry. IDC Energy Insights 1–5
4. Saybani MR (2011) Anomaly detection and prediction of sensors faults in a refinery using data mining techniques and fuzzy logic. Sci Res Essays 6(27):5685–5695

5. Singh R, Ogaji S (2006) Computational Intelligence in Fault Diagnosis. Springer, London, London
6. Patri P, Panangadan AV, Chelmis C, California S, Randall G (October 2014) Predicting failures from oilfield sensor data using time series shapelets. In: SPE annual technical conference, Netherlands, pp 27–29
7. Isa D, Rajkumar R, Woo KC (2007) Pipe error detection using support vector machines. In: 6th WSEAS international conference on circuits, systems, electronics, control & signal processing, Cairo, Egypt, pp 29–31
8. Orrù PF, Zoccheddu A, Sassu L, Mattia C, Cozza R, Arena S (2020) Machine learning approach using MLP and SVM algorithms for the fault prediction of a centrifugal pump in the oil and gas industry. Sustainability 12(11):4776
9. Tewari S, Dwivedi UD, Biswas S (2021) A novel application of ensemble methods with data resampling techniques for drill bit selection in the oil and gas industry. Energies 14(2):432
10. Cunnigham P, Cord M, Delaney SJ (2008) Supervised learning, machine learning techniques for multimedia. In: Cunningham P, Cord M (eds), Springer-Verlag Berlin, Heidelberg
11. Gupta MR, Chen Y (2011) Theory and use of the EM algorithm. Found. Trends® Sig Process 4(3):223–296
12. Platt JC (1998) Sequential minimal optimization: a fast algorithm for training support vector machines. Adv Kernel Methods Support Vector Learn 208:1–21
13. Karush W, W A (2008) Karush–Kuhn–Tucker conditions. Optimization 1–3
14. Platt JC (1998) Fast training of support vector machines using sequential minimal optimization. Adv Kernel Methods 185–208
15. Polikar R (2006) Ensemble based systems in decision making. Circuits Syst Mag IEEE 6(3):21–45
16. Breiman L (1996) Bagging predictors. Mach Learn 24(2):123–140
17. ZhiYuan C, Selere O, Seng NLC Equipment failure analysis for oil and gas industry with an ensemble predictive model. Preprint at https://arxiv.org/abs/2012.15030

Classification of Oil Palm Fresh Fruit Bunches (FFB) Based on Its Maturity Colour Using Convolutional Neural Network (CNN) Approach

Stephenie Carmeline Valentine, Rayner Alfred, Florence Sia Fui, Mohd Shamrie Sainin, and Zamhar Iswandono

Abstract It is very crucial to identify the maturity of Fresh Fruit Bunches (FFB) of oil palm to ensure the appropriate timing is set for harvesting them and thus maintaining a good quality of oil palm production. In Malaysia, Oil Palm Industry is one of the biggest industries and has contributed a large amount of the country's income. Therefore, it is important to check the maturity/ripeness of oil palm fruits before harvesting it. Currently, the grading of FFB based on its maturity is done manually by human and sometimes human can make mistake in performing the inspections. Hence, this paper proposes an automatic grading system of oil palm FFB using a deep learning approach, specifically by using Convolutional Neural Network (CNN). Most previously published works dealt with binary classification (Two classes classification). However, this work investigates the effectiveness of applying a deep learning approach to perform a multi-class classification task. In addition to that, not many works dealt with optimizing the deep learning model in classifying oil palm fresh fruit bunches (FFB) based on its maturity colour. Thus, this work will also evaluate the performance of the proposed CNN model with optimized training parameters that includes the number of epoch and batch size. Based on the results obtained, the number of epochs and the batch size have a great influence on the performance of the CNN models. For instance, having a high number of epoch's value of 100 and bigger batch size's value (e.g., 64), the CNN model can produce higher performance in term of accuracy.

S. C. Valentine (✉) · R. Alfred · F. S. Fui · M. Shamrie Sainin
Creative Advanced Machine Intelligence Research Centre, Faculty of Computing and Informatics, Universiti Malaysia Sabah, Kota Kinabalu, Sabah, Malaysia
e-mail: bi18110134@student.ums.edu.my

R. Alfred
e-mail: ralfred@ums.edu.my

F. S. Fui
e-mail: florence.sfs@ums.edu.my

M. Shamrie Sainin
e-mail: shamrie@ums.edu.my

Z. Iswandono
Higher Colleges of Technology, Abu Dhabi, United Arab Emirates
e-mail: zismail@hct.ac.ae

© The Author(s), under exclusive license to Springer Nature Singapore Pte Ltd. 2023
D.-K. Kang et al. (eds.), *Proceedings of the 9th International Conference on Computational Science and Technology*, Lecture Notes in Electrical Engineering 983,
https://doi.org/10.1007/978-981-19-8406-8_46

Keywords Oil palm · Fresh fruit bunches · Multi-class classification · Convolutional neural network · Deep learning approach

1 Introduction

Malaysia is one of the largest palm oil producers in the world. In 2020, Malaysia has produced 19.14 million tonnes of palm oil [6]. Malaysia also produced and exported 41% of world production and 47% of world exports [5]. Hence, the production of high-quality oil palm fruits is very important because most import countries of oil palm fruit are very concerned about its quality [14]. The oil palm industry is also one of the biggest contributors to the country's economic development. Therefore, the maturity identification of Fresh Fruit Brunches (FFB) is very necessary in order to produce high-quality oil palm fruit thus helping the country in increasing income for this industry. However, there are some problems faced while trying to produce good quality fruits. The main problem faced in the industry of palm oil production is the accuracy in distinguishing or classifying oil palm fruits according to its maturity [4]. Presently, the grading or classification of FFB is done manually by qualified graders (grading staffs of the mill or an agent appointed by the mill who has the capability and experience in the grading of FFB) [12]. Basically, the grading will be done in palm mills and graders need to follow the standard inspections given by the authorities. Unfortunately, this method of classification (manual grading) has some limitations. One of the biggest limitations is the human's ability to do examinations [19]. Different human will have different grading classification and they can sometimes make mistakes too while examining. Disputes can occur between the seller and the grader regarding the maturity and quality of the oil palm fruit as the maturity of FFB will be assessed through its color. As human beings, we view things differently and also similar to graders, they can also contradict with each other in terms of color classification of FFB.

Most previously published works are dealing with binary classification (Two classes classification). Thus, this work will investigate the effectiveness of applying a deep learning approach to perform the multi-class classification task. In addition to that, not many works are dealing with optimizing the deep learning model in classifying oil palm fresh fruit bunches (FFB) based on its maturity. Thus, this work will also investigate the effects of optimizing the parameters of the CNN models such as the number of epoch and the batch size. The expected output of this work will be an optimized prototype that can be used to perform the classification of the maturity of Fresh Fruit Bunches (FFB) of oil palm and thus overcome the limitations of manual grading system. Section 2 provides the detail of the literature review or related works. Section 3 presents the proposed approach, Sect. 4 presents the results and discussion and finally Sect. 5 concludes this paper.

2 Related Works

Much research on classification or grading system have been done since it is becoming a future trend because it has high potential for the study of new approaches [13]. There are several efforts done by some researchers to ensure the efficiency of the grading or classification of oil palm FFBs by using different approaches, such as Convolutional Neural Network (CNN) [18]. The study aimed to classify FFBs into two categories which are categorized into ripe and unripe bunches. They have trained the deep CNN to classify 628 images and the result showed that the classification has 98% accuracy. A grading system of oil palm FFB ripeness has been developed using supervised machine learnings that includes Artificial Neural Network (ANN), K-Nearest Neighbours (KNN), and Support Vector Machine (SVM) [3]. Basically, the processes involved are shown in Fig. 1. The performance of CNN depends on the number of training data and the number of layers used in the CNN model [9]. It was suggested that using AlexNet, a pre-trained CNN was suitable for classification tasks where a huge amount of data was not available and also for tasks in which higher processing time was not an issue. The CNN algorithm is considered as a good approach for the FBBs grading system [18]. There are other related projects that include the grading of maturity of oil palm fruits based on visible and NIR Band [17], palm oil fresh fruit bunch ripeness grading identification using colour features with ANN approach [16], and evaluations of oil palm FFB maturity degree using Multiband Spectrometer with ANN approach [20].

However, there are gaps based on the studies conducted by those researchers. Some researchers used ANN model and others to grade FFBs while most used CNN models. Based on the reviews, deep learning approach using CNN is the most suitable method for classifying FFBs ripeness. But there are some enhancements that need to be done to increase the efficiency of the grading system. For example, the research by [2] has already achieved 98% accuracy but it only has two categories of ripeness. Another class of ripeness can be added which is the overripe FFB. In other words, this work deals with a multi-class classification and focuses especially on optimizing the training parameters such as number of epoch and batch size to improve the classification of the oil palm Fresh Fruit Bunches (FFB) based on its maturity colour using a deep learning approach.

In this study, CNN will be used for the multi-class classification task. A research was done by Anupama Jawale and Ganesh Magar in 2019 to compare the image classification techniques which are Binary and Multiclass using CNN and SVM [10]. What they did was basically comparing CNN and SVM in terms of accuracy of classification images. Based on their findings, SVM is a better choice of algorithm for binary image classification with 80.95% accuracy while for CNN the accuracy was only 75%. But for Multiple Class classification, CNN shows higher accuracy than SVM. Similarly, the CNN model is an effective and most suitable solution for multi-class images classification task where the Precision, Error, and Accuracy rates are 90.81%, 4.25%, and 95.07% respectively [7].

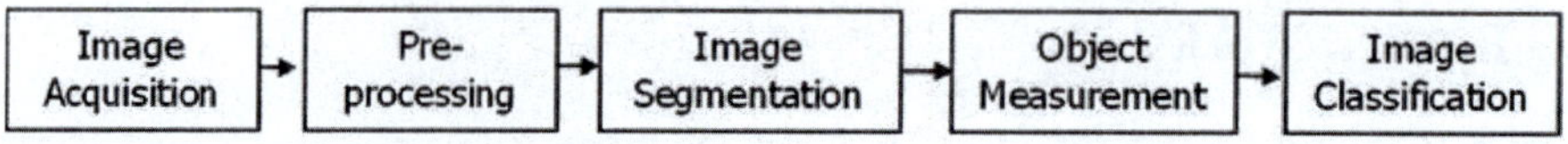

Fig. 1 Common image processing system configuration

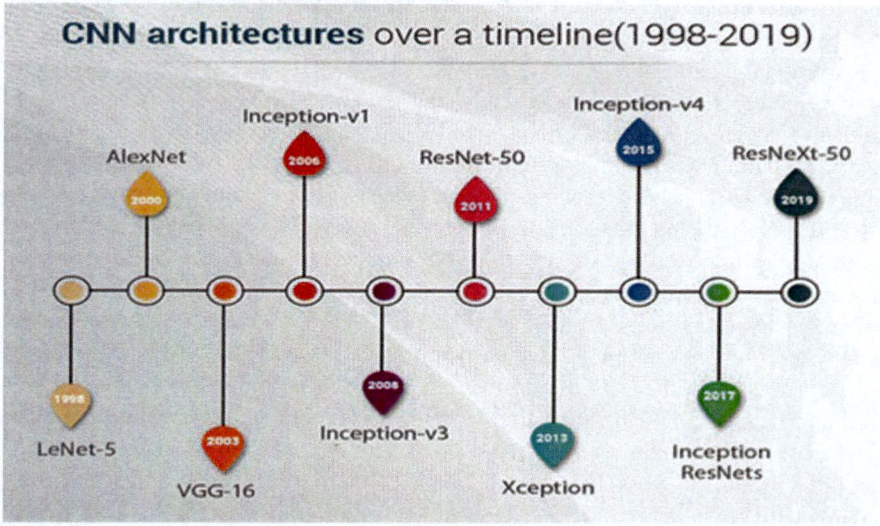

Fig. 2 CNN architectures over a timeline (1998–2019) [1]

In order to improve the classification of FFBs based on its maturity colour using a deep learning approach, optimizing the training parameters has to be done. Therefore, a suitable optimizer should be chosen. There are several choices of optimizers such as Stochastic Gradient Descent (SGD), Momentum, RMSProp, Adam, AdaGrad, and AdaDelta. The effectiveness of each optimizer depends on the CNN architectures. Figure 2 shows the CNN architectures over a timeline from 1998–2019 [1]. Adaptive Moment Estimation (Adam) optimizer is often used as the default optimizer in machine learning. This optimizer has several advantages, including the fact that it is simple to build and requires minimal tuning, that it is computationally efficient, memory efficient, and has low memory needs, and that Adam performs well in instances with huge datasets and high parameter values. The best combination of CNN Architecture-Optimizer is the VGG16-Adam combination. Several optimizers were used for VGG16 Architecture Network including NAG, AdaGrad, RMSProp, Adam, Adamax, and Nadam [11]. The Adam optimizer was found to be the most stable optimizer. It is because it has high results and low variance between learning rates. Three optimizers were compared which were Adam, RMSProp, and SGD and it was found that Adam was the most often used optimizer with the optimal learning rate between 0.00005 and 0.0001 [15].

3 Methods

3.1 Datasets

In this work, images of FFBs with its actual ripeness level are acquired from Kaggle. The total images acquired for the dataset are 413 and the images are stored in a folder named "Dataset" and they are separated into three sub-folders which are "Training", "Testing", and "validation". For each sub-folders there will be another three folders named "Ripe", "Unripe", and "Overripe". The name of the sub-folders indicates the actual ripeness level of the FFB images. The dataset is retrieved using the `flow_from_directory` method. The target size (128, 128), batch size (32), classes ('Overripe', 'Ripe', 'Unripe') and class mode 'categorical' are defined respectively. The datasets are split into three sets which are training set, testing set, and validation set.

For the image pre-processing, the FFN images are manually cropped to remove unneeded parts of the image such as other objects so that only the image of FFB is kept. Next, image data augmentation is used to expand the training, testing, and validation dataset in order to improve the performance of the proposed deep learning model to generalize. `ImageDataGenerator` class which is supported by the Keras library will be used for image data augmentation. The augmentation techniques that are used in this work includes shifting, rotations, zooming, and flipping. This class also provides a real-time data augmentation where it generates augmented images even when the CNN model is still in the training stage. Figures 3, 4 and 5 show the sample images of FFBs for each class or ripeness level. Each class will have 10 sample images.

Fig. 3 Overripe FFBs images

Fig. 4 Ripe FFBs images

Fig. 5 Unripe FFBs images

3.2 *Designing the Convolutional Neural Network (CNN) Architecture*

CNN has three main layers and every layer has its own function. These layers are the convolution layer and pooling layer, which will deal with the feature extraction while the fully connected layer is used for classification task or predicting the output. Hence, image of FFB will be used as the input and CNN model is expected to classify the ripeness level of the FFB based on the given image.

Convolutional Layer Convolutional layer accepts an RGB picture or an output from another layer. In order to produce a feature map, the incoming data is converted into pixel values. This map represents low-level features such as edges and curves. A sequence of extra layers of convolution may be used to find higher level unique characteristics. The image is modelled as a volume with the dimensions Width × Height × Depth (Fig. 6).

1. The number of filters is a configurable parameter (k). More filters are required in a more sophisticated model. To create an activation feature map, depth 3 convolutional filters are dragged over the screen.

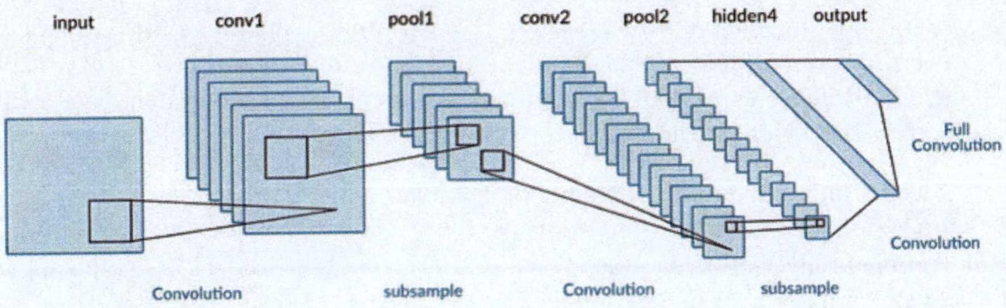

Fig. 6 Architecture of CNN

2. Depending on the application, the spatial scope of the filter (width and height) can be changed. The most common filter sizes are 3×3, 5×5, and 7×7. When a $3 \times 3 \times 3$ filter is applied to a $32 \times 32 \times 3$ picture, the resulting image is $30 \times 30 \times 1$. This is due to one pixel being deleted from the image's left, right, top, and bottom.
3. If S = 1, the kernel or filter window will be slid by one pixel by default. Additionally, the window may be moved by more than one pixel at a time. The stride is defined as the number of pixels slid. Because more sites are bypassed, larger strides result in less overlap between the receptive fields and the smaller feature map. As a result, the feature map is smaller than the input.
4. By injecting a number of zeros along the image's border, padding is employed to keep the dimension of the feature map the same as the input. As a result, it's also known as zero padding. The number of zeros added is determined by the size of the convolution filter. If the filter size is set to 3, the number of zeros added will be one, according to the formula.

Pooling Layer A down sampling layer is placed after the activation layer to lower the spatial dimension while maintaining the depth constant. A 2×2 filter is frequently used to the input in order to generate an output based on the pooling type. Maximum or average pooling can be used, in which case the maximum or average value from each filter-delimited subarea is used. There are no pooling options. The pooling layer lowers the feature map's length and breadth but not its depth. As a consequence, the number of parameters and weights, as well as the training time and computing cost, are decreased. In addition, overfitting is avoided. When a model achieves 100% or 99% accuracy on the training set but only 50% accuracy on the test data, this is referred to as overfitting. Dropout layers, which remove a random set of activations by changing the value to 0, can be used to circumvent this issue. Dropout is a pattern generalisation function that discovers several representations of the same pattern.

Fully Connected Layer The Fully Connected layer identifies extremely high-level characteristics that are closely linked with a certain item or class. A fully connected layer is fed a set of characteristics for categorizing photographs that do not take into consideration the spatial structure of the images as input. Most of the models have two totally connected layers. The output of the fully connected layer is a 1D vector created by flattening the preceding pooling layer's output. Flattening is the technique of converting a three-dimensional volume into a one-dimensional vector.

The initialization of the model can be made by specifying the model as a sequential model and these layers are added.

1. 1 × Convolution layer of 32 channel of 3 × 3 kernel
2. 1 × Max pooling layer of 2 × 2 pool size
3. 1 × Dropout layer of 0.25 value
4. 1 × Convolution layer of 32 channel of 3 × 3 kernel
5. 1 × Max pooling layer of 2 × 2 pool size
6. 1 × Dropout layer of 0.25 value
7. 1 × Convolution layer of 32 channel of 3 × 3 kernel
8. 1 × Max pooling layer of 2 × 2 pool size
9. 1 × Dropout layer of 0.25 value.

The ReLU (Rectified Linear Unit) is also added as the activation to each layer to ensure that negative values are not passed to the next layer. After creating all the convolution layer, flatten layers are added to make the multidimensional input to become a one-dimensional and also for the transition from the convolution layer to the fully connected layer. Next, the dense layers are added as shown below:

1. 1 × Dense layer of 64 layers, using ReLU as activation
2. 1 × Dropout layer of 0.25 value
3. 1 × Dense Softmax layer of 3 units.

The ReLU activation is used for the dense layer of 64 units and for the dense layer that has the softmax activation, and only 2 units are used since there are only 3 classes to predict which are ripe, unripe, and overripe. The model is finally done after the creation of the softmax layer. Next, the Adam Optimizer is used it is proven to have the highest accuracy when applied with CNN model. The optimizer is also defined with the learning rate of 0.01. After that, the model will be complied using `model.compile()` function. In the training process, the number of epochs is set to 100 with 64 batch size. Figure 7 outlines the coding used for creating the CNN model. The `Model.summary()` function is used to summarize the model that has just been created. In Fig. 8, there are three layers of convolution layer, three layers for max pooling layer, and one fully connected layer. Every layer has dropout layer and the activation layer used are ReLU and Softmax in the last layer.

In this work, the CNN model parameters settings are varied in order to investigate which setting produces the highest accuracy performance. Basically, the parameters that are varied are the epochs (e.g., 25, 50, 75 and 100) and batch size (e.g., 32, 64) while the other settings will be constant including:

- Height = 128
- Width = 128
- Learning rate = 0.01
- Adam optimizer
- CNN Layers

```
Model = Sequential()
#Conv-relu-pool
Model.add(Conv2D(32, (3, 3), input_shape=(IMAGE_SIZE, IMAGE_SIZE, 3), activation="relu"))
Model.add(MaxPooling2D(pool_size=(2, 2)))

Model.add(Conv2D(32, (3, 3), activation="relu"))
Model.add(MaxPooling2D(pool_size=(2, 2)))
Model.add(Dropout(0.25))

Model.add(Conv2D(64, (3, 3), activation="relu"))
Model.add(MaxPooling2D(pool_size=(2, 2)))
Model.add(Dropout(0.25))

Model.add(Flatten())

Model.add(Dense(units= 64, activation="relu"))
Model.add(Dropout(0.5))

Model.add(Dense(units =3, activation = 'softmax'))

Model.compile(optimizer="adam", loss="categorical_crossentropy", metrics=["accuracy"])


h=Model.fit(training_set,epochs=100,batch_size=64, validation_data=val_set, shuffle=True)
Model.save("C:/Users/USER/Downloads/FYP DAE/FFB-CNN.h5")
```

Fig. 7 Codes for creating the CNN model

Table 1 Evaluation results of the trained CNN models

CNN model	Epochs	Batch size	Train Acc	Test Acc	Valid Acc
1	25	32	0.936	0.864	0.7391
2	50	32	0.954	0.268	0.652
3	75	32	0.967	0.878	0.783
4	100	32	0.967	0.878	0.826
5	25	64	0.921	0.878	0.783
6	50	64	0.961	0.780	0.826
7	75	64	0.967	0.902	0.652
8	100	64	0.978	0.927	0.867

- Dropout Regularization
- Activation Layer including ReLU and Softmax.

4 Results and Discussion

Based on the results shown in Table 1, it shows that by using Batch Size of 64 and Epochs of 100, the constructed CNN model 8 can provide the highest result which is 0.978 for Training Accuracy, 0.927 for Testing Accuracy, and 0.867 for Validation Accuracy. Figure 9 depicts the comparison of training, testing and validation accuracies for all the CNN models.

```
Model: "sequential_3"
_________________________________________________________________
 Layer (type)                Output Shape              Param #
=================================================================
 conv2d_9 (Conv2D)           (None, 126, 126, 32)      896

 max_pooling2d_9 (MaxPooling  (None, 63, 63, 32)       0
 2D)

 dropout_11 (Dropout)        (None, 63, 63, 32)        0

 conv2d_10 (Conv2D)          (None, 61, 61, 32)        9248

 max_pooling2d_10 (MaxPoolin  (None, 30, 30, 32)       0
 g2D)

 dropout_12 (Dropout)        (None, 30, 30, 32)        0

 conv2d_11 (Conv2D)          (None, 28, 28, 64)        18496

 max_pooling2d_11 (MaxPoolin  (None, 14, 14, 64)       0
 g2D)

 dropout_13 (Dropout)        (None, 14, 14, 64)        0

 flatten_3 (Flatten)         (None, 12544)             0

 dense_6 (Dense)             (None, 64)                802880

 dropout_14 (Dropout)        (None, 64)                0

 dense_7 (Dense)             (None, 3)                 195

=================================================================
Total params: 831,715
Trainable params: 831,715
Non-trainable params: 0
_________________________________________________________________
```

Fig. 8 Summary of CNN model

5 Conclusion

This work investigated the effectiveness of applying a deep learning approach to perform a multi-class classification task to classify the degree of ripeness for the oil palm FBBs (e.g., overripe, ripe, unripe) based on its maturity. The number of epochs and the batch size of the CNN model influenced the performance of the CNN models. Having high number of epochs' value of 100 and bigger batch size value (e.g., 64), the CNN model can produce higher performance in term of accuracy. For future works, there are several suggestions that can be made to ensure that the output of this work can be further improved. These suggestions include using larger dataset and also using a Cross Validation to help reducing overfitting of the model.

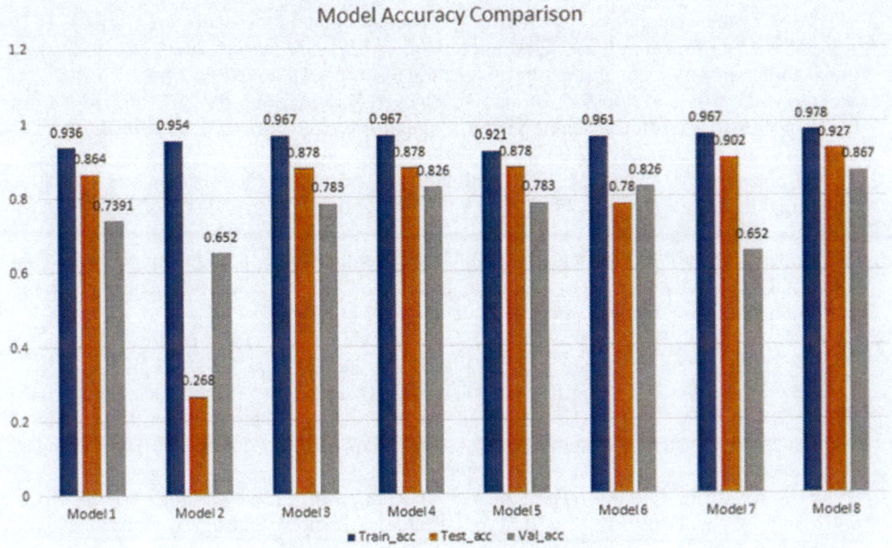

Fig. 9 Comparison of training, testing and validation accuracies for all the CNN models

References

1. CNN architectures over a timeline (1998–2019). https://www.aismartz.com/blog/cnn-architectures/
2. FFB grading guideline. http://www.mjmpom.com/ffb-grading-guideline/
3. Alfatni MSM, Shariff ARM, Bejo SK, Saaed OMB, Mustapha A (2018) Real-time oil palm FFB ripeness grading system based on ANN, KNN and SVM classifiers. IOP Conf Ser: Earth Environ Sci 169:012067. https://doi.org/10.1088/1755-1315/169/1/012067
4. Bensaeed OM, Shariff AM, Mahmud AB, Shafri H, Alfatni M (2014) Oil palm fruit grading using a hyperspectral device and machine learning algorithm. IOP Conf Ser: Earth Environ Sci 20:012017. https://doi.org/10.1088/1755-1315/20/1/012017
5. Chaudhari D, Waghmare S (2022) Machine vision based fruit classification and grading—a review. In: Kumar A, Mozar S (eds) ICCCE 2021. Springer, Singapore, pp 775–781
6. Chu MM. Malaysia's palm oil board cuts 2021 production outlook. https://www.nasdaq.com/articles/malaysias-palm-oil-board-cuts-2021-production-outlook-2021-09-07
7. Ezat WA, Dessouky MM, Ismail NA (2020) Multi-class image classification using deep learning algorithm. J Phys: Conf Ser 1447(1):012021. https://doi.org/10.1088/1742-6596/1447/1/012021
8. huseinzol05: Huseinzol05/Malaya: natural language toolkit for Bahasa Malaysia. https://malaya.readthedocs.io/. https://github.com/huseinzol05/malaya
9. Ibrahim Z, Sabri N, Isa D (2018) Palm oil fresh fruit bunch ripeness grading recognition using convolutional neural network. J Telecommun Electron Comput Eng 10:109–113
10. Jawale A (2019) Comparison of image classification techniques: binary and multiclass using convolutional neural network and support vector machines. INFOCOMP J Comput Sci 18(2):28–35. https://infocomp.dcc.ufla.br/index.php/infocomp/article/view/618
11. Kandel I, Castelli M, Popovič A (2020) Comparative study of first order optimizers for image classification using convolutional neural networks on histopathology images. J Imaging 6(9). https://doi.org/10.3390/jimaging6090092

12. Makky M (2016) A portable low-cost non-destructive ripeness inspection for oil palm FFB. Agric Agric Sci Procedia 9:230–240. https://doi.org/10.1016/j.aaspro.2016.02.139. International conference on food, agriculture and natural resources, IC-FANRes 2015
13. Makky M, Soni P. Development of an automatic grading machine for oil palm fresh fruits bunches (FFBS) based on machine vision. Comput Electron Agric 93:129–139. https://doi.org/10.1016/j.compag.2013.02.008
14. Mba OI, Dumont MJ, Ngadi M (2015) Palm oil: processing, characterization and utilization in the food industry—a review. Food Biosci 10:26–41. https://doi.org/10.1016/j.fbio.2015.01.003
15. Podgorelec V, Pečnik p, Vrbančič G (2020) Classification of similar sports images using convolutional neural network with hyper-parameter optimization. Appl Sci 10(23). https://doi.org/10.3390/app10238494. https://www.mdpi.com/2076-3417/10/23/8494
16. Sabri N, Ibrahim Z, Syahlan S, Jamil N, Mangshor NNA (2018) Palm oil fresh fruit bunch ripeness grading identification using color features. J Fundam Appl Sci 9:563–579
17. Saeed OMB, Sankaran S, Shariff ARM, Shafri HZM, Ehsani R, Alfatni MS, Hazir MHM (2012) Classification of oil palm fresh fruit bunches based on their maturity using portable four-band sensor system. Comput Electron Agric 82:55–60. https://doi.org/10.1016/j.compag.2011.12.010
18. Saleh A, Liansitim E (2020) Palm oil classification using deep learning. Sci Inf Technol Lett 1(1):1–8. https://doi.org/10.31763/sitech.v1i1.1. https://pubs2.ascee.org/index.php/sitech/article/view/1
19. Shabdin MK, Shariff ARM, Johari MNA, Saat NK, Abbas Z (2016) A study on the oil palm fresh fruit bunch (FFB) ripeness detection by using hue, saturation and intensity (HSI) approach. IOP Conf Ser: Earth Environ Sci 37:012039. https://doi.org/10.1088/1755-1315/37/1/012039
20. Yarak K, Witayangkurn A, Kritiyutanont K, Arunplod C, Shibasaki R (2021) Oil palm tree detection and health classification on high-resolution imagery using deep learning. Agriculture 11(2). https://doi.org/10.3390/agriculture11020183

Copy-Move Forgery Detection Using Canny Edge Detector and SIFT-Based Blob Analysis

Bashir Idris, Lili N. Abdullah, Mohd Taufik Abdullah Selimun, and Alfian Abdul Halin

Abstract This work presents a copy-move forgery detection method based on blobs detection from canny edge-response images, followed by SIFT (Scale Invariant Feature Transform) key points analysis. The SIFT key points are extracted from grayscale images, which are then clustered according to their distance from within generated blobs of that image. Different key points in different blobs are ignored (unique non-duplicated key points) while the same key points in different blobs are matched. The assumption made here is that these duplicate regions are highly indicative of forgery. The proposed method is tested on the MICC-F220, MICC-8Multi, and CoMoFoD public benchmark datasets. The results demonstrate that copy-move forgery detection can be reliably performed on geometrically transformed images, i.e., rotated, scaled, combined and post-processed images such as JPEG compression, blurred, noise added, and contrast adjusted. Our method is also robust on multiple copy-move forgeries.

Keywords Canny edge detector · SIFT · Blobs · Copy-move · Forgery · Copy-move forgery

---

B. Idris · L. N. Abdullah (✉) · M. T. A. Selimun · A. A. Halin
Faculty of Computer Science and Information Technology, Universiti Putra Malaysia, 43400 Serdang, Selangor, Malaysia
e-mail: liyana@upm.edu.my

B. Idris
e-mail: gs54656@student.upm.edu.my

M. T. A. Selimun
e-mail: taufik@upm.edu.my

A. A. Halin
e-mail: alfian@upm.edu.my

© The Author(s), under exclusive license to Springer Nature Singapore Pte Ltd. 2023
D.-K. Kang et al. (eds.), *Proceedings of the 9th International Conference on Computational Science and Technology*, Lecture Notes in Electrical Engineering 983,
https://doi.org/10.1007/978-981-19-8406-8_47

1 Introduction

In image forensics, forgery detection in digital images has become a hot topic within the research community [1]. This in part is due to the vast improvements in the capabilities of image editing tools. For this reason, the profound legitimacy and trustworthiness of images are at stake, especially in major image application areas such as forensic investigation, criminal examination, journalism, surveillance system, law enforcement, insurance claim and medical imaging [2, 3]. Instead of the primary purpose of images for ascertaining visual confidence, especially during court judgements, they are now used as tools for fraud, false presentations, and political propaganda [4].

The two main approaches for image forgery detection include active and passive approaches [5]. The active approach requires an already pre-coded signature or watermarks, which, when decoded, can reveal manipulation of the image [6]. Passive techniques, which are arguably more useful, function without prior knowledge of the image [7]. Block-based matching and key point-based techniques are two different ways to implement the passive approach. Due to the numerous block comparisons needed, the block-based approaches are computationally expensive despite being non-affine transformation independent [8]. Key point-based techniques determine the most important or prominent pixels in an image. Hence, they are said to be effective at detecting forgeries in rotated, scaled, or even JPEG compressed images [8]. Since key points are described using specific features, the disadvantage of comparing blocks in the block-based method is now reduced to only comparing key point descriptors.

As for the bigger domain of image forgery types, the two most common are splicing and copy-move. In splicing, two opposite images are combined to form one representation of the images [9]. On the other hand, copy-move forgery is when a portion of an image is copied and pasted into another region of the same image [10, 11]. However, because the copied-pasted region is from the same image, the characteristic of the copied portion is interwoven with the background region. This makes copy-move forgery difficult to detect without proper forensic measures. The above reasons motivated this research to focus on passive detection of copy-move forgeries.

Scale and rotation independent features are required for reliable forgery detection since forgeries often include transformations. As a result, key point-based algorithms are more suited for forgery detection since they create fewer feature vectors by selecting and identifying high-entropy key points [12]. However, one issue with key point-based methods is that the copied regions are sparsely covered by matched key points. If only a small portion in the image covers the copied regions, the region may be completely missed. Key point-based algorithms also ineffectively detect a forgery in smooth regions since they are more sensitive to textured regions (where most of the detected key points are in high contrast regions [12]). Sharp images can guarantee the detection of adequate key points to match, while smooth regions sparsely allow a small number of key points to be detected. In textured areas, filtering algorithms usually require selecting the salient feature points to reduce outlier formation [11].

In our work, the intuition is to ensure key points are generated from any image under consideration. Due to this, we propose to apply the canny edge detector as the initial pre-processing step. The canny operator is known to suppress noise while exposing edges. This in turn allows for key points in regions to be generated. We also include SIFT (Scale Invariant Feature Transform) and blobs analysis (hence the term CannySIFT + Blob) for the final forgery detection process.

2 Related Works

Copy-move forgery (CMF) is the simplest and consequently most common type of image forgery, which has made it one of the most studied areas in forgery detection [13]. CMFs are normally detected either using block- or key point-based approaches.

Block-based approaches are complicated since all features are exhaustively searched. This CMF detection (CMFD) approaches however can be improved where FMT, DWT, DCT, SVD, HOG, and Zernike Moment are some of the proposed improvements in terms of features for block-based methods [1, 14, 15]. Fridrich et al. [16] method divides the image into small overlapping chunks and extracts DCT coefficients as features. The retrieved features were lexicographically sorted and compared for forgery detection. This approach uses 64 quantized DCT coefficients per block and an exhaustive search to detect duplicate regions. Popescu and Farid [17] employed PCA to reduce feature vector dimensions. Each block is lexicographically sorted and compared for duplicates. Luo et al. [18] partition an image into small, overlapped chunks, analyze their similarity, and locate repeated portions. The block is divided into four sub-blocks, and the feature vectors are the average of the red, green, and blue channels. His approach detected Gaussian blur, additive noise, and JPEG pictures. Mahdian and Saic [19] replaced overlapping blocks with blur-invariant moments relevant to blur deterioration, additive noise, and arbitrary contrast shift, and used PCA to minimize the dimension of RGB feature vectors. Forgeries were detected using a kd-tree nearest neighbour search. The approach works fine but does not handle post-processing operations [19]. Lin et al. [20] partition the input image into equal-sized overlapping chunks as integers. Radix sort compares block feature vector numbers. Huang et al. [21] created a method for forgery detection that uses the discrete cosine transform (DCT) and clusters 16 low-frequency DCT values in a row vector for feature matching. Ardizzone et al. [22] examined features in a triangle to improve block-based forgery detection. Shape, triangular content, and geographical features are used to match triangles. The approach was not affected by lossy compression and additive noise. However, it cannot detect affine-transformed forgeries. Hayat et al. [23] suggest using DCT with DWT to extract and reduce features. The correlation coefficient between row vectors is used to detect duplicated blocks. Multiple forgeries decrease the method's effectiveness. Parveen et al. [24] proposed a method that separates greyscale images into 8 overlapping blocks and extracts features using DCT. K-means clusters and radix sorts the blocks. The proposed method detects forged parts from digital photos but fails with geometrical attacks. Gani and Qadir

[25] proposed a CMF approach employing DCT. DCT is used to extract features from blocks and generate feature vectors using cellular automata. Matched feature vectors with Kd-tree nearest-neighbor searches. The approach improves post-processing, however the weakness of block-based against geometric attach like rotation and scaling and complexity remains the same. Block-based approaches identify straightforward copy-move forgery but fail when geometrical modifications like scaling, rotation, and combination occur around the forged region. Block-based approaches are computationally expensive, so key point-based methods were developed.

Key point-based techniques do not require blocking the image [4]. They use high-entropy image key points [1]. The image's key points are descriptor related. Identifying these descriptors identifies copy-move forgeries. The Scale Invariant Feature Transform (SIFT) is a prominent CMF detection method [15]. Both SIFT [7] and SURF [26] extract local image features. Local features are computed with a reduced magnitude, reducing their computational cost. Amerini et al. initially employed the SIFT descriptor to detect repeated copy-move forgeries using g2NN matching. SIFT and AHC decrease false positives. In a subsequent study, the authors introduce the J-Linkage clustering approach to improve matching [27]. Pan and Lyu suggested a SIFT-based approach for duplicate identification. The SIFT invariance to geometric attack detects rotational duplication. The approach fails to detect forgeries with blurring and flipping, especially in smooth sections of the image [28]. Zhu et al. [29] devised a method for obtaining ORB key points from a Gaussian pyramid. Similar ORB key points are matched. This method works for the geometric attack but takes longer for high-resolution images and requires an outlier reduction algorithm (like RANSAC) to eliminate false matches. Xu et al. [30] used SURF to locate duplicated areas. It compares SURF descriptors from two randomly assigned groups. Rotation, resizing, blurring, and white gaussian were identified effectively. Silva et al. [31] proposed a method using SURF descriptor for feature extraction and applied clustering to reduce the complexity of the technique. The method was able to detect some forms of forgeries. However, it performs poorly in the presence of JPEG compression and post-processing attack. Recently, key point-based methods are used for detecting images with various attacks. Alberry et al. [7] suggested a method for copy-move forgery detection using SIFT and FCM. The approach detects geometric attacks like rotation scaling and multiple CM forgeries, but its post-processing is poorly robust. Yang et al. [32] employed a modified SIFT-based detector to find image key points. The method used smooth regions and a novel key-points distribution strategy to evenly distribute key points across the image. An improved SIFT descriptor detects copy-move forgeries. Rotation is detected, but affine modifications are not. Niyishaka and Bhagvati [11] developed an approach using blobs and BRISK. The approach discovers blobs and matches the detected BRISK features inside them. Forgery is recognized by detecting the same BRISK features in different blobs. The strategy improves detection but needs improvement. Because of Sobel edge detection. The Sobel edge detector is noise-sensitive and rarely produces a good edge. This study builds on the method in [11] by using a canny edge detector to find good edges, detect SIFT features within blobs and match duplicates once.

3 Proposed Method

The proposed CMFD technique follows both the block-based and key point-based constraints. Our CMFD approach utilizes edge detectors and SIFT blobs. Our method is loosely based on [11], where the authors generate their edge using the Sobel operator, and then utilize the edge to extract BRISK key points. Similar key points of opposing blobs are then matched for forgery detection. From our observations, Sobel edge detection can be affected by Gaussian smoothing [33, 34]. However, we concur that edge images can be useful. To cater for issue with Gaussian noise, we opted for the Canny edge detected instead. Furthermore, we did not go with BRISK since SIFT gives superior performance for images containing geometric transformations [5]. Our method has six steps: preprocessing, canny edge detection, SIFT detection, blob detection, key point extraction and matching, and forgery detection. It uses MICC-F220, MICC-Multi, and CoMoFoD pictures (Fig. 1).

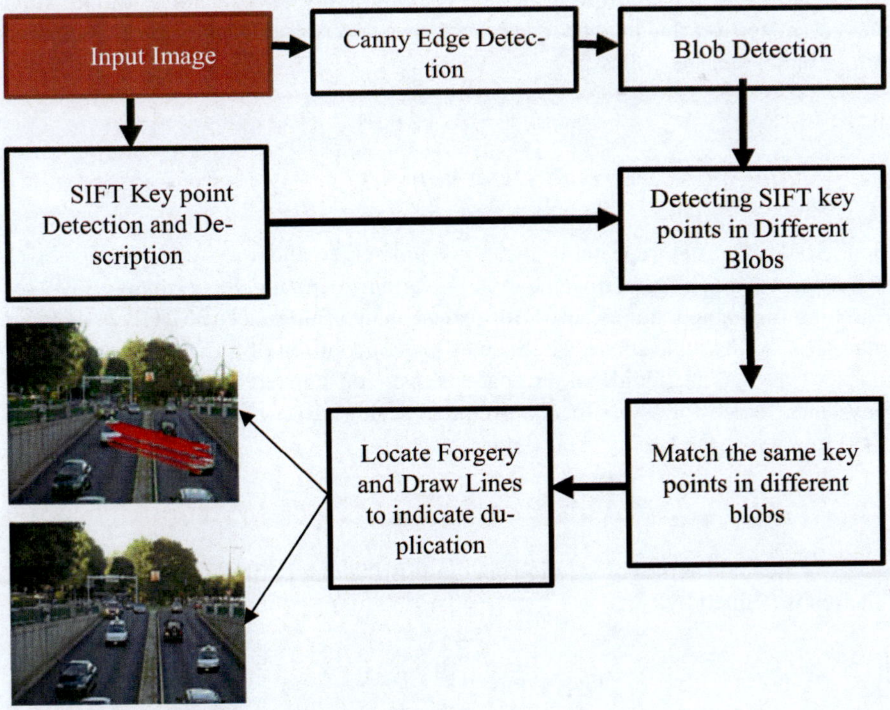

Fig. 1 Flow diagram of the proposed method

4 Preprocessing

To enable batch processing, we followed the scaling advice in [11]. The canny edges were detected on grayscale images, which is consistent with the work in [35]. Blob detection allows matches between similar blobs as duplicates. Edge detection is conducted before blob detection because the foreground images contain more information while the dark background appears less informative. Edge detection divides objects into distinct blobs. Each detected region is a separate blob. A good edge detector suppresses the image and extracts the best information.

5 SIFT Feature Generation

Distinctive features of local image patches are extracted using the SIFT algorithm. SIFT features are invariant to scale and rotation and robust to noise, illumination changes, distortion, and image viewpoints. Lowe describes SIFT key points detection in four major steps as follows:

5.1 Scale-Space Extrema Detection

First, SIFT identifies repeatable locations and scales under different perspectives of the same object. Detecting locations invariant to image scale change involves searching for stable features across all scales, using scale space [36]. Given an input image I (x, y), its scale space is defined as the convolution of a variable-scale Gaussian, G (x, y,). The initial colour space conversion converts the RGB image I to grayscale. Search for stable local extrema in scale space to find SIFT key points. I (x, y) in scale-space L (x, y, σ) is defined as:

$$L(x, y, \mathrm{s}) = G(x, y, \mathrm{s}) * I(x, y) \tag{1}$$

where * is denoted as the convolution operation in the x and y directions. The Gaussian function is defined by:

$$\mathrm{G(x, y, s)} = \frac{1}{2\pi\sigma^2} e \frac{\mathrm{x}^2 + \mathrm{y}^2}{\sigma^2} \tag{2}$$

Where σ is used as the scale space factor for extracting key points that are invariant to scale and orientation. It uses the scale-space extrema of DoG, convolved with *D (x, y, σ) image, and computed as a difference of two nearby scales separated by k.*

$$D(x, y, \sigma) = [G(x, y, k\sigma) - G(x, y, \sigma)] * I(x, y) = L(x, y, k\sigma) - L(x, y, \sigma) \tag{3}$$

To maintain a fixed blurred image per octave, we choose a specific value of k. It facilitates the generation of the same DoG images. Once we obtain the DoG images. Then, key points are extracted which are considered the local minima or maxima of the DoG images. Each pixel in these DoG images is compared with eight neighbors within the same scale and nine other similar nearby pixels in each of the neighboring scales. If the pixel is the highest or lowest among the compared pixels, then the pixel is considered the candidate key point.

The DoG function, on the other hand, has a close approximation to the scale-normalized Laplacian of Gaussian, $\sigma^2\nabla^2 G$. Normalization with a factor of σ^2 is required to obtain a true scale-invariance image [36]. It was found in an experimental comparison that the maxima features compared to a range of other possible images and minima of $\sigma^2\nabla^2 G$ produce the most stable image functions, such as the gradient, Hessian, or Harris corner function [37].

5.2 *Key Point Localization*

Scale-space Extrema detection generates an excessive number of candidates' key point, including unstable ones. To be more accurate, key points are filtered such that only stable key points are kept. After identifying the key point candidate through neighborhood pixel comparison, a thorough fit to the neighboring data is performed to ensure proper placement, scale, and the ratio of principal curvatures. This information permits data points with low contrast or that are poorly located along an edge to be eliminated.

6 Blob Detection

Blob detection is used to locate and mark image areas that have different characteristics from the surrounding areas, such as brightness or colour. Two of the most popular blob detectors are the Laplacian of Gaussian (LoG) operator [11, 38, 39] and the Difference of Gaussian (DoG) operator [11, 36, 38]. The LoG computes the second order of the gaussian filter through Eq. (4):

$$LoG(x, y) = \frac{1}{\pi\sigma^4}\left[1 - \frac{x^2 + y^2}{2\sigma^2}\right]\mathrm{e}^{-(x+y)/2\sigma^2} \tag{4}$$

where σ denote the gaussain filter standard deviation. The blob is determined at the maximum response of the LoG magnitude. But the LoG's limitation is its sensitivity to noise and complexity in computing the second derivative of an image. However, the LoG is also determined by the approximation of the difference of two gaussian (DoG) computed as (Fig. 2):

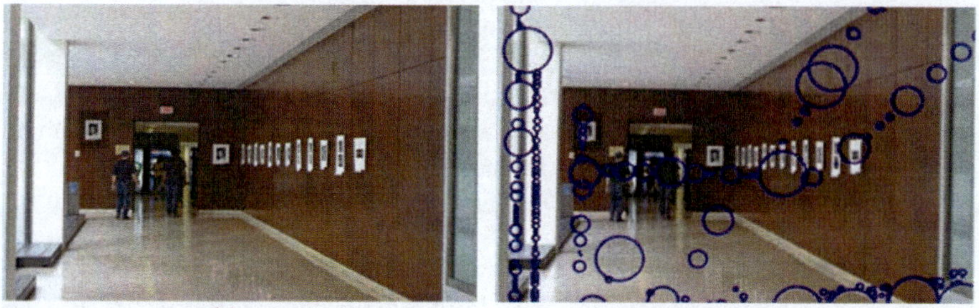

Fig.2 **a** original image **b** blob detected image

$$(DoG = g(x, y, \sigma k)g(x, y, \sigma) \quad (5)$$

where σ denote the gaussian filter standard deviation and (x, y, σ) denote the gaussian filter, while k stands as the scale variable. The blob is detected at the scale-space extrema of the DOG at multiple scales with a relative radius of $\sqrt{2\sigma}$, as shown in Fig. 3.

6.1 Advantage of Using Blobs in Copy-Move Forgery

This research aims to develop a robust algorithm that can use blobs to detect copy-move forgery. Research has been conducted over the decade for blob detection. But the commonly applied blob detection is thresholding. The thresholding-based approach assumes that the blobs' intensity has significant differences from that of the background [40]. However, this assumption is often invalid and can lead to the detection of false positives on the detected object. Instead of using thresholding directly, the researcher uses thresholding in the preprocessing step, with a variety of filters such as erosion, dilation, opening and closing. Then the intensity-based features are further generated from the preprocessed images. However, such filters are limited to blob detection because they can only be effective when the objects to be detected are large and few. Alternatively, other filters based on space transformation theory are used, where raw images are mapped to a space where properties and patterns of blobs are recognized. Such application includes Radial Symmetry [41], SIFT [36], SURF [42] and BRISK [43]. In recent times, researchers also detect blobs based on used space theory [42], where the application of the Laplacian of Gaussian (LoG) [42, 44] is proposed. However, the drawback of LoG in detecting symmetrical blobs rationally has led to extended LoG-based solutions, these include generalized Laplacian of Gaussian (gLoG) [39], Difference of Gaussian (DoG) [36, 45] and Hessian Difference of Gaussian (HDoG) [46]. This research applied DoG for its blob detection. A blob can be detected by applying the DoG function on an image *I (x, y)* of size M*N. Blobs are bright regions on the dark background Fig. 2.

```
Algorithm for Image forgery detection with canny edge detector, SIFT and
Blob;
Input I: RGB image
Output: Detected forgery
    Define the input image I(x,y);
    Gi ← Compute grayscale image from I(xy);
    Ci ← Apply canny edge detector to Gi;
    D(x,y) ← compute the distance using equation 6
    Blobs( x_b, y_b), r) ← extract edge blob features and spatial coordinates
    from canny edge image Ci;
    SIFT (x_k, y_k) ← extract SIFT features and spatial coordinates from Gi;
    For blob in blobs do
        sameBlob =empty
        for Keypoints in SIFT do
            if D ≤ r then
                sameBlob.append(Keypoints);
            else
                continue;
            end if
        end for
        if sameBlob !=empty then
            Blobs = Blobs - sameBlobs
            matches = bf.knnMatch(Blobs, sameBlobs);
            match_ratio = 0.43;
            matches = 0;
            for m, n, in matches do
                if m.distance < n.mach_ratio then
                    matches +=1;
                else
                    continue;
                end if
            end for
            If matches ≥2 then
                Print('there is match');
                Indicate matches by drawing lines against corresponding
    similar
                key points in different blobs;
            else
                continue;
            end if
        else
            continue;
        end if
    end for
```

Fig. 3 Algorithm for CMFD using canny edges, SIFT and blobs

In blob detection, the blobs detected are scale independent since they are resulted from a scale-space. The radius of each blob in the scale space is approximately $\sqrt{2\sigma}$ [11]. We employ the use of blob because the image can be affected with forgery in a small and large proportions; forgery in small proportions can be detected by small blobs while large forge areas can be detected by large blobs.

In copy-move forgery detection, the computational cost is reduced since the pixel comparison in blobs is done only when a blob contains a key point associated with the forgery. It makes all identical blobs remain idle while blobs of different characteristics are compared. On the contrary, in the block-based approach, for an image $I(x, y)$, if the image is divided into M*N overlapping fixed-size of blocks β with a step size slide n and number of blocks N_b given in [11]. The number of blocks is directly proportional to the size of the image. Therefore, as the size of the image increases the number of blocks also increases. The complexity is also subjected to the number of blocks to process since block-based compares all blocks to detect the affected ones.

6.2 *Identifying SIFT Interest Points Located Within the Same Blob*

As in [11], we identifySIFT the interest points within the same blob by extracting the 2D spatial coordinates for each blob $B(x_b,, y_b\ \sigma)$ and SIFT keypoints $K(x_k, y_k)$. The (x_b, y_b) as the blob's centre and sigma (σ) corresponding to the standard deviation of the Gaussian kernel are used to detect the blob. Using $\sqrt{2\sigma}$ as the approximate radius of each blob from its centre to the location of the keypoints. Computed from $D(x, y)$:

$$D(x, y) = \sqrt{(x_k - x_b)^2 + (y_k - y_b)^2} \tag{6}$$

Since key points located in the same blobs are not match. It increases the chances of reducing pair of blobs matching during the copy-move forgery detection. The blob containing the original image and its corresponding duplicate are matched in copy-move forgery detection, thereby limiting the number of false matches drastically.

7 Feature Matching and Forgery Detection

After feature description and blob detection, next is matching the key points appearing in various blobs. The SIFT features obtained associated with a 128-dimensional vector that behaves as a descriptor for that specific feature is invariant to rotation and illumination changes. Hence, in a given a test image I, a set of key-points $K = (k_1, k_2, \ldots, kn)$ and the corresponding descriptors $D = (d_1, d_2, \ldots, d_n)$ are extracted. For each key-point k_i in image I, the distance in the Euclidean space of all the other $(n - 1)$ keypoints are calculated. Then the nearest neighbour with the minimum distance is detected as the best candidate key-point of k_i, for each keypoint located within the blobs, the L2 norm was used in getting the distance to the two nearest neighbours. If the distance D of the nearby pixels is by d_1 and d_2 and threshold $Th \in (0, 1)$, a match is confirmed if $d_1/d_2 < Th$. We adopted the threshold value as examined experimentally in [11] i.e., 0.43, which seems to provide the best accuracy.

8 Experimental Results

The MICC-F220 and MICC-8Multi datasets were used to evaluate scaling and rotation performance and multiple duplicates, while the CoMoFoD dataset tests post-processing operations on copy-move forgeries, respectively.

8.1 *Description of Datasets and Metrics Used for the Evaluation*

MICC-F220 dataset composes of 220 images (110 originals and 110 tampered pairs). The dataset is used to evaluate copy-move forgery with geometrical transformation operations such as scaling and rotation. A combination of multiple copy-move images from MICC-8Multi and CoMoFoD datasets are used to evaluate multiple copy-move detection performance. The evaluation take note of rotation angle at various degree (θ = 10, 20 30 and 40), while the scale at each orientation $S(x, y)$ is considered, $S = (1.2x, 1.2y), (1.3x, 1.3y, (1.4x, 1.2y)$ as well as the combination of scaling and rotation $SR = \theta(x, y) = 10(1.2x, 1.2y)$ and $20(1.4x, 1.2y)$. We also employ images from CoMoFoD datasets to test our method on post-processing performance. Specifically, JPEG compression, and contrast adjustment are tested. These parameters are detailed in [11].

We assess the performance of the algorithm using the well-known image processing metrics known as the confusion matrix. With the forged image determined as the positive class, True positive *(TP)*, False Positive *(FP)*, False Negative *(FN)* and True Negative *(TN)* are defined as follows:

- TP: Images that are forged and are correctly detected forged.
- FP: Images that are original and are wrongly detected as forged.
- FN: Images that are forged and are wrongly detected as the original.
- TN: Images that are original and correctly detected as the original.

After obtaining the, TP, FP, FN and TN, True Positive Rate (TPR), False Positive Rate (FPR), Precision, Recall, and F-measure can be calculated. TPR: is the proportion of correctly classified forged images, FPR: is the proportion of original images that are misclassified, Precision is the chance that a forged image is detected, and recall is the percentage of the forged image that are successfully detected. The harmonic mean of precision and recall is known as an F – measure. If at least two pair matches are detected between two image regions, we consider the image to be forged. The following equations depict the above evaluation metrics:

$$TPR = TP/TP + FN \tag{7}$$

$$FPR = FP/FP + TN \tag{8}$$

$$Precision = TP/TP + FP \tag{9}$$

$$Recall = TP/TP + FN \tag{10}$$

$$\text{F - Measure} = 2(\text{P/R}) \tag{11}$$

8.2 Results

The major problem of copy-move forgery is the inability of key points to detect a forgery in smooth regions, especially when geometric transformations and post-processing operations are used. We show the result of our forgery detection approach using the proposed **CannySIFT + Blob** method; the performance is shown in Fig. 3 using images from MICC-F220 and the CoMoFoD dataset. Various geometric and postprocessing attacks are considered. Multiple copy-move attach was also tested from a combination of images from both MICC-F220 and CoMoFoD datasets.

We consider separately the effect of rotation scaling and their combination as characterized in MICC-F220 and JPEG compression, blur, noise addition and contrast adjustment as characterized in CoMoFoD datasets. We employ the same procedure in using datasets as used in [11] and compare our results with their method. Table 1 and Fig. 4, show a comparison of copy-move forgery under various transformation attacks. Figure 4 shows matching performance based on true positive rate and false positive rate defined in Table 1. The proposed method has improved the performance to 100% in terms of TPR for the rotation effect. In terms of scaling effect, the proposed method maintained similar TPR but resulted a zero FPR (no false alarm). The proposed method also detected forgery with both rotation and scaling effect at 90% TPR and no false alarm. The performance of the proposed method in terms of post-processing operations is lower than the previous method in TPR but reliably has low FPR (low false alarm).

Based on the two datasets used in the analysis, we focus on various attacks such as geometrical transformation and post-processing operations. Figure 5 shows the

Table 1 Result of forged images on rotation, scaling, rotation + scaling, multiple copy-move, and post-processing operation

Methods	*TPR*	*FPR*
Rotation		
blob + brisk [11]	0.95	0.09
Proposed method	1	0.09
Scaling		
blob + brisk [11]	0.9	0.09
Proposed method	0.90	0
Scaling + rotation		
blob + brisk [11]	0.86	0.09
Proposed method	0.90	0
Multiple copy-move		
blob + brisk [11]	0.83	0.06
Proposed method	0.875	0
JPEG compression		
blob + brisk [11]	0.95	0.1
Proposed method	0.94	0.05
Blur, noise, contrast adj		
blob + brisk [11]	0.96	0.16
Proposed method	0.93	0

result of the various attacks on the images. To further validate the result in Fig. 7, on whether there is a statistically significant improvement in the proposed method from the method used in [11], we employ the used of statistical test known as the chi-square t-test (X^2) or the goodness of fit test [46]. On which the calculated value of X^2 obtained is compared with the critical value of X^2 to accept or reject the null hypothesis. From Fig. 7, we hypothesized that (Figs. 6 and 8):

H_0: There is no statistically significant improvement between the method used in [11] and the proposed method on the geometric transformation and post-processing operations on the forge images.
H_1: There is a statistically significant improvement between the two methods.

We computed the Chi-square t-test (X^2) from the result in Fig. 7 using a contingency table (Table 2) and find the calculated X^2 and critical X^2 values (Table 3). The values in bold are the total values computed in each raw to be used in the computation. See [47] for the detail procedure.

Since the Table 2 has two techniques to measure with 3 different values (P, R, and F-M), we computed the degree of freedom from the Eq. (10):

$$X_{\upsilon}^{2} = (\mathrm{h}-1)(\mathrm{k}-1) \tag{12}$$

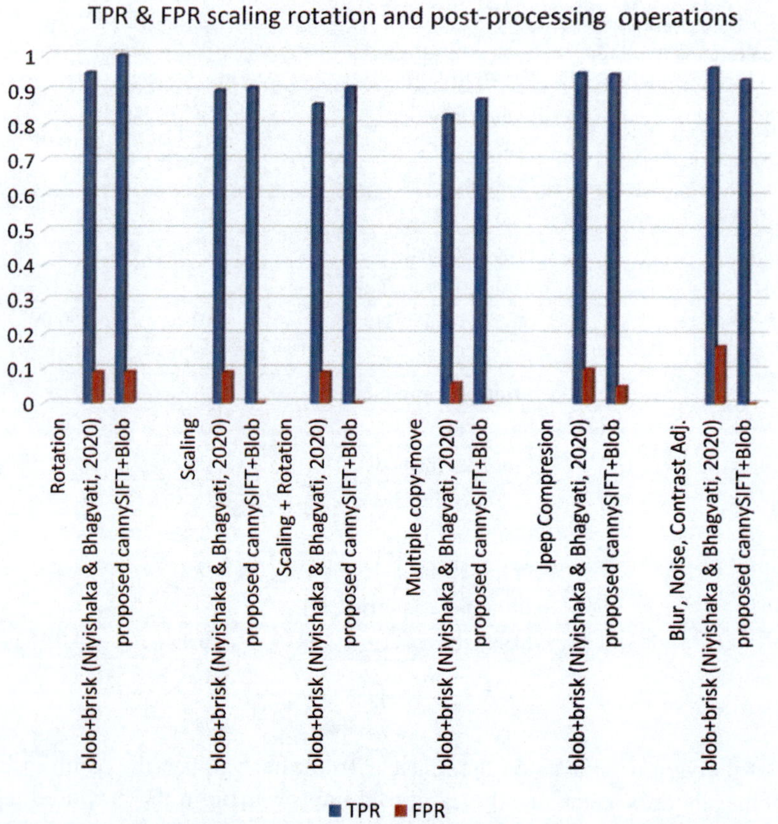

Fig. 4 Result of forged images on rotation, scaling, rotation + scaling, multiple copy-move, and post-processing operation

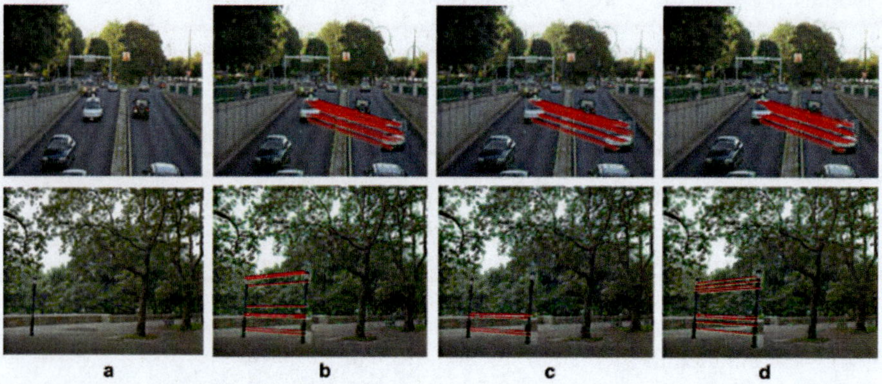

Fig. 5 Detected forgeries with various rotation and scale factors. **a** Original images, **b** scaled (1.2, 1.2), **c** scaled (1.3, 1.3), **d** scaled (1.4, 1.2)

a b c d e

Fig. 6 Detected forgeries with various rotation and scale factors. **a** Original image. **b** Rotated (10°). **c** Rotated (20°). **d** Rotated (30°) and **e** rotated (40°)

a b c

Fig. 7 Multiple copy_move forgeries detection using MICC-8multi dataset. **a** The original images. **b** The copy-move forged images. **c** The detected multiple copy-move forgery (top) and CoMoFoD (bottom) benchmark datasets

where X_v^2 is the degree of freedom, h is the number of rows and k is the number of columns. Therefore ((2 − 1) (3 − 1) = 2) degree of freedom using Eq. (12). We obtained a critical X^2 value for 2° of freedom at 5% (p = 0.05) significance level (Table 3). The values of the expected frequency (E) are generated from total rows and columns in Table 2 using Eq. (11).

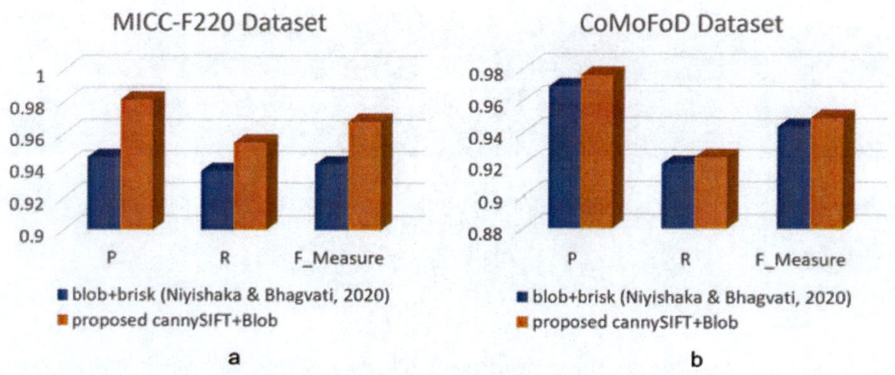

Fig. 8 Comparison of the proposed method and the blob + brisk [11] method

Table 2 Precision, recall and F-measures for MICC-F220 datasets

Methods	P	R	F-M	**Total**
[11]	0.944954	0.936364	0.940639	**2.821957**
Proposed method	0.981308	0.954545	0.967742	**2.903595**
Total	**1.926262**	**1.890909**	**1.908381**	**5.725552**

Table 3 Statiscally significant improvement between the method in [10] and the proposed method

Datasets	Calculated X^2	Critical X^2 at 5% significance level
MICC-F220	94.18576	5.991
CoMoFoD	0.000182197	5.991

$$E = \frac{row\,total - column\,total}{grand\,total} \tag{13}$$

The calculated X^2 value is then obtained using the summation of the square of the difference between the observed frequency (O_i) and the expected frequency (E_i) using Eq. (12).

$$X^2 = \sum \frac{(O_i - E_i)^2}{E_i} \tag{14}$$

where $\boldsymbol{O_i}$ is the observed frequency values (actual values in P, R and F-M) and $\boldsymbol{E_i}$ is the expected values obtained using Eq. (13)

To test the above hypothesis, we compare the calculated X^2 value and critical X^2 values in Table 3. If the Calculated value is large, the null hypothesis is rejected and vice vasa. The result for both MICC and CoMoFoD are presented in Table 3 simultaneously.

As indicated in Table 3, the calculated X^2 value is greater than the critical X^2 value (94.18576 > 5.991). We, therefore, reject the null hypothesis. This evidence indicated that there is a statistically significant improvement in the proposed method based on the geometrical transformation attacks (Fig. 6) in test data. While on the CoMoFoD dataset, the calculated chi-square X^2 value is less than the critical X^2 value (0.000182197 < 5.991). We, therefore, accept the null hypothesis. This evidence indicated that there is no statistically significant improvement in the proposed method based on the post-processing attack on CoMoFoD dataset in Fig. 8b.

9 Conclusion

This paper presents a method based on a Canny edge detector and SIFT algorithm using blobs as a clustering method for detecting forged regions. The main contribution of this work is the enhancement of blob detection by using a strong edge detection filter as well as detecting strong key points located within blobs. Because the Sobel operator is noise-sensitive and does not take steps to reduce noise, which affects the detection of a good edge and results in inadequate key point localization, the Sobel operator is replaced by a canny edge detector. On the other hand, the canny edge operator emphasized the strong edges while suppressing the weak ones using non-maximum suppression and hysteresis. In this study, features from the query image are extracted and described using SIFT. Proximate features are separated using blob, and similar key points in various blobs are matched using L2 norm to identify copy-move forgeries. Blobs are identified using the DoG algorithm. It is a region within an image that differs in properties such as brightness or color.

Therefore, applying a canny edge detector to the suspected image, enhances blob localization. By comparing the SIFT key points from different detected blobs, copy-move forgery is detected. The same key points in separate blobs are matched, but different key points in different blobs are automatically ignored throughout the matching process. By ignoring opposite key points in different blobs, the number of blob matching decreases, hence reducing the complexity of finding the forgery. The results indicated that our method could identify forgery in the presence of geometric transformation and post-processing operations and has outperformed the state-of-the-art method that used blob in forgery detection. The result of the two methods is further validated to test the level of significant improvement from the proposed method and the method in [10], using t-test statistics. It shows a significant improvement in the MICC-F220 dataset while in CoMoFoD, there is no significant improvement. In the future work, the post-processing operation in the forged images should be improved.

References

1. Hegazi A, Taha A, Selim MM (2021) An improved copy-move forgery detection based on density-based clustering and guaranteed outlier removal. J King Saud Univ Comput Inf Sci 33(9):1055–1063. https://doi.org/10.1016/j.jksuci.2019.07.007
2. Alipour N, Behrad A (2020) Semantic segmentation of JPEG blocks using a deep CNN for non-aligned JPEG forgery detection and localization. Multimed Tools Appl. https://doi.org/10.1007/s11042-019-08597-8
3. Asghar K, Habib Z, Hussain M (2017) Copy-move and splicing image forgery detection and localization techniques: a review. Aust J Forensic Sci 49(3):281–307. https://doi.org/10.1080/00450618.2016.1153711
4. Prakash CS, Panzade PP, Om H, Maheshkar S (2019) Detection of copy-move forgery using AKAZE and SIFT keypoint extraction. Multimed Tools Appl 78(16):23535–23558. https://doi.org/10.1007/s11042-019-7629-x
5. Wang D, Gao T (2020) An efficient USM sharpening detection method for small-size JPEG image. J Inf Secur Appl 51:102451. https://doi.org/10.1016/j.jisa.2020.102451
6. Zheng L, Zhang Y, Thing VLL (2019) A survey on image tampering and its detection in real-world photos. J Vis Commun Image Represent 58(December):380–399. https://doi.org/10.1016/j.jvcir.2018.12.022
7. Alberry HA, Hegazy AA, Salama GI (2018) A fast SIFT based method for copy move forgery detection. Futur Comput Inform J 3(2):159–165. https://doi.org/10.1016/j.fcij.2018.03.001
8. Manu VT, Mehtre BM (2018) Copy-move tampering detection using affine transformation property preservation on clustered keypoints. Signal, Image Video Process 12(3):549–556. https://doi.org/10.1007/s11760-017-1191-7
9. Destruel C, Itier V, Strauss O, Puech W (2018) Color noise-based feature for splicing detection and localization. In: 2018 IEEE 20th International workshop on multimedia signal process, MMSP 2018, pp 1–6. https://doi.org/10.1109/MMSP.2018.8547093
10. Ismail M, Kanwal N (2018) A review block based copy move forgery detection techniques. Int J Comput Sci Mob Comput 7(4):205–212
11. Niyishaka P, Bhagvati C (2020) Copy-move forgery detection using image blobs and BRISK feature. Multimed Tools Appl. https://doi.org/10.1007/s11042-020-09225-6
12. Wang XY, Li S, Liu YN, Niu Y, Yang HY, li Zhou Z (2017) A new keypoint-based copy-move forgery detection for small smooth regions. Multimed Tools Appl 76(22):23353–23382. https://doi.org/10.1007/s11042-016-4140-5
13. Bilal M, Habib HA, Mehmood Z, Saba T, Rashid M (2020) Single and multiple copy-move forgery detection and localization in digital images based on the sparsely encoded distinctive features and DBSCAN clustering. Arab J Sci Eng 45(4):2975–2992. https://doi.org/10.1007/s13369-019-04238-2
14. Christlein V, Riess C, Jordan J, Riess C, Angelopoulou E (2012) An evaluation of popular copy-move forgery detection approaches. IEEE Trans Inf Forensics Secur 7(6):1841–1854. https://doi.org/10.1109/TIFS.2012.2218597
15. Warif NBA et al (2016) Copy-move forgery detection: survey, challenges and future directions. J Netw Comput Appl 75:259–278. https://doi.org/10.1016/j.jnca.2016.09.008
16. Fridrich J, Soukal D, Lukáš J (2003) Detection of copy-move forgery in digital images. In: Proceedings of digital forensic research workshop, Cleveland, pp 1163–1168. https://doi.org/10.1109/ICMLA.2015.137
17. Popescu AC, Farid H (2004) Exposing digital forgeries by detecting duplicated image regions. Technology report, TR2004-515, no. 2000, pp 1–11. Department Computer Science, Dartmouth College, Hanover. Retrieved from http://os2.zemris.fer.hr/ostalo/2010_marceta/Diplomski_files/102.pdf
18. Luo WQ, Huang JW, Qiu GP (2007) Robust detection of region-duplication forgery in digital image. Jisuanji Xuebao/Chinese J Comput 30(11):1998–2007
19. Mahdian B, Saic S (2008) Blind authentication using periodic properties of interpolation. IEEE Trans Inf Forensics Secur 3(3):529–538. https://doi.org/10.1109/TIFS.2004.924603

20. Lin HJ, Wang CW, Kao YT (2009) Fast copy-move forgery detection. WSEAS Trans Signal Process 5(5):188–197
21. Huang Y, Lu W, Sun W, Long D (2011) Improved DCT-based detection of copy-move forgery in images. Forensic Sci Int 206(1–3):178–184. https://doi.org/10.1016/j.forsciint.2010.08.001
22. Ardizzone E, Bruno A, Mazzola G (2015) Copy-Move forgery detection by matching triangles of keypoints. IEEE Trans Inf Forensics Secur 10(10):2084–2094. https://doi.org/10.1109/TIFS.2015.2445742
23. Hayat K, Qazi T (2017) Forgery detection in digital images via discrete wavelet and discrete cosine transforms. Comput Electr Eng 62:448–458. https://doi.org/10.1016/j.compeleceng.2017.03.013
24. Parveen A, Khan ZH, Ahmad SN (2019) Block-based copy–move image forgery detection using DCT. Iran J Comput Sci 2(2):89–99. https://doi.org/10.1007/s42044-019-00029-y
25. Gani G, Qadir F (2020) A robust copy-move forgery detection technique based on discrete cosine transform and cellular automata. J Inf Secur Appl 54:102510. https://doi.org/10.1016/j.jisa.2020.102510
26. Karami E, Prasad S, Shehata M (2017) Image matching using SIFT, SURF, BRIEF and ORB: performance comparison for distorted images. Retrieved from http://arxiv.org/abs/1710.02726
27. Amerini I, Ballan L, Caldelli R, Del Bimbo A, Del Tongo L, Serra G (2013) Copy-move forgery detection and localization by means of robust clustering with J-Linkage. Signal Process Image Commun 28(6):659–669. https://doi.org/10.1016/j.image.2013.03.006
28. Pan X, Lyu S (2010) Region duplication detection using image feature matching. IEEE Trans Inf Forensics Secur 5(4):857–867. https://doi.org/10.1109/TIFS.2010.2078506
29. Zhu Y, Shen X, Chen H (2015) Copy-move forgery detection based on scaled ORB, no. 2699. https://doi.org/10.1007/s11042-014-2431-2
30. Xu B, Wang J, Liu G, Dai Y (2010) Image copy-move forgery detection based on SURF. In: Proceedings—2010 2nd international conference on multimedia information networking and security, MINES 2010, pp 889–892. https://doi.org/10.1109/MINES.2010.189
31. Silva E, Carvalho T, Ferreira A, Rocha A (2015) Going deeper into copy-move forgery detection: Exploring image telltales via multi-scale analysis and voting processes. J Vis Commun Image Represent 29:16–32. https://doi.org/10.1016/j.jvcir.2015.01.016
32. Yang B, Sun X, Guo H (2018) A copy-move forgery detection method based on CMFD-SIFT, no. 1800, pp 837–855. https://doi.org/10.1007/s11042-016-4289-y
33. Othman Z, Rafiq M, Kadir A (2009) Comparison of canny and sobel edge detection in MRI images, pp 133–136
34. Kim D (2013) Sobel operator and canny edge detector, pp 1–10. Retrieved from http://www.egr.msu.edu/classes/ece480/capstone/fall13/group04/docs/danapp.pdf
35. Lorencin I, Anđelić N, Španjol J, Car Z (2019) Using multi-layer perceptron with Laplacian edge detector for bladder cancer diagnosis. Artif Intell Med 102(May):2020. https://doi.org/10.1016/j.artmed.2019.101746
36. Lowe DG (2004) Distinctive image features from scale-invariant keypoints. Int J Comput Vis 60(2):91–110
37. Mikolajczyk K (2002) Detection of local features invariant to affine transformations: application to matching and recognition. Most, p 171
38. Lindeberg T (1996) Scale-space: a framework for handling image structures at multiple scales. Computing 1–12
39. Kong H, Akakin HC, Sarma SE (2013) A generalized Laplacian of Gaussian filter for blob detection and its applications. IEEE Trans. Cybern. 43(6):1719–1733. https://doi.org/10.1109/TSMCB.2012.2228639
40. Wu KCQ, Merchant F (2010) Microscope image processing. Elsevier
41. Loy G, Zelinsky A (2003) Fast radial symmetry for detecting points of interest. IEEE Trans Pattern Anal Mach Intell 25(8):959–973. https://doi.org/10.1109/TPAMI.2003.1217601
42. Bay H, Ess A, Tuytelaars T, Van Gool L (2008) Speeded-up robust features (SURF). Comput Vis Image Underst 110(3):346–359. https://doi.org/10.1016/j.cviu.2007.09.014

43. Leutenegger S, Chli M, Siegwart RY (2011) BRISK: binary robust invariant scalable keypoints. In: Proceedings of the IEEE international conference on computer vision, pp 2548–2555. https://doi.org/10.1109/ICCV.2011.6126542
44. Witkin AP (1983) Scale-space filtering, vol 2. Morgan Kaufmann Publishers, Inc.
45. Mikolajczyk K, Schmid C (2004) Scale and affine invariant interest point detectors. Int J Comput Vis 60(1):63–86. https://doi.org/10.1023/B:VISI.0000027790.02288.f2
46. Zhang M et al (2016) Efficient small blob detection based on local convexity, intensity and shape information. IEEE Trans Med Imaging 35(4):1127–1137. https://doi.org/10.1109/TMI.2015.2509463
47. Yeh A (2000) More accurate tests for the statistical significance of result differences, p 947. https://doi.org/10.3115/992730.992783
48. Johnson JW (2018) Adapting mask-RCNN for automatic nucleus segmentation, pp 734–745. https://doi.org/10.1007/978-3-030-17798-0
49. Tareen SAK, Saleem Z (2018) A comparative analysis of SIFT, SURF, KAZE, AKAZE, ORB, and BRISK. In: 2018 International conference on computing, mathematics and engineering technologies, pp 1–10. https://doi.org/10.1109/ICOMET.2018.8346440

An Optimization for Access Point Placement in Indoor Communication

Bashar Alathari, Mohammad R. Radi, Mohammed Falih Kadhim, Nabeel Salih Ali, and Zaid Abdi Alkareem Alyasseri

Abstract Access Points (APs) in wireless networks have gained significant attention in several prospective such as estimating, propagating the Radio Frequency (RF), managing, and operating the network, and several challenges pose a critical manner, especially in indoor localization like obstacles, moving objects, the different materials building, and signal interference. According to previous issues, selecting an optimum location for APs has attracted researchers to find low-cost localization methods and approaches for indoor communication as it significantly affects the networks' coverage, operation, and management. In this article, a problem investigation analysis and evaluation are conducted inside the Department of Electrical Engineering, Faculty of Engineering, at the University of Kufa by using the combination of the received signal levels (RSL) and the effective service coverage area for APs that installed on both sides of the department building. Each AP has different points in several rooms inside the building (17 points for the left and 18 points for the right side). Furthermore, this research proposed a new optimized design that provides high coverage area and service quality coverage in signal and connection. The evaluation results show the highest RSL value achieved, which is − 47 dBm, and

B. Alathari · M. R. Radi · M. F. Kadhim · N. S. Ali · Z. A. A. Alyasseri (✉)
Information Technology Research and Development Centre, University of Kufa, P.O. Box (21), Kufa, Najaf Governorate, Iraq
e-mail: zaid.alyasseri@uokufa.edu.iq

B. Alathari
e-mail: bashar.alathari@uokufa.edu.iq

M. R. Radi
e-mail: mohammad@uokufa.edu.iq

M. F. Kadhim
e-mail: mohammed.kadhim@uokufa.edu.iq

N. S. Ali
e-mail: nabeel@uokufa.edu.iq

Z. A. A. Alyasseri
ECE Department Faculty of Engineering, University of Kufa, Najaf, Iraq

College of Engineering, University of Warith Al-Anbiyaa, Karbala, Iraq

© The Author(s), under exclusive license to Springer Nature Singapore Pte Ltd. 2023

D.-K. Kang et al. (eds.), *Proceedings of the 9th International Conference on Computational Science and Technology*, Lecture Notes in Electrical Engineering 983,
https://doi.org/10.1007/978-981-19-8406-8_48

the lowest RSL reaches − 87 dBm. Hence, the presented wireless signal coverage and performance have been analyzed and shown disheartening results in covering the entire building because it is on barrier two walls. Moreover, the far-away AP distance and the presence of the walls have weakened the transmitted signal, which caused the poor quality of service.

Keywords Indoor communication · AP location · Indoor network · Wireless access point · Received signal level · Service coverage area

1 Introduction

In the higher education sector, most universities seek to boost their educational level by providing optimal infrastructure for learning to students on the university campus [1]. Hence, many universities employ wireless Wi-Fi networks to allow students to use educational and learning services with reliable internet connections [2]. Wi-Fi technology plays a significant role in utilizing wireless internet connections in different buildings because it offers a reliable internet connection based on features such as high bandwidth, ease of installation, unlicensed band, and economic infrastructure [3–5]. Besides, it can be used and implemented for several devices, which are widely used in our digital era, such as smartphones and tablets [6, 7]. Proper coverage of the Wi-Fi network can be gained by the appropriate deployment of access points (APs) to serve enough indoor users due to its higher attenuation, which can signal pass through walls, obstacles, and objects that result in a dramatic impact on providing the total coverage for the end-users [8, 9]. Therefore, many researchers aim to estimate the best distribution approaches for AP devices in the targeted area to reduce costs and interferences and avoid dead zone, which leads to coverage gaps, especially for indoor communication [8, 10].

On the other hand, different engineering and architectural aspects should be leveraged in modern converged campus networks from a design standpoint. These principles include resiliency, modularity, and hierarchy in university campus networks. The performance of a network is influenced by user satisfaction that is related to network services that are provided [11]. In addition, the WLAN campus has several primary characteristics that a network should have to enhance its performance and simultaneously lower costs, such as efficiency, stability, safety, ease of installation, maintenance, and management [12, 13]. Wireless networks can cover the signals to the places where there are difficult to the arrangement of wires [13]. Managing a network includes making changes as the network grows to carry more traffic or reach more users and troubleshooting the network when things go wrong, or performance is not as desired [11].

Authors attempted to deploy indoor services to manage, operate the network, and estimate the Radio Frequency (RF) propagation characteristics when deploying the access points (AP) in wireless networks besides trying to avoid indoor placement challenges, which can be summarized by signal interference, moving objects, obstacles, and different building materials.

According to [13], the authors use a heuristic approach to specify locations for reference node installation and improve signal performance for single-floor and multi-floor buildings. The results lack detailed analysis [14]. In addition, a hybrid approach is proposed by [15] to implement the signal coverage of AP devices and their position coordination. Combining the Greedy algorithm with simulated Annealing suggests that utilizing such methods would regard the cost function. Meanwhile, [16] suggests a model for optimal AP placement using Particle Swarm Optimization (PSO), which considers the wall's attenuation.

Moreover, it lacks consideration of the attenuation and effects of different walls. Moreover, the authors in [17] present a solution to the AP placement problem based on a mathematical representation. The solution proposed an algorithm to recommend better locations for the APs and increase the whole network performance with the best accuracy that allows testing the algorithm in any actual or simulated environment. Likewise, a Genetic algorithm (GA) is developed by [18] to find the optimal global solution and solve the placement problem. The proposing method ignores the effects of obstacles in different buildings.

This paper addresses several issues and challenges concerning network infrastructure several software tools were used to conduct the study, such as the Edraw Max tool to design and document the network topologies, and NetSpot software is used to measure and evaluate the reliability of the existing network signals. After identifying several issues associated with the network, the paper proposes an upgrading network infrastructure to accommodate the network users and provide an optimal access point location.

In this paper, several objectives are addressed (1) To identify the various obstacles and factors encountered by conducting a questionnaire based on several hypotheses to propose an optimized design for access point location for Indoor Communication. Likewise, (2) To evaluate the existing indoor wireless for signals strengths and connection quality by NetSpot software analyzer. Finally, (3) To present an optimized design based on the network analyzer results that provide a larger coverage area and better service quality through relocating and adding APs.

2 Materials and Methods

To understand the research methodology, different sections are set in this section to elaborate on the study objectives, such as network model and tools, problem investigation, and performance evaluation, and two evaluation metrics, namely network coverage area and network service quality.

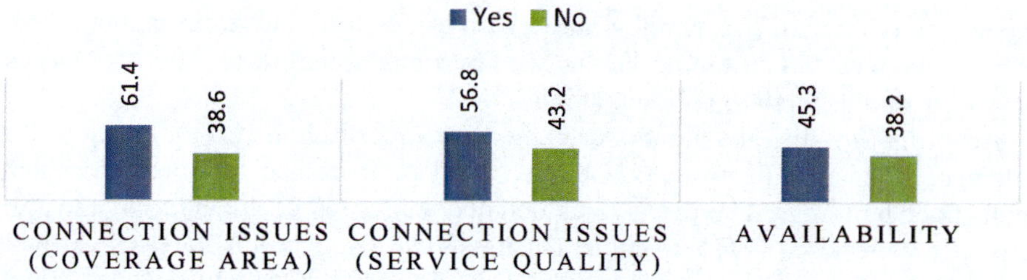

Fig. 1 Percentage of respondents concerning connection quality questions

Fig.2 RSL and distance for AP1 and AP2 on the left side

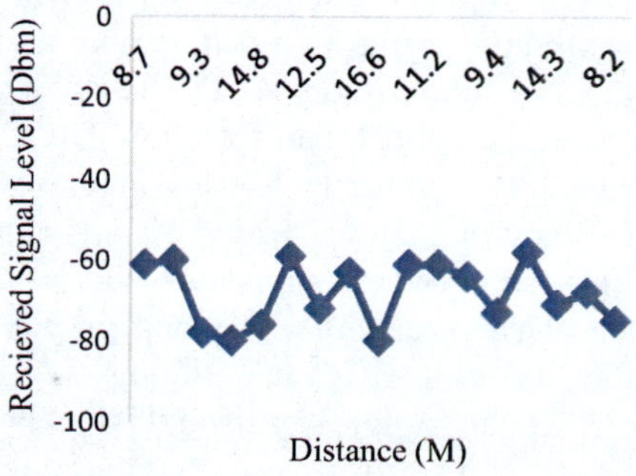

2.1 *Network Model and Tools*

This section documents the current network topology of the engineering faculty using the Edraw Max program, as shown in Fig. 2. The Fig. 2 also shows the Department of Electrical Engineering map for the engineering faculty as a testbed. The existing network services are described, checked, and evaluated in terms of signal quality and service availability. The existing network topology is documented via Edraw Max software (Fig. 1).

2.2 *Problem Investigation*

A questionnaire has been used to collect data that describe an observed population. The aim is to investigate the current network characteristics (services) by showing the direct correlation between network design and the users and the University's demands, as well as to identify the various obstacles and factors that would be encountered to set optimal indoor access points network.

Regarding the questionnaire, firstly, we considered respondents from employees who work in the Department of Electrical Engineering at the faculty of engineering. Secondly, the questions in this questionnaire mainly were related to the influential factors of this study, such as availability, network coverage area, network service quality, and documentation. The questionnaire was divided into six sections with 27 questions in total. Thirdly, five experts have performed validation and verification by interviewing and consulting. In this stage, the questions were revised and polished many times to clarify the questionnaire to the respondents. Fourthly, distribution was conducted in two ways, namely, online, and complex copy. The advantage of the online survey is the ease of access via a link shared through email, social networks, etc. Finally, an analysis was conducted after data collection.

2.3 *Performance Evaluation*

As a case study, we have addressed several issues and challenges concerning network infrastructure in the engineering faculty at the University of Kufa. We have documented, checked, examined, and evaluated the existing network characteristics in terms of network signal quality and availability of services. Several software tools were used to conduct the study, such as the Edraw Max tool to design the network topologies and NetSpot software to measure and evaluate the existing network signal.

Network Coverage Area

The surrounding environment has a significant effect on the wireless network signal. Reflection, absorption, distortion, or obstruction are signal experiences for propagation information for signals in general. Besides, collecting whole propagation data between each floor in the building is crucial to understand the propagation inside the building and determine the propagation path between floors. Hence, various wall materials have various attenuation amounts within the same other properties of the wall in addition to the passing signal frequency [14]. Therefore, indoor propagation achieves power that is attenuated by the wall; this power will be expressed as the formula below [19]:

$$P_r = P_t G_t G_r \left(\frac{\lambda}{4\pi}\right)^2 d^{-\alpha} 10^{\frac{X_\varphi}{10}} \prod_{m=1}^{M} |\Gamma_m|^2 \tag{1}$$

- where Pr is the power of the AP transmission,
- Gt is the transmitting antenna's gain,
- Gr is the receiving antenna's gain,
- d is the distance between the AP and the user,
- α is the distance power,
- φ is the standard deviation of the shadowing as a regular log,
- Γ, m and M respectively refer to the transmission coefficients after the m-th wall is passed by the direct propagation path and the wall's number [20].

To measure the network coverage area's performance for the building APs, Receive Signal Level (RSL) is determined by the NetSpot tool. RSL is used to measure the level of power of an antenna that is achieved. The higher the amount of RSLs, the stronger the obtained signal. RSL is calculated mathematically based on Eq. (1). The Wi-Fi Analyzer measures and analyzes the surrounding Wi-Fi network via measuring the received signal power in dBm. Therefore, the analyzer can determine the quality of the signal and network saturation information in 5 s.

Network Service Quality

To achieve optimal service delivery, the wireless network coverage from the AP must be measured. NetSpot is used to provide information and signal strength about the network. The tool can quickly determine and analyze nearby networks. In the survey mode, it can measure the strength and weaknesses of the network signal [21].

The color indicator shows different colors in Table 1, indicating signal strength that denotes the difference between AP and user.

RSL and adequate service area coverage were carried out in indoor spaces of the buildings inside the Department of Electrical Engineering, part of the faculty of Engineering. The building plan showing measurements can be noticed in Fig. 3. From Fig. 3, there are 2 AP in the main halls of the building. The first AP was mounted on the side of the building. Users can be found anywhere within the building.

RSL is calculated according to the distance between the APs and is conducted to measure the status of a wall barrier between the AP and the end-user in actual circumstances. The measurements were determined using an AP device and a virtual server on a laptop for setting the Tx rate. From the end of the tools, the measurement

Table 1 The colorful color indicators indicate signal strength

Color indicators	Strength of signal (dBm)	Quality of Signal
	-40 to -50	Excellent
	-50 to -60	Good
	-60 to -85	Less
	-85 to -100	Poor

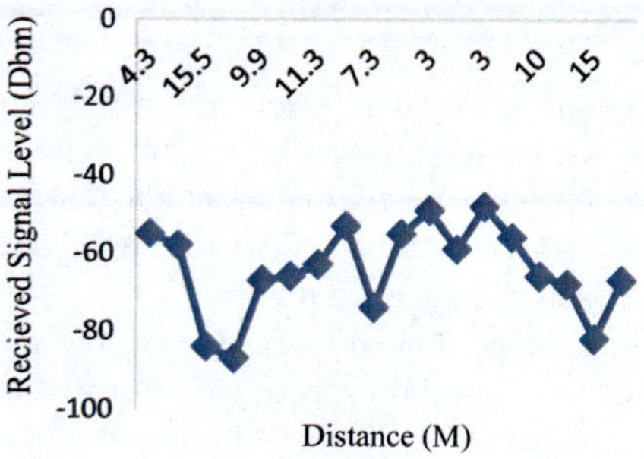

Fig. 3 RSL and distance for AP1 and AP2 on the right side

phase is started after confirming the AP is connected to the university networks, then turning on the NetSpot tool, which automatically reveals results in the RSL form.

Therefore, the NetSpot tool was used to examine the signal coverage area in the indoor area of the Department of Electrical Engineering, faculty of engineering, University of Kufa. The measurement results are analyzed to set several recommendations to improve the quality of the network and the indoor coverage. Based on the previous observation, the received signal is measured for signal coverage area at similar measurement points. The results show that Wi-Fi quality is greatly affected by the wall barriers around the APs.

3 Results and Discussions

In this section, the network availability analysis is discussed, besides the present results of the performance evaluation for the network connection measured by calculated coverage area and the quality of network signal.

3.1 Network Availability Issues Analysis

Both hard and electronic copies of the questionnaire were delivered to the target respondents. The questionnaire also involved 230 employees at the University of Kufa. The collected data were analyzed with the statistical package of Microsoft Excel. Data analysis was conducted to identify which network characteristics, such as availability, Quality of Service, redundancy, and fault tolerance, can influence the network performance. Figure 2 presents the percentage of respondents concerning connection quality questions.

Figure 1 illustrates respondents' answers to three questionnaire questions to assess the quality of the network connections and signal strengths in the targeted Department of the faculty. From the first bar chart, which represents questions about signal coverage area, (38.6%) of the respondents acknowledged that the network connection is not stable, meaning there is a problem with the network coverage in the department indoor area. The second bar chart shows that (43.2%) said "Yes" when asked a set of questions concerning whether they have issues connecting to the network inside their rooms (services quality connection issue). The last chart shows that (38.2%) of the respondents have issues connecting to the network in general.

3.2 Performance Evaluation

As we mentioned, the performance evaluation for the network connection can be measured by the calculated coverage area and the quality of the network signal. The

Table 2 RSL for AP1 on the left side

Point No.	Distance	RSL (dB)
1	8.7	– 61
2	7.4	– 60
3	9.3	– 78
4	15.2	– 80
5	14.8	– 76
6	5.5	– 59
7	12.5	– 72

received signal value was a metric to evaluate the signal coverage area measured earlier at the measurement points. In contrast, the NetSpot tool is used to measure the quality of the signal coverage area based on the number of wall barriers passed by the AP signal.

Received Signal Level Measurement

Within both sides of the corresponding Department, 17 points were distributed over eight rooms to measure the signal quality for AP1 and AP2 on the left side, as shown in Tables 2 and 3. While on the right side, where AP1 and AP2 are located along with 18 points distributed over eight rooms, as shown in Tables 4 and 5. The APs results are presented in the same tables (e.g., Tables 2, 3, and 4), respectively. Then the measurement results are sorted by distance to the APs position, the closest to the furthest.

According to the results shown in Fig. 2, the measurement reading of RSL decreased significantly at some points. The maximum value of RSL obtained is – 58 dBm, and the lowest is beyond – 80 dBm. While Fig. 6 shows how RSL values have decreased significantly at some point. The maximum RSL was – 47, whereas the lowest RSL reached – 87 dBm. Decreasing values of RSL started to appear when the user distance to the AP was increased. According to Eq. (1), the value of RSL

Table 3 RSL for AP2 on the left side

Point No.	Distance	RSL (dB)
8	9.5	– 63
9	16.6	– 80
10	6	– 61
11	11.2	– 61
12	6.3	– 64
13	9.4	– 73
14	7	– 58
15	14.3	– 71
16	13.3	– 68
17	8.2	– 75

Table 4 RSL for AP1 on the right side

Point No.	Distance	RSL (dB)
1	4.3	– 55
2	7.4	– 58
3	15.5	– 84
4	16	– 87
5	9.9	– 67
6	13	– 66
7	11.3	– 63

Table 5 RSL for AP2 on the right side

Point No.	Distance	RSL (dB)
8	3.7	– 53
9	7.3	– 74
10	6.3	– 55
11	3	– 49
12	5.3	– 59
13	3	– 48
14	5	– 56
15	10	– 66
16	12	– 68
17	15	– 82
18	13	– 67

is influenced by distance factors and dampness and shadowing of the wall. On the other hand, a few points have similar distances registered with different RSL values. This phenomenon occurred as a wall blocked one user while two blocked another (Figs. 4 and 5).

Results for APs Services coverage area in the Network Model

The AP1 coverage area was measured by switching off AP2 and performing a scenario to AP2 on each side of the network model, which is the Department, according to Figs. 4 and 5. Besides, users' route in measurement depends on Figs. 4 and 5 and the results presented via NetSpot Software that can be seen in Fig. 6. A heat map featuring different colors is presented in Fig. 6 to describe the strength of the received signal and correspond to service quality. An excellent service quality denotes the dark green color, while Light green refers to good service quality, and yellow indicates lower service quality for users located further from APs on both sides of the area.

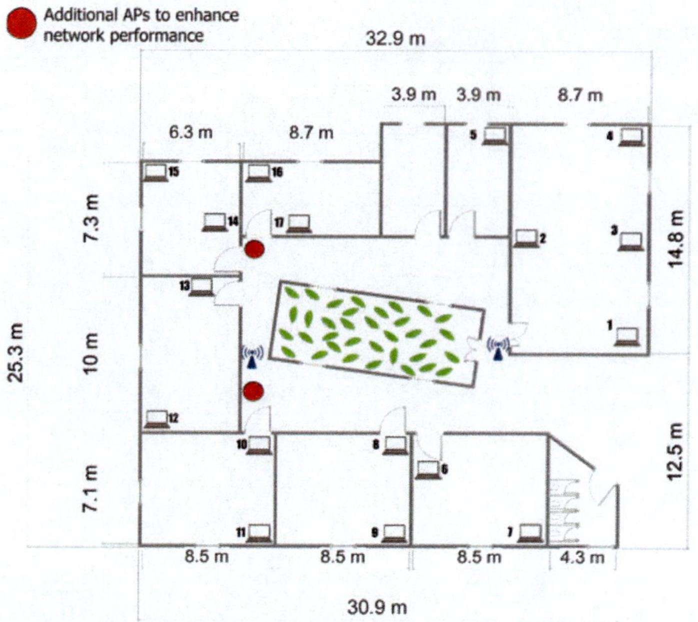

Fig. 4 New optimized design for the left

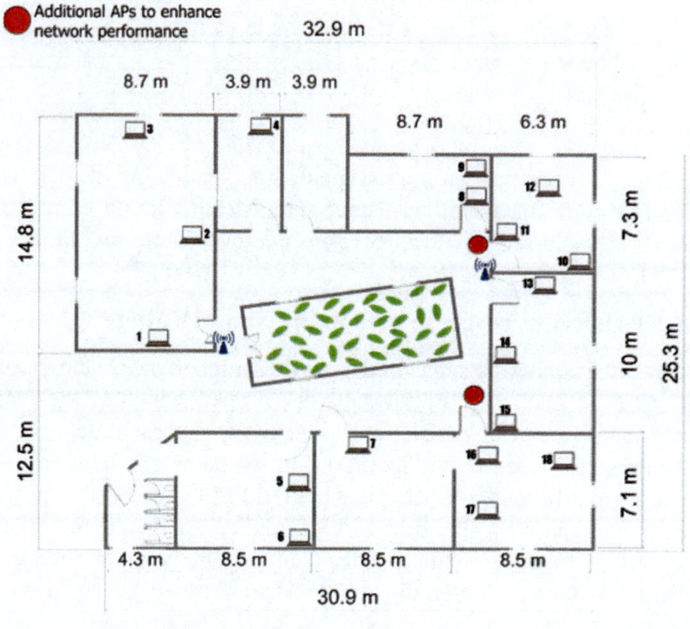

Fig. 5 New optimized design for the right

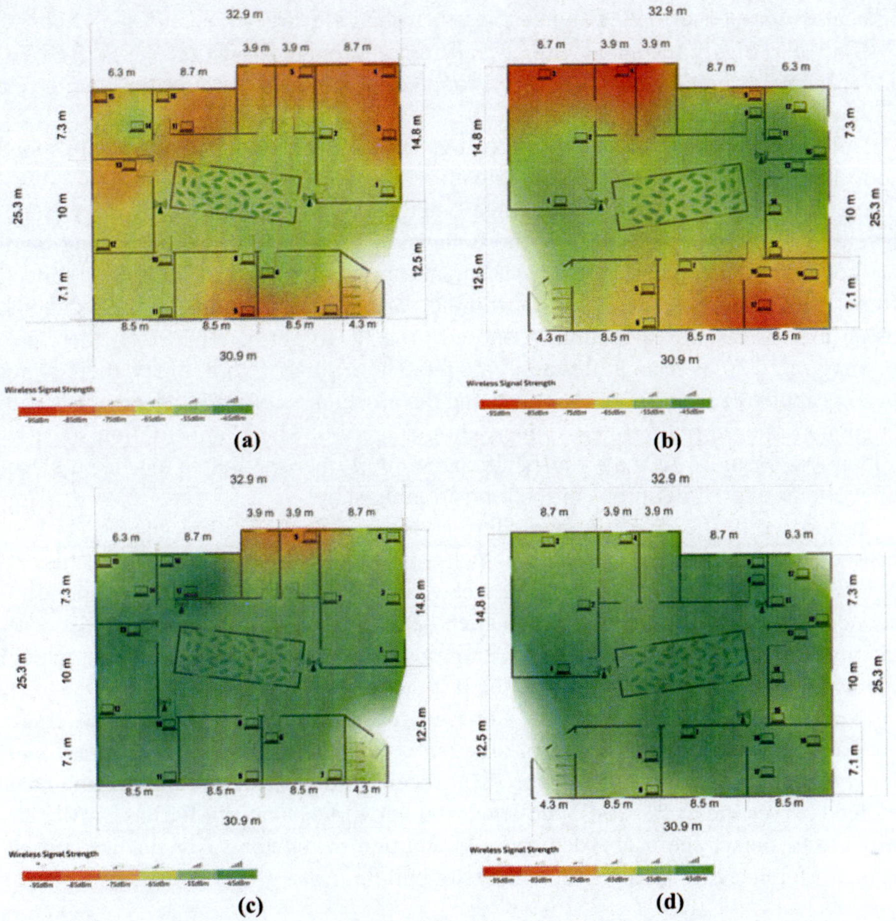

Fig. 6 A heat map coloring describing the received signal strength before and after. **a** Signal coverage area for APs on the left, **b** signal coverage area for APs on the right, **c** signal coverage area for APs on the left, **d** signal coverage area for APs on the right

3.3 *Optimized Design of the Proposing Network Model.*

Network equipment might need to be re-positioned to ensure decent signal coverage within the indoor area. Likewise, one access point is placed on each side of the Department (network model) to be a new-optimized design after the distribution is done (see Fig. 5).

In the beginning, the entire site plan of the targeted region must be drawn in detail by using any of the corresponding Software more professionally. In addition, all the positions of the current APs, that provide the signal should be illustrated.

Receive Signal Level (RSL) for the targeted region is determined using a network service quality tool (see Sect. 3.2), such as the NetSpot tool, which provides colored indicators for measuring the signal strength around the targeted area (Figs. 4 and 5). As shown in Fig. 5, the spots with a weak signal level are shown orange. Moreover, never being covered is shown with red canbe easily recognized to be covered in the following stages.

To achieve optimal signal coverage in general, all the weak signals or bare spots must be minimal or eliminated along with a minimized number of additional APs. Therefore, a procedure with three stages is proposed to fulfill the optimal coverage level in a general approach, and these stages are Measuring, Repositioning, and Compensation. The measuring stage is the same process as the RSL mentioned above. Reposition stage is achieved by examining various combinations of AP locations using the Circle Pack algorithm proposed in [22]. To get better RSL, if there are no uncovered areas found after applying reposition stage several times, there is no need to apply the Compensation stage, and the procedure will end. Indeed, this stage is applied after applying the Compensation stage too. The Compensation stage is adding one or more APs after a specific amount of uncovered area has taken place after attempting all iterations of the repositioning stage.

In this work, the coverage procedure above was applied till all bare spots had been covered, as shown in Fig. 6, which presents a heat map featuring different colors to describe the strength of the received signal and which also correspond to service quality to compare before and after the optimal design of the proposed model is implemented. All stages of the coverage procedure are applied as shown, which dramatically affects the results after the last measuring stage.

According to Fig. 6, all points shown in A and B concerning signal coverage area and the strength of the received signal were optimized to be excellent, and good signal quality depends on the dBm received. Thus, the proposed optimized method serves the expansion of colleges and network scalability for the University of Kufa to satisfy the user's demands. In addition to reducing APs devices, costs, and interference effects inner the university building.

3.4 Comparison of the State of the Art

A comparison with state-of-the-art methods was conducted to evaluate APs services coverage area in the network model proposed by researchers. Table 6 presents a comparison associated with the state-of-the-art methods that shed light on their coverage area and service quality in the proposed network model.

Referring to Table 6, different studies have been attempted to investigate the indoor Wi-Fi communication to measure the signals and the strength of APs localization inside the buildings, such as [15], which support a coordinate system and graphic representation that is solved by using particle swarm optimization (PSO) optimization method to facilitate the AP placement algorithm. Moreover, [16] solve the problem of optimal AP placement for the indoor positioning system. Besides [23], which conducting to optimal APs localization in indoor infrastructure. As a case study, the proposed method was deployed in the natural world environment, particularly at

Table 6 Comparison of the proposed method with the state-of-the-art

Refs	APs localization	Coverage area	Services quality
[14]	×	✓	×
[15]	✓	✓	×
[16]	✓	✓	×
[17]	×	✓	✓
[23]	✓	✓	×
[24]	✓	✓	×
[25]	✓	✓	×
[26]	×	✓	✓
[27]	×	✓	✓
[28]	×	✓	×
	×	✓	×
Proposed	✓	✓	✓

Sunway University. The results show that DCCLA has produced the highest accuracy with well-equipped wireless access points. Also, the study presented in [24] that selecting reliable APs positioning for WLAN network uses a quasi-entropy function to measure the signal discrimination quantification between distinct locations. As well as in [25], which presents a low-cost method optimal AP location algorithm by two-step measurement, the results have shown that the conducted method achieved optimum AP location and enhanced the network coverage area. While only [28] proposed the PSO method to improve the network coverage by (40%).

On the other hand, some works have been conducted to assess the network service quality by [17] using a Genetic Algorithm to enhance the quality of the indoor navigation system. Likewise, the study [26] provides statistical results to evaluate large-scale Wi-Fi network performance on the Baghdad University College of Arts campus, and finally, [27] introduces performance measurements for the AP network based on RSL and service coverage Area criteria. The measurement results have shown areas that have low RSL. Thus, the current study aims to identify the access point performance in Indoor Wi-Fi communication. RSL and service coverage area parameters are used to distinguish dead zones around the campus besides selecting an optimal campus APs localization by optimization method to reduce APs devices, costs, and interference effects inner the building.

4 Conclusions and Future Work

Achieving good wireless coverage within an indoor environment can be challenging, as it requires good planning and designing using the right tools and equipment. This paper discussed the importance of APs localization in an indoor environment at the

Department of Electrical Engineering, faculty of engineering adopted as a case study to evaluate the network performance.

The evaluation is conducted by investigating the reasons influencing the network performance through a survey questionnaire. In terms of performance evaluation, the network connection is measured by examining coverage area and service quality for network signal and connections. The evaluation results show that the highest RSL obtained is − 47 dBm, and the lowest RSL reaches − 87 dBm. Hence, the presented Wireless network coverage and performance have been analyzed and showed disheartening results in covering the entire building. As a result, an optimization for the repositioning of AP devices has been proposed to enhance the coverage area and overall wireless network performance by adding additional APs devices. A comparison with state-of-the-art methods was conducted to evaluate APs services coverage area in the network model in Table 6.

The optimized design provides a higher coverage area and service quality. Furthermore, it has increased the received power, reduced the dead zones, and achieved a higher coverage area at the lower possible implementation costs. For the plan, the outdoor area is examined and analyzed to verify the results evaluated by the network analyzer.

References

1. Sriram B, Sarrayrih MA (2016) An analytical study on user satisfaction on Wi-Fi services: a case study on Omani undergraduate environment. Indian J Sci Technol 1–9
2. Kadhim MF, Ali NS, Al-Khammasi S (2018) Multi-phase methodology for proposing a high performance switched campus network: University of Kufa case study. J Eng Appl Sci 13(16):6700–6707
3. Amano T, Higashino T (2019) Outdoor Wi-Fi RSSI map construction based on crowd-sourcing and simulation. In: 2019 IEEE International conference on pervasive computing and communications workshops (PerCom Workshops). IEEE, pp 443–444
4. Karnık J, Streit J (2016) Summary of available indoor location techniques. IFAC-PapersOnLine 49:311–317
5. Abdulkareem KH, Azmi MS, Ali NS, Al-Mhiqani MN, Alsharqi H (2018) Systematic Phases for proposing a new model of qualifications gap based on network technician data. J Soft Comput Decis Support Syst 1:27–35
6. Puspitasari NF, Al Fatta H, Wibowo FW (2015) Implementation of greedy and simulated annealing algorithms for wireless access point placement. In: 2015 3rd International conference on artificial intelligence, modelling and simulation (AIMS). IEEE, pp 165–170
7. Al-Khammasi S, Alhelal D, Ali NS (2018) Energy efficient cluster based routing protocol for dynamic and static nodes in wireless sensor network. TELKOMNIKA (Telecommun Comput Electron Control) 16(5):1974–1981
8. Shareef OA, Abdulwahid MM, Mosleh MF, Abd-Alhameed RA (2019) The optimum location for access point deployment based on RSS for indoor communication. In: International conference on modelling and simulation (UKsim2019), vol 20, pp 2.1–2.6
9. Hussain TH, Salman AA (2011) Indoor access point optimization using differential evolution. In: Proceedings of the second Kuwait conference on e-services and e-systems, pp 1–8
10. Nurminen H, Dashti M, Piché R (2017) A survey on wireless transmitter localization using signal strength measurements. Wireless Commun Mobile Comput

11. Abdulkareem KH, Azmi MS, Ali NS, Nasser M (2016) Factors affecting qualifications gap for network technicians: Baghdad universities case study. Int J Recent Trends Eng Res 5–11
12. Oppenheimer P (2004) Top-down network design. Cisco Press
13. Rattanaopas K, Boonchuay K (2017) A high-performance network virtualization architecture with bandwidth guarantees. In: 2017 International Electrical Engineering Congress (iEECON). IEEE, pp 1–4
14. Du X, Yang K (2016) A map-assisted Wi-Fi AP placement algorithm enabling mobile device's indoor positioning. IEEE Syst J 11:1467–1475
15. Voronov RV (2017) The problem of optimal placement of access points for the indoor positioning system. Вестник Санкт-Петербургского университета. Серия 10. Прикладная математика. Информатика. Процессы управления 13(1):61–73
16. Alsmady A, Awad F (2017) Optimal Wi-Fi access point placement for RSSI-based indoor localization using genetic algorithm. In: 2017 8th International conference on information and communication systems (ICICS). IEEE, pp 287–291
17. REMCOM Inc. (2009) The wireless insite user's manual. Version 2.6.3, Romcom Inc., 315s, Allen St., Suite 416 State College, pa16801
18. Rappaport TS (2002) Wireless communication principles and practice. Prentice-Hall, USA
19. Gunantara N, Hendrantoro G (2013) Multi-objective cross layer optimization for selection of cooperative path pairs in multihop wireless ad hoc networks. J Commun Softw Syst 9(3)
20. Singh R, Kumar S (2018) A comparative study of various wireless network monitoring tools. In: 2018 First international conference on secure cyber computing and communication (ICSCCC). IEEE, pp 379–384
21. Collins CR, Stephenson K (2003) A circle packing algorithm. Comput Geom 25(3):233–256
22. Toh C, Lau SL (2016) Indoor localisation using existing Wi-Fi infrastructure—a case study at a university building. In: 2016 22nd International conference on virtual system and multimedia (VSMM). IEEE, pp 1–5
23. Fang SH, Lin TN (2010) Accurate indoor location estimation by incorporating the importance of access points in wireless local area networks. In: 2010 IEEE global telecommunications conference, GLOBECOM 2010. IEEE, pp 1–5
24. Abdulwahid MM, Al-Ani OAS, Mosleh MF, Abd-Alhmeed RA (2019) Optimal access point location algorithm based real measurement for indoor communication. In: Proceedings of the international conference on information and communication technology, pp 49–55
25. Habeeb HS, Al_Yasiri HA, Al_Nakkash AH (2020) Performance evaluation of a campus hotspot network in Baghdad. IOP Conf Ser: Mater Sci Eng 881(1):012137. IOP Publishing
26. Gunantara N, Sudiarta PK, Prasetya AA, Dharma A, Antara IG (2018) Measurements of the received signal level and service coverage area at the IEEE 802.11 access point in the building. J Phys: Conf Ser 989(1):012014. IOP Publishing
27. Srivastava A, Vatti R, Deshpande V, Patil J, Nikte O (2018) Coverage improvement of IEEE 802.11 n-based campus wide wireless LANs. In: 2018 International conference on advances in communication and computing technology (ICACCT). IEEE, pp 126–129
28. Abdulwahid MM, Al-Hakeem MS, Mosleh MF, Abd-alhmeed RA (2020) Investigation and optimization method for wireless AP deployment based indoor network. IOP Conf Ser: Mater Sci Eng 745(1):012031. IOP Publishing

The Hybrid of WOABAT-IFDO Optimization Algorithm and Its Application in Crowd Evacuation Simulation

Hamizan Sharbini, Roselina Sallehuddin, and Habibollah Haron

Abstract This paper proposes a new hybrid of nature inspired optimization algorithm (IFDO-WOABAT) based on the latest optimization algorithm namely Improved Fitness Dependent Optimization (IFDO) with Whale-Bat Optimization algorithm (WOABAT). The hybrid is essential to overcome the inaccuracy in searching optimal path when dealing with many agents in conjunction with exploration and exploitation element in WOABAT signify the process of searching behaviour and optimizing the speed value of agent. The performance of the new hybrid optimization algorithm is verified using standard classical test function and further evaluated with other four renowned optimization algorithms and the results showed that it is better in most cases compared with the existing algorithms. Ultimately, the algorithm's performance also has been tested in crowd simulation evacuation that involves a different number of agents and with/without obstacle scenario. The conducted experiment reveals promising results and signify effectiveness in minimizing the evacuation time.

Keywords Hybrid whale-bat with improved fitness dependent optimization · Nature inspired optimization · Crowd evacuation simulation

H. Sharbini (✉) · R. Sallehuddin · H. Haron
School of Computing, Universiti Teknologi Malaysia, 81310 Johor Bahru, Malaysia
e-mail: hamizan6@live.utm.my

R. Sallehuddin
e-mail: roselina@utm.my

H. Haron
e-mail: habib@utm.my

H. Sharbini
Faculty of Computer Science and Information Technology, Universiti Malaysia Sarawak, 94300 Kota Samarahan, Malaysia

© The Author(s), under exclusive license to Springer Nature Singapore Pte Ltd. 2023

D.-K. Kang et al. (eds.), *Proceedings of the 9th International Conference on Computational Science and Technology*, Lecture Notes in Electrical Engineering 983,
https://doi.org/10.1007/978-981-19-8406-8_49

1 Introduction

In recent trend of crowd evacuation simulation system, there is a need to optimized using different optimization algorithms (hybrid). This is to help finding the best location of the exit door of evacuation area. Some may say that shortest distance towards exit is crucial, but optimization can help decide to find the best or near optimal solution (exit) [1]. The coordination between subgroups within the crowd, and uncertainty in crowd movement due to obstacle's position led to the studies of several soft computing (SC) based for developing intelligent approaches to govern and simulate crowd motion [2]. In the context of optimization issues, a traditional evacuation system indicates only a fixed direction and may mislead people to a dangerous or wrong place. However, only a few studies are dedicated towards this system [3]. The optimization is crucial for retrieving a minimal evacuation time [4]. Optimization algorithms under nature-inspired metaheuristic algorithms has been in research domain and some of the most efficient has been widely proposed in past literature. As coined by [5], they explained the term 'No Free Lunch Theorems for Optimization' to indicate that there will always in a need for latest developed algorithms provided that the proposed algorithm is better than previously and much more efficient in terms of performance wise. Traditional algorithms might have problems to cater for real-world problems (high non-linear problem) because of the impossibilities to have every single solution for the problems in terms of time, cost, and space (search space). Thus far, it is recommended that metaheuristic is said to be more suitable to cater for such predicaments [6]. The organization of this paper is as follows: Sect. 2 will describe about the related works. Section 3 introduces on the proposed method. Section 4 discusses on experiment and simulation analysis and results. Finally in Sect. 5 is describing the conclusion and future work.

2 Related Works

The current emergency research shows emergency simulation is important for mitigation to avoid risk of human life, but not considered the interaction of human i.e., human to obstacles, social interaction with each other that mostly takes place during emergency (optimizing planning behaviour) to shorten the period for this post emergency evacuation time. There is still an issue on how to define the right rules in terms of interactions of the individual speeds that clearly will show the behaviour of agent (particle) of the swarm (crowd) [7].

Another most notable optimization algorithm under nature inspired is the Ant Colony Optimization. This algorithm has been used to solve the evacuation routing problem, which is based on the organization of the evacuee's space–time paths within a hierarchical directed network. The results showed that it can reducing evacuation time within the scope of the experiment, total travel distance and congestion. There will be a need to further improve in future research that focus on improving the space–time paths of evacuees [8]. In another work done by [1], the implementation of Cellular Automata integrates with Fuzzy Logic, Kth Nearest Neighbour's (KNN) has been used to simulate evacuation scenario. They have not incorporated an element to resemble panic situation. Another example of nature inspired algorithm is the Modified Artificial Bee Colony (MOABC) done by [9]. The results from their work have shown that the time for seeking shelter from open area to close area was reduced. The drawback of this work is they still need to model conceptualization in regards of the area characteristics, topography, and agent's behaviour other than to include other factors namely as traffic and risks along evacuation paths. The latest optimization algorithm such as Improved Fitness Dependent Optimization (IFDO) still cannot deal with searching accuracy when the search agent is more than five. Therefore, it is a need to use large number of search agents to enhance the accuracy of the algorithm with more space and time by further hybrid with other popular nature-inspired algorithms, such as Whale-Bat (WOABAT) algorithm [10].

3 Proposed Hybrid of IFDO-WOABAT Algorithm

3.1 Evaluation and Experiment Setting on Function Optimization

As usual, the evaluation is normally done for an algorithm to get better comparison. There are two ways for evaluation; (i) by average and standard deviation result and (ii) by comparison of hybrid WOABAT-IFDO with other selected nature inspired metaheuristic algorithms. Based on the first point, the algorithm's performance or evaluation is tested as function optimization problem that use the Classical Benchmark testing. The proposed hybrid WOABAT- IFDO based on a group of 19 classical benchmark test functions. Next, for the second evaluation it is crucial to make a comparison between the proposed WOABAT-IFDO with other selected nature inspired algorithm namely as IFDO, FDO and PSO [10] and WOABAT. As it is important that an algorithm to avoid being trapped in local optimal solution and to get to global optimum solution. Figure 1 the flowchart of WOABAT-IFDO respectively. The testing has been done using MATLAB2020(b) in Windows 10 operating system. The processor is AMD Ryzen 5 4600H with 16 GB RAM. The size of dimension for testing is set to 10, whilst the algorithm is tested 30 times for 500 iterations, 30 scout bee.

Table 1 shows the results of mean and standard deviation for each tested algorithm. It shows that the hybrid algorithm WOABAT-IFDO obtained quite a promising performance compare with other algorithms on almost every benchmark function [10] and shows the same result with addition of WOABAT experiment included. The results showed that the proposed hybrid algorithm obtained most of the minimum results over 19 test functions. The mean value depicts the convergence speed and

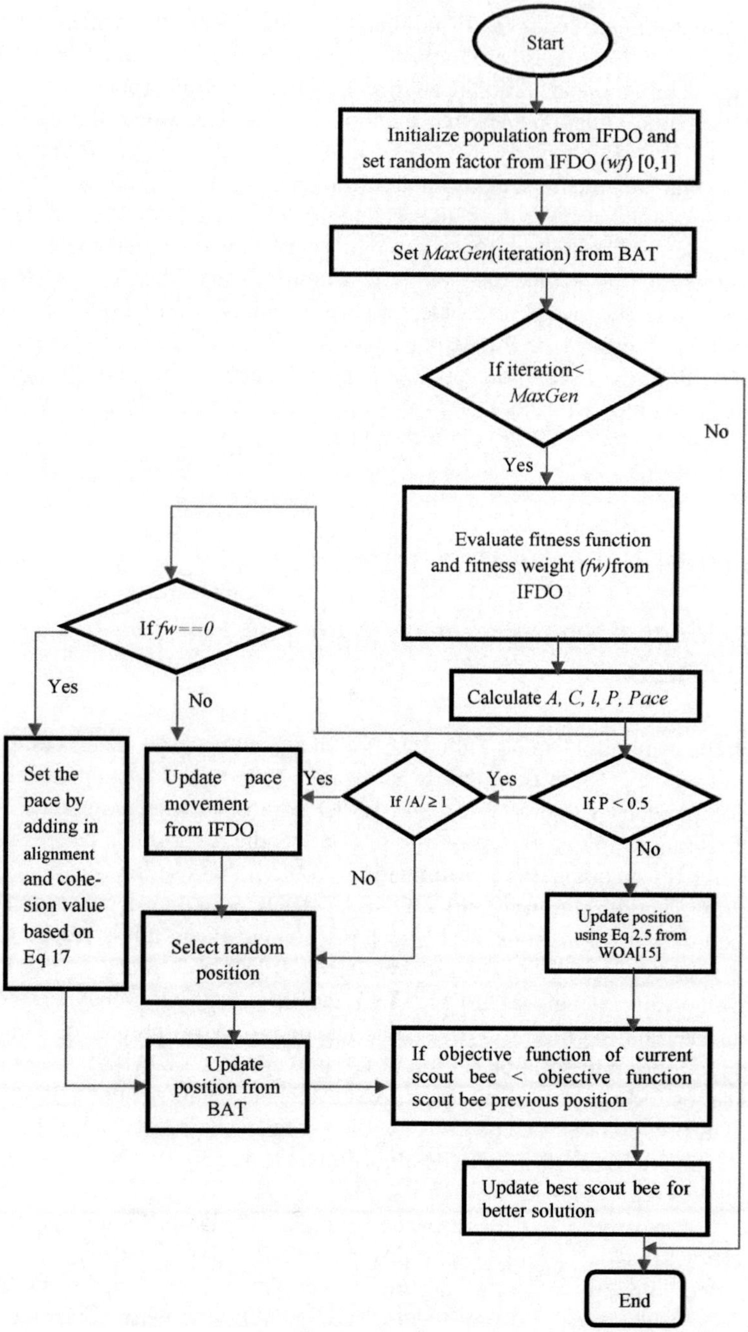

Fig. 1 The proposed hybrid WOABAT-IFDO flowchart

accuracy of the optimization function of an algorithm. Meanwhile, the standard deviation reflects the dispersion and stability of an algorithm. Thus, the proposed hybrid algorithm WOABAT-IFDO has better optimization accuracy, convergence speed and stability.

A quantitative measurement metrics were adapted for the test function details namely unimodal, multimodal, and composite standards as a feature to measure specific algorithm outcome. As for unimodal benchmark, this is used for testing the exploitation stage and for convergence of an algorithm (targeting on single optimum). On the other hand, multimodal is used for testing the exploration stage and avoidance in local optima condition. Multimodal benchmark functions mean it has multi-ideal or optimal solution for example global and local optimal solutions. As mentioned before, an algorithm should be able to prevent local optimum solutions. Instead, it should get to converge towards global optimum solution. As for the composite benchmark, it comprises of diverse and different region of search landscape and tend to have a very big local optimum. The mean value from the result of test function shows the speed on convergence rate and the accuracy of the proposed algorithm. Based on Tables 1 and 2, the unimodal standard functions which range from TF1 to TF7, the result of hybrid WOABAT + IFDO in TF7 is comparable with IFDO and FDO. However, TF1 and TF4 showed poor performance in comparison with the selected algorithms. For multimodal standard which range from TF8 to TF13, the test function result of TF8, TF9 and TF11 showed that the hybrid WOABAT + IFDO outperformed the other competitor algorithm. Moreover, for the composite standard functions range from T14 to T19, the result of TF15, TF17, TF18 and TF19 for hybrid WOABAT + IFDO showed better performance than other selected algorithms. Generally, the proposed hybrid of WOABAT and IFDO has the capability to exploit the search space with improved exploration stage and overall can move in the direction of optimality and avoid local optima.

3.2 Crowd Evacuation Simulation Experiment

The hybrid WOABAT-IFDO algorithm can be applied in many ways either to optimize other crowd model or as the optimizer for a path planning of agent for evacuation simulation. For this paper, it will encompass on the later purpose. The environment of simulation is developed using MATLAB R2020b under Windows 10 operating system. The simulation will have few scenarios which include obstacle and non-obstacles condition. The parameter for the simulations is detailed as in Tables 3 and 4 which is based on the map area focusing on the near main entrance that normally done for exhibition event location that can be accessed via Borneo Convention Centre Kuching website [11]. The walking speed is based on general adult speed and the radius size are based on the data that has been published in literature [12, 13].

From the optimization performance experiment from Tables 1 and 2, the fitness that has been assigned for this evacuation shows better evacuation of agents during evacuation procedure. The agent's position and direction are impacted with different

Table 1 Comparison of performance on selected optimization algorithms (FDO and IFDO)

Test function	IFDO_WOABAT		FDO		IFDO	
	Ave	Std	Ave	Std	Ave	Std
*TF1	4.88E − 06	1.73E − 06	7.47E − 21	7.26E − 19	5.38E − 24	2.74E − 23
TF2	0.0058	0.001	9.39E − 06	6.91E − 06	0.534345844	1.62026
TF3	1.08E − 05	4.65E − 06	8.55E − 07	4.40E − 06	2.88E − 07	6.90E − 07
*TF4	0.0518	0.041	6.69E − 04	0.0024887	2.60E − 04	9.11E − 04
TF5	4.263	1.1094	23.501	59.7883701	1.94E + 01	3.31E + 01
TF6	5.61E − 06	2.06E − 06	1.42E − 18	4.75E − 18	4.22E + 06	8.15E − 09
TF7	0.5253	0.2892	0.544401	0.3151575	5.68E − 01	3.14E − 01
TF8	**− 2.14E + 306**	1.03E + 307	− 228,520	206,684.91	− 2.92E + 06	2.24E + 05
TF9	**0.0011**	3.88E − 04	14.56544	5.202232	1.35E + 01	6.66E + 00
TF10	0.003	4.37E − 04	4.00E − 15	6.38E − 16	5.18E − 15	1.67E − 15
TF11	**1.17E − 05**	5.86E − 06	0.568776	0.1042672	0.525690405	8.90E − 02
TF12	2.24E − 04	8.96E − 04	19.83835	26.374228	1.81E + 01	2.57E + 01
TF13	0.0121	0.0151	10.2783	7.42028	4.10E + 09	1.50E − 05
TF14	1.1634	0.5266	3.79E − 07	6.32E − 07	2.68E − 07	4.68E − 07
TF15	**2.41E − 16**	4.48E − 16	0.001502	0.0012431	4.03E − 16	9.25E − 16
TF16	− 1.0316	4.82E − 07	0.006375	0.0105688	9.14E − 16	3.61E − 16
TF17	**0.3979**	1.87E − 07	23.82013	0.2149425	2.38E + 01	1.24E − 01
TF18	**3.000**	1.95E − 05	222.9682	9.96E − 06	2.24E + 02	2.68E − 05
TF19	**− 3.8112**	0.1961	22.7801	0.0103584	3.15E + 01	1.32E − 03

The bolded values signifies the best values for the test functions listed

Table 2 Comparison of performance on selected optimization algorithms (FDO and IFDO)

Test function	IFDO_WOABAT		WOABAT		PSO	
	Ave	Std	Ave	Std	Ave	Std
*TF1	4.88E – 06	1.73E – 06	1.92E – 08	2.30E – 08	4.20E – 18	1.31E – 17
TF2	0.0058	0.001	0.0062	9.79E – 04	0.003154	0.009811
TF3	1.08E – 05	4.65E – 06	0.0155	0.085	0.001891	0.003311
*TF4	0.0518	0.041	4.41E – 04	6.69E – 05	0.001748	0.002515
TF5	4.263	1.1094	1.2156	2.77	63.45331	80.12726
TF6	5.61E – 06	2.06E – 06	1.45E – 08	2.22E – 08	4.36E – 17	1.38E – 16
TF7	0.5253	0.2892	0.0029	0.0095	0.005973	0.003583
TF8	**– 2.14E + 306**	1.03E + 37	– 39.3375	1.45E – 14	– 7.10E + 11	1.20E + 12
TF9	**0.0011**	3.88E – 04	5.6381	8.1308	10.44724	7.879807
TF10	0.003	4.37E – 04	12.6424	5.42E – 15	0.280137	0.601817
TF11	**1.17E – 05**	5.86E – 06	0.2816	0.227	0.083463	0.035067
TF12	2.24E – 04	8.96E – 04	9.75E – 10	1.24E – 09	8.57E – 11	2.71E – 10
TF13	0.0121	0.0151	3.37E – 09	4.40E – 09	0.002197	0.004633
TF14	1.1634	0.5266	2.7558	3.8086	150	135.4006
TF15	**2.41E – 16**	4.48E – 16	0.0011	0.0037	188.1951	157.2834
TF16	– 1.0316	4.82E – 07	6.42E + 03	1.85E – 12	263.0948	187.1352
TF17	**0.3979**	1.87E – 07	5.8444	3.18E – 15	466.5429	180.9493
TF18	**3.000**	1.95E – 05	8.4	10.9846	136.1759	160.0187
TF19	**– 3.8112**	0.1961	– 3.8628	8.09E – 14	741.6341	206.7296

The bolded values signifies the best values for the test functions listed

Table 3 The environment of evacuation setting under BCCK selected hall layout

Input	Value/number
Walking speed (adult) Radius size (agent)	1.47 m/s (14.77)* 0.2 (2)*
Area	49.7 × 57.2 m (main populated area)
Exit $(X)_e$	1
Exit width	9 m (90)
Obstacle $(X)_n$	1–15
Particles (a)	Range from 50 to 1000

Table 4 Model parameters for evacuation

Parameter	Maxgen	Fmin	Fmax	b_woa	lb = min(Area)	p_sigma	C_obs	Stepsize
Value	3000	0	2	0.001	Populated area	0.2	0.1	14.77

scenario of obstacles involve and position of exit. The impact from obstacle is derived from bat algorithm [14] as expressed in Eq. (1) below:

$$cost(p,q) = exp\left(-\left(\frac{(p_x - q_{x)})^2}{(\sigma_{px} + \sigma_{qx})^2}\right) + \left(\frac{(p_y - q_{y)})^2}{(\sigma_{py} + \sigma_{qy})^2}\right)\right) \tag{1}$$

The px, qx shows the position of agent (individual p), whilst the qx, qy shows the position of certain dynamic and static obstacle that represent the setting in the environment. The sigma denoted the agent size and area of obstacle's coverage. The overall cost function that corresponds as fitness (Fobj) is formulated based on Eq. (2) [14] below:

$$F_{obj}(p) = C_{obs} X \underset{o \in O}{max}(cost(p, O)) + \frac{1}{cost(p, g)} \tag{2}$$

The O shows the static and dynamic obstacles (if any exist), whilst g denotes exit position. Cobs means the value of obstacle's weight coefficient; the higher the value of Cobs, the bigger the influence of the obstacle has towards agents going to the exit way. From IFDO evaluation, the following expression Eq. (3) [10] for search landscape is expressed as follows:

$$nl = \frac{lB}{2 * PI} \tag{3}$$

The n is to denote which scout (agent) involves in deciding the alignment and cohesion [10]. The IFDO moved the (scout) agents' population randomly during initialization stage. For scout that maintain its position during its global best in predetermined time, it will move randomly until the movement get the better position. The range of search landscape is being reduced (shrink) based on WOA algorithm [15] as to allow for any scout (agent) new position can be decided better in more focus range between current position and targeted new position. The time for evacuation is analyzed and compared with different number of agents and obstacles using algorithms namely WOABAT, BAT, WOA, FDO, IFDO and PSO and the output of simulation as shown in Fig. 2.

4 Results and Discussions

Figure 3a shows that the proposed hybrid WOABAT-IFDO algorithm in evacuation simulation with no obstacle in population setting from 200 to 1000 agents. The mean average of time taken shows that the WOABAT-IFDO is 54.75 s and 124.4 s therefore is faster compared to other algorithms respectively. The time is gradually increase when the number of agents is higher, but overall WOABAT-IFDO marked

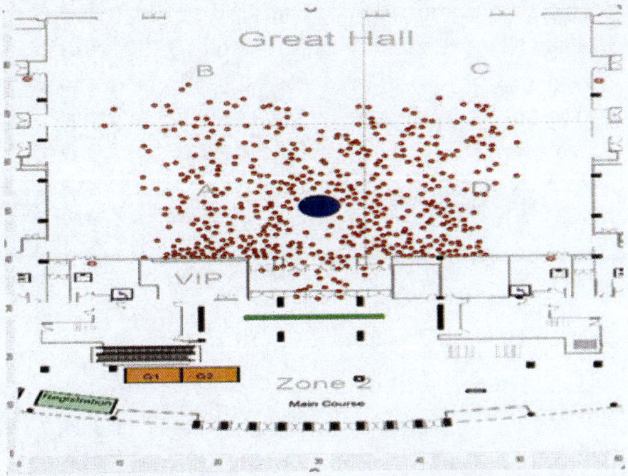

Fig. 2 Example of the evacuation scenario based on 200–1000 agents with 1 obstacle. The simulation is repeatedly run with other different scenario such as no obstacle condition and with 2, 6 and 15 obstacle(s) with red particle denotes the agents and green area denotes the exit

the fastest evacuation time. Figure 3b shows that the proposed hybrid WOABAT-IFDO algorithm in evacuation simulation with one (1) obstacle in population setting from 200 to 1000 agents. The mean average of time taken shows that the WOABAT-IFDO is faster compared to other algorithms which is 54.89 s and 116.6 s respectively, whilst gradually increased with the increasing number of total agents (Table 5).

Figure 3c shows that the proposed hybrid WOABAT-IFDO algorithm in evacuation simulation with two (2) obstacles in population setting from 200 to 1000 agents. The mean average of time taken for 200 population size of WOABAT-IFDO is 55.38 and 113 s. gradually increased with the increasing number of total agents. This time of evacuation during 400 population shows slightly faster evacuation time that is 64.64 s compared to one (1) obstacle that showed 66.09 s condition due to influence of more than one obstacle. Figure 3d shows that the proposed hybrid WOABAT-IFDO algorithm in evacuation simulation with six (6) obstacles in population setting from 200 to 1000 agents. The mean average of time taken for 200 population size of WOABAT- IFDO is 60.13 s whilst for 1000 population is 118 s. The trending shows it is gradually increased with the increasing number of total agents. The theory of many obstacles can elevate evacuation time does not portray in this condition of square shape obstacle even the quantity of obstacle is higher compared to previous condition.

Finally, in Fig. 3e, it shows that the proposed hybrid WOABAT-IFDO algorithm in evacuation simulation with fifteen (15) obstacles in population setting from 200 to 1000 agents. The mean average of time taken for 200 population size of WOABAT-IFDO is 57 s whilst for 1000 population is 108 s. The time shows increasing trend and faster than two, one and non-obstacle conditions in 600–1000 population, but slightly slower time for 200 and 400 number of total agents with same condition of obstacle. The time taken for evacuation simulation from the proposed hybrid of WOABAT-IFDO is less than other selected algorithms.

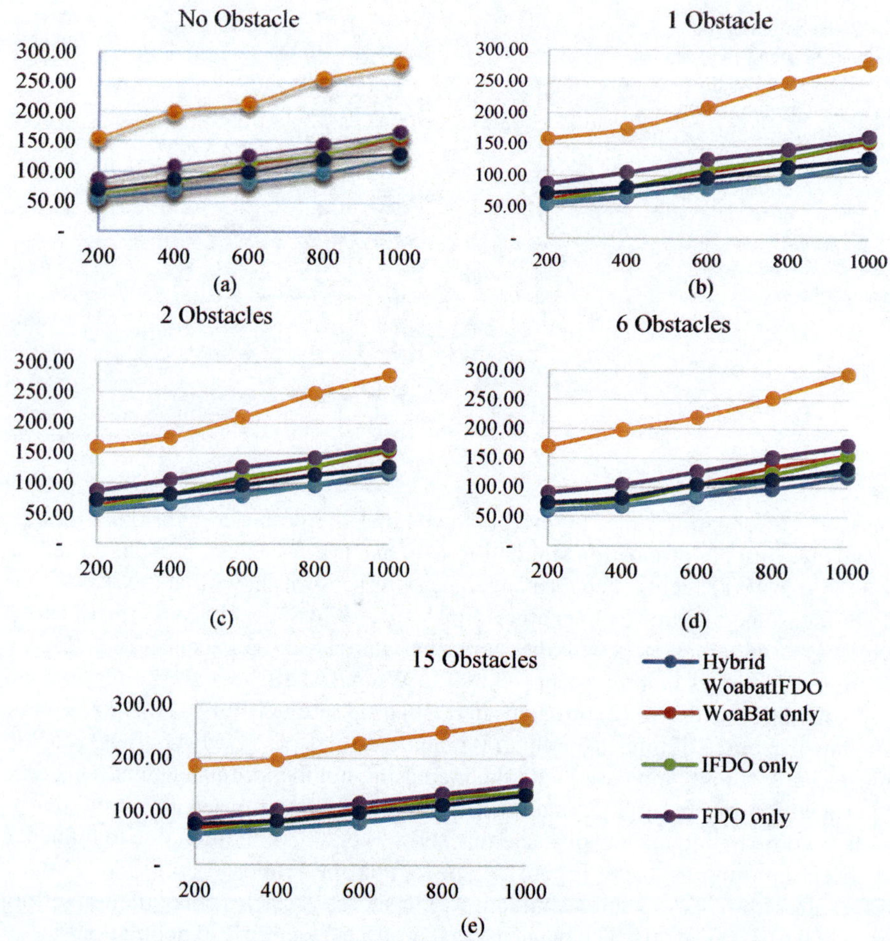

Fig. 3 a-e The evacuation time versus agent graph for 200–1000 agents' various scenarios

From the simulation results, it is observed that the evacuation time taken for hybrid WOABAT-IFDO is less than the rest of the algorithm. This is due to the alignment and cohesion updated as the particles seems more aligned and move coherently towards the exit (back to the nature of IFDO). At the same time, the use of WoaBat for exploration and exploiting the point and updating current position contribute to better searching and faster global convergence, hence reduce the time for evacuation. For single WOABAT, it still has exploration and exploitation capability but searching behaviour haven't been enhance as particle still wandering to find exit way. In this simulation, whilst for single IFDO, it is slightly slower than WOABAT algorithm since the particles only focus on group alignment and coherent value. Whale algorithm from simulation observation there is also a behaviour of last particle wandering as the coherency works better when in swarm or with other particles nearby to help

Table 5 The results of mean evacuation time for 200–1000 agents and selected algorithms. WOABAT-IFDO is demonstrating majority minimal evacuation time compared with others (in bold)

Particles	Algo/scene	No obstacle	1 obstacle	2 obstacles	6 obstacles	15 obstacles
200	**Woabat_IFDO**	**54.75**	**54.89**	**55.38**	**60.13**	**57.00**
	WoaBat only	72.36	62.00	59.20	70.60	67.80
	IFDO only	62.00	56.33	61.89	73.18	57.27
	FDO only	85.73	86.27	88.27	90.36	84.18
	BAT only	55.78	55.88	57.57	61.88	57.50
400	**Woabat_IFDO**	**67.45**	**66.09**	**64.64**	**68.00**	**66.67**
	WoaBat only	83.73	81.11	81.27	81.20	81.82
	IFDO only	81.73	80.60	80.18	76.36	80.00
	FDO only	108.27	104.73	106.45	104.14	101.80
	BAT only	70.75	67.88	65.17	69.20	67.83
600	**Woabat_IFDO**	**81.18**	**86.30**	**84.80**	**84.60**	**81.80**
	WoaBat only	109.91	105.73	106.18	104.00	104.25
	IFDO only	114.00	109.73	107.67	105.13	98.22
	FDO only	125.50	125.90	126.36	126.60	115.50
	BAT only	84.00	79.50	82.80	87.60	79.29
800	**Woabat_IFDO**	**97.75**	**96.38**	**95.13**	**97.43**	**96.89**
	WoaBat only	130.20	127.67	126.80	134.20	126.40
	IFDO only	129.50	128.67	126.00	122.17	123.40
	FDO only	144.88	141.40	140.40	150.80	133.80
	BAT only	99.89	98.00	97.86	112.13	98.89
1000	**Woabat_IFDO**	**124.40**	**116.60**	**113.00**	**118.00**	**108.00**
	WoaBat only	155.20	153.60	144.40	154.00	152.40
	IFDO only	162.20	157.80	156.00	153.00	138.40
	FDO only	166.20	162.20	165.40	171.20	151.20
	BAT only	125.60	120.00	116.40	125.33	110.25

The bolded values signifies the best values for the test functions listed

moving towards the exit way. As for Bat algorithm, in average this algorithm can search faster but the movement is unrealistic as if the bat 'jump' into the possible solution which cannot be consider as real behaviour in evacuation procedure. Nevertheless, for PSO, the particles searching behaviour which is equal to the local search behaviour tends to wander (trapped in local minima), thus contributes to slow convergence (slow getting to the exit/goal). The movement seems not realistic as it depicts a discrete movement behaviour [16].

5 Conclusions and Future Work

In this paper, the new WOABAT-IFDO algorithm has been proposed to boost the accuracy of searching process by enhancing the IFDO with hybridizing it with other latest nature-inspired algorithms WOA-BAT optimization algorithm. The hybrid is the concept of complementing in both of exploitation and exploration stage from WOABAT as well as embedding the element of alignment and cohesion as new pace value in IFDO. The proposed WOABAT-IFDO is being tested based on Classical Benchmark to test the performance of the algorithm and against other closely related nature-inspired algorithms. Extensively, the WOABAT-IFDO is applied to crowd evacuation simulation and the results showed efficient level in reducing the evacuation time with no- obstacle and obstacle condition. Although the results are quite competitive with BAT and IFDO in few experiments simulation, the hybrid WOABAT-IFDO is still notably well accepted as an efficient algorithm among all based on the percentage differences. The research project to optimize crowd evacuation model by incorporating with other evacuation model such as social force model with further validation on crowd evacuation time in future simulation work.

Acknowledgements This research was supported by Ministry of Higher Education (MOHE) through Fundamental Research Grant Scheme FRGS/1/2019/ICT02/UTM/02/13 and Research Management Centre (RMC), UTM.

References

1. Muhammed DA, Saeed SAM, Rashid TA (2016) A simulation model for pedestrian crowd evacuation based on various AI techniques. In: Revue d'Intelligence Artificielle 33(4):283–292. https://doi.org/10.18280/ria.330404
2. Mohd Ibrahim A, Venkat I, Subramanian KG, Khader AT, Philippe DW (2016) Intelligent evacuation management systems. ACM Trans Intell Syst Technol 7(3):1–27. https://doi.org/10.1145/2842630
3. Purva G, Saxena A (2018) Evacuation plans and simulations for crowd egress—a review. Int J Adv Eng Manag Sci 4(8):637–639. ISSN: 2454-1311. https://doi.org/10.22161/ijaems.4.8.10
4. Niyomubyeyi O, Pilesjö P, Mansourian A (2019) Evacuation planning optimization based on a multi-objective artificial bee colony algorithm. ISPRS Int J Geo-Inf 8(3):110. https://doi.org/10.3390/ijgi8030110
5. Wolpert DH, Macready WG (1997) No free lunch theorems for optimization. IEEE Trans Evol Comput 1(1):67–82
6. Fang ZX, Zong XL, Li QQ, Li QP, Xiong SW (2011) Hierarchical multi-objective evacuation routing in stadium using ant colony optimization approach. J Transp Geogr 19(3):443–451
7. Aljedaie AM, Hijji B, Alrashidi B, Alkhassawneh HM, Yadav K (2019) Swarm intelligence algorithms in concepts models, 31(2):217–222
8. Fausto F, Adolfo RO, Erik C, Ángel GA, Marco PC (2020) From ants to whales: metaheuristics for all tastes. Artif Intell Rev. https://doi.org/10.1007/s10462-018-09676-2
9. Laizhong C et al. (2018) A smart ABC algorithm with distance-fitness-based neighbor search and its application. Future Gen Comput Syst 89:478–493
10. Muhammed DA, Saeed SAM, Rashid TA (2020) Improved fitness-dependent optimizer algorithm. Access IEEE 8:19074–19088

11. BCCK, Borneo Convention Center Kuching. www.bbck.com.my. Accessed 7 June 2020
12. Seah et al. (2018) Agent-based modelling and simulation. In: Singapore management university report
13. Wang S et al. (2018) A multi-species ABC algorithm and its application for crowd simulation. IEEE Access. https://doi.org/10.1109/ACCESS.2018.2886629
14. Yu T et al (2019) Modeling and simulation of evacuation based on bat algorithm. IOP Conf Ser Earth Environ Sci 2:67 032017
15. Mohammed HM, Umar SU, Rashid TA (2019) Systematic and meta-analysis survey of whale optimization algorithm. Comput Intell Neurosci. https://doi.org/10.1155/2019/8718571
16. Peichuan T, Chihming C, Yingping C (2014) PSO-based evacuation simulation framework. In: IEEE Congress on Evolutionary Computation, pp 1944–1950

Cyberbully Detection Survey: Malay-English Glossary of Cyberbullying Incidents

Marina Md. Din, Fiza Abdul Rahim, Rina Md. Anwar, Asmidar Abu Bakar, and Aliza Abdul Latif

Abstract This article presents the results of a survey conducted to elicit keywords or phrases relating to cyberbullying incidents in both English and Malay languages commonly used in Malaysian society. The keywords or phrases can be utilized as a Malay and English cyberbullying glossary in the development of an auto-detection cyberbullying tool. A set of questionnaires were distributed among 329 respondents ages 15–30 years in Malaysia in the form of an online survey over one-and-a-half-month starting 1 November 2021. This study was conducted to test the items' reliability using Cronbach's alpha values. There are three (3) Sections to this questionnaire; Sect. 1 is about the demographics of the respondents, Sect. 2 is related to Cyberbullying, and Sect. 3 shows a few scenarios that might be or might not be a cyberbullying incident. Findings from the analyses showed that 447 words were collected, and all of these were later grouped into five (5) categories 5; Intellectual, Physical Appearance, Insulting/Offensive, Intimidating and Others. Making offensive comments or doing insulting posts was the most cyberbullying form made by the bullies. Five (5) popular words or phrases were identified as the common cyberbullying content in Malaysian society.

M. Md. Din (✉) · F. A. Rahim · R. Md. Anwar · A. A. Bakar · A. A. Latif
Institute of Informatics and Computing in Energy (IICE), Universiti Tenaga Nasional, Putrajaya Campus, Jalan Ikram-Uniten, 43000 Kajang, Selangor, Malaysia
e-mail: marina@uniten.edu.my

F. A. Rahim
e-mail: fiza.abdulrahim@utm.my

R. Md. Anwar
e-mail: mrina@uniten.edu.my

A. A. Bakar
e-mail: asmidar@uniten.edu.my

A. A. Latif
e-mail: aliza@uniten.edu.my

F. A. Rahim
Universiti Teknologi Malaysia, Jalan Sultan Yahya Petra, 54100 Kuala Lumpur, Malaysia

© The Author(s), under exclusive license to Springer Nature Singapore Pte Ltd. 2023

D.-K. Kang et al. (eds.), *Proceedings of the 9th International Conference on Computational Science and Technology*, Lecture Notes in Electrical Engineering 983,
https://doi.org/10.1007/978-981-19-8406-8_50

Keywords Cyberbullying · Cyberbullying detection · Text extraction · Machine learning

1 Introduction

With the recent growth of social media and online communication, cyberbullying has become a significant issue. Cyberbullying is any type of harassment involving the dissemination of offensive material (such as text, images, comments, rumors, and private information) about a person to make them feel bad [1]. Cyberbullying is a type of bullying that occurs using information and communication technologies such as social media, smartphones, and the Internet. In Malaysia, social networks are the most used platform for cyberbullying, accounting for 71% compared to the global average of 65% [2]. This is supported by U-Report Malaysia [3], which reported that 80% of cyberbullying occurs on social networks, with 47% using Facebook, and 36% using Instagram.

Flaming, online harassment, cyberstalking and exclusion are examples of cyberbullying [4]. Verbal abuse is allegedly the most common type of cyberbullying among adolescents [5, 6]. As of January 2022, approximately 44% of social media users in the United States [7] and 34% of social media users globally [8] experience cyberbullying, making it an international issue [9]. A Pew Research Center study found that cyberbullying had occurred to 59% of American teenagers [10].

According to a study by IPSOS [2], global cyberbullying awareness is 75%, while in Malaysia is 85%. 17% of the parents worldwide say that their children have been cyberbullied, while 23% in Malaysia. Comparitech.com conducted a survey to measure cyberbullying and discovered that Malaysia ranked sixth out of 28 countries and surprisingly second in Asia in 2018 [11]. According to U-Report Malaysia [3], 44% of the cyberbullying victims were between the ages of 15 and 19, while 28% were between the ages of 20 and 24. This has demonstrated that Malaysian youngsters are the ones who are primarily bullied. In Malaysia, a cyberbullying incident led to suicide, in which the abuser threatened to make the victim's personal photos on social media. This frightening incident is one of many examples of serious issues in cyberbullying incidents.

2 Related Works in Cyberbullying Prevention

Numerous programs have been put in place to address or stop cyberbullying events. Increasing awareness of cyberbullying through events and activities is one of the most common contemporary strategies for dealing with it.

Anti-bullying initiatives such as the Anti-Bullying Campaign [12], #DareToShout [13], and Nickelodeon's Together for Good anti-bullying initiative [14], were aimed at high school and college students. They sought to prevent bullying by teaching children and students about it and offering solutions, such as contacting Malaysia's Talian Kasih (care hotline) [15].

Numerous routes exist for reporting bullying incidents; one of them is Facebook's Bullying Prevention Hub [16]. Teens, parents, and educators can access services if they need assistance with bullying-related difficulties. These materials, which explain online bullying [17] and how to stop it, are offered by Non-Governmental Organizations (NGOs), governmental organizations, or private blogs. Users can also find instructions on Facebook on reporting and blocking posts.

The Instagram help center outlines how to report these issues. Offensive or improper activity on the social media platform Instagram can lead to the user being blocked [18] and can result in the user's account being suspended. The website recently implemented shadow banning, which restricts publishing or commenting on a post) as a tool to combat cyberbullying [19].

Twitter on the other hand, offers detailed instructions on how to report offensive or threatening behaviour [20]. A single report can have many Tweets packed.

In Malaysia, Cybersecurity Malaysia (CSM) has launched the CyberSAFE program to educate the Malaysian public on cyber safety and the threats they may face online. CyberSAFE is a platform where cyberbully victims can lodge a complaint to the Cyber999 Help Centre via email by sending an email to cyber999@cybersecurity.my or mycert@mycert.org.my. The public can also call the hotline 019-2665850 or use the Cyber999 mobile application on their mobile phones [21].

A measure in combating and reducing cyberbullying perpetration is needed whenever the bullies are trying to carry out a cyberbully attack. This includes to detect a cyberbullying posting by utilizing machine learning algorithms. There are four features used in detection outlined by Salawu et al. [22]; content-based, sentiment-based [23], user-based, and network-based features. Work done by [24] used a machine learning approach for text classification in cybercrime detection. [25–27] utilizing deep learning for textual extraction (content-based) in their work. Interestingly, work done by [17, 23, 28] developed hybrid models integrating content based, user based and network-based features to achieve the ultimate result in cyberbully detection. Most works have examined cyberbullying in English-language texts.

To stop or at least lessen future instances of cyberbullying, all the measures are essential and ought to be maintained. There are lists of keywords or words regarded as unsuitable that are used as part of the vocabulary in cyberbullying to decide whether any postings can be considered cyberbullying or not. A software solution appropriate for this multilingual situation will assist in identifying these words in Malaysia, where many internet users publish comments on social media in both English and Malay. It may operate independently or as a component of other mobile or web-based applications. The main function is to detect and filter potential cyberbullying communications, and at the same time, it may help the authorities act before any real cyberbullying occurs. Before that, a collective of cyberbully words and phrases is needed as a source for any detection tool. Our study intends to produce a glossary and propose a suitable algorithm to detect cyberbullying incidents in Malaysian society based on English and Malay.

3 Methodology

An opinion survey [29] was used to obtain the required information from the public. Most of the questions were close ended; only three questions were open-ended.

3.1 Questionnaire Design

The questionnaire was initially composed of 35 questions, 6 of which had between 6 and 20 items, but it underwent a few adjustments as a consequence of conversations with the research team members. 32 questions were close-ended, and the choices of answers were either nominal or ordinal. The questionnaires were divided into 5 Sections. There are:

i. Section A: Demographic Details (7 questions)
ii. Section B: Cyberbullying Prevention Measure (5 questions)
iii. Section C: Cyberbullying Experience (5 questions)
iv. Section D: Cyberbully Words/Phrases (3 questions), and
v. Section E: Scenario (1 question with 15 scenarios).

Two experts reviewed the questions based on their structure, suitability, language used, grammar, and relevance. Prior to its actual distribution, a pilot test was performed on the initial version of the questionnaire involving 40 identified respondents, ages 17 to 28 with the purpose of identifying problems concerning the questions' clarity. The questionnaire was distributed online to 40 identified respondents, ages 17 to 28. The idea of giving the survey to the identified respondents is that any doubts or issues raised can be recorded by the researcher and discussed later. As a result of the pilot test, several potential problems regarding the clarity of the questions that might affect the participants' understanding were identified and rectified. Every suggestion is valuable, even though not all the recommendations are considered due to some valid reason. The following Table 1 list all improvements made to the questionnaire.

The final questionnaire eventually had 30 questions and was broken up into sections A, B, and C. Section A comprised eight questions (Q1–Q8) that asked about demographic information. Q2–Q8 are compulsory, while Q1 is optional for the respondent to supply the information. However, it is very helpful to the researcher for a reward/honorarium purpose to be awarded to the respondents. Questions in this section collected the following information from the respondents.

Table 1 List of improvements to the questionnaire

No.	Section	Item (question)	Remark	Actions taken
1	A	2	To put age range rather than let the respondent supply the information	No action was taken since the researcher would like to know the exact age of respondents so it would be easier to check the age range of respondents. The targeted respondents are aged 15 to 30 years old
2	A	7	Add one option i.e., "Others", to the list and allow a respondent to spell the option	No action was taken since the researcher would like to limit the options. Currently, there are already seven on the list
3	A	Suggested to add no. 8	To add language usage of the respondent	This question is not relevant because this questionnaire is about cyberbullying. People might use any language when cyberbullying. However, the researcher adds 1 question requesting the respondent to provide an email address
4	C	9 and 10	These two questions can be omitted since all statements are clearly in their suitable category, i.e., item 9 is under insulting/offensive situation, and item 10 is under an intimidating situation	The researcher agrees with this. Removal of these questions as it was already under the cyberbully category
5	C	9 sub-question 14	Revise the Malay phrase using the right verb i.e. *Anda mempunyai tenaga positif yang luar biasa umpama bunga yang mekar*	It is updated accordingly
6	D	11	Suggest highlighting the **Malay** word/phrases	It is updated accordingly
7	D	12	Suggest highlighting the **English** word/phrases	It is updated accordingly

(continued)

Table 1 (continued)

No.	Section	Item (question)	Remark	Actions taken
8	D	13	Suggest highlighting the **Mixed Malay-English** word/phrases	It is updated accordingly
9	E	14 Scenario 1	To use the right Malay word, i.e., *dianggap*	It is updated accordingly
10	E	14 Scenario 1–4	Suggest enlarging the image/figure	All four figures are resized accordingly
11	E	14 Scenario 13	To use the right Malay word, i.e., *disalahertikan*	It is updated accordingly
12	General comment		Need to include a definition for each section	Elaboration of each section for a clear definition and purpose of it is included
13	General comment		Provide consistency for each question in terms of the structure of each question	The questions are rephrased to provide consistency
14	General comment		Suggest reducing the number of scenarios from 18 to 15 only	Removal the ones that are a cyberbully post
15	General comment		Suggest grouping the questions and limiting the number of sections	Regroup the questions into 3 Sections

- Email Address (Q1)
- Gender (Q2)
- Age (Q3)
- Race (Q4)
- Religion (Q5)
- Education level (Q6)
- Residential status (Q7) and
- Social media platform of choice (Q8)-Respondents are allowed to choose more than one social media of their choice.

Section B of the questionnaire consisted of 9 questions that were arranged according to the cyberbullying-related matters. Details of the questions are as follows:

- Cyberbullying prevention measures (Q9 with six items)
- Cyberbullying format if one has come across cyberbullying on social media platforms (Q10 with six items)
- Language one witnessed the cyberbully post/s (Q11 with three items)
- To rate the internet slang or terms that one might think are considered cyberbullying (Q12 with 20 items)
- State examples of Malay words/phrases that one knows in use as a sign of cyberbullying (Q13 – open-ended question)

- State examples of English words/phrases that one knows in use as a sign of cyberbullying (Q14 – open-ended question)
- State examples of Mixed Malay-English words/phrases that one knows in use as a sign of cyberbullying (Q15 – open-ended question).

Section C of the questionnaire consisted of 15 questions that were arranged as scenarios equipped with images on social media posting that might be categorized as cyberbully or non-cyberbully. These postings were obtained from 3 popular social media like Instagram, Twitter and Facebook. Prior to setting up the questionnaire, the researcher has already grouped/labeled these scenarios into two (2) categories, i.e., cyberbully and non-cyberbully, and kept the record aside. This part aimed to get mutual agreement and understanding between the respondents and researchers in identifying each scenario, either cyberbully or non-cyberbully.

3.2 *Procedure*

Microsoft Forms was chosen as a platform to obtain responses from respondents for its accessibility almost using any web browser or mobile device, easiness of observing real-time results as they're submitted, the usefulness of its built-in analytics to evaluate responses, and export results to Excel for additional analysis. Two surveys were conducted with different respondents: Pilot Test and the Actual Survey.

Pilot Test

As discussed in section A, a pilot test on the initial version of the questionnaire was conducted with 40 identified participants of various ages. The survey had 35 questions and took an average of 23 min to complete. The survey was distributed on 28 September 2021 and ran for three days. The researcher recorded any issues raised during the questionnaire answering session to be discussed later.

Actual Survey

Following the completion of the amendments based on feedback from the pilot test and expert reviews, the actual survey was conducted starting on 1 November 2021 and lasted until the middle of December 2021, with 30 questions asked. This exercise yielded a total of 329 respondents, who took an average of 15 min and 39 s to complete the survey. Improvements made after implementing the pilot test resulted in a shorter response time. This time saving enabled the researcher to obtain a relatively large number of respondents. The initial target was 200 respondents, including the pilot test. However, given that the survey was only open for two months, the response was overwhelming and unexpected.

4 Results and Analysis

Detailed results and analysis of the actual survey are as follows:

4.1 Demographics Information Section

About 188 of the respondents were female, and 133 respondents were males. For this survey, we targeted young adolescents between 15 and 30 years old. According to the results, the most common age group was between 20 and 23 years old, with approximately 156 respondents, with 16–19 years old coming in second.

There were five distinct races: Malay, Chinese, Indian, Arab and others. Most respondents were Malays, about 53.0% with 170 respondents. 53.6% of the respondents, which is about 179 are Muslim. One respondent identified as Sikh. With 202 respondents, most respondents had a degree education. There were only four postgraduate respondents. In the survey, there were two options for residence status: local and foreign. We received 317 local respondents for this study, with 12 foreign respondents.

The final question in this section was about the respondents' preferred social media sites [30]. Seven social media platforms were on the list, and respondents were allowed to select more than one. Instagram is the most frequently used social media platform by 232 respondents. For this study, the preferred dataset sources were Instagram, Twitter, and Facebook. Even though Telegram was among the top three (3) in the ranking, the researcher did not use it as the source of this study due to the privacy concerns of personal account holders. As a result, the researcher chose Facebook as the next highest social media platform in the ranking. Table 2 shows the profiles of respondents.

4.2 Cyberbullying Information Section

The primary goal of this paper is to produce cyberbullying glossaries that can be used in the next phase of this study. This was accomplished by asking the respondents to answer three (3) open-ended questions related to words or phrases used in cyberbullying. Respondents were allowed to answer these questions with as many words/phrases as they wanted. All the glossaries were grouped into five categories with a specified scale used to analyze the results. Because it involves numerical analysis, each category was assigned a scale to calculate the number of occurrences in each category.

The researcher collected 179 Malay words, 87 English words and 182 mixed English-Malay words/phrases from the survey. Table 3 shows the detailed results of each category of the glossaries with its dedicated scale.

Table 2 Profiles of respondents

Characteristics	Frequency	Percentage
Gender		
Female	194	58.97
Male	135	41.03
Age		
12–15	3	0.91
16–19	117	35.56
20–23	156	47.42
24–27	45	13.68
28–32	8	2.43
Race		
Malay	176	53.0
Indian	112	35.5
Chinese	32	9.66
Arab	1	0.62
Others	8	1.25
Religion		
Islam	179	53.6
Hindu	104	32.4
Buddha	31	9.7
Christian	10	3.1
Others	4	1.4
Sikh	1	
Education level		
Degree	202	62.93
Diploma	29	9.03
STPM/foundation/matriculation	26	8.1
SPM	32	9.97
Secondary School	28	8.73
Postgrad	4	1.25
Residential status		
Local	317	98.8
International	12	1.2
Social media platform of choice		
Instagram	232	
Telegram	141	
Facebook	119	
WhatsApp	119	

(continued)

Table 2 (continued)

Characteristics	Frequency	Percentage
Twitter	115	
Snapchat	68	
LinkedIn	31	

Table 3 Detailed results of each category

Words/phrases		Malay	English	Mixed Malay-English
Category	Scale	Total		
Intellectual	1	10	7	18
Physical appearance	2	20	14	39
Insulting/offensive	3	124	59	100
Intimidating	4	13	6	11
Others	5	12	1	14
Total		179	87	182

4.3 Scenarios Section

This survey section seeks respondents' opinions on the scenarios using the 5 Likert scales. A total of 15 scenarios were chosen randomly from three (3) mainstream social media platforms: Instagram, Twitter and Facebook. They were labeled Cyberbully (CB) or Non-Cyberbully (NCB) and surreptitiously kept to the researcher. Scenarios 1, 2, 3, 4, 8, 10, 11, 14, and 15 were labelled as CB, while scenarios 5, 6, 7 9, 12, and 13 were labeled NCB. The results were based on the Cronbach Alpha level. Detailed analysis of 15 scenarios are as follows:

Cyberbully

There were nine (9) scenarios under this category that showed body shaming, insulting, cursing, expressing anger, and those intended to enliven the environment. Table 4 shows the reliability statistics with Cronbach's alpha value was 0.966, indicating a high level of internal consistency because of all recommended values. The high value of alpha of 0.966 indicated the items being measured are highly correlated to each other. The survey was designed to measure overall experience of cyberbully-ing and with the result of 0.966, it able to measure the constructs well. Questions posed in the survey are appropriately worded and are reliable in measuring experienc-es of cyberbullying of the respondents.

Non-Cyberbully

Under this category, there were six (6) scenarios evaluated. The Cronbach's alpha value was 0.853, as shown in Table 5, although slightly lower than the CB category, but it was considered good internal consistency. This result may be reflected from the

Table 4 Reliability statistics for CB category scenarios

Reliability statistics		
Cronbach's alpha	Cronbach's alpha based on standardized items	No. of Items
0.966	0.966	9

Table 5 Reliability statistics for Non-CB category scenarios

Reliability statistics		
Cronbach's alpha	Cronbach's alpha based on standardized items	No. of items
0.853	0.854	6

lower number of items being measured in the Non-CB categories, 6 items compared to 9 in CB category.

The results from (i) and (ii) are in line with the views of researchers, and it reached a mutual agreement between researchers and the public.

5 Threats to Validity

While every care has been taken to ensure the reliability of the information gathered, its representativeness cannot be guaranteed as the data were obtained from the sampled population. However, the following measures have been taken to mitigate the possible threats to data validity. Before the dissemination of the questionnaire, a pilot test on construct validity was conducted to ensure that the possible respondents had a common understanding of the questions. As a result, a few questionnaires were rearranged and rephrased for clarity. Two questions were also removed. Furthermore, there was no technical jargon or terms in the questionnaire that were difficult to understand and may cause misunderstanding.

The questionnaires were reviewed for external validity by two experts: one from Universiti Tenaga Nasional to check on the linguistic component as well as the construction of each question. Simultaneously, another expert from Universiti Teknologi Malaysia provided a quantitative and qualitative review of each question. Respondents from a variety of backgrounds were able to complete the questionnaire because it was available online. Regarding internal validity, it can be shown in this study that the results produced directly from the raw data were analyzed using just basic descriptive statistics. The survey method and questionnaire structure have also been thoroughly described to enable the study to be repeated by other researchers.

6 Conclusion

The results shared in this study serve as evidence that cyberbullying incidents occur not just in English but also other languages. With social media now becoming the norm and most people having access to the Internet and smartphones, the risk of becoming a victim of cyberbullying is increasing. Our results showed that incorporating context in the form of scenarios might assist in generating cyberbullying-related words or phrases, which may contribute to the detection of cyberbullying. This work can be extended to develop models to detect cyberbullying words or phrases discovered in this study. In future studies, other features, such as demographic profiles or behavioral characteristics, could also be considered, which might be beneficial to employ predicting algorithms for improving detection accuracy.

Acknowledgements Information presented in this paper constitutes a part of the research titled An Empirical Study on Malay-English Glossary for Detecting Cyberbullying Incidents in Various Social Media Platforms funded by BOLD2021 Grant, Universiti Tenaga Nasional.

References

1. An official website of the United States government. Retrieved from https://www.stopbullying.gov/bullying/what-is-bullying. Accessed 02 July 2022
2. IPSOS (2018). Accessed 02 July 2022
3. U-Reporters. Retrieved from https://malaysia.ureport.in/opinion/3444/. Accessed 02 July 2022
4. Watts LK, Wagner J, Velasquez B, Behrens PI (2017) Cyberbullying in higher education: a literature review. Comput Hum Behav 69:268–274
5. Jun W (2020) A study on the cause analysis of cyberbullying in Korean adolescents. Int J Environ Res Public Health 17(13):4648
6. Abaido GM (2020) Cyberbullying on social media platforms among university students in the United Arab Emirates. Int J Adolesc Youth 25(1):407–420
7. Cyberbullying: twenty crucial statistics for 2022, Security.org Team. Retrieved from https://www.security.org/resources/cyberbullying-facts-statistics/. Accessed 02 July 2022
8. Kuadli J (2021) 25 Shocking cyberbullying statistics you should know eye-opening cyberbullying statistics, Legaljobs. Retrieved from https://legaljobs.io/blog/cyberbullying-statistics/. Accessed 02 July 2022
9. Park MA, Golden KJ, Vizcaino-Vickers S, Jidong D, Raj S (2021) Sociocultural values, attitudes and risk factors associated with adolescent cyberbullying in East Asia: a systematic review. Cyberpsychol: J Psychosoc Res Cyberspace 15(1)
10. Anderson M (2018) A majority of teens have experienced some form of cyberbullying. Justice Research and Statistics Association
11. Cook S (2022) Cyberbullying facts and statistics for 2018–2022, Comparitech. Retrieved from https://www.comparitech.com/internet-providers/cyberbullying-statistics/. Accessed 26 July 2022
12. R.E.A.L. Education Group. Anti-bullying campaign. Retrieved from https://real.edu.my/anti-bullying-campaign/. Accessed 02 July 2022
13. Quah E (2018) Beauty queen takes on cyberbullying | The Star. Accessed 02 Jul 2022
14. Nickelodeon announces its anti-bullying campaign in Malaysia, The Sun Daily (2017)

15. Talian Kasih 15999 (2022) Ministry Info, Ministry of Women Family and Community Development. Retrieved from https://www.kpwkm.gov.my/kpwkm/index.php?r=portal/full&id=NGtVYXZIMjRqM3diWmV2K005Mm9Hdz09. Accessed 02 July 2022
16. Put a stop to bullying, Facebook. Retrieved from https://web.facebook.com/safety/bullying?_rdc=1&_rdr. Accessed 02 July 2022
17. Al-Garadi MA, Varathan KD, Ravana SD (2016) Cybercrime detection in online communications: the experimental case of cyberbullying detection in the Twitter network. Comput Hum Behav 63:433–443. https://doi.org/10.1016/j.chb.2016.05.051
18. How to report things. Help center, Instagram. Retrieved from https://help.instagram.com/. Accessed 02 July 2022
19. Cost B (2019) Instagram users can now 'shadow ban' cyberbullies. Tech, Instagram. Retrieved from https://nypost.com/2019/10/02/instagram-users-can-now-shadow-ban-cyberbullies/
20. Report abusive behavior. Help center, Twitter. Retrieved from https://help.twitter.com/en/safety-and-security/report-abusive-behavior. Accessed 02 July 2022
21. CyberSAFE Gallery, CyberSAFE Malaysia. Retrieved from https://www.cybersafe.my/en/. Accessed 02 Jul 2022
22. Salawu S, He Y, Lumsden J (2020) Approaches to automated detection of cyberbullying: a survey. IEEE Trans. Affect Comput 11(1):3–24.https://doi.org/10.1109/TAFFC.2017.2761757
23. Im TL, San PW, On CK, Alfred R, Anthony P (2013) Analysing market sentiment in financial news using lexical approach. In: 2013 IEEE conference on open systems (ICOS), pp 145–149
24. Kumari S, Saquib Z, Pawar S (2018) Machine learning approach for text classification in cybercrime. In: 2018 Fourth international conference on computing communication control and automation (ICCUBEA), pp 1–6. https://doi.org/10.1109/ICCUBEA.2018.8697442
25. Aind AT, Ramnaney A, Sethia D (2020) Q-bully: a reinforcement learning based cyberbullying detection framework. In: 2020 International conference for emerging technology (INCET), pp 1–6. https://doi.org/10.1109/INCET49848.2020.9154092
26. Banerjee V, Telavane J, Gaikwad P, Vartak P (2019) Detection of cyberbullying using deep neural network. In: 2019 5th International conference on advanced computing and communication systems (ICACCS), pp 604–607. https://doi.org/10.1109/ICACCS.2019.8728378
27. Al-Ajlan MA, Ykhlef M (2018) Optimized twitter cyberbullying detection based on deep learning. In: 2018 21st Saudi computer society national computer conference (NCC), pp 1–5. https://doi.org/10.1109/NCG.2018.8593146
28. Rosa H, Carvalho JP, Calado P, Martins B, Ribeiro R, Coheur L (2018) Using fuzzy fingerprints for cyberbullying detection in social networks. In: 2018 IEEE international conference on fuzzy systems (FUZZ-IEEE), pp 1–7. https://doi.org/10.1109/FUZZ-IEEE.2018.8491557
29. Kitchenham BA, Pfleeger SL (2008) Personal opinion surveys. In: Guide to advanced empirical software engineering. Springer, pp 63–92
30. Chatzakou D et al. (2019) Detecting cyberbullying and cyberaggression in social media. ACM Trans Web 13. https://doi.org/10.1145/3343484

VANET Performance Evaluation for Malaysian Urban Federal Highway

Nurshahrily Idura Ramli, Arief Fahmi Mohd Fuad, Mohd Faisal Ibrahim, and Mohd Izani Mohamed Rawi

Abstract The recently deployed Vehicular Ad-Hoc Network (VANET) system for the vehicular road users can improve communication amongst drivers through early traffic congestion warnings, route suggestions, and emergency alerts. Regardless of how advantageous it may seem to many urban areas around the world, because the technology has not yet been completely implemented in Malaysia, it is unknown whether or not the implementation of VANET are viable or practical. Given that Vehicle-to-Roadside (V2R) communication's transmission range is only 500 m at most, it must be used in close proximity to a road or highway where vehicles are moving in fast mobility. This study investigates the viability of VANET using a variety of methods, taking into account the requirement for VANET to be operational in urban highway environment. In order to establish RSU placement guidelines and determine the best deployment strategy for RSUs along the Federal Highway in the Klang Valley, Malaysia, this study examines the viability of VANET utilising computer-aided software simulation tools. It primarily focuses on the section of the highway that is in close proximity to solid structures such as high rise buildings. This study also establishes the framework for simulations of the VANET environment for the Federal Highway, including the modelling of vehicular traffic, the establishment of the RSU, and the performance of the VANET. The results indicate that positioning RSUs in front of obstructions, with sufficient antenna power levels, along with lower vehicle density and velocity, could lead to higher packet throughput and lower latency.

Keywords VANET · V2R · Urban highway · Simulation

N. I. Ramli (✉) · A. F. M. Fuad · M. F. Ibrahim · M. I. M. Rawi
Faculty of Computer and Mathematical Sciences, Universiti Teknologi MARA,
Shah Alam 40450, Selangor, Malaysia
e-mail: idura@uitm.edu.my

© The Author(s), under exclusive license to Springer Nature Singapore Pte Ltd. 2023

D.-K. Kang et al. (eds.), *Proceedings of the 9th International Conference on Computational Science and Technology*, Lecture Notes in Electrical Engineering 983,
https://doi.org/10.1007/978-981-19-8406-8_51

1 Introduction

1.1 Background Study

Modern technology has drastically changed how people live. Information and communication technologies, or ICTs, are a part of every aspect of our life. Information technology has generated vast amounts of urban data that reveal information on the environment and infrastructure of cities. For example, VANET technology enables data exchange between moving vehicles and infrastructure while adamantly delivering data even at high speeds.

One of the most urbanised nations in South-East Asia is Malaysia, which has one of the highest rates of urbanisation in the region. Urban and urban-like areas were already home to 77.16% of the nation's inhabitants in 2020, a sharp increase from the meagre 34.2% in 1980 [1]. The fastest rate of urbanisation in Malaysia is in Kuala Lumpur, and both urbanisation and motorization were significant, particularly in the Klang Valley [2] where the first road was constructed. The Malaysian Federal Highway in Klang Valley is now one of the busiest roads utilised by commuters that contributes to daily major traffic jams [3].

The effect from the integration of contemporary telecommunications and information technology into the existing transportation systems are the Intelligent Transportation Systems (ITS), which is mainly applying the VANET technology. These systems, which are generally made up of road networks and mobile vehicles, are primarily designed to address significant issues including passenger safety and traffic congestion [4], however it is also diverted to other information dissemination applications. Thus, Vehicle-to-vehicle (V2V) and vehicle-to-Infrastructure (V2I) which includes V2R communication paradigms, has emerged as the two fundamental communication paradigms that form the basis of such data communication system in VANET [5] allowing connectivity between vehicles as well as the Roadside Unit (RSU) that are placed alongside highways as well as arterial roads.

Considering that the dependability and advantages of VANET in Malaysia have not been fully established, such extensive deployment is now unlikely to be a practical option. In essence, thorough planning and testing of RSU deployment schemes is required to provide effective V2R connection for highway scenarios in Malaysia's Klang Valley. Hence, this study seeks to investigate the viability and effectiveness of V2R communications, emphasizing on RSU deployment schemes and network connectivity in urban locations like the Klang Valley, Malaysia, in light of the critical need to employ this technology, particularly for urban highway applications.

1.2 VANET Communication Architecture

In theory, VANET is not required to adhere to any particular architecture or topology. A VANET ecosystem is composed of numerous communication architectures.

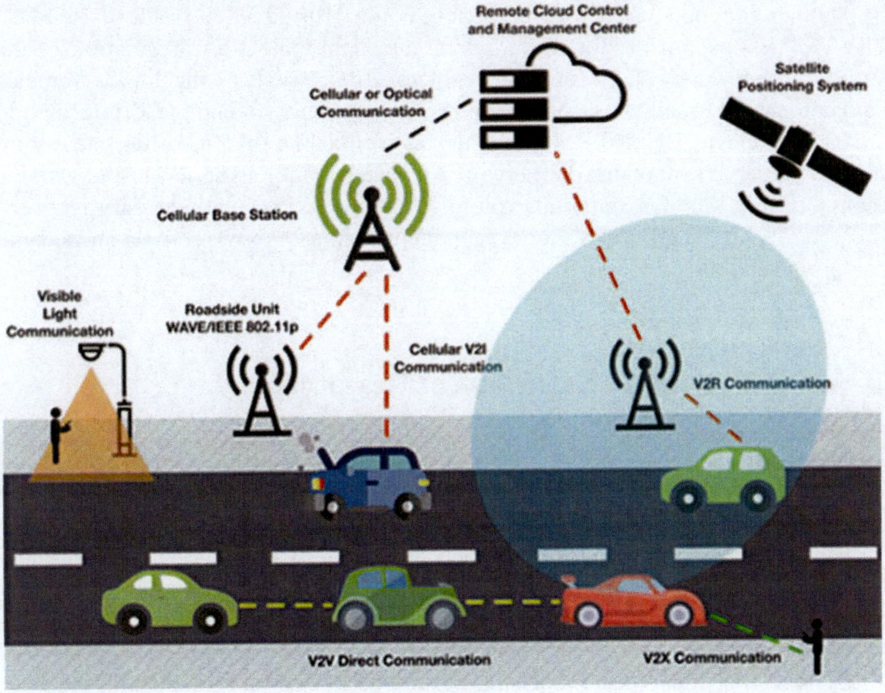

Fig. 1 VANET ecosystem

The Vehicle-to-Everything (V2X) architecture includes many of these communication architectures such as the Vehicle-to-Building (V2B), Vehicle-to-Grid (V2G), Vehicle-to-Pedestrian (V2P) and Vehicle -to-Infrastructure (V2I), Vehicle-to-Vehicle (V2V) and notably Vehicle-to-Roadside (V2R) communication. Figure 1 shows an example of the VANET ecosystem using a combination of the many communication architectures.

The RSU and other VANET Components On-Board Units (OBU) and Road-Side Units are the two types of DSRC devices that are utilised in the VANET for communication. Each vehicle has an OBU, which combines a transmitter located within the vehicle with a computing device. The OBU accesses the wireless channel using an omnidirectional antenna that is likewise installed on each vehicle.

The RSU on the other hand are stationary devices that are mounted road side. The RSU is similar to an OBU in that it has a transceiver, antenna, processor, and sensors. They are strategically placed along the road in order to provide services to vehicles. It is a core component of the VANET network which integrates mobile vehicles to the rest of the framework [6]. They serve as relay points, and also be used to deliver lower-priority information such as commercial advertisements and featured media content [7].

Another common VANET commponent is the Trusted Main Authority (TMA). The TMA is responsible for the infrastructure deployment and regulation of the VANET framework [8]. A common regulatory body such as the Motor Vehicles Division (MVD) maintains control by registering and validating the information of road vehicles, OBU and RSU identities as well as the road user IDs into secure centralized servers off of the road network to prevent malicious actions to the VANET ecosystem [9]. Another important role of TMAs is to execute necessary recovery and security measures should any component in the VANET ecosystem become compromised [10].

2 VANET Highway Environment Researches

Recently, research on VANET is growing due to their tremendous capabilities. However, the installation of RSUs that can cater for a large geographical area can be very costly. Many related works revolves around the study of the connectivity problem for vehicles travelling on highways, where there is only one entrance and exit and numerous RSUs are evenly spaced out along the route. This includes strategic RSU placements, hence includes connectivity and interference issues and their effects VANET's performance.

A connectivity analytical model was developed by Mousa et al. [11] for the positioning of RSUs in various scenarios, the model can be used to evaluate the impact of key factors on the highway connectivity rate. Kim et al. [12] have proposed a new approach on how effectively to deploy RSUs to leverage their spatiotemporal coverage under a small budget. Consideration of an advanced RSU distribution mechanism that uses a combination of three different methodologies were proposed by applying static RSU placement on certain locations, taking advantage of the public transportation as well as utilizing fully governable automobiles owned by the local authorities. Wang et al. [13] embraces the concept of centrality in the social network to the deployment of the RSU to assess the significance of the RSU location for site-wide deployment, while Guerna et al. [14] produced a taxonomy of RSU-based IoV deployment analyses, similarly reviewing important RSU placement approaches, highlighting their strengths and limitations.

The works mentioned above as well as other related VANET highway communication mostly examines the placements of RSUs in a typical sparse density highway and rarely on urban highways as such the Federal Highway in Malaysia. RSU placements are measured in distance to each other as well as to other moving vehicles without emphasizing the existence of solid structures or buildings that may obstruct the V2R communications. Moreover VANET performance analysis and parameters used are distinct to each scenarios and study, in which findings of these relevant studies vary hence are only used as guidelines and references.

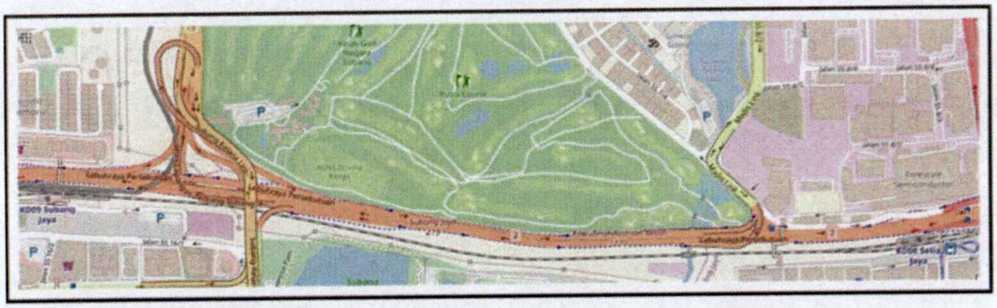

(a) Imported map view from OpenStreetMap

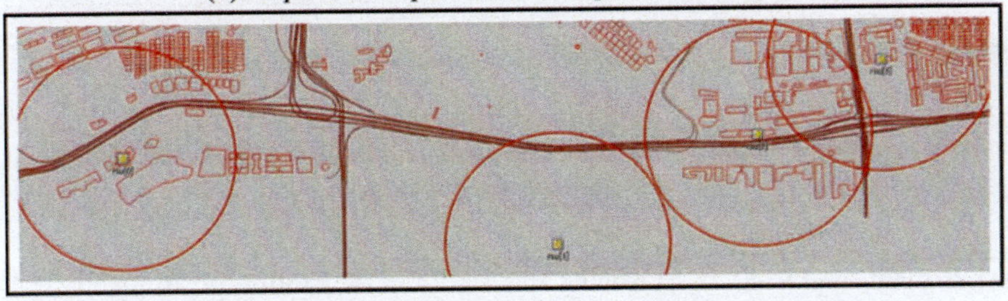

(b) Imported map view from OMNet++

Fig. 2 Imported map of the federal highway

3 VANET Simulation Setup

A network simulator, a vehicular mobility simulator, and an integration framework are the necessary simulation tools that are commonly used in VANET analysis project, hence utilized in this study. The experiments were carried out in the OMNeT++ 5.6.1 network simulator using SUMO 1.8.0 as the vehicular mobility simulator and Veins 5.1 as the networking framework for RSU scenarios.

VANET Federal Highway Topology The Federal Highway segment that was used for the experiments had a perimeter that stretched from 3.0796°N, 101.5830°E to 3.0922°N, 101.6228°E. The simulated stretch had a total length of 4.6 km. This 4.6 km stretch of highway features the most substantial structures, including high buildings, ramps, monuments, and billboards. In the Klang Valley, it is also noted to be one of the Federal Highway's most crowded sections. Figure 2 depicts the OpenStreetMap's aerial view and the imported map linked with OMNet++ with RSU placements for the stated section of the Federal Highway.

VANET Experiment Setup Four experiments were set to determine multiple parameters that might impact the performance of VANET based on the RSU deployment scheme utilizing the basic configuration of the RSU as explained below.

RSU Configuration Parameter RSUs that are utilised in the simulations are specifically set up to meet the diverse requirements of each experiment. The RSU is gen-

(a) First location (b) Second location (c) Third location

Fig. 3 RSU Placements

erally set to operate on IEEE 802.11p DSRC standard on frequency of 5.850–5.925 GHz on an omnidirectional type antenna. The transmission radius is et to 500 m and the transmission power is tested from 5 mV to 20 mW. The bitrate is 6 Mbps.

Experiment 1—Obstacle Existence Impact on V2R Communications Experiment 1 is carried out to evaluate the network efficiency of V2R communication concerning the existence obstacles of solid structured buildings in the area. RSUs were imposed in three possible locations along the Federal Highway at three key sites of the most commonly congested stretch that is also packed with high-rised solid structured buildings to measure the impact on the signal transmission in V2R communication.

Based on Fig. 3, three positions were chosen at three separate locations to place the RSU. The marked rsu[0] is positioned in the back of the buildings, rsu[1] is in the middle, and rsu[2] is positioned in front of the solid structure. Figure 3 displays all three of the RSU placement plans in three separate locations represented in Google Street View and its representation in OMNet++.

At 3.083357°N, 101.584035°E, close to a few shopping malls and a commuter train station, is where the first RSU positioning site placed, as shown in Fig. 3a.

Similarly in Fig. 3b, the second location for RSU positioning is at 3.083547°N, 101.609408°E at a row of shop lots and factories along the Federal Highway, opposite a commuter train station. Here, the impact of the RSU position were evaluated also based on 3 positions at the shop lots.

The third and final location decided for RSU positioning, as depicted in Fig. 3c, is at the Icon City high-rised building complex; (3.086060°N, 101.614303°E), which is close to the interchange from Federal Highway to Lebuhraya Damansara-Puchong (LDP) which is another busy highway.

The impact of nearby obstacles is measured using three data sets obtained from VANET simulation experiments; the first variable is the number of received Beacon Safety Messages (Number of Received Base Safety Messages), the number of Received WAVE Short Messages (Number of Received WAVE Short Messages) and the total number of lost packets (Number of Total Lost Packets).

Experiment 2: RSU Antenna Power Level Impact on V2R Communications The second experiment is carried out to evaluate the RSU's antenna power level in relation to the network performance of V2R communication. A similar configuration of RSU as in Table 1 were utilized in experiment 2, however only 1 location is used and the antenna power level was set to 5mw, 10mW, 15mW, 20mW for each run.

Experiment 3: Vehicular Velocity Impact on V2R Communication The third experiment is conducted to measure the network performance of V2R communication concerning the velocity of the vehicles. The simulation conducted using one vehicle and tested using velocity or speed of 80 kmph, 140 kmph and 200 kmph.

Experiment 4: Vehicular Density Impact on V2R Communication The final experiment is conducted to measure the network performance of Vehicle-to-Roadside (V2R) communication concerning the density of the vehicles. In this experiment, the simulation on velocity or speed of 80 kmph, and evaluating the effects for 20, 40 and 60 vehicles were performed.

4 Performance Evaluation Metrics

Multiple parameters were obtained from the simulations conducted in OMNeT++. These parameters are the number of Received Base Safety Messages (receivedB-SMs), Number of Received WAVE Short Messages (receivedWSMs), the number of Total Lost Packets (totalLostPackets) and, the Average Packet Latency (avgPacket-Latency).

The overall received beacon messages throughput of the RSU can be calculated by dividing the number of Received Base Safety Messages against the duration of the simulation (simTime) in seconds, as shown in Eq. 1.

$$bsmThoughput = \frac{(\sum receivedBSMs) * packetSize * 8)}{simTime} \tag{1}$$

The overall Number of Received WAVE Short Messages throughput of the RSU can be determined using a similar equation, by dividing the number of Received WAVE Short Messages against the duration of the simulation in seconds (simTime), as shown in Eq. 2.

$$wsmThoughput = \frac{(\sum receivedWSMs) * packetSize * 8)}{simTime} \tag{2}$$

Another performance metric that is used to evaluate the effectiveness of the RSU placement scheme is the packet loss rate. This performance metric can be calculated using the total number of lost packets divided by the simulation time duration in seconds (simTime) and multiplied by 100%, as shown in Eq. 3.

$$packetLossRate = \frac{totalLostPackets}{\sum receivedWBSMs + \sum receivedWSMs} * 100\% \quad (3)$$

The final performance metric that is used in this project is the Average Packet Latency (avgPacketLatency). As shown in Eq. 4, this metric is calculated by subtracting the Vehicular Node Arrival Time (nodeArrivalTime) in seconds from the RSU Packet Arrival Time (rsuArrivalTime) and divided by the total number of all packets (totalPacketCount).

$$avgPacketLatency = \frac{rsuArrivalTime - nodeArrivalTime}{totalPacketCount} \quad (4)$$

5 Results and Analysis

5.1 Experiment 1—Obstacle Existence Impact on V2R Communications

The first experiment's findings demonstrated that V2R connectivity performs best when RSUs are placed in front of obstacles. This has the effect of minimising packet loss and allowing the RSU antenna to fully transmit its signal to nearby vehicles. RSUs that were placed in the middle on the other hand, will experience performance deterioration and higher packet loss. RSUs placed behind the obstacles performs the worst, receiving far fewer messages and experiencing more packet losses. This performance negation is a result of obstacle shadowing, which prevents radio signals from propagating through solid structures like buildings [15].

Figure 4a shows the Base Safety Messages Throughput for RSU Placement 1, 2 and 3 which recorded the highest in RSU placement 1 that is positioned in front at 11187.12 bits per second. For RSU Placement 2, the highest is also the RSU placed in front at 21132.72 bits per second, similarly for RSU Placement 3 that achieved 17992.80 bits per second.

A similar outcome shown in Fig. 4b was produced in the WAVE Short Messages Throughput for experiment 1, which also recorded the highest value for RSUs positioned in front of the building for all placement schemes. For RSU Placement 1, the highest throughput is at 1579.60 bits per second. The highest value for RSU Placement 2 is at 1834.80 bits per second, and 2041.60 bits per second for RSU Placement 3.

As for the rate of packet loss for the first experiment, Fig. 4c shows that the RSUs positioned at the back of the buildings suffered from the highest values compared to RSUs positioned in the middle or front of the buildings. In RSU Placement 1, the highest value of packet loss is 24.76% for the RSU positioned at the back. For RSU

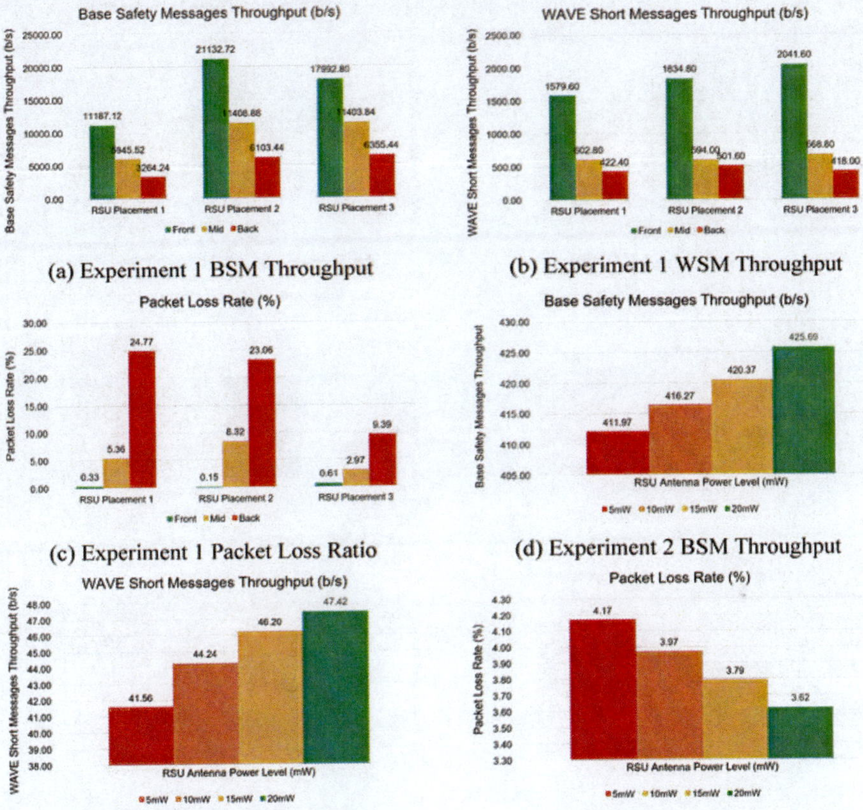

(a) Experiment 1 BSM Throughput

(b) Experiment 1 WSM Throughput

(c) Experiment 1 Packet Loss Ratio

(d) Experiment 2 BSM Throughput

(e) Experiment 2 WSM Throughput

(f) Experiment 2 Packet Loss Ratio

Fig. 4 Results for experiment 1 and experiment 2

Placements 2 and 3, the same position also resulted in the highest rates at 23.06% and 9.39% respectively.

Experiment 2: RSU Antenna Power Level Impact on V2R Communications Experiment 2 results obtained show that V2R connectivity is at maximum performance when the RSUs were set to operate at maximum power which is 20 mW. This meant that the RSU antenna can fully propagate its signal to nearby vehicles and minimize packet loss. RSUs that operate at lower power levels will suffer from performance degradation and increased packet loss. The RSU will perform the worst with significantly decreased messages received and a higher number of packet losses when its antenna power level is set to the minimum level, 5 mW.

As shown in Fig. 4d, the throughput of the BSM packets is recorded lowest at 411.97 bits per second, when the RSU is operating at 5 mW, followed by 416.27 bits per second at 10 mW, 420.37 bits per second at 15 mW, and the highest value at 425.69 bits per second at 20 mW. Similarly in Fig. 4e, for WAVE Short Messages

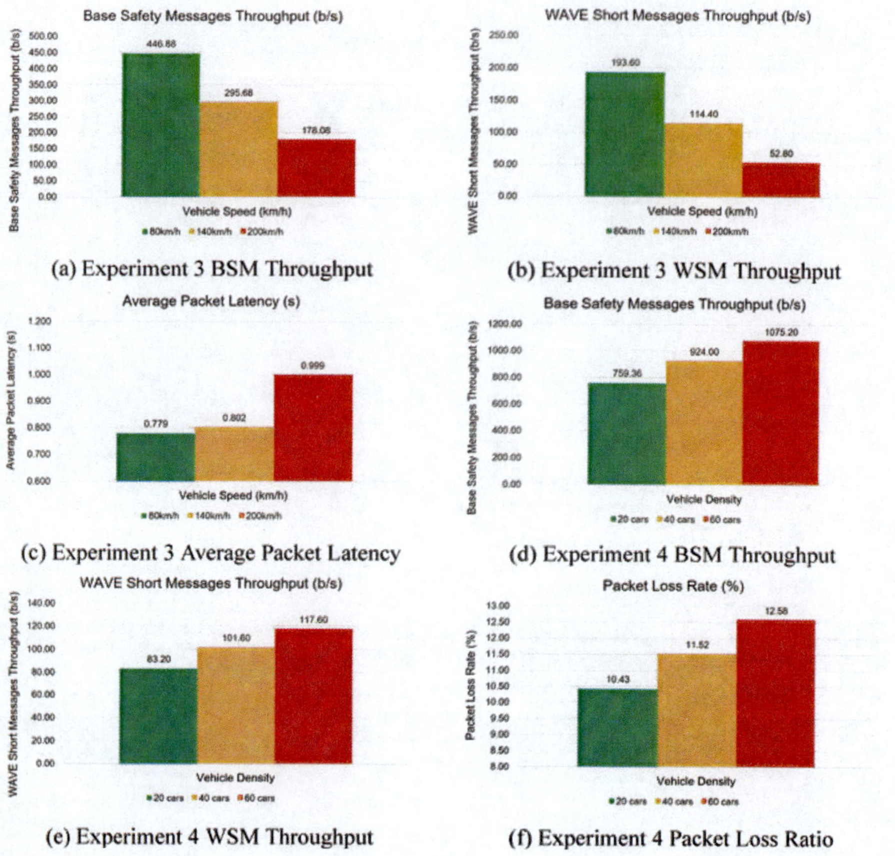

(a) Experiment 3 BSM Throughput

(b) Experiment 3 WSM Throughput

(c) Experiment 3 Average Packet Latency

(d) Experiment 4 BSM Throughput

(e) Experiment 4 WSM Throughput

(f) Experiment 4 Packet Loss Ratio

Fig. 5 Results for experiment 3 and experiment 4

Throughput evaluations, the lowest is at 5mW antenna power with 41.56 bits per second. At 10 mW, the value increases to 44.24 bits per second, which increases to 46.20 bits per second at 15 mW and gets to the highest value of 47.42 bits per second at 20 mW.

Finally, as the power level of the RSU antenna rises, the packet loss rate exhibits a decreasing trend as depicted in Fig. 4f. The maximum loss rate, 4.17%, is experienced while the RSU is only using 5 mW. This number falls to 3.97% at 10 mW, 3.79% at 15 mW, and 3.62% at 20 mW, where it is at its lowest.

Experiment 3: Vehicular Velocity Impact on V2R Communication The results obtained from the third experiment showed that when the vehicles are travelling at 80 km/h, the resultant throughput is the highest at 446.88 bits per second. At 140 km/h, the value drops down to 295.68 bits per second and drops further to only 178.08 bits per second at 200 km/h. Figure 5a illustrates the results obtained.

The outcomes for WAVE Short Messages Throughput also showed a similar pattern as clearly depicted in Fig. 5b, whereby at 80km/h vehicular speeds, the throughput value is at the highest with 193.60 bits per second. This value declines to 114.40 bits per second when the vehicle is moving at 140 km/h and decreases to the lowest point of 52.80 bits per second at 200 km/h. Alternatively, the average packet latency showed a directly proportional relationship to the speed of the vehicle, as shown in Fig. 5c When the vehicle travelled at 80km/h, the average packet latency is lowest at 0.779 s, followed by 0.801 s at 140km/h. The average packet latency almost reached 1 s when the car travelled at 200 km/h.

The findings clearly demonstrate that V2R connectivity is at maximum performance when the velocity of the vehicle is low. This meant that the vehicle node stays in the RSU antenna radius for longer intervals which provide a larger communication window between the RSU and the vehicle. In contrast, vehicles that travel at higher speeds will have less time to stay in the RSU communication radius, therefore decreasing the communication window with the RSU. As a result, the number of packets received by the RSU is reduced while the packet latency increased. From a usability standpoint, higher vehicular mobility speeds might lead to less effectiveness in V2R communications [16].

Experiment 4: Vehicular Density Impact on V2R Communication In the final experiment conducted to measure the vehicular density impact on V2R communications, the data shown in Fig. 5d shows an upward trend directly proportional to the number of vehicles. At 20 cars, the value of Base Safety Messages Throughput is 759.36 bps, followed by 924 bps at 40 cars, and reaches the highest point of 1075.20 bps at 60 cars. As for the WAVE Short Messages Throughput values, the pattern shown in Fig. 5e is also similar to Fig. 5d. The value is lowest at 83.20 bps for 20 cars, which rises to 101.60 bps for 40 cars and increases further to 117.60 bps for 60 cars. In terms of the packet loss rate for experiment 4 illustrated in Fig. 5f, the value increases in correlation to the density of the vehicles. The lowest loss rate is 10.43% at 20 cars, which increases slightly to 11.52% at 40 cars, and reaches 12.58% at 60 cars.

Due to larger packet collision in V2R connections, the RSU dropped even more packets as the number of cars increased. Additionally, each vehicle poses as an obstruction for the others, which raises the possibility of higher packet loss. Another aspect to take into account is that as the number of vehicles on the road increases, the communications system is also subjected to additional noise, which results in increased packet loss. In conclusion, if the traffic density is too high, the V2R network may become ineffective for critical situations and emergencies.

6 Conclusion

The best possible scenario for RSU deployment is clearly the one that is furthest from obstacles, however, in close proximity in the locations tested, and due to the effects of obstacle shadowing, RSUs deployed in front of obstacles performs best in terms of V2R connectivity. This outcome also highly dependent and correlated to the RSU's antenna power level, in which operating at a higher power would increase the performance of RSU's signal propagation to nearby vehicles, hence minimizing packet loss.

Analyzing the network performance of the V2R communications system in respect to the RSU deployment and configurations, it can be concluded that the V2R connectivity is optimal in situations with lower vehicular velocity and density within the RSU antenna coverage.

From a real-world standpoint, the results obtained from this study can be served as a viable resource in future actual implementations of a V2R communications system for the Federal Highway. Although the deployment scheme of the RSUs is conducted using computer-aided simulations, the RSUs use accurate physical configuration parameters, and GPS coordinates used in this project are based on OpenStreetMap and street-level imagery from Google Street View. Therefore, town planners or network architects that are planning to develop a full-scale VANET system for the Federal Highway can use these coordinates and configurations for the most optimal RSU deployment scheme in the real world.

Acknowledgements The authors gratefully acknowledge the support from the Ministry of Higher Education, Malaysia in providing the Fundamental Research Grant Scheme (FRGS) (Project Number: FRGS/1/2018/ICT03/UITM/02/3) research grant.

References

1. The Star Online, Ismail BH (2021) Malaysian Future is Urban. https://www.thestar.com.my/news/focus/2021/09/19/malaysian-future-is-urban
2. Aldukali Salem I, Almselati Riza Atiq, Rahmat OK, Jaafar Othman (2011) An overview of urban transport in Malaysia. Soc Sci 6:24–33
3. IKalbana P (2022) Nation's first highway in bad shape. https://www.nst.com.my/news/exclusive/2019/03/472118/nations-first-highway-bad-shape. Last accessed 25 July 2022
4. Susan S, Finson R (2013) Intelligent transportation systems. Institute of Transportation Studies, Research Reports, Working Papers, Proceedings qt3hh2t4f9, Institute of Transportation Studies, UC Berkeley
5. Trullols O, Fiore M, Casetti C, Chiasserini CF, Barcelo Ordinas JM (2010) Planning roadside infrastructure for information dissemination in intelligent transportation systems. Comput Commun 33(4):432–442. https://doi.org/10.1016/j.comcom.2009.11.021
6. Kim D, Velasco Y, Wang W, Uma RN, Hussain R, Lee S (2017) A new comprehensive RSU installation strategy for cost-efficient VANET deployment. IEEE Trans Veh Technol 66(5):4200–4211. https://doi.org/10.1109/TVT.2016.2598253
7. Filippini I, Malandrino F, Dán G, Cesana M, Casetti C, Marsh I (2012) Non-cooperative RSU deployment in vehicular networks. In: 2012 9th Annual conference on wireless on-demand network systems and services, WONS 2012, pp 79-82. https://doi.org/10.1109/WONS.2012.6152243
8. Silva CM, Masini BM, Ferrari G, Thibault I (2017) A survey on infrastructure-based vehicular networks. Mobile Inf Syst. https://doi.org/10.1155/2017/6123868
9. Huang D, Misra S, Verma M, Xue G (2011) PACP: an efficient pseudonymous authentication-based conditional privacy protocol for VANETs. IEEE Trans Intell Transp Syst 12(3):736–746. https://doi.org/10.1109/TITS.2011.2156790

10. Ashritha M, Sridhar CS (2015) RSU based efficient vehicle authentication mechanism for VANETs. In: Proceedings of 2015 IEEE 9th international conference on intelligent systems and control, ISCO 2015. https://doi.org/10.1109/ISCO.2015.7282299
11. Mousa RJ et al (2021) J Phys Conf Ser 1804 012104
12. Wang Jing, Zhang Huyin, Tang Xing, Hao Sheng (2019) Social-aware routing for cognitive radio-based vehicular ad hoc networks. Int J Distrib Sens Netw 15:155014771986638. https://doi.org/10.1177/1550147719866389
13. Guerna A, Bitam S, Calafate CT (2022) Roadside unit deployment in internet of vehicles systems: a survey. Sensors 22:3190. https://doi.org/10.3390/s22093190
14. Carpenter SE (2014) Obstacle shadowing influences in VANET safety. In: Proceedings—international conference on network protocols, ICNP, pp 480–482. https://doi.org/10.1109/ICNP.2014.72
15. Alwan MH, Ramli KN (2018) Performance evaluation for high speed vehicle in VANET. Int J Appl Eng Res. 13(10):7937-7941. ISSN 0973-4562. https://www.ripublication.com/ijaer18/ijaerv13n1091.pdf
16. Harkat Y, Amrouche A (2019) Vehicle density, vehicle speed and packet interarrival time analysis in IEEE 802.11p EDCA based VANETs. In: 2018 International conference on signal, image, vision and their applications, SIVA 2018. https://doi.org/10.1109/SIVA.2018.8661070

Design and Development of Maiji Mountain Grottoes Immersive Virtual Reality Learning Environment

Ying Cao, Giap Weng Ng, and Hui Wang

Abstract This study attempted to take Maiji Mountain Grottoes as an example to design an immersive virtual reality learning environment serving for the field of cultural heritage or archaeology. Cultural heritage needs to be protected and not easy to arrive, which brings difficulties and challenges to learners. Virtual reality learning environment can provide immersive experience for users without worrying about time and space and allow flexibly switch between macro and micro world to learn more intuitively and efficiently. Based on the constructivist learning principles, this study designed a learning-guide model for immersive virtual reality learning environment, which not only dynamically display the 3D knowledge map and guide users to explore interactively in the immersive virtual reality environment, but also intelligently recommend the learning path according to the real-time feedback of learning performance and learning behavior. This paper will focus on the key points of the design of the immersive virtual reality learning environment and put forward the plan of future work.

Keywords Virtual reality · Head-mounted display · Knowledge map

1 Introduction

With the advent of the metaverse era, the application of virtual reality technology continues to extend to various fields including education. Taking Maiji Mountain Grottoes as an example, this study designed an immersive virtual reality learning environment for the field of cultural heritage or archaeology. There are many grottoes

Y. Cao · H. Wang
College of Arts and Science, Beijing Union University, 197, Beitucheng West Rd., Haidian District, Beijing, China
e-mail: cao_ying_di20@iluv.ums.edu.my; ying19810403@sina.com

G. W. Ng (✉)
Faculty of Computing and Informatics, University Malaysia Sabah. Jalan UMS, 88400 Kota Kinabalu, Sabah, Malaysia
e-mail: nggiapweng@ums.edu.my

© The Author(s), under exclusive license to Springer Nature Singapore Pte Ltd. 2023
D.-K. Kang et al. (eds.), *Proceedings of the 9th International Conference on Computational Science and Technology*, Lecture Notes in Electrical Engineering 983,
https://doi.org/10.1007/978-981-19-8406-8_52

with high historical research value in China, among which the most famous Dunhuang Mogao Grottoes, Longmen Grottoes, Yungang Grottoes and Maiji Mountain Grottoes are listed in the world cultural heritage list. Strengthening the protection of grottoes and personnel training is of great significance not only to China, but also to the world. Therefore, this study took grottoes archaeology as an example to explore the design of immersive virtual reality learning environment in this field.

Sometimes the limitations of objective conditions during cultural heritage or archaeology learning such as geographical location, narrow space, and pressure on cultural heritage protection may challenge instructional activities. Moreover, the knowledge points in these fields usually have complex connections through time and space. For example, explaining both the grotto feature of a certain era and the grotto changes through time lapse needs comparative learning. Although compare with learning on site through much moving, watching pictures or videos in classroom solves these problems more easily, obviously the latter lacks the sense of intuitive experience. How to use head-mounted display to meet these needs while saving time and cost and how to establish learning-guide model for users in immersive virtual reality learning environment is the focus of this study.

2 Related Works

Since the advent of Virtual Reality technology, it had been widely used in all walks of life and brought great changes to the world, especially in the field of education. The research showed that students' motivation to learn will increase if they can be visually engaged in the class. This goal can be achieved by developing immersive learning environment with visualized 3D models based on virtual reality technology. Users were not passive observers but participated in immersive virtual reality learning environment as active participants, which allowed the development of the learning paradigm based on exploration. In recent years, more and more researchers used Virtual Reality technology to develop educational applications such as serious games or learning environment, hoping to trigger motivation to engage in learning.

2.1 Virtual Reality Technology Background

Virtual reality technology, which is characterized by immersion and interaction, can provide beneficial experiences to travel through time and space with the sense of immersion. One of the most critical values of virtual reality technology to the existing technology was the sense of immersing in the virtual world built around users [1]. It is generally believed that learners are not easy to be disturbed by the outside world and can be more absorbed in instructional contents in an immersive virtual reality learning environment. Virtual worlds become an excellent opportunity for educators to implement learner-centered pedagogies where students took an active role to do activities which would be impossible to carry out due to the environments that were difficult to access, dangerous, or expensive in real life [2, 3]. At the same time, due to the interaction of virtual reality technology, it brings more opportunities for learners to explore actively, and the learning process may change from traditional passive input to learner-centered mode. Interactive experience was the preferred method of learning and training, because it balanced cost, technological development, immersion and the possibilities that users had of learning and improving their skills [4]. In addition, the interactivity of virtual reality is based on multimedia, which can affect learners' visual, auditory and other senses [5], and provide more personalized choices for learners with different learning behavior preferences.

2.2 *Virtual Reality for Education*

Instructional design is one of the most critical aspects of the transactions that go with the learning process in terms of the pedagogy. If instructional tasks and activities are designed by an inappropriate pedagogical approach may lead that learning environment fail to meet learners' needs. However, related studies indicated that a scarcity of studies which had a clear theoretical pedagogical model to guideline instructional design of virtual reality educational application [6–12]. Learning technologists and experts seem to lack expertise in how to use virtual reality technology to design learning solutions. In addition, there are few studies on the potential and challenges of instructional design strategies for virtual reality technology, which may affect the final learning performance with immersive virtual reality learning environment.

Based on the organizational pedagogical categories developed by Kebritchi and Hirumi [13], constructivism relates closely to experiential and discovery learning, and which adds the construction of personal meaning by the learner as an important step [2]. In constructivist learning, learners construct or make sense of experience in ways that are personally meaningful, and gain knowledge and understanding through interact and conduct activity, that is consistent with the user-centered learning environment built by virtual reality which encourage learners to explore freely and achieve learning results. A growing body of literature [6, 7, 10, 14] concluded that researchers and educators could apply constructivist principles in the design of virtual reality learning environments to improve learning performance. However, although some studies have adopted approaches based on constructivism, there wasn't a very clear model to guide the design of immersive virtual reality learning environment. How to design and develop an immersive virtual reality learning environment guided by constructivist learning principles needs to be further investigated.

2.3 *Virtual Reality for Cultural Heritage or Archaeology*

Due to relatively few cases of virtual reality technology applied to in grottoes archaeology learning, related studies in cultural heritage or archaeology field in recent three years were analyzed, some of which were shown in Table 1. These studies can be roughly divided into two categories. The first is mainly to use virtual reality technology to build a virtual environment for cultural heritage, sometimes combine with museums, face the public to provide a platform for virtual navigation tourism and broad knowledge dissemination [15–20]. The second is mainly to provide students with a virtual cultural heritage learning environment by combining virtual reality technology with instructional content for the purpose of learning [4, 21–23].

This study mainly aims at designing an immersive virtual reality environment of grotto archaeology for learners. Based on learning from previous relevant studies, we explored the following attempts.

Due to the complex distribution of Maiji Mountain Grottoes, the design of tracking navigation and space–time tunnel, which can cross space constraints through the flexible switching between macro and micro perspectives to construct geographical cognition, and span time changes to support comparative learning between different dynasties.

According to the characteristics of rich and relatively scattered knowledge points, it designed learning-guide model based on the constructivist learning principle, which use the dynamically generated of 3D knowledge map to guide learners to understand and construct knowledge combined with their own experience and recommend personalized learning paths due to the analysis and operation of performance feedback and interactive behavior.

To fully meet the preferences of users with different learning styles (visual learners and auditory learners) for knowledge acquisition channels, both annotation and audio forms of knowledge are superimposed in the interactive learning process. These functions are detailed in the later chapters.

3 Method and Tools

In this study, HTC Vive was selected as the head-mounted display for immersive experience and interaction. The development of interactive scenes and the learning-guide model including 3D knowledge map was mainly completed by Unity 3D. Based on Linux operating system, MySQL database was used to store relevant resource information. The implement of logical operation and interface tools were mainly coded by C#. The architecture of this immersive virtual reality learning environment is shown in Fig. 1.

The main objective of this study is to design an immersive virtual reality environment with learning-guide model for grottoes archaeology and other related fields. In order to meet the characteristics of the rich and complex knowledge points in

Table 1 Some representative studies of virtual reality for cultural heritage or archaeology

Author/year	Summary
Bozorgi et al. [15]	This study described the data collection, 3D processing, and next steps of using 3D scanning and immersive virtual reality technologies to document the Ganjali Khan Complex in Iran
Bozzelli et al. [16]	One case of this study was about using HTC Vive to explore the Hera II Temple of Paestum, which can enhance motivation and activate the learning process faster
Cantatore et al. [17]	This study created a virtual environment containing the actual state of conservation and historic and technical information by using 360° images of Cryptoporticus in an archaeological site of Egnatia in Italy
Chan et al. [21]	This study took the Pantheon in Rome as a representative case to explore the use of immersive virtual reality technologies for teaching architectural history and evaluate the learning effectiveness
Checa et al. [4]	Two teaching methodologies, learning by immersive virtual reality tours and viewing video renderings environments of a fifteenth-century Spanish town called Briviesca were compared in this study
Farooq et al. [22]	This study reconstructed the Bannu rail system, cultural heritage in Pakistan for learning purposes by virtual reality technology and compared with traditional methods to evaluate learning outcome
Ghani et al. [24]	This research studied on the effect of immersion towards place presence through feedbacks from three levels of virtual reality system of a mosque building in Kampung Hulu, Melaka in Malaysia
Kim et al. [18]	This study used virtual reality technology to display the construction, completion, and excavation of Muryeong Royal tomb in Korea, including navigation, interaction, immersion, reality, content and community
Kyrlitsias et al. [23]	This study analyzed the experience and attitude of learning a digital reconstruction of the archaeological site of Choirokoitia through two different system, immersive virtual reality and desktop computer
Pecci et al. [19]	This study used virtual reality for an inaccessible cultural heritage, a famous chapel containing a cycle of frescoes by Pietro Antonio Ferro, one of the most important painters of theseventeenth century in Basilicata
Puig et al. [20]	This study focused on using virtual reality technology to disseminate the knowledge of La Draga Neolithic settlement in archaeological museum and analyzed the interaction patterns and learning experience
Zilles et al. [25]	This study compared and evaluated two groups' (archeologists and VR experts) experience of using interactive archaeometry tools in Itapeva Rocky Shelter in Brazil virtual reality environment

these fields, both knowledge information and learning performance database were designed. The former was mainly to use to store basic information, correlation, assessment, and virtual reality scene index of all the knowledge points, and the latter was mainly used to store the learning per and feedback of users, including assessment results, operation duration, repetition times and other information. At the same time, convenient interface tools were developed for teachers or professionals to

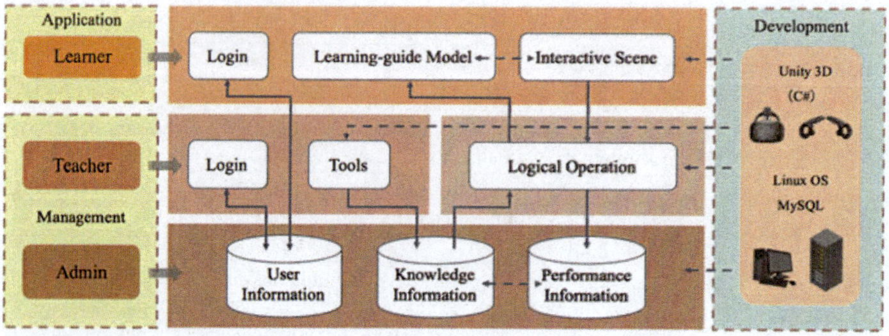

Fig. 1 Schematic diagram of the system architecture

dynamically maintain the knowledge database. The main interactive functions in this immersive virtual reality learning environment will be described in the next chapter.

4 Design and Develop

Main functions of this Maiji Mountain Grottoes immersive virtual reality learning environment will be introduced from two aspects in this chapter, including some important interactive scenes and the learning-guide model designed for these fields.

4.1 Interactive Scenes

In the process of learning in cultural heritage or archaeology fields, it usually needs much movement at historic sites, and Maiji Mountain Grottoes is no exception. Tracking navigation was designed to help learners complete the point-to-point path planning and perceive spatial position (Fig. 2). When one target grotto is selected through this navigation, it will recommend the optimal path through logical operation (Fig. 2, left). The implementation of this function is mainly based on recast navigation, including triangulation algorithm, A-Star algorithm and funnel algorithm. Moreover, when learners interact in one virtual reality scene, who can also view the current position from a macro perspective in real time to get clearer understanding of the spatial location in the immersive environment (Fig. 2, right).

In addition to tracking navigation, the historical dynasties of the grottoes also marked on the global map (Fig. 3). Superimposing time information on the spatial position may help learners construct space–time connections, which is usually used for learning in cultural heritage or archaeology fields.

To take the advantages of immersive virtual reality technology, not only the superposition of space–time information as mentioned above, but also other interaction

Fig. 2 Path planning and real-time position

Fig. 3 Spatiotemporal information superposition

scenes based the space–time relationship was designed. For example, it is usually necessary to compare and study the grottoes in different dynasties to better understand the development and changes of the style of Buddha statues. On one hand, if learners were to study on site, they need to move many times between different grottoes. On the other hand, if learners were to study in the classroom through traditional method, they can only use pictures, video, or other multimedia materials, lacking an immersive experience. The space–time tunnel (part of which is shown in Fig. 4) designed by using virtual reality technology combines the advantages of the above two methods and provides a sense of immersion free from time and space constraints.

Additionally, learners can also observe the details from a micro perspective without worrying about the threat to the protection of the statue. The information of knowledge can be learned not only by annotation but also by synchronized audio for both visual learners and auditory learners (Fig. 5).

With the changes of historical dynasties and the switching from macro perspective to micro perspective, immersive virtual reality environment provides more possibilities for learning.

Fig. 4 Part of the space–time tunnel

4.2 Learning-Guide Model

As mentioned before virtual reality technology can provide an immersive learning experience and make cross space–time learning possible, especially in cultural heritage or archaeology fields. However, compared with the traditional learning environment, when leaners construct knowledge in virtual reality scenes, they need to change the method from teacher led "teaching" to independent exploration "learning". This study attempted to propose a learning-guide model to present knowledge structure by 3D knowledge map and recommend learning path by logical operation.

3D Knowledge Map. Knowledge points in cultural heritage or archaeology fields are usually rich and complex, taking Maiji Mountain Grottoes as an example, part of the knowledge structure defined by teachers or professionals is shown in Fig. 6 (left). The relationship between these knowledge points can be divided into three categories as below. (i) Sequential relationship: the mastery of one knowledge point restricts the learning of another one, as the solid line with arrow shown in Fig. 6 (left). (ii) Inclusion relationship: one knowledge point may contain some sub knowledge points, as the dotted line with arrow shown in Fig. 6 (left). (iii) Association relationship: there may be an associational relationship between knowledge points, such as the intersection of space and time, the dotted line without arrow shown in Fig. 6 (left). Based on the relationship between knowledge points, it is projected into the virtual reality space through logical operation to form 3D knowledge map (Fig. 6, right).

Once the knowledge structure is imported into the system, 3D knowledge map can be automatically generated through this method, which can not only present the relationship between knowledge points, but also avoid the overlap of the spatial position of knowledge point, part of the codes is shown in Fig. 7. Learners can view each knowledge point and enter the interactive scene for learning by controlling the trigger.

Fig. 5 Interaction for details by HTC Vive

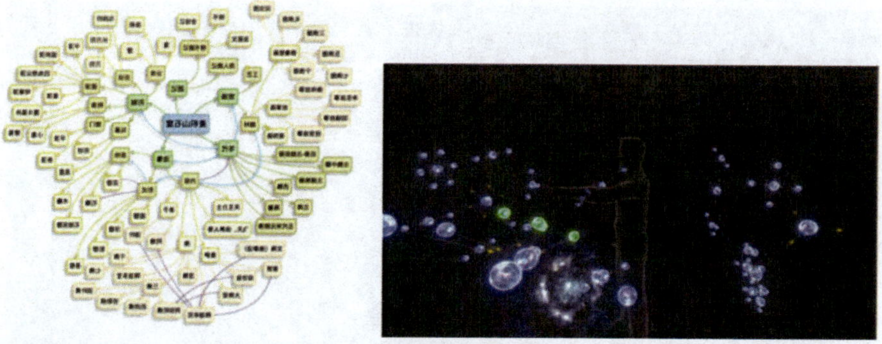

Fig. 6 Knowledge structure and 3D knowledge map of Maiji Mountain Grottoes

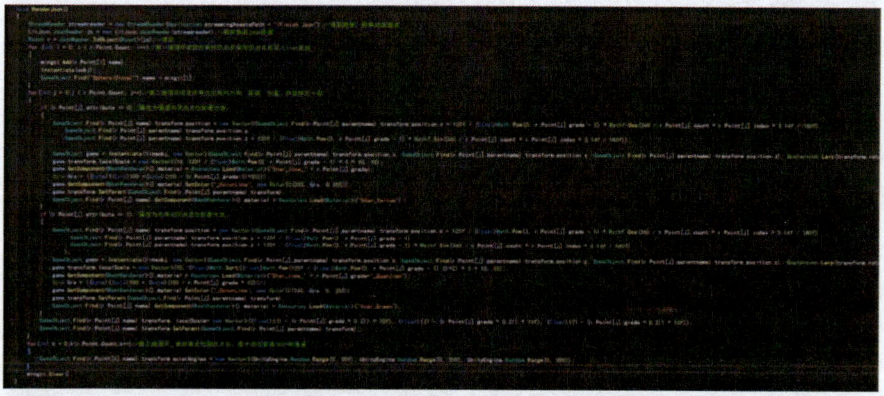

Fig. 7 Part of the codes for the generation of 3D knowledge map

Learning Path Recommendation. The interaction scene of each knowledge point includes test, the result of which is used as the main reference for the recommendation of learning path. The calculation formula of *Score(A)* is as follows:

$$Score(A) = \sum_{i=1}^{n}(k_i * w_i) \Big/ \sum_{i=1}^{n}(w_i)$$

Score(A) represents the test result of knowledge point A, k_i represents the score of question i of knowledge point A, and w_i represents the weight of question i. Besides, the learner's behavior parameters also take into account, such as the number n of tests and the duration t of completing the test. When $n > N_A$ or $t > T_A$, it will calculate n_A or t_A as below.

$$n_A = 1 - \left[\frac{n - N_A}{n}\right]$$

$$t_A = 1 - \left[\frac{t - T_A}{t}\right]$$

$$E = Score(A) * s_w + t_A * t_w + n_A * n_w$$

$$s_w + t_w + n_w = 1$$

Teachers or professionals can dynamically adjust N_A and T_A according to the difficulty of knowledge point A. S_w, n_w and t_w represent weights of *Score(A)*, n_A and t_A respectively. When the mastery degree E of knowledge point a is less than E_A, which is also defined by the teachers or professionals. Some test content and result are shown in Fig. 8.

Based on the relationship between knowledge points, the learning path is recommended for learners through iterative operation and displayed in the 3D knowledge map (Fig. 9).

In this immersive virtual reality learning environment, 3D knowledge map can be generated dynamically according to different knowledge structures. At the same time, teachers or professionals can flexibly adjust parameters such as N_A, T_A or E_A according to the difficulty of knowledge points. If it is proved that the system can

Fig. 8 Test content and result show in 3D knowledge map

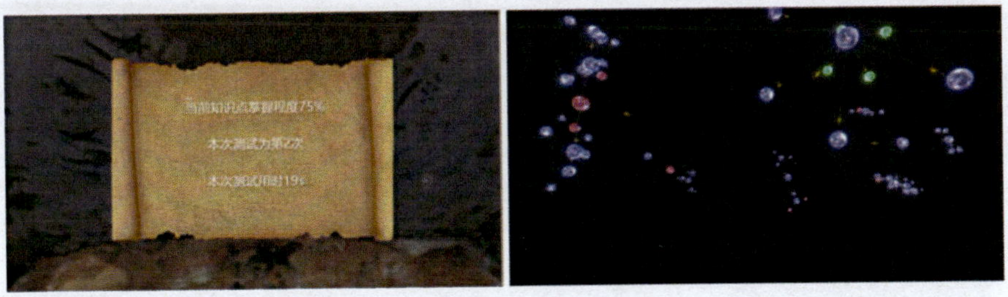

Fig. 9 Test result feedback and learning path recommendation

improve learning performance through evaluation, this learning-guide model may be applied in relevant cultural heritage or archaeology fields in the future.

5 Conclusion and Outlook

In this study, taking Maiji Mountain Grottoes as an example, some attempts based on virtual reality technology are made, such as the learning-guide model and spatiotemporal interaction scenes of immersive learning environment. At present, the main functions of the system have been designed and developed, then a small-scale pre-evaluation involved a few teachers and students will be carried out after testing. Next, the system will be optimized based on the feedback collected, and the learning performance will be evaluated through comparing with the traditional teaching mode. If the system shows better learning performance through evaluation, it is possible to replicate this design in the future and may provide richer immersive virtual reality environment in cultural heritage and archaeology fields for learners.

References

1. Oyelere SS, Bouali N, Kaliisa R et al. (2020) Exploring the trends of educational virtual reality games: a systematic review of empirical studies. Smart Learn Environ 7:31. https://doi.org/10.1186/s40561-020-00142-7
2. Johnston E, Olivas G, Steele P, Smith C, Bailey L (2018) Exploring pedagogical foundations of existing virtual reality educational applications: a content analysis study. J Educ Technol Syst 414–439
3. Papanastasiou G, Drigas A, Skianis C et al. (2019) Virtual and augmented reality effects on K-12, higher and tertiary education students' twenty-first century skills. Virtual Reality 23:425–436. https://doi.org/10.1007/s10055-018-0363-2
4. Checa D, Bustillo A (2020) Advantages and limits of virtual reality in learning processes: Briviesca in the fifteenth century. Virtual Reality 24:151–161. https://doi.org/10.1007/s10055-019-00389-7
5. Melo M, Coelho H, Gonçalves G et al (2022) Immersive multisensory virtual reality technologies for virtual tourism. Multimedia Syst 28:1027–1037. https://doi.org/10.1007/s00530-022-00898-7
6. Chen JC, Huang Y et al (2020) Developing a hands-on activity using virtual reality to help students learn by doing. J Comput Assist Learn 36:46–60
7. Ferguson C, van den Broek L, van Oostendorp H (2020) On the role of interaction mode and story structure in virtual reality serious games. Comput Educ
8. Fowler C (2015) Virtual reality and learning: where is the pedagogy? Br J Edu Technol 46(2):412–422
9. Patterson T, Han I (2019) Learning to teach with virtual reality: lessons from one elementary teacher. TechTrends 63:463–469
10. Pellas N, Mystakidis S, Kazanidis I (2021) Immersive virtual reality in K-12 and higher education: a systematic review of the last decade scientific literature. Virtual Reality 25:835–861
11. Stefan L, Moldoveanu F, Gheorghiu D (2016) Evaluating a mixed-reality 3D virtual campus with big data and learning analytics: a transversal study. J E-Learn Knowl Soc 12(2):42–54

12. Zhang XH, Shi YX, Bai H (2021) Immersive virtual reality physical education instructional patterns on the foundation of vision sensor. J Sens 2021. Article ID: 7752447
13. Kebritchi M, Hirumi A (2008) Examining the pedagogical foundations of modern educational computer games. Comput Educ 51:1729–1743
14. Wu J, Guo R, Wang Z, Zeng R (2019) Integrating spherical video-based virtual reality into elementary school students' scientific inquiry instruction: effects on their problem-solving performance. Interact Learn Environ 29(3):496–509
15. Bozorgi K, Lischer-Katz Z (2020) Using 3D/VR for research and cultural heritage preservation: project update on the virtual Ganjali Khan project. Preser Digit Technol Cult 49(2):45–57. https://doi.org/10.1515/pdtc-2020-0017
16. Bozzelli G, Raia A, Ricciardi S, Nino M, Barile N et al. (2019) An integrated VR/AR framework for user-centric interactive experience of cultural heritage: the arkaevision project. Digit Appl Archaeol Cult Heritage e00124. https://doi.org/10.1016/j.daach.2019.e00124
17. Cantatore E, Lasorella M, Fatiguso F (2020) Virtual reality to support technical knowledge in cultural heritage. The case study of cryptoporticus in the archaeological site of Egnatia (Italy). Int Arch Photogram Remote Sens Spat Inf Sci 44:465–472
18. Kim S, Ko H, Hong JY et al (2019) A study on experience contents of Baekje Muryeong Royal tomb using virtual reality. J Ambient Intell Hum Comput. https://doi.org/10.1007/s12652-019-01506-9
19. Pecci A, Campanile I (2020) The Chiesa della Madonna dei Mali (Ferrandina, Italy) in VR: virtual reality for inaccessible cultural heritage in peripheral areas. IOP Conf Ser: Mater Sci Eng 949(1):012080. IOP Publishing
20. Puig A, Rodríguez I, Arcos JL et al. (2020) Lessons learned from supplementing archaeological museum exhibitions with virtual reality. Virtual Reality 24:343–358. https://doi.org/10.1007/s10055-019-00391-z
21. Chan, CS., & Bogdanovic, J. & Kalivarapu, V. (2022). Applying immersive virtual reality for remote teaching architectural history. Educ Inf Technol 27, pp. 4365–4397. https://doi-org.ezproxy.ums.edu.my/https://doi.org/10.1007/s10639-021-10786-8
22. Farooq U, Rabbi I, Akbar S et al. (2022) The impact of design on improved learning in virtual worlds: an experimental study. Multimed Tools Appl 81:18033–18051. https://doi.org/10.1007/s11042-022-12593-w
23. Kyrlitsias C, Christofi M, Michael D, Banakou D, Ioannou A (2020) A virtual tour of a hardly accessible archaeological site: the effect of immersive virtual reality in user experience, learning and attitude change. Front Comput Sci 2, article. no. 23. https://doi.org/10.3389/fcomp.2020.00023
24. Ghani I, Rafi A, Woods P (2020) The effect of immersion towards place presence in virtual heritage environments. Pers Ubiquit Comput 24:861–872. https://doi.org/10.1007/s00779-019-01352-8
25. Zilles Borba E, Corrêa AG, de Deus Lopes R et al. (2020) Usability in virtual reality: evaluating user experience with interactive archaeometry tools in digital simulations. Multimed Tools Appl 79:3425–3447. https://doi.org/10.1007/s11042-019-07924-3

A Diagnostic Method for Detecting Tomato Leaf Diseases Based on a Deep Learning Approach

Zizie Vanessa Singkui, Rayner Alfred, Florence Sia Fui, Januarius Gobilik, Ervin Gubin Moung, Zamhar Iswandono, Azrina Abd Aziz, Nasreen Badruddin, and Micheal Drieberg

Abstract Nowadays, the agricultural area has been enhancing every day due to the innovation that makes farming far more effective and efficient in the production of crops. The Industrial Revolution 4.0 (IR4.0) plays a major role in ensuring that Smart Farming concept can be realized. For instance, a plant diagnostic tool using a deep learning approach can be developed to monitor the growth of plant or to investigate the health of the plant based on its leaf conditions. This study aims to investigate the feasibility of optimizing the deep learning algorithm to diagnose the quality of tomato crops using an optimized Convolutional Neural Network (CNN) model. In other words, this paper investigates the effects of varying the parameter settings

Z. V. Singkui · R. Alfred (✉) · F. S. Fui · J. Gobilik · E. G. Moung
Creative Advanced Machine Intelligence Research Centre, Faculty of Computing and Informatics, Universiti Malaysia Sabah, Kota Kinabalu, Sabah, Malaysia
e-mail: ralfred@ums.edu.my

Z. V. Singkui
e-mail: bi18110120@student.ums.edu.my

F. S. Fui
e-mail: florence.sfs@ums.edu.my

J. Gobilik
e-mail: jgobilik@ums.edu.my

E. G. Moung
e-mail: ervin@ums.edu.my

Z. Iswandono
Higher Colleges of Technology, Abu Dhabi, United Arab Emirates
e-mail: zismail@hct.ac.ae

A. A. Aziz · N. Badruddin · M. Drieberg
Department of Electrical and Electronic Engineering, Universiti Teknologi PETRONAS, Seri Iskandar, Perak, Malaysia
e-mail: azrina_aaziz@utp.edu.my

N. Badruddin
e-mail: nasreen.b@utp.edu.my

M. Drieberg
e-mail: mdrieberg@utp.edu.my

© The Author(s), under exclusive license to Springer Nature Singapore Pte Ltd. 2023

D.-K. Kang et al. (eds.), *Proceedings of the 9th International Conference on Computational Science and Technology*, Lecture Notes in Electrical Engineering 983,
https://doi.org/10.1007/978-981-19-8406-8_53

used in the CNN architecture on the prediction accuracy of the tomato diseases. The objective of this project is to formulate a deep learning model, specifically a CNN model, that can be used to learn tomato leaf images to detect tomato diseases and also to optimize the proposed CNN model by utilizing the CNN parameters such as number of epochs, batch size and learning rate. The methodology used in this work can be divided into three main tasks which are Data Capturing and Fusion, Data Modelling and Model Assessment. Based on the obtained results, the combination of all three parameters has a great influence on the performance accuracy of the proposed deep learning algorithm. For instance, the highest performance accuracy of the CNN model could be obtained by setting the parameters as follows; Epoch = 500, Batch Size = 16, Learning Rate = 1e − 1, accuracy of 95.54% for the training dataset and 65.0% for the test dataset.

Keywords Tomato leaf disease · Classification · Convolutional neural network · Deep learning approach

1 Introduction

The agricultural area has been enhancing every day due to the innovation that makes farming far more effective and efficient in the production of crops. The Industrial Revolution 4.0 (IR4.0) plays a major role in ensuring that Smart Farming concept can be realized [1–3]. For instance, a plant diagnostic tool using a deep learning approach can be developed to inspect the health of the crops by analyzing the leaf of the crops and indirectly monitor the growth of plant by investigating the health of the plant based on its leaf conditions [4, 5]. Plant diagnostic tool is essential to ensure the growth and the production of crops can be monitored closely. With the plant diagnostic tool, farmers can easily monitor the crops easily just by reading the data gained. Thus, making decision process can be done easily to manage the growth of the crops.

Smart farming can be defined as the use of technology and data discoveries to improve complex production frameworks [2, 6]. It concentrates on obtaining and applying the data, how the data obtained can be used actively. In other words, the implementation of smart farming focuses on accessing and applying the data, and on how the data coolected can be used smartly. Thus, it will overcome critical issues such as low crop yield, lack of ability to manage calamities, variation within farms and regions and produce wastage [5, 7]. The plant diagnostic tool can be used in smart farming, and it can incorporate any deep learning approaches to learn data captured from the field for monitoring the growth of the crops and also to detect disease of the plants. For instance, the neural network algorithm as well as its subtypes (e.g., Convolutional Neural Network (CNN)) have been found to be very successful in categorizing tasks for observing diseases in plants [8–11] as well as in human [12]. In this work, a tomato plant will be used as the main crops to be investigated due to the availability of the dataset. Tomato fruit has been considered as a domestic need [13].

The aim of this work is to design an optimized deep learning solution to diagnose the quality of plant being examined. In order to improve or maintain the quality of the tomato crops production, the tomato plant tree needs to be monitored and observed from the beginning when the tomato seeds are being sow. After the tomato seedlings has grown, the leaf of the plant can be inspected and monitored, and data will be captured. Based on the quality of the leaf of the tomato plant, farmers will be able to model the deep learning approach to detect any tomato diseases encountered. For example, bacterial spot and early blight are several tomato diseases that can be caused by fungus infection. Some of the factors that influence the performances of the deep learning approaches include limited annotated datasets, symptom representation, covariate shift, image background, image capture conditions, symptom segmentation, symptom variations, selection of initial values for the hyperparameter settings [14, 15]. The most common type of deep learning used was the CNN algorithm and it was able to produce acceptable accuracies. However, in each of the works, they used different settings to perform the modelling. Thus, this work is partly designed to investigate the effects of varying the values of the settings used in CNN architecture (e.g., number of epochs, batch size and learning rate) on the prediction accuracy of the tomato diseases. Section 2 provides the detail of the literature review or related works. Section 3 presents the proposed approach, Sect. 4 presents the results and discussion and finally Sect. 5 concludes this paper.

2 Related Works

Deep learning is a feature of artificial intelligence, which replicates the functioning of the human mind with regard to data processing and the creation of patterns of decision making. It is also an artificial intelligence machine learning subcategory which includes methods that can develop from unmodified or processed data without supervision, which is also referred to as deep neural education or deep neural network. Based on the features of what deep learning are capable of, it is a waste for not optimizing the use of deep learning.

The VGG convolutionary neural network can be considerd as the most successful system architecture for a deep learning approach [16–20]. Based on this high level of performance, it is evident that convolutionary neural network is best suited for automated identification and treatment of plant diseases through the analysis of simple foliage images. In order to enhance this direction, a much wider range of training data should be obtained from a number of sources from various geographical regions, the conditions of cultivation and the trends and patterns of images acquired. The proposed deep learning approach has shown its high potential, so it is a matter of quantity and quality of existing data to develop the method and to make it broader (in terms of plant species and diseases that can be identified) and more robust in real cultivation conditions.

In this work, CNN has been chosen because it conducts feature extraction automatically, resulting in a more reliable and computationally efficient, and sparing specialists from the time-consuming operation of extracting features which often produces inaccurate output [21]. Deep learning methodologies can be used to develop CNN frameworks for plant disease identification and treatment using simple image datasets of affected and unaffected plants [22]. A few other studies have been performed whereby the deep learning concepts have been used for digital image-based plant stress phenotyping [23].

Hyperspectral data [24, 25], soil and environment data [26] and high definition of plant photographs [27] were used for quality assessment in all plants. The two most often utilized methods in machine learning to evaluate the consistency of paddy rice plant in these works were the SVM and the CNN algorithms. In measuring the quality of paddy rice, CNN was considered more significant than SVM algorithm [27]. On the other hand, a combination of the classic artificial neural networks and SVM was applied to predict the lack of nitrogen in rice plants [28].

Neural network algorithm and its kind of variants (e.g., CNN) are shown to be extremely efficient in the classification of plant disease monitoring tasks [29–31]. For example, the artificial neural network (ANN) outperformed the fuzzy classification (FC) and SVM algorithms [31], and the calibrated CNN model still demonstrated good classification ability in a small-scale sample set and was chosen as the best classification model over the DT, k-NN, and SVM [25]. However, CNN model requires many samples for training purposes [32]. There are a few factors that can affect the output result for plant disease recognition such as limited annotated datasets, symptom representation, covariate shift, image background, image capture conditions, symptom segmentation, symptom variations and finally the selection of initial values for the hyperparameter settings of the CNN model [14, 15, 33].

There are several tools developed especially to diagnose plant's health. A web-mobile phone application has been developed that uses deep learning algorithm for tomato leaf classification [34]. The FCNNs deep learning algorithm is used for removing the background of tomato leaf photos obtained during difficult outdoor circumstances. Based on their findings, FCNNs was able to accurately elimate the background of the tomato leaf photos collected in the field. They have also shown that semantic segmentation systems can be used for instance segmentation for specifically single example of an item if the target occupies the whole picture, even though additional comparable but less noticeable items are present.

TutaAbsoluta DeepNet was introduced in which it relies on CNN models trained using ImageNet dataset [21]. `Tuta absoluta` is a common pest which attacks tomato plant that can cause a huge loss for farmers and the proposed solution was expected to overcome the problem of limited human labours to perform manual inspections. Based on the experiment conducted, VGG16 was able to produce the best performance accuracy, 91.9% on 85:15 ratio for training and testing dataset.

A comparison has also been conducted using various performance indicators (e.g., accuracy, precision, sensitivity, specifying, F-score and AUC) among CNN architecture for classifying tomato plant diseases that includes AlexNet, GoogleNet, Inception V3, ResNet 18 and ResNet 50 [35]. FGoogleNet has the highest percentage for accuracy, precision, sensitivity, specifying, F-score and AUC. As demonstrated, all of them exhibited highly relevant performances for every test, only with GoogleNet implementation attaining greater percentages. On the other side, Inception V3 produced the poorest performance compared to other approaches despite the enormous number of layers. To conclude the study, each model in this study seemed to be able to identify diseases in leaf of tomato plant, but with the GoogleNet with 22 layers achieving 99.72% classification of tomato diseases by using transfer learning training mechanism, which is highly statistically significant.

Resnet were also used for extracting core illness features of tomato disease [36]. A separate study was conducted on creating a template so that it can be used to categorize the type of tomato by using CNN [37]. The accuracy of effectiveness obtained was 93.0%, in which 5266 images were taken from Kaggle dataset for testing the proposed model. From the dataset, 70% of the dataset wwas used to train the CNN model while the other 30% of the dataset were used as validation set.

Another investigation was conducted that proposed a CNN model that was able to learn images which were extracted in Full-Color and Grey-scale images [38]. Based on the results obtained, the CNN model produced better performance with 99.84% classification accuracy when using the Full-Color model, while only 95.54% classification accuracy when using the Grey-Scale model. There were 9000 images dataset of tomato leaves were taken and the classification of dataset were divided into 6 classes, which were Bacterial Spot, Early Blight, Healthy, Septoria Leaf Spot, Leaf Mold and Yellow Leaf Curl Virus.

3 Methods

The methodology that has been chosen in this can be divided into three phases namely, Data Collection, Building CNN Model and Data Modeling, Model Optimization and Model Assessment.

3.1 Data Collection

The PlantVillage Tomato Leaf dataset used in this work is gathered from the Kaggle website [39]. There are 9 classes of tomato leaf corresponding to 9 tomato diseases. These leaf classes include the Target Spot (See Fig. 1), Mosaic Virus (See Fig. 2), Yellow Leaf Curl Virus (See Fig. 3), Backterial Spot (See Fig. 4), Early Blight (See Fig. 5), Late Blight (See Fig. 6), Healthy (See Fig. 7), Leaf Mold (See Fig. 8), Septoria Leaf Spot (See Fig. 9) and Spider Mites (See Fig. 10).

Fig. 1 Target spot tomato disease (1404 images)

Fig. 2 Mosaic virus tomato disease (373 images)

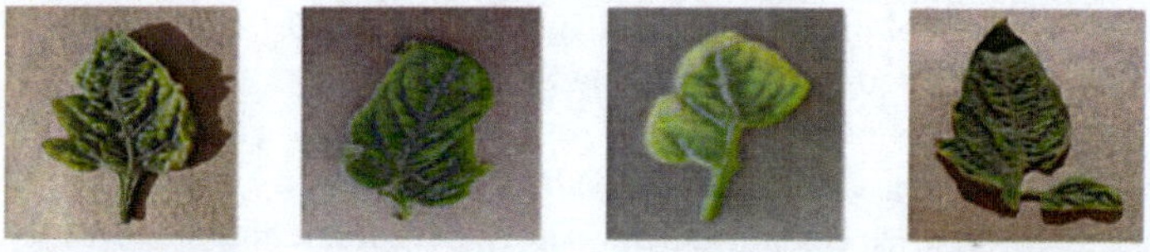

Fig. 3 Yellow leaf tomato disease (3209 images)

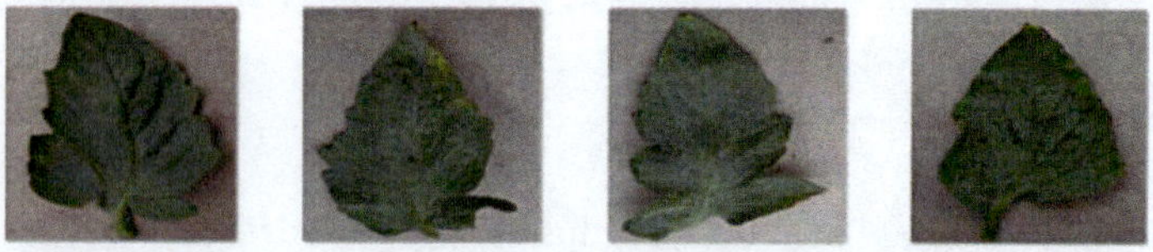

Fig. 4 Backteria spot tomato disease (2127 images)

Fig. 5 Early blight tomato disease (1000 images)

Fig. 6 Late blight tomato disease (1909 images)

Fig. 7 Healthy tomato (1591 images)

Fig. 8 Leaf mold tomato disease (952 images)

Fig. 9 Septoria leaf tomato disease (1771 images)

Fig. 10 Spider mites tomato disease (1676 images)

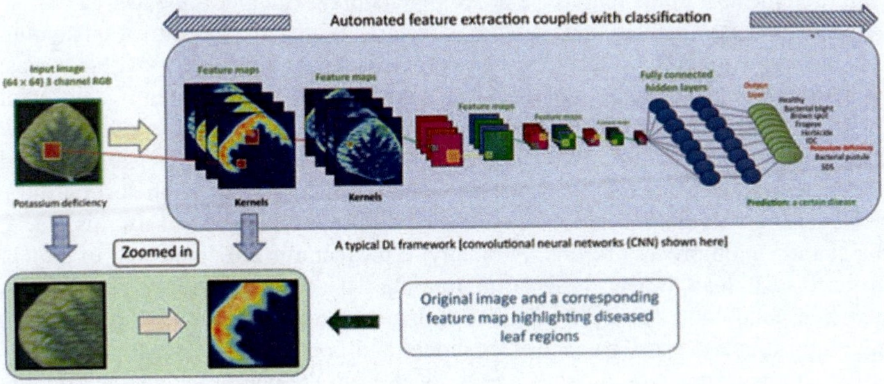

Fig. 11 CNN model

```
#CNN with softmax activation layer in last output layer
model = models.Sequential([
    resize_and_rescale,
    data_augmentation,
    layers.Conv2D(32, kernel_size = (3,3), activation='relu', input_shape=input_shape),
    layers.MaxPooling2D((2, 2)),
    layers.Conv2D(64,  kernel_size = (3,3), activation='relu'),
    layers.MaxPooling2D((2, 2)),
    layers.Conv2D(64,  kernel_size = (3,3), activation='relu'),
    layers.MaxPooling2D((2, 2)),
    layers.Conv2D(64, (3, 3), activation='relu'),
    layers.MaxPooling2D((2, 2)),
    layers.Conv2D(64, (3, 3), activation='relu'),
    layers.MaxPooling2D((2, 2)),
    layers.Conv2D(64, (3, 3), activation='relu'),
    layers.MaxPooling2D((2, 2)),
    layers.Flatten(),
    layers.Dense(64, activation='relu'),
    layers.Dense(n_classes, activation='softmax'),
])
```

Fig. 12 Construction of CNN model

3.2 Building CNN Model and Data Modeling

Figure 11 depicts the CNN modela and Fig. 12 outlines the sequential method on constructing the CNN model. The CNN model has 6 layers of convolution and various activation functions such as ReLU have been added. Below is the CNN code and the summary of the model. A total of 15 layers are available within the CNN. The Conv2D layer performs math operations on the convolution between the input image and the MxM size filter. After, this feature map will be sent to another layer to learn some other features of the input image.Then, Max Pooling layer gets the largest element from the feature map. Average pooling calculates the average of the elements in the predefined image sections. Total pooling calculates the total amount of items in the predefined sections. The Fully Connected layer, along with neurons, consists of weights and biases and is used to connect neurons between two different layers. These layers are usually placed before the output layer to form the last few layers of the CNN architecture. The input image of the previous layer is smoothed and added to the FC layer. Next, the flattened vector passes through several more FC layers, where math function operations are normally performed. In this phase, the classification process begins. Normally, if all features are connected to the FC layer, this can lead to overfitting of the training dataset. Overfitting occurs when a particular model works very well with training data and, when used with new data, adversely affects the performance of the model. A dropout layer is used to resolve this issue. At this level, some neurons are removed from the neural network during the training process, reducing the size of the model. Finally, one of the most important parameters of the CNN model is the activation function. The sigmoid and softmax functions are recommended for binary classification of the CNN model, and softmax

16. Chang X, Wu J, Yang T, Feng G (2020) Deepfake face image detection based on improved vgg convolutional neural network. In: 2020 39th Chinese control conference (CCC). IEEE, pp 7252–7256
17. Cheng S, Zhou G (2020) Facial expression recognition method based on improved vgg convolutional neural network. Int J Pattern Recogn Artif Intell 34(07):2056003 (2020)
18. Khan MSM, Ahmed M, Rasel RZ, Khan MM (2021) Cataract detection using convolutional neural network with vgg-19 model. In: 2021 IEEE World AI IoT congress (AIIoT). IEEE, pp 0209–0212
19. Radha M, Fonseca P, Moreau A, Ross M, Cerny A, Anderer P, Long X, Aarts RM (2021) A deep transfer learning approach for wearable sleep stage classification with photoplethysmography. NPJ Digital Med 4(1):1–11
20. Ye H, Han H, Zhu L, Duan Q (2019) Vegetable pest image recognition method based on improved vgg convolution neural network. J Phys Conf Ser 1237:032018
21. Mkonyi L, Rubanga D, Richard M, Zekeya N, Sawahiko S, Maiseli B, Machuve D (2020) Early identification of tuta absoluta in tomato plants using deep learning. Sci Af 10:e00,590
22. Ferentinos KP (2018) Deep learning models for plant disease detection and diagnosis. Comput Electron Agric 145:311–318
23. Singh AK, Ganapathysubramanian B, Sarkar S, Singh A (2018) Deep learning for plant stress phenotyping: trends and future perspectives. Trends Plant Sci 23(10):883–898
24. Nutini F, Confalonieri R, Crema A, Movedi E, Paleari L, Stavrakoudis D, Boschetti M (2018) An operational workflow to assess rice nutritional status based on satellite imagery and smartphone apps. Comput Electron Agric 154:80–92
25. Zhang L, Sun H, Rao Z, Ji H (2020) Hyperspectral imaging technology combined with deep forest model to identify frost-damaged rice seeds. Spectrochimica Acta Part A Mol Biomol Spectro 229:117973
26. Padarian J, Minasny B, McBratney A (2019) Using deep learning to predict soil properties from regional spectral data. Geoderma Reg 16:e00198
27. Liu C, Xu C, Liu S, Xu D, Yu X (2017) Study on identification of rice false smut based on cnn in natural environment. In: 2017 10th International congress on image and signal processing, biomedical engineering and informatics (CISP-BMEI). IEEE, pp 1–5
28. Sethy PK, Barpanda NK, Rath AK, Behera SK (2020) Nitrogen deficiency prediction of rice crop based on convolutional neural network. J Ambient Intell Humanized Comput 11(11):5703–5711
29. Harris C, Trisyono YA (2019) Classifying, detecting, and predicting infestation patterns of the brown planthopper in rice paddies. In: 2019 18th IEEE international conference on machine learning and applications (ICMLA). IEEE, pp 246–251
30. Lakmal D, Kugathasan K, Nanayakkara V, Jayasena S, Perera AS, Fernando L (2019) Brown planthopper damage detection using remote sensing and machine learning. In: 2019 18th IEEE international conference on machine learning and applications (ICMLA). IEEE, pp 97–104
31. Narmadha R, Arulvadivu G (2017) Detection and measurement of paddy leaf disease symptoms using image processing. In: 2017 International conference on computer communication and informatics (ICCCI). IEEE, pp 1–4
32. Kamrul, MH, Paul P, Rahman M (2019) Machine vision based rice disease recognition by deep learning. In: 2019 22nd International conference on computer and information technology (ICCIT). IEEE, pp 1–6,
33. Barbedo JG (2018) Factors influencing the use of deep learning for plant disease recognition. Biosystems Eng 172:84–91
34. Ngugi LC, Abdelwahab M, Abo-Zahhad M (2020) Tomato leaf segmentation algorithms for mobile phone applications using deep learning. Comput Electron Agric 178:105788
35. Maeda-Gutiérrez V, Galvan-Tejada CE, Zanella-Calzada LA, Celaya-Padilla JM, Galván-Tejada JI, Gamboa-Rosales H, Luna-Garcia H, Magallanes-Quintanar R, Guerrero Mendez CA, Olvera-Olvera CA (2020) Comparison of convolutional neural network architectures for classification of tomato plant diseases. Appl Sci 10(4):1245

36. Zhang Y, Song C, Zhang D (2020) Deep learning-based object detection improvement for tomato disease. IEEE Access 8:56607–56614
37. Alajrami MA, Abu-Naser SS (2020) Type of tomato classification using deep learning. Int J Acad Pedagogical Res (IJAPR) 3(12)
38. Ashqar BA, Abu-Naser SS (2018) Image-based tomato leaves diseases detection using deep learning
39. Emmanuel TO (2018) Plantvillage dataset. https://www.kaggle.com/datasets/emmarex/plantdisease
40. Aradhya AM, Ashfahani A, Angelina F, Pratama M, de Mello RF, Sundaram S (2022) Autonomous CNN (autocnn): a data-driven approach to network architecture determination. Inf Sci
41. Dileep P, Das D, Bora PK (2020) Dense layer dropout based cnn architecture for automatic modulation classification. In: 2020 National conference on communications (NCC). IEEE, pp 1–5
42. Kandel I, Castelli M (2020) The effect of batch size on the generalizability of the convolutional neural networks on a histopathology dataset. ICT Express 6(4):312–315